IEEE Technology Update Series

KU-384-860

Fuzzy Logic Technology and Applications

IEEE Technology Update Series

Fuzzy Logic Technology and Applications

Robert J. Marks II
EDITOR

PREFACE BY
Lotfi A. Zadeh

IEEE Technical Activities Board

A Selected Reprint Volume under the sponsorship of the Products Council of the IEEE Technical Activities Board

The Institute of Electrical and Electronics Engineers, Inc.
New York City, New York

Abstracting is permitted with credit to the source. Libraries are permitted to photocopy beyond the limits of U.S. Copyright law for private use of patrons those articles in this volume that carry a code at the bottom of the first page, provided the per-copy fee indicated in the code is paid through the Copyright Clearance Center, 222 Rosewood Drive, Danvers, MA 01923. For other copying, reprint, or republications permission, write to the IEEE Copyright Manager, IEEE Service Center, 445 Hoes Lane, P.O. Box 1331, Piscataway, NJ 08855-1331. All rights reserved. Printed in the USA. Copyright © 1994 by The Institute of Electrical and Electronics Engineers, Inc.

Second Printing

Library of Congress Cataloging-in-Publication Data

Fuzzy logic technology and applicatons / Robert J. Marks II, editor
  ; IEEE Technical Activities Board.
      p.   cm. -- (IEEE technology update series)
    "A selected reprint volume under the sponsorship of the Products
Council of the IEEE Technical Activities Board."
    "IEEE catalog number: 94CR0101-6"--T.p.verso.
    Includes bibliographical references and indexes.
    ISBN 0-7803-1383-6
    1. Automatic control. 2. Fuzzy systems. 3. Logic, Symbolic and
mathematical. 4. Systems engineering. I. Marks, Robert J.
II. Institute of Electrical and Electronics Engineers. Technical
Activities Board. III. Series.
TJ213.F89 1994
629.8--dc20   94-14291
                                                             CIP

IEEE Technical Activities Board

445 Hoes Lane
Piscataway, New Jersey 08855-1331

Robert J. Marks II
Editor-in-Chief

Associate Editors

Surendra Bhatia	Thomas P. Caudell	Jai J. Choi
Mohamed El-Sharkawi	Toshio Fukuda	Stamatios Kartalopoulos
Thomas F. Krile	Dr. Ichiro Masaki	Dr. Hiroyuki Mori
Edgar Sanchez-Sinencio	Francis A. Spelman	Colin Wiel
Sinclair Yee		

Robert T. Wangemann
Executive Editor

Tania Skrinnikov, Managing Editor
Harry Strickholm, Technical Editor
John Vitale, Technical Editor
Jayne Cerone, Administrative Editor
Lois Pannella, Administrative Editor
Patricia Thompson, Administrative Editor
Mark A. Vasquez-Jorge, Administrative Editor
Ann Burgmeyer, Production Editor
Janet Romano, Art Director

FOREWORD

Fuzzy systems and neural networks have emerged as important new technologies for consumer and industrial applications. In recognition of the growing volume of research and interest in this field, the IEEE Technical Activities Board (TAB) and the TAB Book Broker Committee selected fuzzy logic as the topic of the second volume in the IEEE Technology Update Series. This Series was developed to furnish readers with up-to-date research and practical applications information in specific fields of interest. "Fuzzy Logic: Technology and Applications" contains applications-oriented material from IEEE conferences and journals for electrotechnology professionals around the world.

For bringing this book to fruition, my thanks go to Robert J. Marks II for serving as Editor-in-Chief and for gathering an outstanding group of Associate Editors and papers in the field of fuzzy logic. Prof. Marks served as the first President of the IEEE Neural Networks Council and continues to be a leader in the field. Conferences sponsored and co-sponsored by the Neural Networks Council provided much of the material for this reprint book.

I am indebted to Prof. Donald M. Bolle, IEEE Vice President-Technical Activities, and Dr. Jan Brown, Chair of the TAB Book Broker Committee, for their guidance in this venture.

I would also like to recognize the diligence and time expended by my staff in fulfilling the task of bringing this book to print. Thanks go to Tania Skrinnikov, Harry Strickholm, John Vitale, Mark A. Vasquez-Jorge, Lois Pannella, Patricia Thompson and Jayne F. Cerone for their editorial input. The production assistance of Ann H. Burgmeyer, Production Editor, and Art Director Janet Romano completed the task.

Robert T. Wangemann
Staff Director - IEEE Technical Activities

CONTENTS

Foreword — vii
Preface L. Zadeh — xvii
Introduction R.J. Marks II — xix
Suggested Additional Reading list — xxi

Section I: Technology

Chapter 1: Overviews — 1

Selected Papers:

Introduction to Fuzzy Models
1.1 "Editorial: Fuzzy Models - What are they, and why?"
   J.C. Bezdek — 3

Consumer Products
1.2 "Application of Neural Networks and Fuzzy Logic to Consumer Products"
   Hideyuki Takagi — 8

Vehicular Technology
1.3 "Fuzzy Logic Technology & the Intelligent Highway System (IHS)"
   Bruno Bosacchi and Ichiro Masaki — 13

Control
1.4 "Twenty Years of Fuzzy Control: Experiences Gained and Lessons Learnt"
   E. H. Mamdani — 19

Computer Vision
1.5 "Fuzzy Set Theoretic Approach to Computer Vision: An Overview"
   Raghu Krishnapuram and James M. Keller — 25

Image Processing and Recognition
1.6 "Fuzzy Sets in Image Processing and Recognition"
   S.K. Pal — 33

Section II: Applications

Chapter 2: Vehicular Technology   **41**

Selected Papers:

2.1 "Trainable Fuzzy and Neural-Fuzzy Systems for Idle-Speed Control"
  L. A. Feldkamp and G. V. Puskorius   43

2.2 "Evaluation of Fuzzy and Neural Vehicle Control"
  Jos Nijhuis, Stefan Neuber, Jurgen Heller, Jochen Sponnemann   50

2.3 "Follow-up Characteristics of a Small Automatic Guided Vehicle System with Fuzzy Control
  "Nobuhiko Yuasa, Mitsuo Shioya and Gunji Kimura   56

2.4 "Self Organizing Fuzzy Logic Control of a Level Control Rig"
  Nader Vijeh   62

2.5 "Fuzzy Logic Anti-Lock Brake System for a Limited Range Coefficient of Friction Surface"
  D.P. Madau, F. Yuan, L.I. Davis, Jr., and L.A. Feldkamp   68

2.6 "Intelligent Cruise Control with Fuzzy Logic"
  Rolf Muller and Gerhard Nocker   74

2.7 "Design of a Rule-Based Fuzzy Controller for the Pitch Axis of an Unmanned Research Vehicle"
  Deepak Sabharwal and Kuldip S. Rattan   81

2.8 "Fuzzy Expert System for Automatic Transmission Control"
  H.G. Weil, G. Probst and F. Graf   88

2.9 "Application of Fuzzy Logic to Shift Scheduling Method for Automatic Transmission"
  S. Sakaguchi, I. Sakai and T. Haga   94

2.10 "Adaptive Traffic Signal Control Using Fuzzy Logic"
  Stephen Chiu and Sujeet Chand   101

2.11 "A Control System of Carburization Using Fuzzy-PID Combined Controller"
  Liya Hou and Zhengqing Wang   107

2.12 "Fuzzy Control for Active Suspension Design"
  Edge C. Yeh and Yon J. Tsao   109

Chapter 3: Robotics   **115**

Selected Papers:

3.1 "Fuzzy Controlled Gait Synthesis for a Biped Walking Machine"
  Luis Magdalena and Feliz Monasterio   117

3.2 "A Fuzzy Logic Force Controller for a Stepper Motor Robot"
  J.G. Hollinger, R.A. Bergstrom and J.S. Bay   123

3.3 "Hierarchical Intelligent Control for Robotic Motion by Using Fuzzy, Artificial Intelligence, and Neural Network"
Toshio Fukuda and Takanori Shibata 129

3.4 "Hierarchical Control for Autonomous Mobile Robots with Behavior-Decision Fuzzy Algorithm"
Yoichiro Maeda, Minoru Tanabe, Morikazu Yuta and Tomohiro Takagi 135

3.5 "Fuzzy Navigation of a Mobile Robot"
Kai-Tai Song and Jen-Chau Tai 141

3.6 "Blending Reactivity and Goal-Directedness in a Fuzzy Controller"
Alessandro Saffiotti, Enrique H. Ruspini and Kurt Konolige 148

3.7 "Fuzzy Logic Based Robotic Arm Control"
Robert N. Lea, Jeffrey Hoblit, and Yashvant Jani 154

3.8 "Robotic Deburring Based on Fuzzy Force Control"
M.H. Liu 160

3.9 "Manipulator for Man-Robot Cooperation (Control Method of Manipulator/Vehicle System with Fuzzy Inference)"
Yoshio Fujisawa, Toshio Fukuda, Kazuhiro Kosuge, Fumihito Arai, Eiji Muro, Haruo Hoshino, Takashi Miyazaki, Kazuhiko Ohtsubo and Kazuo Uehara 168

Chapter 4: Motors, Servos, and Drives   175

Selected Papers:

4.1 "Adaptive Fuzzy Control of High Performance Motion Systems"
E. Curruto, A. Consoli, A. Raciti and A. Testa 177

4.2 "Fuzzy Algorithm for Commutation of Permanent Magnet AC Servo Motors without Absolute Rotor Position Sensors"
Dong-II Kim, Jin-Won Lee and Sungkwun Kim 184

4.3 "Fuzzy Logic-Based Control of Flux and Torque in AC-Drives"
Wilfried Hofmann and Michael Krause 190

4.4 "Adaptive Fuzzy Techniques for Slip-Recovery Drive Control"
L.E. Borges da Silva, G. Lambert-Torres, V. Ferreira da Silva and K. Nakashima 196

4.5 "Fuzzy Controller for Inverter Fed Induction Machines"
Sayeed A. Mir, Donals S. Zinger and Malik E. Elbuluk 204

4.6 "A Fuzzy Current Controller for Field-Oriented Controlled Induction Machine by Fuzzy Rule"
Seong-Sik Min, Kyu-Chan Lee, Jhong-Whan Song and Kyu-Bock Cho 212

Chapter 5: Power Systems — 219

Selected Papers:

5.1 "A Fuzzy Knowledge-Based System for Bus Load Forecasting"
H.G. Lambert-Torres, L.E. Borges da Silva, B. Valiquette, H. Greiss and D. Mukhedkar — 221

5.2 "A Symptom-Driven Fuzzy System for Isolating Faults"
Jiann-Liang Chen, Ronlon Tsai and Huan-Wen Tzeng — 229

5.3 "Comparison of Fuzzy Logic Based and Rule Based Power System Stabilizer"
Juan Shi, L. H. Herron and A. Kalam — 233

5.4 "Analysis of Power System Dynamic Stability via Fuzzy Concepts"
Pei-Hwa Huang — 239

Chapter 6: Industry Applications — 245

Selected Papers:

6.1 "Application of Neuro-Fuzzy Hybrid Control System to Tank Level Control"
Tetsuji Tani, Shunji Murakoshi, Tsutomu Sato, Motohide Umano and Kazuo Tanaka — 247

6.2 "Minimization of Combined Sewer Overflows Using Fuzzy Logic Control"
Sheng-Lu Hou and N. Lawrence Ricker — 253

6.3 "Identification and Analysis of Fuzzy Model for Air Pollution - An Approach to Self-Learning Control of CO Concentration"
Kazuo Tanaka, Manabu Sano and H. Watanabe — 261

6.4 "Self-Learning Fuzzy Modeling of Semiconductor Processing Equipment"
Raymond L. Chen and Costas J. Spanos — 267

6.5 "Range Tests Made Fuzzy: An Alternate Perspective on the Built-in-Test of Real Time Embedded Systems"
Jennifer D. Brown and George J. Klir — 274

6.6. "Fuzzy Seam-Tracking Controller"
Yoshito Sameda — 280

6.7 "Neural Network Based Decision Model Used for Design of Rural Natural Gas Systems"
W. Pedryczk, J. Davidson and I. Goulter — 285

6.8 "Application of Fuzzy Control System to Hot Strip Mill"
Naoki Sato, Noriyuki Kamada, Shuji Naito, Takashi Fukushima and Makoto Fujino — 293

6.9 "A Fuzzy Classification Technique for Predictive Assessment of Chip Breakability for Use in Intelligent Machining Systems"
J. Fei and I.S. Jawahir — 298

6.10 "A Rule-Based Fuzzy Logic Controller for a PWM Inverter in Photo-Voltaic Energy Conversion Scheme"
Rohin M. Hilloowala and Adel M. Sharaf   304

6.11 "Fuzzy Control of Wire Feed Rate in Robot Welding"
Satoshi Yamane, Guillermo Alzamora and Takefumi Kubota   312

Chapter 7: Electronics   317

Selected Papers:

7.1 "Autonous Navigation of a Mobile Robot Using Custom-Designed Qualitative Reasoning VLSI Chips and Boards"
Francois G. Pin, Hiroyuki Wantanabe, Jim Symon and Robert S. Pattay   319

7.2 "Fuzzy Control of an Industrial Robot in Transputer Environment"
Jarmo Franssila and Heikki N. Koivo   325

7.3 "Fuzzy Logic Approach to Placement Problem"
Rung-Bin Lin and Eugene Shragowitz   331

7.4 "A Fuzzy Algorithm for Multiprocessor Bus Arbitration"
Robert T. Tran, Timothy R. Slator and Anaikuppam R. Marudarajan   337

Chapter 8: Sensors   343

Selected Papers:

8.1 "Improving Dynamic Performance of Temperature Sensors with Fuzzy Control Technique"
Wang Lei and Volker Hans   345

8.2 "Multi-Sensor Integration System Based on Fuzzy Inference and Neural Network for Industrial Apprication"
Toshio Fukuda, Koji Shimojima, Fumihito Arai and Hideo Matsuura   347

8.3 "A Fuzzy Logic Approach for Handling Imprecise Measurements in Robotic Assembly"
H.B. Gurocak and A. de Sam Lazaro   353

Chapter 9: Aerospace   361

Selected Papers:

9.1 "Space Shuttle Attitude Control by Reinforcement Learning and Fuzzy Logic"
Hamid R. Berenji, Robert N. Lea, Yashvant Jani, Pratap Khedkar, Anil Malkani and Jeffrey Hoblit   363

9.2 "Intelligent Control of a Flying Vehicle Using Fuzzy Associative Memory System"
T. Yamaguchi, K. Goto, T. Takagi, K. Doya and T. Mita   369

9.3 "Development and Simulation of an F/A-18 Fuzzy Logic Automatic Carrier Landing System"
Marc Steinberg — 380

Chapter 10: Communications — 387

Selected Papers:

10.1 "An RLS Fuzzy Adaptive Filter with Application to Nonlinear Channel Equalization"
Li-Xin Wang and Jerry M. Mendel — 389

10.2 "Model-Reference Neural Color Correction for HDTV Systems Based on Fuzzy Information Criteria"
Po-Rong Chang and C. C. Tai — 395

10.3 "Classified Vector Quantization Using Fuzzy Theory"
Ferran Marques and C.C. Jay Kuo — 401

Chapter 11: Bioengineering — 409

Selected Papers:

11.1 "Fuzzy Control of Blood Pressure During Anesthesia with Isoflurane"
R. Meier, J. Nieuwland, S. Hacisalihzade, D. Steck and A. Zbinden — 411

11.2 "Real-Time Fuzzy Control of Mean Arterial Pressure in Post-surgical Patients in an Intensive Care Unit"
Hao Ying, Michael McEachern, Donald W. Eddleman and Louis C. Sheppard — 418

11.3 "Fuzzy ARTMAP Neural Network Compared to Linear Discriminant Analysis Prediction of the Length of Hospital Stay in Patients with Pneumonia"
Philip H. Goodman, Vassilis G. Kaburlasos and Dwight D. Egbert — 424

11.4 "Fuzzy Classification of Heart Rate Trends and Artifacts"
Dean F. Sittig, Kei-Hoi Cheung and Lewis Berman — 430

Chapter 12: Image Processing and Recognition — 441

Selected Papers:

12.1 "Application of the Extended Fuzzy Pointing Set to Coin Grading"
P.A. Laplante, D. Sinha and C.R. Giardina — 443

12.2 "Region Extraction for Real Image Based on Fuzzy Reasoning"
Kohi Miyajima and Toshio Norita — 447

12.3 "A Fuzzy Approach to Scene Understanding"
Weijing Zhang and Michio Sugeno — 455

Chapter 13: Pattern Recognition   461

Selected Papers:

13.1 "Recognition of Facial Expressions Using Conceptual Fuzzy Sets"
Hirohide Ushida, Tomohiro Takagi and Toru Yamagugchi   463

13.2 "Qualitative/Fuzzy Approach to Document Recognition"
Hiroko Fujihara and Elmamoun Babiker   469

13.3 "Fuzzy Artificial Network and its Application to a Command Spelling Corrector"
N. Imasaki, T. Yamaguchi, D. Montgomery and T. Endo   476

13.4 "A New Similarity Measurement Method for Fuzzy-Attribute Graph Matching and its Application to Handwritten Character Recognition"
G.M.T. Man and J.C.H. Poon   482

13.5 "Automatic Target Recognition Fuzzy System for Thermal Infrared Images"
Christiaan Perneel, Michel de Mathelin and Marc Acheroy   486

13.6 "A Vowel Recognition Using Adjusted Fuzzy Membership Functions"
Sung-Soon Choi and Kyung-Whan Oh   493

Chapter 14: Management   501

Selected Papers:

14.1 "Dynamics and Fuzzy Control of a Group"
Kenji Kurosu, Tadayoshi Furuya, Masaaki Nakamura, Hiroshi Utsunomiya and Mitsuru Soeda   503

14.2 "The Fuzziness Index for Examining Human Statistical Decision-Making"
Sumiko Takayanagi and Norman Cliff   509

14.3 "Linking the Fuzzy Set Theory to Organizational Routines: A Study in Personnel Evaluation in a Large Company"
Alessandro Cannavacciuolo, Guido Capaldo, Aldo Ventre and Giuseppe Zollo   515

Chapter 15: General and Multi-Discipline   521

Selected Papers:

15.1 "An Electronic Video Camera Image Stabilizer Operated on Fuzzy Theory"
Yo Egusa, Hiroshi Akahori, Atsushi Morimura and Noboru Wakami   523

15.2 "Fuzzy Logic Based Banknote Transfer Control"
Masayasu Sato, Tohru Kitagawa, Takehito Sekiguchi, Keisuke Watanabe and Masao Goto   531

15.3 "Electrophotography Process Control Method Based on Neural Network and Fuzzy Theory"
Tetsuya Morita, Mitsuhisa Kanaya, Tatsuya Inagaki, Hisao Murayama and Shinji Kato   537

15.4 "Fuzzy Logic Implementation of Intent Amplification in Virtual Reality"
John Dockery and David Littman ... 543

15.5 "On-Line Analysis of Music Conductor's Two-Dimensional Motion"
Zeungnam Bien and Jong-Sung Kim ... 549

15.6 "Multilevel Database Security Using Information Clouding"
Sujeet Shenoi ... 556

15.7 "Active Control of Broadband Noise Using Fuzzy Logic"
Oscar Kipersztok ... 562

Author Index ... 569

Subject Index ... 571

Editor's Biography ... 575

PREFACE

Lotfi A. Zadeh

The past few years have witnessed an explosive growth in the number and variety of papers dealing with the applications of fuzzy logic. A consequence of the wide variety of applications is a wide dispersal of the sources of publication. By assembling and structuring a representative collection of applications in a single volume, Professor Robert Marks II, the Editor, and the Associate Editors of "Fuzzy Logic: Technology and Applications" have performed a valuable service.

Although some of the earlier controversies regarding the applicability of fuzzy logic have abated, there are still influential voices which are critical and/or skeptical. Some take the position that anything that can be done with fuzzy logic can be done equally well without it. Some are trying to prove that fuzzy logic is wrong. And some are bothered by what they perceive to be exaggerated expectations. That may well be the case but, as Jules Verne had noted at the turn of the century, scientific progress is driven by exaggerated expectations.

To view the claims and the counterclaims in a proper perspective, it is necessary, first, to clarify what is meant by fuzzy logic. There is a need for such a clarification because the label fuzzy logic is used in two different senses.

In a narrow sense, fuzzy logic, FLn, is a logical system which aims at a formalization of approximate reasoning. In this sense, FLn is an extension of multivalued logic. However, the agenda of FLn is quite different from that of traditional multivalued logics. In particular, such key concepts in FLn as the concept of a linguistic variable, canonical form, fuzzy if-then rule, fuzzy quantification, fuzzification and defuzzification, predicate modification, truth qualification, the extension principle, the compositional rule of inference and interpolative reasoning, among others, are not addressed in traditional systems. This is the reason why FLn has a much wider range of applications than traditional logical systems.

In its wide sense, fuzzy logic, FLw, is fuzzily synonymous with fuzzy set theory, FST, which is the theory of classes with unsharp boundaries. FST is much broader than FLn and includes the latter as one of its branches. What is important to realize is that any field X can be fuzzified -- resulting in fuzzy X -- by replacing the concept of a crisp set in X by a fuzzy set. This is the genesis for fuzzy arithmetic, fuzzy mathematical programming, fuzzy probability theory, fuzzy decision analysis, fuzzy control, fuzzy neural network theory, fuzzy topology, etc. The question is: What is accomplished by fuzzifying X? The answer is (a) greater generality; and (b) enhanced ability to deal with real-world problems, especially in the realms of control, probability theory, decision analysis and, more generally, those fields in which crisp models are unrealistic or there is an opportunity to exploit the tolerance for imprecision to achieve higher MIQ (Machine Intelligence Quotient) and/or lower cost.

In arguing about fuzzy logic, it is necessary to recognize that, at this juncture, the label fuzzy logic, FL, is used most frequently in its wide sense, mainly because as a label fuzzy logic is more euphonious and self-explanatory than fuzzy set theory. This is the perspective in which the contents of "Fuzzy Logic: Technology and Applications" should be viewed.

Today, most of the applications of fuzzy logic are based -- implicitly rather than explicitly -- on the use of a subset of FL which might be called the calculus of fuzzy rules (CFR). As its name suggests, CFR is concerned with the generation and processing of fuzzy if-then rules. The importance of CFR derives from the fact that it is basically a language which serves to describe and analyze imprecise dependencies -- dependencies which do not lend themselves to representation via differential equations, difference equations, algebraic equations or other conventional techniques. What is important to recognize is that CFR does not replace the conventional techniques. Rather, it adds to them a body of concepts and techniques which can be used effectively in the conception, analysis, design and construction of systems which in their entirety or in part involve imprecisely defined or imprecisely known dependencies or commands. The wide range of applications covered in this volume suggest that such systems are ubiquitous. The controversies surrounding fuzzy logic will fade away when this becomes a widely accepted view.

"Fuzzy Logic: Technology and Applications" is certain to find an appreciative audience and contribute importantly to the awareness of ways in which fuzzy logic can be applied in the solution of real-world problems. The editors of the volume deserve our thanks and congratulations.

INTRODUCTION

Lord Kelvin argued in 1885 that "Heavier than air flying machines are impossible." In 1923, Nobel Laureate Robert Milikin claimed, "There is no likelihood man can ever tap the power of an atom." Henry M. Warner of Warner Brothers fell victim to his own thinking inertia in 1927 when he muttered "Who the hell wants to hear actors talk?" Today, a US President quoted saying "Sensible and responsible women do not want to vote" as Grover Cleveland did in 1905 would be committing political suicide. These are but a few examples of resistance to a paradigm shift -- a revolutionary alteration in a way of thinking.

The term 'fuzzy', first introduced by Lotfi Zadeh in 1965 [1], invokes similar responses. Consider the following recently published comment.

"The image of (fuzzy control) which is portrayed is of the ability to perform magically well by the incorporation of 'new age' technologies of fuzzy logic, neural networks, expert systems, approximate reasoning, and self organization in the dismal failure of traditional methods. This is pure, unsupported claptrap which is pretentious and idolatrous in the extreme, and has no place in the scientific literature" [2].

Mamdani [3] counters that statements such as this emanate from a *'cult of analyticity'*. What is the truth? Are fuzzy systems 'claptrap' or are they, as others claim, a revolutionary technology? Can't other technology be used to do the same thing fuzzy systems do [4]? Isn't the lack of stability assurance a hindrance to the use of fuzzy control [5]? Doesn't a dimensional explosion of rules prohibit the use of fuzzy control of multivariate systems? These questions are examples of concerns over fuzzy systems that largely remain unanswered. Nevertheless, these and other open questions have not been a hindrance in attempts to apply fuzzy technology to certain problems. One of the purposes of this book is to address the utility of fuzzy systems by looking at the evidence, i.e., the applications, of fuzzy systems. One indisputable fact is that fuzzy systems have been applied to a wide spectrum of engineering applications and, in certain important cases, work quite well.

The research and development activity in fuzzy systems have been increasing rapidly. The vitality of the field can be seen in the recent increase in publication and patent activity [6]. Figure 1 shows the publication activity in fuzzy systems over the last few years. In 1991, over 1400 papers dealt with the topic. A better gauge of fuzzy system application and implementation is in the United States patent activity shown in Figure 2. The number of fuzzy patents issued in 1992 exceeds the sum total of all previously issued fuzzy system patents.

The papers in this volume were chosen from hundreds of fuzzy papers published in IEEE conference records over the last few years. Except for overview manuscripts, each paper deals with an application of fuzzy systems. The papers were chosen by a distinguished pool of Associate Editors. In general, the Associate Editors are neither specialists nor advocates of fuzzy systems. They are, rather, disinterested practitioners in an applications area where fuzzy systems can be used. Papers were thus chosen in accordance to the worth of their proposed application.

Robert J. Marks II
Editor

SCANNING THE BOOK

Engineering specialties are either problems looking for solutions (e.g. robotics, power engineering, industrial applications, vehicular technology) or solutions looking for problems (e.g. signal processing, neural networks, computer science). Fuzzy systems fall into the latter category. The section titles in this book fall into the former. Indeed, most of the section titles are the names of IEEE Societies wherein problems seek solutions. Section 2, for example, contains papers appropriate for the IEEE Vehicular Technology Society, Section 3 the IEEE Robotics and Automation Society, Section 5 the IEEE Power Engineering Society, etc. The proposed applications across numerous IEEE specialties is evidence of the widespread utility of and interest in fuzzy systems.

The first section of this volume contains surveys and *Overviews* of various areas of fuzzy systems. The first paper, by James C. Bezdek, is reprinted from the inaugural issue of the *IEEE Transactions on Fuzzy Systems*. Bezdek directly addresses many of the misconceptions about fuzzy logic using fundamental examples. His 'evolution of new technology' curve nicely describes the perception of fuzzy logic throughout the last three decades. It is a curve traveled by the evolution of a number of technologies, including information theory in the 1950's, optical computing in the 1970's, neural networks in the 1980's and virtual reality in the 1990's.

Takagi provides an overview of the application of fuzzy logic to appliances including washing machines, air conditioners, vacuum cleaners, microwave ovens, clothes dryers, electric fans, refrigerators and rice cookers. The vast majority of these products are produced by Japanese corporations who, historically, are singularly responsible for the reduction of fuzzy logic to practice. Fuzzy logic has also found an important place in the Intelligent Highway System. Bosacchi and Masaki provide a broad overview. One of the most popular applications of fuzzy logic is to control. Mamdami, who pioneered fuzzy control, reflects on two decades of its development. Finally, Krishnapuram and Keller present an overview of fuzzy computer vision and Pal reviews fuzzy image processing and recognition. Both of these papers contain extensive bibliographies.

The papers in the *Vehicular Technology* section explore applications of fuzzy systems to such diverse vehicular technology applications such as speed and cruise control, guidance, automatic transmission control, traffic signal control, carburetor control and suspension design. Anti-lock braking is both a nonlinear and time variant problem. Madau, Yuan, Davis and Feldkamp show that fuzzy systems provide a viable approach to the problem.

Fuzzy systems offers solutions to many of the control problems encountered in *Robotics* both in arm manipulation and navigation. Control of nonlinear time variant *Motors, Servos and Drives* can be quite difficult. Fuzzy approaches are shown through experimentation and prototyping to be a solution. In *Power Systems*, fuzzy expert systems for forecasting and fault isolation are tested with actual power system data. Power system stabilization using fuzzy methods is both simulated and applied to a four machine study system. Papers in *Industry Applications* deal with application of fuzzy systems to diverse problems, including tank level control, sewer overflow control, air pollution monitoring, semiconductor manufacturing, built-in testing, welding control, natural gas system design, machining, and photo-voltaic energy conversion.

In *Electronics*, Pin & Watanabe discuss the use of a custom-designed VLSI chip in the control of a car. Franssila & Koivo control a robot using fuzzy transputer circuitry. Use of fuzzy logic in VLSI circuit design and adaptive multiprocessor arbitration protocol is also presented. Fuzzy systems are used to increase the performance of certain *Sensors*. Improved performance is verified through a series of extensive experiments.

In the *Aeronautics* section, Berenji et.al. contrast conventional and fuzzy control of the Space Shuttle. Fuzzy control and automatic carrier landing of flying vehicles is also addressed. *Communications* applications of fuzzy logic include HDTV, vector quantization and nonlinear channel equalization.

Bioengineering has potentially a number of interesting fuzzy applications, including blood pressure control, hospital stay forecasting and heart rate trend classification. LaPlante, Sinha and Giardina present an interesting technique of fuzzy *Pattern Recognition* where coin quality is graded. Region extraction and scene understanding are also discussed in this section. *In Image Recognition*, facial expressions, documents, spelling, handwriting, thermal infrared images and vowels are recognized using fuzzy systems.

The uncertainties of *Management* suggest application of fuzzy logic. A fuzzy model is proposed to simulate the activities of a group. Fuzzy models are also proposed for human decision-making and personnel evaluation.

The last section is entitled *General and Multidisciplinary*. It contains papers who belong in a section of their own. Egusa et.al. discuss the operation of video camera stabilization based on fuzzy reasoning. Banknote transfer control by fuzzy methods is proposed by Sato et.al. Fuzzy solutions are also proposed for electrophotography process control, virtual reality intent amplification, database security clouding, and simulation of the motions of an orchestra conductor.

FOR FURTHER INFORMATION:

There are a number of resources available on fuzzy systems.

• Videos:

A number of videos covering fuzzy systems are available from the *IEEE Educational Activities Board* (800 678-IEEE). A wonderful video conference proceedings covering the *1993 International Conference on Fuzzy Systems* (FUZZ-IEEE) and the *1993 IEEE International Conference on Neural Networks* is **Fuzzy Logic and Neural Networks: Clips from the Field (FUZZ-IEEE '93)**, San Francisco, March 1993.

It nicely covers a number of fascinating applications of fuzzy systems including self tuning fuzzy systems, and helicopter and robotics control.

Also available from EAB is a series of topical tutorials. Each runs about two hours.

- ○ **Introduction to Fuzzy Set Theory and Fuzzy Logic: Basic Concepts and Structures** by Enrique Ruspini, SRI
- ○ **International Fuzzy Logic: Advanced Concepts and Structures,** Lotfi Zadeh, UC/Berkeley
- ○ **Information Processing With Fuzzy Logic,** Pierro Bonissone, General Electric
- ○ **Fuzzy Logic and Neural Networks for Control Systems,** Hamid R. Berenji, NASA Ames Research Center
- ○ **Fuzzy Logic and Neural Networks for Pattern Recognition,** James C. Bezdek, University of West Florida
- ○ **Fuzzy Logic for Neural Networks for Computer Vision,** James Keller, University of Missouri

All of these tapes are sponsored by the *IEEE Neural Networks Council*.

• Journals:

Journals devoted primarily to fuzzy logic include

- ○ *IEEE Transactions on Fuzzy Systems*, IEEE
- ○ *Fuzzy Sets and Systems*, North Holland
- ○ *International Journal of Approximate Reasoning*, Elsevier

- SOFT (Society of Fuzzy Technology) Journal, Japan

Papers dealing with fuzzy systems also commonly appear in

- *IEEE Transactions on Neural Networks*
- *IEEE Transactions on Systems, Man and Cybernetics*

• Books:

As in any new field, a plethora of books on fuzzy systems are available. Below is a partial list. The first few chapters of the book by Klir and Folger provide a particularly good introduction to fuzzy logic. Advanced concepts are built from fundamental axioms. The IEEE Press book by Bezdek contains classical papers in the development of fuzzy pattern recognition. Included are seminal papers by Zadeh and Ruspini. A nice survey of fuzzy systems applications is in Terano, Asai and Sugeno.

- J.C. Bezdek, **Pattern Recognition with Fuzzy Objective Function Algorithms**, Plenum Press, New York, (1981).

- J.C. Bezdek, **Analysis of Fuzzy Information**, CRC Press, Boca Raton, Fla, (1985).

- J.C. Bezdek, editor, **Fuzzy Models for Pattern Recognition**, IEEE Press, (1992).

- D. Dubois and H. Prade, **Fuzzy Sets and Systems, Theory and Application**, Academic Press, New York, (1979).

- A. Kandel, **Fuzzy Mathematical Techniques with Applications**, Addison Wesley, Reading, Mass (1982).

- A. Kaufmann, **Introduction to the Theory of Fuzzy Subsets**, Academic Press, New York, (1975).

- A. Kaufmann and M.M. Gupta, **Introduction to Fuzzy Arithmetic: Theory and Applications**, Von Nostrand Reinholt, New York.

- G.J. Klir & T.A. Folger, **Fuzzy Sets: Uncertainty, and Information**, Prentice Hall, (1988).

- B. Kosko, **Neural Networks and Fuzzy Systems: A Dynamical Systems Approach to Machine Intelligence**, Prentice Hall, (1992).

- D.K.D. Majumder, **Fuzzy Mathematical Approach to Pattern Recognition**, John Wiley, New York, (1968).

- C.V. Negoita, **Expert Systems and Fuzzy Systems**, Benjamin/Cummings, Menlo Park, CA (1985).

- W. Pedrycz, **Fuzzy Control and Fuzzy Systems, 2nd extended ed.,** New York: Wiley, (1993).

- K.J. Schmucker, **Fuzzy Sets, Natural Language Computations, and Risk Analysis,** Computer Science Press, Rockville, MD (1983).

- T. Terano, K. Asai and M. Sugeno, **Fuzzy Systems Theory and its Applications,** Academic Press, (1992).

- H.J. Zimmermann, **Fuzzy Set Theory and Its Applications,** Second Edition, Kluwer Academic Publishers, (1991).

- **Societies and Conferences:**

Fuzzy activities in IEEE are under the *IEEE Neural Networks Council* (NNC). The NNC has, as members, a number of IEEE Societies. Other international fuzzy professional organizations include the *International Fuzzy Systems Association* (IFSA) and the *North American Informations Processing Society* (NAFIPS). Each sponsors a fuzzy conference.

The largest conference devoted solely to fuzzy systems is *The International Conference on Fuzzy Systems* (a.k.a. FUZZ-IEEE). Dates and location of this annual conference are listed in the *IEEE Technical Activities Guide* (TAG) and in the *IEEE Spectrum*.

REFERENCES

° 1. L.A. Zadeh, "Fuzzy Sets", Information and Control, Vol. 8, pp. 338-353 (1965); reprinted in J.C. Bezdek, editor, **Fuzzy Models for Pattern Recognition,** IEEE Press, 1992.

° 2. "On fuzzy control ... and fuzzy reviewing", *IEEE Control Systems,* Vol. 13, no. 3, pp. 5-7 (June 1993).

° 3. E.H. Mamdani, "Twenty Years of Fuzzy Control: Experiences Gained and Lessons Learnt", **IEEE International Conference on Fuzzy Systems** (FUZZ-IEEE) 1993, p. 339-reprinted in this volume.

° 4. Possibly, but one can also unwisely analyze the frequency response of linear time-invariant circuits without the use of complex numbers. Fuzzy logic has the advantage of providing a match between technology design and the way we think and, in this sense, is a technology choice. A more proper question might be, 'can other technology more simply be used to do the same thing fuzzy systems do?' An example was voiced to the author by Enrique Ruspini.

° 5. There is currently much attention being focused on the question of fuzzy system stability. Some recent results can be found in

> L.X. Wang, "Stable adaptive fuzzy control of nonlinear systems", **IEEE Transactions on Fuzzy Systems,** vol. 1, pp. 146-155, (1993).
>
> H. Kang, "Stability and control of fuzzy dynamic systems via cell-state transitions in fuzzy hypercubes", **IEEE Transactions on Fuzzy Systems,** vol. 1, pp. 267-279, (1993).

° 6. R.J.. Marks II, "Intelligence: Artificial Versus Computational", **IEEE Transactions on Neural Networks** (September, 1993).

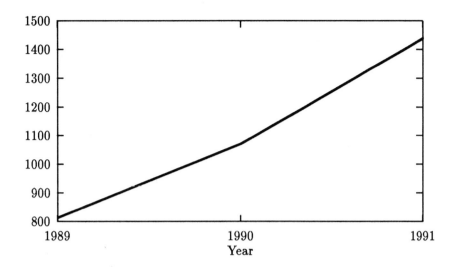

Figure 1: This data was obtained from the INSPEC *(Information Service for Physics and Engineering Communities)* data base compiled by the IEE and the IEEE. The INSPEC data base contains titles, authors and abstracts from over 4000 journals and is augmented with entries of books, reports and conference records. It is focused on the fields of physics, electrical engineering, computer science and electronics. Over one million entries have been logged into INSPEC since 1989. Contents are updated monthly. Searches for key words are performed over titles, authors, journal titles, and abstracts. At this printing, the data base for 1992 was not complete.

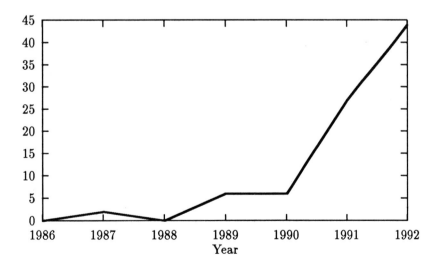

Figure 2: US Patent data was obtained from CASSIS *(Classification for Search Support Information System)*. Searches are also performed over titles and abstracts.

SECTION I: Technology

Chapter 1: Overviews

Editorial
Fuzzy Models—What Are They, and Why?

IT is my great pleasure to welcome you to the inaugural issue of the IEEE TRANSACTIONS ON FUZZY SYSTEMS. Many of you are probably quite familiar with the basic ideas underlying fuzzy sets and systems that utilize fuzzy models. However, there will also be readers who are looking at the contents of this issue, wondering what it's all about. For this latter group, the papers in this first issue may seem bewildering, for they are quite technical, and none are tutorial in nature. Consequently, this preface is divided into two parts. Section I contains a brief introduction to the basic ideas of our science, in hopes that this glimpse will help you understand the articles, but perhaps more importantly, pique your interest in fuzzy models. Section II gives a short précis and historical perspective of the field, and how it has led to the creation of this journal.

I. FUZZY MODELS

Fuzzy sets are a generalization of conventional set theory that were introduced by Zadeh is 1965 as a mathematical way to represent vagueness in everyday life [1]. The basic idea of fuzzy sets is easy to grasp. Suppose, as you approach a red light, you must advise a driving student when to apply the brakes. Would you say, "Begin braking *74 feet* from the crosswalk"? Or would your advice be more like, "Apply the brakes *pretty soon*"? The latter, of course; the former instruction is too precise to be implemented. This illustrates that precision may be quite useless, while vague directions can be interpreted and acted upon. Everyday language is one example of ways vagueness is used and propagated. Children quickly learn how to interpret and implement fuzzy instructions ("go to bed *about* 10"). We all assimilate and use (act on) fuzzy data, vague rules, and imprecise information, just as we are able to make decisions about situations which seem to be governed by an element of chance. Accordingly, computational models of real systems should also be able to recognize, represent, manipulate, interpret, and use (act on) both fuzzy and statistical uncertainties.

Fuzzy interpretations of data structures are a very natural and intuitively plausible way to formulate and solve various problems. Conventional (crisp) sets contain objects that satisfy *precise properties* required for membership. The set of numbers H from 6 to 8 is crisp; we write $H = \{r \in \Re \mid 6 \leq r \leq 8\}$. Equivalently, H is described by its *membership* (or characteristic, or indicator) *function* (MF), $m_H : \Re \mapsto \{0, 1\}$, defined as

$$m_H(r) = \begin{cases} 1; & 6 \leq r \leq 8 \\ 0; & \text{otherwise} \end{cases}$$

The crisp set H and the graph of m_H are shown in the left half of Fig. 1. Every real number (r) either is in H or is not.

Since m_H maps all real numbers $r \in \Re$ onto the two points (0,1), crisp sets correspond to two-valued logic: is or isn't, on or off, black or white, 1 or 0. In logic, values of m_H are called truth values with reference to the question, "Is r in H?" The answer is yes if and only if $m_H(r) = 1$; otherwise, no.

Consider next the set F of real numbers that are *close to 7*. Since the property "close to 7" is fuzzy, there is *not* a *unique* membership function for F. Rather, the modeler must decide, based on the potential application and properties desired for F, what m_F should be. Properties that might seem plausible for this F include (i) normality ($m_F(7) = 1$), (ii) monotonicity (the closer r is to 7, the closer $m_F(r)$ is to 1, and conversely), and (iii) symmetry (numbers equally far left and right of 7 should have equal memberships). Given these intuitive constraints, either of the functions shown in the right half of Fig. 1 might be a useful representation of F. m_{F1} is discrete (the staircase graph), while m_{F2} is continuous but not smooth (the triangle graph). One can easily construct a MF for F so that *every* number has some positive membership in F, but we would not expect numbers "far from 7," 20 000 987 for example, to have much! One of the biggest differences between crisp and fuzzy sets is that the former always have unique MF's, whereas every fuzzy set has an infinite number of MF's that may represent it. This is at once both a weakness and a strength; uniqueness is sacrificed, but this gives a concomitant gain in terms of flexibility, enabling fuzzy models to be "adjusted" for maximum utility in a given situation.

In conventional set theory, sets of real objects, such as the numbers in H, are equivalent to, and isomorphically described by, a unique membership function such as m_H. However, there is no set-theory equivalent of "real objects" corresponding to m_F. Fuzzy sets are always (and only) *functions*, from a "universe of objects," say X, into [0,1]. This is depicted in Fig. 2, which illustrates that the fuzzy set is the *function* m_F that carries X into [0,1].

As defined, *every* function $m : X \mapsto [0, 1]$ is a fuzzy set. While this is true in a formal mathematical sense, many functions that qualify on this ground cannot be suitably interpreted as realizations of a conceptual fuzzy set. In other words, functions that map X into the unit interval *may* be fuzzy sets, but *become* fuzzy sets when, and only when, they match some intuitively plausible semantic description of imprecise properties of the objects in X. There are many good texts and monographs that describe various aspects of fuzzy sets and models; for example, interested readers may consult [2]–[14].

One of the first questions asked about this scheme, and the one that is still asked most often, concerns the relationship of fuzziness to probability. Are fuzzy sets just a clever disguise

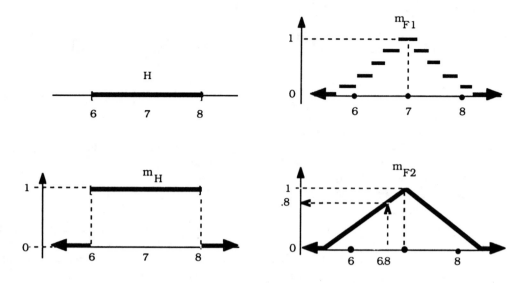

Fig. 1. Membership functions for hard and fuzzy subsets of $\Re$.

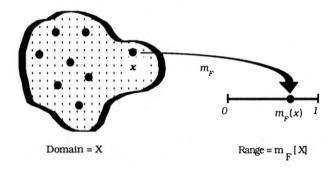

Fig. 2. Fuzzy sets are membership functions.

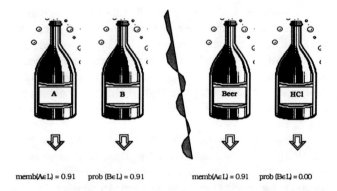

Fig. 3. Bottles for the weary traveler—disguised and unmasked!

for statistical models? Well, in a word, NO. Perhaps an example will help.

Example 1: Let the set of all liquids be the universe of objects, and let fuzzy subset L = {all *potable* (="suitable for drinking") liquids}. Suppose you had been in the desert for a week without drink and you came upon two bottles, A and B, marked as in the left half of Fig. 3 (memb = "membership," and prob = "probability").

Confronted with this pair of bottles and given that you must drink from the one that you choose, which would *you* choose to drink from first? Most readers familiar with the basic ideas of fuzzy sets, when presented with this experiment, immediately see that while A could contain, say, swamp water, it would not (discounting the possibility of a Machiavellian fuzzy modeler) contain liquids such as hydrochloric acid. That is, a *membership* of 0.91 means that the contents of A are "fairly similar" to perfectly potable liquids (pure water). On the other hand, the *probability* that B is potable = 0.91 means that over a long run of experiments, the contents of B are expected to be potable about 91% of the trials; and the other 9%? In these cases the contents will be unsavory (indeed, possibly deadly)—about one chance in ten. Thus most subjects will opt for a chance to drink swamp water, and will choose bottle A.

Another facet of this example concerns the idea of *observation*. Continuing then, suppose that we examine the contents of A and B, and discover them to be as shown in the right half of Fig. 3; that is, A contains beer, while B contains hydrochloric acid. *After* observation then, the membership value for A will be unchanged, whilst the probability value for B clearly drops from 0.91 to 0.0.

Finally what would be the effect of changing the numerical information in this example? Suppose that the membership and probability values were both 0.5—would this influence your choice? Almost certainly it would. In this case many observers would switch to bottle B, since it offers a 50% chance of being drinkable, whereas a membership value this low would presumably indicate a liquid unsuitable for drinking (this depends, of course, entirely on the MF of the fuzzy set L).

In summary, Example 1 shows that these two types of models possess philosophically different kinds of information: fuzzy memberships, which represent similarities of objects to imprecisely defined properties, and probabilities, which convey information about relative frequencies. Moreover, interpretations about and decisions based on these values also depend on the actual numerical magnitudes assigned to particular objects and events. See [15] for an amusing contrary view.

Another common misunderstanding about fuzzy models over the years has been that they were offered as *replacements* for crisp (or probabilistic) models. To expand on this, first note from Figs. 1 and 2 that every crisp set is fuzzy, but not conversely. Most schemes that use the idea of fuzziness use it in this sense of *embedding*; that is, we work at preserving the conventional structure, and letting it dominate the output whenever it can, or whenever it must. Another example will illustrate this idea.

Example 2: Consider the plight of early mathematicians, who knew that the Taylor series for the real (bell-shaped) function $f(x) = 1/(1 + x^2)$ was divergent at $x = \pm 1$ but could not understand why, especially since f is differentiable infinitely often at these two points. As is common knowledge for any student of complex variables nowadays, the *complex* function $f(z) = 1/(1 + z^2)$ has poles at $z = \pm i$, two purely imaginary numbers. Thus, the complex function, which is an embedding of its real antecedent, cannot have a convergent power series expansion anywhere on the boundary of the unit disk in the plane; in particular at $z = \pm 0i \pm 1$, i.e., at the real numbers $x = \pm 1$. This exemplifies a general principle in mathematical modeling: given a real (seemingly insoluble) problem; enlarge the space, and look for a solution in some "imaginary" superset of the real problem; finally, specialize the "imaginary" solution to the original real constraints.

In Example 2 we spoke of "complexifying" the function f by embedding the real numbers in the complex plane, followed by "decomplexification" of the more general result to solve the original problem. Most fuzzy models follow a very similar pattern. Real problems that exhibit nonstatistical uncertainty are first "fuzzified," some type of analysis is done on the larger problem, and then the results are specialized back to the original problem. In Example 2 we might call the return to the real line decomplexifying the function; in fuzzy models, this part of the procedure has come to be known as defuzzification. Defuzzification is usually necessary, of course, because even though we instruct a student to "apply the brakes pretty soon," in fact, the brake pedal must be operated crisply, at some real time. In other words, we cannot admonish a motor to "speed up a little," even if this instruction comes from a fuzzy controller—we must alter its voltage by a specific amount. Thus defuzzification is both natural and necessary. Example 2 illustrates that this is hardly an idea that is novel; instead, we should regard it as a device that is useful.

Example 3: As a last, and perhaps more concrete, example about the use of fuzzy models, consider the system shown in Fig. 4, which depicts a simple inverted pendulum free to rotate in the plane of the figure on a pivot attached to the cart. The control problem is to keep the pendulum vertical at all times by applying a restoring force (control signal) $F(t)$ to the cart at some discrete times (t) in response to changes in both the linear and angular position $(x(t), \theta(t))$ and velocity $(\dot{x}(t), \dot{\theta}(t))$ of the pendulum. This problem can be formulated many ways. In one of the simpler versions used in conventional control theory, linearization of the equations

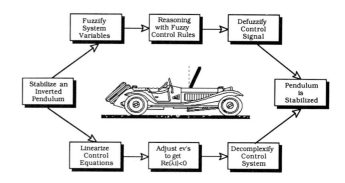

Fig. 4. Conventional and fuzzy solutions to real control problems found by embedding them in "imaginary" supersets.

of motion results in a model of the system whose stability characteristics are determined by examination of the *real parts* of the eigenvalues $\{\lambda_i\}$ of a 4×4 matrix of system constants. The lower track in Fig. 4 represents this case. It is well known that the pendulum can be stabilized by requiring $\text{Re}(\lambda_i) < 0$, as shown in the middle of the lower track. This procedure is so commonplace in control engineering that most designers don't even think about the use of imaginary numbers to solve real problems, but it is clear that this process is exactly the same as was illustrated in Example 2—a real problem is solved by temporarily passing to a larger, imaginary setting, analyzing the situation in the superset, and then specializing the result to get the desired answer.

The upper track in Fig. 4 depicts an alternative solution to this control problem that is based on fuzzy sets. This approach to stabilization of the pendulum is also well known, and yields a solution that in some ways is much better; e.g., the fuzzy controller is much less sensitive to changes in parameters such as the length and mass of the pendulum [16]. Note again the embedding principle: fuzzify, solve, defuzzify, control.

The point of Example 3? Fuzzy models aren't really that different from more familiar ones. Sometimes they work better, and sometimes not. This is really the only criterion that should be used to judge any model, and there is much evidence nowadays that fuzzy approaches to real problems are often a good alternative to more familiar schemes. This is the point to which our discussion now turns.

II. NOTES ON THE EVOLUTION OF FUZZY MODELS AND THE IEEE TRANSACTIONS ON FUZZY SYSTEMS

Why an IEEE TRANSACTIONS ON FUZZY SYSTEMS? And further, why done under the aegis of the IEEE Neural Networks Council (NNC)? There are several compelling answers. First, the enormous success of commercial applications which are at least partially dependent on fuzzy technologies fielded (in the main) by Japanese companies has led to a surge of curiosity about the utility of fuzzy logic for scientific and engineering applications. Over the last five or ten years fuzzy models have supplanted more conventional technologies in many scientific applications and engineering systems, especially in control systems and pattern recognition. A recent *Newsweek* article indicates that the Japanese now hold thousands of patents on fuzzy devices used in applications as

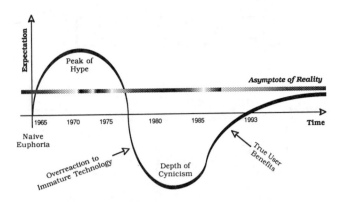

Fig. 5. Evolution of new technologies—time axis specialized to fuzzy models.

diverse as washing machines, TV camcorders, air conditioners, palm-top computers, vacuum cleaners, ship navigators, subway train controllers, and automobile transmissions [17]. It is this wealth of deployed, successful applications of fuzzy technology that is, in the main, responsible for current interest in the subject area.

Since 1965, many authors have generalized various parts of subdisciplines in mathematics, science, and engineering to include fuzzy cases. However, interest in fuzzy models was not really very widespread until their utility in fielded applications became apparent. The reasons for this delay in interest are many, but perhaps the most accurate explanation lies with the salient facts underlying the development of any new technology, which is succintly captured in Fig. 5.

The horizontal axis if Fig. 5 is time, and the vertical axis is expectation—whose expectation? Well, usually, of the people who pay for development of the technology; but here I encourage you to interpret this axis in a much broader sense, for utility is, of course, in the eye of the user. The crucial part of Fig. 5 is the *asymptote of reality*, which bounds the delivery of the technology to a much lower expected value than early users project for it. The years shown along the time axis pertain to fuzzy models, and are, of course, approximate at best (with the exception of the initial one). When you look at this figure, you may enjoy deleting these years, and substituting your favorite new technology for the one illustrated. Each technology has its own evolution, and not all of them follow the pattern suggested by Fig. 5 (but you may be surprised to see how many do!). For example, try putting dates and identifying the people and events associated with, say, computational neural networks (which has an atypical, bimodal graph!); or artificial intelligence; or fractals; or complex numbers; and so on.

Every new technology begins with naive euphoria—its inventor(s) are usually submersed in the ideas themselves; it is their immediate colleagues that experience most of the wild enthusiasm. Most technologies are overpromised, more often than not simply to generate funds to continue the work, for funding is an integral part of scientific development; without it, only the most imaginative and revolutionary ideas make it beyond the embryonic stage. Hype is a natural handmaiden to overpromise, and most technologies build rapidly to a peak of hype. Following this, there is almost always an overreaction to ideas that are not fully developed, and this inevitably leads to a crash of sorts, followed by a period of wallowing in the depths of cynicism. Many new technologies evolve to this point, and then fade away. The ones that survive do so because someone finds a good use (= true user benefit) for the basic ideas. What constitutes a "good use"? For example, there are now many "good uses" in real systems for the complex numbers, as we have seen in Examples 2 and 3, but not many mathematicians thought so when Wessel, Argand, Hamilton, and Gauss made imaginary numbers sensible from a geometric point of view in the later 1800's. And in the context of fuzzy models, of course, "good use" corresponds to the plethora of products alluded to above.

Interest in fuzzy systems in academia, industry, and government is also manifested by the rapid growth of national and international conferences. A short list of conferences includes NAFIPS (North American Fuzzy Information Processing Society), annual, 1982–present; IFSA (International Fuzzy Systems Association), biannual, 1985–present; NASA, Fuzzy Logic and Neural Networks, biannual, 1988–present; MCC, Industrial Conference on Fuzzy Systems, 1992; and, finally, the IEEE International Conference on Fuzzy Systems, annual 1992–present. This latter set of conferences is organized and sponsored by the IEEE NNC.

As noted above, successful applications of fuzzy models have gained great visibility through commercial applications in Japan. MITI in Japan started LIFE (Laboratory of Industrial Fuzzy Engineering) in 1988 with an annual budget of about $24 000 000 (U.S. dollars) for seven years. Japan now has a very large and active professional society (SOFT, the Japanese Society for Fuzzy Theories), which publishes a technical journal in Japanese. Other professional societies include IFSA (International Fuzzy Systems Association); NAFIPS (North American Information Processing Society); and societies in Korea, India, and China that have an estimated combined membership of about 10 000. Most of the European countries also have very active societies, and there are several journals and newsletters associated with various organizations throughout Europe and Asia.

And what about neural nets—why the NNC? There has been, in the last five years, a large and energetic upswing in research efforts aimed at synthesizing fuzzy logic with computational neural networks (CNN's). There are several reasons for this. The marriage of fuzzy logic with CNN's has a sound technical basis, because these two approaches generally attack the design of "intelligent" systems from quite different angles. CNN's are essentially low-level, computational algorithms that (sometimes) offer good performance in dealing with sensor data used in pattern recognition and control. On the other hand, fuzzy logic is a means for representing, manipulating, and utilizing data and information that possess nonstatistical uncertainty. Thus, fuzzy methods often deal with issues such as reasoning on a higher (semantic or linguistic) level than CNN's. Consequently, the two technologies often complement each other, CNN's supplying the brute force necessary to accommodate and interpret large amounts of sensor data; and fuzzy logic providing a structural framework that utilizes and exploits these low-level results. There are also many ways to use either technology as a "tool" within the framework of a model based on the other. For example, the CNN is well known for its ability to represent functions. The basis of every fuzzy model is the membership function. So, a natural application of CNN's in fuzzy models is to provide good approximations to the membership functions that are essential to the success of any fuzzy approach. Broadly speaking, then we may characterize efforts at merging these two technologies as (i) fuzzification of conventional CNN architectures and models and (ii) the use of CNN's as tools in fuzzy models. One need look no further than the September 1992 issue of the IEEE TRANSACTIONS ON NEURAL NETWORKS to see evidence of this marriage, which is a special issue of TNN containing 19 papers on precisely this topic. References [18]–[24] are a sampler of books and articles that articulate or illustrate various aspects of this evolving relationship.

There are several major journals devoted to fuzzy systems: the *SOFT Journal* (Japan), *Fuzzy Sets and Systems* (North Holland), and the *International Journal of Approximate Reasoning* (Elsevier). It is appropriate that the IEEE create a flagship publication in the area of fuzzy systems, to collect and publish the best research on this important new technology. The IEEE TRANSACTIONS ON FUZZY SYSTEMS will publish only the highest quality technical papers in the theory, design, and application of fuzzy sets and systems that use them. Readers are encouraged to submit papers which disclose significant technical knowledge, exploratory developments, and applications of fuzzy systems. Emphasis will be given to engineering systems and scientific applications. The TRANSACTIONS will also contain a letters section, which will include information of current interest, as well as comments and rebuttals submitted in connection with published papers. Representative applications areas include, but are not limited to, the following aspects of fuzzy systems:

1) Fuzzy estimation, prediction, and control
2) Approximate reasoning
3) Intelligent systems design
4) Machine learning
5) Image processing and machine vision
6) Pattern recognition
7) Fuzzy neurocomputing
8) Electronic and photonic implementations
9) Medical computing applications
10) Robotics and motion control
11) Constraint propagation and optimization
12) Civil, chemical, and industrial engineering applications

Well, this has been a much longer preface than most journals ever contain! All that is left to do is thank everyone who has helped make the TRANSACTIONS a reality. First and foremost, that includes the thousands of researchers who have developed the field to a point where this journal is well justified. There are, of course, far too many people who had an active hand in starting TFS for me to recognize each one individually. However, it is appropriate to say that what has been done would have been quite impossible without the able and professional help of the people on (and behind) the various publication boards and NNC committees that helped mold the journal into a reality. A few persons should be specifically mentioned. First, Bob Marks, Pat Simpson, Russ Eberhart, and Toshio Fukuda, who had the vision to lead the IEEE Neural Networks Council toward its decision to sponsor this journal. Second, the associate editors and advisory board, who really did almost all of the hard work in getting this first issue to press, Without them, none of this would have been possible. And, finally, Chris Ralston and the staff at IEEE Publishing Services should be credited for accounting for many of the tedious details that go unnoticed when things work.

I hope you enjoy reading this inaugural issue, and that you find its contents useful and illuminating. Your suggestions on how to make this journal more valuable for the academic, industrial, and governmental communities are both welcome and appreciated—please let me know how we can improve it.

Jim Bezdek, *Founding Editor*
February 1993

REFERENCES

[1] L. A. Zadeh, "Fuzzy sets," *Information and Control*, vol. 8, pp. 338–352, 1965.
[2] G. Klir, G. Folger, and T. Folger, *Uncertainty and Information.* Englewood Cliffs, NJ: Prentice-Hall, 1988.
[3] D. Dubois and H. Prade, *Fuzzy Sets and Systems: Theory and Applications.* New York: Academic Press, 1980.
[4] H. Zimmermann, *Fuzzy Set Theory—and its Applications*, 2nd ed. Boston: Kluwer, 1990.
[5] A. Kaufmann and M. Gupta, *Introduction to Fuzzy Arithmetic: Theory and Applications.* New York: Van Nostrand Reinhold, 1985.
[6] K. Schmucker, *Fuzzy Sets, Natural Language and Computation.* Rockville, MD: Computer Science Press, 1984.
[7] V. Novak, *Fuzzy Sets and Their Applications.* Bristol: Adam Hilger, 1986.
[8] M. Smithson, *Fuzzy Sets Analysis for Behavioural and Social Sciences.* New York: Springer-Verlag, 1986.
[9] A. Kandel, *Fuzzy Mathematical Techniques with Applications.* Reading, MA: Addison-Wesley, 1986.
[10] J. C. Bezdek, *Pattern Recognition with Fuzzy Objective Function Algorithms.* New York: Plenum, 1981.
[11] A. Kandel, *Fuzzy Techniques in Pattern Recognition.* New York: Wiley Interscience, 1982.
[12] R. Di Mori, *Computerized Models of Speech Using Fuzzy Algorithms.* New York, Plenum, 1983
[13] S. K. Pal and D. K. Dutta Majumder, *Fuzzy Mathematical Approach to Pattern Recognition.* New York: Wiley, 1986.
[14] J. C. Bezdek and S. K. Pal, *Fuzzy Models for Pattern Recognition.* Piscataway, NJ: IEEE Press, 1992.
[15] P. Cheeseman, "An inquiry into computer understanding," *Comp. Intell.*, vol. 4, pp. 57–142, 1988 (with 22 commentaries/replies).
[16] C. C. Lee, "Fuzzy logic in control systems: Fuzzy logic controllers" (parts I and II), *IEEE Trans. Syst., Man, Cybernet.*, vol. 20, no. 2, pp. 404–435, 1990.
[17] "The future is fuzzy," *Newsweek*, May 1990.
[18] Y. H. Pao, *Adaptive Pattern Recognition and Neural Networks.* Reading, MA: Addison-Wesley, 1989.
[19] B. Kosko, *Neural Networks and Fuzzy Systems: A Dynamical Approach to Machine Intelligence.* Englewood Cliffs, NJ: Prentice-Hall, 1991.
[20] J. M. Keller and D. J. Hunt, "Incorporating fuzzy membership functions into the perceptron algorithm," *IEEE Trans. Pattern Anal. Mach. Intell.*, vol. 1, pp. 693–699, 1985.
[21] L. Hall et al., "A comparison of neural network and fuzzy clustering techniques in segmenting magnetic resonance images of the brain," *IEEE Trans. Neural Networks*, vol. 3, pp. 672–682, Sept. 1992.
[22] G. A. Carpenter, S. Grossberg, and D. B. Rosen, "Fuzzy ART: Fast stable learning and categorization of analog patterns by an adaptive resonance system," *Neural Networks*, vol. 4, no. 6, pp. 759–772, 1991.
[23] P. Werbos, "Neurocontrol and fuzzy logic—Connections and designs," *Int. J. Approx. Reasoning*, vol. 6, no. 2, pp. 185–200, 1992.
[24] J. C. Bezdek, "Computing with uncertainty," *IEEE Communications Magazine*, vol. 30, no. 9, pp. 24–36, 1992.

APPLICATION OF NEURAL NETWORKS AND FUZZY LOGIC TO CONSUMER PRODUCTS [1]

Hideyuki TAKAGI *
Central Research Laboratories,
Matsushita Electric Industrial Co., Ltd.

Abstract

This paper describes how neural networks and fuzzy logic have been applied to consumer products. First, the background behind why both technologies have been applied to this field is described. Second, briefly overview of the fusion technology of neural networks and fuzzy logic is given. As a good example of the R&D process, the application of neural nets to the design and tuning of fuzzy system is introduced. In Sections 4 and 5, applications of both technologies are categorized into four cases. They are: (1) neural networks being used to automate the task of designing and fine-tuning the membership functions of fuzzy systems, (2) both fuzzy inference and neural network learning capabilities provided separately, (3) neural networks work as correcting mechanisms for fuzzy system, (4) neural networks cascaded (serially) with fuzzy systems, and (5) neural networks used to customize the standard system according to each user's preferences and individual needs. Finally, the new trend that aims at the realization of adaptive systems for the user is discussed. As examples of the trend, four consumer products that apply learning capability of neural networks for the user are introduced.

1 Introduction

1.1 Background: Applying Fuzzy Logic

The use of fuzzy technology has rapidly spread in the realm of consumer product design in order to satisfy the following requirements: (1) to develop control systems with nonlinear characteristics and decision-making systems for controllers, (2) to cope with an increasing number of sensors and exploit the larger quantity of information, (3) to reduce development time, (4) to reduce costs associated with incorporating the technology into the product. Fuzzy technology can satisfy these requirements for the following reasons.

Nonlinear characteristics are realized in fuzzy logic by partitioning the rule space, by weighting the rules, and by the nonlinear membership functions. Rule-based systems compute their output by combining results from different parts of the partition, each part being governed by separate rules. In fuzzy reasoning, the boundaries of these parts overlap, and the local results are combined by weighting them appropriately. That is why the output of a fuzzy system is a smooth, nonlinear function.

In decision-making systems, the target of modeling is not a control surface but the person whose decision-making is to be emulated. This kind of modeling is outside the realm of conventional control theory. Fuzzy reasoning can tackle this easily since it can handle qualitative knowledge (e.g. linguistic terms like 'big' and 'fast', and rules of thumb) directly. In most applications to consumer products, fuzzy systems do not directly control the actuators, but determine the parameters to be used for control. For example, they may determine washing time in washing machines, or if it is is the hand or the image that is shaking in a camcoder, or they compute which object is supposed to be the focus in an autofocus system, or they determine the contrast optimal for watching television.

A fuzzy system encodes knowledge at two levels : knowledge which incorporates fuzzy heuristics, and the knowledge that defines the terms being used in the former level. Due to this separation of meaning, it is possible to directly encode linguistic rules and heuristics. This reduces the development time since the expert's knowledge can be directly built in.

Although the developed fuzzy system may have complex input-output characteristics, as long as the mapping is static during the operation of the device, the mapping can be discretized and implemented as a memory lookup on simple hardware. This reduces the costs involved in incorporating the knowledge into the device.

1.2 Background: Combination with Neural Networks

One reason for incorporating neural networks into the above process is that the Japanese consumer strongly requires more intelligent and more sensitive appliances with finer capabilities. Manufacturers address this by increasing the number of sensors and the amount of information available to the device. So fuzzy technology is a natural choice for solving this problem.

However, this increase in information leads to a higher complexity of design that cannot be fully tackled by fuzzy reasoning. Neural networks are being introduced to speed up the development of a complex rulebased system. Also, knowledge that cannot be made explicit can be handled by these nets. The first product whose development used neural networks appeared on the market in December 1990. Moreover, there is a further need to have adaptive systems which can be tailored to the user's needs and preferences. Neural networks are expected to be one way to realize this learning capability. This is why both the methodologies of neural nets and fuzzy logic have been combined recently.

2 Status of Fusion Research of Neural Networks and Fuzzy Logic

2.1 Overview

The earliest research in fusing neural networks and fuzzy logic started in 1974. The main thrust of this research was to use fuzzy logic in neurophysiology. Fuzzy languages, fuzzy entropy, and fuzzy automata were used to model neurons and hearing-vestibular

---

*The author is a Visiting Industrial Fellow at the University of California, Computer Science Division, Berkeley, CA 94720 USA. Tel: +1-510-642-8015, Fax: +1-510-642-5775, e-mail: takagi@cs.berkeley.edu

[1] Sections 4 and 5 will appear in [1] as the selected paper from the First Int'l Workshop on Industrial Applications of Fuzzy Control and Intelligent Systems, which was held on November 21-22, 1991 in Texas.

nerves and to analyze the nervous system. This has been largely abandoned now.

In the late eighties, especially in 1988, the fusion of neural nets and fuzzy logic began anew. Research has increased dramatically during the nineties. Some of the issues addressed in this research are automatic design and tuning of membership functions, knowledge acquisition and representation, fuzzy cognitive maps, clustering, pattern recognition etc. [2]. From the viewpoint of practical relevance and the quantity of research, the first of these issues is the major one.

2.2 Design Support of Fuzzy Systems

In this section, we discuss the developmental path from research in design support to concrete products.

One of the features of fuzzy logic is that it separates the logic from the fuzziness of the terms used. The logic is usually expressed by if-then type of rules, and the fuzziness is captured by defining suitable membership functions for the terms used. Therefore, it is easy to tackle knowledge or skills which can be expressed as rules using qualitative (or linguistic) terms. On the other hand, designing the membership functions (the meaning of the terms) remained difficult and had to be done by trial and error in order to optimize performance. The first research that addressed this difficulty was "NN-driven Fuzzy Reasoning" [3] [4].

First, NN-driven fuzzy reasoning determines the number of rules by clustering the data for designing fuzzy systems. Using this clustered data, a neural network decides on a multidimensional, nonlinear membership function, and this network is then used as a generator of the membership function. One contribution of this research was to introduce neural networks into the design process of fuzzy systems. Secondly, this membership function is designed completely at one stroke, rather than separately along each input axis.

NN-driven fuzzy reasoning can acquire the skill from a human expert automatically. This was demonstrated by the swing-up pole-balancing system designed using this approach [5]. First, a human manually controlled the cart so that the pole was swung up and kept balanced. The demonstration data so collected implicitly included the knowledge of the expert. Extracting this knowledge is difficult from the conventional knowledge engineering standpoint. NN-driven fuzzy reasoning automatically acquired the knowledge, and resulted in a fuzzy system which could duplicate the skill. From a practical viewpoint, this was an important achievement because it reduced the design time of a fuzzy system by using neural networks.

For practical feasibility, the next step is to further reduce the development time. Since designing membership functions is the central theme of the idea and takes the most time, a simplification was proposed to reduce this time [6]. Here, the earlier multidimensional function was decomposed into several one-dimensional functions, which were much easier to design. The error between the designed fuzzy system and the actual data depends on the parameters which define these one-dimensional membership functions. These parameters can then be tuned to minimize the error, in a manner similar to backpropagation learning of weights in a neural network.

The fuzzy system designed as above was applied to the operation of a washing machine, which appeared on the market in early 1991. Following this, similar ideas were used for other kinds of equipment. Details of such techniques and products are presented in Section 4.

Matsushita Electric	air conditioner washing machine vacuum cleaner rice cooker kerosene fan heater electric thermo pot microwave oven induction heating cooker
Sanyo	washing machine cloth drier microwave oven desk type electric heater electric carpet electric fan kerosene fan heater rice cooker
Hitachi	rice cooker kerosene fan heater washing machine vacuum cleaner
Sharp	refrigerator washing machine rice cooker kerosene fan heater forced-flue kerosene fan heater
Mitsubishi	kerosene fan heater induction heating cooker
Toshiba	kerosene fan heater washing machine
Fujitsu General	kerosene fan heater
Corona	kerosene fan heater
Toyotomi	kerosene fan heater

Table 1: Consumer products in Japan using Neural Networks technology (as of September 20, 1991)

3 Overview of Applications of Neural Networks and Fuzzy Logic

In the context of consumer products, neural networks and fuzzy logic have been put to use in the following ways.

1. Neural Nets as Development Tools for Fuzzy Systems
2. Independent Use of Neural Nets and Fuzzy Logic
3. Neural Nets as Correcting Mechanisms for Fuzzy System
4. Cascade Combination of Neural Nets and Fuzzy System
5. Learning User Preferences.

The last way above is a recent trend which uses neural nets to customize the standard system according to each user's preferences and individual needs. The learning capability is being used for adaptation here.

As of September 1991, there were 14 consumer products using neural networks and fuzzy logic in the Japanese market. These are listed in Table 1.

Besides Japan, Korean companies have also been applying these technologies in some of their products. The Korean exhibition [7] displayed an air-conditioner (by Goldstar), washing machines (by Goldstar, Samsun and Daewoo), a kerosene fan heater (by Samsun), and an electric microwave oven range (by Goldstar).

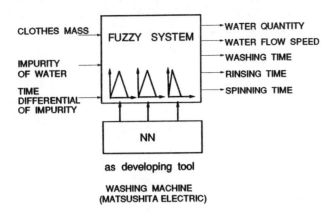

Figure 1: A neural net determines membership functions.

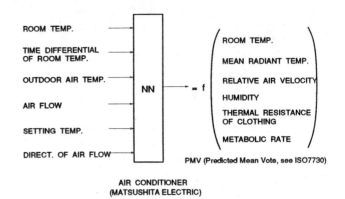

Figure 2: Non-linear mapping from 6-D space to 6-D space

4 Non-Consumer-Trainable Applications

4.1 Development Tools

Fuzzy logic can encode expert knowledge directly using rules. This uses linguistic (qualitative) labels. However, the precise (quantitative) definitions of these linguistic labels are difficult to obtain, and are often tuned manually, depending on performance. Neural network technology can automate this process of determining the membership function parameters, and can substantially reduce development time while increasing performance. Initial research [3] [4] to this end used partitioning in multi-dimensional spaces. Alternatively, one can use simple-shapes (e.g. triangles) for membership functions, and partition each dimension separately [6]. The latter trades off performance in exchange for faster training.

Consider the inference system in a washing machine, which is shown in Figure 1. It uses three inputs, so a fuzzy partitioning of each of the three one-dimensional input spaces has to be done, and triangular shapes are used to do this. For example, the input variable *Clothes-Mass* can take the values *light*, *medium* and *heavy*. So three triangles are used to partition the one-dimensional space for this input variable. A similar process is used on the other inputs.

Performance will depend on how this partitioning is done. In other words, the shape and positions of these triangles will affect the output of the system. The partitioning is completely specified by the centers of the triangles and their widths (they are symmetric about their centers). Since performance depends on the parameters of the partitioning, an optimization method is used to determine these so that performance is maximized. Both the fuzzy inference and the optimization process is similar to the way a neural net operates.

Such an approach is being applied to the development of consumer products by the companies of the Matsushita Electric Group.

4.2 Independent Type

This approach applies both neural networks and fuzzy logic, but in an unrelated fashion.

One characterization of bodily comfort is that the heat produced by our bodies is absorbed by the surroundings at the same rate, so that we do not accumulate heat or feel cold. The International Standard Organization defines PMV (Predictive Mean Vote) [8] in ISO-7730 as an index of comfort. This is a mathematical function of six input variables, shown on the right side in Figure 4.2. If we can compute this function's value, then we can control the air-conditioning so as to achieve maximum comfort levels.

However, some of these 6 input variables such as metabolic rate and the thermal resistance of the user's clothing, are difficult to detect by using sensors. Moreover, it would be extremely user-unfriendly to ask the user to input these values periodically. In contrast, the variables shown on the left of Figure 4.2 are easy to calculate using sensors.

If we can discover and implement the mapping from the 6 variables on the left to the 6 on the right (a six-dimension to six-dimension mapping), we can estimate the value f of the PMV index using sensor inputs only. This mapping introduces some error since the 6 variables on the left do not completely capture the information provided by the right-hand-side variables. As long as this error in f is within the tolerances allowed for it, we can still use this mapping as a practical method.

The problem now is to discover and calculate this nonlinear relationship between the two sets of variables. Neural nets can handle such a task well, if they are provided with training data (samples of input-output pairs). The input data is the vector of 6 variables on the left and is sensed. The values of the six on the right are calculated manually (this is the desired output corresponding to the input data). Then the room environment is changed to get another sample (this yields another input-output data pair). Several such samples are used to train the network, which learns the relationship between one set of variables and the other. Now we have a way of finding the value of the PMV index (output of the neural net), and this is used to control the air-conditioner [9]. This is a nice application of the nets' mapping ability, and is being used in the air-conditioners marketed by Matsushita Electric Company.

4.3 Corrector Type

The application of fuzzy systems technology to Japanese consumer products began in 1990. The next phase involved incorporating more sensor data so as to make the control smoother, sensitive and more accurate. This complicates the task as the input space increases in dimension. In this approach, the neural net handles the larger set of sensor inputs and corrects the output of the fuzzy system (which was designed earlier for the old set of inputs). A complete redesigning of the fuzzy system is thus avoided. This causes a substantial saving in development time (and cost), as redesigning the membership functions becomes more difficult as the number of inputs increases.

Figure 3 shows the schematic underlying the Hitachi washing machine [10]. The fuzzy system shown in the upper part was part of the first model. Later, an improved model incorporated extra information by using a neural net as shown. The neural net pro-

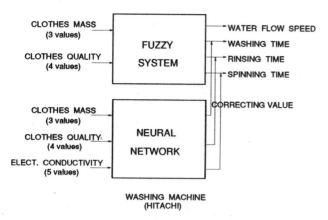

Figure 3: A neural net corrects the output of a fuzzy system

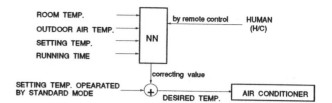

Figure 4: A user-trained network modifies a standard control

vided corrections to the values output by the fuzzy controller. The additional input (fed only to the net) is Electrical Conductivity, which is used to measure the opacity/transparency of the water. To train the neural net, the desired correction is used. This value is the difference between the desired output and what the fuzzy system outputs. Sanyo uses a similar approach in its washing machine, although some of the inputs/outputs are different [11].

Sanyo has also applied similar methods in the design of its conventional-cum-microwave oven [12]. In the older model, the fuzzy system used the temperature and quantity of the food, and the user-set heater temperature, to compute the heating time, power level to be used, and the boiling point of the food. The newer version eliminates the variation in heating due to changes in the temperature outside the oven, since seasonal changes in temperature do affect the final food temperature slightly. The extra input can be incorporated using a correcting neural network.

4.4 Cascade Combination Type

Another way to combine the two technologies is to connect them up serially, where one does a part of the task and then passes its result to the other. In the Sanyo electric fan [13], the fan must rotate towards where the user is, and this requires calculating the direction of the remote controller.

Three infrared sensors in the fan's body detect the strengths of the signal from the remote controller. This information is to be used to compute the direction of the remote control. First, the distance to the remote is calculated by a fuzzy system; then this distance figure and the ratios of sensor outputs are used by a neural network to compute the required direction. The latter calculation is done by a neural net because neither mathematical models nor fuzzy reasoning were successful for doing this computation. The final product has an error of $\pm 4°$ as opposed to the $\pm 10°$ error of statistical regression methods [14].

5 Consumer-Trainable Neural Networks

In the examples given in the previous section, all the neural networks were trained during the design phase. They were then implemented and built into the product, such that the user did not have the option to train them any more. Now we consider products in which the end-user can exploit the trainability of these nets. Of the four products described below, information about the last two may be incomplete because their news releases did not reveal detailed technical information regarding the learning capability.

5.1 Reducing Pre-Heating Time

A product known as a kerosene fan heater is used extensively in Japan to heat the house. The burner burns vaporized oil and this requires pre-heating the burner before it can start heating its surroundings. This pre-heating phase takes considerable time. It would be advantageous if the heater could predict when it was going to be turned on, so that it could start pre-heating accordingly. This prediction depends on the usage pattern which varies from one home to another. Sanyo uses a neural network to learn this usage pattern [15].

Earlier models of this device used a strong, separate heater (600 Watts) to speed up the pre-heating. The model using the neural network to learn the user's pattern took the same time to pre-heat, but reduced the energy consumption to a half, because of its predictive ability.

The on/off switching of the pre-heating is controlled by a conventional 3-layer feed-forward neural network with 5 hidden units. The three inputs to it are current time (CT), the time elapsed since switching it on (ΔT), and the previous day's lowest temperature (LT). CT is required to distinguish usage variation between mornings and evenings. ΔT affects heating power required in the future. Seasonal variation is captured by LT.

Since users can train the network, they may misuse it, causing dangerous accidents. Therefore the manufacturer must pay special attention to consumer safety in the face of misuse. This applies especially to products like the fan heater, which uses combustion. Sanyo has avoided this risk by letting the neural net control only the pre-heating, but not the firing. In the worst case, bad training samples may cause higher energy consumption or inconvenience (longer heating time), but there will be no dangerous consequences of bad training.

This Sanyo heater also uses a fuzzy controller for regulating temperature, and another fuzzy rulebase to estimate the room size.

5.2 Correction According to Personal Preference

The air-conditioner described in Section 4.2 has been augmented by a neural network to further fine-tune the equipment to the user's personal preferences. In the older version, one neural net computed the nonlinear mapping as described before, but this may not be the same as the optimal value for that user. This output is now corrected by a signal produced by the second network, which has been trained by the user. The corrected value is then used in the control of the air-conditioner. The second (user-trained) net uses four inputs from sensors — room temperature, outdoor temperature, temperature setting, and running time of the air-conditioner. In addition, the user pushes +/− buttons on the remote control to make adjustments for personal comfort. These are an indication of his/her preferences and act as desired output values in supervised training. The output is a correction applied to the output of the first network. See Figure 4 for the schematic diagram.

Standard learning algorithms such as backpropagation are time-consuming and may use too much memory, so the above product uses newer and more simplified methods for learning [16]. Current and future trends require the development of new and simpler models to suit the product specifications.

5.3 Adjusting the Control Program to the User's Environment

In the case of the fan heater, the characteristics of the space to be heated up, such as size and material (wood or concrete) affect the time needed to heat it up. Ordinarily, the user sets a firing time, keeping in mind the time to warm up. But in the new Sharp heater, the user just has to specify when the room should be warm. The manufacturer cannot set a predesignated time to start pre-heating or firing, since that would depend on the room being warmed up. The factory sets some average standard time only, which is modified by a neural net. Since such equipment (once bought) is generally used in the same room, the neural net learns the heating time by observing it the first few times, and then controls the switching accordingly [17]. So differences in the temperature curve due to room material and size are compensated for.

5.4 Predicting Pre-Cooling Time

Another instance of learning usage patterns is found in a refrigerator marketed by Sharp [18]. A higher frequency of door opening/closing of the refrigerator door causes an increase in temperature inside and causes a degradation in the quality of frozen food, which must be kept below $-18°C$. The food still remains frozen, so it is hard to detect the degradation in quality. The refrigerator therefore tries to predict the time when the user increases frequency of door openings, and compensates by pre-cooling the insides. Both neural nets and fuzzy technology are used to solve this problem.

6 Requirement for Future Systems

The introduction of fuzzy and neural technology into consumer products from 1990 onwards has led to the introduction of user-trainable networks. The trend is evolving towards more user-friendly and intelligent equipment. The adaptive ability of neural networks makes them invaluable for this purpose.

Having learning functionality in the product is a major improvement, but it alone is not sufficient. Consumer products are targeted towards a wide range of non-specialized users, so incorrect training by them may lead to deterioration in performance, and in the worst case, may be dangerous. Therefore, better safety precautions also need to be built in. One way to handle this requirement is to let the logic ensure strict safety, and let the neural network optimize performance within these safe limits.

One of the realization is to use both implicit and explicit knowledge. It is easy for explicit knowledge to describe rules for the safety of system. The implicit knowledge of neural networks can handle the adaptability for the user. Future advances would involve combining this with high-level cognitive processing, leading to flexible knowledge processing involving fuzzy logic, neural nets, and conventional AI, paving the path to truly intelligent systems.

References

[1] H. Takagi, "Cooperative system of neural networks and fuzzy logic and its application to consumer products", (ed. J. Yen and R. Langari) Industrial Applications of Fuzzy Control and Intelligent Systems, Van Nostrand Reinhold (will be published in 1993)

[2] H. Takagi, "Fusion technology of fuzzy theory and neural networks — survey and future directions", 1st Int'l Conf. Fuzzy Logic & Neural Networks (IIZUKA'90), pp.13-26 (July, 1990)

[3] H.Takagi & I.Hayashi, "Artificial_neural_network-driven fuzzy reasoning", Int'l Workshop on Fuzzy System Applications (IIZUKA'88), pp.217-218 (Aug., 1988)

[4] H.Takagi & I.Hayashi, "NN-driven fuzzy reasoning", Int'l J. Approximate Reasoning (Special Issue of IIZUKA'88), Vol. 5, No.3, pp.191-212 (1991)

[5] I. Hayashi, H. Nomura and N. Wakami, "Artificial_neural_network-driven fuzzy control and its application to the learning of inverted pendulum system", 3rd IFSA Congress, pp.610-613 (Aug. 1989)

[6] H. Nomura, I. Hayashi, and N. Wakami, "A self-tuning method of fuzzy control by descent method", 4th IFSA Congress, Vol. Engineering, pp.155-158 (1991)

[7] The 1st Joint Conf. & Exhib. on Artificial Intelligent, Neural Networks, and Fuzzy Systems (Nov. 1991)

[8] P. O. Fanger, "Thermal Comfort — Analysis and Application in Environmental Engineering", McGraw-Hill (1970)

[9] M. Saito, M. Naka, K.Yoshida, & I. Akamine, "Estimation of thermal comfort by neural network", Japanese Assoc. of Refrigeration Annual conf., pp.125-128 (1990) (*in Japanese*)

[10] "Neuro & Fuzzy logic automatic washing machine and fuzzy logic drier", Hitachi News Release, No.91-024 (Feb. 26, 1991) (*in Japanese*)

[11] "Neuro & Fuzzy logic automatic washing machine ASW-50v2", Sanyo News Release (Feb. 29, 1991) (*in Japanese*)

[12] "Two types of Sensor - Oven Range", Sanyo News Release (July 8, 1991) (*in Japanese*)

[13] "Electric fan series in 1991", Sanyo News Release (Mar., 14, 1991) (*in Japanese*)

[14] "New trend in consumer electronics: combining neural networks and fuzzy logic", Nikkei Electronics, No.528, pp.165-169 (1991.5.25) (*in Japanese*)

[15] K. Morito, M. Sugimoto, T. Araki, T. Osawa, and Y. Tajima, "Kerosene fan heater using fuzzy control and neural networks ⟨CFH-A12JD⟩", Sanyo Technical Review, Vol. 23, No. 3, pp.93-100 (1991) (*in Japanese*)

[16] M. Naka, T. Shida, K. Yoshida, & I. Akamine, "Application of neural network for air-conditioner's control", Tech. Report of IEICE, HC91-37, pp.9-16 (Dec. 1991) (*in Japanese*)

[17] "Developing 9 type and 19 Kerosene heaters, 1991", Sharp News Release, No.3-032 (June 11, 1991) (*in Japanese*)

[18] "Slim 66: Marketing a new refrigerator which door opens right/left side freely", Sharp News Release, No.3-053 (Sept. 5, 1991) (*in Japanese*)

Fuzzy Logic Technology & the Intelligent Highway System (IHS)

Bruno Bosacchi
AT&T Bell Laboratories - Princeton NJ 08542-0900

&

Ichiro Masaki
General Motors Research Laboratories - Warren MI 48090

Abstract - **This paper presents an overview of the applications of Fuzzy Logic technology (FLT) to the Intelligent Highway System and, more specifically, to one of its essential components, the Intelligent Vehicle (IV). In particular, status and trends of the effort are highlighted by reviewing and discussing some representative papers from two recent conferences on the subject (Intelligent Vehicle '92 [IV'92], Tokyo, 1991, and Intelligent Vehicle '93 [IV'93], Detroit, 1992).**

I. Introduction

Fuzzy Logic Technology (FLT) has been enjoying a remarkable popularity over the past few years, thanks to an abundance of commercial applications which have fueled the FLT development with technical challenges and financial incentives. However, with the exception of a few major projects, most of the commercial successes so far concern relatively simple problems in the home appliances and consumer electronics markets. To maintain its momentum, FLT may need to scale up its effort to more challenging tasks, in markets lucrative enough to justify and support the required R&D effort.

In this respect, that complex of efforts and technologies which in this paper we group under the names of Intelligent Highway System (IHS) and Intelligent Vehicle (IV) appears as a promising field. The IV/IHS technical challenges will be discussed later on. As for its financial reward, we will just mention a few projections on two of its most important sectors: the automotive electronics and the communication products and services associated with the flow of information in the intelligent highway network. The automotive electronics is the fastest growing segment of the electronics technology. According to forecasts by BPA, its volume in the year 2000 will be higher than the volume of the entire electronics industry in the year 1980! And according to Leading Edge Reports, the world automotive electronic market, which was $2 billion in 1980 and $27 billion in 1990, is projected to reach $61 billion in 1995! As for the IHS communication products and services, AT&T estimates that the US market alone may reach a total volume of $5 million in the next 5 years, and $212 billion over the next 20 years! [1]

In this paper we attempt to provide a picture of the FLT potential in the IV/IHS market by discussing status and trends of the field in the light of papers presented at two recent conferences specifically devoted to the subject: Intelligent Vehicle '91 [IV'91], held in Tokyo, in Nov.1991, and Intelligent Vehicle '92 [IV'92], held in Detroit in June 1992. We first present an overview of the IV/SHS effort (Sect.2), and then discuss the potential role of FLT (Sect.3). Finally, in Sect. 4, we highlight some specific FLT applications by reviewing representative papers from IV'91 and IV'92.

II. Overview of the IV/IHS effort [2]

We denote under the umbrella name of IV/IHS a wide range of electronic-based systems and technologies aimed at assisting the driver and providing new and better ways to improve traffic efficiency and safety. This entails the realization of a well managed system of "intelligent drivers in intelligent vehicles on intelligent highways". This ubiquitous request for "intelligence" translates into concepts such as human-like behavior in the car automatic operations, and human-friendly interface between the driver on one side, and the vehicle and highway system on the other. Indeed these concepts, as shown, for example, by the Trilby plan of General Motors or the Delphi prediction of the University of Michigan,[3] are a powerful driver of the IV/IHS effort.

The pursuit of the IV/IHS objectives requires the complex interaction of a high number of players (drivers, highway authorities, car manufacturers, communication companies, commercial fleets, etc.). As a consequence, the effort is most effectively managed at the national level, with national programs such as the IVHS (Intelligent Vehicle Highway System) in the US, PROMETHEUS in Europe, and AMTICS in Japan.

For the sake of discussion, we can distinguish three main components in the IV/SHS effort: vehicle control, information, and overall system management. (In the IVHS initiative of the US, these components closely pattern the Advanced Vehicle Control System, the Advanced Driver Information System, and the Advanced Traffic Management System, respectively).

The vehicle control component is concerned with the use of advanced control engineering techniques and "intelligent" control technologies to assure the passengers a safe and comfortable ride. The task also includes the communication of signals across the interface vehicle/highway, to provide the vehicle with the information needed by the implementation of intelligent control functions and autonomous navigation. Problems such as gear shift control, cruise control, automatic system control, collision avoidance, autonomous navigation, etc. belong to this effort. Of course, some of these functions already exist in some form aboard to-day cars. However, the IV/IHS thrust is towards the introduction of a considerable level of "intelligence", which translates mainly in the requirement of "human-like" features in the control operations.

The information component aims at using a broad spectrum of wireless communication technologies (cellular, satellite, short range beacon, etc.) to establish a complete two way-communication (audio, video and data) between the passengers inside the car and the world outside. It is basically the implementation at the car level of the Personal Communication Network concept). An important role is played by information delivery technologies, such as image processing, voice recognition, display systems, etc., and the drive towards more "intelligence", in this effort, translates mainly in the requirement of "human-friendly" interfaces.

Finally, the overall management component aims at the centralized system management of the highway operations, gathering, in traffic management centers, information from the vehicles and from a highway network of video motion detectors, image processing systems, character recognition devices, etc., in interaction with highway dispatch centers (police, ambulances, tow-trucks, etc.). It includes the monitoring of traffic conditions, the analysis of flow patterns and traffic composition to map out traffic strategies, the redirection of vehicles to reduce traffic congestion, the rapid response to emergencies and highways accidents, the identification of vehicles, the charging of tolls, the detection of traffic rule violations, etc. Here again, though many of the above task already exist in an elementary form, the drive is towards more "intelligence". This translates mainly in the need of system optimization and quick decisions in complex and ill-defined situations.

III. THE ROLE OF FLT IN THE IV/IHS EFFORT

In this Section we present a general discussion of the role of FLT in the IV/IHS effort, as it emerges by our reading of the 27 papers presented at IV'91 and IV'92 which, on a total of 102 contributions, were somehow related to FLT. (These papers are listed in Appendix A) [4]

The discussion of Sect.2 has shown that the IV/IHS effort encompasses tasks ranging from straightforward control engineering problems, to that complex interplay of control, communication and decision which characterizes the information and management systems in which the driver-vehicle system is embedded. This wide range of tasks offers an ideal testground for FLT, which, in the management of complex control and decision problem has not only its main objective, but even its original motivation. [5]

Vehicle control engineering includes many functions which are characterized by complex and nonlinear behavior and cannot be completely described by mathematical models. The potential of FLT for dealing with such cases is well known. Therefore, vehicle motion control is a natural target for FLT, with its ability to capture the heuristic "rules of thumb" ordinarily used by human operators. In particular, FLT is intended to provide more human-like behavior and better performance than conventional PID controllers in terms of frequency and smoothness of change in the controlling variables. This point is well documented by the commercial success of FLT in the home appliances and consumer electronics applications, and is object of strong interest in the automotive industry, as can be seen from the affiliation of the authors of the papers of App.A.

On the other hand, automotive control problems tend to be more complex than those of home appliances and consumer electronics, and are further complicated by considerations of safety, reliability, etc. Therefore more fundamental work is probably needed to compensate for the ad-hoc nature of fuzzy rule design and establish procedures to estimate the completeness and consistency of the rules and the stability of the solutions. The move of FLT towards control problems of higher complexity is also accompanied by some interesting trends, such as the emergence of VLSI hardware implementations of fuzzy logic microcontrollers and the integration of FLT with other "intelligent" technologies such as Neural Networks, Genetic Algorithms, Object Oriented Programming, etc. For example, several IV'91 & IV'92 papers integrate FLT and Neural Networks to elicit membership functions, build knowledge bases, implement fuzzy operations, etc.

At a more subtle level of control engineering, FLT looks promising in dealing with the vaguer field of human-vehicle relationship through the modeling of the human recognition, judgement and decision processes which occur during driving operations. Due to the trend towards human-like behavior, which conventional engineering concepts cannot realistically model, this is a field of increasing importance. A few papers address the possibility of using FLT to introduce in the control process variables such as driver's feelings, predictions and intentions, rather than simple reliance on the "mechanical" input from sensors.

Traditionally, the IV conferences have been mostly focused on the Intelligent Vehicle component of the IHS effort, i.e., on the vehicle control and the management of the control signals across the vehicle/highway interface. So it may not be surprising that only a few among the IV'91 & IV'92 papers address the application of FLT to the information and management systems in which the driver-vehicle system is embedded, leaving the impression of a strong unbalance in favour of the control component. This unbalance, however, is probably real, and mirrors the general situation in which FLT finds itself. In spite of the extensive and impressive literature on the unique effectiveness of FLT for decision and information processing problems in ill-defined situations, when processing of heuristic information is required, or when human interpretation plays a relevant role, significant practical applications still wait for development. Perhaps, given its commercial importance, the IV/IHS field may provide the decisive push.

IV. SPECIFIC IV/IHS APPLICATIONS OF FLT

Since space requirements preclude a detailed review of all the papers of Appendix A, we have grouped them in Table I according to specific IV/IHS tasks, with the tasks approximately listed in order of increasing complexity, from straightforward control engineering to autonomous navigation and overall system management. In the review which follows, we limit our attention to a few representative contributions which illustrate and support the discussion of Sect.3, with a certain emphasis on contributions from IV'92, since a review of the papers of IV'91 has already been given.[6]

Table 1. Automotive Applications of FLT

IV/IHS Application	FLT Papers
Engine Diagnostics & Control	#17
Anti-Lock Braking System	#4
Active Suspension System	#5, #19
Integrated Chassis Control	#1
Cruise Control	#13, #24
Gear Shift Control	#2, #14, #15
Automatic Steering Control	#7, #21- #23
Electronic Brake System	#18
Vision Enhancement System	#25
Collision Avoidance System	#11, #12
Automobile Tracking Problem	#6
Autonomous Navigation System	#8- #11, #18, #21- #23, #25
Traffic Signal Control	#20

As an example of advantageous application of FLT in a straightforward control problem, we consider paper #17, which describes the implementation and testing of a fuzzy controller (FC) for idle speed, an important parameter to optimize stability, fuel consumption, power output, drivability and exhaust emission of an engine. This paper reports that, when compared to a conventional PID controller, and to an Optimal Linear Quadratic (OLQ) regulator, both of which use mathematical models, the FC performs much better than the first, and only slightly below the second. It is further claimed that FC should outperform the OLQ in the long run, due to its superior robustness in adjusting to the unavoidable drift in engine parameters, due to wear, manufacturing tolerances, nonlinearities, operating conditions, etc., which affect the optimality of the tuning.

At a more sophisticated level of control engineering, it is claimed that FLT can outperform conventional control not only in terms of reduced overshoot and undershoot, and less frequent and smoother changes in the variables under control, but also in terms of adaptive behavior and "intelligent" operational mode. These features, for example, are observed in the cruise control systems discussed in paper #13 and #24. The first reports on an FLT constant speed cruise controller which since 1987 has been used as standard equipment on certain Japanese models. Since the traffic congestion on Japanese road does not easily allow constant speed, the controller has been being modified to allow adaptive control, i.e., the vehicle speed is automatically adjusted by a fuzzy estimate of the optimal distance from the vehicle running ahead. The second describes an intelligent human-like cruise control which optimizes speed and distance with respect to preceding vehicles, taking also into account specific driving and weather conditions.

Increasing further the sophistication level, various attempts are being made to introduce intelligent "human-like" behavior in the implementation of the control operation. A good example is provided by the problem of automatic gear shift control, discussed in several papers (#2, #14, #15). The engineering of automatic transmission is by now a mature field. Yet, it is a common experience of automatic transmission car drivers that gears do not always shift properly under certain driving conditions. For example, when the driver lets up on the accelerator to navigate a slight downgrade while driving an uphill slope, the gear may automatically shift, even when, seeing the uphill ahead, a "human" driver would avoid needless shifting. The same situation is encountered on a winding road.

To introduce human-like behavior in the automatic transmission problem one needs to "input" the driving environment in the control system. In paper #14 this is achieved using a multiplicity of inputs (car speed, throttle opening, engine speed, engine negative pressure, shift position and brake signals) to generate IF...THEN rules which take into account common sense experience and driver intentions. Papers #2 and #15 take a different approach inferring the driving environment from the pattern of accelerating behavior of the driver. In particular, they assume that car speed, displacement and rate of displacement of the accelerator pedal sufficiently characterize the driving environment. In paper #2, data related to the driving pattern (accelerator inputs, speed changes, brake and steering input, and road conditions) of a large number of drivers on a large variety of roads are recorded and statistically analyzed to extract distinctive features which are then used to create fuzzy logic (FL) rules and membership functions. During operation, the vehicle speed, the accelerator displacement and displacement rate are used as inputs of the fuzzy inference scheme, whereas the output is computed as the minimum value (logical product) of the grade obtained from their respective membership functions. This result is compared with the value of the corresponding parameter predetermined by the statistical analysis and, with the help of the FL rules, a decision is made as to whether the vehicle is traveling in conditions requiring the switch to a more appropriate gear shifting pattern. In paper #15, the control knowledge of experience drivers is used to create a Fuzzy Controller (FC) with a desirable gear shifting behavior. This FC is realized with a neural network chip which is used to represent membership functions, implement rules and, through a novel learning algorithm, to adjust the weight of each rule, thus achieving self-tuning capabilities.

The role of FLT in the management of signals across the interface vehicle/highway, necessary to implement "intelligent" and "human-like" control operation is abundantly exemplified in the large majority of papers. The unique ability of FLT to deal with the interpretation of fuzzy signals is used in a variety of tasks, from automatic steering control, cruise control, collision avoidance, etc. up to the complex interplay of information and control of system operations such as autonomous navigation.

A typical example is provided by paper #7, which reports on a fuzzy controller for human-like automatic steering control, using fuzzy inferences through a fuzzy LSI chip and real-time self-adjustment of the membership functions. Human like operation in automatic steering control is also pursued in papers 21, 22, and 23. Paper 21 even attempts to improve on human-like behavior by eliminating human fatigue! The system has been designed to substitute human drivers and increase duration and difficulty of durability tests during the development of new models.

A second example for dealing with fuzzy signals is provided by paper #27 on the vision system of an autonomous vehicle. This system analyzes 2D images to derive scene descriptions which are used to control the vehicle as it moves in a 3D environment. Deriving scene description requires searching and matching the extracted features against the features of the model base. This is a very complex task due to the feature instabilities caused by changing viewpoints. Paper #27 proposes using viewpoint invariance relations (VIRs) on line segments, such as parallelism, collinearity, proximity, as the primitives on which to base the analysis of the segmentation results. Whereas in a noise-free image the extraction of VIR's is relatively straightforward, in the actual practice the task is strongly complicated by the uncertainties of the image formation process and by the imprecision of the information. For example, a pair of parallel segments characterizing a scene may only appear as "nearly" parallel. The paper proposes to use FLT to model these uncertainties. For example, the parallelism of a pair of "approximately parallel" segments is quantitatively characterized by the angle 0 between their directions, and the fuzzy set "approximate parallelism" (AP) is represented by the membership function u(AP)=0. A method ("anchoring & adjusting") is then introduced to aggregate the various pieces of fuzzy information.

A final example on the management of fuzzy signals, which also anticipates a prototype situation in any future traffic information and management system, is the development of a fuzzy map-matching algorithm for autonomous navigation, discussed in paper #16. In general, the vehicle location is obtained by processing inputs from positioning sensors and is then related to a geographic database. Due to sensor errors and database imperfections, the map-matching process involves many ambiguities and a resulting vagueness in the determination of the vehicle location in the road network. The fuzzy algorithm proposed in #16 uses FL to manage this kind of problems.

We conclude the review with two other examples of FLT application to information and traffic management problems. Paper #20 develops an adaptive distributed approach to traffic signal control, in which the signal timing parameters at a given intersection are adjusted as functions of the local traffic conditions and of the signal timing parameters at adjacent intersections. The signal timing at an intersection is defined by three parameters: cycle time, phase split, and offset. Fuzzy decision rules are employed to adjust these parameters based only on local information. According to the author, this approach provides a fault-tolerant, highly responsive traffic management system, whose effectiveness has been demonstrated through a simulation of the traffic flow in a network of controlled intersections.

Finally, in paper #18, a discussion is given of an exploratory study on the application of fuzzy control to emergency braking in the framework of ISIS, a French interactive road information display system. ISIS is meant to supplement roadside signs with IR beacons which broadcast data, instructions and warnings to passing vehicles. In particular, ISIS aims at ensuring observance of speed limits and stop signs, even "dismissing" the driver who does not collaborate to slow down the vehicle or bring it to a complete stop. Preliminary results on experimental vehicles are reported as encouraging.

V. CONCLUSIONS

We conclude this review with some broad summarizing remarks. The papers of IV'91 and IV'92 provide a good overview of the challenges and prospects, and a reasonably updated picture of status and trends of the FLT application to the the IVHS field (For more recent developments, the readers are invited to attend the 1993 edition of the conference, Intelligent Vehicles '93, which will be held Tokyo, in July '93). IVHS is a challenging and potentially rewarding field for FLT, for both control and management/decision aspects. For the latter, interesting application are beginning to appear, but its enormous potential is still virtually untapped. As for the control aspects, the volume itself of the effort, well documented by the IV'91 & IV'92 contributions, clearly shows that the FLT community has the challenge of scaling up its effort from the relatively simple successes in the consumer electronics and home appliances markets to the more demanding applications in the automotive field, where issues of safety and reliability are of paramount importance. An area where fundamental work is urgently needed is the development of convenient criteria to assess stability, robustness, and reliability of the solutions, to avoid the extensive simulations presently used to deal with this problem.

ACKNOWLEDGEMENTS

The authors are indebted to several people at AT&T Bell Labs and General Motors Research and Environment Staff, in particular S.R.Nagel and G.G.Dodd, for their interest and encouragement.

REFERENCES

[1] AT&T Focus, Nov. 1992, p.18

[2] For a review, see, for example, the Special Issue of IEEE Trans. on Vehicular Technology, **40**, #1, (Feb. 1991).

[3] The University of Michigan: Delphi IV, Vol.2, Technology (TECH-63A), p.172-173

[4] Information on the Proceedings of IV'91 and IV'92 can be obtained from Ichiro Masaki. See also "Vision-Based Vehicle Guidance", ed by I.Masaki (Springer-Verlag, 1991)

[5] L.A.Zadeh Inf. and Control **8**, 338-352 (1965)

[6] B.Bosacchi & I.Masaki in Science of Artificial Neural Networks, ed. by D.W.Ruck, SPIE Proc. **#1710**, p. 686, (1992)

APPENDIX A

List of FLT papers presented at IV'91 & IV'92

IV'91

1. Y.Uyeda (OMRON, Japan) *Improvement of Chassis Control by Fuzzy Controller*

2. H.Takahashi (Nissan Motor Co., Japan) *A Method of Detecting the Driving Environment Using Fuzzy Reasoning*

3. M.J.Patyra (Carnegie Mellon University, USA) *Implementation of Fuzzy Operations with Neural Networks*

4. P.M.Basehore, J.T.Yestrebsky, K.A.Tucker (American NeuraLogix, Inc., USA) *Innovative Architecture and Theoretical Considerations Yield Efficient Fuzzy Logic Controller VLSI Design*

5. L.A.Feldkamp, G.V.Puskorius, L.I.Davis, F.Yuan (Ford Motor Co., USA) *Decoupled Kalman Training of Neural and Fuzzy Controllers for Automotive Systems*

6. J.R.Jang (University of California, Berkeley, USA) *A Self-Learning Fuzzy Controller with Application to Automobile Tracking Problems*

7. Y.Hashimoto, T.Shigematsu, K.Ohnishi, N.Ohta (Toyota Motor Co., Toyota Central R&D Labs., Inc., Japan) *Application of Fuzzy Algorithm to Automatic Steering Control*

8. T.Hessburg, M.Tomizuka (University of California, Berkeley, USA) *A Fuzzy Rule Based Controller for Automotive Vehicle Guidance*

9. E.H.Ruspini, D.C.Ruspini (SRI International, & NASA Ames Research Center, USA) *Autonomous Vehicle Motion Planning Using Fuzzy Logic*

10. L.Huang, W.Kao, H.Oshizawa, M.Tomizuka (ZEXEL USA Co. & University of California at Berkeley, USA) *A Fuzzy Map-Matching Algorithm for Automotive Navigation Systems*

11. R.N.Lea, Y.Jani, M.G.Murphy, M.Togai (NASA, LinCom Co., University of Houston, Togai InfraLogic Inc., USA.) *Design and Performance of a Fuzzy Logic Based Vehicle Controller for Autonomous Collision Avoidance*

12. R.Fujioka (OMRON, Japan) *An Anticollision System Concept*

13. A.Takayama, A.Hirako (Isuzu Advanced Engineering Center, Ltd., Japan) *Adaptive Cruise Control According to Optimal Distance*

14. I.Sakai, S.Sakaguchi, T.Haga, M.Togai (Honda R&D & Togai InfraLogic,Inc., Japan) *Shift Scheduling Method of Automatic Transmission Vehicles Using Fuzzy Logic*
15. Y.Dote, K.Hayashi, J.Nasu, M.Strefezza, A.Takayama, A.Hirako (Muroran Institute of Technology & Isuzu Advanced Engineering Center, Ltd., Japan) *Neuro Fuzzy Transmission Control for Automobile*
16. B.S.Widmann, G.R.Widmann (GM Hughes Aircraft Co., USA) *Implementation of Object Oriented Constructs for Use in Fuzzy Logic Classifiers*
17. M.Abate (Centro Ricerche FIAT - Orbassano, Italy) *An Application of Fuzzy Logic to Engine Control*
18. J.P.Aurrand-Lions, L.Fournier, P.Jarri, M. de Saint Blancard, E.Sanchez (University of Aix-Marseille II & IMT, Peugeot S.A., France) *Application of Fuzzy Control for ISIS Vehicle Braking*

IV'92

19. E.C.Yeh, H.J.Tsao (National Tsing Hua University, R.O.C.) *Fuzzy Control for Active Suspension Design*
20. S.Chiu (Rockwell International, U.S.A.) *Adaptive Traffic Signal Control Using Fuzzy Logic*
21. K.Ohnishi, J.Komura, T.Ishibashi (Toyota Motor Corporation, Japan) *Development of Automatic Driving System on Rough Road - Realization of Highly Reliable Automatic Driving System*
22. T.Shigematu, Y.Hashimoto, T.Watanabe (Toyota Motor Corporation, Japan) *Development of Automatic Driving System on Rough Road - Automatic Steering Control by Fuzzy Algorithm*
23. N.Ooka, T.Tsuboi, H.Oka (Nippondenso Co., Ltd., Japan) *Development of Automatic Driving System on Rough Road - Fault Tolerant Structure for Electronic Controller*
24. R.Muller, G.Nockler (Daimler-Benz AG, Germany) *Intelligent Cruise Control with Fuzzy Logic*
25. B.S.Widmann, G.R.Widmann (GM Hughes Aircraft Co., USA) *Fuzzy Logic Object Oriented Classes*
26. S.Baocheng, L.Xihui, Z.F.Zhang (Chinese Academies of E.I. & of SINICA) *A New Approach to Fuzzy Control by Using Neural Networks*
27. S.Toh (Arizona State U., U.S.A.) *Extracting Viewpoint Invariance Relations Using Fuzzy Sets*

Twenty years of Fuzzy Control: Experiences Gained and Lessons Learnt

E. H. Mamdani
Dept. of Electronic Engineering
Queen Mary & Westfield College
Mile End Road
London E1 4NS
UK

Abstract—This paper is based on 20 years of involvement with Fuzzy control since its first inception. The main lesson one learns from this involvement is that process control can now be considered a multi-paradigm discipline, involving not just the dynamic system theory but also decision-making approaches gleaned from the field of artificial intelligence. Experience has shown that such approaches may provide better ways of implementing controllers than using conventional approaches based on dynamic system theory.

I. Introduction

This paper examines fuzzy control from its inception to its recent wide-spread acceptance in industrial use in order to reflect upon matters such as: its earlier dismissal, the reasons for its appeal, the nature of scientific research and engineering innovations. This paper only deals with fuzzy control and not other applications of fuzzy sets. Nor is the intention here to provide a retrospective look at fuzzy control. The fuzzy control area – for it is a whole area judging by the explosion of recent research into fuzzy control and neural networks, self-organising control, genetic algorithms and so on – is still continuing to expand with new results appearing regularly. (Such is the advance of this new work that any list of references to it runs the risk of errors caused by omitting important works of major significance.) Nevertheless, it is worth learning some lessons after some 20 years since the idea first appeared.

The dominant position of analytic control theory prevented fuzzy control from being taken seriously until its increasing application in Japan. Many of the arguments against fuzzy control were framed in the language of that dominant theory and also countered in that language. This paper argues that fuzzy control and allied other techniques such as self-organising fuzzy control, neural networks, genetic algorithms and so on, provide an alternate paradigm to the analytic control theory. This paradigm consists of non-analytic approaches to control and is based on decision-making approaches from artificial intelligence.

II. A brief historical look

Fuzzy control [1] came out of a doctoral research work that was concerned with the application of learning techniques to process control. At that time learning and adaptive control systems were well established research topics. There were already many results in these areas within the established control systems theory. Our work, however, was drawing on the emerging learning techniques from the field of artificial intelligence and pattern recognition: rule–based systems, protocol analysis, neural networks and statistical pattern recognition. In a sense this was one of the first studies to investigate the application of AI to control. This is nowadays sometimes referred to as intelligent control but this paper proposes the alternative reference to it as non-analytic control for reasons that will be given in the sequel.

The main part of the study concentrated upon the use of a supervised Bayesian learning system for learning the state–action statistics to control a laboratory prototype steam engine from the control behaviour of a human operator. The state of the steam engine (the boiler pressure and the engine speed) was periodically displayed to the human operator on a teletype. The human responded to this by typing the changes in the inputs (the boiler heat and the engine throttle) he wanted made and this action was implemented by the computer. At the same time the displayed state and the human action were noted by the Bayesian learning system algorithm which proceeded to learn the state–action statistics. After the learning control algorithm had converged, it was used on its own to control the steam engine. We noted that its behaviour was consistently inferior to that of the human. As experimenters our feeling was that we could tell a machine better in more abstract terms how to behave than just showing how as operators we behaved at each instance on observing the state of the engine.

The above should not be read to mean that learning controllers can never perform well. There is no doubt that we could have persevered and improved the behaviour of the learning controller. However, it was felt that a rule–based approach such as had been demonstrated in studies like DENDRAL [2] would be more effective. Indeed this was also tried, but the LISP systems of that period were primitive and slow. It was Zadeh's paper [3] published at that time which persuaded us to use a fuzzy rule–based approach. Between reading and understanding Zadeh's paper and having a working controller took a mere week and it was "surprising" how easy it was to design a rule–based controller. Other workers also have commented on the "surprising" ease with which a fuzzy controller can be implemented. This surprise reflects that results are different from what conventional wisdom would lead one to expect. That in turn begs the question as to what that conventional wisdom is. The answer seems to be that conventional control theory and the way it is taught suggests that controller design requires a good analytical treatment and any other technique such as one borrowed from AI would need to be sufficiently complex and involved to match the apparent sophistication involved in that analytical theory. Therefore, the use of a few simple fuzzy control rules is counter to that intuition.

On reflection it becomes clear that our approach to control was based on a different view of control (that based on the AI paradigm) than the conventional view of control (that based on the analytical control theory). The latter was at that time a very successful discipline and thus all powerful. So much so that any discussion on the problem of control could only be expressed in the terminology of that established paradigm. This culture clash was at the very heart of the early criticisms of fuzzy control. With the hindsight of twenty years and the industrial success of fuzzy control with us we can examine the arguments that took place then.

III. THE CULTURE CLASH OF TWO PARADIGMS [4] [5]

It was the insight by research workers during the '40s and the '50s that many dynamical systems can be mathematically modelled using differential equations which led to the foundations of control theory. With the addition of transform theory, this provided an extremely powerful means of studying (i.e. both analysing and designing) control systems. First as the theory of servo-mechanisms and later as control systems theory, the discipline blossomed and continued to do so until the '70s. These were such heady days that often the area was called simply systems theory to indicate its definitiveness. The theory provided many powerful insights about control systems and although not all of these needed to be expressed using the body of mathematics of the area, it became mandatory to do so in keeping with definitive high ground that the paradigm soon staked for itself. Unfortunately, in too many instances and for a variety of reasons this approach could not be sustained because many systems are simply not amenable to the full force of mathematical analysis as dictated by the control theory.

A. *The limitations of control systems theory*[1]

There were attempts to apply control theory to other systems, that is systems whose behaviour cannot be derived from first principles of physics and described as differential equations: systems such as management systems, economic systems even telecommunications systems. Such systems have many of the properties of industrial processes such as inertia, momentum, damping, time lags, oscillations and other forms of instability and so on; however, there is no intrinsic and immutable physical law of dynamics that governs their behaviour. In economic systems for example, various parameters of the economy are measured such as inflation rate, unemployment, money growth and so on and the goal of control is to manipulate fiscal and monetary variables in order to maintain the measured variables at some desired values. Similarly in telecommunications networks, the purpose of network management is to measure the network performance in terms of quality of service metrics and to maintain these at some desired values (sometimes decreed by regulations and / or contracts) by taking control actions such

---

[1] In passing let it be said that in no other discipline is it more difficult to distinguish an engineer from a scientist. For many, science is a precursor to engineering. This is a misguided view of the relationship between science and engineering. Engineering is related to technology and in human civilisation, technology came long before science. It may be permissible to say that engineering is concerned with creating technology by making use of the best available science, but this is a far cry from saying that technology is the application of science.

In control engineering journals, many of the papers had nothing to do with technology or engineering whatsoever. They were increasingly mathematical and often only about mathematics. Zadeh has commented many times that it was this trend that motivated him to come up with Fuzzy sets theory. The year of his first paper was 1965 so the origins of the trend were fully established in the mid-sixties.

as reconfiguration of the network. The behaviour of these type of systems is governed by humans who interact with them to make them work. That is to say, it is the human behaviour with respect to these systems that produces their dynamic behaviour. While it may be tempting to describe such systems by discovering a mathematical law that fits their input / output behaviour, the long term validity of any such law cannot be guaranteed because it is ultimately dependent on humans who may change their behaviour suddenly and in an unpredictable way.

There are yet other systems – process control systems no less – for which the mathematical law of system behaviour is difficult to derive from basic physical principles because the physical processes taking place are too complex. Cement kilns and Steel making are the best examples of such systems. Again it may be possible to discover the mathematical law governing their behaviour by an input output analysis. However, as with human systems mentioned above, the validity of that law would be limited. Indeed such systems have more in common with human systems than with classical linear dynamic systems.

When faced with such difficult systems, control engineers have always invoked the ultimate catch phrase: non-linearity. This is used in such an all encompassing way that it has never made much sense; for it is unclear where non-linearity ends and non-mathematical (non-analytic to be more accurate) begins. Non-linearity ought only to refer to those systems which have to be described by non-linear mathematical equations resulting in a degradation of analyticity; however, it is frequently applied to any non-analytic system. This discussion has been introduced here to point to an important attribute that has come about within control theory: the *cult of analyticity*. This cult had much to do with early criticisms of fuzzy control resulting in a slow acceptance of the approach outside Japan. The criticism centred around the lack of stability analysis of fuzzy logic controllers.

B. Argument concerning system stability

Some of the insight on systems in control theory concerned the "stability" of the systems. It is worth noting that those well trained in the discipline will conjure up poles and zeroes in the s-plane and certain properties of matrices in order to properly discuss stability. To put it another way, it is often difficult for a control systems practitioner to define stability to a non-practitioner without using the mathematical theory. It is in this context that the main criticism of fuzzy control has to be understood - that it does not provide stability analysis. The answer provided at the time against this criticism was that in order to carry out such an analysis it was first necessary to have a reliable mathematical model of the system, and complex systems such as cement kilns (the first industrial application which used fuzzy control [6]) cannot be accurately modelled mathematically. This argument continued but the matter rested there with the implied suggestion that fuzzy control was best suited only for those complex systems such as cement kilns that could not be modelled mathematically. To-day, given the large number of other industrial applications of fuzzy control in Japan, one needs to modify that view considerably. To do so we need to examine two issues: firstly, the requirement for stability analysis from an industrial perspective; and secondly, the problem of control as viewed from within the AI paradigm.

C. Industrial requirement for stability analysis

Industry has never put forward a view that mathematical stability analysis is a necessary and sufficient requirement for the acceptance of a well designed control system. That is merely the view that control system scientists wished to put forward, but it has never gained currency outside academic circles. It is noteworthy that control system area never developed a comprehensive and prescriptive design methodology like we have for Information System Design such as SSADM. Control system methodologies are at best, recommendations on how to use a set of computer based analytic tools. They do not provide an industry approved standard for a structured step by step approach for the analysis and design of a system. An industrially accepted methodology that included stability analysis as an essential validation step will have a very limited use as so few systems would be amenable to that approach.

Prototype testing is more important than stability analysis; stability analysis by itself can never be considered a *sufficient* test. Moreover, in any practically useful methodology, a stability analysis step would need to be made a desirable but an optional step; it cannot be a *necessary* step. Control system scientists have continued to stress stability analysis with great vigour, but the argument is in fact disingenuous and should not have been treated by fuzzy logic control workers with the seriousness with which it was. Stability as understood by a control system scientist is not the property of the process being controlled but an operational definition expressed in the mathematical language of analytic control theory. It turns out on reflection that fuzzy control is one of the techniques that forms part of an alternative non-analytic control theory. Stability is still an important issue but a different way has to be found to study it. In the final analysis all one may be able to do is to build prototypes for the purpose of approval certification. That is a well tried and tested approach used in industry and there is no reason why it may not suffice with control systems as well.

IV. NON-ANALYTIC CONTROL SYSTEMS - AN ALTERNATE SYSTEM DESIGN PARADIGM

The problem of control is also one of decision making, viz.: given the observation of the state of the process to decide from encoded knowledge what action to take. Knowledge based systems and, in particular rule-based approaches, are ideally suited for such a decision making task. On the one hand one may use deep knowledge, perhaps in the form of qualitative models along with qualitative reasoning for such a decision making. On the other, a simple rule–base derived from the experience of a process operator can also be used. The only requirement is that decision making is automated so that the KBS system does not simply give advice on what action is best suited but actually goes ahead and implements the decision without checking with an operator. Control systems based on knowledge based systems, neural networks, genetic algorithms are all viable alternatives to those derived from conventional control theory.

The fuzzy controller is one such simple rule-based control system. The knowledge used for this does not derive only from an expert operator. It can also come from the designer of the system. Some of the knowledge can be based on the understanding of the behaviour of the class of dynamic systems in general to which the particular system belongs, and this can be refined further to suit the particular system if need be. Indeed if one looks at the rules in a fuzzy controller, one finds three sets of rules: a set dealing with providing a rapid response to large errors (rise time rules); a set for preventing overshoots (damping rules); and a set maintaining errors near to zero (steady state rules).

Whatever knowledge (intuition) that is available about dynamic systems, but which does not need to be expressed only mathematically can be encoded as rules. There is much knowledge on dynamic systems that can only be expressed using the mathematical language of control system theory; but there is equally much knowledge about particular dynamic systems that definitely cannot be expressed in any mathematical way. Such knowledge can be used in rule–bases but not in the conventional control paradigm. For some kind of knowledge a mathematical expression would be much more cumbersome even if it existed than a rule–based expression. Thus often a rule–based non-analytic control paradigm may in fact produce better controllers than the analytic control theory paradigm. It must now be stated clearly and forcefully that the goal of pursuing research on any kind of non-analytic controller is first and foremost to prove that the technique is applicable, and not just to find ways of designing controllers for only those systems for which the conventional theory cannot be used. Even if a controllers can be designed using the conventional analytic theory, a designer may still prefer to use the non-analytic approach because the resulting controller may be easy to design; it may be easy to implement in hardware; and / or, it may result in a more robust controller. (A robust controller is less susceptible to system parameter changes or to noise.) Recent industrial experience has shown that all these three criteria apply to fuzzy logic based controllers [7].

A. The advantages of using fuzzy control

One of the main advantages of using a fuzzy approach is that fuzzy logic provides the best technique for knowledge representation that could possibly be devised for encoding knowledge about continuous (analogue[2]) variables. This can be briefly explained with reference to the figure below. Assume the system to have two measured variables and one control action variable. The controller will be composed of a set of control rules with two antecedents and one consequent.

The antecedents of each fuzzy rule describes a fuzzy region in the state-space (see figure below). By so doing one effectively quantizes an otherwise continuous state-space so that it is covered by a finite number of such regions (and consequently fuzzy rules). There is a specific fuzzy action associated with each such fuzzy rule (fuzzy region). During the decision making process (control process) each point in the state-space is differently affected by the actions associated with all the fuzzy regions in whose footprint the point falls. The fuzzy aggregation rule and the defuzzification method then yield a specific action for that point. As the point moves, the action also changes smoothly. This means that an effectively quantized representaion of the state-space nevertheless yields a smooth action surface over the state-space.

---

[2] "analogue" is not a synonym of "continuous" in English. Such usage in electrical engineering owes much to the primacy of control system theory and involves a convoluted reasoning: "analogue" is contrasted with "digital"; digital quantities are discrete; the opposite of discrete is continuous; hence etc.

In effect given only the actions associated with the centre point of overlapping rectangular regions of the state–space, fuzzy logic algorithm provides the action for any individual point in the state–space. Obviously all the other values are interpolated from the available finite sets of values by a method which depends upon the choice of the mathematical expression used for the various fuzzy logic connectives (AND, OR, ELSE etc.) as well as the choice of the defuzzification method. It is agreed by the workers in fuzzy control that one should limit the choice of mathematical expressions to those that would result in a smooth action surface. The result is that the control can be specified compactly with as few rules as possible.

This form of knowledge representation is applicable in any rule–based application dealing with continuous variables, not just for feed–back control. Many of the consumer products in Japan using fuzzy logic use it as a feed–forward decision maker. The washing machine is a case in point. Having introduced a new sensor for measuring the "turbidity" of the water, the manufacturers needed to introduce the associated intelligence to respond to this new information. It would be pointless to display the turbidity value to the user of a washing machine for it would not make much sense on how to interpret the information so conveyed and what to do with this new information. Because the engineers discovered what useful information this new sensor conveyed (how the articles being washed were soiled and how dirty they were), and thus what to do with it (adjust the washing time), this information could be encoded as rules. Furthermore as the information was itself "analogue", fuzzy control approach was the best knowledge representation method.

It is instructive to note that in most of its industrial applications (both in consumer products and in industrial systems) the fuzzy controller comprises only a few rules (around 7 to 15). It represents a tiny amount of intelligence being embedded into the overall system. In this respect it is analogous to the use of tiny electric motors in applications such as automobile rear view mirrors and wind-up mechanisms of even cheap cameras. Often there is more technical detail in the sensors and actuators of the overall system than in the controller itself. This shows that one of the problems with fuzzy control is that it is percieved by some (both control system and AI scientists) as too simple. While this factor provides technological strength it is also seen as a scientific weakness [3]. Outside Japan, where intellectual sophistication is revered, fuzzy control has been seen as scientifically weak and this is another strong reason why its is only belatedly being commercially exploited there. In Japan its technical strengths have been most apparent and it is not at all surprising that large scale exploitation has first occurred in Japan.

V. CONCLUSIONS

This paper has concentrated mainly on the humble rule–based fuzzy controller. The experience gained with it over the past several years has shown that fuzzy control may often be a preferred method of designing controllers for dynamic systems even if traditional methods can be used. Experience has shown that a fuzzy controller is often more robust than a PID controller in the sense that it is less susceptible to noise and system parameter changes. Furthermore, a fuzzy controller may also be easier to design and, thanks to the fuzzy chips and so on, easier to implement. Its application is not limited to dynamic system control. Its main merit is that it gives the most efficient knowledge representation method that can be devised for rule–based systems that deal with continuous variables. It can be used wherever such knowledge exists. For example, in mathematical programming there is often more specific domain related problem knowledge available than can be expressed in the framework of the mathematical model.

Its delayed exploitation outside Japan teaches several lessons. This paper has argued that there are two main reasons for this: the tradition of intellectualism in engineering research in general and the cult of analiticity within control engineering research in particular. The former is responsible for continuing to produce a number of unscientific distortions in engineering research, albeit in the name of science. For example, journal editors and conference organis-

---

[3] To continue with the theme of the first footnote, the aims of scientific enquiery and technological implementation are not the same. Science delights in complexity as it discovers ever more subtle phenomena. Technology abhors complexity as it tends to render the product in question more difficult to use and maintain. In order to make use of advanced scientific concepts and techniques into products, it is necessary to encapsulate the complexity of a subsystem within a black box both functionally and physically. Functionally, the modules and subsystems that contain complexity must not be allowed to impose their subtleties on other modules that make up the system. Physically the complex parts must be tightly sealed so that when they fail they can be replaced as a whole and not require in situ maintenance. The purpose of the black box is to hide the scientific complexities in order to make the product usable. Designers of many products have not been able to resist showing the beauties of the complex science to the user.

ers have been heard to assert that they would like to see more application papers *even if this means lowering the quality*. Quality according to what criteria? Should one not question these criteria themselves? Evidence suggests that the quality of engineering as seen in all manners of applied work including the technology in the end products is of a higher a standard than anything seen in run-of-the-mill journal papers. Secondly, the authors of technical papers feel obliged to introduce a large amount of symbolism (again in the name of good science) even to the point of obscuring the main message[4].

The main lesson learnt is that the subject of automatic control can no longer be the exclusive preserve of conventional analytic control theory. An alternative paradigm of non-analytic control systems now exists. The advantage of the analytic approach is that it gives an abstract view of system dynamics which applies to a whole variety of linear systems and many non-linear systems as well. The problem is that often more is known about a particular system than can be expressed analytically and this knowledge may thus not be captured effectively. The alternative paradigm is extremely well suited to express special characteristics and knowledge as rule–bases. However, at the current state of research, the more abstract knowledge about system dynamics is not well captured.

New research should be aimed at strengthening the concepts and techniques about system dynamics available to this new paradigm. To achieve this it would prove useful to study the powerful control system concepts that already exist within the conventional theory and then see if they can be redefined from the mathematical language of that theory to be meaningful within the new paradigm (for example, by introducing qualitative modelling techniques). Some workers have suggested that a synthesis of the two paradigms is what should be sought. Aims of synthesising theories are fine as ideals, but in practice they lead to artificial shoe-horning of one set of concepts within the framework of an alien theory. It is far better to accept that control engineering has now become a multi–paradigm discipline.

REFERENCES

1. Mamdani, E. H. and Assilian S. "An Experiment with in Linguistic Synthesis with a Fuzzy Logic Controller", *Int. J. Man-Machine Studies,* vol. 7 pp 1-13, 1975

2. R. K. Lindsay, B.G. Buchanan, E. A. Feigenbaum, J. Lederber *Applications of Art. Int. for Organic Chemistry: The DENDRAL PROJECT* New York, McGraw-Hill, 1980

3. Zadeh, L. A. "Outline of a new approach to the analysis of a complex system and decision processes", *IEEE Trans. SMC,* vol. 3, pp 28, 1973

4. T. S. Kuhn, *The Structure of Scientific Revolutions* Chicago, 1962

5. Brian Easlea, *Liberation and the Aims of Science*, Chatto & Windus (Sussex University) 1973.

6. Holmblad, L.P. and Ostergaard, J. J. , "Control of a Cement Kiln by Fuzzy Logic", *Fuzzy Information and Decision Processes* (Ed M.M. Gupta et al) North Holland 1982

7. Sugeno, M. (Ed) *Industrial Application of Fuzzy Control,* North-Holland, 1985

---

[4]It is possible to think of a paper on, let us say, fuzzy control that can either be written as plainly as possible (a really good scientific practice) or with as much symbolism as possible (the more involved Greek symbols that can be used the better the scientific impact!!). A good heuristic for potential referees to use is that if an author has anything of substance to say in a paper than he would not wish to obfuscate it with excessive symbolism.

FUZZY SET THEORETIC APPROACH TO COMPUTER VISION: AN OVERVIEW

Raghu Krishnapuram and James M. Keller
Department of Electrical and Computer Engineering
University of Missouri-Columbia
Columbia, Missouri 65211

1. Introduction

Computer vision is the study of theories and algorithms for automating the process of visual perception. This involves tasks such as noise removal, smoothing, and sharpening of contrast (low-level vision); segmentation of images to isolate objects and regions and description and recognition of the segmented regions (intermediate-level vision); and finally interpretation of the scene (high-level vision). Uncertainty abounds in every phase of computer vision. Some of the sources of this uncertainty include: additive and non-additive noise of various sorts and distributions in low-level vision, imprecisions in computations and vagueness in class definitions in intermediate-level vision, and ambiguities in interpretations and ill-posed questions in high-level vision. The use of multiple modalities as a means of overcoming some of the limitations imposed by a single image is receiving increased attention, but the use of more than one source of information raises new questions regarding how the complementary and supplementary information should be combined, how redundant information should be treated and how conflicts should be resolved.

Traditionally, probability theory was the primary mathematical model used to deal with uncertainty problems in computer vision. More recently, both Dempster-Shafer belief theory and fuzzy set theory have gained popularity in modeling and propagating uncertainty in imaging applications. While both probability theory and belief theory are important frameworks for this field, the purpose of this paper is to give an overview of the fuzzy set theoretic approach to computer vision. We discuss the applications of fuzzy set theory in computer vision in the areas of image modeling, preprocessing, segmentation, boundary detection, object/region recognition, and rule-based scene interpretation. Although pattern recognition methods are used in computer vision for object labeling, fuzzy-set-theoretic methods meant specifically for pattern recognition are outside the scope of this paper.

2. Images as Fuzzy Subsets

An image is a function $f: R^n \to R^m$ where normally n is 2 and m is 1 (intensity) or 3 (color). If $m = 1$, then the value $f(x,y)$ is called the gray level of pixel (x,y); if $m>1$, then $f(x,y)$ is referred to as a feature vector. In order to apply the rich assortment of fuzzy set theoretic operators to an image, the gray levels (or feature values) must be converted to membership values. Let X denote the domain of the digital image. Then a fuzzy subset of X is a mapping $\mu_f: X \to [0,1]$, where the value of $\mu_f(x,y)$ (i. e., the membership) depends on the original feature vector $f(x,y)$. The calculation of membership functions is central to the application of fuzzy set theory, just as the calculation of conditional probability density functions or basic probability assignments are crucial in the use of probabilistic or Dempster-Shafer belief models.

There are many methods of transforming pixel feature vectors into membership functions. In the case of gray scale images, most researchers have used S functions (or functions resembling them) when there are only two regions (object and background), and combinations of S functions and Π functions (or suitable generalizations) for multiple regions [1-6].

A fuzzy subset of a digital image of size $M \times N$ is characterized by a membership function array $\mu_f(x_{ij})$, $1 \leq i \leq M$, $1 \leq j \leq N$. In a series of papers [7-9], Rosenfeld studied the geometry and topology of fuzzy subsets of the digital (i.e., image) plane. These properties were later generalized by Dubois and Jaulent [10]. Many of the basic geometric properties of image regions and relationships among regions can be generalized to fuzzy subsets. Rosenfeld has extended the theory of these fuzzy subsets to include the topological concepts of connectedness, adjacency and surroundedness, extent and diameter, and convexity.

Pal et al have defined several indices that measure the fuzziness of fuzzy subsets μ_f of an image [11-13]. These include the linear and quadratic indices of fuzziness, and entropy. These indices are maximum when all the membership values in μ_f are equal to 0.5, and minimum when all membership values are either 0 or 1 (i. e., when μ_f is crisp). Hence, they can be used to measure the "goodness" of the end result of a fuzzy-set-based algorithm.

3. Low-Level Vision

Pal and King [1,2,4] use functions resembling the S function as the basic membership functions for both contrast enhancement and smoothing. The contrast enhancement is achieved by applying an intensification operator to the fuzzy subset of an image. The intensification operator is a function that moves the membership values closer to 0 or 1 depending on whether the membership value is less than or greater than 0.5 respectively. They show that the enhancement process reduces the indices of fuzziness. The intensified memberships are converted back to intensity values using an inverse membership function. De et al propose a similar technique by defining a more flexible sinusoidal intensification operator [3]. One disadvantage of these enhancement methods is that there is no simple way to pick the values of the parameters that are used in the membership functions and in the intensification operators. Different values give rise to very different enhanced images.

Following Nakagawa and Rosenfeld [14], Pal et al have applied min and max operations on membership values in the neighborhood of each pixel to produce smoothing or edge detection [12,15]. Other approaches to edge detection using fuzzy set methods can be found in [16,17].

4. Segmentation and Region Representation

Image segmentation is one of the most critical components of the computer vision process. Errors made in this stage will impact all higher level activities. Therefore, methods which incorporate the uncertainty of object and region definition and the faithfulness of the features to represent various objects (regions) are desirable. The first connection of fuzzy set theory to computer vision was made by Prewitt [18] who suggested that the results of image segmentation should be fuzzy subsets rather than crisp subsets of the image plane. The result of a fuzzy segmentation is a fuzzy partition of $X = \{x_1, x_2, \ldots, x_n\}$, where each of the x_j represents the features (such as grey level value) of a pixel, defined as follows.

For an integer C, $2 \leq C \leq n$, a $C \times n$ matrix $U = [u_{ik}]$ is called a fuzzy C-partition of X whenever the entries of U satisfy the constraints:

$$\sum_{i=1}^{C} u_{ik} = 1, \; u_{ik} \in [0,1] \text{ for all } k, \text{ and } \sum_{k=1}^{n} u_{ik} > 0 \text{ for all } i.$$

Column j of the $C \times n$ matrix U represents membership values of x_j in the C fuzzy subsets of X. Row i of U exhibits values of a membership function u_i on X where $u_{ik} = u_i(x_k)$ denotes the grade of membership of x_k in the ith fuzzy subset of X. The objective of fuzzy segmentation is to convert image feature values into class membership numbers. Furthermore, if we define a uniformity predicate $P(\mu_{ij})$ such that it assigns the value true or false to the sample point x_k based on its membership value (for example, $P(\mu_{ij}) = 1$ if $\mu_{ij} \geq \mu_{kj}$ for all k, and it is 0 otherwise), we will have paralleled crisp segmentation.

In the two-class (object and background) case, this problem reduces to computing the membership function μ for the object, since the membership function for the background can be simply taken to be $1-\mu$ to satisfy the conditions for a fuzzy partition. Pal et al achieve this by using the intensity value as the feature, and assuming that μ is an S function (or an approximation thereof). To estimate optimum values of the parameters of the S function, they plot the index of fuzziness of the resulting fuzzy subset as a function the parameters and pick the values corresponding to the global minimum of the plot [11,19]. This method can be extended to the case of multiple objects by choosing all the local minima instead of just the global minimum. Another measure that may be minimized is the compactness (or a similar measure called the index of area coverage) of the fuzzy subset representing the object. This approach has also been used by Pal and Rosenfeld in a two class problem to find the best choice of S function parameters [20,13]. Another interesting approach that uses statistical methods to perform a fuzzy segmentation may be found in [21].

Probably the most popular method of assigning multi-class membership values to pixels, for either segmentation or other processing, is to use the fuzzy C-means (FCM) algorithm [22-29]. The FCM algorithm attempts to cluster feature vectors by searching for local minima of the following objective function:

$$J_m(U,V) = \sum_{u=1}^{n} \sum_{i=1}^{C} u_{ik}^m \|x_k - v_i\|_A^2 \quad 1 \leq m < \infty,$$

where U is a fuzzy C-partition of X, $\|*\|_A$ is any inner product norm, $V = \{v_1, v_2, \ldots, v_C\}$ is a set of cluster centers, $v_i \in R^d$, and $m \in [1,\infty)$ is the membership weighting exponent. Cluster center v_i is regarded as a prototypical member of class i, and the norm measures the similarity (or dissimilarity) between the feature vectors and cluster centers. It is shown in [23] that for $m>1$ under the assumption that $x_k \neq v_i$ for all i, k, (U, V) may be a local minimum of J_m only if

$$u_{ik} = \left(\sum_{j=1}^{C} \left(\frac{\|x_k - v_i\|_A^2}{\|x_k - v_j\|_A^2} \right)^{2/(m-1)} \right)^{-1} \text{ and } v_i = \frac{\sum_{k=1}^{n} u_{ik}^m x_k}{\sum_{k=1}^{n} u_{ik}^m} \text{ for all } i, k.$$

The algorithm defined by looping iteratively through the above conditions is known as the FCM algorithm. The inner product norm $\|*\|_A$, or its replacement by more general distance metrics $d^2(x_k, v_i)$ controls the final shape of the clusters generated by the FCM: hyperspherical, hyperellipsoidal, linear subspace, etc.

In terms of generating membership functions for later processing, the fuzzy C-means has several advantages. It is unsupervised, that is, it requires no initial set of training data; it can be used with any number of features and any number of classes; and it distributes the membership values in a normalized fashion across the various classes based on "natural" groupings in feature space. However, being unsupervised, it is not possible to predict ahead of time what type of clusters will emerge from the fuzzy C-means from a perceptual standpoint. Some of the resulting fuzzy subsets may be disconnected. Also, the number of classes must be specified for the algorithm to run. Finally, iteratively clustering features for a 512×512 resolution image can be quite time consuming. Approximations and simplifications have been introduced to ease this computational burden [25].

The fuzzy C means algorithm is effective only when the clusters are hyperspherical and approximately equal in size. When this is not the case, other variations of this algorithm may be used. For example, the Gustafson-Kessel (G-K) algorithm is obtained by using the scaled Mahalanobis distance in the fuzzy C means algorithm [23]. The GK algorithm works well when the clusters are hypersherical or when the clusters are in linear subspaces of the original feature space (i. e., when the clusters are linear, planar, and hyperplanar). Another variation due to Gath and Geva uses a distance measure derived from maximum likelihood estimation methods [30-32]. This algorithm works well even when the clusters are ellipsoidal in shape and unequal in size.

In general situations, the a priori setting of the number of classes is not always possible, especially in segmentation of natural scenes. In such cases, certain cluster validity measures may be used to find the optimum number of clusters. This is done as follows. A clustering procedure such as the FCM is run first. After the algorithm converges, certain validity measures are computed for the resulting fuzzy partition. This process is repeated for increasing values of C, computing validity measures in each run, until a maximum (or minimum, as the case may be) of the validity measure is found. The C value corresponding to this situation is taken to be the optimum value. Several validity criteria have been suggested in the literature. These include partition coefficient, classification entropy, proportion coefficient, total within class distance of clusters, total fuzzy hypervolume of clusters, and partition density of clusters [23,30,33,34].

The boundaries of regions after segmentation is performed are often not very smooth. Also, sometimes several tiny spurious clusters also result. These effets can be remedied by using clean up procedures such as shrink-expand. Rosenfeld et al. have developed geometrical operations on fuzzy image subsets, including shrinking and expanding [35,36].

Skeletons of the object regions are often used as compact representation of the shape of the boundary. When all points in an object belong to the skeleton to varying degrees, it may be called a fuzzy skeleton. Several fuzzy skeletonization algorithms developed based on this idea [35,37,38].

5. Boundary Detection and Representation

Boundary detection is another approach to segmentation. In this approach, an edge operator is first used on the image to detect edge elements. The edge elements so detected are considered to be part of the boundaries between various objects or regions in the image. Pal and King describe a simple algorithm which can classify and approximate boundary segments in terms of lines and arcs [39]. The boundaries are sometimes described in terms of analytical curves such as straight lines, circles, and other higher degree curves.

Variations of the FCM algorithm can be used to detect (or fit) straight lines to edge elements. This is achieved by initializing the FCM with C linear prototypes rather than C centers. Each linear prototype consists of a point (which acts as cluster center) and a parameter defining the orientation of the cluster. The fuzzy covariance matrix of each cluster may be used to define its orientation since its principal eigenvector gives the direction of maximum variance of the cluster [23]. The C prototypes are updated in each iteration. Several algorithms have been developed for the detection of lines based on this idea [40-43]. We have shown that the Gustafson-Kessel algorithm is also very effective for the detection of lines or linear clusters[44]. As mentioned earlier, one problem with these algorithms is that the number of clusters needs to be specified. In the line detection case, one way to overcome this is to specify a relatively high value of C and then merge compatible clusters after the algorithm converges [44,45]. In this implementation, two (or more) clusters were considered compatible if i) their orientation was the same, ii) the line joining their centers had the same orientation as the clusters and iii) the cluster centers were not more than 4 principal eigenvalues apart.

The FCM algorithm may be generalized further to detect non-linear clusters such as circles and ellipses. The Fuzzy C Shells (FCS) algorithm due to Dave uses a prototype consisting of the radius and location of the center for each cluster. A novel distance measure is defined, and it is shown that using this distance, the center and the radius of the C clusters may be estimated in each iteration by solving two non-linear equations [46]. The use of Newton's method is suggested to solve the equations, which can be time-consuming. To overcome this disadvantage, Krishnapuam et al have suggested a new approach which results in simple equations for the estimation of the center and the radius of the C clusters [47]. Dave et al have also devised an algorithm that can be used to fit ellipses to image data [48]. Krishnapuram et al have extended this method to the case of quadric curves and surfaces [70].

6. Object Recognition and Region Labeling

The area of computer vision concerned with assigning meaningful labels to regions in an image can be thought of as a subset of pattern recognition. There is a large amount of research in the use of fuzzy set theory in pattern recognition. Here we will only discuss a few approaches for object recognition in image analysis.

We first present two techniques that view the labeling problem as an aggregation of evidence problem. The evidence can be derived from several sensors (for example, color), several distinct pattern recognition algorithms, different features, or the combination of image data with non-image information (intelligence). The support for a labeling decision may depend on supports for (or degrees of satisfaction of) several different criteria, and the degree of satisfaction of each criterion may in turn depend on degrees of satisfaction of other sub-criteria, and so on. Thus, the decision process can be viewed as a hierarchical network, where each node in the network "aggregates" the degree of satisfaction of a particular criterion from the observed evidence. The inputs to each node are the degrees of satisfaction of each of the sub-criteria, and the output is the aggregated degree of satisfaction of the criterion. Thus, the labeling problem reduces to i) determining the structure of the aggregation network to be used, ii) determining the nature of the connectives at each node of the network, and iii) computing the input supports (degrees of satisfaction of criteria) based on observed features. The structure of the aggregation network depends on the problem at hand [49]. The connectives used at each node of the network are based on fuzzy union, fuzzy intersection, or compensative operators (such as generalized mean or the γ-model) [49,50]. Krishnapuram and Lee developed a learning procedure so that both the type of connective at each node, as well as the parameters associated with the connective can be learned from training data [49,51]. This method not only partitions the image into connected components of similar properties, but also labels these components. In other words, it produces both a segmentation and a region recognition simultaneously, while capturing an abstract model of the decision making process.

The fuzzy integral is another numeric-based approach which we have used for both segmentation and object recognition [52-54]. It also uses a hierarchical network of evidence sources to arrive at a confidence value for a particular hypothesis or decision. The difference from the proceeding method is that besides this directly supplied objective evidence, the fuzzy integral utilizes information concerning the worth or importance of the sources in the decision making process. The fuzzy integral relies on the concept of a fuzzy measure which generalizes probability measure in that it does not require additivity, replacing it with a weaker continuity condition. A particularly useful set of fuzzy measures is due to Sugeno [55]. The fuzzy integral is interpreted as an evaluation of object classes where the subjectivity is embedded in the fuzzy measure. In our applications, the integral is evaluated over a set of information sources (sensors, algorithms, features, etc.) and the function being integrated supplies a confidence value for a particular hypothesis or class from the standpoint of each individual source of information. In comparison with probability theory, the fuzzy integral corresponds to the concept of expectation. The fuzzy integral values provide a different measure of certainty in the classification than posterior probabilities. Since the integral evaluation need not sum to one, lack of evidence and negative evidence can be distinguished.

In the methods discussed above, one needs to compute membership values in different classes from observed feature data. Several methods can be used for this purpose. One approach is to run the FCM algorithm on the training data to estimate the prototypes which can then be used to compute membership values. Recently, we have used normalized histograms of the feature values generated from training data to estimate the particular membership functions [51,52]. This has the advantages that it does not force any particular shape to the resultant distributions, can be extended to deal with multiple features instead of gray level alone, and can easily accommodate the addition of new classes.

The above techniques, as well as many other fuzzy pattern recognition algorithms, are numeric feature-based procedures. On the other hand, fuzzy logic, and in general possibility theory, is inherently set-based, and so, offers the potential to manipulate higher order concepts. Keller et al have used linguistic weighted averaging of possibility distributions [56,57] to generate object confidence from a combination of feature level results and harder-to-quantify values relating to range and motion. Rough estimates of object range and motion were used to construct trapezoidal possibility distributions which were averaged, using alpha-level set methods [58], with similar trapezoidal numbers formed from the output of fuzzy pattern recognition algorithms such as the fuzzy k-nearest-neighbors [59]. A scaling technique was also developed to actually turn the averaging procedure into a confidence fusion methodology overcoming the spreading inherent in fuzzy arithmetic [57].

In [60], normalized histograms of color components of images of beef steaks were used directly in a linguistic approximation scheme to assess the degree-of-doneness of the steak. It was felt that because of the large amount of uncertainty inherent in food processing, the entire distribution of color (primarily in the red and brown regions) was important for class recognition. Note that this is conceptually distinct from those techniques described earlier which used normalized histograms of training data to calculate membership numbers for particular instances of the domain variable. Here, the object (a steak image) is represented by a group of fuzzy sets (various color histograms) and a set-based nearest prototype algorithm was used to assign class labels and confidences.

7. Modeling Relationships and Reasoning

Rule-based systems have gained popularity in computer vision applications, particularly in high level vision activities. In guiding the choice of parameters for low level algorithms, a vision knowledge base may have a rule such as
> IF the range is LONG, THEN
> the object detection window size is SMALL.

If LONG and SMALL are modeled by possibility distributions over appropriate domains of discourse, then fuzzy logic offers numerous approaches to translate such rules and to make inferences from the rules and facts modeled similarly. Nafarieh and Keller [60] designed a fuzzy logic rule based system for automatic target recognition which contained the above rule and approximately 40 other such rules.

Most fuzzy logic inference is based on Zadeh's composition rule. This generalizes traditional modus ponens which states that from the proposition
> P_1: If **X** is **A** Then **Y** is **B**
> and P_2: **X** is **A**,

we can deduce **Y** is **B**. If proposition P_2 did not exactly match the antecedent of P_1, for example, **X** is **A'**, then the modus ponens rule would not apply. However, in [62], Zadeh extended this rule if **A**, **B**, and **A'** are modeled by fuzzy sets. In this case, **X** and **Y** are fuzzy variables [62] defined over universes of discourse U and V respectively. The proposition P_1 concerns the joint fuzzy variable (**X,Y**) and is characterized by a fuzzy set over the cross product space $U \times V$. This relation is "composed" with the input relation **X** is **A'**, the result being projected onto the V universe of discourse, providing the meaning for the output **Y** is **B'**. The flexibility of this form of reference lies in the methods of translating proposition P_1 and in the particular form of composition of fuzzy relations chosen.

Besides changing the way in which P_1 is translated into a possibility distribution, methods involving truth modification have been proposed [63,64]. The proposition **X** is **A'** is compared with **X** is **A**, and the degree of compatibility is used to modify the membership function of **B** to get that for **B'**. Keller et al describe a rule-based system in [61] which utilizes a new inference technique based on truth value restriction. This method is shown to outperform most methods of fuzzy logic inference when the inputs were exponentially defined functions of the antecedent clause (VERY LONG, MORE-OR-LESS LONG, etc.) [64].

To ease the computational burden of performing modus ponens inferences with fuzzy sets, and to preserve the generalization capability, Keller et al have introduced neural network architectures to accomplish the fuzzy logic inferences. These architectures can be trained on multiple conjunctive or disjunctive antecedent clause rules and can actually store several compatible rules in one structure, providing a natural method of conflict resolution [65-67].

Spatial relationships between regions in an image play an important role in scene understanding. Humans are able to quickly ascertain the relationship between two objects, for example "B is to the right of A", but this has turned out to be a somewhat illusive task for automation. The determination of spatial relationships is critical for higher level vision processes (based on artificial intelligence) involved in, for example, autonomous navigation, medical diagnosis, or more generally, scene interpretation. When the objects in a scene are represented by crisp sets, the all-or-nothing definition of the subsets actually adds to the problem of generating such relational descriptions. It is our belief that definitions of spatial relationships based on fuzzy set theory, coupled with a fuzzy segmentation will yield realistic results. While this activity is embryonic, we note that in [68], initial models for spatial relationships of fuzzy regions are proposed, and in [69], fuzzy logic inference regarding propagation of spatial and temporal constraints is investigated.

8. Conclusions

The use of fuzzy set theory is growing in computer vision as it is in all intelligent processing. The representation capability is flexible and intuitively pleasing, the combination schemes are mathematically justifiable and can be tailored to the particular problem at hand from low level aggregation to high level inferencing, and the results of the algorithms are excellent, producing not only crisp decisions when necessary, but also corresponding degrees of support.

There is much work left to be done at all levels of computer vision. One area of particular need is the calculation and subsequent use of (fuzzy) features from the output of fuzzy segmentation algorithms. Fuzzy set methods can also be used to solve the correspondence problem in stereo and motion. More research is also necessary in high level vision processes. Fuzzy set theory offers excellent potential for describing and manipulating object and region relationships, thereby assisting with scene interpretation. Finally, we believe that possibility distributions should be the model for the interface between (1) the human and the vision system and (2) high level vision subsystem and mid or low level vision processes.

9. References

1. S.K. Pal, and R.A. King. "Image enhancement using smoothing with fuzzy sets," *IEEE Transactions on System, Man, and Cybernetics*, Vol. SMC-11, 1981, pp. 494-501.
2. S.K. Pal, "A Note on the Quantitative measure of image enhancement through fuzziness", *IEEE Transactions on Pattern Analysis and Machine Intelligence*, vol. PAMI-4, no. 2, 1982, pp. 204-208.
3. T.K. De and B.N. Chatterji, "An Approach to a generalized technique for image contrast enhancement using the concept of fuzzy set", *Fuzzy Sets and Systems*, vol. 25, 1988, pp. 145-158.
4. S.K. Pal, and R.A. King. "Histogram equalization with S and π functions in detecting x-ray edges", *Electronics Letters*, Vol. 17, 1981, pp. 302-304.
5. J. Keller, H. Qiu, and H. Tahani, "The fuzzy integral in image segmentation", *Proceedings of NAFIPS Workshop*, New Orleans, June 1986, pp. 324-338.
6. R. Sankar, "Improvements in image enhancement using fuzzy sets", *Proceedings of NAFIPS Workshop*, New Orleans, June 2-4, 1986, pp. 502-515.
7. A. Rosenfeld, "Fuzzy digital topology", *Information and Control*, 40, 1979, pp. 76-87.
8. A. Rosenfeld, "On connectivity properties of gray scale pictures", *Pattern Recognition*, 16, 1983, pp. 47-50.
9. A. Rosenfeld, "The fuzzy geometry of image subsets", *Pattern Recognition Letters*, 2, 1984, pp. 311-317.
10. D. Dubois and M.C. Jaulent, "A general approach to parameter evaluations in fuzzy digital pictures", *Pattern Recognition Letters*, Vol. 6, 1987, pp. 251-259.
11. S.K. Pal, R.A.King and A.A. Hishim, "Automatic grey level thresholding through index of fuzziness and entropy", *Pattern Recognition Letters*, vol. 1, 1983, pp. 141-146.
12. S.K. Pal, "A measure of edge ambiguity using fuzzy sets", *Pattern Recog Letters*, vol. 4, 1986, pp. 51-56.
13. S.K. Pal and A. Ghosh, "Index of area coverage of fuzzy image subsets and object extraction", *Pattern Recognition Letters*, vol. 11, 1990, pp. 831-841.

14. Y. Nakagawa, and A. Rosenfeld, "A note on the use of local min and max operators in digital picture processing," *IEEE Transactions on System, Man and Cybernetics*, Vol. SMC-8, 1978, pp. 632-635.

15. S.K. Pal, and R.A. King. "On edge detection of x-ray images using fuzzy sets," *IEEE Transactions on Pattern Analysis and Machine Intelligence*, Vol. PAMI-5, 1983, pp. 69-77.

16. M.M. Gupta, G.K. Knopf, and P.N. Mikiforuk, "Edge Perception Using Fuzzy Logic", in *Fuzzy Computing: Theory Hardware and Applications*, North Holland, 1988.

17. Huntsberger, and M. Desclazi, "Color edge detection", *Pattern Recognition Letters*, 3, 1985, 205.

18. J.M. Prewitt, "Object enhancement and extraction", in *Picture Processing and Psychopictorics*, B.S. Lipkin and A. Rosenfeld (Eds.), Academic Press, New York, 1970, pp. 75-149.

19. C.A. Murthy and S.K. Pal, "Fuzzy thresholding: mathematical framework, bound functions and weighted moving average technique", *Pattern Recognition Letters*, vol. 11, 1990, pp. 197-206.

20. S.K. Pal and A. Rosenfeld, "Image enhancement and thresholding by optimization of fuzzy compactness", *Pattern Recognition Letters*, vol. 7, 1988, pp. 77-86.

21. J. T. Kent and K.V. Mardia, "Spatial classification using fuzzy membership models", *IEEE Transactions on Pattern Analysis and Machine Intelligence*, vol. 10, no. 5, 1988, pp. 659-671.

22. J.C. Dunn, A fuzzy relative of the Isodata process and its use in detecting compact well-separated clusters, *Journal Cybernet* 31(3), 1974, pp. 32-57.

23. C. Bezdek, *Pattern Recognition with Fuzzy Objective Function Algorithms*, Plenum Press, New York, 1981.

24. T. Huntsberger, C. Jacobs, and R. Cannon, "Iterative fuzzy image segmentation", *Pattern Recognition*, vol. 18, 1985, pp. 131-138.

25. R. Cannon, J. Dave and J. Bezdek, "Efficient implementation of the fuzzy c-means clustering algorithm," *IEEE Transactions on Pattern Analysis Machine Intelligence*, Vol. 8, No. 2, 1986, pp. 248-255.

26. T. Huntsberger., "Representation of uncertainty in low level vision", *IEEE Transactions on Computers*, Vol. 235, No. 2, 145, 1986, p. 145.

27. R. Cannon, J. Dave, J.C. Bezdek, and M. Trivedi, "Segmentation of a thematic mapper image using the fuzzy c-means clustering algorithm," *IEEE Transactions on Geographical Science and Remote Sensing*, Vol. 24, No. 3, 1986, pp. 400-408.

28. J. Keller and C. Carpenter, "Image Segmentation in the Presence of Uncertainty," *International Journal of Intelligent Systems*, vol. 5, 1990, pp. 193-208.

29. M.M. Trivedi and J. Bezdek, "Low-Level segmentation of aerial images with fuzzy clustering, *IEEE Transactions on Systems, Man, and Cybernetics*, vol. SMC-16, No. 4, 1986, pp. 589-598.

30. I. Gath and A.B. Geva, "Unsupervised Optimal Fuzzy Clustering", *IEEE Transactions on Pattern Analysis Machine Intelligence*, vol. PAMI-11, no. 7, July 1989, pp. 773-781.

31. R. Krishnapuram and A. Munshi, "Cluster-Based Segmentation of Range Images Using Differential-Geometric Features", to appear in *Optical Engineering*, October 1991.

32. J. Keller and Y. Seo, "Local fractal geometric features for image segmentation", to appear in *International Journal of Imaging Systems and Technology*, Vol. 2, 1990, pp. 267-284.

33. M.P. Windham, "Cluster validity for the fuzzy c-means clustering algorithm", *IEEE Transactions on Pattern Analysis and Machine Intelligence*, vol. PAMI-4, no. 4, 1982, pp. 357-363.

34. E. Backer and A.K. Jain, "A clustering performance measure based on fuzzy set decomposition", *IEEE Transactions on Pattern Analysis and Machine Intelligence*, vol. PAMI-3, no. 1, 1981, pp. 66-75.

35. S. Peleg and A. Rosenfeld, "A mini-max medial axis transformation, *IEEE Transactions on Pattern Analysis and Machine Intelligence*, Vol. PAMI-3, 1981, pp. 208-210.

36. C.R. Dyer and A. Rosenfeld, "Thinning operations on grayscale pictures," *IEEE Transactions on Pattern Analysis and Machine Intelligence*, Vol. PAMI-1, 1979, pp. 88-89.

37. S.K. Pal "Fuzzy sketetonization of an image", Pattern Recognition Letters, vol. 10, 1989, pp. 17-23.

38. S. K. Pal and L. Wang, "Fuzzy Medial Axis Transformation (FMAT): Redundancy, Approximation and Computational Aspects", *Proceedings of the International Fuzzy Systems Association Congress*, Brussels, 1991, volume on *Engineering*, pp. 167-170.

39. S.K. Pal, R.A. King, and A.A. Hashim, "Image description and primitive extraction using fuzzy sets", *IEEE Transaction on Systems, Man, and Cybernetics*, vol. SMC-13, No. 1, 1983, pp. 94-100.

40. J. Bezdek and I.M. Anderson, "An Application of the c-varieties clustering algorithms to polygonal curve fitting", *IEEE Transactions on Systems, Man, and Cybernetics*, vol. SMC-15, No. 5, 1985, pp. 637-641.

41. R. Dave, "Use of the adaptive fuzzy clustering algorithm to detect lines in digital images", *Proceedings of the Intelligent Robots and Computer Vision VIII*, vol. 1192, no. 2, 1989, pp. 600-611.

42. J. Bezdek, C. Cordy, R. Gunderson and J. Watson, "Detection and characterization of cluster substructure", *SIAM Journal Applied Mathematics*, Vol. 40, 1981, pp. 339-372.

43. M. Windham, "Geometrical fuzzy clustering algorithms", *Fuzzy Sets and Systems*, vol. 10, 1983, pp. 271-279.

44. R. Krishnapuram and C.-P. Freg, "Algorithms to detect linear and planar clusters and their applications", *Proceedings of the IEEE Conference on Computer Vision and Pattern Recognition*, Hawaii, June 1991, pp. 426-431.

45. R. Krishnapuram and C.-P. Freg, "Fitting an unknown number of lines and planes to image data through compatible cluster merging", to appear in *Pattern Recognition.*

46. R. Dave, "Fuzzy Shell-Clustering and applications to circle detection in digital images", *International Journal of General Systems*, vol. 16, No. 4, 1990, pp. 343-355.

47. R. Krishnapuram, O. Nasraoui, and H. Frigui, "Fuzzy C Shells: A New Approach", submitted to the *IEEE Transactions on Neural Networks, Special Issue on Fuzzy Systems.*

48. R. N. Dave, "Adaptive C-shells clustering", *Proceedings of the North American Fuzzy Information Processing Society Workshop*, Columbia, Missouri, 1991, pp. 195-199.

49. R. Krishnapuram and J. Lee "Fuzzy-Connective-Based Hierarchical Aggregation Networks for Decision Making", to appear in *Fuzzy Sets and Systems*, March 1992.

50. H.J. Zimmermann and P. Zysno "Decisions and evaluations by hierarchical aggregation of information", *Fuzzy Sets and Systems*, vol.10, no.3, 1983 pp. 243-260.

51. R. Krishnapuram and J. Lee, "Fuzzy-Compensative-Connective-Based Hierarchical Networks and their Application to Computer Vision", to appear in *Journal of Neural Networks.*

52. Tahani, "The generalized fuzzy integral in computer vision," Ph.D. dissertation, University of Missouri - Columbia, 1990.

53. H. Qiu and J. Keller, "Multispectral segmentation using fuzzy techniques," *Proceedings of the NAFIPS Workshop*, Purdue University, May 1987, pp. 374-387.

54. H. Tahani and J. Keller, "Information fusion in computer vision using the fuzzy integral", *IEEE Transactions on System, Man and Cybernetics*, vol. 20, no. 3, 1990, pp. 733-741.

55. M. Sugeno, "Fuzzy measures and fuzzy integrals: A survey", in *Fuzzy Automatic and Decision Processes*, North Holland, Amsterdam, 1977, pp. 89-102.

56. J. Keller, G. Hobson, J. Wootton, A. Nafarieh, and K. Luetkemeyer, "Fuzzy confidence measures in midlevel vision," *IEEE Transactions on System, Man and Cybernetics*, Vol. SMC-17, No. 4, 1987, pp. 676-683.

57. J. Keller and D. Jeffreys, "Linguistic computations in computer vision", *Proceedings of the NAFIPS Workshop*, Vol. 2, Toronto, 1990, pp. 432-435.

58. W. Dong, H. Shaw and F. Wang, "Fuzzy computations in risk and decision analysis", *Civil Engineering Systems*, vol. 2, 1985, pp. 201-208.

59. J. Keller, M. Gray, and J. Givens, "A fuzzy k-nearest neighbor algorithm," *IEEE Transactions on System, Man, and Cybernetics*, vol. 15, 1985, pp. 580-585.

60. J. Keller, D. Subhanghasen, K. Unklesbay, and N. Unklesbay, "An approximate reasoning technique for recognition in color images of beef steaks", *International Journal General Systems*, Vol. 16, 1990, pp. 331-342.

61. A. Nafarieh and J. Keller, "A fuzzy logic rule-based automatic target recognizer", *International Journal of Intelligent Systems* to appear, Vol. 6, 1990, pp. 295-312.

62. L. Zadeh, "The concept of a linguistic variable and its application to approximate reasoning", *Information Sciences*, Part 1, Vol. 8, pp. 199-249; Part 2, Vol. 8, pp. 301-357; Part 3, Vol. 9, pp. 43-80, 1975.

63. I. B. Turksen and Z. Zhong, "An approximate analogical reasoning approach based on similarity measures", *IEEE Transactions on Systems, Man and Cybernetics*, vol. 18, 1988, pp. 1044-1056.

64. A. Nafarieh and J. Keller, "A new approach to inference in approximate reasoning", *Fuzzy Sets and Systems*, vol. 41, 1991, pp. 17-37.

65. J. Keller and H. Tahani, "Backpropagation neural networks for fuzzy logic", *Information Sciences*, to appear 1992.

66. J. Keller and R. Yager, "Fuzzy logic inference neural networks", *Proceedings of the SPIE Symposium on Intelligent Robots and Computer Vision VIII*, 1989, pp. 582-591.

67. J. Keller and H. Tahani, "Implementation of conjunctive and disjunctive fuzzy logic rules with neural networks", *International Journal of Approximate Reasoning*, to appear 1991.

68. J. Keller and L. Sztandera, "Spatial relations among fuzzy subsets of an image", *Proceedings of the First International Symposium on Uncertainty Modeling and Analysis*, College Park, MD, 1990, pp. 207-211.

69. S. Dutta, "Approximate spatial reasoning: Integrating qualitative and quantitative constraints", *International Journal of Approximate Reasoning*, Vol. 5, 1991, pp. 307-331.

70. R. Krishnapuram, H. Frigui and O. Nasraoui, "New fuzzy shell clustering algorithms for boundary detection and pattern recognition", *Proc. of the SPIE Conf. on Intelligent Robots and Computer Vision*, Boston, Nov. 1991.

Fuzzy Sets in Image Processing and Recognition

Sankar K. Pal[*], Senior Member, IEEE
Software Technology Branch/PT4
NASA Johnson Space Center
Houston, Texas 77058, USA.

Abstract: The various aspects of image processing and analysis problems where the theory of fuzzy sets has so far been applied are addressed along with their relevance and applications. The possibility of making fusion of the merits of fuzzy set theory, neural network theory and genetic algorithms for improved performance is discussed. Some contours of future directions of research are outlined. A list of representative references is also provided.

Introduction

The task of pattern recognition by a computer can be viewed as a transformation from the measurement space M to the feature space F and finally to the decision space D, i.e.,

$$M \rightarrow F \rightarrow D.$$

When the input pattern is a gray tone image, the measurement space involves some important processing tasks such as enhancement, filtering, noise reduction, segmentation, contour extraction and skeleton extraction, in order to derive salient features from the image pattern. This is what is basically known as *image processing*. The ultimate aim is to use data contained in the image to enable the system to understand, recognize and interpret the processed information available from the image pattern. Such a complete image recognition/interpretation system is called a *vision system* which may be viewed as consisting of three levels namely, low level, mid level and high level.

The relevance of fuzzy set theory in pattern recognition problems has adequately been addressed in the literature [1-4]. It is seen that the concept of fuzzy sets can be used at the feature level in representing an input pattern as an array of membership values denoting the degree of possession of certain properties and in representing linguistically phrased input features, at the classification level in representing multi-class membership of an ambiguous pattern, and in providing an estimate (or a representation) of missing information in terms of membership values. In other words, fuzzy set theory may be incorporated in handling uncertainties (arising from deficiencies of information available from a situation; the deficiencies may result from incomplete, imprecise, ill-defined, not fully reliable, vague, contradictory information) in various stages of a pattern recognition system. While the application of fuzzy sets in cluster analysis and classifier design was in the process of development, an important and related effort in fuzzy image processing and recognition was evolving more or less in parallel with the aforesaid general developments. This evolution was based on the realization that many of the basic concepts in image analysis, e.g., the concept of an edge or a corner or a relation between regions, do not lend themselves well to precise definition. The present paper will address the development of this branch of pattern recognition in the light of fuzzy set theory.

A gray tone image possesses ambiguity within pixels due to the possible multi-valued levels of brightness in the image. This indeterminacy is due to inherent vagueness rather than randomness. Incertitude in an image pattern may be explained in terms of grayness ambiguity or spatial (geometrical) ambiguity or both. Grayness ambiguity means "indefiniteness" in deciding whether a pixel is white or black. Spatial ambiguity refers to "indefiniteness" in the shape and geometry of a region within the image.

Conventional approaches to image analysis and recognition [5-7] consist of segmenting the image into meaningful regions, extracting their edges and skeletons, computing various features/properties (e.g., area, perimeter, centroid etc.) and primitives (e.g., line, corner, curve etc.) of and relationships among the regions, and finally, developing decision rules/grammars for describing, interpreting and/or classifying the image and its subregions. In a conventional system each of these operations involves crisp decisions (i.e., yes or no, black or white, 0 or 1) to make regions, features, primitives, properties, relations and interpretations crisp.

Since the regions in an image are not always crisply defined, uncertainty can arise within every phase of the aforesaid tasks. Any decision made at a particular level will have an impact on all higher level activities. A recognition (or vision) system should have sufficient provision for representing and manipulating the uncertainties involved at every processing stage; i.e., in defining image regions, features, matching, and relations among them, so that the system

---

[*] *On leave from the Electronics & Communication Sciences Unit, Indian Statistical Institute, Calcutta 700035.*

retains as much of the "information content" of the data as possible. If this is done, the ultimate output (result) of the system will possess minimal uncertainty (and unlike conventional systems, it may not be biased or affected as much by lower level decision components).

For example, consider the problem of object extraction from a scene. Now, the question is "How can one define exactly the target or object region in a scene when its boundary is ill-defined?" Any hard thresholding made for the extraction of the object will propagate the associated uncertainty to subsequent stages (e.g., thinning, skeleton extraction, primitive selection,) and this might, in turn, affect feature analysis and recognition. Consider, for example, the case of *skeleton extraction* of a region through medial axis transformation (MAT). The medial axis transformation of a region in a binary picture is determined with respect to its boundary. In a gray tone image, the boundaries are not well defined. Therefore, errors are likely, (and hence further increase uncertainty in the system), if we compute the MAT from the aforesaid hard-segmented version of the image.

Thus, it is convenient, natural and appropriate to avoid committing ourselves to a specific (hard) decision (e.g., segmentation/ thresholding, edge detection and skeletonization), by allowing the segments or skeletons or contours to be fuzzy subsets of the image, the subsets being characterized by the possibility (degree) to which each pixel belongs to them. Similarly, for describing and interpreting ill-defined structural information in a pattern, it is natural to define primitives (line, corner, curve etc.) and relations among them using labels of fuzzy sets. For example, primitives which do not lend themselves to precise definition may be defined in terms of arcs with varying grades of membership from 0 to 1 representing their degree of belonging to more than one class. The production rules of a grammar may similarly be fuzzified to account for the fuzziness (impreciseness) in physical relation among the primitives; thereby increasing the generative power of a grammar for syntactic recognition of a pattern.

Basic principles and operations of image processing and recognition in the light of fuzzy set theory are available in [3]. In the following sections we will be describing the different aspects of image processing and analysis where the theory of fuzzy sets has so far been found to be useful. These include (i) development of image processing operations for weakening the strong commitments (as is done in the case of conventional techniques) in extracting them as ordinary subsets, (ii) providing image ambiguity/information measures and quantitative evaluation, (iii) computing fuzzy geometrical properties, (iv) describing/representing uncertainties in various operations, (v) using neural nets and genetic algorithms and (vi) their applications. A list of references representing these research areas has also been provided.

Image Definition

An image X of size MxN and L levels can be considered as an array of fuzzy singletons, each having a value of membership denoting its degree of brightness relative to some brightness level ℓ, $\ell = 0, 1, 2, \ldots L-1$. In the notation of fuzzy sets, we may therefore write $X = \{\mu_X(p)/p, p \in X\}$ where $\mu_X(p)$, $0 \leq \mu_X(p) \leq 1$ denotes the grade of possessing some property μ (e.g., brightness, edginess, smoothness) or of belonging to some subset (e.g., object, skeleton or contour) by a pixel or a point p. In other words, a fuzzy subset of an image X is a mapping μ from X into [0, 1]. For any point $p \in X, \mu(p)$ is called the degree of membership of p in μ. (Note that an ordinary subset of X can be regarded as a fuzzy subset for which μ takes on only the values 0 and 1.)

One may use either global or local information of an image in defining a membership function characterizing some property. For example, brightness or darkness property can be defined only in terms of gray value of a pixel whereas, edginess, darkness or textural property need the neighborhood information of a pixel to define their membership functions. Similarly, positional or coordinate information is necessary, in addition to gray level and neighborhood information to characterize a dynamic property of an image. Again, the aforesaid information can be used in a number of ways (in their various functional forms), depending on individuals opinion and/or the problem at hand, to define a requisite membership function for an image property or an image subset.

Fuzzy Geometry, Entropy and Ambiguity Measures

In a series of papers Rosenfeld has explained the concept of fuzzy geometry of a gray image which is a generalization of many crisp properties of and relations among regions in an image. These extensions include topological concepts of connectedness, adjacency and surroundness, starshapedness and convexity, area, perimeter, compactness, height, width, extent, diameter etc. Their various properties have also been defined. Papers [8-18] describe these concepts. Some more geometrical properties e.g., length, breadth, index of area coverage, adjacency, major axis, minor axis, center of gravity and density have recently been defined in [19, 20]. The faster methods of computation of these parameters using the co-occurrence matrix are also reported in [20]. These measures quantify ambiguities in the geometry (spatial domain) of an image.

The concept of fuzzy sets has also been found to be useful in providing image information measures [21-29]. These include index of fuzziness, correlation, global entropy, local (higher order) entropy and hybrid entropy. These measures may be used to describe grayness ambiguity of an image. For example, higher order entropy of order r gives a measure of the average amount of difficulty in making a decision whether any subset of pixels of size r possesses an image property or not. Hybrid entropy, on the other hand, represents an amount of difficulty in deciding whether a pixel possesses a certain property or not by making a prevision on its probability of occurrence. (It is assumed here that the fuzziness occurs because of the transformation of the complete white (0) and black pixels (1) through a degradation process; thereby modifying their values to lie in the intervals [0, 0.5] and [0.5, 1] respectively.)

Image Processing Operations and Quantitative Evaluation

It has been possible to develop some operations on fuzzy sets (e.g., shrinking and expanding, thinning, splitting and merging, measures of fuzziness, dilation, concentration and intensification) which can be used effectively in processing an image pattern. These applications include image enhancement, filtering, edge detection, smoothing, thinning and skeleton extraction, and quantitative analysis. A gray scale thinning algorithm is described in [30] based on the concept of fuzzy connectedness [8] between two pixels such that the dark regions in an image can be thinned without ever being explicitly segmented. Paper [31] demonstrates the formulation of a basic processing scheme, involving fuzzification of the image space, performing some operations and then defuzzification, which has been implemented for enhancement of an image (such as smoothing and contrast enhancement) using the max, min and INT operators along with S and π functions [32]. It has also been shown that the fuzziness in an image decreases with its enhancement and how it can be measured quantitatively with the concept of fuzzy entropy and index of fuzziness. An extension of this concept to enhance the contrast among various ill-defined regions using multiple applications of π and $(1-\pi)$ functions has been described in [33, 34] for edge detection of x-ray images. The edge detection operators involve max and min operations. A recent attempt to use relaxation (iterative) algorithms for fast image enhancement using various orders of fuzzy S functions along with its convergence property is made in [35]. Readers may consult the articles [3, 36-50] for further reference on fuzzy set theoretic image processing techniques, information measures (including color image), their applications and quantitative indices for image analysis.

The problem of evaluation of image quality forms another research area where the fuzzy set theory finds significant applications [26-28, 40-42, 45-47] because of the fuzziness of the human senses. An application of a relaxation type pixel classification procedure using fuzzy measures and fuzzy integrals has recently been reported [51] for subjective evaluation of printed color images.

Segmentation and Object Extraction

The problem of image segmentation plays a key role for its recognition analysis and description. It can be performed in two ways, namely gray level (histogram) thresholding and pixel classification using global and/or local information of an image space. Papers [52, 53] typically represent these two aspects on fuzzy segmentation. A histogram thresholding technique in providing both fuzzy and nonfuzzy versions by minimizing the grayness ambiguity (global entropy, index of fuzziness, index of crispness) and geometrical ambiguity (fuzzy compactness) of an image has been described in [52]. It uses different S type membership functions to define fuzzy "object regions" and then selects the one which is associated with the minimum value of these ambiguity measures. (Note that the nonfuzzy thresholds obtained automatically can then be used in defining cross-over points of S functions or π functions [31] for enhancing contrast among various regions.) The mathematical framework of the algorithm including the selection of S functions, its band width and bounds is available in [54]. Many other measures of image ambiguity e.g., local and conditional entropy [24, 55], fuzzy correlation [56], index of area coverage [19], adjacency [20] may similarly be used. Use of the aforesaid grayness and spatial ambiguity measures has recently been made in automatic selection of an appropriate enhancement (nonlinear) function for an unknown image [46].

Paper [53], on the other hand, shows an application of c-means fuzzy clustering techniques (the theory of which is adequately described in [1, 4] for segmentation of a color image by pixel classification. It involves coarse segmentation using thresholding techniques followed by a fine segmentation which uses the fuzzy c-means algorithm for assigning those pixels which remain unclassified after the coarse segmentation. An evaluation criterion is also defined on the basis of probability of error. The coarse-fine strategy reduces the enormous computational burden of the c-means algorithm. Some recent results on image segmentation using the fuzzy c-means algorithm and the conventional c-means algorithm with fuzzy integral and geometrical properties as features are reported in [20, 57-61].

It is to be mentioned here that the aforesaid developments on image processing operations are mainly based on the applications of fuzzy operators, properties and mathematics. Image processing and recognition based on the theory of approximate reasoning (i.e., based on "if-then" rules) should constitute another field of research in the near future.

Membership Function Evaluation

The problem of determining the appropriate membership function in image processing drew the attention of many researchers. Consider for example, the problem of gray level thresholding [52] using S functions. If there is a difference in opinion in defining an S function, the concept of spectral fuzzy sets [62] can be used to provide soft decisions (a set of thresholds along with their certainty values) by giving due respect to all the opinions. The bounds for S type functions have also been defined recently based on the properties of fuzzy correlation [63] so that any function lying in the bounds would give satisfactory segmentation results. It therefore demonstrates the flexibility of fuzzy algorithms. Xie and Bedrosian [23] have also made attempts in determining membership functions.

Primitive Extraction

As described in the introduction, a vision system should have sufficient provisions for representing the uncertainties involved at every stage so that it retains as much as possible the information content of the input image for making a decision at the highest level. The ultimate output of the system will then be associated with least uncertainty and, unlike conventional systems, it will not be biased or affected very much by the lower level decisions. There have been several attempts in this line to extract fuzzy primitives (or features) from fuzzy edge and segmented outputs of image regions for shape analysis, matching and recognition. For example, consider the article [64] which considers the fuzzy c-means algorithm for segmentation of color images and shows the methods of performing the tasks of edge detection and shape matching within the domain of fuzzy sets derived during the segmentation phase. An interpretation of the shape parameters of triangle, rectangle and quadrangle in terms of membership for "approximate isosceles triangles", "approximate equilateral triangles" and "approximate right triangles" and so on has been made for their classification. The other earlier attempts for extraction of fuzzy primitives (e.g., vertical, horizontal and oblique lines, sharp, fair and gentle curves, corner, cornerity, symmetry) or in providing similarity measures between triangle, rectangle, quadrangle for shape analysis and syntactic classification are available in [3, 65-76]. Recognition algorithms for skeletal maturity from x-ray images and for brain cell images using fuzzy grammars and primitives are also described in [75, 76].

Attempts have also been made [20, 77] in extracting the fuzzy skeleton of a region from its fuzzy segmented version which is obtained by a gray level thresholding technique [52]. Here, the membership of a pixel for the subset skeleton can be computed with respect to the ε edge (edge points of object after which its class membership value is less than or equal to ε, $0 \leq \varepsilon \leq 1$) of the object region. The skeletons produced by these methods do not depend much on the boundary selection. A fuzzy medial axis transformation (FMAT) based on fuzzy disks has recently been formulated [78, 79] in this regard which provides fuzzy skeletons of the higher intensity regions (and also the exact representation of the image) *without requiring* any kind of segmentation or thresholding. (Note that the skeleton or MAT of a region in a binary picture is determined with respect to its boundary [5, 6].) The definition of the FMAT involves natural extensions (generalization) of the concepts of maximal disk, union, inclusion and symmetry for an ordinary set to a fuzzy set. Some other attempts in this regard are reported in [80-82].

Polygonal approximation, linear splining, curve fitting or segmentation of plane curves constitutes another important area [5, 7, 83-85] in machine vision. An application of the c-varieties fuzzy clustering algorithm for providing good polygonal approximations (in the mean-squared error sense) of planar data sets is described in [86]. The utility of this approach, called boundary fit program, is demonstrated through several numerical examples. The use of fuzzy clustering techniques in detecting ellipses or circles in two dimensional data is recently reported in [87, 88].

Use of Neural Networks

Currently, a large number of researchers are concentrating on making fusion of the merits of both fuzzy set theory and neural network theory for intellectual and material gain in the field of computer and system science. The readers may refer to the articles [89-91] which demonstrate such an application for image segmentation, scene labeling and object extraction problems. Use of Hopfield type neural nets and self-organizing nets has also been made [92-94] for the object extraction problem. The concept of fuzzy processing (e.g., possibility of belonging to more than one class or property) is of value if the import of the decision propagates into a network of other related decisions. The knowledge of the other possibilities should not be ignored or discarded, particularly when these are available. For example, the fuzzy geometrical properties (described above) of a pattern can be used as features for learning the network parameters. The fuzzy segmented version, fuzzy edge detected version or fuzzy skeleton of an image may also be used along with their degrees (values) of ambiguities for the purpose of network training and its recognition.

A good review explaining the merits of combining these two technologies and indicating the future direction of research is available in [4]. The current state of the art of the aforesaid fusion appears to be directed mainly towards the use of the fuzzy set concept in NN, the use of NN theory in usual fuzzy algorithms and the development of fuzzy neurons. This proceedings will definitely contain many articles along this line.

Use of Genetic Algorithms (GAs)

Genetic algorithms [95], another new technology, have also been attempted recently to make them useful in pattern recognition problems involving adaptive and optimization processes. GAs differ from many conventional search algorithms in the following ways. They consider many points in the search space simultaneously, not a single point, and therefore have less chance of converging to local optima. They deal directly with strings of characters representing the parameter sets, not the parameters themselves. They use probabilistic rules to guide their searching process instead of deterministic rules.

In handling uncertainty in pattern analysis, GAs may be helpful in determining the appropriate membership functions, rules and parameter space, and in providing a *reasonably suitable* solution. For this purpose, a suitable fuzzy fitness function needs to be defined depending on the problem. Fuzziness may also be incorporated in the encoding process by introducing a membership function representing the degree of similarity/closeness between the chromosome parameters (strings). For example, consider a scene analysis problem where the relations amongst various segments (or objects) may be defined in terms of fuzzy labels such as close, around, partially behind, occluded, etc. Given a labelling of each of the segments the degrees to which each relationship fits each pair of segments can be measured. These measures can be combined to define an overall fuzzy fitness function. Given this fitness function, the relations amongst objects, and the relations amongst classes to which the objects belong, a genetic algorithm searches the space to find the best solution in determining a class to be associated most appropriately to each object. An approach based on genetic algorithm for scene labelling is reported in [96].

Some recent attempts in applying the GAs for classification, segmentation, primitive extraction and vision problems are reported in [97]. The basic idea is to use the GA to search efficiently the hyperspace of parameters in order to maximize some desirable criteria. For example, consider the task of extracting fuzzy medial axis transformation (FMAT) [79] of an image which involves enormous computation and it is not guaranteed even if the resulting output provides a compact minimal set for image representation. Searching based on GAs may be helpful in this case. It is to be mentioned here that the GAs are computationally expensive. Moreover, one should be careful in selecting the initial population and the recombination operators.

Conclusions

The different aspects of image processing and analysis (low and mid level vision) problems where the theory of fuzzy sets has so far been applied are discussed along with their applications. These include enhancement, edge detection, thinning, segmentation, object extraction, skeleton extraction, primitive extraction, information and ambiguity measures, curve fitting, and use of neural learning and GAs. The relevance of fuzzy sets in weakening the strong commitments, as is done in the case of conventional operations, and in managing uncertainty in a recognition system is explained. Some contours of future directions of research are also outlined. Application of fuzzy image processing and recognition techniques in space autonomous research [98, 99] is in progress at the Johnson Space Center.

Acknowledgements: This article was written while the author held an NRC-NASA Senior Research Associateship at the Johnson Space Center, Houston, Texas. He is also grateful to Dr. R.N. Lea for his interest in this work.

References

[1] J.C. Bezdek, Pattern Recognition with Fuzzy Objective Function Algorithm, Plenum Press, NY, 1981.
[2] A. Kandel, Fuzzy Techniques in Pattern Recognition, Wiley Interscience, NY, 1982.
[3] S.K. Pal and D. Dutta Majumder, Fuzzy Mathematical Approach to Pattern Recognition, John Wiley & Sons (Halsted), N.Y., 1986.
[4] J.C. Bezdek and S.K. Pal, Fuzzy Models for Pattern Recognition, IEEE Press, N.Y., 1992 (to appear).
[5] A. Rosenfeld and A.C. Kak, Digital Picture Processing, 2nd ed., Academic Press, N.Y., 1982.

[6] R.C. Gonzalez and P. Wintz, <u>Digital Image Processing</u>, 2nd ed., Addison-Wesley Pub. Co., Reading, Massachusetts, 1987.
[7] D. Marr, <u>Vision</u>, W.H. Freeman, San Francisco, 1982.
[8] A. Rosenfeld, "Fuzzy digital topology, <u>Inform. Control,</u> vol. 40, pp. 76-87,1979.
[9] A. Rosenfeld, "The fuzzy geometry of image subsets", <u>Patt. Recog. Lett.</u>, vol. 2, pp. 311-317, 1984.
[10] A. Rosenfeld, "Fuzzy graphs", in [18], pp. 75-95.
[11] R. Lowen, "Convex fuzzy sets", <u>Fuzzy Sets and Systems</u>, vol. 3, pp. 291-310, 1980.
[12] L. Janos and A. Rosenfeld, "Some results on fuzzy (digital) convexity", <u>Pattern Recognition</u>, vol. 15, pp. 379-382, 1982.
[13] A. Rosenfeld, "On connectivity properties of grayscale pictures", <u>Pattern Recognition</u>, vol. 16, pp. 47-50, 1983.
[14] A. Rosenfeld, "The diameter of a fuzzy set", <u>Fuzzy Sets and Systems</u>, vol.13, pp. 241-246, 1984.
[15] A. Rosenfeld and S. Haber, "The perimeter of a fuzzy set", <u>Pattern Recognition</u>, vol. 18, pp. 125-130, 1985.
[16] A. Rosenfeld, "Distance between fuzzy sets", <u>Patt. Recog. Lett.</u>, vol. 3, pp. 229-233, 1985.
[17] A. Rosenfeld, "Fuzzy rectangles", <u>Patt. Recog. Lett.</u>, vol. 11, pp. 677-679, 1990.
[18] L.A. Zadeh, K.S. Fu, K. Tanaka and M. Shimura, (eds.), <u>Fuzzy Sets and Their Applications to Cognitive and Decision Processes</u>, Academic Press, New York, 1975.
[19] S.K. Pal and A. Ghosh, "Index of area coverage of fuzzy image subsets and object extraction", <u>Patt. Recog. Lett.</u>, vol. 11, pp. 831-841, 1990.
[20] S.K. Pal and A. Ghosh, "Fuzzy geometry in image analysis", <u>Fuzzy Sets and Systems</u>, (to appear).
[21] A. De Luca and S. Termini, "A definition of a nonprobabilistic entropy in the setting of fuzzy set theory", <u>Inform. Control</u>, Vol. 20, pp. 301-312, 1972.
[22] W.X. Xie and S.D. Bedrosian, "An information measure for fuzzy sets", <u>IEEE Trans. Syst., Man and Cyberns.</u>, vol. SMC-14, pp. 151-156, 1984.
[23] W. X. Xie and S.D. Bedrosian, Experimentally derived fuzzy membership function for gray level images", <u>J. Franklin Inst.</u>, vol. 325, pp. 154-164, 1988.
[24] N.R. Pal and S.K. Pal, "Entropy: a new definition and its applications", <u>IEEE Trans. Syst. Man and Cyberns.</u>, vol. SMC-21, no. 5, October 1991 (to appear).
[25] N.R. Pal and S.K. Pal, "Higher order fuzzy entropy and hybrid entropy of a set", <u>Inform. Sci.</u>, vol. 61, no. 3, pp. 211- 231, June 1992 (to appear).
[26] N.R. Pal, <u>On Image Information Measures and Object Extraction,</u> Ph. D. Thesis, Indian Statistical Institute, Calcutta, March 1990.
[27] W.X. Xie, "An information measure for a color space", <u>Fuzzy Sets and Systs.</u> vol. 36, pp 157-165, 1990.
[28] S.K. Pal, "Fuzzy geometry, entropy and image information", (Invited Talk), <u>NASA Conference Publication 10061,</u> February 1991, <u>(Proc. Second Joint Technology Workshop on Neural Networks and Fuzzy Logic,</u> NASA, Johnson Space Center, Houston, Texas, April 11-13, 1990), vol. II, pp. 211-232.
[29] S.K. Pal, "Fuzziness, image information and uncertainty management in pattern recognition" (invited paper), <u>J. Sci. Industrial Research</u>, (in press).
[30] C.R. Dyer and A. Rosenfeld, "Thinning algorithms for gray scale pictures", <u>IEEE Trans. Patt. Anal. Mach. Intell.,</u> vol. PAMI-1, pp. 88-89, 1979.
[31] S.K. Pal and R.A. King, "Image enhancement using smoothing with fuzzy set", <u>IEEE Trans. Syst. Man & Cyberns.,</u> vol. SMC-11, pp. 494-501, 1981.
[32] L.A. Zadeh, "Calculus of fuzzy restrictions', in [18], pp. 1-39.
[33] S.K. Pal and R.A. King, "Histogram equalisation with S and π functions in detecting x-ray edges", <u>Electronics Letters,</u> vol. 17, no. 8, pp. 302-304, 16th April 1981.
[34] S.K. Pal and R.A. King, "On edge detection of x-ray images using fuzzy set", <u>IEEE Trans. Pattern Anal. Machine Intell.,</u> vol. PAMI-5, pp. 69-77, 1983.
[35] H. Li and H.S. Yang, "Fast and reliable image enhancement using fuzzy relaxation technique", <u>IEEE Trans. Syst. Man & Cyberns</u>, vol. SMC-19, pp. 1276-1281, 1989.
[36] Y. Nakagowa and A. Rosenfeld, "A note on the use of local max and min operations in digital picture processing", <u>IEEE Trans. Syst., Man Cyberns.</u> , vol. SMC-8, pp. 632-635, 1978.
[37] R. Jain, "Application of fuzzy sets for the analysis of complex scenes", in <u>Advances in Fuzzy Set Theory and Applications,</u> M.M. Gupta Ed., North Holland, Amsterdam, 1979, pp. 577-587.
[38] S.K. Pal and R.A. King, "Image enhancement using fuzzy set", <u>Electronics Lett.,</u> vol. 16, pp. 376-378, 1980.
[39] V. Goetcherian, "From binary to gray tone image processing using fuzzy logic concepts", <u>Pattern Recognition</u>, vol. 12, pp. 7-15, 1980.

[40] S.K. Pal, "A note on the quantitative measure of image enhancement through fuzziness", IEEE Trans. Patt. Anal. and Machine Intell. , vol. PAMI-4, pp. 204-208, 1982.
[41] T.L. Huntsberger and M.F. Descalzi, "Color edge detection", Patt. Recog. Lett., vol.3, pp. 205-209, 1984.
[42] S.K. Pal, "A measure of edge ambiguity using fuzzy sets", Patt. Recog. Lett., vol. 4, pp. 51-56, 1986.
[43] R. Jain and S. Haynes, "Imprecision in computer vision", Computer, vol. 15, pp. 39-48, 1982.
[44] J. Keller, G. Hobson, J. Wootton, A. Nafarieh and K. Luetkemeyer, "Fuzzy confidence measures in midlevel vision", IEEE Trans. Syst., Man Cyberns. , vol. SMC-17, pp. 676-683, 1987.
[45] M.K. Kundu and S.K. Pal, "A note on gray level-intensity transformation: effect on HVS thresholding", Patt. Recog. Lett., vol. 8, no. 4, pp. 257-269, 1988.
[46] M.K. Kundu and S.K. Pal, "Automatic selection of object enhancement operator with quantitative justification based on fuzzy set theoretic measure", Patt. Recog. Lett., vol. 11, pp. 811-829, 1990.
[47] S.K. Pal, "Fuzzy tools for the management of uncertainty in pattern recognition, image analysis, vision and expert system", Int. J. Syst. Sci., vol. 22, pp. 511-549, 1991.
[48] S.K. Pal and N.R. Pal, "Higher order entropy, hybrid entropy and their applications", Proc. INDO-US Workshop on Spectrum analysis in one and two dimensions, Nov 27-29, 1990, New Delhi, NBH Oxford Publishing Co., New Delhi (to appear).
[49] H. Tahani and J. Keller, "Information fusion in computer vision using the fuzzy integral", IEEE Trans. Syst., Man Cyberns. , vol. SMC- 20, pp. 733-741, 1990.
[50] A. Nafarieh and J. Keller, "A fuzzy logic rule-based automatic target recognizer", Int. J. Intell. Systs., vol. 6, pp. 295-312, 1991.
[51] K. Tanaka and M. Sugeno, "A study on subjective evaluations of printed color images", Int. J. Approximate Reasoning, vol. 5, pp. 213-222, 1991.
[52] S.K. Pal and A. Rosenfeld, "Image enhancement and thresholding by optimization of fuzzy compactness", Patt. Recog. Lett., vol. 7, pp. 77-86, 1988.
[53] Y.W. Lim and S.U. Lee, "On the color image segmentation algorithm based on the thresholding and fuzzy c-means techniques", Pattern Recognition, vol. 23, pp. 935-952, 1990.
[54] C.A. Murthy and S.K. Pal, "Fuzzy thresholding: mathematical framework, bound functions and weighted moving average technique", Patt. Recog. Lett., vol. 11, pp. 197-206, 1990.
[55] N.R. Pal and S.K. Pal, "Object-background segmentation using new definitions of entropy", IEE Proceedings-E, vol. 136, pp. 284-295, 1989.
[56] S.K. Pal and A. Ghosh, "Image segmentation using fuzzy correlation", Inform. Sci., vol. 62, (to appear).
[57] T.L. Huntsberger, C.L. Jacobs and R.L. Cannon, "Iterative fuzzy image segmentation", Pattern Recognition, vol. 18, pp. 131-138, 1985.
[58] R. Cannon, J. Dave, J.C. Bezdek and M. Trivedi, "Segmentation of a thematic mapper image using the fuzzy c-means clustering algorithm", IEEE Trans. Geographical Sci. and Remote Sensing, vol. 24, pp. 400-408, 1986.
[59] M. Trivedi and J.C. Bezdek, "Low level segmentation of serial images with fuzzy clustering", IEEE Trans. Syst., Man Cyberns. , vol. SMC-16, pp. 580-598, 1986.
[60] J. Keller, D. Subhangkasen and K. Unklesbay, "Approximate reasoning for recognition in color image of beef steaks", Int. J. General Syst., vol. 16, pp. 331-342, 1990.
[61] B. Yan and J. Keller, "Conditional fuzzy measures and image segmentation", Proc. NAFIPS' 91, University of Missouri-Columbia, May 14-17, 1991, pp. 32-36.
[62] S.K. Pal and A. Dasgupta, "Spectral fuzzy sets and soft thresholding", Inform. Sci., (to appear).
[63] C.A. Murthy, S.K. Pal and D. Dutta Majumder, "Correlation between two membership functions", Fuzzy Sets and Systems, vol. 17, pp. 23-38, 1985.
[64] T.L. Huntsberger, C. Rangarajan and S.N. Jayaramamurthy, "Representation of uncertainty in computer vision using fuzzy sets", IEEE Trans. Comp., vol. C-35, pp. 145-156, 1986.
[65] E.T. Lee, "Proximity measures for the classification of geometric figures", J. Cyberns., vol. 2, pp. 43-59, 1972.
[66] E.T. Lee, "Shape-oriented chromosome classification", IEEE Trans. Syst., Man Cyberns. , vol. SMC-5, pp. 629-632, 1975.
[67] M.G. Thomason, "Finite fuzzy automata, regular fuzzy language and pattern recognition", Pattern Recognition, vol. 5, pp. 383-390, 1973.
[68] E.T. Lee, "The shape oriented dissimilarity of polygons and its application to the classification of chromosome images", Pattern Recognition, vol. 6, pp. 47-60, 1974.
[69] D. Dutta Majumder and S.K. Pal, "On fuzzification, fuzzy language and fuzzy classifier", Proc. IEEE 7th Int. Conf. Cyberns. and Soc., Washington, D.C., 1977, pp. 591-595.

[70] B.B. Chaudhuri and D. Dutta Majumder, "Recognition of fuzzy description of sides and and symmetries of figures by computer", Int. J. Syst. Sci., vol. 11, pp. 1435-1445, 1980.

[71] L. Vanderheydt, F. Dom, A. Dosterlinck and H. Van den Berghe, "Two-dimensional shape decomposition using fuzzy set theory applied to automated chromosome analysis", Pattern Recognition, vol. 13, pp. 147-157, 1981.

[72] E.T. Lee, "Fuzzy tree automata and syntactic pattern recognition", IEEE Trans. Patt. Anal. and Machine Intell., vol. PAMI-4, pp. 445-449, 1982.

[73] S.K. Pal, R.A. King and A.A. Hashim, "Image description and primitive extraction using fuzzy set", IEEE Trans. Syst., Man Cyberns., vol. SMC-13, pp. 94-100, 1983.

[74] S.A. Kwabwe, S.K. Pal and R.A. King, "Recognition of bones from x-ray of the hand", Int. J. Syst. Sci., vol. 16, pp. 403-413, 1985.

[75] A. Pathak and S.K. Pal, "Fuzzy grammars in syntactic recognition of skeletal maturity from x-rays", IEEE Trans. Syst., Man and Cyberns., vol. SMC-16, pp. 657-667, 1986.

[76] S.K. Pal and A. Bhattacharyya, "Pattern recognition technique in analyzing the effect of thiourea on brain neurosecretory cells", Patt. Recog. Lett., vol. 11, pp. 443-452, 1990.

[77] S.K. Pal, "Fuzzy skeletonization of an image", Patt. Recog. Lett., vol. 10, pp. 17-23, 1989.

[78] S.K. Pal and A. Rosenfeld, "A fuzzy medial axis transformation based on fuzzy disks", Patt. Recog. Lett., (in press).

[79] S.K. Pal and L. Wang, "Fuzzy Medial Axis Transformation (FMAT): Redundancy, Approximation and Computational Aspects", Proc. IFSA' 91 Congress, Brussels, July 7-12, 1991, pp. 167-170.

[80] S. Peleg and A. Rosenfeld, "A min-max medial axis transformation", IEEE Trans. Patt. Anal. Mach. Intell. vol. PAMI-3, pp. 208-210, 1981.

[81] E. Salari and P. Siy, "The ridge-seeking method for obtaining the skeleton of digital images", IEEE Trans. Syst., Man Cyberns., vol. SMC-14, pp. 524-528, 1984.

[82] J. Serra, Image Analysis and Mathematical Morphology, Academic Press, London, 1982, 1988.

[83] E.L. Hall, Computer Image Processing and Recognition, Academic Press, N.Y., 1978.

[84] T. Pavlidis, Structural Pattern Recognition, Springer-Verlag, N.Y., 1977.

[85] R. Duda and P. Hart, Pattern Classification and Scene Analysis, Wiley Interscience, NY, 1973.

[86] J.C. Bezdek and I.M. Anderson, "An application of the c-varieties clustering algorithms to polygonal curve fitting", IEEE Trans. Syst. Man & Cyberns, vol. SMC-15, pp. 637-641, 1985.

[87] R. Dave, "Fuzzy shell-clustering and applications to circle detection in digital images", Int. J. General Syst., vol. 16, pp. 343-355, 1990.

[88] R. Dave and K. Bhaswan, "Adaptive fuzzy c-shells clustering", Proc. NAFIPS' 91, University of Missouri-Columbia, May 14-17, 1991, pp. 195-199.

[89] L.O Hall, A. Bensaid, L. Clarke, R. Velthuisen and J.C. Bezdek, "Segmentation of MR images with fuzzy and neural network techniques", IEEE Trans. Neural Networks,1992, (in press).

[90] A. Ghosh, N.R. Pal and S.K. Pal, Self-organization for Object Extraction using Multilayer Neural Network and Fuzziness Measures, IEEE Trans. Neural Networks, (communicated).

[91] R. Krishnapuram and J. Lee, Fuzzy-Compensative -- Connected-Based Hierarchical Networks and Their Applications to Computer Vision, J. Neural Networks, (to appear).

[92] A. Ghosh, N.R. Pal and S.K. Pal, "Object background classification using Hopfield type neural network", Int. J. Patt. Recog. and Artificial Intell., (communicated).

[93] A. Ghosh, N.R. Pal and S.K. Pal, "Object extraction using a self-organizing neural network", Proc. Int. Symp. on Intell. Robotics (ISIR), Bangalore, India, January 1991, pp. 686-697.

[94] A. Ghosh, N.R. Pal and S.K. Pal, "Image segmentation using neural networks", Biological Cyberns., (to appear).

[95] D.E. Goldberg, Genetic Algorithms in Search, Optimization, and Machine Learning, Addison-Wesley Publishing Co, Reading, 1989.

[96] C. Ankenbrandt, B. Buckles and F. Petry, "Scene recognition using genetic algorithms with semantic nets", Patt. Recog. Lett., vol. 11, pp. 285-293, 1990.

[97] Proc. Fourth Int. Conf. on Genetic Algorithms (eds. R.K. Belew and L.B. Booker), Univ. of California, San Diego, July 13-16, 1991.

[98] S.K. Pal, "Uncertainty management in space station autonomous research: pattern recognition perspective", Inform. Sci., (accepted).

[99] R.N. Lea and Y. Jani, "Fuzzy logic in autonomous orbital operations", Int J. Approxt. Reasoning, (to appear).

SECTION II: Applications

Chapter 2: Vehicular Technology

Trainable Fuzzy and Neural-Fuzzy Systems for Idle-Speed Control

L. A. Feldkamp and G. V. Puskorius
Research Laboratory, Ford Motor Company
Suite 1100, Village Plaza
23400 Michigan Avenue
Dearborn, Michigan 48124
lfeldkam@smail.srl.ford.com

Abstract— We describe the use of a neural-network based procedure to train fuzzy or hybrid neural-fuzzy systems as vehicle idle-speed controllers. Simulation with a nonlinear model containing a significant delay is used, and we attempt to simulate the effects of realistic sampling and controller update frequencies. The present treatment may be regarded as a step toward on-line training with an actual system. The fuzzy system has a parameterized form similar to that described previously, allowing us to use methods identical to those we use for training neural networks. We illustrate the results of training by imposing various torque disturbances and showing the controller actions and the response of the system.

I. Introduction

The control of the speed of an automobile's engine under idle conditions is an old problem, but one that continues to be interesting and important. Modern passenger-car engines are usually controlled by computer to yield a combination of performance, economy, and emissions that is far superior to that of engines controlled by earlier methods. The whole of engine control is fertile ground for the application of new techniques employing neural networks and fuzzy logic. The idle-speed control (ISC) problem seems a logical first application of these methods, since ISC is comparatively isolated from other aspects of the overall engine control problem. At the same time, idle-speed control is a challenging problem, as it represents an attempt to regulate, against both anticipated and unanticipated disturbances, a highly nonlinear system operating well away from its optimal region. Because of time delays, the system state vector is incompletely known. Furthermore, the length of such delays can vary with time, making an analytical treatment very difficult. Finally, though the system itself may be regarded as continuous, information from it is available only in sampled form and control may be exercised only at discrete times.

Traditional approaches to the problem involve building system models whose parameters can be determined by a series of measurements. Even if the model is chosen to have nonlinearities, the controller design has usually been based on linearization of the model about an appropriate operating point. Delay elements in the model can either be explicitly ignored, as in PID control, or can be treated by posing the controller in state-variable form, in terms of both externally observed system states and appropriately defined internal states. Since the latter are not directly observable, it is asserted that they can be reconstructed by an *observer*, such as a Kalman filter. It is probably accurate to say that most production idle-speed controllers do not employ this last sophistication.

Recent papers [1] present a different approach using a fuzzy proportional-integral (PI) controller. The authors made use of simulation studies for a large part of the design process, using experimentation with an actual system largely for verification. The two-dimensional input space was divided into 64 regions, each of which corresponds to a rule of a max-min fuzzy system. Adequate performance was observed when all but 16 rules had been eliminated. Comparison to an LQ optimal controller derived by one of the authors [2] suggested that the fuzzy controller was slightly inferior. The difference between these should not be overinterpreted, however, since the LQ controller had the benefit of more information (the manifold pressure in addition to the speed error) and was subjected to optimization in terms of a performance function, a process that does not seem to have been applied in the design of the fuzzy controller.

A full approach to the ISC problem would involve simulation studies followed by on-line training of both fuzzy and neural controllers. In the present paper we describe simulation-based training of fuzzy controllers. We make use of a representative model that incorporates nonlinearity and time delay. Because we view the model as a convenient means of preparing to deal with the physical system, we attempt to simulate some of the harsh reality that may be anticipated. For example, we impose restrictions on the quality and timeliness of information supplied to the controller and on when control actions can be computed and supplied to the system. We carry out training on the fuzzy system using the same procedure that we have applied to the training of both neural and fuzzy systems for other problems, such as vehicle active suspension [3] and anti-lock braking [4]. During training, we minimize a quadratic performance function, as in conventional optimal control. We discuss the controller performance that is obtained when such fuzzy systems are provided inputs from fixed or trained preprocessing layers. Whether fixed or trained, the preprocessing layer is recurrent, i.e., it uses both current and past information from the system being controlled. Elsewhere [5] we consider in greater detail the efficacy of recurrent neural architectures for the ISC problem.

We have organized the remainder of this paper as follows. Section 2 contains a description of the engine model under idle conditions and details the way we interact with the model during the training process. In Section 3 we describe various controller architectures and in Section 4 consider the process of training. In Section 5 we discuss the results of training and make comparisons to results for a conventional control architecture and relate this work to that on recurrent neural networks. In Section 6 we provide some final comments.

II. ENGINE MODEL

The dynamic engine model employed in this study was derived from steady-state engine map data and empirical information by Powell and Cook [6], with revisions as described by Vachtsevanos et al. [7]. The engine model parameters are for a 4-cylinder fuel injected engine. The model is a two-state, two-input system. The states are manifold pressure P in kPa and engine speed N in rpm; the control inputs are throttle angle θ and spark advance δ in degrees. Disturbances act on the engine in the form of load torques T_d in N-m.

The evolution of the system is described by the following set of coupled equations:

$$\dot{P} = k_P(\dot{m}_{ai} - \dot{m}_{ao}), \ k_P = 42.40$$
$$\dot{N} = k_N(T_i - T_L), \ k_N = 54.26$$
$$\dot{m}_{ai} = (1 + 0.907\theta + 0.0998\theta^2)g(P)$$
$$g(P) = \begin{cases} 1 & P < 50.6625 \\ 0.0197(101.325P - P^2)^{\frac{1}{2}} & P \geq 50.6625 \end{cases}$$
$$\dot{m}_{ao} = -0.0005968N - 0.1336P$$
$$+ 0.0005341NP + 0.000001757NP^2$$
$$m_{ao} = \frac{\dot{m}_{ao}(t-\tau)}{120N}, \ \tau = \frac{45}{N}$$
$$T_i = -39.22 + 325024 m_{ao} - 0.0112\delta^2$$
$$+ 0.000675\delta N(2\pi/60) + 0.635\delta$$
$$+ 0.0216 N(2\pi/60) - 0.000102 N^2 (2\pi/60)^2$$
$$T_L = (N/263.17)^2 + T_d .$$

For notational simplicity, we have suppressed explicit dependence on time in these equations, except in the case of the delayed $\dot{m}_{ao}$. The time delay τ is treated here as a lumped quantity that represents the effect of the time between induction of the fuel mixture into a particular cylinder and the corresponding power stroke. The value of τ given above corresponds to an induction-power lag of 270 degrees.

Controller design with the neural network methods used here depends almost entirely on observing the output behavior of the system for actively determined input patterns. Since the above differential equations exhibit "stiff" behavior in certain regions of operation, a backward Euler scheme with a step size of 1 ms is used. Control commands are throttle angle θ in the range 5–25 degrees and spark advance δ in the range 10–45 degrees. Measurements of manifold pressure P and engine speed N are assumed to be available 4 times per engine revolution, with the current engine position being calculated from the engine speed. Control commands are computed at 20 ms intervals but are applied twice per engine revolution at fixed values of crank angle. With the assumed timing, crank rotation of at least 90 degrees (20 ms at 750 rpm) occurs before a measurement can be reflected in an applied control command. Measured manifold pressure P_m is corrupted by Gaussian noise of zero mean and variance 1 kPa^2, while measured engine speed N_m is corrupted by Gaussian noise of zero mean and variance of 6.25 rpm^2. Note that while control actions are computed at uniform time intervals, state variables are measured and control is asserted uniformly in engine position. Disturbances, ranging from 0 to 61 N-m, may begin and end at any time within the resolution implied by the 1 ms update interval for the differential equations. This means that changes in disturbance input occur asynchronously with the application of control and the measurement of system outputs. The range of applied disturbances is relatively large, apparently about twice as large as considered in [1].

The primary goal of ISC is to maintain a constant engine speed while the system experiences unobserved disturbances. For the engine model described above, a set point of 750 rpm is desired. The two controls affect engine speed rather differently. The throttle command θ has a large dynamic range (thousands of rpm), but its effect is delayed by a time inversely proportional to engine speed (for a steady state engine speed of 750 rpm, the delay is 60 ms). On the other hand, the spark advance δ has an immediate effect on the engine speed, but its dynamic range is roughly an order of magnitude smaller than that of the throttle command. In addition to maintaining engine speed at 750 rpm, we must respect secondary criteria. For example, we would like to maximize fuel efficiency while simultaneously minimizing vehicle vibrations. A means of increasing fuel efficiency is to operate the engine with spark angle advanced. For the engine model described above, the timing that results in maximum brake torque (MBT) for a steady-state speed of 750 rpm is found to be 30.9 degrees of spark advance. However, the spark advance must be retarded relative to this value to leave room to respond to torque load disturbances. We choose, somewhat arbitrarily, the nominal value $\delta = 22.9$ degrees (8 degrees retarded from MBT) to allow rapid system response to torque disturbances while maintaining a reasonable level of fuel efficiency.

III. CONTROLLER ARCHITECTURES

We have previously described [8] our procedure for tuning or training max-min fuzzy and hybrid neural-fuzzy systems. In brief, we parameterize membership functions in terms of analytic functions and replace MAX and MIN operations with soft approximations. This results in a computational structure that is completely differentiable and hence can be optimized (trained) in the same way as a neural network. For convenience, we made the following minor changes: 1) the analytic form for soft minimum given in [8] was replaced by the form used by Berenji [9], $smin(x,y) = \frac{xe^{-kx}+ye^{-ky}}{e^{-kx}+e^{-ky}}$, which maintains order invariance when generalized to more than two arguments. 2) The output membership functions were taken to be singletons with weighted average defuzzification. This modification simplifies and speeds up the training process.

We experimented with several possibilities for inputs to the fuzzy system. In the first we constructed estimates of speed error proportional and integral terms. This is the approach taken by Abate and Dosio [1]. After training several different fuzzy systems, we concluded that these inputs contain insufficient information for satisfactory control of the present system. We next supplied the fuzzy system with five inputs: speed error, integral of speed error, difference in speed error between consecutive time steps, manifold pressure, and difference in pressure between time steps. We term this PID/PD information. We derived these five inputs, as described in [5], by means of a preprocessing layer of seven recurrent nodes with a prespecified connection pattern and weights. This fixed preprocessing layer is also used for the training of a conventional PID/PD linear controller. Finally, we allowed the weights of the preprocessing layer, initialized to PID/PD values, to be trained along with the parameters of the fuzzy system.

The consequences of dealing with more than two or three inputs to a fuzzy system, whether trained or not, poses some difficulties that have nowhere been adequately explored. In our view, at least three approaches can be taken. The first is to construct a "fuzzy lookup table" by fully tiling the relevant portion of the input space using, for example, Gaussian input membership functions. The latter could be held fixed and only the output membership functions, perhaps taken as singletons, allowed to be trained. This approach could require hundreds or thousands of rules (here n^5, where n determines the resolution along each dimension) and a correspondingly large number of trainable parameters. Assuming a limited range on the input functions, indexing could be used so as to involve only a small fraction of the rules in calculation of each control action, but even this number would be significant, of order 2^5 or 3^5.

A second approach would be to use intuitive understanding of the problem to create a much smaller number of rules, many of which might involve a subset of the input variables. This procedure has been shown to work well in many cases, mostly simpler problems possessing limited dynamics or involving a single control action.

Our present approach is to train a system with relatively few rules (here 24), counting on the flexibility of the input membership functions to make up for the sparse coverage of the input space. Each rule of the fuzzy system has two (singleton) outputs. After defuzzification, the outputs of the fuzzy system feed sigmoidal nodes which explicitly limit the controller outputs θ and δ to their allowed ranges.

IV. TRAINING

We employ the same two-step procedure that we use to train neural controllers. In the first step, we train a neural network to identify the input-output characteristics of the unknown dynamical system. Then the trained identification network is used to provide estimates of the dynamic derivatives of plant outputs with respect to the trainable controller parameters. A training algorithm based upon a decoupled extended Kalman filter (DEKF) is employed for training both the identification network and the fuzzy controller, as described in [3, 10, 11, 12]. On the basis of considerable experience, we feel that the training of both feedforward and recurrent neural networks as well as fuzzy systems by DEKF algorithms generally leads to superior results with less total training time and fewer presentations of training data than does training by pure gradient descent.

Training of controllers is performed in an indirect fashion by using the identification network to model the input-output behavior of the system. In this scheme, the desired control signals are not known, but rather must be inferred indirectly through a specification of the system's desired behavior. This is provided by a subjectively and empirically determined cost function. For the ISC problem, we choose a quadratic cost function consisting of four terms. The first component penalizes deviations of engine speed from the desired value, here 750 rpm. The remaining

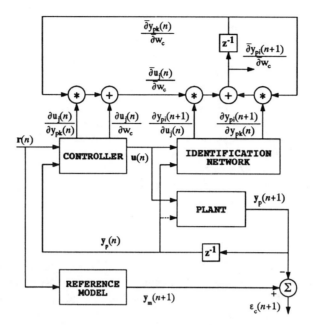

Figure 1: Recurrent derivative structure for controller training. This figure assumes that there are no internal feedback connections within either the identification or control networks, although there are external recurrent connections. The vertical lines emanating from the identification and control networks denote derivatives that are computed by backpropagation. The *total* partial derivatives that this recurrent derivative structure produces are used with the error vector $\varepsilon_c(n+1)$ to update the weights of the controller.

terms discourage certain behaviors of the control signals. We penalize deviations of the spark advance from 22.9 degrees for reasons discussed above. The third and fourth terms penalize large changes in throttle and spark angle between two successive control time steps. In this way we implement a "smoothness" constraint [13] that tends to inhibit oscillatory behavior in the controls for dynamical systems with significant internal time delays. The contribution to the cost function at time step n is given by

$$C(n) = \frac{1}{2} \left(\beta_1 (750 - N(n+1))^2 + \beta_2 (22.9 - \delta(n))^2 + \beta_3 (\theta(n-1) - \theta(n))^2 + \beta_4 (\delta(n-1) - \delta(n))^2 \right),$$

where the empirical weighting factors are $\beta_1 = 1.6 \times 10^{-7}$ for $N_m(n+1) > 750$ and $\beta_1 = 2.4^{-7}$ for $N_m(n+1) \leq 750$; $\beta_2 = 3.25 \times 10^{-7}$; $\beta_3 = 10^{-3}$; and $\beta_4 = 1.63 \times 10^{-5}$.

A critical step in training of controllers by gradient methods is the proper computation of derivatives of plant and controller outputs with respect to the trainable controller parameters. The computation of these derivatives is guided by two observations: 1) the evolution of the system state is defined recursively in terms of the previous state; 2) the computed control signals are defined recursively as a function of the measured system state and/or output, which is itself a function of the previous state and previously applied control signals. Hence the derivatives of the system state and controller outputs with respect to the trainable controller parameters should likewise be defined recursively in terms of the derivatives from previous time steps. The temporal evolution and computation of the *total* partial derivative of a component of plant output with respect to a controller parameter is illustrated in the sensitivity circuit of Figure 1. The derivatives computed can then be used by gradient methods such as dynamic backpropagation [14] or Kalman-filter-based algorithms to update the controller's trainable parameters.

A neural identification network consisting of a single layer of 8 completely interconnected (recurrent) nodes with linear activation functions was trained to model the input-output characteristics of the plant. The network has four inputs, consisting of the measured system state, P_m and N_m, and the control inputs, θ and δ. The output of the network is a prediction of the system state 20 ms into the future. The feedback weights of linear recurrent nodes are constrained to be less than unity in magnitude by treating them as the outputs of bipolar sigmoidal functions. This tends to make the training process computationally stable for linear recurrent nodes. The control inputs and torque disturbances were varied through their entire ranges during training of the identification networks. Since the torque disturbances are not observable, the prediction of system state by the identification network will be in error, particularly during large transitions in torque disturbance. However, we only require that on average the derivatives of system state with respect to controller parameters be correct.

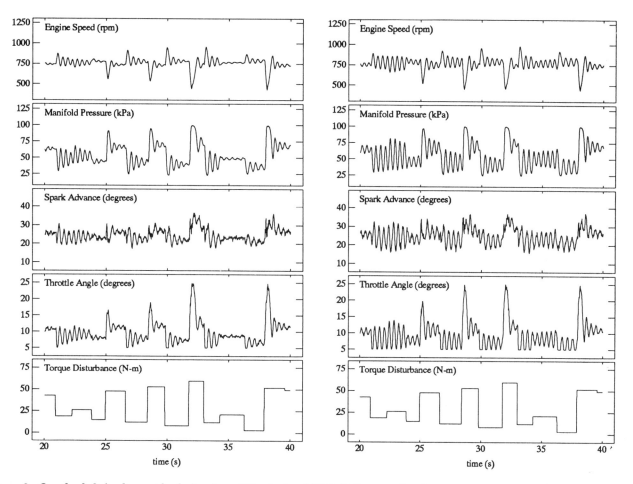

Figure 2: On the left is shown the behavior of the trained PID/PD controller for a test sequence of load disturbances, shown in the bottom panel. On the right is shown the performance of the (unmodified) controller for an altered plant with an offset in the measurement of manifold pressure, as described in the text.

V. Results

Results for a trained PID/PD linear controller are given in Figure 2 for a sequence of load disturbances independent from but generated in the same way as the disturbance pattern used in training. The response to both small and large disturbances reflects the unavoidable induction-power delay as well as delays in data acquisition and application of control. With this controller, the engine exhibits considerable oscillation in its response to torque disturbances.

The results obtained for a fuzzy system fed by the fixed preprocessing layer are superior to that of the PID/PD controller, but are not nearly as good as that obtained when the weights of the preprocessing layer are trained along with the fuzzy system. Figure 3 shows the performance of this hybrid controller. After the engine speed recovers from each disturbance, the response is fairly well damped. Additional testing with closely spaced disturbances disclosed no tendency to instability. In response to long periods of constant disturbance, the speed remains nearly constant (with very small fluctuations due to simulated measurement noise) at approximately the target value of 750 rpm.

As a measure of controller robustness, we altered the plant equations and introduced a systematic offset (-10 kPa) in the measurement of manifold pressure. The constants k_P and k_N are set to 38.16 and 59.69, respectively. The intake mass air equation is changed to $\dot{m}_{ai} = (1.2 + 0.907\theta + 0.12\theta^2)g(P)$. The expression for load torque is given by $T_L = (N/236.85)^2 + T_d$. The performance of the PID/PI controller on this altered plant, shown on the right side of Figure 2, is not very satisfactory. Though the performance of the hybrid controller on the altered plant (right side of Figure 3) is not as good as on the unaltered plant (greater ringing), it is clearly superior to that of the PID/PD controller.

As a further test, we repeated the test of the hybrid controller with a different altered plant (offset of +10 kPa, $k_P = 59.69$, $k_N = 38.16$, $\dot{m}_{ai} = (0.8 + 0.907\theta + 0.08\theta^2)g(P)$, and $T_L = (N/300)^2 + T_d$). The response in

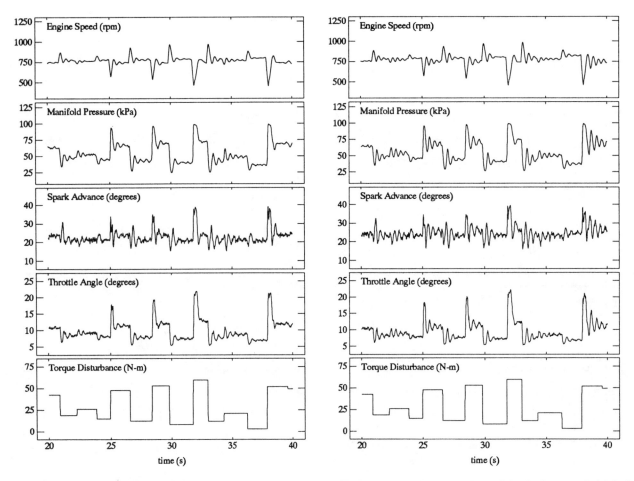

Figure 3: Illustration (left) of the behavior of the neural-fuzzy hybrid controller for the original plant and (right) for the altered plant and measurement offset.

engine speed was almost as good as for the original plant, i.e., better than for the first altered plant.

VI. Discussion

Control of the plant defined here seems to be a difficult problem. We found that the training process proceeded more slowly than similarly executed training for simpler plants. We doubt that a purely intuitive approach to fuzzy control would have much success in dealing with the model presented here. (Relaxing the key stipulations of our problem statement, as by neglecting the induction-power delay, makes the problem **much** easier.) To the extent that the model and supplementary conditions used in this exercise reflect the complexity of an actual system, it is not surprising that production ISC systems, which tend to contain relatively simple control architectures, sometimes exhibit interesting behavior.

This work confirms the evident principles that a good controller architecture must 1) be able to extract the required information from the available data and 2) have sufficient representational power to form good control actions when this information has been extracted. With the same fixed PID/PD preprocessor, a fuzzy system performed better than a linear controller, presumably due to greater representational power. Similarly, when preceding the same fuzzy system architecture, a trainable preprocessing layer gave rise to a better controller than did a fixed preprocessor, undoubtedly because it was providing better information. In parallel work, we have used a more extensive recurrent layer to produce even better control for this same plant. In [5] we present results obtained with a recurrent hidden layer of 8 nodes feeding 2 recurrent output nodes. Nearly equivalent performance can be obtained without recurrence in the output layer, either by single sigmoidal nodes or by a simple fuzzy system. This supports our view that a properly trained recurrent preprocessing layer can recover useful state information that may not be present in temporal combinations, such as PID, that are constructed *a priori*.

Acknowledgments: We thank our colleague B. K. Powell for helpful discussions of idle-speed control and other aspects of engine dynamics. We are grateful to S. S. Farinwata and Prof. G. Vachtsevanos of Georgia Institute of Technology for stimulating discussions and for providing details of the revised engine model.

REFERENCES

[1] M. Abate and N. Dosio (1990). Use of Fuzzy Logic for Engine Idle Speed Control. *SAE Paper 900594*, 107–114; M. Abate (1991). An Application of Fuzzy Logic to Engine Control. In *Proceedings of the Fuzzy and Neural Systems and Vehicle Applications '91 Conference*, November 8–9, 1991, Tokyo, Japan.

[2] M. Abate and V. Di Nunzio (1990). Idle Speed Control Using Optimal Regulation. *SAE Paper 905008*, 87–96.

[3] L. A. Feldkamp, G. V. Puskorius, L. I. Davis, Jr., and F. Yuan (1991). Decoupled Kalman Training of Neural and Fuzzy Controllers for Automotive Systems. *Proceedings of the Fuzzy and Neural Systems and Vehicle Applications '91 Conference*, November 8–9, 1991, Tokyo, Japan; L. A. Feldkamp, G. V. Puskorius, L. I. Davis, Jr., and F. Yuan (1992). Neural Control Systems Trained by Dynamic Gradient Methods for Automotive Applications. In *Proceedings of the 1992 International Joint Conference on Neural Networks* (Baltimore 1992) vol. II 798–804.

[4] L. I. Davis, Jr., G. V. Puskorius, F. Yuan, and L. A. Feldkamp (1992). Neural Network Modeling and Control of an Anti-Lock Brake System. *Proceedings of Intelligent Vehicles '92* (Detroit 1992) 179–184.

[5] G. V. Puskorius and L. A. Feldkamp (1993). Automotive Engine Idle Speed Control with Recurrent Neural Networks. Submitted to the *American Control Conference, 1993*.

[6] B. K. Powell and J. A. Cook (1987). Nonlinear Low Frequency Phenomenological Engine Modeling and Analysis. *Proceedings of the 1987 American Control Conference*, vol. 1, 336–340; J. A. Cook and B. K. Powell (1988). Modeling of an Internal Combustion Engine for Control Analysis. *IEEE Control Systems Magazine*, vol. 8, No. 4, 20–26.

[7] G. Vachtsevanos, S. S. Farinwata and H. Kang (1992). A Systematic Design Method for Fuzzy Logic Control With Application to Automotive Idle Speed Control. To appear in *Proceedings of the 31st IEEE Conference on Decision and Control* (Tuscon, AZ 1992).

[8] F. Yuan, L. A. Feldkamp, L. I. Davis, Jr., and G. V. Puskorius (1992). Training a Hybrid Neural-Fuzzy System. In *Proceedings of the 1992 International Joint Conference on Neural Networks* (Baltimore 1992) II-739–II-744; L. A. Feldkamp, G. V. Puskorius, F. Yuan, and L. I. Davis, Jr. (1992). Architecture and Training of a Hybrid Neural-Fuzzy System. In *Proceedings of the 2nd International Conference on Fuzzy Logic and Neural Networks* (Iizuka, Japan 1992) 131–134.

[9] H. Berenji (1992). Basic Concepts of Fuzzy Control. Tutorial in the *IEEE International Conference on Fuzzy Systems* (San Diego 1992).

[10] G. V. Puskorius and L. A. Feldkamp (1991). Decoupled Extended Kalman Filter Training of Feedforward Layered Networks. In *International Joint Conference on Neural Networks* (Seattle 1991), vol. I, 771–777. New York: IEEE.

[11] G. V. Puskorius and L. A. Feldkamp (1992). Recurrent Network Training with the Decoupled Extended Kalman Filter Algorithm. In *Proceedings of the 1992 SPIE Conference on the Science of Artificial Neural Networks* (Orlando 1992).

[12] G. V. Puskorius and L. A. Feldkamp (1992). Model Reference Adaptive Control with Recurrent Networks Trained by the Dynamic DEKF Algorithm. In *Proceedings of the 1992 International Joint Conference on Neural Networks* (Baltimore 1992).

[13] M. I. Jordan (1989) Generic Constraints on Underspecified Target Trajectories. In *International Joint Conference on Neural Networks* (Washington D.C. 1989), vol. I, 217–225. New York: IEEE.

[14] K. S. Narendra and K. Parthasarathy (1990). Identification and Control of Dynamical Systems Using Neural Networks. *IEEE Transactions on Neural Networks* 1, no. 1, 4–27; ibid. (1991). Gradient Methods for the Optimization of Dynamical Systems Containing Neural Networks. *IEEE Transactions on Neural Networks* 2, no. 2, 252–262.

Evaluation of Fuzzy and Neural Vehicle Control

Jos Nijhuis, Stefan Neußer,
Lambert Spaanenburg
Institut für Mikroelektronik Stuttgart
Allmandring 30A, 7000 Stuttgart-80

Jürgen Heller, Jochen Spönnemann
Institut für Steuerungstechnik
der Universität Stuttgart
Seidenstraße 36, 7000 Stuttgart-1

– **abstract**– *In the devlopment of autonomous mobile robots, control strategies must be formulated for collision avoidance, obstacle round-about, navigation and path planning. Where analytical methods have difficulties in translating sensor measurements into control directives, alternative approaches may provide a solution. This paper presents a neural and fuzzy solution to the collision avoidance problem of an automated guided vehicle (AGV). The advantages and problems of each approach are evaluated.*

1. Introduction

The application of an automated guided vehicle (AGV) in flexible manufacturing systems entails such probrem areas as navigation, path planning and collision avoidance. Within the last twenty years different navigation systems reaching from wire guidance to vision based systems have been developed [1], [2]. Looking at todays industrial applications one will find only a few wireless AGV-systems compared to the standard wire guided systems. All of those wireless systems use artificial landmarks in the surrounding in order to determine the vehicle position and orientation [3], [4]. Systems with no artificial marks have not yet reached an acceptable performance for application in industry. However, due to their enormous flexibility in terms of user-programming they offer large appeal for automated guided vehicles.

Another major problem in applying wireless guided vehicles in industrial systems is safety. The problem of collision avoidance either with the surrounding (or even more critical with humans) is currently solved by using mechanical bumpers. Only a few systems use non-contact methods for obstacle detection, e.g. ultrasonic or optical sensors. The performance of ultrasonic sensors strongly depends on the surface of the obstacle and/or other interference frequencies like ventilators and they produce misleading data if there are oncoming vehicles. Where optical sensors are concerned: they are either too expensive, not accurate enough or very much susceptible to dirt. Even more critical is the integration of the sensor system in the vehicle control. A well-performed control has to distinguish between different situations like travelling along a wall, in narrow alleys, in curves or omnidirectional movements, all in the presence of oncoming traffic. If possible the accuracy in sensor measurement should also be considered. An analytical approach to this problem leads to a lot of modelling difficulties. Therefore two methods based on neural networks and fuzzy logic respectively have been developed.

Section 2 describes the AGV and its sensor characteristics. A neural network solution is presented in Section 3. Next, Section 4 shows a fuzzy logic solution to the collision avoidance problem. Both approaches are compared in Section 5. Finally Section 6 draws some conclusions.

2. The Automated Guided Vehicle (AGV)

The described strategy for collision avoidance has been integrated in the control system of the automated guided vehicle called FLEXL [5]. Figure 1 shows the kinematic with two drives for the position and one drive for the orientation that can all be separately actuated. The environment is observed by means of one infrared scanner.

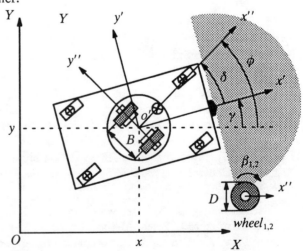

Figure 1 *Kinematic concept of the automated guided vehicle called FLEXL with (o,x,y) the world coordinate system, (o',x',y') the AGV coordinate system and (o',x'',y'') the rotation core coordinate system. The shaded semicircle indicates the area that is observed by the scanner.*

Considering a certain world coordinate system the kinematic equations of the vehicle are given by:

$$\dot{x} = \frac{D}{4}(\dot{\beta}_1 + \dot{\beta}_2)\cos(\gamma + \delta) \quad (1)$$

$$\dot{y} = \frac{D}{4}(\dot{\beta}_1 + \dot{\beta}_2)\sin(\gamma + \delta) \quad (2)$$

$$\dot{\gamma} = \frac{D}{2B}(\dot{\beta}_1 - \dot{\beta}_2) - \dot{\delta} \quad (3)$$

For obstacle detection a newly developed 180° infrared scanner is used that is located at the front of the AGV. Figure 1 shows the position and scanning direction of the scanner with respect to the AGV. The semicircle is divided into sectors of 3°. For each sector the scanner delivers an 8 bit digital value that represents the distance to an object in this sector.

Both neural and fuzzy controller should, based on the scanner information $s_1 \cdots s_{62}$, current speed v and heading direction δ, decelerate the AGV when continued traveling would lead to a collision. The diagram of the AGV control system is shown in Figure 2

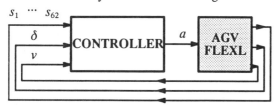

Figure 2 *The control system diagram.*

3. Neural Network based collision avoidance

Traditional control methodologies fall short as they are often unable to infer relevant information for control decisions from the sensory input data. Artificial neural networks can provide a solution to this problem as neural networks are capable of learning complex, highly non-linear relations between input and output even if these relations are not explicitly known, i.e. the network

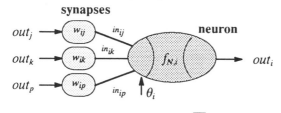

$$out_i(t) = f_{No,i}(\mathbf{w}_i \cdot \mathbf{out}_i(t) - \theta_i) = f_{No,i}(\sum_{\forall j \in IN_i} w_{ij} * out_j(t) - \theta_i)$$

$$f_{No,i}(x) = \frac{1}{1 + e^{-x}}$$

Figure 3 *The basic elements in a neural network: neurons and synapses.*

will learn to extract the essential sensor information. The basic elements of a neural network are synapses and neurons (see Figure 3). The complete neural network is built with these two basic elements. The actual behavior is determined by the synapse weights w_{ij} and the neuron thresholds θ_i. The design of a neural controller involves the following three basic steps:

1. Selection of controller inputs/outputs.
2. Selection of the network topology.
3. Definition of a training set.

ad. 1) Additional decoder stages may be used to adjust the input and output signals to the network requirements and to simplify the task that should be learned. The number of input and output neurons will correspond with respectively the number of input and output signals. In our case the input and output signals are perfectly suitable for direct use by the network, i.e. no additional processing (other than a simple linear scaling) between the infrared sensor and the controller is required.

ad. 2) Inspired by [6], [7], a multi-layer feedforward (MLF) neural network is used (see Figure 4). The net-

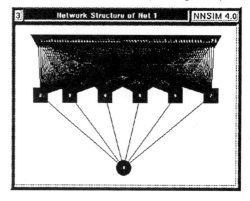

Figure 4 *The MLF-network that is used to implement the collision avoidance controller. The 66 inputs are present for the sensor information. The output neuron in the bottom layer delivers the desired break signal for the AGV.*

work has one input neuron for each sector of the scanner, e.g. 62 inputs. Two additional input neurons are used to receive the actual speed of the AGV, and the direction in which the AGV is heading with respect to the AGV coordinate system (e.g. angle δ in Figure 1). The output neuron delivers the deceleration for the AGV. The actually required number of hidden neurons has to be determined by experimentation. Note that a surplus of hidden neurons will not hamper the learning process.

ad. 3) The conventional error back-propagation learning rule [8] is used to adjust the network behavior to the desired control function, e.g. adjust the synapse weights and neuron thresholds. During learning the output of the network is compared with a desired or target

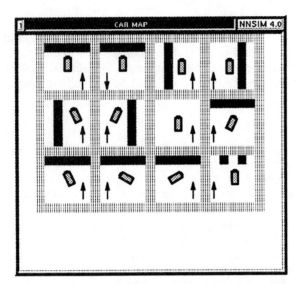

Figure 5 *The situations in the training set. The arrow indicates the heading direction of the AGV that is used for each particular training situation. The dark areas represent obstacles.*

response. Rather than explicitly defining a control action for each possible input combination, a limited set of 12 training examples is used (see Figure 5). It is assumed that the generalization capabilities of the neural network will guarantee a correct operation in situations that are not explicitly trained. During each learn step the AGV is placed in a randomly selected training situation. The corresponding sensor values are determined and together with the pre-defined heading direction and speed used as network inputs. The required output value or target deceleration can be easily calculated on forehand for each of the training situations.

Experiments showed that after ~ 500 learn cycles the neural controller was capable of halting the AGV when necessary. An example of a test situation is shown in Figure 6.

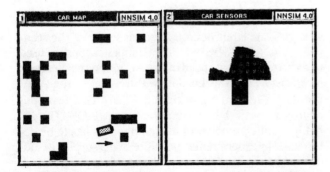

Figure 6 *An example of a test environment with random distributed obstacles (left screen). The right screen shows the actual range of the scanner. The arrow indicates the heading direction.*

4. Fuzzy Logic based collision avoidance

A paper by Zadeh [9] in 1965 spurred researchers all over the world to start working on the fuzzy theory itself and its application [10]. The use of fuzzy logic in control and closed-loop controlled systems in combination with sensor data processing showed that in many cases it is easier to deal with linguistic variables and if ... then rules than to develop a mathematical model covering all aspects of the real world. The design of a fuzzy logic controller involves the following three basic steps:

1. Determination of the linguistic variables.
2. Selection of if ... then rules.
3. Determination of membership functions.

ad. 1) The infrared scanner detects obstacles within a semicircle with a radius of 3.5m. However, the vehicle is able to move omnidirectional so that only obstacles in a part of the semicircle could lead to collisions whereas others are detected by the sensor but are not critical for this situation. Depending on the travelling direction of the vehicle a collision corridor is defined by d_c and the position of an obstacle in the corridor by d_o. Hence, the ratio $c = d_o/d_c$ is used as a linguistic variable. Small values of c indicate that an obstacle is right in front of the vehicle and a fast deceleration has to start, whereas large values of c indicate that the situation is not so dangerous. It is obvious that the distance d to the obstacle and the velocity v of the AGV will contribute to the strategy as well. The output of the fuzzy logic controller, e.g. the deceleration necessary to stop safely, will be taken as the fourth linguistic variable.

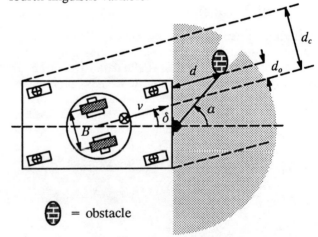

Figure 7 *The scanner and the definition of a dangerous zone d_c.*

ad. 2) The next step involves the determination of a membership function for each variable and a set of if ... then rules. Using the terms PS (positive small), PM (positive medium), PL (positive large), NS (negative small), NM (negative medium), NL (negative large), IN (within the collision corridor), OUT (outside the collision corridor) the following set of if ..then rules are used:

1. Rule: if(d = PS) and (v = PM) and (c = OUT) then (a = NS)
2. Rule: if(d = PS) and (v = PL) and (c = OUT) then (a = NM)
3. Rule: if(d = PS) and (v = PS) and (c = IN) then (a = NS)
4. Rule: if(d = PS) and (v = PM) and (c = IN) then (a = NM)
5. Rule: if(d = PS) and (v = PL) and (c = IN) then (a = NL)
6. Rule: if(d = PL) and (v = PL) and (c = OUT) then (a = NS)
7. Rule: if(d = PL) and (v = PM) and (c = IN) then (a = NS)
8. Rule: if(d = PL) and (v = PL) and (c = IN) then (a = NM)

ad. 3) For each of the four linguistic variables it should be specified how their actual value can be expressed in terms of PL, PM, PS, NS, NM, NL, OUT and IN, i.e. their degree of membership (D.o.M) of each term. For each variable membership functions should be defined. The evaluation results of if ... then rules (and thus the behavior of fuzzy controller) depends strongly on the actual shape of the membership functions. Experiments are needed to find an 'optimal' set. Figure 8 shows the membership functions for the four linguistic variables in our example.

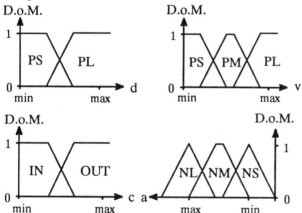

Figure 8 *The membership functions for the linguistic variables d, c, v, a.*

The behavior of the fuzzy controller is defined by the membership functions and the if ... then rules. During operation of the fuzzy controller, each rule is evaluated using the max–prod inference method. Figure 9 gives an example of this inference method for all the rules. After the evaluation of all rules, eight fuzzy values for the deceleration variable *a* are obtained. One defuzzified value is determined by means of the center of gravity method (see Figure 9).

With the shown membership functions and the if ... then rules the AGV halts when nearby obstacles make a collision free traveling impossible.

5. A Comparison and Combination

The previous two sections described how neural networks and fuzzy logic can solve the collision avoidance problem. The quality of the obtained controllers, with respect to the generated deceleration signal, is the same in both approaches. Differences between the fuzzy and neural network approach occur during the development and testing phase of the controller. At this stage each approach has its own distinct advantages.

- **Handling of complex sensor data**: Neural networks learn from examples, thereby automatically extracting relevant information from complex sensor data. The fuzzy logic solution requires an explicit transformation of the sensor data into one or more linguistic variables. Such a transformation be may difficult to find, especially when the input data is composed of the information from several, cheap (i.e. noisy) sensors. Furthermore, a change of the number of sensors or the sensor characteristics can by easily learned by the neural network whereas the fuzzy logic approach requires the construction of a new transformation rule in order to get reliable linguistic variables.

- **Expert knowledge needed**: Both approaches involve steps that require expert knowledge. Compared to the conventional system–theoretical approaches, this knowledge is more problem related and less control theory related. During the neural network controller design experience is required during the development of the training set as it determines the quality of the controller. The selection of the network topology causes normally no problems as for most control problems a standard MLF–topology can be used. The fuzzy logic approach requires experience during the determination of the linguistic variables, the rules and the membership functions. No structured design theory exists to guided the designer during these stages.

- **Understandability**: The distributed decision making by the neural network is very difficult to understand. It is hardly possible to identify individual neurons and synapses that handle a certain part of the controller behavior. Rather can rules be identified. The fuzzy logic solution is directly based on a set of understandable if ... then rules. Single rules can be pointed out and altered when a particular part of the controller behavior is not satisfactionary.

- **Realization efforts**: Traditional IC–design approaches are more familiar with the integration of a number of if ... then rules than with the integration of many densely interconnected neurons and synapses. This has led to a number of single chip hardware realizations of fuzzy logic controllers [11], whereas the hardware for neural controllers has not completely left the experimental stage [12]. Actual applications of neural networks are usually software emulated on standard hardware.

Combining the neural network and fuzzy logic approach may result in a system that is capable of learning

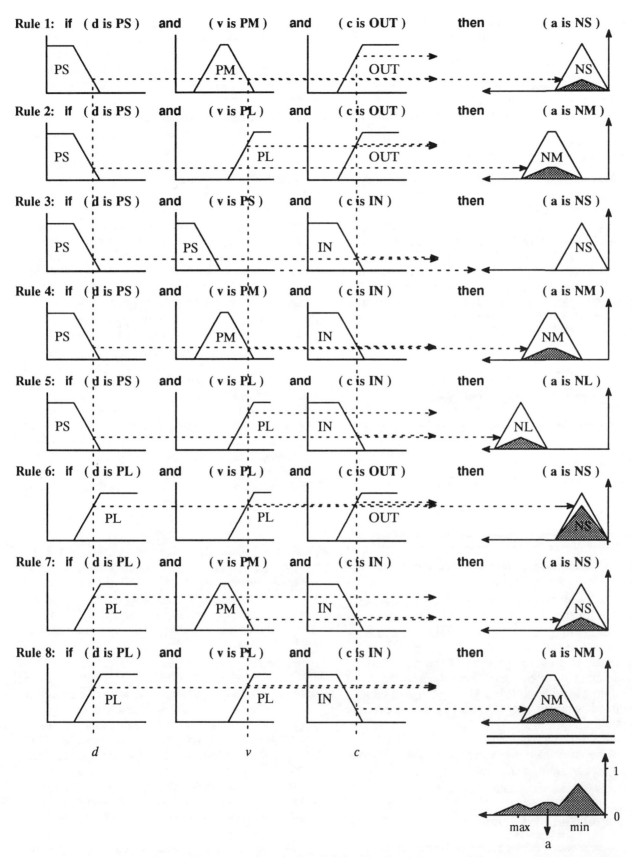

Figure 9 *Fuzzy inference with max–prod method and defuzzification by means of the center of gravity method.*

complex input/output relations whereas the knowledge is represented by understandable if ... then rules. The neural network can be used to: a) generate the linguistic variables, i.e. extract relevant information for complex sensor data (NN1 in Figure 10) [13] and/or b) to generate the fuzzy rule base (NN2 in Figure 10) [14], [15].

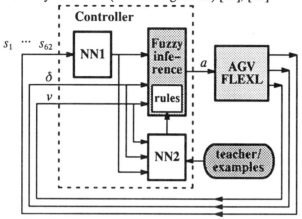

Figure 10 *The mixed neural/fuzzy control system diagram.*

By using a rule base to store the controller knowledge rather than the neural network itself, an expert can easily evaluate and add to the control behavior.

6. Conclusions

Experiments showed that the collision avoidance problem can be successfully tackled by both neural networks and fuzzy logic. Both approaches have the advantage that almost no control specific knowledge is needed. Neural network controllers are easier to design, whereas the operation of the fuzzy logic controller is more understandable, i.e. individual rules can be adjusted so optimize certain parts of the controller behavior.

A next step would be the use of a mixed neural/fuzzy system to realize the merits of both approaches in combination. Further research is needed in this direction.

7. References

[1] M. Jantzer, "Bahnverhalten und Regelung Fahrerloser Transportsysteme ohne Spurbindung", ISW Forschung und Praxis, Springer Verlag, Berlin, Germany, (in German), Band 82, 1990

[2] U, Rembold, "Autonome Mobile Roboter", *Robotsysteme*, (in German), Vol. 4, No. 1, pp. 17-26, 1988

[3] U. Wiklund, U. Andersson and K. Hyyppä, " AGV Navigation by Angle Measurements", (in: *Proceedings of the 6th International Conference on Automated Guided Vehicle Systems*), IFS Ltd and authors, pp. 199-212, 1988.

[4] H. van Brussel, C.C. van Helsdingen and K. Machiels, "FROG - Free Ranging on Grid: New Perspectives in Automated Transport", (in: *Proceedings of the 6th International Conference on Automated Guided Vehicle Systems*), IFS Ltd and authors, pp. 223-232, 1988.

[5] G. Pritschow and J. Heller, "Advanced Sensor-Guided Motion Control for Mobile Robots", (in: *Information Processing in Autonomous Mobile Robots*, ed. G. Schmidt), Springer-Verlag, Berlin, Germany, pp. 201-215, 1991.

[6] F. Tuijnman and B.J.A. Kroese, "Neural Networks for Collision Avoidance between Autonomous Mobile Robots", (in: *Intelligent Autonomous Systems*, eds. T. Kanade, F.C.A. Groen and L.O. Hertzberger), IOS, Amsterdam, The Netherlands, Vol. 2, pp. 407-416, 1989.

[7] J.A.G. Nijhuis, B. Hofflinger, S. Neusser, A. Siggelkow and L. Spaanenburg, "A VLSI Implementation of a Neural Car Collision Avoidance Controller", (in: *Proceedings International Joint Conference on Neural Networks*), Seattle, WA, Vol. 1, pp. 493-499, 8-12 July, 1991.

[8] D.E. Rumelhart, G.E. Hinton and R.J. Williams, "Learning Internal Representations by Error Propagation", (in: *Parallel Distributed Processing: Foundations*, eds. D.E. Rumelhart and J.L. McClelland), MIT Press, Cambridge, MA, Vol. 1, pp. 318-362, 1986.

[9] L.A. Zadeh, "Fuzzy Sets", *Information and Control*, Vol. 8, No. 3, pp. 338-353, June 1965.

[10] H-J. Zimmermann, *Fuzzy Sets Theory – and its Applications*, Kluwer-Nijhoff Publishing, Boston, MA., 1990.

[11] T. Wolf, "Fuzzy, die Revolution aus japanischen High-Tech-Tempeln", *MC*, (in German), pp. 44-49, March 1991.

[12] L. Spaanenburg, J.A.G. Nijhuis, S. Neußer and A. Siggelkow, "Digital Hardware für neuronale Netze", *HMD*, (in German), No. 159, pp. 87-109, May 1991.

[13] S. Nakanishi and T. Takagi, "Pattern Recognition by neural networks and fuzzy Inference", (in: *Proceedings of the International Conference on Fuzzy Logic*), Iizuka, Japan, pp. 183-186, 20-24 Juny, 1990.

[14] I. Enbutsu, K. Baba and N. Hara, "Fuzzy Rule Extraction from a Multilayered Neural Network", (in: *Proceedings of International Joint Conference on Neural Networks*), Seattle, WA, Vol. II, pp. 461-465, 8-12 July, 1991.

[15] W. Eppler, "Implementation of Fuzzy Production Systems with Neural Networks", (in: *Parallel Processing in Neural Systems and Computers*, eds. R. Eckmiller, G. Hartmann and G. Hauske), North-Holland, Amsterdam, The Netherlands, pp. 249-252, 1990.

FOLLOW-UP CHARACTERISTICS OF A SMALL AUTOMATIC GUIDED VEHICLE SYSTEM WITH FUZZY CONTROL

Nobuhiko YUASA , Mitsuo SHIOYA , Gunji KIMURA

Dept. of Electrical Engineering Faculty of Technology
Tokyo Metropolitan University
1-1, Minamiohsawa, Hachiouji-shi, Tokyo, 192-03, JAPAN

ABSTRACT

This paper describes a fuzzy control of a small automatic guided vehicle system. The purpose of this system is to maintain continuously constant distance between a forward running vehicle and a following vehicle. For this system, we propose a new dynamic tuning method of a scaling factor to improve the running performance of the automatic guided vehicle system. The simulation and experimental results verify the validity of the proposed method.

1. INTRODUCTION

Recently, the researches and the developments on the fuzzy controls have been popular in Japan [1][2]. This paper describes the fuzzy control of a small automatic guided vehicle(AGV) system constructed by a motor drive unit and an ultrasonic sensor. The purpose of this system is to maintain continuously constant distance between the forward running vehicle and the following vehicle, and a fuzzy control is applied to the system to improve the follow-up performance of the AGV [3].

The performance of the fuzzy control system depends on the control parameters,

Keywords
Fuzzy control, Automatic guided vehicle, Follow-up characteristics, and Scaling factor.

membership functions, control rules, scaling factors and other factors included in the system. In order to improve the performance of the system, it is necessary that these parameters are optimum [4]. In conventional fuzzy control, the scaling factors of the input and the output membership functions are treated as the constant values obtained experimentally. Besides, the fuzzy controller is constructed by the linguistic control rules, which are composed of the fuzzy variables and the fuzzy sets. And the fuzzy control rules have been derived by modeling a human's driving actions.

This paper proposes new dynamic tuning method of a scaling factor, which adopts a non-linear function in order to optimize the running performance of the AGV. The characteristics of the transient response are improved by new method, and the performance of the follow-up control is better than that of the conventional fuzzy control. Furthermore the performance by applying the fuzzy control is compared with that by the PID(Proportional, Integral, and Derivative) control. It is cleared that the fuzzy control is superior to the PID control.

2. SYSTEM CONSTRUCTION

Figure 1 shows a configuration of a used AGV system and figure 2 shows a distance measurement system using an ultrasonic sensor. This system consists of a forward running vehicle, an AGV(following vehicle) and a processing unit, using a personal computer to transmit and receive many control data between them. The AGV has the rear wheels driven by a stepping motor unit to control the vehicle velocity. An ultrasonic sensor is put on in front of the AGV to measure the distance between itself and the forward running vehicle. In the distance measurement system using the ultrasonic sensor, the distance between the AGV and the forward running vehicle is measured by counting the pulse generated by an oscillator. The forward running vehicle has only the rear wheels to control its velocity.

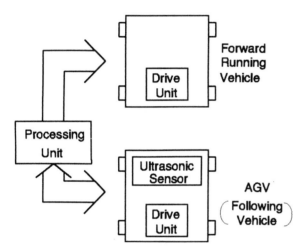

Figure 1 System Configuration

3. CONTROL METHOD

3.1 Outline of Control Method

As the AGV system has non-linear characteristics for variations of the vehicle velocity and the distance between itself and the forward running vehicle, the AGV has to vary its running performance according to these variations. Then, in order to compare the control characteristics, the fuzzy control method and the PID control method are applied to the AGV.

The simulation and the experimental running tests are performed as follows; At first, both the AGV and the forward running vehicle are at standstill. At the second state, the forward running vehicle runs at the velocity of 50[mm/sec], and after the passage of 12[sec] the forward running vehicle will speed up at 70[mm/sec]. The rear AGV follows up to the forward running AGV. The sampling time of the controllers are 0.1[sec] respectively.

3.2 PID Control

A structure of PID controller is shown in figure 3 and the block diagram of its controller is shown in figure 4. The PID controller consists of a combination of each control element as shown figure 3. The transfer functions of the rear AGV and the forward running vehicle is approximated to $1/s$. In this system a necessary torque depends on the induced torque by moment of inertia and the friction of the AGV. Then it is not necessary to apply an adjustable speed control. The slips of the tires are omitted, because of the negligible effect.

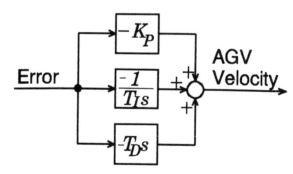

Figure 3 Structure of PID Controller

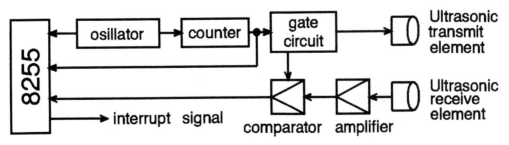

Figure 2 Ultrasonic Distance Measurement System

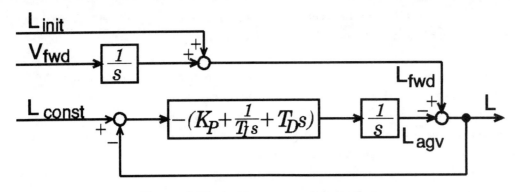

Figure 4 Block Diagram of PID Control

3.3 Fuzzy Control

The membership functions and a block diagram are shown in figure 5 and figure 6 respectively. The fuzzy control rules are shown in table 1. The meanings of the fuzzy labels are as follows:

NB : Negative Big
NS : Negative Small
ZO : Zero
PS : Positive Small
PB : Positive Big

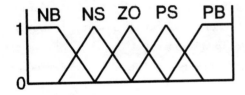

Figure 5 Membership Functions

And the specifications on the fuzzy inference are shown in table 2.

The membership functions are selected to be triangular and trapezoidal forms for calculation. The fuzzy controller consists of two input variables and one output variable, and estimates the AGV velocity and the distance between the AGV and the forward running vehicle. Then the controller infers and commands the appropriate change of velocity on the AGV.

Table 1 Control Rules of Fuzzy Inference

		ΔE				
		NB	NS	ZO	PS	PB
	NB	ZO	ZO	NS	NB	NB
	NS	ZO	ZO	NS	NS	NB
E	ZO	PS	PS	ZO	NS	NS
	PS	PB	PS	PS	ZO	ZO
	PB	PB	PB	PS	ZO	ZO

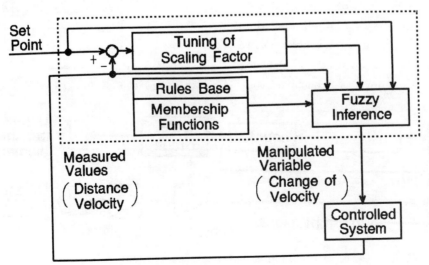

Figure 6 Block Diagram of Fuzzy Control

Table 2 Specifications of Fuzzy Inference

Control Period	0.1 [sec]
Rule Form	Production Rule of 2 inputs and 1 output
Rule Numbers	25 Rules
Form of Fuzzy Inference	Min/Max Operation
Form of Resultant Conclusion	Method of the Center of Gravity
Linguistic Values	5 Levels
Membership Functions	Continuous Function used the Triangular and the Trapezoidal Forms
Inputs	Velocity of the AGV and Relative Distance
Output	Change of Velocity of the AGV

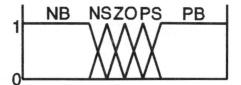

Figure 8 Membership Function
(Small Scaling Factor)

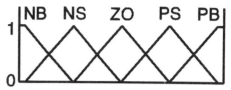

Figure 9 Membership Function
(Large Scaling Factor)

4. TUNING OF SCALING FACTOR

A new dynamic tuning method of the scaling factor using arctangent function is adopted to improve the running performance of the AGV system, because there are several appropriate shapes of the membership functions according to the conditions of the system. The scaling factor function is shown in figure 7.

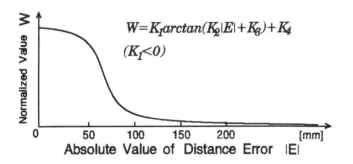

Figure 7 Scaling Factor Function

By using the arctangent function, the shapes of the membership functions are modified to correct the distance error. The dynamic tuned scaling factor is applied to the input variable of the distance error. The shapes of the varied membership functions are shown in figure 8 and figure 9. Figure 8 shows the membership functions when the distance error is large, and means that the gain of the system is large. Figure 9 shows the membership functions when the distance error is small, and means that the gain of the system is small.

To improve the control characteristics, the normalized input variable by the non-linear scaling factor is introduced by evaluating the control response at the real time. The normalizing function is defined as

$$W(t) = K_1 arctan(K_2|E(t)|+K_3)+K_4 \quad \cdots \quad (1)$$

where $W(t)$ is the normalized value of distance error, $E(t)$ is the distance error, and K_1, K_2, K_3, and K_4 are constant values.

5. SIMULATION AND EXPERIMENTAL RESULTS

The simulation results by the fuzzy control are shown in figure 10 to figure 12.

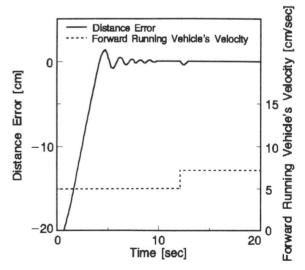

Figure 10 Characteristic of Response
(Small Scaling Factor)

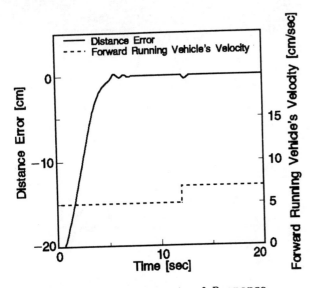

Figure 11 Characteristic of Response
(Large Scaling Factor)

Figure 10 and figure 11 show the characteristics of the response at the small constant value of scaling factor and the large one respectively. Figure 12 shows the characteristic of the response on the tuned scaling factor. On the way, the simulation results by the PID control and the fuzzy control are shown in figure 13 and figure 14. Figure 13 and figure 14 show the characteristics of the distance and the velocity respectively. The simulation and the experimental results by the fuzzy control are shown in figure 15.

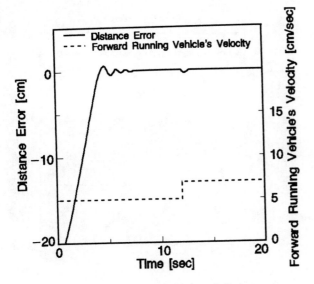

Figure 12 Characteristic of Response
(Tuned Scaling Factor)

For a constant value of the scaling factor as shown in figure 10 and figure 11, it is difficult to improve together a rise time and an overshoot at the same time. But by the application of the dynamic tuning of the scaling factor, a rise time and an over shoot are improved together as shown in figure 12. It is clarified that the PID control is inferior to the fuzzy control as shown in figure 13 although the PID controller is regulated optimally. For the structure of the PID controller, the values of the coefficient K_P, T_P and T_D included in the control elements are 0.4, 13.3, and 0.2 respectively.

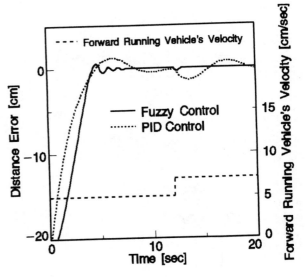

Figure 13 Characteristics of Distance

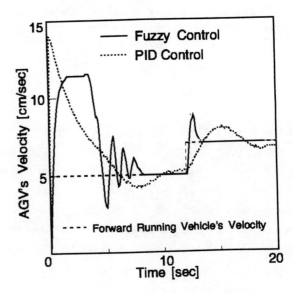

Figure 14 Characteristics of Velocity

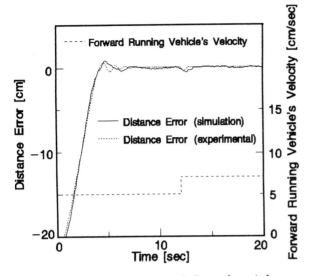

Figure 15 Simulation and Experimental Results

The experimental results have clarified that the new fuzzy control method is useful for the purpose of maintaining continuously constant distance of the AGV system. In the experimental result as shown in figure 15 a small fluctuation of error is occurred by the sensor, but the simulation result and the experimental result show similar characteristics of the response. Table 3 shows the evaluated transient response by each control method.

To evaluate the results of the Fuzzy controls and the PID control, we introduce an evaluation function that is given by,

$$PI = \int t|E(t)|dt \qquad (2)$$

where PI means a performance index, $E(t)$ expresses the distance error.
The evaluation function consists of an integral of the time multiplied by the absolute value of the distance error so as to estimate importantly the duration, while the error is causing. The period of evaluation is 0[sec] to 20[sec]. The performance indexes by each control method are shown in table 4, and show that the tuned fuzzy control is prefer than above mentioned other control methods.

Table 3 Evaluation of Each Control Method

Evaluation Scale	Fuzzy Scaling Factor			PID
	Small	Large	Tuned	
Rise Time [sec]	4.2	5.5	4.2	4.6
Overshoot [cm]	1.3	0.3	0.7	1.2
Osillatory Duration [sec]	9.8	7.3	7.4	23.1

Table 4 Performance Index

Control Methods	PI (ratio)
Fuzzy (Tuned Scaling Factor)	1.00
Fuzzy (Large Scaling Factor)	1.07
Fuzzy (Small Scaling Factor)	1.08
PID	1.95

6. CONCLUSION

This paper proposes a new dynamic tuning method of scaling factor which is non-linear function. It is found that the new dynamic tuning method of scaling factor is very useful for the AGV control. And new fuzzy control method has the better characteristics of response than those of the conventional fuzzy control method and the PID control method.

References

[1] M.Suzuki, M.Shioya, M.Nomura, "Driving of An Automated Guided Vehicle System with Fuzzy Control", National Convention Record of IEE JAPAN, p.6-92, March(1990)

[2] M.Nomura, M.Suzuki, M.Shioya, "An Automated Guided Vehicle SYSTEM With Fuzzy Control", National Convention Record IEE JAPAN -Industry Applications Society-, pp.I36-40, August(1990)

[3] N.Yuasa, M.Shioya, G.Kimura, "Driving of A Small Automated Guided Vehicle System with Fuzzy Control", National Convention Record of IEE JAPAN, p.6-94, April(1991)

[4] M.Hayashi, "A Tuning Method of Fuzzy Control System", 6th Fuzzy System Symposium, Tokyo, JAPAN, pp.189-192, September (1990)

Self Organizing Fuzzy Logic Control of a Level Control Rig

Nader Vijeh[1]
Department of Engineering Sciences
University of Exeter
Exeter, Devon, UK

Abstract - Self Organizing Fuzzy Controllers are able to generate and modify their Control Policy based on a given Performance Criteria. This ability allows these controllers to be used in applications were the knowledge to control the process does not exist or the process is subject to changes in its dynamic characteristics. The Self Organizing Controller (SOC) described here is applied to a Level Control Rig and the results are presented.

Introduction

The simple fuzzy controllers, with a fixed set of rules, are used successfully when the control protocol is available in a textual format and the process under control does not exhibit dynamic changes due to aging or variation in the set-point. The problem of rule acquisition manifests itself particularly when the rules have to be obtained from a human operator whose knowledge is based on experience and he applies them in an intuitive manner.

A possible solution to these problems is a Self-Organizing Controller (SOC) which is capable of generating and modifying the Control Protocol by a learning process based on measuring performance. The controller described here is based on that presented by Procyk and Mamdani [3].

The learning process of the SOC is comprised of algorithms that allow the controller to assess its own performance on the basis of a set of predetermined performance rules. At every sampling point the controller uses a Performance Decision Table to determine the performance in terms of Error and Change in Error variables. If the performance is unsatisfactory, then either an existing rule, leading to the present poor performance, is modified or a new one generated and the redundant rules are deleted. The block diagram below represents the conceptual structure of the SOC.

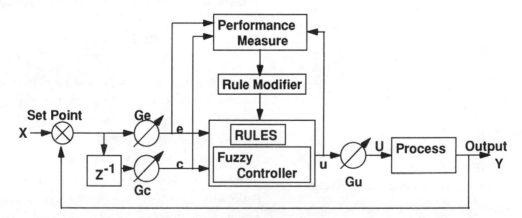

Fig 1. Conceptual Block Diagram of SOC

The controller can be viewed as a hierarchical rule based system, where a set of fixed performance rules are used to modify a second set of control rules. As can be seen from the block diagram, the lower half of the structure is essentially identical to that of a simple fuzzy controller. The performance rules represent the acceptable performance of the controller. In the case of a single-input single-output (SISO) process an assumption is made that the process is monotonic and an increase in the input will eventually cause an increase in the output.

---

[1]Author can be reached at: Advanced Micro Devices Inc., 901 Thompson Place, M/S 45, Sunnyvale, CA 94088. Email: nader.vijeh@amd.com.

Protocol modification Using a Performance index

Table 1. Performance Decision Table

		-6	-5	-4	-3	-2	-1	0	1	2	3	4	5	6
	-6	6	6	6	6	6	6	6	0	0	0	0	0	0
	-5	6	6	6	6	6	6	6	3	2	2	0	0	0
	-4	6	6	6	6	6	6	6	5	4	2	0	0	0
	-3	6	5	5	4	4	4	4	3	2	2	0	0	0
	-2	6	5	4	3	2	2	2	0	0	0	0	0	0
E	-1	5	4	3	2	1	1	1	0	0	0	0	0	0
R	-0	4	3	2	1	0	0	0	0	0	0	0	0	0
R	+0	0	0	0	0	0	0	0	0	0	-1	-2	-3	-4
O	+1	0	0	0	0	0	0	-1	-1	-1	-2	-3	-4	-5
R	+2	0	0	0	0	0	0	-2	-2	-2	-3	-4	-5	-6
	+3	0	0	0	-2	-2	-3	-4	-4	-4	-4	-5	-5	-6
	+4	0	0	0	-2	-4	-5	-6	-6	-6	-6	-6	-6	-6
	+5	0	0	0	-2	-2	-3	-6	-6	-6	-6	-6	-6	-6
	+6	0	0	0	0	0	0	-6	-6	-6	-6	-6	-6	-6

CHANGE IN ERROR

The SOC modifies the control protocol according to the performance decision table. The desired performance is described in a per sample basis and is therefore a local criterion as opposed to a global performance one such as Integral Square of Error (ISE). Although this table can be tuned to a specific process, minor changes in this table were found to cause insignificant changes in the controller performance.

The non-zero entries in this table represent cases where the performance is not satisfactory and a modification to the rules is necessary. The upper left hand and lower right hand corners of the table present situations where there is a large error and the output is moving away from the set-point. The lower left hand and upper right hand corners contain zeros as these are the situations where there is a large error but the output is moving towards the set-point rapidly. The middle region of the table represents zero error and stationary output, therefor no modification to the rules is necessary.

The region of table containing zeros, stretching from the lower left hand corner to the upper right hand corner, represents the desired trajectory of the process output.

The rule modification algorithm operates on previously stored states of the controller. The computer program stores up to 40 previous Error (e), Change in Error (c) and the Controller output (u) instances.

If the normalized error and change in error in a given sample point (n) are given by e_n and c_n, then the performance obtained from the above table is given by:
$p_n = P[e_n, c_n]$.

Where, P stands for the Performance decision table. If p_n is not zero, then a delay factor (d) is used to retrieve a previous state of the controller from the delay buffer. The three values extracted are: e_{n-d}, c_{n-d} and u_{n-d}. The delay in reward factor (**d**) depends on the delays in the process and is generally equal to one for a first order system.

A new rule is then formed by generating *antecedents* that correspond to e_{n-d} and c_{n-d} and a *consequence* that corresponds to $u_{n-d} + p_n$. This rule is then added to the control protocol and redundant rules with similar antecedents are deleted. Note that if no rule had existed for the **n-d** sample, then the corresponding change in control action (u_{n-d}) would be zero and the new rule would simply contain the consequence corresponding to p_n.

Self-Organizing Control Algorithm

Before describing details of the SOC algorithm it is necessary to clarify the manner in which rules are stored internally. Rules are made-up of fuzzy subsets with a fixed shape. In general these are vectors containing a single maximum value of 10 with two other values of 7 and 3 surrounding the maximum. The exceptions to this are boundary values and error subsets NZ and PZ. It is therefore possible to store each rule as a set of three numbers (e_m, c_m and u_m), corresponding to the position of the maximum value in the vector. Tables 2 and 3 describe the shape of Fuzzy Variables referenced here. For example the rule: *If* E is PM *and* C is NB *then* U is ZE can be stored as three integers: 12, 1 and 7.

The fuzzy subsets for each rule is then reconstructed using a *Fuzzification* procedure which simply recreates each subset in the shape of a vector, as specified in the following tables.

Table 2. Normalized ERROR (e)

Fuzzy Sets	-6	-5	-4	-3	-2	-1	-0	+0	+1	+2	+3	+4	+5	+6
NB	10	7	3	0	0	0	0	0	0	0	0	0	0	0
NM	3	7	10	7	3	0	0	0	0	0	0	0	0	0
NS	0	0	3	7	10	7	3	0	0	0	0	0	0	0
ZE-	0	0	0	0	3	7	10	0	0	0	0	0	0	0
ZE+	0	0	0	0	0	0	0	10	7	3	0	0	0	0
PS	0	0	0	0	0	0	0	3	7	10	7	3	0	0
PM	0	0	0	0	0	0	0	0	0	3	7	10	7	3
PB	0	0	0	0	0	0	0	0	0	0	0	3	7	10
Index	1	2	3	4	5	6	7	8	9	10	11	12	13	14

Table 3. Normalized Change in Error (c) and Change in Control Action (u)

Fuzzy Sets	-6	-5	-4	-3	-2	-1	0	+1	+2	+3	+4	+5	+6
NB	10	7	3	0	0	0	0	0	0	0	0	0	0
NM	3	7	10	7	3	0	0	0	0	0	0	0	0
NS	0	0	3	7	10	7	3	0	0	0	0	0	0
ZE	0	0	0	0	3	7	10	7	3	0	0	0	0
PS	0	0	0	0	0	0	3	7	10	7	3	0	0
PM	0	0	0	0	0	0	0	0	3	7	10	7	3
PB	0	0	0	0	0	0	0	0	0	0	3	7	10
Index	1	2	3	4	5	6	7	8	9	10	11	12	13

A procedure known as *Quantification* is used to transform the inferred consequence fuzzy subset into a single value. The procedure used here is the 'Mean of Maxima', which is described in [3] as well as other related literature.

The overall algorithm for the SOC can be described as follows; At every sampling point **n**:

(a) The Error is computed by subtracting the process output (**y**) from the desired set-point (**x**):

$$e_n = y_n - x_n$$

(b) The Change in Error (**c**) is computed as:

$$c_n = e_n - e_{n-1}$$

(c) These values are then multiplied by their corresponding Gain factors and are scaled into values in the range 1 to 14 for error and 1 to 13 for change in error.

$$e^o_n = f(e_n \cdot G_e)$$
$$c^o_n = f(c_n \cdot G_c)$$

(d) These values are used to find the performance index, p^o, from performance decision table **P**:

$$p^o_n = P[e^o_n, c^o_n]$$

if p^o_n is not zero then create a rule with the following elements:

$$e' = e^o_{n-d}$$
$$c' = c^o_{n-d}$$
$$u' = u^o_{n-d} + p^o_n$$

(e) A consequence vector, U, is initialized with all members being zero. Next steps (f) through (h) are repeated for all the rules in the protocol.

(f) For rule, i, in the control protocol, the three numbers (e^i_m, c^i_m and u^i_m) are used to generate fuzzy subsets, E^i, C^i and U^i, using the Fuzzification procedures, ϕ for error and ϕ' for change in error and change in control action:

$$E^i = \phi(e^i_m)$$
$$C^i = \phi'(c^i_m)$$
$$U^i = \phi'(u^i_m)$$

(g) If p^o_n is not zero then move to step (h). if Rule i is dissimilar to the rule generated by step (d) then continue to step (h), else delete rule i and repeat this procedure for the next rule in the protocol.

(h) Then Membership Grades (μ) of the consequence vector, U, are computed from:

$$\mu_U(u) = \text{Max}_u \{\mu_U(u), \text{Min}(\alpha_i, \mu_{U_i}(u))\},$$

where: $\alpha_i = \text{Min}\{\mu_{E_i}(e_0), \mu_{C_i}(c_0)\}$.

(i) If p^o_n is not zero then add the new rule to the protocol and repeat step (h) for this rule.

(j) Quantify the consequence fuzzy subset, U, thus generated into value u^o_n and store the values e^o_n, c^o_n and u^o_n in the First-in First-out (FIFO) delay buffer for future use by the rule modification algorithm.

(k) Compute the controller output from:

$$u_n = u_{n-1} + G_u \cdot u^o_n$$

SOC Applied to a Level Control Rig

The controller was applied to a Level Control Rig in order to examine the controller behavior in dealing with a real plant and demonstrate the learning capability of the controller in dealing with changes in the characteristics of the process under control. The rig used for these experiments was located at the University of Exeter Chemical Engineering Laboratories and consisted of a liquid flow pipe network and a system of pneumatic valve control lines. This rig was used to conduct experiments with a variety of control algorithms and exhibits inherent non-linear characteristics. Experiments with the conventional PID controllers were found to have difficulty performing satisfactorily at different operating points, once these are tuned for a specific set-point [6].

The block diagram below shows the general set-up for these experiments. The liquid pipe network consisted of an open tank (T) of approximately one cubic meter capacity. The drain from the bottom of the tank leads to an electric pump (P) as shown in the diagram. A pneumatically operated control valve (V) adjusts the flow of water downstream from the valve. If the valve is fully closed, then the water is returned to the tank. If the valve is open, then a proportion of the flow is passed through an orifice plate and is then forced up a distance of over two meters.

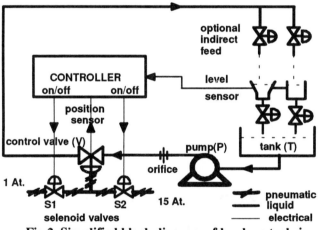

Fig 2. Simplified block diagram of level control rig

The flow can be directed towards either a narrow cylindrical vessel of approximately 50 cm height or a vessel of similar height, but a variable cross section, ranging from about 2 cm at the bottom to 20 cm at the top. The outflow from these vessels can be altered by means of a manually operated valve.

The position of the control valve can be adjusted by the sequential control of two electrically driven solenoid-pneumatic valves (S1 and S2). These valves are linked in series arrangement as shown and are supplied with pressurized air at approximately 15 psig at one side and atmospheric pressure at the other side.

The position of the control valve is sensed through a transducer and is fed to one of the analog inputs of the controller. Two of the digital outputs from the controller are used to turn on or off the solenoid valves, responsible for the lowering or raising the control valve.

A second transducer is responsible for measuring the level of the water in vessels and its output is also fed to one of the analog inputs of the controller. A third analog input is connected to a potentiometer and can be used to adjust the desired set-point for the water level.

The overall characteristics of the rig correspond to a real application and as such exhibit typical inherent non-linearities and delays.

Controller modification

The basic SOC algorithm described earlier is modified in order to set the position of the pneumatically operated control valve.

This modification consisted of a simple on-off control loop that would take the SOC output as the desired set-point for the control valve. The control valve would then be raised or lowered and its position monitored until the desired set-point is reached. To raise the valve, the solenoid-pneumatic valve S1 is closed and S2 is opened. To lower the control valve the opposite procedure is performed. Once the desired valve position is reached both solenoid valves are closed.

Experimental Results

Cylindrical vessel

A number of experiments with the cylindrical vessel were performed in order to demonstrate the learning behavior and overall response of the SOC. Figures below show the results of one of the experiments performed with the cylindrical vessel and a direct feed. The Sampling interval for this experiment was set to 2 seconds. Delay in reward parameter was set to 2, in order to account for long delays inherent in the system. The first graph shows the water level in the vessel versus the desired set-point and the second graph shows the controller output (Control valve position).

The controller is started with no rules and several changes to the set-point are made. The graph of the output (next page) shows a good performance by the controller, after the initial learning period. The control rules generated by the SOC are also shown on the next page.

Other experiments [6] with this setup showed the SOC to be relatively insensitive to minor changes to the performance rules. Also, as noted, the controller is started with no rules and is able to adapt to the process relatively quickly, without causing the process to become unstable.

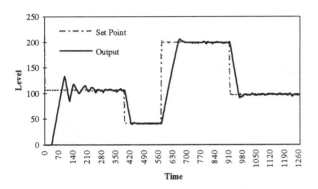

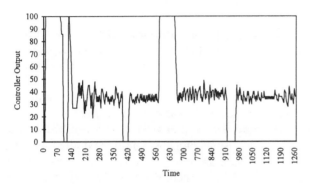

Rule:	If E	and C	then U
1	NB+	PB-	PB
2	NB	ZE	PB
3	NB	PS	PB
4	NS+	PB-	NB
5	PS	PB-	NB
6	PB-	PB	NB
7	PB-	PM-	NB+
8	ZE+	NB+	PB
9	NS	NB	PB
10	NM	NB	PB
11	ZE+	PB	NB
12	ZE-	NM+	PB-
13	ZE-	ZE	PS-
14	ZE+	ZE	ZE
15	ZE+	PS	NM+
16	ZE+	NS	NS+
17	NS+	PS	NS
18	ZE-	NB	PB

Variable Cross Section Vessel

Experiments with the variable cross section vessel demonstrate the ability of the controller to deal with a process with variable characteristics. The results of one experiment is shown in the following graphs. The process characteristics in this case depended on the set-point. It was observed that the controller had to adjust the control policy for a change in the set-point and as the result the initial response to a change in the set-point was poorer than the cylindrical vessel. The total number of rules generated in this case was 20.

For this experiment the sampling interval was set to 2 seconds and the delay in reward parameter was set to 2.

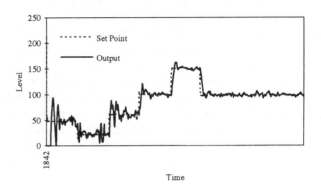

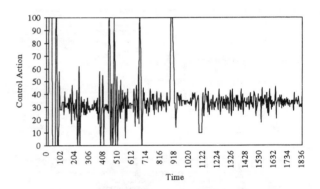

Conclusions and Recommendations for future work

The learning behavior of the SOC is based upon using a set of performance rules that describe the desired performance of the controller in terms of the deviation of the process output from a desired set-point. This learning method generates rules that are produced with a reward mechanism based on a local performance measure. There are no mechanisms for producing rules that result in an optimal form of control.

The controller can be viewed as a hierarchical decision making system, where the decision rules are generated by a second set of fixed performance rules. Further work is needed to explore the possibility of extending this hierarchical structure to include a higher level set of rules that would be used to generate the local performance measure on the basis of a more global performance criterion. These criteria could be based on describing the optimal performance of the controller based on the controller output and the process output.

This implies rules with multiple antecedents, referring to a number of previous states of the system and the consequence would be the changes made to the performance index.

The controller described here was implemented in the form of a simple microprocessor based system with 64K byte of memory and an 8 bit Analog to Digital converter. This system can be implemented in a very cost effective manner and applied to a variety of applications.

Acknowledgments

I would like to thank Professor John O. Flower for the opportunity to carry out this research at the University of Exeter and for his invaluable advice.

References

[1] King, P. J., Mamdani,E.H. (1977). "The Application of Fuzzy Control Systems to Industrial Processes", Automatica, Vol.13, 235-242.
[2] Mamdani,E.H. (1974). "Application of Fuzzy Algorithms for the control of a Dynamic Plant", Proc. IEE, Vol.121.
[3] Procyk,T.J., Mamdani,E.H. (1979). "A linguistic Self- Organising Process Controller", Automatica Vol.15, 15- 30.
[4] Tong,R.M. (1977). "A Control Engineering Review of Fuzzy Systems", Automatica, Vol.13, 559-569.
[5] Tong,R.M., Beck,M.B., Latten,A. (1980). "Fuzzy Control of Activated Sludge Wastewater Treatment Process", Automatica Vol. 16, 659-701.
[6] Vijeh, N. (1988). "Microprocessor Engineering Aspects of a Fuzzy Logic Self-Organizing Controller", Ph.D. Thesis, Department of Engineering Science, University of Exeter, Exeter, Devon, UK.
[7] Zadeh,L.A. (1965). "Fuzzy Sets", Information and Control, Vol.8, 338-353.
[8] Zadeh,L.A. (1973). "Outline of a New Approach to the Analysis of Complex Systems and Decision Processes", IEEE Transaction on Systems, Man and Cybernetics, SMC- 1, 28-44.
[9] Zadeh,L.A. (1976). "A Fuzzy Algorithmic Approach to the definition of Complex or Imprecise Concepts", Int. Journal of Man-Machine Studies, Vol.8, 249-291.

Fuzzy Logic Anti-Lock Brake System for a Limited Range Coefficient of Friction Surface

D. P. Madau, F. Yuan, L. I. Davis, Jr., and L. A. Feldkamp
Research Laboratory, Ford Motor Company
Suite 1100, Village Plaza, 23400 Michigan Avenue
Dearborn, Michigan 48124
davis@smail.srl.ford.com

Abstract— The use of fuzzy logic has recently gained recognition as an approach for quickly developing effective controllers for higher-order, nonlinear time-variant systems. This paper describes the preliminary research and implementation of a fuzzy logic controller to control wheel slip for an anti-lock brake system. The dynamics of braking systems are highly nonlinear and time-variant. Simulation was used to derive an initial rule base which was then tested on an experimental brake system. The rules were further refined by analysis of the data acquired from vehicle braking maneuvers on a surface with high coefficient of friction. The robustness of the fuzzy logic slip regulator was further tested by varying operating conditions and external environmental variables.

I. Introduction

When braking force is applied to a rolling wheel, it will begin to slip; that is, the wheel circumferential velocity V_{whl} will be less than the vehicle velocity V_{veh}. Slip is defined as the difference between vehicle velocity and wheel circumferential velocity, normalized to vehicle velocity:

$$\lambda = \frac{V_{veh} - V_{whl}}{V_{veh}}.$$

If sufficient braking force is applied, the wheel will "lock up," that is, slide without turning at all. A locked wheel has no lateral stability and usually slides with much less friction than a wheel with $\lambda < 1.0$. The relationship between slip, vehicle velocity, and the coefficient of friction μ is complicated and changes with different surfaces. Figure 1 is a plot of typical μ-slip functions, both lateral and longitudinal. The lateral coefficient of friction is greatest at zero slip. Lateral friction provides lateral stability, the ability to steer and control the direction of the vehicle. The longitudinal coefficient of friction is zero at zero slip and typically exhibits a peak in μ at some intermediate value of slip. For most surfaces, as braking force is increased longitudinal μ increases with slip until a point is reached where μ decreases for increasing slip. If braking force is not quickly reduced at this point, the reduction in road force leads to a rapid increase in slip and eventual lockup. Anti-Lock Brake Systems (ABS) sense when this point has been reached and reduce braking force so that lockup is avoided. The two curves shown in Figure 1 for longitudinal coefficient of friction are typical of hard surfaces. It would appear that maintaining slip at the value of λ which gives the peak value of μ would be ideal. Unfortunately, the position of the peak varies for different surfaces and speeds (for example, the curve may not have an extremum at all until $\lambda = 1.0$). Most control strategies define their performance goal as maintaining slip near a value of 0.2 throughout the braking trajectory. This represents a compromise between lateral stability, which is best at zero slip, and maximum deceleration, which usually peaks for some value of slip between 0.1 and 0.3. The goal of ABS control then becomes the regulation of slip to a known and desired level.

In terms of practical ABS design, there are many complicating factors. For example, the amount of brake torque at the wheel is nonlinearly related to the temperature of the brake linings. The viscosity of the brake fluid is affected by the temperature of the brake fluid and affects the rate at which the brake pressure can be increased and decreased. Also, the anti-lock brake system must handle external disturbances such as variations in the adhesive force between the road and tire due to changes in road surfaces, loading, and steering, and variations in the frictional force due to irregularities in the road surface. External variables must be considered, including anomalies in the wear patterns of the braking components which may increase noise in the wheel velocity sensor, variations in the adhesion of the brake linings, and brake hysteresis. The control algorithm must accommodate variations in tire inflation pressure and tire wear patterns (as well as possible use of a mini-spare tire). Tolerances on the sensor wheel and tooth errors can significantly affect the accel-

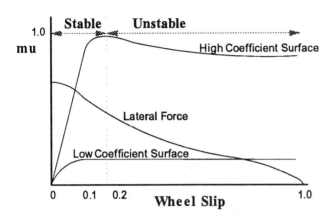

Figure 1: Coefficient of Friction **mu**, as a function of wheel slip.

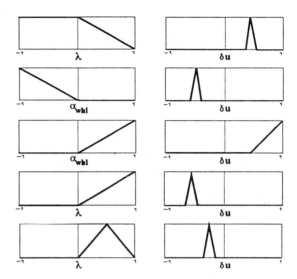

Figure 2: Initial ABS Rule Set. Input membership functions are shown on the right, with corresponding output functions on the left.

eration calculations. In summary, the internal ABS plant dynamics are difficult to model accurately and tend to be nonlinear, time-varying, and complicated by the inclusion of higher order terms.

In this paper, we describe the development of a fuzzy logic controller for anti-lock braking. Fuzzy logic is being recognized as a useful tool in developing robust controllers for higher-order, nonlinear, time-varying systems. We begin with a description of the initial controller design via simulation and proceed to follow the design process through practical system considerations and development issues, test results, and adjustment of the control strategy. Although the development has not been carried through as far as a production system, the resulting controller shows great promise.

II. INITIAL FUZZY LOGIC ABS CONTROLLER DEVELOPMENT

An initial fuzzy logic controller was developed using a simple simulation model for the braking process. By observing simulated trajectories we were able to propose, test, and evaluate rules corresponding to particular strategies. Once a reasonable set of rules had been established, their corresponding membership functions were adjusted to improve results. Evaluations were made on the basis of statistical comparisons of tests performed while varying road surface μ, sensor noise, initial conditions, etc. The simulations of ABS control were performed using a simple, one-wheel, straight-line model as described in [1]. The system model was initialized to a standard set of conditions and then the dynamics were simulated according to what brake output the controller produced at each step. The model maintained separate state variables for the "actual" system and an estimate of the state variables for deriving sensor readings to be provided to the controller. Simulation time steps for the model were ten times finer than the controller cycle times. At each controller cycle, the model made available to the controller three values: estimates of vehicle speed, and estimates of wheel speed and acceleration. These estimates included noise and integration error accumulated during the simulated trajectory.

The resulting initial set of rules is pictured in Figure 2. The input variables used were slip λ, as calculated from estimated vehicle and wheel speed, and estimated wheel acceleration α_{whl}. There was only one antecedent for each rule. The output δu, was chosen to be *change* in braking force in anticipation of the use of hydraulic actuation. These rules resulted from simulating many braking trajectories and adjusting for minimum stopping distance. The rules and the control they represent was presented as a generic starting point, with important details such as scaling of the inputs and output to the parameters of the vehicle and braking system remaining to be established.

III. PRACTICAL ABS CONTROLLER DEVELOPMENT

The inputs necessary to implement the rules in Figure 2 are, for each wheel, wheel velocity, wheel acceleration, and slip. The only available inputs to the ABS controller that are actually measured are the wheel velocities. All

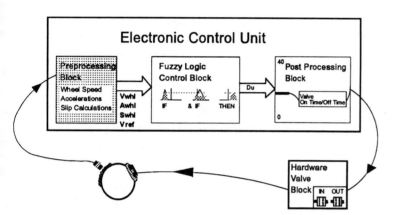

Figure 3: Control Block Diagram

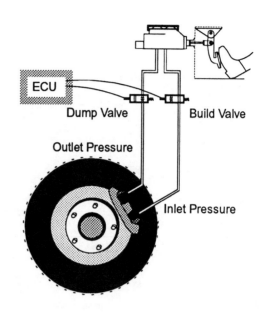

Figure 4: ABS Hydraulic System

other potential inputs are derived from these measurements. The wheel velocities are provided by a wheel speed sensor at each of the four wheels. A wheel speed sensor generates a signal whose frequency is directly proportional to the velocity of the wheel V_{whl}. The computed derivative of the wheel speed is used to estimate the angular acceleration of the wheel α_{whl}. The velocity of the vehicle V_{veh} is necessary in calculating the individual wheel slip λ according to Eq. 1. However, the determination of the vehicle velocity can become a very uncertain task if all four wheels are slipping at once. Generally an estimate of V_{veh}, called the reference velocity V_{ref}, is calculated using certain assumptions. Under normal (non-ABS) conditions, V_{ref} for each wheel is the average of the non-driven wheel velocities. As ABS operation begins for a wheel, its V_{ref} starts at the prevailing average and then is decelerated at a fixed rate. Any time the wheel's velocity exceeds its V_{ref}, V_{ref} is allowed to track the wheel velocity. The value of V_{ref} continues to be updated in this manner until the wheel leaves ABS mode. The amount of slip for each wheel is calculated using Eq. 1, but with V_{ref} instead of V_{veh}.

The fuzzy logic controller consists of three blocks (Figure 3): the preprocessing block, the fuzzy logic control block, and the postprocessing block. The preprocessing block handles the hardware interface and calculates the necessary variables for the fuzzy logic control block. The fuzzy logic control block fuzzifies the inputs and maps them to a rule base to determine the control output δu.

The following five rules were used to drive the fuzzy logic kernel:

1. If λ is Pos. Small then δu is Pos. Small
2. If α_{whl} is Neg. Large then δu is Neg. Med.
3. If α_{whl} is Pos. Large then δu is Pos. Large
4. If λ is Pos. Large then δu is Neg. Large
5. If λ is Pos. Medium then δu is Neg. Small

The fuzzy logic kernel utilizes standard max-min fuzzy inference. The center of area method was used to generate the control output δu.

The ABS hydraulic system components include an inlet valve and an outlet valve located at each wheel used to regulate the pressure at the caliper (Figure 4). For the simulation, there was just one output, change in brake force; for the real brake system, we must derive two output signal rates, one for each valve. In the build pressure state, the inlet valve is open and the outlet valve is closed, allowing for pressure to build at the caliper. The dump pressure state opens the outlet valve while closing the inlet valve, which allows for pressure reduction. The hold pressure constant state closes both valves, maintaining a constant pressure. The outlet valve orifice size is twice that of the inlet valve to accommodate the lower pressures at the outlet valve. Note that the inlet pressure at the build valve is unknown and varies as a function of the driver pedal effort. This leads to some uncertainty as to the rate of increase of brake pressure. It is possible for the

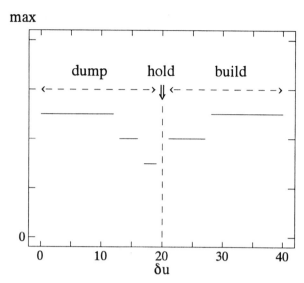

Figure 5: Brake Output Scaling: Build, hold, and dump rates

outlet pressure to be estimated if μ can be estimated, under the assumption that the control algorithm is operating in the optimal braking slip range. We did not, however, include μ estimation in our strategy. To accommodate integer math used by the microcontroller, the output was scaled to a domain from 0 to 40, such that 20 is equivalent to no change in brake pressure (see Figure 5).

When the output is negative (i.e., less then 20) the controller commands a specific dump pressure whose rate increases as the output approaches zero. A positive output specifies a build pressure rate whose rate increases as the output approaches 40. These rates are not interpolated linearly from the output but are derived from a lookup table in the postprocessing block. The postprocessing lookup table maps the output from the fuzzy logic control block δu to a specific build or dump rate as indicated in Figure 5. Each rate specifies an on time and a period for the valve. Normally, the adjustment of the table of these rates would be determined from bench testing of the various valve flow rate capabilities in conjunction with the hydraulic system. Instead, we developed the postprocessing block of the fuzzy logic ABS from in-vehicle testing.

IV. Development Issues

The implementation of the fuzzy logic algorithm in software is limited by constraints that are embedded in the base architecture design for the ABS system. The loop time is constant and fixed at 5 ms. During the loop time, the software must preprocess the inputs, determine the control action, drive the outputs for all four wheels, check for hardware fail-safe, and communicate to the serial data acquisition system. Initially, processing the fuzzy logic algorithm took far in excess of the 5 ms loop time. By reducing resolution in the defuzzification portion of the max-min inference and in-lining the code, we were eventually able to squeeze the algorithm time to within the necessary 5 ms constraint.

In order to record the values of controller variables during vehicle braking tests, an internal serial data link was used to communicate to external data acquisition equipment. The link could read four data words located in memory at a maximum rate of 9600 baud within the 5 ms loop time. As a result of this time constraint, the data that could be collected largely consisted of the reference velocity, one of the individual wheel speeds, the wheel valve signal, and the output of the fuzzy controller signal δu.

The performance of the initial fuzzy controller was far from optimal. The system did not attempt to take control of the wheel on the first cycle until it was deep into slip. In order to recover, the decrease in brake pressure necessary to reaccelerate the wheel resulted in an excessive loss of brake fluid. In a closed loop hydraulic configuration there exists the potential for depleting the total energy of the system more rapidly than the internal ABS pump motor can compensate. As a result, the inlet brake pressure may not be able to provide enough energy to allow the wheel to climb up the stable side of the $\mu - \lambda$ curve and reach the optimal braking force. This results in the cycling of the build and dump valves at a greater frequency at the beginning of the stop than later in the stopping maneuver. We observed this to be the case for the initial rule set provided from the simulation studies. Since most of the energy in the system was used in recovering the initial wheel's velocity, the length of time spent in build pressure

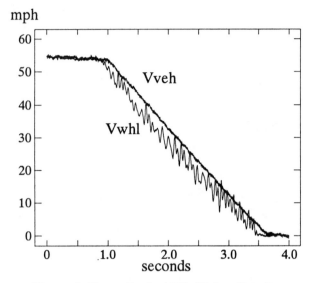

Figure 6: Fuzzy Logic ABS: High μ Results

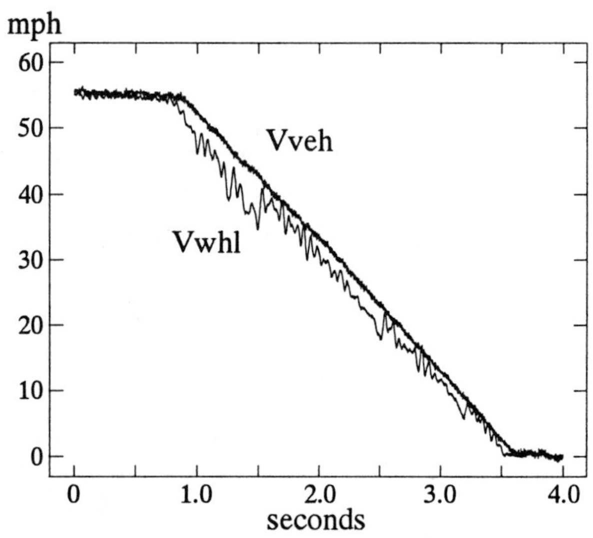

Figure 7: Conventional ABS: High μ Results

state near the end of the stop is significantly greater than at the beginning. The total deceleration rate for the stop on a high μ surface was 8.36 m/s^2 (= 0.853 G). The theoretical optimal deceleration rate for a vehicle on high μ is 9.8 m/s^2 (= 1.0 G). Although we do not expect to achieve the theoretical result (for high μ, zero degrees pitch, straight line stop), we expect to see deceleration rates greater than 8.82 m/s^2 (0.9 G).

Since dumping brake pressure quickly over a short time interval severely limits the ability of achieving optimal deceleration, the output scaling was modified to decrease the dumping of brake pressure. After this modification, it was still evident that the system did not try to control the initial slip of the wheel until it was deep within slip. On the other hand, the build and dump frequencies were similar during the entire stop, leading to the conclusion that the system energy was better preserved during the stop. Initial vehicle stability was still an issue due to the deep first cycle of the wheels.

The next stage of development was to adjust the trigger points that initiate ABS control to enable the control algorithm to react more quickly to a pending lockup situation. With this modification, wheel slip was better maintained within desired limits throughout the stop. Although this did not improve the deceleration rate very much (8.62 m/s^2 = 0.88 G), the stability of the vehicle was increased.

Another consideration during the testing was the temperature of the brake linings. During repeated stops, the temperature of the brake linings will increase, thus affecting the ability for the brake to displace any additional energy and thereby reducing braking potential. Of concern is that at a certain point the brake pads would begin to burnish which would further decrease the maximum amount of brake torque available. In observing the performance of the control from initial stop to several stops later there was evidence that the initial brake stop displayed a greater force on the wheels than the final stops. On the other hand, we noted that the control of the hot brakes was actually better: with less force able to be applied to the wheel, stabilizing the wheel required less effort. This was an indication that the algorithm tended to allow pressure to build too fast. With hot brakes, the algorithm regulated the slip of the wheel to a desired level without going deep into the unstable region.

Finally, after slight modification to the postprocessing valve timing, the following results were achieved on a high μ surface. The average deceleration rate for the entire stop was 8.91 m/s^2 (= 0.91 G). The wheel slip cycled around 0.9 of V_{ref}. As the vehicle velocity decreased the wheel was allowed to cycle deeper (around 0.8). The algorithm appeared to gain better control of the wheel as time increased. The wheel oscillations were kept to a minimum near the end of the stop, which increased the stability of the vehicle.

V. Results: High and Low μ

The main design intent for this first set of experiments was preventing the wheels from locking and decreasing the stopping distance. The results as recorded with an external data acquisition system were very promising, as shown in Figure 6. Results for a conventional ABS are shown in Figure 7. On the high μ surface the performance of the fuzzy ABS system was quite comparable to that of the production ABS system. The wheels spend most of the time around the reference velocity. In a straight-line stop, the fuzzy system decelerated the vehicle at a slightly higher rate than that of a production system.

A characteristic of this fuzzy logic approach to ABS is that no attempt is made to estimate the overall coefficient of friction. None of the rules directly utilize the

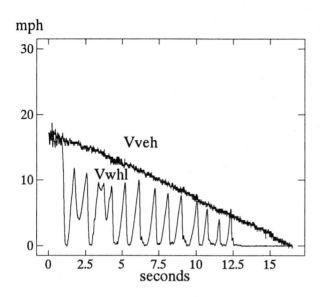

Figure 8: Fuzzy Logic ABS: Low μ Results

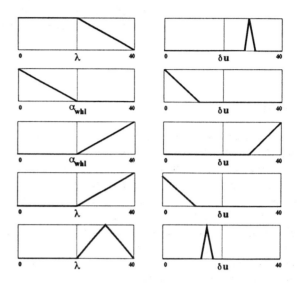

Figure 9: Final Rule Set for ABS

value for μ; therefore it was interesting to test the robustness of the fuzzy logic controller with a lower coefficient of friction test surface. The initial results for a low μ surface are shown in Figure 8. Although the controller was surprisingly functional, the performance naturally lacked features that further tuning undoubtedly would provide. The control algorithm cycled the wheels over a greater slip range than for the high μ case. The wheels spent too much time deep in slip which limited the steering ability of the vehicle. Further analysis of the controller output indicated that the difficulty of slip regulation was not necessarily the result of the fuzzy control but rather the limited resolution of the postprocessing block. Since the build valve on time was limited, the controller could not request a slow enough pressure build for the valve to keep the wheel from locking while maximizing the available frictional forces and maintaining wheel stability. Further work with a better valve resolution is warranted.

VI. Discussion

The intent of this work was to establish an initial practical understanding of the ability of fuzzy logic to control a system with nonlinear dynamics and variable parameters, specifically, an anti-lock brake system. Simulation was used to establish an initial rule base for use in the fuzzy logic controller. With minor tuning of the rules and postprocessing block for the high μ surface, results achieved were comparable to that of a production ABS controller. When tested on a low μ surface, the untuned performance was found to be surprisingly functional. The controller cycled the wheels over a wider slip range and consequently the steering ability was not very good; but it was not clear whether this was due to any deficiency in the strategy or merely a result of a limitation in the valve response. Further simulation suggests that low μ performance can be improved without degrading performance at high μ by adding additional rules.

One observation is clear from our experience: fuzzy logic supplies a framework for quick development of prototype controllers without detailed understanding of the internal plant dynamics. This is apparent both in the process of refining the controller and in the initial development of a rule set. Comparison of the final rule set (Figure 9) with the initial rule set shows how close one may come to a final controller through simulation. Further work will include addition of rules to handle various braking maneuvers, road surfaces, and refinement of the valve resolution time.

References

[1] L. I. Davis, Jr., G. V. Puskorius, F. Yuan and L. A. Feldkamp (1991). Neural Network Modeling and Control of an Anti-Lock Brake System. *Proceedings of the Intelligent Vehicle '92 Conference*, June 29 – July 1, 1992, Ypsilanti, Michigan.

Intelligent Cruise Control with Fuzzy Logic

Rolf Müller, Gerhard Nöcker
Daimler-Benz AG, Forschung und Technik
W-7000 Stuttgart 80, Postfach 80 02 30
Phone (49) 711-17-92870, Fax (49) 711-17-94194
email-x400: c=de;a=dbp;p=daimlerbenz;ou1=tei;s=mueller;g=rolf

Abstract

This paper describes a fuzzy based intelligent cruise control which maintains speed and distance from preceding vehicles. We have developed a modular fuzzy controller that takes into account the specific driving and weather situation as well as the individual driver's needs. This intelligent cruise control system has been implemented in a test-vehicle equipped with an infrared distance sensor and a drive-by-wire actuator system that allows accelerating and braking. Experiments prove the effectiveness of our approach and show the human-like behaviour of our intelligent cruise control system.

1. Introduction

Ordinary cruise control systems for passenger cars are becoming less and less meaningful because the increasing traffic density rarely makes it possible to drive at a preselected speed. Advanced systems controlling both speed and distance to preceding vehicles - so called Autonomous Intelligent Cruise Control (AICC) systems - are currently under investigation in the European traffic research programm PROMETHEUS. This paper describes the Daimler-Benz approach.

Longitudinal control has been under investigation for at least 20 years and a lot of theoretical research work has been done [CrP 70, WoN 71, Har 74, FeC 77, Ack 78, Shl 78, Swi 78]. Because of sensor problems and high costs only a few experimental vehicles have been built. Daimler-Benz started with longitudinal control in 1980. First experiments with a linear PID-controller were not very successful because the controller was not able to handle the noisy sensor data, resulting in a very jerky driving behaviour. The next step was a nonlinear soft controller that operated on an ordinary cruise control and an automatic braking system [Nöc 90]. This approach was successful and we made some 50000 km of driving trials.

However, in order to achieve high customer acceptance an intelligent cruise control system has to perform similarly to an experienced human driver. Therefore, it is necessary to adjust the following distance and the control dynamics according to the individual driver's needs and the specific driving situation and weather conditions.

Applying fuzzy logic to intelligent cruise control seems to be an appropriate way to achieve this human behaviour, because driver's experience can be transformed easily into rules [AFJ+91, TaH 91]. Based on this idea we have developed a modular fuzzy controller for AICC.

2. Vehicle Equipment

Our test vehicle is a Mercedes-Benz 300E equipped with a 5-beam infrared distance sensor. The sensor runs at a frequency of 904 nm with a maximum distance of 150 m and covers 7.5 degrees. Other sensors measure the speed, the acceleration and the steering angle of the vehicle. Driver's actions are monitored as well by measuring the actions on pedals and switches. The distance sensor is connected to an Intel386 control computer via an IEEE interface. The other sensors are connected via CAN-bus.

Each beam of the infrared sensor delivers, every 100ms, the distance to a detected target, the relative speed between the host vehicle and the target and information about the size of the target. Thus, every tick of the clock we receive a set of five not necessarily consistent distance values, relative speeds and target size information.

The distance to a preceding vehicle and the relative speed between the two vehicles have to be determined from these data by a filtering procedure, the so called tracking filter. The tracking filter has to cut out disturbances from on-coming vehicles and from stationary obstacles which are not on the track.

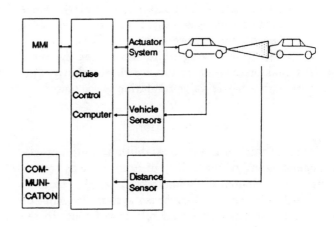

Figure1: Control Loop

Figure 9: *Results on image sequences*

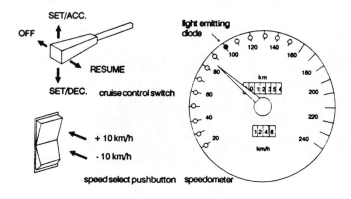

Figure 2: Man-Machine-Interface

The vehicle is controlled by a drive-by-wire actuator system that carries out either acceleration or deceleration control with a fast underlying control-loop. The control computer and the actuator system are wired by a CAN-bus. A block diagram of the car is shown in Figure 1.

3. Function and Operation

Our AICC-system works as follows:

First, the driver has to choose a desired travelling speed. This can be done either by the SET/ACC., SET/DEC. and RESUME commands of the cruise control switch or by a SPEED SELECT pushbutton (see Figure 2). Maximum speed is displayed on the speedometer by a light emitting diode as long as the AICC mode is switched ON. This lamp always reminds the driver of the preselected maximum speed.

As long as no target vehicle is detected by the tracking filter the AICC system provides an ordinary *speed control* function, driving at the speed set. If a target vehicle is detected the AICC system switches automatically to *distance control*, driving with the same speed as the target vehicle at an "optimal distance". If the target vehicle accelerates over the preselected maximum speed or leaves the lane the AICC system switches automatically to speed control.

The AICC system is not designed as a collision avoidance system but as a driver support system for driving on overcrowded roads and highways. The driver is always responsible for the task of driving and has to deal with emergency situations. The AICC system can be overriden by the driver at any time by braking or accelerating. When the driver leaves the pedals the system switches back to automatic control. The automatic accelerations and decelerations of the actuator system are limited from -2.5 m/sec² to 1.5 m/sec². If a maximum deceleration of -2.5 m/sec² is not enough in a specific driving situation, the driver is warned by a buzzer in order to brake with more force.

4. Basic Distance Control with Fuzzy Logic

Our first step was the development of a basic distance controller with two inputs and one output. One input is the distance error *de*, i.e. the percentage distance error to the reference distance. The reference distance is determined by a well-known rule of thumb - half the speed of the controlled vehicle measured in metres. The second input is the relative speed *rs* between the two vehicles. The output of the controller is the nominal value for acceleration or deceleration *ac* for the drive-by-wire system.

For each of the two inputs we defined seven equidistant triangular-shaped membership functions ranging from negative large to positive large (see Figure 3).

NVL: negative very large   PS: positive small
NL : negative large   PM: positive medium
NM: negative medium   PL : positive large
NS : negative small   PVL: positive very large
Z : Zero

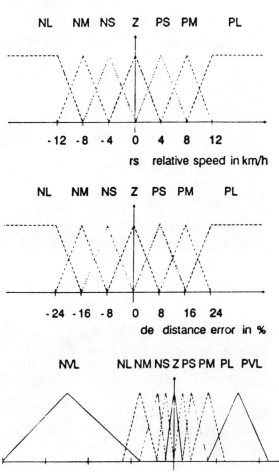

Figure 3: Membership functions for controller inputs and output

In order to obtain very smooth control we provided nine membership functions for the output variable. As described before the value for the drive-by-wire system has to be between -2.5 m/sec² and 1.5 m/sec². This fact has to be taken into account by the definition of the output membership functions and the choice of the defuzzification method. If we use for example the centroid method for defuzzification the centroid for the leftmost (NVL) and rightmost (PVL) membership function must be at -2.5 m/sec² and 1.5 m/sec² respectively. This ensures that all output values are in the desired range.

The rule set shown in Figure 4 reflects the experience of the driver and takes into account special features of the sensor data.

zone is accomplished by the eight Z's in the center of the rule base. With this dead zone we achieve smooth control behaviour while following a preceding vehicle with distance error and relative speed close to zero.

For the realization of the logical AND which is implicitly assumed for all rules in the rule base we tried the mimimum-operator, the algebraic product and the γ-operator [Zim 91]. For the defuzzification we tested the center-of-gravity and the Sugeno methods.

In order to evaluate the performance of the fuzzy controller we used, in a first step, 3-dimensional control surfaces that can be easily interpreted by a control engineer. Figure 5 shows the control surface for the fuzzy controller described in Figures 3 and 4 with the algebraic product for the logical AND, the bounded sum for the OR-combination of all active rules and the Sugeno-method for defuzzification.

	de	too close -> danger!				too far		
rs		NL	NM	NS	Z	PS	PM	PL
	NL	NVL	NVL	NVL	NL	NM	NS	NS
rs < 0 -> danger!	NM	NVL	NL	NM	NS	Z	Z	Z
	NS	NL	NM	NS	Z	Z	Z	Z
	Z	NM	NS	Z	Z	Z	PS	PS
rs > 0	PS	NS	Z	Z	Z	Z	PM	PL
	PM	NS	Z	Z	PS	PM	PL	PVL
	PL	NS	Z	Z	PS	PL	PVL	PVL

Figure 4: Rule base for fuzzy controller

As an example of how driver's experience has been incorporated consider the case when a vehicle cuts in, causing a negative distance error. If the relative speed is greater than zero, i.e. the preceding vehicle is driving faster than we are, an experienced driver wouldn't brake hard but would maintain speed or brake only slightly. This behaviour is reflected by the nine rules in the lower left corner of the rule base in Figure 4. On the other hand consider the case when the target vehicle changes lane, causing a positive distance error with respect to a new target vehicle. In order to reduce the distance error an experienced driver would profit from a slightly or moderately higher speed with respect to the new target rather than accelerating sharply. We take account of this behaviour by the block of Z's in the upper right corner of the rule base.

In order to make the controller robust to measuring noise we provide a dead-zone close to the operating point where distance error and relative speed are nearly zero. The

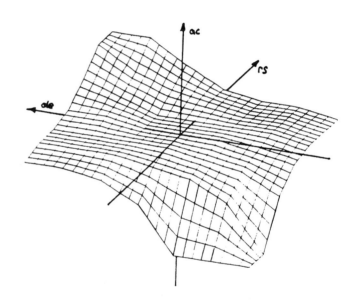

Figure 5: Control surface

In a second step we tested the controller by software simulations of certain driving situations such as cutting-in and cutting-out of the target vehicle, driving towards a slower target vehicle and so on.

Figure 6 gives the simulation plot of a cutting-in-manoeuvre. It shows a simulation over 25 seconds. Initially, both vehicles drive at a speed of 100 km/h and have a distance of 50 m, i.e. the controller is in the operating point. After 2 seconds a new target vehicle cuts in with a speed of 110 km/h and a distance of 40 m. Although this results in a negative distance error the controller does not accelerate. This is because the relative speed is positiv, i.e. the target vehicle is driving faster than the controlled vehicle. After 5 seconds the distance error becomes 0 due to the higher speed of the target vehicle. Now, the controller begins to accelerate reaching the same speed as the target vehicle after 8 seconds without any overshoot.

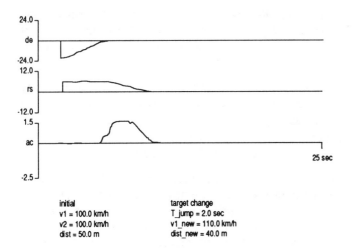

Figure 6: Simulation plot of a cutting-in manoeuver

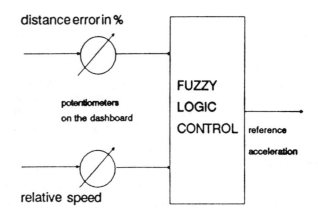

Figure 7: On-line modification of the fuzzy controller

The experiences gained during the fine tuning of the fuzzy controller can be summarized as follows:

- During the modification of membership functions and rules our experience was roughly the same as described in [Zhe 92]. In particular, we often exploited the fact that the control surface can be changed in certain portions without affecting the rest of the control surface.

- Using the algebraic product for the logical-AND and the bounded sum for the OR-combination of all active rules leads to a smoother control surface than using the elementary MIN-MAX-method.

- There was no significant difference between the center-of-gravity and Sugeno methods in the relevant portions of the control surface.

- The fine tuning of the fuzzy controller could be done on a more intuitive level without the knowledge of an exact mathematical model.

For the implementation of the fuzzy controller in our test vehicle we used two different methods. The first method was implementation via an off-line-generated look-up table describing the control surface. As a second approach we used two commercial fuzzy development tools which provide C-routines that can be linked to the control loop. Both methods worked well, causing no running time problems on the Intel386 control computer with respect to the clock rate of 100ms.

In order to influence the behaviour of the controller on-line during real test drives we multiplied the input data by factors that can be set by potentiometers on the dashboard (see Figure 8). With these scaling factors, we are able to carry out linear compression and stretching of the universe of discourse of the input membership functions. Thus, we can increase or decrease the influence of the two input parameters continuously.

5. Advanced Distance Control with Fuzzy Logic

The results of experiments with the basic controller show a good performance in terms of rise time, overshoot and settling time.

However, it turned out that these are not the criteria important to human drivers. They compared the system with their own driving behaviour and criticized the inflexibility of the system in certain situations. Mainly, the following points had to be improved:

- The definition of the reference distance was not always realistic. The "optimal distance" should not depend solely on the speed of the car but also on weather conditions (rain, ice,...) and - to a certain extent - on the individual driver's behaviour (sporty, neutral, comfortable,...).

- The accelerating and braking behaviour must be adapted to the driving situation (straight road, winding road, ...) and again to the individual driver's behaviour.

Based on these requirements we extended our basic distance controller. This resulted in a modular fuzzy controller consisting of two fuzzy blocks (see Figure 8).

The first fuzzy block determines the value for the "optimal distance". Here, the actual speed, the weather conditions and the individual driver's behaviour are taken into account. Weather conditions are derived from data about the outside temperature, wiper actions and the friction of the wheels. In a first stage driver's behaviour can be adjusted by a potentiometer. In a second stage, we will determine the driver's mood by evaluating the actions on the brake and accelerator pedal. The output of the first fuzzy block is combined with the actual measured distance to achieve the distance error that serves as an input for the second fuzzy block.

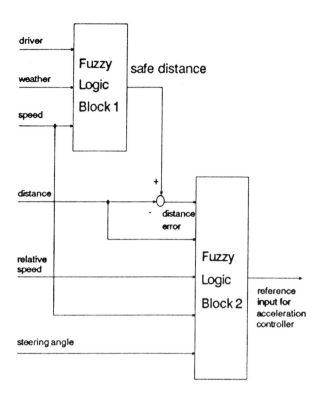

Figure 8: Block diagram of advanced distance controller

The second fuzzy block is an extension of the basic distance controller described in Section 4. In addition, we take into account the speed, distance and steering angle. We use the steering angle in order to detect whether we are in a curve or not. Thus, we can avoid powerful acceleration in a curve which could cause unsafe lateral acceleration. Furthermore, both the steering angle and the speed are analyzed to classify the momentary driving situation. For stop-and-go driving situations it is sensible to use the absolut distance as an additional controller input. As in the basic controller the output is the reference value for the drive-by-wire system.

6. Results of Experiments

The advanced distance controller was realised using the commercial fuzzy development tools. Each of the two fuzzy blocks was itself refined into two fuzzy modules so that no fuzzy module has more than three inputs. The whole fuzzy controller is desbribed by 4 fuzzy modules with a total number of about 200 rules. Although this fuzzy system is very complex, the fine tuning can be done efficiently exploiting the modular structure of the controller.

Figure 9 gives measurements made on a highway. The target vehicle drove at a speed of about 88 km/h. After 17 seconds we halved the reference distance by the potentiometer. Thus, the test vehicle accelerated at a rate of 0.4 m/sec² and closed up. When it came too close to the target vehicle the brake was applied for a short time (see last diagram at 30-33 sec). After a little overshoot of about 4 m the reference distance was reached (see first diagram at 40 sec).

One reason for the little overshoot was that the target vehicle decreased its speed to 84 km/h during the manoeuvre. After 65 seconds we switched back to the initial reference distance by using the potentiometer. As the fourth diagram shows the test vehicle braked very hard and uncomfortably when approaching the target vehicle. This behaviour could be easily changed to a more moderate behaviour by changing a few rules.

7. Conclusions

In this paper we have shown the effectiveness of fuzzy control to achieve a near-human behaviour in an automotive control task.

A basic fuzzy controller was rapidly designed by incorporating human driving experience and behaviour. The fine tuning was able to be done without the knowledge of the exact mathematical model of the vehicle and in a manner that supports human intuition. The basic fuzzy controller was easily adapted to user-specific requirements using a modular technique. Test drives show the reliability and robustness of the fuzzy controller.

In the next step we will use fuzzy logic for data fusion. Here, the goal is to obtain more information out of the existing sensors and to make the information more reliable.

In future, the AICC systems will be improved by establishing communication links to roadside information systems, which provide further information about weather conditions and traffic flow. Moreover, local traffic control systems will be enabled to influence the speed of the vehicle, avoiding traffic congestions by green-wave systems.

8. Acknowledgements

The authors thank Georg Geduld for his technical support in realization and programming and during the test-drives.

9. References

[Ack 78] F. Ackermann, "Erste Erfahrungen mit Abstandsregelung und automatischem Bremsen bei Einsatz von Abstandswarngeräten in Kraftfahrzeugen," in Entwicklungslinien in Kraftfahrzeugtechnik und Straßenverkehr Forschungsbilanz 1978, TÜV Rheinland, pages 250-255, 1978

[AFJ+91] J.P. Aurrand-Lions, L. Fournier, P. Jarri, M. de Saint Blanchard and E. Sanchez, "Application of Fuzzy Control for ISIS Vehicle Braking," in Proc. of the IEEE Roundtable Discussion on Fuzzy and Neural Systems, and Vehicle Applications, paper #32, 1991

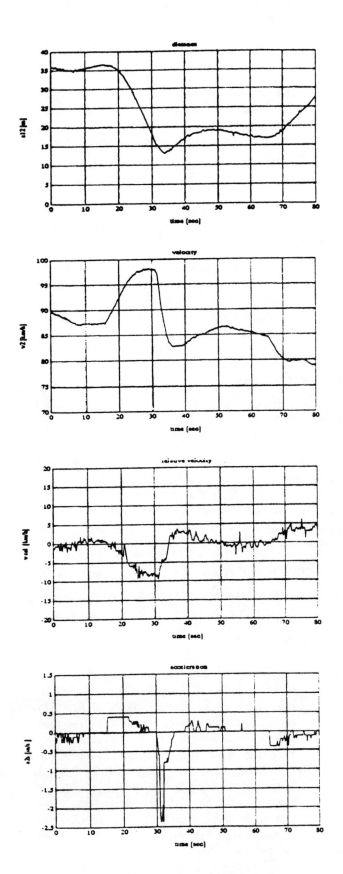

Figure 9: Measurements of real test drives

[CrP 70] D. Crow and R. Parker, "Automatic Headway Control. An Automatic Vehicle Spacing System," SAE Paper 700086, 1970

[FeC 77] R.E. Fenton and P.M. Chu, "On Vehicle Automatic Longitudinal Control," in Transportation Science, Vol. 11 No.1, pages 73-91, 1977

[Har 74] E. Hartwich, "Entwurf eines automatischen Abstandsreglers für ein Straßenfahrzeug," in Regelungstechnik und Prozeßdatenverarbeitung 22, Heft 7, pages 204-212, 1974

[Nöc 90] G. Nöcker, "Abstandsregelung," in VDI Berichte Nr. 817, pages 327-337, 1990

[Shl 78] S.E. Shladover, "Longitudinal Control of Automated Guideway Transit Vehicles Within Platoons," in Transactions of the ASME, Journal of Dynamic Systems Measurements and Control; Vol. 100, pages 302-310, 1978

[Swi 78] R. Swik, "Auslegung einer digitalen Abstandsregelung für Kraftfahrzeuge," Regelungstechnik Heft 9, pages 300-304, 1978

[TaH 91] A. Takayama and A. Hirako, "Adaptive Cruise Control According to Optimal Distance," in Proc. of the IEEE Roundtable Discussion on Fuzzy and Neural Systems, and Vehicle Applications, paper #25, 1991

[WoN 71] B. Wocher and J. Nier, "Automatische Abstandshaltung zwischen Kraftfahrzeugen als Teilproblem der automatischen Verkehrssteuerung," in Bosch Technische Berichte 3 Heft 6, pages 250-255, 1971

[Zhe 92] L. Zheng, "A Practical Guide To Tune Of Proportional and Integral (PI) Like Fuzzy Controllers," in Proc. of the IEEE International Conference on Fuzzy Systems, pages 633-640, 1992

[Zim 91] H.-J. Zimmermann, "Fuzzy Set Theory and its Applcations," Kluwer Academic Publishers, Boston/Dordrecht/London, 1991

Design of a Rule-Based Fuzzy Controller for the Pitch Axis of an Unmanned Research Vehicle

Deepak Sabharwal and Kuldip S. Rattan
Department of Electrical Engineering
Wright State University
Dayton, OH 45435

Abstract

A control scheme based on fuzzy logic is proposed in this paper. The objective is to design a fuzzy controller for the pitch axis of an unmanned research vehicle. The inputs to the controller are the error and the change in error. A set of fuzzy rules that makes the controller look like a proportional-plus-derivative controller is obtained and simulated on $Matrix_x$, a software package for the analysis and design of control systems. The fuzzy controller was found to give better results than those obtained from the existing analog controller.

1 Introduction

Fuzzy control research was started by Mamdani's [6] pioneering work in 1974 which was motivated by Zadeh's two papers on fuzzy algorithms [7] and linguistic analysis [8]. Since then, there have been numerous applications of fuzzy control, e.g., control of warm water processes [9], sinter plant [5], heat exchanger [10] and other industrial processes [11]. Until now, the trend has been to apply fuzzy control to complex processes where it is difficult to obtain a proper transfer function. The controller is designed to model a human operator's actions and can accept a linguistic input and return a 'crisp' or defuzzified output which is used to control the system. The main advantage of a fuzzy controller is that it is able to accept linguistic input. The input is classified into various categories (membership functions) like small, large, very large, etc. A set of rules determines the category of the output based on the category of the input. The rules are obtained intuitively from some knowledge of the system that we want to control. For example, if the error is small, the output is small; if the error is large and change in error is medium, the output is small, etc. The weighted average of the membership values is used to obtain the actual output. This is called the center of area (COA) method of defuzzification.

In this paper, the use of a fuzzy controller for controlling the pitch axis of the unmanned research vehicle (URV) is described. The URV is an experimental aircraft developed by WL/FIGL to study flight control concepts. A mathematical model containing split surfaces and linearized equations of motion for the longitudinal direction axes is used [4]. The URV model for the flight condition CRUISE-MACH = 0.118 at 1500 ft. is given by

$$\underline{\dot{X}}(t) = A\underline{X}(t) + B\underline{\delta}(t)$$
$$\underline{y}(t) = C\underline{X}(t)$$

where

$$\underline{X}(t) = [\alpha \ \theta \ q]^T$$
$$\underline{\delta}(t) = [\delta_{eL} \ \delta_{eR}]^T$$

$$A = \begin{bmatrix} -2.632 & 0 & 1 \\ 0 & 0 & 1 \\ -39.63 & 0 & -3.382 \end{bmatrix}$$

$$B = \begin{bmatrix} 0.0073 & 0.0073 \\ 0 & 0 \\ 0.5118 & 0.5118 \end{bmatrix}$$

$$C = \begin{bmatrix} 0 & 1 & 0 \\ 0 & 0 & 1 \end{bmatrix}$$

δ_{eL} and δ_{eR} are the left elevator and right elevator inputs, respectively and α, θ and q are the angle of attack, pitch angle and the pitch rate outputs, respectively.

The block diagram of the existing control law for the URV is shown in Figure 1.

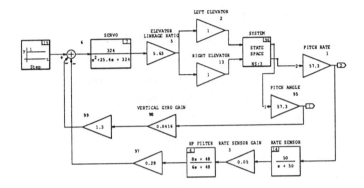

Figure 1: *Existing Control System.*

The existing controller for the pitch axis was replaced by a fuzzy controller. The fuzzy controller as shown in figure 2 is designed to obtain a response twice as fast as that of the existing controller. The performances of the original and the fuzzy controller are compared. The design of the controller is explained in Section 2, the implementation is described in Section 3, the simulation results are described in Section 4 and the conclusions are drawn in Section 5.

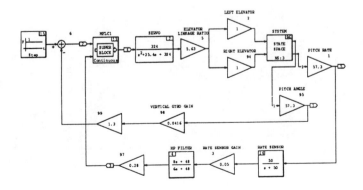

Figure 2: *Pitch Axis Control Using Fuzzy Controller.*

2 Design of a Fuzzy Controller

A fuzzy controller consists of three main parts: fuzzification, table of control rules and defuzzification. Fuzzification is done using membership functions. Membership functions form the heart of any fuzzy controller. These functions determine the assignment of a certain value of input to a particular category. The degree of assignment is quantified as the membership value. The membership values of the inputs determine which control rule(s) can be applied which in turn decides the output generated or the nature of controlling action. Thus, the shape of membership functions used and the limits for each category directly affect the controlling action. Generally, fuzzy variables are associated with exponential, trapezoidal or triangular types of membership functions.

In this paper, fuzzification was performed using triangular functions. The input range -1 to +1 was divided into 7 categories: *NB* (negative big), *NM* (negative medium), *NS* (negative small), *ZO* (zero), *PS* (positive small), *PM* (positive medium) and *PB* (positive big). The ranges of the membership functions were chosen to satisfy :

$$\sum_j \mu_{\tilde{A}_j}(e) = 1 \quad and \quad \sum_l \mu_{\tilde{B}_l}(de) = 1$$

where $\tilde{A} = \{A_j\}$ and $\tilde{B} = \{B_l\}$ are collections of fuzzy subsets over E and DE respectively. e (error) is some element of E and similarly, de (change in error) is some element of DE. This convention was adopted by Langari [1]. Consider a segment of the universe of discourse E that includes the interval $[E_j, E_{j+1}]$ as shown in Figure 3. It can be shown that to meet the above requirements, for each j, $R_j(.)$ and $L_{j+1}(.)$ intersect E at precisely the points E_{j+1} and E_j respectively, and likewise, for each l, $\acute{R}_l(.)$ and $\acute{L}_{l+1}(.)$ must similarly intersect DE at precisely the points DE_{l+1} and DE_l.

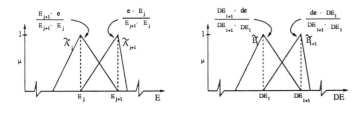

Figure 3: Membership Functions.

This assumption reduces the number of variables required to describe different categories. Using the membership functions, the membership of e and de in the various categories is obtained. The membership functions have been chosen in such a manner that any input can belong to a maximum of two categories. As a result, at any instant a maximum of four rules apply:

$$R_{j,l}: \text{ if } e(t) \text{ is } \tilde{A}_j \text{ and } de(t) \text{ is } \tilde{B}_l \text{ then } u(t) \text{ is } \tilde{C}_{j,l}$$

and so on for $R_{j+1,l}, R_{j+1,l+1}$ and $R_{j,l+1}$. The instantaneous value of the control action is given by

$$\begin{aligned} u(t) &= \frac{\sum_i \mu_i U_i}{\sum_i \mu_i} \\ &= \frac{\mu_{j,l} U_{j,l} + \mu_{j+1,l} U_{j+1,l} + \mu_{j+1,l+1} U_{j+1,l+1} + \mu_{j,l+1} U_{j,l+1}}{\mu_{j,l} + \mu_{j+1,l} + \mu_{j+1,l+1} + \mu_{j,l+1}} \end{aligned} \quad (1)$$

where
$\mu_{j,l} = [\text{membership of } e \text{ in } \tilde{A}_j][\text{membership of } de \text{ in } \tilde{B}_l]$
Note that the *product* has been used here instead of *min* and is essential for obtaining a closed form expression for $u(t)$ [1]. The procedure is illustrated in Figure 4.

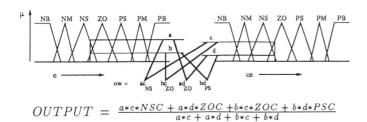

$$OUTPUT = \frac{a*c*NSC + a*d*ZOC + b*c*ZOC + b*d*PSC}{a*c + a*d + b*c + b*d}$$

Figure 4: Output Generation.

The above expression gives:

$$\begin{aligned} u(t) = &\frac{e\ DE_l[U_{j,l+1} - U_{j+1,l+1}] + e\ DE_{l+1}[U_{j+1,l} - U_{j,l}]}{(E_{j+1} - E_j)(DE_{l+1} - DE_l)} \\ &+ \frac{de\ E_{j+1}[U_{j,l+1} - U_{j,l}] + de\ E_j[U_{j+1,l} - U_{j+1,l+1}]}{(E_{j+1} - E_j)(DE_{l+1} - DE_l)} \\ &+ \frac{E_{j+1}[DE_{l+1} U_{j,l} - DE_l U_{j,l+1}] + E_j[DE_l U_{j+1,l+1} - DE_{l+1} U_{j+1,l}]}{(E_{j+1} - E_j)(DE_{l+1} - DE_l)} \\ &+ \frac{e\ de[U_{j,l} - U_{j,l+1}] + e\ de[U_{j+1,l+1} - U_{j+1,l}]}{(E_{j+1} - E_j)(DE_{l+1} - DE_l)} \end{aligned} \quad (2)$$

It can be seen from eqn(2) that a fuzzy controller acts like a piecewise linear controller. If the last two terms of equation (2) are ignored, then we can say that the fuzzy controller behaves like a PD controller. The difference is the piecewise linearity which should be exploited to extract full advantage from the controller. In a conventional PD controller, the gains K_p and K_D remain the same in all regions of operation, whereas in a fuzzy controller the gains for e and de vary depending on the ranges in which these values lie at any instant. This property is used to choose the various ranges of the membership functions.

2.1 Some Guidelines For Choosing The Ranges

1. It can be shown that for second order systems, the overshoot can be suppressed by using a small K_p initially and a large K_p when the output gets close to the setpoint. Hence, the ranges for the membership functions for error e should be chosen to achieve this. Often, it is seen in literature that the membership function for zero (ZO) category is kept very narrow. This results in small ($E_{j+1} - E_j$) which gives large K_p around zero.

2. Also, the ranges should ensure that K_p increases continuously as error decreases from 1 (normalized) to 0. If variation in K_p is not monotonic, the response will look jerky. A rule of thumb is to use E_j's so that the successive $\Delta E_j = E_{j+1} - E_j$ decrease as e decreases to 0. On the other hand, it is required that the successive $\Delta U_j = U_{j+1} - U_j$ increase. In other words,

$(PS - ZO) > (PM - PS) > (PB - PM)$

The above expression validates the fact that we need coarse control away from zero and fine control around zero.

3 Implementation of the Fuzzy Controller

The implementation for simulation purposes was done using blocks available in the SystemBuild module of $Matrix_x$.

3.1 Fuzzification

The FLC consists of an error block, a change in error block, a gain scheduler block, multipliers and summers. The error (e) and change in error (de) blocks are used to obtain the membership values of e and de in various categories. The type of membership function being used is triangular. The slope of the triangular functions is determined by the limits of each category.

The entire range of inputs is divided into seven categories each. The individual ranges are determined by the user and can be changed in a macro in $Matrix_x$. Depending on the range in which the error and change in error values lie (which is tested using the logical operator blocks), weights of the outputs are computed with the algebraic expression block using the expression $w = \frac{e - E_j}{E_{j+1} - E_j}$ and $w' = \frac{E_{j+1} - e}{E_{j+1} - E_j}$. Similar operations are performed for de as shown in Figure 4.

Since there are seven categories for the error and change in error, seven weights are calculated from the error and change in error blocks, and hence there are a total of 49 rules. Of the seven weights calculated, only two weights are non-zero for both the error and the change in error blocks and the rest are zero. Thus, finally there will be only four rules applicable.

3.2 Defuzzification

The defuzzified output is obtained using a process of matrix multiplication.

		\multicolumn{7}{c}{ERROR (e)}
		NB	NM	NS	ZO	PS	PM	PB
CHANGE OF ERROR (ce)	NB	ZO	PS	PM	PB	PB	PB	PB
	NM	NS	ZO	PS	PM	PB	PB	PB
	NS	NM	NS	ZO	PS	PM	PB	PB
	ZO	NB	NM	NS	ZO	PS	PM	PB
	PS	NB	NB	NM	NS	ZO	PS	PM
	PM	NB	NB	NB	NM	NS	ZO	PS
	PB	NB	NB	NB	NB	NM	NS	ZO

$e = \text{ref - output}$

$ce = \frac{\text{old e - new e}}{\text{sampling period}}$

Figure 5: *The Control Rules.*

The table of control rules given in Figure 5 is stored as a 7x7 matrix with centers of output categories as individual elements. The row matrix (1x7) consisting of weights from the error block is multiplied by the table of control rules stored in the gain scheduler and the resulting 1x7 matrix is multiplied by the 7x1 column matrix obtained from the weights from the change in error block to give a single crisp value. This is the defuzzified output.

4 Simulation Results

The ranges for the membership functions of the fuzzy controller used are given by:

$E1 = -4/3$ $DE1 = -1$ $NB = -7/6$
$E2 = -2/3$ $DE2 = -2/3$ $NM = -1$
$E3 = -1/6$ $DE3 = -1/3$ $NS = -2/3$
$E4 = 0$ $DE4 = 0$ $ZO = 0$
$E5 = 1/6$ $DE5 = 1/3$ $PS = 2/3$
$E6 = 2/3$ $DE6 = 2/3$ $PM = 1$
$E7 = 4/3$ $DE7 = 1$ $PB = 7/6$

These values result in a small K_p (around 1) near the extremes and a large K_p (around 4 to 6) close to the setpoint. As a result the response with the fuzzy controller is twice as fast as the response with the existing controller. The gain of the existing controller was doubled to do a fair comparison between the different control strategies. Now both controllers had similar speeds but the fuzzy controller resulted in a **much smoother** response. The output with the fuzzy controller is monotonically increasing while that of the analog controller results in regions of negative $\dot{\theta}$. The responses are shown in Figures 6 and 7.

The control surface of the fuzzy controller was also plotted. It was observed that if the above guidelines are followed the surface is smooth. Otherwise, there are ridges and peaks in between. This is responsible for the output being uneven and not smooth. The control surface for the above ranges is shown in Figure 8.

5 Conclusions

The use of a fuzzy controller for pitch axis control of an aircraft is presented in this paper. Triangular membership functions are chosen and it is shown that under certain assumptions, the fuzzy controller reduces to a piecewise linear PD controller. This result is used to obtain some rough guidelines for designing the fuzzy controller. It is shown that a fuzzy controller can lead to a better response as compared to the existing controller.

References

[1] Gholamreza Langari, *A Framework for Analysis and Synthesis of Fuzzy Linguistic Control Systems*, Ph.D. thesis, University of California at Berkeley, December 1990.

[2] M. Sugeno, *Industrial Applications of Fuzzy Control*, North-Holland, 1985.

[3] M. Sugeno, "An Introductory Survey of Fuzzy Control," *Information Sciences* 36, 59-83, 1985.

[4] Kuldip S. Rattan, "Evaluation of Control Mixer Concept for Reconfiguration of Flight Control System", *NAECON Proceedings*, Dayton, Vol. 1:pp 560-569, May, 1985.

[5] D. A. Rutherford and G. A. Carter, "A heuristic adaptive controller for a sinter plant," *Proceedings of the 2nd IFAC Symposium on Automation in Mining, Mineral and Metal Processing*, Johannesburg, R.S.A., Sept. 1976.

[6] E. H. Mamdani, "Applications of fuzzy algorithms for control of simple dynamic plant," *Proc. IEE* 121(12):1585-1588 (1974)

[7] L.A. Zadeh, "Fuzzy algorithm," *Information and Control* 12:94-102 (1968).

[8] L.A. Zadeh, "Outline of a new approach to the analysis of complex systems and decision processes," *IEEE Trans. Systems Man Cybernet* SMC-3:28-44 (1973).

[9] W. J. M. Kickert and H. R. Van Nauta Lemke, "The application of fuzzy set theory to control a warm water process," *Automatica* 12(4):301-308 (1976).

[10] J. J. Ostergaard, "Fuzzy logic control of a heat exchanger process," *Fuzzy Automata and Decision Processes* (M. M. Gupta, G. N. Saridis and B. R. Gaines, Eds.), North-Holland, 1977.

[11] P. J. King and E. H. Mamdani, "The application of fuzzy control systems to industrial processes," *Automatica* 13(3):235-242 (1977).

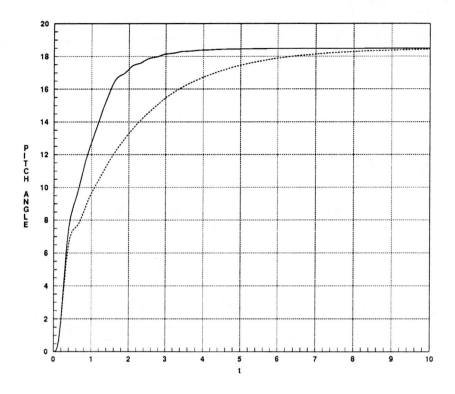

Figure 6: Computer Simulation of Pitch Angle Output for Fuzzy Controller (solid line) and Existing Controller.

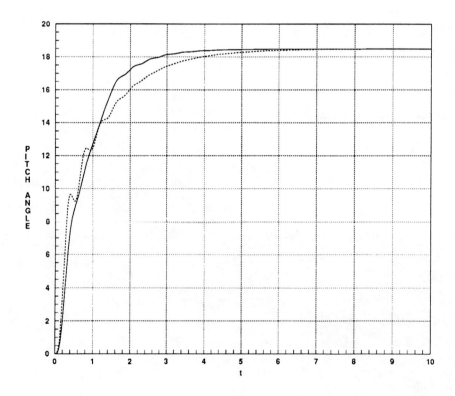

Figure 7: Computer Simulation of Pitch Angle Output for Fuzzy Controller (solid line) and Existing Controller with Double Gain.

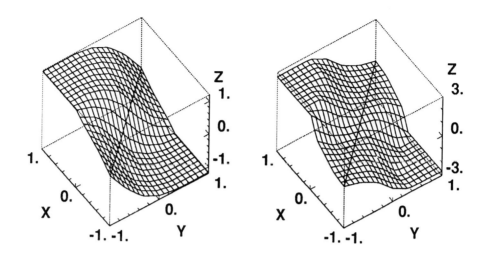

Figure 8: The Control Surfaces.

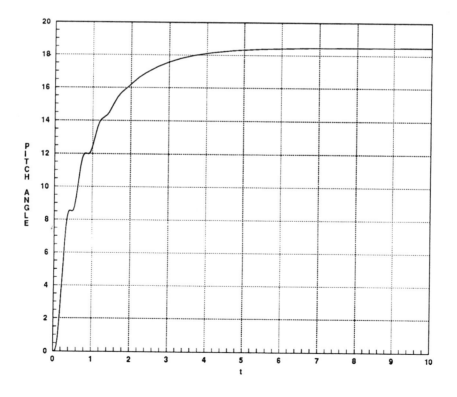

Figure 9: Computer Simulation of Pitch Angle Output for Fuzzy Controller with Uneven Control Surface.

FUZZY EXPERT SYSTEM FOR AUTOMATIC TRANSMISSION CONTROL

H.-G. Weil, G. Probst, F. Graf

Siemens AG ZFE ST SN 42   AT22T412
D-8000 Siemens AG Munich 83, Germany
e-mail: weil1@smaragd.zfe.siemens.com

Abstract - This paper presents an intelligent electronic transmission control system using both fuzzy logic and conventional methods. The controller is part of the drive train, therefore requiring that special attention is paid to the system's security. We propose two different methods to priorize the controller outputs and show how these methods can be realized within fuzzy control theory. Next we show in an application how rules can be derived as a mixture of human expert knowledge and dynamic process models. The controller is tested in our simulation environment. The results are compared to those obtained by using conventional shift logic.

Keywords: Fuzzy Control, Shift Pattern Adaptation, Automatic Transmission, System Security

1. Introduction

Over the last few years the number of vehicles equipped with an automatic transmission have increased substantially in Europe. Beneath the mechanical system needed for the shift operation there has to be an intelligent control module, which changes gears based on the observations of the environment. Therefore more recently a trend to electronic transmission control units (ECU) can be observed. The advantage of such systems is, that an intelligent environment dependent behaviour can be implemented cost-effectively. This paper deals with the development of a ECU based on a fuzzy expert system. It can deal with varying conditions such as a changing environment due to the road gradient, changing driver's intention and mechanical limitations of the different components of the drive train.

In the next chapter the problems of current automatic transmissions are discussed. The models of the driver, vehicle and environment used in the simulation are explained in chapter 3. Chapter 4 discusses the structure of our fuzzy controller, how the rules for the fuzzy rulebase are obtained and how the membership functions are designed. Special attention is given to the aspect of setting priorities within a fuzzy controller. In chapter 5 simulation results are shown. The last chapter gives a short conclusion and outlook to our future work.

2. Problems with current automatic transmission

The majority of automatic transmission controllers marketed today are designed to control the planetary gear type transmission with torque converter. Conventional transmission control systems determine the required transmission gear ratio with a manually selected pretermined shift pattern. The shift pattern consists of charcteristic curves, which describe when a shift operation takes place according to the measured vehicle speed and throttle opening (fig. 1). There are at most three different shift patterns available in today's automatic transmissions (e.g. sporty, defensiv, winter). With these shift patterns it is impossible to optimally adapt the transmission gear ratio to all possible driving conditions. In addition the shift patterns are not adapted to the driving conditions automatically but must be selected manually.

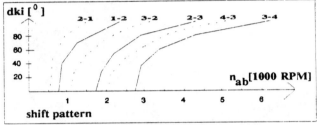

Fig. 1   Shift pattern for an automatic transmission

Only the transmission output speed n_{ab} and the throttle opening dki are considered when determining a suitable gear position (fig. 1). That is not enough to deal with a changing environment (hillside), changing vehicle load or detecting drivers intention (sporty, defensive). Therefore automatic adaptation to the driving conditions can hardly be implemented. A typical case of the weakness of todays automatic transmissions is shown in the following example. When driving down "steep incline" the driver reduces the throttle opening and possibly actuates the brakes. The intention is to drive in a gear as low as possible to utilize the increased braking torque of the combustion engine. The transmission control system does quite the opposite. The shift pattern in figure 1 shows that the reduction of the throttle opening leads to a higher gear position. That means the braking torque of the engine is reduced and therefore the mechanical brakes have to compensate by absorbing more energy. This behaviour is not desirable for vehicles.

An intelligent control module should take these effects into consideration. It therefore needs additional information about the environment, which means additional sensors. Modern cars are equipped with a lot of sensors for different control systems, eg cruise control, engine management systems, electronic anti lock brakes, etc. In the fuzzy ECU we only use sensors already available on these vehicles.

3. Basic Vehicle Simulation

For the development of intelligent control modules a powerful simulation testbed for testing and optimization is useful. Our testbed contains three different parts, one for modeling the driver's behaviour, one for the vehicle's dynamic differential equations and one for a realistic modeling of the changing environment.

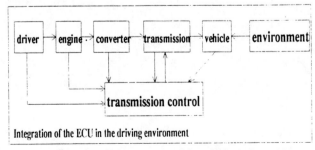

Fig. 2   Fuzzy transmission control and the vehicle environment

The basic simulation model and the environment in which the fuzzy controller interacts with these modules is shown in figure 2. Figure 4 shows the underlying dynamic model of the vehicle.

3.1. Modeling the drive train and the vehicle

The driver can operate the vehicle in two different ways. He can change the throttle opening, which results in a changing motor torque, and he can activate the brakes. In our simulation the drive train is subdivided into the combustion engine, the torque converter and the transmission. The engine and the torque converter are both modeled via characteristic curves (fig. 3). In the case of the engine these curves show the dependence of the motor torque on the throttle opening and the engine rotational speed. In case of the torque converter they show the dependence of the converter torque M_W on the motor torque M_{Mot} and the rotational speed of the motor n_{Mot} and turbine wheel n_{Tur}.

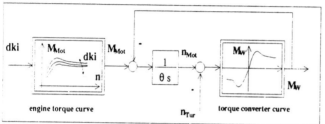

Fig. 3   Modeling the engine and the torque converter

The converter torque is used as the input of the next module of the power train, the transmission module. The transmission has to transform the converter torque to the actually required driving torque. This is done by changing the transmission ratio p_s. The output of the transmission module is therefore a torque dependent on the actual gear position and the converter torque. It is transmitted through the rear axis differential p_{Diff} to the driving wheels. Finally without slipping wheels and additional braking forces this leads to a force F_b acting to the vehicle's chassis.

$$M_{Mot}(t) = f(dki(t), n_{Mot}(t)), \qquad (1)$$

$$F_b(t) = M_w(t) \frac{p_s(t)\, p_{Diff}}{r_{wheel}} \qquad (2)$$

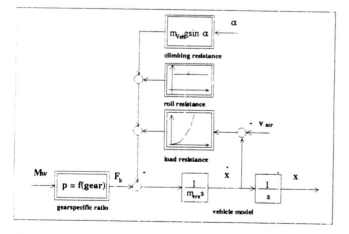

Fig. 4   Block diagram of the transmission and vehicle

There are several different forces acting against the driving force [18]. Some of them can be modeled and/or measured satisfactorily such as the air-resistance or the rolling resistance. Others such as the climbing resistance are difficult to measure, especially without additional sensors. These forces are treated as disturbances in the dynamic vehicle model shown in figure 4.

$$F_L = c_w A_q \frac{\rho_L}{2} v^2_{Veh} \qquad \text{air resistance} \qquad (3)$$

$$F_R = \mu\, m_{Veh}\, g \cos\alpha \qquad \text{rolling resistance} \qquad (4)$$

$$F_{incl} = m_{Veh}\, g \sin\alpha \qquad \text{climbing resistance} \qquad (5)$$

In addition to the already mentioned values the changing vehicle mass m_{Veh} must be taken into account. In extreme cases there is a difference of 100% and more between the actual and the normal mass. The unknown values must be considered when developing an intelligent ECU.

4.   Structure of the rule based fuzzy controller

The fuzzy controller consists of three modules. In the *fuzzification module* the measured states are transformed to probabilities of linguistic states. An *inference system* working with predetermined rules calculates probabilities for the linguistic output states. These rules determine the behaviour of the controller in the closed loop system. The transformation from linguistic states to output signals for process control is done by the *defuzzification module* [7]. Each of the modules will now be looked at more closely.

Figure 5 shows the overall structure of the fuzzy ECU. The controller receives seven input signals and calculates three output signals. One output signal for load detection, one for driver classification and one shift signal indicating whether a shift is allowed or not.

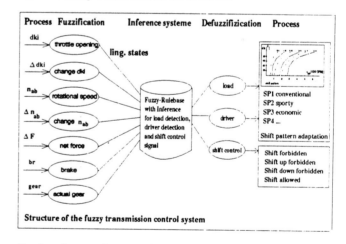

Fig. 5   Inputs and outputs of the fuzzy controller

Both the load and driver signals are used to automatically select a situation dependent shift pattern. The shift pattern adaptation to the driver and the load is shown in the following example. With increasing load or a sporty driver there is an automatic selection of shift patterns with flatter characteristic curves. Selecting a gear with these curves leads to a delay in shift up or may even result in a shift down of the transmission. Therefore the engine is running at higher speed. A behaviour very well suited to the situation described above.

The shift signal helps to survey mechanical limitations of the engine and the gearbox. For example if there are four available gear positions, there is no possibility to shift up when driving in the fourth gear. Just as often a

Fuzzification

The *fuzzifier* transforms the measured variables to the linguistic states. This is done by defining membership functions (MSF) for each linguistic state. With a MSF the connection between a physical state with respect to a probability of membership to a linguistic state is specified. For each input several MSFs are defined. The interaction of measurement, the MSF and the linguistic states is shown in figure 6. For example the drive shaft revolution varies from 0 to 6000 RPM. We chose five linguistic states *very low, low, medium, high* and *very high*. These are sufficient to describe the vehicle's velocity in linguistic terms. The probability for the different linguistic states for n_{ab}=1000RPM can be seen in figure 6. The fuzzification is done for each input state using different fuzzy sets for each.

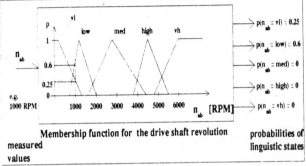

Fig. 6 Fuzzification using membership functions

In our fuzzy transmission control system we use the following sensor signals as input: the throttle opening sensor, the brake switch, the drive shaft revolution, the engine revolution and the gear position sensor. Some of these signals are used to calculate the relative change in the input signals. In particular these are the sum of all known forces, the change of the throttle opening and the change of the transmission output speed or vehicle acceleration.

The number of MSFs and their shape varies. In our implementation we use between three and five MSFs for one variable with triangular or rectangular shapes. The advantage is that with these MSFs time efficient code can be generated. The two different shapes are necessary to represent the nature of the real world. The triangle is used for real "fuzzy" values, the rectangular shape for "crisp" values. In some cases a mixture of both fuzzy and crisp representation is necessary. For example the rotational speed allowed for a special gear position is very well known due to limitations of the construction. Therefore "crisp" MSFs are used to describe these intervals. On the other hand the linguistic state *low* is not as well defined and so a "fuzzy" representation seems more realistic. Therefore a triangular MSF is used to desrcibe these "fuzzy" states.

Inference system

Based on the probabilities of the different linguistic states there are inference rules specifying the desired closed loop behaviour. The inference rules are IF ... THEN constructs. The IF-part refers to the linguistic input states. The THEN-part concerns the linguistic output states. Our fuzzy transmission controller consists of different rules. The rules are derived from human expert's knowledge and from the vehicle model.

The rules include different linguistic expressions in the IF part and at least one linguistic output in the THEN part. The fuzzy logic expressions are combined via fuzzy operators. Analogous operators to conventional logic AND and OR are used. The fuzzy AND is calculated as the minimum, the fuzzy OR as the maximum of the contributed fuzzy expressions.

The output of each rule is a probability for at least one linguistic output variable. To generate an output signal these variables have to be condensed. In the fuzzy transmission controller this is done by a modified MAX-DOT inference method [17]. Each MSF belonging to an output state is weighted with the related probability. All MSFs of one output are drawn in a common diagram (figure 7). The defuzzification uses this diagram to calculate the output signal.

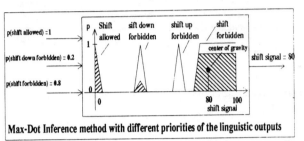

Fig. 7 MAX-DOT inference for condensing data in a common diagram

The fuzzy controller calculates three output signals. Therefore rules must be found which describe the desired shift behaviour in linguistic terms.

Load detection

There are three different load states used in the fuzzy controller: *mountain, plain, valley*. The rules for load detection are derived from the net force which is shown figure 5. The measured acceleration together with the mathematical model allows us to calculate a reference force. This reference force is compared to the actual force minus the known parts of the driving resistances. The resulting force ΔF is a direct measurement of the vehicle load including additional climbing and air resistance. One simple rule for detecting inclinations is:

IF ΔF=positive AND br=not pressed THEN load=mountain.

In addition there are some more heuristic rules describing the behaviour of a normal driver. For example if one is driving with maximal throttle opening and the vehicles velocity is not at maximum and not accelerating, then there must be an additional load acting against the driving force. A rule describing these assumptions is:

IF dki=very_big AND n_{ab}!=very_big AND Δn_{ab}!=positive
  THEN load=mountain.

Driver detection with throttle opening signal

The driver is classified as being either *defensive, medium* or *sporty*. The classification depends on the change of throttle opening and the moving average of the absolute changes Δdki. If the moving average rises it means that the driver is demanding a more sporty shift behaviour and vice versa.

Shift signal with different priorities of the linguistic outputs

The shift signal has four discrete states: *shift allowed, shift down forbidden, shift up forbidden, shift forbidden*. System security is affected by the shift operation and so it is important to give priorities to special rules or states. For example one rule indicates *shift allowed*, another *shift forbidden*. The final decision should be *shift forbidden* and therefore a priority mechanism must be defined.

We have implemented two different mechanisms to guarantee a priority controlled behaviour of the fuzzy controller. The first approach uses the area of the MSFs. Highly priorised outputs are assigned MSFs with relatively big areas. Therefore given the used defuzzification method, the outputs with the highest priority dominates (figure 7).

The second approach uses a special defuzzification method. Not only the center of gravity of the most probable rule for one linguistic output is used for defuzzification, but all weighted areas are taken into account. Therefore the influence of an output increases if more rules supporting the same output exist.

The rules are quite simple. If one is driving in first gear, a shift down is forbidden. A second rule ensures that the engine revolutions will be in the correct interval when changing the gear. Another rule indicates a situation when the vehicle is driving down a hill. The intention is to use the engine braking torque in addition to the normal brake for keeping the vehicle speed low. Therefore the gear position should be as small as possible. Shifting up as a consequence of a reduced throttle opening must therefore be blocked.

Defuzzification

As can be seen in figure 1 the last module of a fuzzy controller is the defuzzification module. This means that the calculated probabilities of different linguistic outputs are transformed to output signals. These signals are then directly fed into the process.

Several different defuzzification methods are described in the literature [17]. For our problem the center of gravity defuzzification method is used. It is applied to the diagrams of figure 1. Not only the envelope of the MSFs but the whole weighted MSFs are calculated. This has two reasons. First it results in very time efficient code. Secondly it ensures that the more rules indicate the same output, the higher is the probability that the output will dominate.

The calculation is done as follows. For every MSF k of a special output j of a rule i both the area A_{jk} and the moment M_{jk} is calculated.

$$A_{jk} = \int MSF_k \tag{6}$$

$$M_{jk} = A_{jk} \, x_s \tag{7}$$

The defuzzification is performed seperately for each output signal by computing the sums of the weighted A_{jk} (6) and M_{jk} (1) of the output of the rules. The weight α_i is the probability of rule i calculated by the inference system. The output signal is computed as the quotient of the sum of the moment divided by the area (10).

$$A_j = \sum_{i=1}^{rules} \alpha_i A_{jk,i} \qquad \text{weighted area} \tag{8}$$

$$M_j = \sum_{i=1}^{rules} \alpha_i M_{jk,i} \qquad \text{weighted moment} \tag{9}$$

$$U_j = \frac{M_j}{A_j} \qquad \text{output } j \tag{10}$$

The chosen defuzzification method is well suited for the described transmission control system. Depending on the shape and the area of the MSFs, the influence of special rules on the output signal can be reduced or amplified. Therefore the effect of "incorrect" rules can be minimized. On the other hand the influence of highly priorized rules can be maximized.

5. Simulation results

The shift behaviour of the fuzzy transmission control was tested in our simulation testbed. Two different driving actions were simulated. The first action shows the effect of load adaptation when driving on roads with changing inclination. The second action demonstrates the effect of shift pattern adaptation on the starting acceleration. The differences between the fuzzy and the conventional controller are discussed in detail.

5.1. Load adaptation of the shift patterns

When driving on roads with changing gradient, the shift behaviour of the ECTS can be shown very clearly. Figure 8 illustrates the underlying road profile.

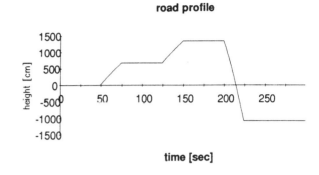

Fig. 8   Road profile

Figure 9 shows the driver's reaction to the changing conditions. When climbing a hill he increases the throttle opening. When driving downhill he decreases the throttle opening and presses the brakes.

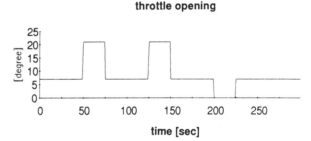

Fig. 9   Changing throttle opening

The vehicle's initial velocity is equal to zero. The diagram for the transmission output speed is shown in figure 10. In this simulation air-, rolling and climbing resistances are taken into account.

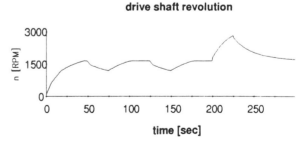

Fig. 10   Rotational speed of the transmission output shaft

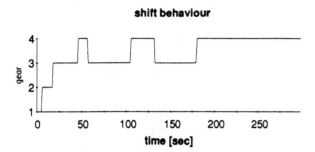

Fig. 11 Shift without load adaptation

Figure 11 shows the shift behaviour of a conventional automatic transmission. The vehicle accelerates. The gear is shifted from the first up to the fourth position. When climbing the first hill the gear is shifted down due to the wider throttle position and the chosen shift pattern of figure 1. When driving down a hill the disadvantages of the transmission control are obvious. Constant speed and reduced throttle opening result in a shifting up of the gear position. Therefore the engine torque is hardly used for braking the vehicle. The mechanical brakes have to compensate this shortcoming.

The shift behaviour of the fuzzy transmission control with load and driver dependent shift pattern adaptation is shown in figure 12. The simulation parameters were left unchanged. An improved shift behaviour can be observed. A faster shift down can be achieved when driving up a hill. This leads to a smaller reduction in vehicle speed due to the higher available engine torque. When driving down a hill the differences are even more obvious. The load adaptation results in a shift down to the second gear. That means the engine braking torque is used optimally. Mechanical limitations are taken into account by the adaptive shift signal of course.

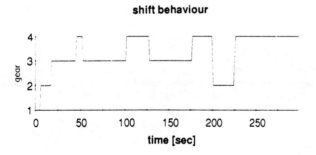

Fig. 12 Shift with load adaptation

5.2. Influence of shift pattern adaptation to the acceleration

When developing automatic transmission controllers a primary design goal is to achieve good vehicle acceleration and less fuel consumption. Load- and/or driver adaptation leads to a shift up delay. This results in a higher acceleration potential. The next figures show some simulation results. The road gradient is positive and constant. The vehicle's initial velocity is zero. The throttle opening is also constant. Without load adaptation there is a straightforward shift from the first to the fourth gear.

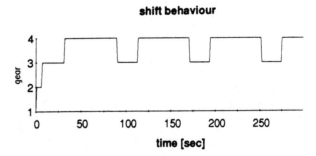

Fig. 13 Shift without load adaptation

When shifting from the third to the fourth gear the driving torque is getting smaller than the sum of all torque resistances. Therefore the velocity decreases after the shift operation and converges to a new stationary value.

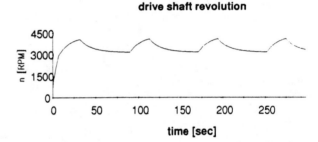

Fig. 14 Rotational speed of the transmission output shaft without adaptation

In this extreme case the reduction of the torque due to the shift up in combination with the reduced velocity of the vehicle result in a cyclic shift behaviour. This behaviour is also well known from manually shifted transmissions. It can be avoided by changing the throttle opening.

The following figures show the simulation results with shift pattern adaptation. An additional torque exists, which acts against the driving torque due to the climbing resistance. The fuzzy controller therefore detects *mountain* and chooses an adequate shift pattern with flatter and possibly translated characteristic curves. This results in a delayed shift up, which can be observed. Another effect is the increasing acceleration due to the lower gear position and the accompanying higher drive shaft torque. The cyclic shift behaviour is avoided by using the fuzzy ECU.

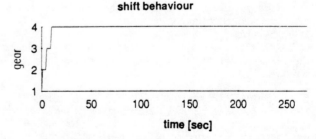

Fig. 15 Shift with shift pattern adaptation

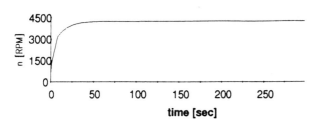

Fig. 16 Rotational speed of the transmission output shaft with shift pattern adaptation

6. Conclusion

A powerful intelligent transmission control system for automatic transmissions is developed using fuzzy logic. Conventional methods are integrated into the fuzzy controller. The shift logic realized with fuzzy logic adapts the shift behaviour according to the changing load and driver's intention. In addition the overall system security is taken into account in the design phase of the fuzzy controller. Incorrect shift behaviour can be avoided.

Future steps involve providing metrics which evaluate the shift quality. At the moment only the number of shift operations and their comprehensive are taken into consideration. A quantitative metric about the quality of the controller is needed. Therefore some functions which evaluate the energy consumption or the torque will be implemented. With such functions the controller can be optimized.

The next step involves implementing the algorithm on our actual microcontroller based transmission controller. This will proof the usefulness of the proposed fuzzy algorithm to changing driving conditions in today's traffic.

7. Bibliography

[1] Beek et al.: Patentschrift DE 3341652 C2 Einrichtung zur Steuerung einer Kupplungs- Getriebeeinheit, Bosch, 1987
[2] P.Y. Glorennec: Adaptive Fuzzy Control. IFSA 1991 Brüssel.
[3] D. Hrovat; W.F. Powers: Power Train Computer Control Systems. IFAC 10th triennial Congress Munich,1987.
[4] M. Ibamoto et al.: Advanced Technology of Transmission Control. Hitachi Review Vol.39(1990),No.5.
[5] Kishi et al.: Fuzzy Control System for automatic Transmission. United States Patent No. 4841815, Nissan Motor Co., 1989.
[6] K.Lorenz et al.: Interactive Engine and Transmission Control. BMW. International Congress on Transportation Electronics. Proceedings (IEEE cat No.88CH2533-8) Dearborn, MI USA, Oct. 1988.
[7] E.H. Mamdani: Application of fuzzy logic to approximate reasoning. IEEE Trans. Comput. 26, 1182-1191. (1977)
[8] P.W. Masding et. al.: Integrated microprocessor control of a hybrid i.c. enging-battery electric automotive power train. Trans Inst MC Vol.12 No.3 1990.
[9] M. Mitschke: Dynamik der Kraftfahrzeuge Band A: Antrieb und Bremsung. Springer Verlag Berlin Heidelberg New York 1982.
[10] Y. Kasai; Y. Morimoto: Electronically controlled continuously variable transmission. International Congress on Transportation Electronics Proceedings 1988 New York.
[11] J.J. Moskwa; J.K Hedrick: Nonlinear Algorithms for Automotive Engine Control. IEEE Control Systems Magazine 1990.
[12] Sakai et al.: Vehicle automatic transmission control system. European Patent Application No 89311976.8, Honda Giken Kogyo Kabushiki Kaisha. (1989)
[13] J.I. Soliman: Trends in automotive microelectronics. 16th International Symposium on Automotive Micro-Electronics, London Univ., England.1987
[14] M. Togai: Applications of Fuzzy Technology to intelligent Control: Present & Future. 33rd Annual Congress of System Control and Information Society, Kyoto, Japan, May 1989.
[15] Li Xue-shi, Qin Gui-he: Simulating test for microprocessor based manual transmission control. International Conference on System Simulation and scientific Computing 1989.
[16] L.A. Zadeh: The concept of a linguistic variable and its application to approximate reasoning. Memorandum ERL-M 411 Berkeley 1973.
[17] H.-J. Zimmermann: Fuzzy Set Theory and its applications.Kluwer Academic Publishers Boston/Dordrecht/London 1991.
[18] A. Zomotor: Fahrwerktechnik: Fahrverhalten. Vogel Buchverlag Würzburg 1987.

Application of Fuzzy Logic to Shift Scheduling Method for Automatic Transmission

S.Sakaguchi, I.Sakai, T.Haga
HONDA R&D Co., Ltd.
1-4-1 Chuo Wako-shi
Saitama 351-01, JAPAN

Abstract-Fuzzy control, which is now widely used in a varied number of fields, has also proved its usefulness in automobile engineering, where it can be used to automatically achieve constant speeds and control anti-lock brake systems. As seen in the case of automatic control of plants, where the operator's strategy is compiled for the purpose, these systems can only be realized after a knowledge base has been established. Knowledge bases have been based on interviews and questionnaires given to operators and their knowledge in how control devices work. This approach was tended to obtain only the most characteristic knowledge and failed to provide satisfactory control results. In view of this problem, we are proposing a way to construct a knowledge base that has the potential for a wide range of applications.

The behavior of human beings is generally regarded as a cycle of recognition of their environment, judgement, and action. This model of behavior consists of two processes: a responsive action devoid of intentional judgement and action based on knowledge and restrictions. Using this model, the authors have constructed a three-layer knowledge base that draws on common sense, knowledge based on experience, and restriction In this way a satisfactory knowledge base is constructed so that the reliability of control devices can be significantly enhanced.

This paper describes how the layers of knowledge have been constructed by applying the fuzzy control to the shift scheduling of automatic transmission cars(AT vehicles), and introduces the field test results of the scheduling system by a fuzzy control device to verify the effectiveness of the aforementioned approach.

I. INTRODUCTION

In recent years, automobiles are designed to have higher performances and more sophisticated functions. As the vehicle performances and functions became higher and more sophisticated, it has become difficult to explain its function without human elements because of the qualitative advancement achieved and the amount of information processed in a modern automobile. In other words, today's automobiles are required to have more sophisticated operational functions and features, in addition to the basic function of being mere by a transportation device in order to respond to more demanding human perceptions. There is also a stronger demand for qualitative improvement of controls that have been considered not so simple to materialize.

On the other hand, for automating complicated controls performed manually by humans, engineers are now placing emphasis on the fuzzy control method which controls objective system by first focusing on human knowledge and then materializing its controlling methods. The fuzzy control attempts to achieve control by incorporating human decision-making abilities into the controller as demonstrated by operators skilled in operating highly complicated systems. In short, the method aims to simulate human behavior using a computer.

For an example, a human operator's control strategy needs to be first established on knowledge base for automating a manufacturing plant.

A conventional approach for composing the knowledge is "questionnaire and interview" methods. This approach, however, tends to extract knowledge of only very specific features and not of overall circumstance. As a result, it is often unsuccessful to materialize the control by merely describing the knowledge, obtained through question and interview methods, in accordance of the rules of "IF...THEN..." which in turn poses a serious problem in designing control systems.

This paper proposes, in consideration of above mentioned problem, a method to construct highly flexible knowledge base and introduces the field test results of the fuzzy control applied to shift scheduling system of vehicle automatic transmission.

II. HUMAN THINKING PROCESS

When in the process of taking action, humans are not conscious of the mathematical model based on which the intended control is implemented. Rather, they roughly recognize the circumstance, then make a judgement and take action. The process can be broken down into four steps : recognition->judgement->decision->action.(Fig. 1) It is easy to describe this series of actions by the production rule of "IF...THEN...". Since this rule itself is regarded as the know-how for the control obtained through experience, practical application of the rule can be considered to be an effective approach to automate the complicated operations performed by humans.

This process reveals a strong linkage of the human action with knowledge they process, the information obtained from the outside world, and object consciousness of how to reflect the intention in their mind to the action. Thus, once the following problems are solved, automatic control by a computer will begin to approximate the flexible operation by humans : how to extract and compile the information in the brains; and how to introduce useful information into the computer.

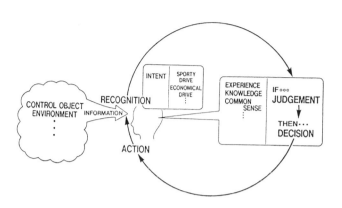

Fig. 1 Human Thinking Process

Another problem is finding how to feed the information about the environment into the control system. This issue relates closely to human's conscious awareness of objectives, as seen in the aforementioned process.

The authors are proposing a way to extract knowledge of human subject and construct a knowledge base through analyzing the human thinking process from recognition to action.

III. DESIGNING CONTOROLLER

A. Compiling knowledge

Human actions are not based on the same levels of decision-making, but are believed to take place at different levels. Fig. 2 depicts the basis of this relationship. In other words, the human actions are divided into two categories, one is the action by reflection and another is the action based on knowledge and restriction. The former is what is called common sense, intuitive, or learned through repeated decision making, by which the action reflectively takes place. The latter corresponds to experiential knowledge used to select a better approach to achieve the intended result. Therefore, the former is a decision making process without going through the intentional judgement process and the latter a decision-making accompanied by the intentional judgement process.

The authors propose to construct a knowledge base using three categories of knowledge : (1) knowledge of common sense that oversees the whole; (2) knowledge of experience that is locally effective or characteristic; and (3) restriction. This will prevent an evasion of knowledge and make the controller to work on the basis of common sense in the normal situations and activate the experience when necessary, thereby improving their reliability to a large extent.

B. Extracting knowledge

For extracting knowledge, knowledge based on experiences can be extracted rather easily from human subject since a human himself can consciously extract the knowledge whenever it is needed, while common sense knowledge is difficult to extract since humans are unconscious of the knowledge. In other words, humans are not normally aware of what they consider common sense. For this reason, questionnaires and interviews method are not appropriate in extracting this knowledge.

1) Abstracting common sense knowledge : Recognizing the problems as described above, the authors decided, instead of extracting knowledge equivalent to common sense knowledge through questionnaires and interviews method, to extract common sense knowledge by asking, "What is the most fundamental requirement of the object of control ?" Consider the water tank model in Fig. 4. A valve on the tank is operated to keep the surface of the water as a target level. In this case, the most fundamental requirement of the object of control is to increase the volume of water when the surface is below the target level and to decrease the volume when the surface is over, or simply:

(1) To open the valve when the surface is below the target level
(2) To close the valve when the surface is over the target level

This knowledge is so basic that it alone cannot control the water level with high precision. And yet, the water level cannot be controlled adequately without this knowledge, though humans are not conscious of it and usually take it for granted.

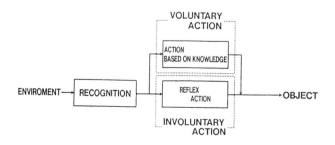

Fig. 2 Human Actions

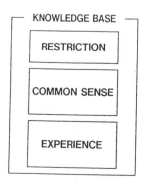

Fig. 3 Configuration of Knowledge Base

In questionnaires and interviews, when asked to explain how this water level is controlled, in most cases, only the knowledge of how to control the level is obtainable in such a characteristic situations.

(1) Adjust the valve opening when the water surface is going to overshoot or undershoot, and
(2) make the valve opening larger if the water surface is below the target level and is still reading.

As this example shows, knowledge obtained through questionnaires and interviews often already contains common sense as a basic component. That is, knowledge is normally extracted not as an overall knowledge but as localized effective knowledge. However, sets of knowledge specific to certain circumstances can hardly control systems satisfactorily. The approach suggested in this paper is effective in extracting the common sense of humans.

2) Abstracting experimental knowledge : As already explained, it is quite easy to extract knowledge that corresponds to experience through questionnaires and interviews. In the water tank model, humans often open or close the valve with the firm intention of bringing the water level to a particular level. Humans would promptly give the following knowledge when asked in a questionnaire or interviews, how to control the water level. Namely, if the level is below the target but is approaching it, they will:

(1) Close the valve if the water surface is rising and approaching with rapid speed to the target level,
(2) close the valve slightly if the water surface is rising slowly to the target level.

In comparison with the common knowledge described above, this knowledge is effective and characteristic in an extremely limited situation. Because of its characteristics, the knowledge can remain with the human consciousness and can easily be extracted as well.

C. Inferring intention

The information that are treated when humans make decision on their action can be classified into two categories: information obtained from external environment and intention closely linked to one's objective. To replace humans with an automatic control system, therefore, one key issue besides the acquisition of knowledge is learning how to incorporate human intentions into the control system. In order to resolve the subject, the authors have first formed human intent based on the information obtained from the human action and environment, and then adopted a multistage fuzzy inference by incorporating its result as input information for the controller.(Fig. 5) This makes it possible to express an antecedent of the knowledge base as a simple description and to express the control rule in short simple phrases, thus leading to improved maintenance and simpler settings. In addition, the separate input information provides a benefit that the control rule does not require rewriting even when the precision of inference is to be upgraded.

D. Overall architecture

As a result of the above study, the fuzzy control system has been composed as an overall architecture as shown in Fig. 6, in which the knowledge base is divided into common sense, experience, and restriction for assessment, and the rules of common sense and experience are referred to as fuzzy inference. The restriction is written as an algorithm

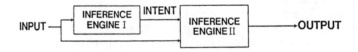

Fig. 5 Multistage Inference

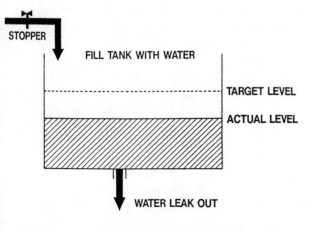

Fig. 4 Water Tank Model

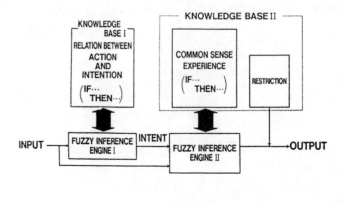

Fig. 6 Overall Architecture

to monitor fuzzy inference output because it is an item that cannot be compromised. Given that the input data contains both the intention and environmental information, a module is added to infer the intention. This architecture seems to have potential for a wide range of applications. This paper adopts the shift scheduling system for AT vehicles to verify the appropriateness of an overall architecture of the system.

IV. FUZZY AT SHIFT SCHEDULING SYSTEM

A. Conventional AT shift scheduling method

As shown in Fig. 7, shift scheduling is usually based on two parameters, vehicle speed and throttle opening. Presently, microprocessors are increasingly used to determine the shift scheduling. In some of the shifting scheduling, command values retrieved from gear shift mapping are automatically adjusted and corrected in detail.

B. Fuzzy AT shift scheduling system

1) Block composition: Fig. 8 shows a fuzzy AT shift scheduling system that consists of an input block, an inference block, a knowledge base and an output block.

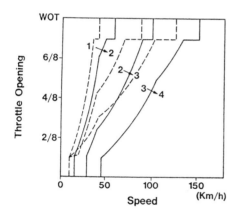

Fig. 7 Gear Shift Map

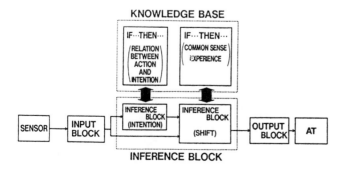

Fig. 8 Fuzzy AT Shift Scheduling System

Input block --- This block receives six input variables --- vehicle speed, throttle opening, engine speed, intake manifold pressure, shift positions and brake signals -- and calculates information necessary for controlling the shifting schedule by performing pre-processing including such statistical processing as averaging and filtering calculations.

Inference block --- This block can be divided into two blocks; intentional inference block and shift position inference block. The intentional inference block decides how much the driver intends to slow down through fuzzy inference using the information from the input block and sets of rules based on the knowledge base. The shift position inference block determines desirable shift positions through fuzzy inference using the information from the input block, results of the intentional inference obtained and sets of rules based on the knowledge base.

Knowledge base --- This block is composed of two sets of knowledge; one used for intentional inference and another used for shift scheduling. In the former, the relationship of the driver's intention to slow down and status of the vehicle is stored as a proposition of the "IF...THEN..." statements. In the latter, the common sense mentioned earlier and driver's control expertise represented as experience are stored as a number of propositions made using "IF...THEN..." statements, where the ambiguity of recognition is expressed by fuzzy sets.

Output block --- This block performs engine over-rev checking of a shift position which is inferred by the inference block and decides the eventual shift position before issuing a command to the actuator. Known restrictions in the human knowledge are thereby considered.

2) Knowledge base

2)-1 Knowledge base for shift scheduling: As explained before, the knowledge for the shift scheduling consists of common sense and experience. For the common sense, it would be most appropriate to adopt the most fundamental characteristic required by the object to be controlled. The authors consider the following two points as the most essential requirements for shift scheduling of automatic transmissions.

(1) Selection of shifting gears according to the vehicle speed.
(2) Shiftings in accordance with the vehicle speeds are variable in accordance with the throttle opening.

These points can be expressed in the form of "IF...THEN..." statements as follows:
(1) If the vehicle speed is high, then use a higher gear, and if the vehicle speed is low, then use a lower gear.
(2) If the throttle opening is large, then use a lower gear, and if the throttle opening is small, then use a higher gear.

These two sets of knowledge have been extracted through a study of characteristics that AT vehicles should have. Together they form an overall knowledge base that works in all driving conditions and agrees with human instinct.

Experiences, under special circumstance where common sense could not be utilized, are responsible for correcting the output power produced by common sense. During uphill driving, for an example, a common sense of "shifting into a higher gear as the vehicle accelerates" is not necessary applicable. In such a case, the driver's experience leads him to use a lower gear. Experience is the basis of this decision.

A list of the concrete knowledge base in the AT shift scheduling system and its membership functions are described in Fig. 9.

2)-2 Knowledge base for intentional inference: Human intentions can be determined only through their actions, since there is no sensor which can read intentions. To predict the driver's intention, it is necessary therefore to clarify the situation (i.e., what action he takes)in which he has an intention to slow down or decides not to slow down. For this reason, the authors created a state transition diagram, Fig. 10 which shows, for an example, that the state changes from the non-existence to the existence of the intention to decrease speed when the throttle is closed, the braking pedal is pressed and the vehicle slows down(arrow A). It is not, however, clear from this diagram whether the intention intensifies even when the pedal is released(arrow B). On the basis of this diagram, the relationship between these actions and changing direction of the intention with the "IF...THEN..." statements

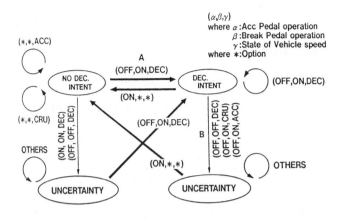

Fig. 10 State Transition Diagram

to make the knowledge base to infer the intention to decrease speed are described. It is basically composed of the following two rules:

(1) Intensifies the intent to decrease the speed when the throttle is closed, the braking pedal is pressed down and the vehicle slows down, and
(2) weakens when the throttle is opened.

In actuality, rules have been written for each of the different situations in which a vehicle is used to better estimate the driver's intention. A list of concrete knowledge base and membership functions are described in the following figure.

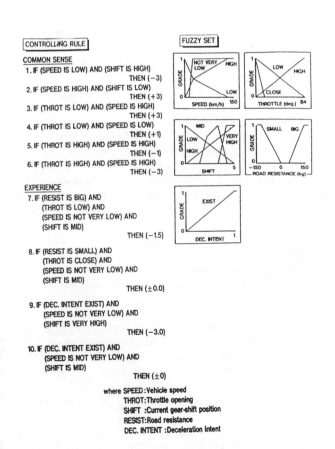

Fig. 9 "IF...THEN..." Rules and Fuzzy Sets(SHIFT POSITION)

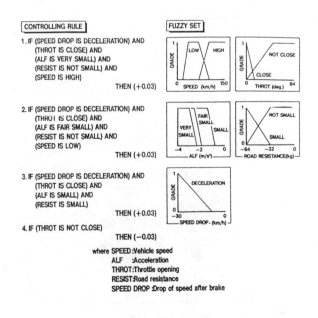

Fig. 11 "IF...THEN..." Rules and Fuzzy Sets (INTENTION)

V. FIELD TEST RESULTS

Field tests on the fuzzy shift scheduling system were conducted on mountainous roads and city streets.

A. Mountainous road test

A mountain road in California was selected as a typical mountainous road and the fuzzy control system (FUZZY) was compared with a conventional automatic drive system (DRIVE).

1) Uphill driving: Fig. 12 shows the uphill driving performances of the two systems, while Fig. 13 shows the frequencies of the shifts used and the number of gear changes. As the figures indicate, when the driver releases the throttle of the conventional system during the uphill driving, the transmission unintentionally shift up creating power insufficiency which in turn generates shift-down. On the other hand, the fuzzy system under the same circumstance will hold the gear appropriately at the third gear, eliminating busy shiftings and improving driveability.

Table I compares the fuel consumption of the two systems to further show the advantage of the fuzzy control system, which seems to be the result of the smaller slip loss in the torque converter arising from the effective utilization of lower gear during uphill driving.

Table I Fuel Consumption

DRIVE	4.43km/ℓ (10.42MPG)
FUZZY	5.00km/ℓ (11.76MPG)

2) Downhill driving: Fig. 14 shows the downhill driving performances of the two systems, while Fig. 15 shows the frequencies of shifting and the number of braking. These figures indicate that the fourth gear is almost always maintained without the effective application of engine braking under the conventional system, while the fuzzy system makes proper down shiftings as if it is responding to the driver's intention to decrease the vehicle speed. Subsequently, the gear is held at the third to apply engine braking, thereby substantially improving the driving stability.

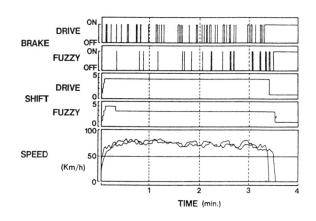

Fig. 14 Comparison of Fuzzy and D-Range III (DOWNHILL DRIVING)

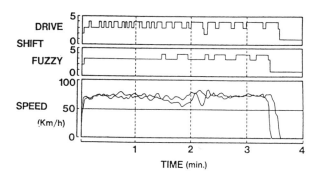

Fig. 12 Comparison of Fuzzy and D-Range I (UPHILL DRIVING)

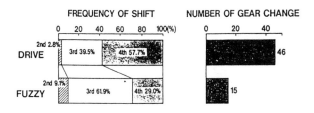

Fig. 13 Comparison of Fuzzy and D-Range II (UPHILL DRIVING)

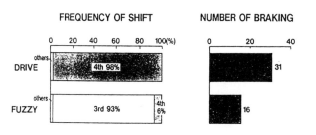

Fig. 15 Comparison of Fuzzy and D-Range IV (DOWNHILL DRIVING)

B. City street test

To re-create typical city street conditions, the authors selected a road in the suburbs of Tokyo and compared the driving performance of the fuzzy control system with that of the conventional AT system. On city streets where there are no extreme slopes, the fuzzy control system makes scheduling, which relies on common sense in the knowledge base, in large part in the same way as the conventional system. This means that the proposed system requires the driver to conduct mostly the same throttle pedal operations as those conducted under the conventional system.

VI. CONCLUSION

Since human behavior can be assessed as a three-layer construction, this paper has proposed to construct a knowledge base for the fuzzy control system by considering common sense, experience and restriction, and to incorporate the human intention into system control as input information.

On the basis of this concept, the authors have developed a system to determine shift scheduling of an automatic transmission.

Field tests have proved that the developed system can realize a shift scheduling which satisfactorily reflects human experiences and intentions. In uphill and downhill driving in particular, the fuzzy system, in comparison with the map method employed by the conventional AT system, shows a significant improvement in the driveability particularly in such areas as riding comfort and firm driving feeling.

In conclusion, this fuzzy control system was actually applied to the production model, 3.2-liter V-6 engine introduced in September of 1992 and has received excellent evaluations in the Japanese market.

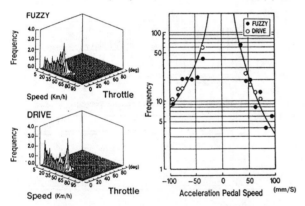

Fig. 16 Accelerator Operation for City Driving

REFERENCES

[1] T. Yamazaki and M.Sugeno, "Fuzzy Control", *System and Control* Vol. 28 No. 10 PP442-446, (July, 1984)
[2] O. Itoh, " Application of Fuzzy Control to Activated Sludge Process" *Preprints of Second IFSA Congress*, PP282-285, (1987)
[3] E. H. Mamdani, "application of Fuzzy Algorithms for Control of Simple Dynamic Plant", *Proc. IEEE*, 121, PP1585-1588, (1974)
[4] Lotfi. A. Zadeh, "Fuzzy Sets", *Information and Control*, Vol. 8, PP338-358 (1965)
[5] S. Yasunobu, "Predictive Fuzzy Control and Application for Automatic Container Crane Operation System", *Preprints of Second IFSA Congress*, PP349-352, (1987)
[6] S. Yasunobu, "A Predictive Fuzzy Control for Automatic Train Operation", *System and Control* Vol. 28 No. 10 PP605-613, (October 1984)
[7] H. Takahashi, "Subjective Evaluation Modeling Using Fuzzy Logic and Neural Network", *Third IFSA Congress*, PP520-523, (1989)
[8] I. Sakai, S. Sakaguchi, T. Haga, "Shift Scheduling Method of Automatic Transmission Vehicles with Application of Fuzzy Logic", *23rd FISITA Congress* Vol. 1, PP343-347, (1990)
[9] I. Sakai, S. Sakaguchi, T. Haga, M. Togai, "Shift Scheduling Method of Automatic Transmission Vehicles Using Fuzzy Logic", *IEEE Roundtable Discussion on "Fuzzy and Neural Systems, and Vehicles Applications"*, paper #27, (1991)

Adaptive Traffic Signal Control Using Fuzzy Logic

Stephen Chiu and Sujeet Chand
Rockwell International Science Center
1049 Camino Dos Rios
Thousand Oaks, CA 91360, USA

Abstract ---- We present a distributed approach to traffic signal control, where the signal timing parameters at a given intersection are adjusted as functions of the local traffic condition and of the signal timing parameters at adjacent intersections. Thus, the signal timing parameters evolve dynamically using only local information to improve traffic flow. This distributed approach provides for a fault-tolerant, highly responsive traffic management system.

The signal timing at an intersection is defined by three parameters: cycle time, phase split, and offset. We use fuzzy decision rules to adjust these three parameters based only on local information. The amount of change in the timing parameters during each cycle is limited to a small fraction of the current parameters to ensure smooth transition. We show the effectiveness of this method through simulation of the traffic flow in a network of controlled intersections.

I. INTRODUCTION

With the steady increase in the number of automobiles on the road, it has become ever more important to manage traffic flow efficiently to optimize utilization of existing road capacity. High fuel cost and environmental concerns also provide important incentives for minimizing traffic delays. To this end, computer technology has been widely applied to optimize traffic signal timing to facilitate traffic movement.

Traffic signals in use today typically operate based on a preset timing schedule. The most common traffic control system used in the United States is the Urban Traffic Control System (UTCS), developed by the Federal Highway Administration in the 1970's. The UTCS generates timing schedules off-line on a central computer based on average traffic conditions for a specific time of day; the schedules are then downloaded to the local controllers at the corresponding time of day. The timing schedules are typically obtained by either maximizing the bandwidth on arterial streets or minimizing a disutility index that is generally a measure of delay and stops. Computer programs such as MAXBAND [1] and TRANSYT-7F [7] are well established means for performing these optimizations.

The off-line, global optimization approach used by UTCS cannot respond adequately to unpredictable changes in traffic demand. With the availability of inexpensive microprocessors, several real-time adaptive traffic control systems were developed in the late 70's and early 80's to address this problem. These systems can respond to changing traffic demand by performing incremental optimizations at the local level. The most notable of these are SCATS [2,3,6], developed in Australia, and SCOOT [3,5], developed in England. SCATS is installed in several major cities in Australia, New Zealand, and parts of Asia; recently the first installation of SCATS in the U.S. was completed near Detroit, Michigan. SCOOT is installed in over 40 cities, of which 8 are outside of England.

Both SCATS and SCOOT incrementally optimize the signals' cycle time, phase split, and offset. The cycle time is the duration for completing all phases of a signal; phase split is the division of the cycle time into periods of green signal for competing approaches; offset is the time relationship between the start of each phase among adjacent intersections. SCATS organizes groups of intersections into subsystems. Each subsystem contains only one critical intersection whose timing parameters are adjusted directly by a regional computer based on the average prevailing traffic condition for the area. All other intersections in the subsystem are always coordinated with the critical intersection, sharing a common cycle time and coordinated phase split and offset. Subsystems may be linked to form a larger coordinated system when their cycle times are nearly equal. At the lower level, each intersection can independently shorten or omit a particular phase based on local traffic demand; however, any time saved by ending a phase early must be added to the subsequent phase to maintain a common cycle time among all intersections in the subsystem. The basic traffic data used by SCATS is the "degree of saturation", defined as the ratio of the effectively used green time to the total available green time. Cycle time for a critical intersection is adjusted to maintain a high degree of saturation for the lane with the greatest degree of saturation. Phase split for a critical intersection is adjusted to maintain equal degrees of saturation on competing approaches. The offsets among the intersections in a subsystem are selected to minimize stops in the direction of dominant traffic flow. Technical details are not available from literature on exactly how the cycle time and phase split of a critical intersection are adjusted. It seems that SCATS does not explicitly optimize any specific performance measure, such as average delay or stops.

SCOOT uses real-time traffic data to obtain traffic flow models, called "cyclic flow profiles", on-line. The cyclic flow profiles are then used to estimate how many vehicles will arrive at a downstream signal when the signal is red. This estimate provides predictions of queue size for different hypothetical changes in the signal timing parameters. SCOOT's objective is to minimize the sum of the average queues in an area. A few seconds before every phase change, SCOOT uses the flow model to determine whether it is better to delay or advance the time of the phase change by 4 seconds, or leave it unaltered. Once a cycle, a similar question is asked to determine whether the offset should be set 4 seconds earlier or later. Once every few minutes, a similar question is asked to determine whether the cycle time should be incremented or decremented by a few seconds. Thus, SCOOT changes its timing parameters in fixed increments to optimize an explicit performance objective.

It is problematic that a specific performance objective will be appropriate for all traffic conditions. For example, maximizing bandwidth on arterial streets may cause extended wait time for vehicles on minor streets. On the other hand, minimizing delay and stops generally does not result in maximum bandwidth. This problem is typically addressed by the use of weighting factors; the TRANSYT optimization program provides user-selectable link-to-link flow weighting, stop weighting factors, and delay weighting factors. A traffic engineer can vary these weighting factors until the program produces a good (by human judgement) compromise solution. Perhaps a performance index should be a function of the traffic condition; it may be appropriate to emphasize an equitable distribution of movement opportunities when traffic volume is low and emphasize overall network efficiency when the traffic is congested. In view of the uncertainty in defining a suitable performance measure, the reactive type of control provided by SCATS, where there is no explicit effort to optimize any specific performance measure, appears to have merit. We believe implementing this type of control using fuzzy logic decision rules can further enhance the appropriateness of the control actions, increase control flexibility, and produce performance characteristics that more closely match human's sensibility of "good" traffic management.

In past work performed by Pappis and Mamdani [4], fuzzy logic was applied to control an intersection of two one-way streets. It was assumed that vehicle detectors were placed sufficiently upstream from the intersection to inform the controller about future arrival of vehicles at the intersection. It is then possible to predict the the number of vehicles that will cross the intersection and the size of the queue that will accumulate if no change to the the signal state takes place in the next N seconds, for N = 1,2,...10. The predicted outcomes are evaluated by fuzzy decision rules to determine the desirability of extending the current state for N more seconds. Each of the possible extensions is assigned a degree of confidence by the rules, and the extension with maximum confidence is selected for implementation. Before the extended period ends, the rules are applied again to see if further extensions are desirable.

Here we apply fuzzy logic to the general problem of controlling multiple intersections in a network of two-way streets. We propose a highly distributed architecture in which each intersection independently adjusts its cycle time, phase split, and offset using only local traffic data collected at the intersection. This architecture provides for a fault-tolerant traffic management system where traffic can be managed by the collective actions of simple microprocessors located at each intersection; hardware failure at a small number of intersections should have minimal effect on overall network performance. By requiring only local traffic data for operation, the controllers can be installed individually and incrementally into an area with existing signal controllers. Each intersection uses an identical set of fuzzy decision rules to adjust its timing parameters. The rules for adjusting the cycle time and phase split follow the same general principles used by SCATS: cycle time is adjusted to maintain a good degree of saturation and phase split is adjusted to achieve equal degrees of saturation on competing approaches. The offset at each intersection is adjusted incrementally to coordinate with the adjacent upstream intersection to minimize stops in the direction of dominant traffic flow. Through simulation of a small network of streets, the distributed fuzzy control system has shown to be effective in rapidly reducing delay and stops.

II. TRAFFIC CONTROL RULES

A set of 40 fuzzy decision rules was used for adjusting the signal timing parameters. The rules for adjusting cycle time, phase split, and offset are decoupled so that these parameters are adjusted independently; this greatly simplifies the rule base. Although independent adjustment of these parameters may result in one parameter change working against another, no conflict was evident in simulations under various traffic conditions. Since incremental adjustments are made at every phase change, a conflicting adjustment will most likely be absorbed by the numerous successive adjustments.

A. Cycle Time Adjustment

Cycle time is adjusted to maintain a good degree of saturation on the approach with highest saturation. We define the degree of saturation for a given approach as the actual number of vehicles that passed through the intersection during the green period divided by the maximum number of vehicles that can pass through the intersection during that period. Hence, the degree of saturation is a measure of how effectively the green period is being used. The primary reason for adjusting cycle time to maintain a given degree of saturation is not to ensure efficient use of green periods, but to control delay and stops. When traffic volume is low, the cycle time must be reduced to maintain a given degree of saturation; this results in short cycle times that reduce the delay in waiting for phase changes. When the traffic volume is high, the cycle time must be increased to maintain the same degree of saturation; this results in long cycle times that reduce the number of stops.

The rules for adjusting the cycle time are shown in Fig. 1 and the corresponding membership functions are shown in Fig. 4. The inputs to the rules are: (1) the highest degree of saturation on any approach (denoted as "highest_sat" in the rules), and (2) the highest degree of saturation on its competing approaches (denoted as "cross_sat"). The output of the rules is the amount of adjustment to the current cycle time, expressed as a fraction of the current cycle time. The maximum adjustment allowed is 20% of the current cycle time. The rules basically adjust the cycle time in proportion to the deviation of the degree of saturation from the desired saturation value. However, when the highest saturation is high and the saturation on the competing approach is low, we can let the phase split adjustments alleviate the high saturation. It should be noted that the "optimal" degree of saturation to be maintained by the controller is only 0.55, whereas SCATS typically attempts to maintain a degree of saturation of 0.9. This discrepancy arises from the method of calculating the maximum (saturated) flow value. We derive the maximum flow value based on a platoon of vehicles with no gaps moving through the intersection at the speed limit, while SCATS uses calibrated, more realistic values.

```
if highest_sat is none
     then cycl_change is n.big;
if highest_sat is low
     then cycl_change is n.med;
if highest_sat is slightly low
     then cycl_change is n.sml;
if highest_sat is good
     then cycl_change is zero;
if highest_sat is high & cross_sat is not high
     then cycl_change is p.sml;
if highest_sat is high & cross_sat is high
     then cycl_change is p.med;
if highest_sat is saturated
     then cycl_change is p.big;
```

Fig. 1. Rules for adjusting cycle time.

B. Phase Split Adjustment

Phase split is adjusted to maintain equal degrees of saturation on competing approaches. The rules for adjusting the phase split is shown in Fig. 2 and the corresponding membership functions are shown in Fig. 4. The inputs to the rules are: (1) the difference between the highest degree of saturation on the east-west approaches and the highest degree of saturation on the north-south approaches ("sat_diff"), and (2) the highest degree of saturation on any approach ("highest_sat"). The output of the rules is the amount of adjustment to the current east-west green period, expressed as a fraction of the current cycle time. Subtracting time from the east-west green period is equivalent to adding an equal amount of time to the north-south green period. When the saturation difference is large and the highest degree of saturation is high, the green period is adjusted by a large amount to both reduce the difference and alleviate the high saturation. When the highest degree of saturation is low, the green period is adjusted by only a small amount to avoid excessive reduction in the degree of saturation.

```
if sat_diff is p.big & highest_sat is saturated
     then green_change is p.big;
if sat_diff is p.big & highest_sat is high
     then green_change is p.big;
if sat_diff is p.big & highest_sat is not high
     then green_change is p.med;
if sat_diff is n.big & highest_sat is saturated
     then green_change is n.big;
if sat_diff is n.big & highest_sat is high
     then green_change is n.big;
if sat_diff is n.big & highest_sat is not high
     then green_change is n.med;
if sat_diff is p.med & highest_sat is saturated
     then green_change is p.med;
if sat_diff is p.med & highest_sat is high
     then green_change is p.med;
if sat_diff is p.med & highest_sat is not high
     then green_change is p.sml;
if sat_diff is n.med & highest_sat is saturated
     then green_change is n.med;
if sat_diff is n.med & highest_sat is high
     then green_change is n.med;
if sat_diff is n.med & highest_sat is not high
     then green_change is n.sml;
if sat_diff is p.sml
     then green_change is p.sml;
if sat_diff is n.sml
     then green_change is n.sml;
if sat_diff is zero then green_change is zero;
```

Fig. 2. Rules for adjusting phase split.

C. Offset Adjustment

Offset is adjusted to coordinate adjacent signals in a way that minimizes stops in the direction of dominant traffic flow. The controller first determines the dominant direction from the vehicle count for each approach. Based on the next green time of the upstream intersection, the arrival time of a vehicle platoon leaving the upstream intersection can be calculated. If the local signal becomes green at that time, then the vehicles will pass through the local intersection unstopped. The required local adjustment to the time of the next phase change is calculated based on this target green time. Fuzzy rules are then applied to determine what fraction of the required adjustment can be reasonably executed in the current cycle. The rules for determining the allowable adjustment are shown in Fig. 3 and the corresponding membership functions are shown in Fig. 4. The inputs to the rules are: (1) the normalized difference between the traffic volume in the dominant direction and the average volume in the remaining directions ("vol_diff"); and (2) the required time adjustment relative to the adjustable amount of time ("req_adjust"), e.g., the amount by which the current green phase is to be ended early divided by the the current green period. The output of the rules is the allowable adjustment, expressed as a fraction of the required amount of adjustment. These rules will allow a large fraction of the adjustment to be made if there is a significant advantage to be gained by coordinating the flow in the dominant direction and that the adjustment can be made without significant disruption to the current schedule.

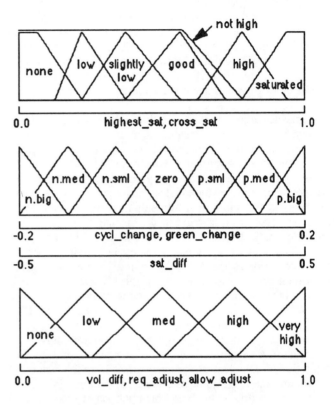

Fig. 4. Membership functions used in rules.

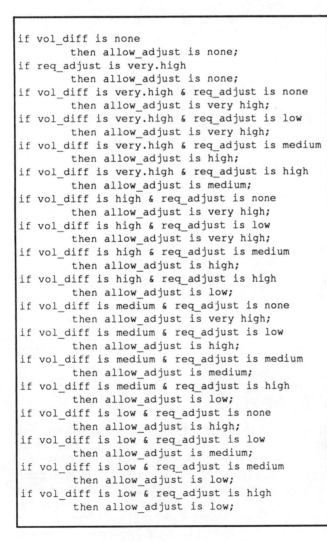

Fig. 3. Rules for adjusting offset.

III. SIMULATION RESULTS

Simulation was performed to verify the effectiveness of the distributed fuzzy control scheme. We considered a small network of intersections formed by six streets, shown in Fig. 5. A mean vehicle arrival rate is assigned to each end of a street. At every simulation time step, a random number is generated for each lane of a street and compared with the assigned vehicle arrival rate to determine whether a vehicle should be added to the beginning of the lane. Some simplifying assumptions were used in the simulation model: (1) unless stopped, a vehicle always moves at the speed prescribed by the speed limit of the street, (2) a vehicle cannot change lane, and (3) a vehicle cannot turn. Vehicle counters are assumed to be installed in all lanes of a street at each intersection. When the the green phase begins for a given approach, the number of vehicles passing through the intersection during the green period is counted. The degree of saturation for each approach is then calculated from the vehicle count and the length of the green period. At the start of each phase change, the controller computes the time of the next phase change using its current cycle time and phase split values. The fuzzy decision rules are then applied to adjust the time of the next phase change according to the offset adjustment rules; the adjusted cycle time and phase split values are used only in the subsequent computation of the next phase change time.

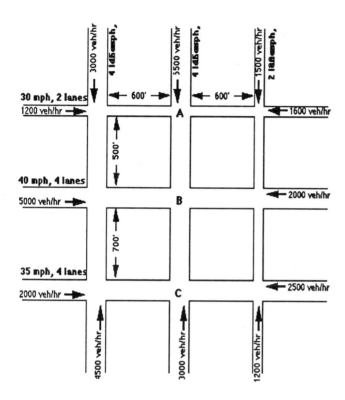

Fig. 5. Network of streets used in simulation.

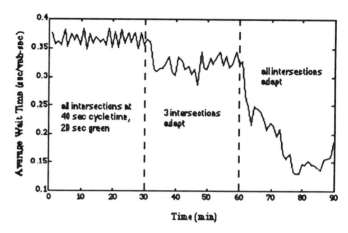

Fig. 6. Average waiting time for the case in which all intersections have an initial cycle time of 40 seconds.

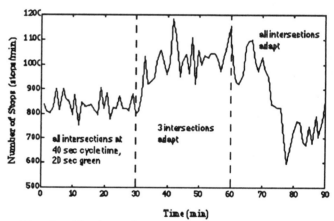

Fig. 7. Number of stops for the case in which all intersections have an initial cycle time of 40 seconds.

Figure 6 shows the average waiting time per vehicle per second spent in the network as a function of time. Figure 7 shows the number of stops per minute encountered by all vehicles. For the first 30 minutes of this simulation, all intersections have a fixed cycle time of 40 seconds, a green duration of 20 seconds, and start their phases at the same time. At the end of 30 minutes, intersections A, B, and C shown in Fig. 5 were allowed to adapt their timing parameters according to the fuzzy decision rules. At the end of 60 minutes, all intersections were allowed to adapt. We see that the improvement in waiting time is minimal when only 3 intersections are adaptive. Furthermore, when only 3 intersections are adaptive, the minor improvement in waiting time was obtained at the expense of greatly increased number of stops. This is because the cycle time chosen by the adaptive intersections (around 20 sec) is widely different from the cycle time for the fixed intersections (40 sec). The mismatch of cycle times resulted in a complete lack of coordination between the adaptive intersections and the fixed intersections, where timing adjustments to facilitate local traffic movement can adversely affect the overall traffic movement. When all intersections were allowed to adapt, all intersections attained similar cycle times (around 20 sec), and significant reductions in both waiting time and number of stops were achieved.

Figures 8 and 9 show the results of a simulation performed using the same sequence of events, but with an initial cycle time of 20 seconds and green duration of 10 seconds for all intersections. In this case, significant reductions in both waiting time and number of stops were achieved even when only 3 intersections are adaptive. This is because the cycle time for the fixed intersections closely matches that chosen by the adaptive intersections. Sharing a common cycle time has enabled the 3 adaptive intersections to have immediate positive effect on overall system performance.

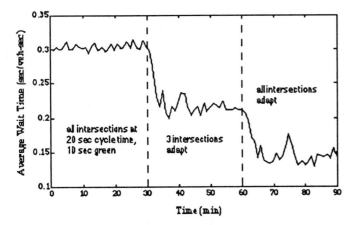

Fig. 8. Average waiting time for the case in which all intersections have an initial cycle time of 20 seconds.

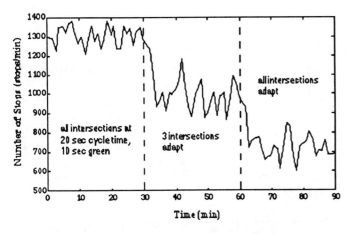

Fig. 9. Number of stops for the case in which all intersections have an initial cycle time of 20 seconds.

IV. CONCLUDING REMARKS

We have investigated the use of fuzzy decision rules for adaptive traffic control. A highly distributed architecture was considered, where the timing parameters at each intersection are adjusted using only local information and coordinated only with adjacent intersections. Although this localized approach simplifies incremental integration of the fuzzy controller into existing systems, simulation results show that the effectiveness of a small number of "smart" intersections is limited if they operate at a cycle time widely different from the rest of the system. In this case, constraining the controller to maintain a fixed cycle time that matches the existing system may provide better overall performance. For the case in which all intersections are adaptive, we need to investigate whether better performance is achieved by constraining all intersections to share a common variable cycle time.

There is much that can be done to further improve the present fuzzy controller, such as including queue length as an input and using trend data for predictive control. The flexibility of fuzzy decision rules greatly simplifies these extensions.

REFERENCES

1. Little, J., Kelson, M. and Gartner, N. (1981). MAXBAND: A Program for Setting Signals on Arteries and Triangular Networks. *Transportation Research Record* 795. National Research Council, Washington, D.C., pp. 40-46.

2. Lowrie, P. (1990). SCATS - A Traffic Responsive Method of Controlling Urban Traffic. Sales information brochure published by Roads & Traffic Authority, Sydney, Australia.

3. Luk, J. (1984). Two traffic-responsive area traffic control methods: SCATS and SCOOT. *Traffic Engineering and Control*, pp. 14-20.

4. Pappis, C. and Mamdani, E. (1977). A fuzzy logic controller for a traffic junction. *IEEE Trans. Syst., Man, Cybern.* Vol. SMC-7, No. 10.

5. Robertson, D. and Bretherton, R. D. (1991). Optimizing networks of traffic signals in real time - the SCOOT method. *IEEE Trans. on Vehicular Technology.* Vol. 40, No. 1, pp. 11-15.

6. Sims, A. (1979). The Sydney Coordinated Adaptive Traffic System. *Proc. ASCE Engineering Foundation Conference on Research Priorities in Computer Control of Urban Traffic Systems*, pp. 12-27.

7. Wallace, C. et. al. (1988). TRANSYT-7F User's Manual (Release 6). Prepared for FHWA by the Transportation Research Center, University of Florida, Gainesville, FL.

A Control System of Carburization Using Fuzzy-PID Combined Controller

Liya Hou
Housei University, Japan

Zhengqing Wang
Beijing University of Aeronautics
and Astronautics, P.R. China

Abstract

A actual control system of carburization, which is nonlinear and contains significant dead-time, is described. Beginning with designing a conventional PID controller, we introduced a new PID-Fuzzy combined controller to get more desired response of system and avoid tuning of PID parameters. The operation of the control system has shown higher carburizing quality and better response of system.

Introduction

Proportional-integral-derivative (PID) controllers are quite popular, and the most widely used as a control strategy in industrial processes. The popularity of PID controllers can be attributed both to their robust performance in a wide range of operating conditions and to their functional simplicity, which allows process engineers to operate them in a simple and straightforward manner. To implement such a controller, three parameters must be determined for the given process proportional gain, integral time constant, and derivative time constant. Often in practice, tuning is carried out by experienced operators using a "trial and error" procedure and some practical rules. This is often a time-consuming and difficult activity, for example when the dynamic process is slow, partly nonlinear, contains significant dead-time or when there are random disturbances acting on the plant. Once tuned, the control performance may later deteriorate because of nonlinear or time-varying characteristics of the process under control. Although PID controllers are common and well-known, they are often poorly tuned. Evidence of this can be found in almost any industrial process.

To overcome this difficulty, several methods have been developed that automatically tune PID controllers. The most well-known method is that of Ziegler and Nichols developed. Their method determines the parameters by observing the gain at which the plant becomes oscillatory and the frequency of this oscillation. A more useful extension of the method allows the determination of the parameters from the observation of the open-loop response of the plant to a step input change. Several other similar simple methods have been developed since, which automatically generate a special input to the process, and by observing its response, they determine the PID parameters. When optimal techniques for designing industrial controllers are introduced in industry, some resistance and difficulties may arise, mostly related to a lack of knowledge about internal mechanisms. They require good models of the controlled processes and the specification of nontrivial design parameters. This seems to explain, for instance, why most of the techniques proposed for tuning PID controllers are empirical.

Interest in the practical application of fuzzy logic control seems to be increasing. Fuzzy logical control techniques have been successfully applied to a wide variety of applications. In many cases, fuzzy logic controllers perform better than conventional controllers and can control systems not feasible with conventional control techniques. Fuzzy logic techniques provide a direct way to translate qualitative and imprecise linguistic statements about control procedures into computer algorithms. These techniques readily allow the implementation of qualitative linguistic statements about control procedures and goals for system performance.

The work described in this paper presents the design, research, and development of a control system of carburization by PID controller / fuzzy-PID combined controller. The carburizing furnace is a nonlinear system which contains a long time lag with dead time. Our development effort began with designing a conventional PID controller, and some expected results was obtained. To get better transition response, a fuzzy-PID combined controllers was introduced. The control system of carburization by fuzzy-PID combined controller has been put into use for two years, and has offered better control quality.

Control System of Carburization by PID Controller

The control system of carburization by PID controller is shown in Fig.1. Controlled variable is carbon potential [c], which is the function of T (furnace temperature), [co] (concentration of CO in furnace atmosphere) and [co$_2$] (concentration of CO$_2$ in furnace atmosphere). By measuring T, [co], and [co$_2$], carbon potential can be calculated according to the equation of calculating carbon potential. There is a constant difference between calculated carbon potential and actual carbon potential. This difference can be corrected by modifying the equation of calculating carbon potential. The input of PID controller is the difference (error) between calculated carbon potential and desired set point value. Manipulated process input, corresponding to the output of PID controller, is the magnitude of kerosene.

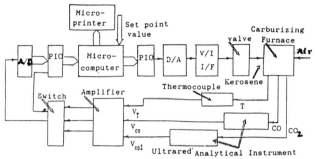

Fig.1 Control System of Carburization by PID controller

To get better carburizing accuracy of parts heat-treated, a model of information processing system, by which the analysis of error propagation can be carried out, is made. This model is shown in Fig.2.

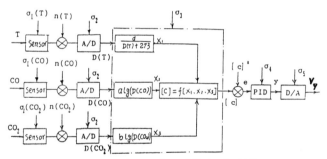

Fig.2 Information Processing of Carburization System by PID Controller

where, σ_1 - sensor error
σ_2, σ_5 - A/D, D/A error respectively
σ_3 - equation error of carburizing carbon potential (called function error)
σ_4 - error of PID discrete algorithm (called algorithm error)
n - noise

Because total error of the system can be devided into several groups, higher control accuracy would be got by decreasing individual error. In this paper, the method to decrease algorithm error is considered. Two PID discrete methods are introduced: back discrete algorithm and bi-linear discrete algorithm. Their pulse transfer functions $D_d(Z)$ are given in Eqn. 1 and Eqn.2 respectively.

For back discrete algorithm,

$$D_d(Z) = Kp + \frac{T_N Kp}{T_I} * \frac{1}{1-Z^{-1}} + \frac{T_D Kp}{T_N} * (1 - Z^{-1}) \qquad (1)$$

For bi-linear discrete algorithm,

$$D_d(Z) = Kp \left[1 + \frac{T_N}{2T_I} * \frac{1+Z^{-1}}{1-Z^{-1}} + \frac{2T_D}{T_N} * \frac{1-Z^{-1}}{1+Z^{-1}} \right] \qquad (2)$$

And the coefficients of error propagation corresponding to back discrete and bi-linear discrete algorithms are given in Eqn.3 and Eqn.4 respectively

$$\xi_d = Kp^2 \left[(1 + \frac{T_N}{T_I})^2 + \frac{2T_D}{T_N} * (1 + \frac{T_D}{T_N}) + n(\frac{T_N}{T_I})^2 \right] \qquad (3)$$

$$\left\{ Kp^2 \left[\frac{n+1}{2}(\frac{T_N}{T_I} - \frac{4T_D}{T_N})^2 + \frac{2n-1}{4}(\frac{T_N}{T_I} + \frac{4T_D}{T_N})^2 + (\frac{T_N}{T_I} + \frac{4T_D}{T_N})^2 \right] + 1 \right.$$

(n is an odd number)

$$\xi_B = \left\{ Kp^2 \left[\frac{n}{2}(\frac{T_N}{T_I} - \frac{4T_D}{T_N})^2 + \frac{2n+1}{4}(\frac{T_N}{T_I} + \frac{4T_D}{T_N})^2 + (\frac{T_N}{T_I} + \frac{4T_D}{T_N}) + 1 \right] \right\}$$

(n is an even number)   (1)

where, Kp - proportional gain
T_I - integral time constant
T_D - derivative time constant
T_N - sample period
n - sample number

The conclusion of comparing the coefficient of error propagation of two PID discrete algorithms is that using back discrete algorithm can decrease algorithm error.

Control System of Carburization by Fuzzy-PID Combined Controller

Basic idea of Fuzzy-PID combined controller is that the error (the difference between the calculated carbon potential and the set point value) is devided into three grades, and control strategy into two models, which is depicted in Fig.3.

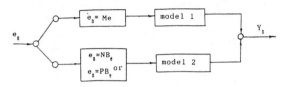

Fig.3 Model of Fuzzy - PID Combined Controller

where, e_n - error
NBe - Negative Big error
Me - Medium error
PBe - Positive Big error
Y_n - output of controller

Model 1 is the conventional PID algorithm, model 2 is the Fuzzy control technique.

The Fuzzy-PID combined control model consists of four control rules, as follows:

Rule 1. If $e_n = Me$
then $Y_n = Y_{n-1} + \Delta Y_n$
$\Delta Y_n = Kp(e_n - e_{n-1}) + K_I e_n + K_D(e_n - 2e_{n-1} + e_{n-2})$

Rule 2. If $e_n = PBe$
then $Y_n = Y_s$
$Y_{n-1} = Y_{n-1} + \Delta Y_{PB}$

Rule 3. If $e_n = NBe$
then $Y_n = Y_B$
$Y_{n-1} = Y_{n-1} - \Delta Y_{NB}$

Rule 4. $\Delta Y_{PB} > \Delta Y_{NB}$

where ΔY_{PB}, ΔY_{NB}, Y_s, and Y_B are empirical adjusted parameters.

Rules 1-3 describe how the control output and its change are adjusted, and rule 4 represents the relationship between ΔY_{PB} and ΔY_{NB} when the error gets different grade. Because increasing carbon potential is easier than decreasing it in carburizing process, the change of the control output should be different.

Fig.4-6 show the response of the system when different ΔY_{NB} (or ΔY_{PB}) is set. If the change of the control output is too big or too small, the transition response of system is not desirable. Only if the change is moderate, the transition response of system is considered to be desirable. Empirically, the change is 5-10% of the control output.

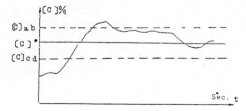

Fig.4 Adjusting magnitude of Y_{n-1} is too small

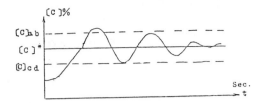

Fig.5 Adjusting magnitude of Y_{n-1} is too big

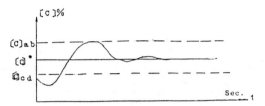

Fig.6 Adjusting magnitude of Y_{n-1} is suitable

Our experiment shows that the selection of ΔY_{PB}, ΔY_{NB}, Y_s and Y_B is easier than the tuning of PID parameters.

Fig. 7 shows the flow chart of Fuzzy-PID combined controller.

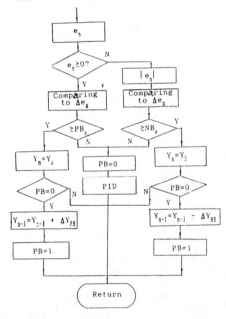

Fig.7 Flow Chart of Fuzzy - PID Combined Controller

The performance of the system using Fuzzy-PID combined controller has improved on that using PID controller. In fact, the control system of carburization using the Fuzzy-PID combined controller has being operated for more than two years in Hui Yang mechanic factory of CHINA.

Conclusions

1. For a nonlinear process, the Fuzzy-PID combined controller can be expected to be very useful.

2. The performance of the system using Fuzzy-PID combined controller is not so sensitive as that using PID controller. The selection of the parameters of our controller is easier than the tuning of the parameters of PID controller.

References

1. E. H. Mamdani, Application of Fuzzy Algorithms for Control of a Simple Dynamic Plant, Proc. Of IEEE, 122(1974)

2. J. G. Ziegler and N. B. Nichols, Optimum Settings for Automatic Controller, Trans. ASME, Vol. 64, (1942).

3. M. Sugeno, ed., Industrial Applications of Fuzzy Controller. Elsevier, New York, (1985).

4. J. G. Bollinger and N. A. Duffie, Computer Control of Machines and Processes. Addison-Wesley, (1988).

5. L. Y. Hou and Z. Y. Ma, A Control System of Carburization Using PID Controller. The First Annual Conference of the Computer Society of China, Xiamen, CHINA, (1984).

Fuzzy Control for Active Suspension Design

Edge C. Yeh* and Yon J. Tsao[†]
*Professor, [†]Graduate Student
Department of Power Mechanical Engineering
National Tsing Hua University, Hsinchu 30043, Taiwan R.O.C.
Phone (886)35-715131 Ext.3742, Fax (886)35-722840, ECYEH@PME.NTHU.EDU.TW

Abstract

A new control algorithm for active suspension design is proposed. In this study, ride comfort will be paid more attention in design procedure. It is suggested that the trajectory of the sprung mass should sustain as smooth as possible for the reason of ride comfort. Thus the controller is designed for this purpose. Fuzzy control concept is employed to deal with such problem based on heuristic reasoning of human intuition by means of qualitative linguistic expressions. Some considerations of selecting fuzzy variables and rules are suggested. Computer simulation is performed to verify the above concept and the fuzzy rules. It is shown that satisfactory performances have been achieved by using such control scheme.

1. Introduction

In recent years, road vehicle suspension design has been received great attention. In order to achieve both good handling and riding comfort, the performance of suspension system plays an important role. Besides the main functions of suspension system in providing vehicle support, stability, and direction control during braking and cornering, it also provides effective isolation from road disturbance. In conventional vehicle with passive suspension, the spring and damper elements can not be dynamically adjusted. So a compromise between ride comfort and handling will occur. Nowadays, research and development on semi-active and active suspension has increased greatly. An active suspension, distinguished from a semi-active suspension which employs variable damping force shock absorber or variable stiffness of spring, uses a hydraulic or pneumatic actuator and is controlled electronically to improve performance.

Many design methods for active suspension have been developed. The application of linear optimal control theory is widely employed by Thompson [1], Wilson et al. [2], ElMadany [3]. By means of evaluation function (or cost function), optimal control is applied to minimize some quantities such as sprung mass acceleration, relative displacemant between sprung mass and unsprung mass, and relative displacement between unsprung mass and road surface. They also represent passenger discomfort, suspension working space, and tire load fluctuations. It is known that if the tire deflection is tolerable and the suspension elements move within its stroke with limited range, the unsprung mass should follow the road surface and moves within the limited stroke as possible as it can for the reason of ride comfort. It is like a soft suspension to isolate the input from road to the sprung mass. From the viewpoint of the optimal control theory, the performance includes the term of the suspension working space to minimize in the linear range of the system. Thus, it is possible that the sprung mass will vibrate due to the unsprung mass moves upward and downward, and the isolation between two mass may be not good. Hedrick and Butsuen [4], Yue, et al. [5] pointed out some fundamental limitations for active suspension design from the viewpoint of frequency domain. The invariant properties had been proposed. It is worth noticed that the tire deflection is large in the frequencies which are near the invariant point. So it should be taken into consideration in design procedure to deal with the tire deflection in such frequencies.

The objective of this study is to apply fuzzy control algorithm to deal with the problem occurs in the active suspension system. Fuzzy set theory has first been proposed by L. A. Zadeh [6]. Nowadays, "fuzzy" concept has been widely used in control system design. As well known the characteristics of fuzzy control are expressed by a number of fuzzy control rules. They are presented in the following form :

IF (fuzzy variable condition)

THEN (controlled variable action),

so the fuzzy control can be nonlinear. With the human intuition, the complex phenomenon can be reduced to some simple rules. And based on the physical insight, decisions can be made to satisfy the desired characteristics of system behavior.

A quarter car model is used in this study for active suspension design in formulating the fuzzy control scheme. Computer simulations are performed to show the performance in time domain and frequency domain using fuzzy control scheme.

2. System Description

The quarter car model is shown in Fig. 1. It is assumed that the tire does not leave the ground, the z_s and z_u can be measured, the nonlinearity behaviors of each element are neglected, and the vehicle sprung mass is considered as a rigid body. Then we can obtain the linear differential equations of the model shown in Fig. 1 as

$$m_s\ddot{z}_s = -k_s(z_s - z_u) - b_s(\dot{z}_s - \dot{z}_u) + f_a, \quad (1)$$

$$m_u\ddot{z}_u = k_s(z_s - z_u) + b_s(\dot{z}_s - \dot{z}_u) - f_a + k_t(z_r - z_u), \quad (2)$$

where

$$f_a \equiv \text{scalar active force.}$$

If we choose the state variables as follow :

$$x_1 = z_s - z_u,$$
$$x_2 = \dot{z}_s,$$
$$x_3 = z_u,$$
$$x_4 = \dot{z}_u.$$

The state equations can be expressed as follow:

$$\dot{x} = Ax + B_1 f_a + B_2 z_r,$$

where

$$A = \begin{bmatrix} 0 & 1 & 0 & -1 \\ -k_s/m_s & -b_s/m_s & 0 & b_s/m_s \\ 0 & 0 & 0 & 1 \\ k_s/m_u & b_s/m_u & -k_t/m_u & -b_s/m_u \end{bmatrix},$$

$$B_1 = \begin{bmatrix} 0 \\ 1/m_s \\ 0 \\ -1/m_u \end{bmatrix} \quad B_2 = \begin{bmatrix} 0 \\ 0 \\ 0 \\ k_t/m_u \end{bmatrix}.$$

The numerical values of the system parameters are given below (adopted from [5]) :

$$m_s = 240 \ kg$$
$$m_u = 36 \ kg$$
$$b_s = 980 \ N \cdot s/m$$
$$k_s = 16000 \ N/m$$
$$k_t = 160000 \ N/m.$$

3. Control Scheme

The main control scheme for active suspension design is divided into two parts:

1. using the relative displacement between two mass as feedback signal to generate control force for balancing the force generated by passive elements;
2. using fuzzy inference rules to regulate the sprung mass trajectory and avoid the excessive tire deflection.

Part I

The purpose of using passive elements with spring and damper in conjuction with the active force actuator is to provide a passive suspension system as a backup system when the power failure occurs for the active force actuator, and to provide a constant force for supporting the sprung mass in the initial state, so that the active actuator needs not to generate force while the vehicle is stopping or parking. But the passive system limits the performance of ride quality, so some parts of the actuator force are used to cancel the dynamic force change generated by passive elements.

From eq.(1), we know that the motion of the sprung mass is dependent on three terms: spring force, damper force, and actuator force. So for the reason of ride quality, if the three forces are balanced, that is, the sum of all forces is zero, then no acceleration occurs for the sprung since no net force is applied on it. And good ride quality can be obtained. In other words, the motion of the sprung mass is dominated by the three forces, so it is desired to design the actuator force to be equivalent to the sum of the spring and damper force to achieve the demand of ride comfort. For this reason, we use **ZSZU**=$Z_s - Z_u$ (the relative displacement between the sprung and unsprung mass) to cancel the spring force and **TCZSZU** (the time change of **ZSZU**) to cancel the damper force. The control laws are shown below:

$$F_s = k_s * \mathbf{ZSZU},$$
$$F_d = (b_s/h) * \mathbf{TCZSZU}.$$

where

F_s represents the part of active force to cancel the spring force;

F_d represents the part of active force to cancel the damper force;

h represents the time interval.

Part II

When the above control laws are applied, it can be found that these laws almost isolate the coupling between the sprung mass and unsprung mass. But there are still some problems should be solved.

First, the acutal damping force is generated from the relative velocity between two masses, not the time difference of the relative displacement of the two masses, so cancellation is not complete and some residues may remain. Therefore, it is desired to use the fuzzy inference process to regulate the sprung mass trajectory for the purpose of ride comfort. Thus, the fuzzy rules (see Table. 1) are developed to provide a smooth trajectory of the sprung mass, where the fuzzy variables shown in Table. 1 are defined as below:

ZS represents the sprung mass position;
TCZS represents the time change of **ZS**;
F_{zs} represents the control force for this rule;
NB,P,... represents the fuzzy sets of these rules, where NB is Negative-Big, P is Positive, and so on.

The membership functions of the fuzzy sets are shown in Fig. 2. Therefore, the total control force f_a is

$$f_a = F_s + F_d + F_{zs}.$$

Second, from the dynamic structure of the suspension, if the sprung mass is isolated from the unsprung mass by the actuator force, the tire body becomes a neutral stable or an unstable system, since no energy dissipating element is connected with the tire if the cancellation in *Part I* is perfect. That is, it becomes a second order system with almost no damping (with only mass and spring). From the discussion in [4], it is shown that there is an invariant point at the special frequency $\omega = \sqrt{k_t/m_u}$, which is coincidentally the natural frequency of the unsprung mass. If we apply the above fuzzy control rules, it will show that the ride quality has improved in all frequencies except the frequencies which are near the special frequency. And the large tire deflection may degrade the road handling performance in such frequencies.

To avoid the effect of large tire deflection, we find that the parameters of spring constant and damping coefficient of the suspension play an important role. From the frequency response of the system, it can be shown that if the spring constant and damping coefficient becomes larger, the tire deflection response will improve greatly in the frequencies near the invariant point. So a stiffer passive system, three times of the spring constant and damper coefficient in this study, is used to avoid large tire deflection in the range around invariant frequency, while in the other frequency range the control force can be determined by the way mentioned previously. Then the stability of the tire motion can be guaranteed to preserve the advantages which the previous control rules provide.

Based on the above discussion, the rules shown in Table. 1 should be modified as below:

If the tire deflection exceeds the limits, then the total control force f_a is changed into

$$f_a = -2 * (F_s + F_d). \quad (3)$$

(because $F_s + F_d$ is used to cancel the spring force and damping force, so $-2 * (F_s + F_d)$ means to add equivalently two times of the spring force and damping force which are generated by the actuator) In this study, it is assumed that the static tire deflection is 1.7cm, and the maximum allowable wheel excursion is 5.7cm. And the tire deflection limits are chosen at 70 percent of the excursion from the static position in this study. That is, the tire deflection limits are set to be 1.19cm above and 2.8cm below the static position.

4. Results

Fig. 3 compares the frequency response of passive system with the fuzzy control active suspension system. The gain of the active system is obtained by the maximum excursion of the steady state response under the sinusoidal input with the magnitude of ±1 cm. It shows that in the low and high frequency range, performance of sprung mass verse road input has been improved greatly. Although the tire defleciton response is little worse than the passive system, but the gain is only higher with about 20 dB and is still small enough to ignore. Due to the invariant property, the frequencies which are near the special frequency $\omega = \sqrt{k_t/m_u}$ have almost the same response as the passive one.

Fig. 4 compares the time response with the Part I and Part II control laws where the sprung mass position is deliberately shifted down 3 cm for the purpose of illustration. It shows the vehicle is passing an irregular road surface which consists of six sinusoidal components as 1, 5, 15 rad/s in the first three seconds and 2, 6, 20 rad/s in the later three seconds. Fig. 4(a) shows the results under the force cancellation scheme in Part I only, and the Fig. 4(b) combines the Part I and the fuzzy control law in Part II. From the two plots, it shows that the actuator force, which is generated only by Part I, can not balance perfectly the force generated by passive elements. And the ride quality is improved greatly by adding the fuzzy control law. This verifies the benefits of fuzzy control in this study.

Fig. 5 shows the ability for avoiding the large tire deflection by the modified control law as a stiffer passive suspension in the large tire deflection range in addition to the normal control law (combining Part I and fuzzy control) for tire deflection within its limits. The road surface is the same as in Fig. 4 except a high frequency disturbance is added for testing the modified control law in the large tire deflection condition. Fig. 5(a) is the results for the passive system. Fig. 5(b) uses only the control law mentioned in Part I, while in Fig. 5(c) the Part I and fuzzy control law both are applied but without using eq.(3). Fig. 5(d) uses all of control laws mentioned in the previos section.

Although the tire deflection is not large in the passive system, it can be seen that the ride quality is worse than the others. The ride quality is improved in the second and the third plot, but the tire deflection is so large that the tire may be to leave the ground. In the fourth plot, the trajectory of sprung mass is a perfect level curve except a small bump due to the high frequency disturbance to result in the satisfactory ride quality. While the tire deflection is little larger than the passive one, it is deemed to be tolrable for achieving good ride quality. And the effectiveness of eq.(3) is thus demonstrated.

5. Conclusion

A fuzzy control scheme for active suspension design has been considered. In the first part, it is supposed that the control force is generated to balance the passive force for the purposes of isolating the sprung mass and achieving good ride quality. In the second part, fuzzy control laws have been applied for regulating the trajectory of sprung mass. In avoiding large tire deflection, the exception condition is also considerd in the control rules. It is shown that great improvements are obtained as compared with the passive one, and the response in time domain and frequency domain are satisfactory by using the above control laws. Although the sprung mass trajectory is somewhat degraded if the tire deflection is large and the extra control law is applied to provide a stiffer passive system, the proposed control scheme can still yield satisfactory ride quality for moderate road surface variation.

References

[1] Thompson, A. G., "An Active Suspension with Optimal Linear State Feedback," Vehicle System Dynamics, Vol.5, 1976, pp. 187-203.

[2] Wilson, D. A., Sharp, R. S. and Hassan, S. A., "The Application of Linear Optimal Control Theory to the Design of Active Automobile Suspensions," Vehicle System Dynamics, Vol.15, 1986, pp. 105-118.

[3] ElMadany, M. M., "Optimal Linear Active Suspensions with Multivariable Integral Control," Vehicle System Dynamics, Vol.19, 1990, pp. 313-329.

[4] Hedrick, J. K. and Butsuen, T., "Invariant Properties of Automotive Suspensions," IMechE, C423/88, 1988, pp. 35-42.

[5] Yue, C., Butsuen, T., and Hedrick, J. K., "Alternative Control Laws for Automotive Active Suspensions," Journal of Dynamic System, Measurement, and Control, Vol.111, 1989, pp. 286-291.

[6] Zadeh, L. A., "Fuzzy Sets," Infor. Contr., Vol.8, 1965, pp. 338-353.

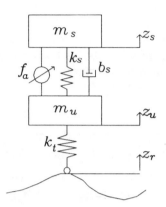

Fig. 1 Quarter car model

Table. 1 Rule table for fuzzy control

		ZS		
	F_{zs}	P	Z	N
TCZS	NB	NB	NB	NB
	NM	NS	NM	NB
	NS	PB	NS	NB
	ZE	PB	ZE	NB
	PS	PB	PS	NB
	PM	PB	PM	PS
	PB	PB	PB	PB
	control force			

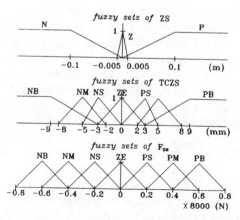

Fig. 2 The membership functions of fuzzy sets

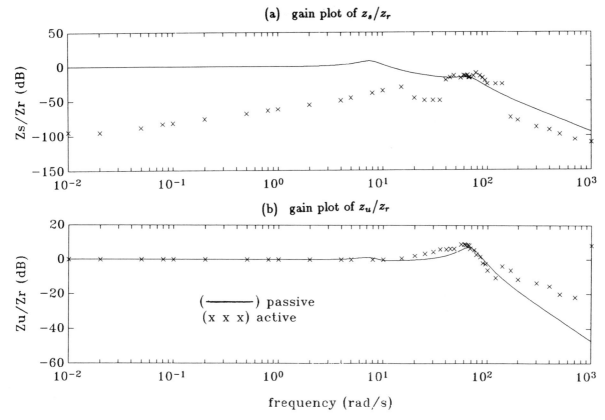

Fig. 3 Frequency response of passive and active system

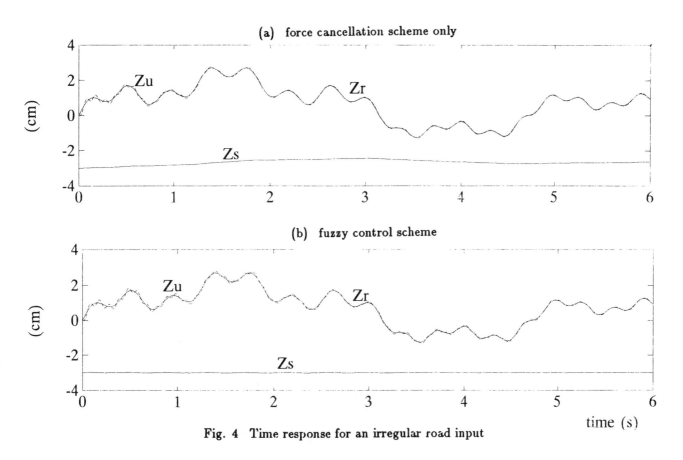

Fig. 4 Time response for an irregular road input

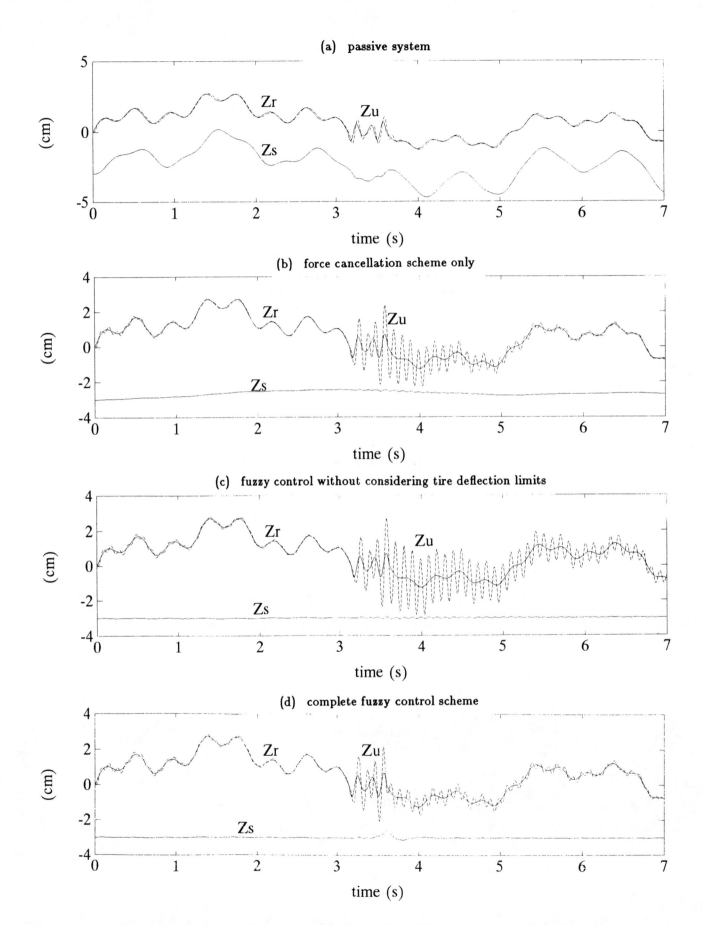

Fig. 5 Time response for an irregular road surface with high frequency disturbance input

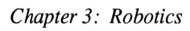

Chapter 3: Robotics

Fuzzy Controlled Gait Synthesis for a Biped Walking Machine *

Luis Magdalena
Dpto. Matemática Aplicada a las Tecnologías de la Información.
llayos@mat.upm.es
Félix Monasterio, PhD
Dpto. Tecnologías Especiales Aplicadas a la Telecomunicación.
fmh@dit.upm.es

ETSI Telecomunicación, Universidad Politécnica de Madrid,
Ciudad Universitaria s/n., E-28040 Madrid, Spain.

Abstract— The control of Biped Walking Machines constitutes a really complex problem that we will focus by using an Intelligent Controller. This controller is based on the Hierarchically Intelligent Control approach, proposed by Saridis, distributing the control in levels according to the principle of Decreasing Precision with Increasing Intelligence. This hierarchy has three basic levels: Organization, Coordination and Execution. In an intermediate level of precision and intelligence we can find the Coordination Level that has a tree structure, with a root called dispatcher and a finite set of subnodes called coordinators. The topic of the paper is the development of the walk coordinator, one of these subnodes. The Walk Coordinator has been designed as a Rule Based System that works with Fuzzy Logic structures. The walk coordinator has two different rule levels we call prescription and compensation levels; this division is based on Prescribed Synergy method introduced by Vukobratovic. The initial rules have been obtained from biomechanical studies of human walking. In addition to the fuzzy controller, some learning modules can be added to the coordinator.

I. INTRODUCTION

The research of legged locomotion mechanisms and machines is primarily concerned with the applications in the rehabilitation of disabled people and with the problems of legged locomotion. The use of walking machines to replace wheeled ones in those situations that are out of its reach is a really interesting task.

The motion of living organisms can be interpreted from the viewpoint of mechanics as a result of changes in equilibrium conditions. The spontaneous motion due to the redistribution of tension in muscle groups modifies the relations between forces, bringing these relations to equilibrium or taking them away from the equilibrium position. Legged locomotion systems represent extremely complex dynamic systems, particularly the anthropomorphic mechanism. Its complexity is the result of the union between a really complex mechanical structure and some control characteristics.

The most important characteristics of biped locomotion systems are:

- An *unpowered degree of freedom* formed between the contact of the foot and the ground surface.

- A certain *repeatability* of movements.

- A permanent change between *double, single and none-support phases* (2, 1, 0 feet in contact with the ground).

- The presence of a *closed kinematic chain* in the double-support phase.

In the field of biped walking robots, a great amount of research have been done from a teorethical point of view, but only a slow progress have been obtained in robot design. Some of these works are only failed projects but others have obtained very interesting results. Different mechanical structures, control strategies, sensors, actuators..., have been proved. The first one was the WL-5 in 1973 [1], a member of the WABOT family, a collection

---
*This project is partially supported by CICYT (Spain) under reference ROB90-0174

of biped robots from Waseda University in Tokyo. The WL-5 was a biped but still static walking machine. In recent years starting at 1984, dynamic walk have been realized by different biped robots. The BIPER-4 of Miura and Shimoyama [6]; new members of WABOT's series, the WL-10RD and the WL-12 [11]; the biped hoping machine by Raibert [9] or the SD-2 by Zheng [15, 16], are examples of this objective.

A. Control Structure

The control system designed for a complex system such as a biped walking machine is expected to have some intelligent functions (e.g. learning or multilevel decision-making). In addition, some practical conditions like lightweight, small dimensions, real time performance, functionality and a small number of sensors, must be imposed to the design. These constraints exclude computationally complex algorithms or long time computation. This kind of controllers take place on *Intelligent Control Theory*. Opposite to control methodologies based on the rigorous mathematical modeling and the analisys of physical processes, these new methodologies demonstrate machine intelligence and decision making.

In the field of Intelligent Control Theory, the *Hierarchically Intelligent Control* approach has been proposed by Saridis [10], distributing hierarchically the control according to the principle of Decreasing Precision with Increasing Intelligence. The three basic levels of control and its role in the control system are:

1. The Organization level that deals with path planning, obstacle crossing and gait selection using high level sensorial information. This level is not considered in this study.

2. The Coordination level has a tree structure with a root called dispatcher and a finite set of subnodes called coordinators [13]. Each coordinator is associated with several devices and will process the operation and data transfer of these devices.

3. The Execution level is the hardware control level.

The main goal of this work is to implement the walk coordinator for a biped walking machine under development at this moment at the Universidad Politécnica of Madrid. This walk coordinator will be a subnode of the Coordinator that constitutes a level of the general control structure, Fig. 1.

B. Mechanical Structure

The biped has two legs with an anthropomorphic morphology, with closed chain kinematics to locate the motors at the hip, near the center of mass. The foot consists on a

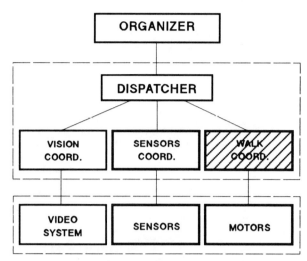

Figure 1: Control Structure.

heel and a big toe. A mathematical simplified model has been obtained to enable computer simulation an test of the basic system.

II. THE WALK COORDINATOR

The control structures and knowledge related with gait synthesis are contained in the walk coordinator at the dination level of the intelligent controller.

The walk coordinator is divided in two different levels we have named prescription and compensation levels; this division is based on the Prescribed Synergy method introduced by Vukobratovic [12]. Each of these two levels has its own rule and data base with different meaning and goal.

The *Prescription level* generates the dynamics for one part of the system. This generated dynamics must ensure functional movements that make possible to maintain the equilibrium condition imposed to the system and to control the unpowered d.o.f formed by the contact between the foot and the ground surface. The movements considered in the generation of fuzzy rules have different bases: biomechanical description of human walk [8]; some gait parameters related with walking speed [2]; the position of the ZMP (Zero Moment Point: the point with respect to which the sum of all moments of active forces is equal to zero) [12].

The *Compensation level* must generate the dynamics for the remaining parts of the system in such a way that the ZMP stays at the position defined by prescribed part and under repeatability conditions. In the 3-D case, the compensation level must generate dynamics for three degrees of freedom because the ZMP defines null moments in the three axes. In the 2-D case, the compensation level must work only with a single degree of freedom because there are moments with only one axis, the one perpendicular to the plane of movement.

The *output variables* of the walk coordinator are position, velocity and acceleration of each powered degree of freedom. These output variables are generated partially in the Prescription level and are completed in the Compensation level. The Prescription level generates some internal variables too; these constitute a part of the input for the Compensation level. These variables are the proposed position, velocity and acceleration for the unpowered degree of freedom and the proposed position for the ZMP.

III. GAIT SYNTHESIS

The only task of the Walk Coordinator is Gait Synthesis, that is performed by means of fuzzy rules. This module has a conventional rule based system architecture. Its inference engine is basically a fuzzy controller.

The acquisition of initial knowledge, knowledge processing and knowledge refinement are the subjects of this paragraph.

A. Initial knowledge

The knowledge acquisition for this problem is really diferent to that realized in expert systems for process control. Knowledge acquisition in such systems is usually accomplished working with human experts that control a similar process. In a sense, we are experts in human walk; every day we control several walking processes, however we have no conscious knowledge on this process. Biomechanical studies must play the role of experts and the knowledge must be obtained from them.

Initial control rules for the walk coordinator have been derived from [5, 8]. These biomechanical studies deal with the *description of human walk* accomplished in 1953 by Saunders, Inman and Eberhart. They consider six determinants for a normal gait: Compass gait, Pelvic rotation, Pelvic tilt, Stance-leg knee flexion, Plantar flexion and Lateral displacement of the pelvis.

Translating these elemental motions on a *sequential description* with relative phase information, we can find: Heel-strike of swinging foot $(0 - 0.15)$; Plain foot on the ground $(0.15 - 0.4)$; Plantar flexion $(0.4 - 0.5)$; Toe-off $(0.5 - 0.6)$; Swinging-leg advance with knee flexion $(0.6 - 0.75)$; Swinging-leg knee extension $(0.75 - 1)$.

Each one of these six sequential phases contains a part (or all) of the determinants of normal gait, and constitutes a global description for the fuzzy controller. In addition, some parameters must be defined to translate this global description into a real one; walking speed and ground characteristics are also needed.

The influence of walking speed on other gait parameters has been studied in [2]. Some conclusions from [2] will be introduced in data bases of the coordinator to modify the meaning of the rules depending on walking speed. The way to modify the meaning of a rule is changing the fuzzyfication and defuzzyfication process (the meaning of fuzzy sets). Changing the meaning of fuzzy sets, an input value has a different fuzzy translation and a fuzzy control output generates new shifted control values.

Both papers report the mean values of some gait parameters and the correlation that can be found between walking speed and them. Gait parameters studied are: cadence, stride length, double support duration, maximum knee flexion on stance and swinging leg, and some other information on coronal plane. In addition there is a direct relation between stride length and compass gait.

B. Knowledge processing

The inference engine of the coordinator is a fuzzy controller [3, 7] with a general structure as follows:

1. A *fuzzyfication interface* that transfers the values of input variables into fuzzy information, assigning grades of membership to each fuzzy set defined for that variable.

2. A *knowledge base* that comprises a dynamic data base, as has been said above, and a fuzzy control rule base. The data base is used in fuzzyfication and defuzzyfication processes. The representation of a fuzzy control rule is that of fuzzy implications

   If $\quad (x_i$ is C_{ij} and $\ldots$ and x_k is $C_{kl})$
   then $\quad y_m$ is D_{mn}

   where x_i is the input variable i, C_{ij} is a fuzzy set belonging to this variable, y_m is an output variable and D_{mn} is a fuzzy set belonging to this variable.

3. The *inference engine* has the capability to infer fuzzy control actions employing fuzzy implications and the rules of inference in fuzzy logic.

4. The *defuzzyfication interface* yields a nonfuzzy control action from an infered fuzzy control action.

The structure of the fuzzy controller allows us to extract the common parts that are equal in prescription and compensation levels (that are 1, 3 and 4) and the only one that differs containing level knowledge. The common parts have a unique implementation.

The coordinator have three different types of *input variables*:

1. Variables defined in the Organization level. They are the general gait parameters Speed, Cadence and Stride length. These variables have a constant value over each step.

2. Status variables received from sensors; these are position, velocity and acceleration for each powered degree of freedom.

3. Status variables not measured but calculated, or estimated from other sensorial information; these are the position, velocity and acceleration of the unpowered degree of freedom and the position of the ZMP.

IV. SIMULATION

A mathematical simplified model has been obtained to enable computer simulation to test the basic system. The simplified model is a five link 2-D structure with two legs and a trunk, each leg has a punctual foot, a knee with a degree of freedom and a hip with another degree of freedom. The model dimensions are shown in Table 1.

Fig. 2 presents variables used in the *Walk Coordinator*, α, and in the *Mathematical Model*, δ.

Relations between both sets of variables are:

$$\begin{aligned}
\delta_1 &= \alpha_1 \\
\delta_2 &= \pi + \alpha_1 - \alpha_2 \\
\delta_3 &= \alpha_1 - \alpha_2 - \alpha_3 \\
\delta_4 &= \pi + \alpha_1 - \alpha_2 - \alpha_3 + \alpha_4 \\
\delta_5 &= \alpha_1 - \alpha_2 + \alpha_5
\end{aligned} \quad (1)$$

that may be represented by

$$\vec{\delta} = [R]\vec{\alpha} + \vec{r} \quad (2)$$

The dynamic equation for the model is represented by the following set of nonlinear equations

$$\tau_\delta = D(\delta)\ddot{\vec{\delta}} + C(\dot{\delta}, \delta)\dot{\vec{\delta}} + \phi(\delta) \quad (3)$$

where τ are generalized forces. Considering (2) we obtain

$$\tau_\alpha = [R]^T \tau_\delta \quad (4)$$

and characterize the presence of the unpowered d.o.f. by

$$\tau_{\alpha_1} = 0. \quad (5)$$

From (3,4,5), the second-order differential equation

$$A_1\ddot{\delta}_1 + A_2(\dot{\delta}_1)^2 + A_3\dot{\delta}_1 + A_4\cos\delta_1 + A_5\sin\delta_1 + A_6 = 0 \quad (6)$$

is obtained and then solved by a numerical method.

Other simplifying assumptions are: Motions in frontal and top-view planes are neglected; Frictional forces at all joints are neglected; Contact between foot and ground is punctual and a large frictional force avoids slipping.

Obviously this is a really hard simplification, but the fuzzy controller will deal with the same kind of problems, and then we think that it can be tested using the model.

Table 1: Model dimensions.

Link	Length (m)	Weight (kg)
Trunk	0.75	40
Thigh	0.50	9
Shank	0.50	6

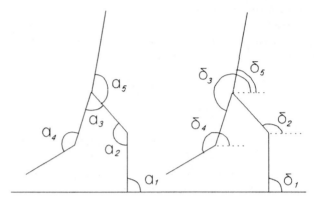

Figure 2: System and model variables.

The simplified model can only perform Compass gait and Stance-leg knee flexion. Plantar flexion has no sense in a punctual foot and must be replaced by knee movements. Pelvic rotation, Pelvic tilt and Lateral displacement of the pelvis are impossible in a 2-D model. This reduces the sequential description joining parts 2 and 3. In the simplified model case, only one degree of freedom must be controlled by Compensation level as we said above. We have selected the one formed between the trunk and the stance-leg, considering that it is the most appropriate to compensate moments from legs.

With the previous scheme, the input and output variables for each level are:

- The *Prescription level receives* from the Organization level:
  the general gait information, such as Walking speed and Stride length;
  the status information, that is, the position, velocity and acceleration of all degrees of freedom including the estimation for the unpowered one and the estimate position of the ZMP.

- The *Prescription level sends* to the Compensation level and to the Actuation level:
  the proposed position, velocity and acceleration for the powered degrees of freedom except for the trunk. It sends to Compensation level only:
  the proposed position, speed and acceleration for the unpowered degree of freedom and the proposed position for the ZMP.

- The *Compensation level receives* -in addition to the information from Prescription- the position, velocity and acceleration of the trunk and the relative phase from the Organization level.

- The *Compensation level sends* to Actuation level the proposed position, velocity and acceleration for the trunk.

Rule base contains fuzzy rules, like this from Prescription level:

If *Phase* is *swinging-leg advance* and *Swinging-foot position* is *normal* then *Knee flexion speed* is *normal positive*

where:
Phase is a variable received from Organization level,
Swinging-foot position is an internal variable,
Knee flexion speed is an output variable.
The linguistic labels *swinging-leg advance*, *normal* and *normal positive* are assigned to different fuzzy sets whose limits are defined in the prescription data base. All fuzzy sets are defined by means of four variable parameters (a,b,c,d): membership function is equal to 0 out of interval (a,d), is equal to 1 in interval $[b,c]$ and have constant slope in intervals $[a,b]$ and $[c,d]$. Parameters a, b, c and d, defining a certain fuzzy set, are not constant and will be calculated when gait parameters change, considering the correlation between the variable that qualifies the fuzzy set and the present Walking speed and Stride length values. These values cannot be changed before the present step has finished.

Some tests have been performed to validate this model, obtaining gait descriptions near to those of biomechanical studies. Fig. 3 and 4 show simulation results. The solid line near the hip is the trajectory of the center of gravity. The Coordination level generates information for the Actuation level each $0.25 sc.$, corresponding with snapshots on figures. Ground marks represents 0.5 m.

Fig. 3 is a sample of a gait that would be called *gait with ground reaction*. This gait may be characterized by initial conditions (speed of the unpowered degree of freedom) not equal to zero, obtained from previous step. This simulation has been realized with a set of *12 Prescription rules* and *5 Compensation rules*.

Fig. 4 presents a *gait without ground reaction*, that is, with initial conditions equal to zero. In this case the conditions to take off the foot are obtained by a sudden extension of rear leg. The second pattern has been obtained by only changing those rules related to the rear knee behavior in the double-support phase.

Both rule sets have successfully passed a 100 walking steps simulation with different stride lengths and cadence.

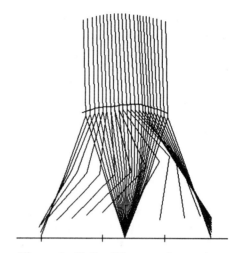

Figure 3: Gait with ground reaction.

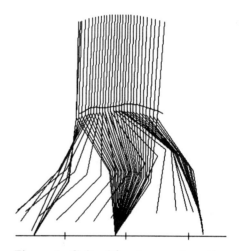

Figure 4: Gait without ground raction.

V. THE LEARNING SCHEME

In addition to its conventional architecture, the coordinator contains two other modules that add some learning capabilities to it:

1. Rule evaluator estimates the payoff for any rule involved at the inference process.

2. Strength modifier adjusts the rule strengths using a function of this rule strength and rule's payoff.

This control architecture is a modified version of the process control architecture proposed by Magdalena and others [4].

The initial knowledge for the fuzzy controllers (Prescription and Compensation) has been obtained from biomechanical studies of human walk. After initial rules have been created, system performance must be improved using learning capabilities. The learning process has an adaptive behavior performed by rule evaluator and strength modifier. Using the results obtained from some proposed action, the rule evaluator must analyze the goodness of the action with different functions depending on variables used for each level. The goodness measure has values in $[-1, 1]$ interval where higher values mean better actions.

After this phase, single contribution to global result must be estimated for each rule to obtain an individual goodness value that constitutes the payoff for that rule. The goodness value is obtained for each rule as a product of global goodness value and an individual influence value. The rule's influence is a value in $[0, 1]$ interval with higher values meaning that the action proposed by rule is closer to control output of coordinator. Each rule has an associated strength value that is always in the $[0, 1]$ interval. This value is modified by the strength modifier in order to get higher values for better rules and viceversa. To do this, we use the following learning function:

$$S(r_i, t+1) = \begin{cases} S(r_i, t) + K \cdot E(r_i) \cdot (1 - S(r_i, t)) & E(r_i) \geq 0 \\ S(r_i, t) + K \cdot E(r_i) \cdot S(r_i, t) & E(r_i) < 0 \end{cases}$$

The meaning of variables in the equation is:
$S(r_i, t)$ is the Strength value for rule i at time t,
K is a constant of learning process that has values in $[0, 1]$,
$E(r_i)$ is the evaluation function (individual goodness) for rule i.

The inference engine selects some rules from all those that it can fire. This selection is made using a function (product at this moment) of the strength value and the truth value of each rule as the possibility measure of selection.

We have said before that the rule evaluator employs different functions to analize the goodness of a proposed action, so we will define those used in simulations. A natural evaluation function for compensation is the compensation error, that is, the ZMP position deviation from the proposed one. Nevertheless, a wide range of possibilities having usually some energetic or security meaning, could be chosen to evaluate the prescription level.

The learning system has been tested previously in a control system now at work in a fossil power plant [4], with good results. The question that remains unanswered is the performance of the system in a hard real time environment. Some tests with a more complex model can answer this question by measuring the decision time of the system.

VI. CONCLUSIONS

A fuzzy control system that implement the Coordinator of an Intelligent Control System have been presented. Learning capabilities have been added to the system. Gait synthesis based on biomechanical initial knowledge have been obtained. At this moment some tests of learning system and others with hard time restriction are under development.

REFERENCES

[1] Kato I. et al., "Information power machine with senses and limbs", *1st. Symposium Theory and Practice of Robots and Manipulators*. Springer-Verlag 1974.

[2] Kirtley C., Whittle M. and Jefferson R., "Influence of Walking Speed on Gait Parameters". *Journal of Biomedical Engineering, vol 7*, pp 282-288. Oct. 1985.

[3] Lee C.C., "Fuzzy Logic in Control Systems: Fuzzy Logic Controller - Parts I and II", *IEEE Transactions on Systems, Man and Cybernetics, vol 20, num 2*, pp 404-435. Mar/Apr. 1990.

[4] Magdalena, L., J.R. Velasco, G. Fernandez and F. Monasterio., "A Control Architecture for Optimal Operation with Inductive Learning", *SICICA-92, IFAC Symposium on Intelligent Componentes and Instruments for Control Applications*, Preprints pp 307-312. 1992.

[5] McMahon T., "Mechanics of Locomotion". *International Journal of Robotics Research, vol 3, num 2*, pp 4-28. 1984.

[6] Miura H. y Shimoyama I., "Dynamic Walk of a Biped", *International Journal of Robotics Research, vol 3, num 2*, pp 60-74. 1984.

[7] Pedrycz W., *Fuzzy Control and Fuzzy Systems*, Research Studies Press Ltd. 1989.

[8] Plas F., Viel E. et Blanc Y., *La Marche Humaine*, MASSON, S.A. Paris, 1984.

[9] Raibert M., *Legged Robots That Balance*, MIT Press, 1986.

[10] Saridis G.N., "Intelligent Robotic Control". *IEEE Transactions Automatic Control, vol 28*, pp 547-557. 1983.

[11] Takanishi A., Ishida M., Yamazaki Y. y Kato I., "The realization of dynamic walking by the biped walking robot WL-10RD", *Proceedings 1985 International Conference on Advanced Robotics.* pp 459-466. 1985.

[12] Vukobratovic M., Borovac B., Suria D. and Stokic D., *Biped Locomotion*, Scientific Fundamentals of Robotics,7, Springer-Verlag, 1990.

[13] Wang F., Kyriakopoulus K.J., Tsolkas A. y Saridis G.N., "A Petri-Net Coordination Model for an Intelligent Mobile Robot", *IEEE Transactions on Systems, Man and Cybernetics, vol 21, num 4*, pp 777-789. Jul/Aug. 1991.

[14] Zadeh L.A. "Fuzzy Sets", *Information and Control, vol 8*, pp 338-353. 1965.

[15] Zheng Y., "Acceleration Compensation for Biped Robots to Reject External Disturbances", *IEEE Transactions Systems, Man and Cybernetics, vol 19, num 1*, pp 74-84. 1989.

[16] Zheng Y. and Shen J., "Gait Synthesis for the SD-2 Biped Robot to Climb Sloping Surface", *IEEE Transactions on Robotics and Automation, vol 6, num 1*, pp 86-96. Feb. 1990.

A FUZZY LOGIC FORCE CONTROLLER FOR A STEPPER MOTOR ROBOT

J. G. Hollinger, R. A. Bergstrom, and J. S. Bay

Bradley Department of Electrical Engineering
Virginia Polytechnic Institute and State University
Blacksburg, Virginia 24061-0111

Abstract

This paper describes a fuzzy logic force controller used to solve the robotic "hard contact" problem. This involves the control of force interactions between a stiff manipulator and an environment with very little compliance. It is demonstrated that this type of force control is well suited for a fuzzy logic controller.

The fuzzy logic force controller was implemented on a MERLIN 6540 Industrial Robot. The design considerations that went into developing a joint-position force controller for a stepper motor robot are discussed, as well as the reasoning behind the selection of the fuzzy membership sets and rule base. Experimental force responses are presented for a step and sinusoid change in desired force when the MERLIN robot was in contact with a hard surface. In addition, results of commanding the robot to maintain a constant force on a surface while following a trajectory tangent to the surface are described.

Introduction

Robotic force controllers can be divided into two categories depending on their mode of implementation. They can be classified as either computed torque or position-based force controllers. A computed torque controller derives the necessary information to control the joint torques from the desired end-effector forces and the dynamic model of the manipulator. In notation this can be expressed as [3, pg. 92]:

$$\tau = I(\Theta)\ddot{\Theta} + H(\Theta,\dot{\Theta}) + G(\Theta) + J^T(\Theta)\Gamma$$

where $H(\Theta,d\Theta/dt)$ and $G(\Theta)$ represent centripetal, Coriolis and gravity effects, $I(\Theta)$ represents the manipulator's inertia matrix and $J^T(\Theta)$ is the transpose of the manipulator's jacobian. Θ is a vector of joint angles, Γ is a vector of the desired end-effector forces and τ is the resulting vector of joint torques. This method is most effective when the dynamics of the manipulator can accurately be determined [2], [6].

Position-based force controllers are most useful for robots without the capability for joint torque control [13]. This method requires the contact force to be sensed. An error signal, proportional to the difference between the desired force and measured force, is fed back to the force controller and used to adjust the position of the manipulator's joints. Through this method, the manipulator can be made to behave like a mass-spring-damper with adjustable stiffness and compliance. This can be accomplished by controlling the Cartesian acceleration of the end-effector based on the sensed contact force.

Elosegui, et al. [2] demonstrated that position-based force control is very sensitive to the compliance of the manipulator-environment connection. Kazerooni, et al. [4] and Wen [13] show that force stability can not be guaranteed for a hard contact: when both the manipulator and environment have very little compliance. Instability can result in this instance because a small change in the end-effector's position will result in a large force reaction between the manipulator and the environment.

The goal of this research is to implement a force controller on a MERLIN 6540 Industrial Robot. The MERLIN is constructed with stepper motors which are stable only in a discrete number of positions, depending on the number of magnetic poles on the rotor and on the microstepping capabilities of the motor controller [12]. A stepper motor will generate as much torque as needed to advance to its commanded position. This makes torque control impossible. As a result, a position-based force controller must be developed for the MERLIN.

PID force control experiments were carried out with the MERLIN robot in contact with a hard surface. Stable responses could only be achieved with extremely low proportional gains, otherwise the end-effector repeatedly lost contact with the surface. Better results were obtained with an empirically designed nonlinear force controller. The nonlinearities attenuated the response of the controller for small force errors and amplified the response for large force errors. The most significant drawback of this controller was that it was difficult to obtain an intuitive feel for adjusting the controller's parameters. This motivated experiments with a fuzzy logic force controller. The benefits of a fuzzy controller are its nonlinear mapping functions and its ability to linguistically associate inputs and outputs [7], [8], [9], [11].

The next section will outline the force control problem for the MERLIN robot. The implementation of the fuzzy logic controller, including the selection of the fuzzy membership sets and the rule base, will then be described. Afterwards, the results of force control experiments and comparisons will be presented.

Force Control Problem

The fuzzy logic force controller has been implemented via joint position control on a MERLIN 6540 Industrial Robot. The MERLIN 6540 is a six degree of freedom articulated robot with a 50 pound payload capacity and a 40 inch reach. Optical encoders mounted on the back of each stepper motor give an axis resolution of 0.0075 degrees per revolution and a Cartesian accuracy of approximately 0.005 inches. The controller is capable of updating the position of all six axes every 4 ms. A wrist mounted force/torque sensor manufactured by Assurance Technologies, Inc. provides force feedback every 10 ms. Near-frictionless contact is achieved by mounting an industrial trackball to the force/torque sensor. This configuration allows only contact forces normal to the surface to be transmitted to the sensor. The rolling freedom of the trackball eliminates tangential forces due to friction.

The inverse kinematics of the MERLIN robot [1] were programmed and used to measure the stiffness of the test surface (a table top). Figure 1 shows the steady state force response when the

MERLIN was programmed to descend on the table in 0.001 inch increments. The staircase-like response is due to the robot only being accurate to 0.005 inches. If the robot is assumed to be infinitely stiff relative to the table, then the table stiffness can be calculated from the plot to be approximately 450 pounds per inch.

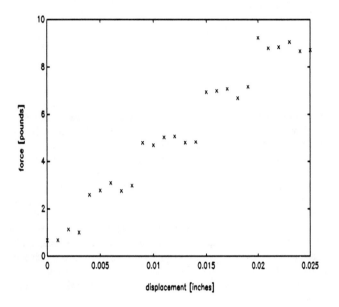

Figure 1. Steady state force response on a table top resulting from the robot descending by 0.001 inch steps.

Figure 1 also shows that a Cartesian based hard contact force controller would be ineffective for this robot because steady state force levels only occur at 2 pound increments. A method of movement that resulted in a higher force resolution would be desirable. This can be achieved by rotating one axis independently of the five remaining axes. The higher force resolution was possible because the position error associated with six axes was reduced to the error of just one axis. The configuration of the robot when it was in contact with most surfaces allowed the elbow axis to be used for this type of force control (the surfaces were in the horizontal plane in front of the robot). The elbow axis was also chosen because it had less mechanical advantage than the shoulder axis. Although rotating the elbow axis independent of the other axes resulted in a component of the end-effector's motion to be tangent to the surface, this component was small because the surface was very hard (450 pounds per inch). Figure 2 shows the steady state force response due to rotating the elbow while keeping the other axes fixed. Steady state force levels occur at 0.25 pound increments. The open loop force response due to a commanded encoder increment of the elbow is shown in figure 3. The force overshoot is due to the motor driver boards only being accurate to four encoder counts. This allowed the stepper motor to deviate from the desired position.

The objective of the fuzzy logic force controller will be to improve the response time of the contact force and maintain force stability when the robot is interacting with hard surfaces. An application of the force controller will be to maintain a four pound contact force while the robot is programmed to explore an unknown surface. This requires force control in the sensed normal of the surface with periodic steps along a trajectory tangent to the surface.

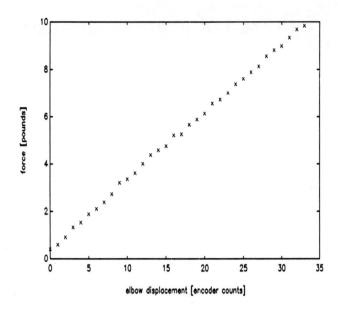

Figure 2. Steady state force response due to elbow increments.

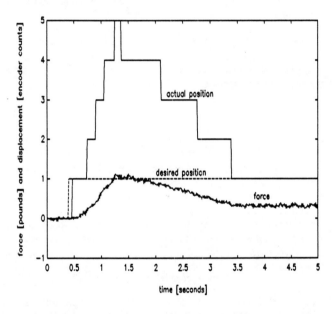

Figure 3. Open loop response due to one encoder count increment of the elbow. The figure shows the relationship between force overshoot and position error.

Fuzzy Logic Controller

The controller uses two inputs. Each input's universe of discourse is portioned into seven membership functions with each function corresponding to a linguistic variable. The variables used are positive and negative large, medium, small, and zero (PL, PM, PS, ZE, NS, NM, NL). The controller treats each measurement as a fuzzy singleton and fuzzifies the value using the fuzzy sets shown in figure 4. Both inputs use the same fuzzy sets for fuzzification. Isosceles triangles were chosen as the membership functions because they simplified the controller's computations (evaluation of the membership functions involves only simple linear equations).

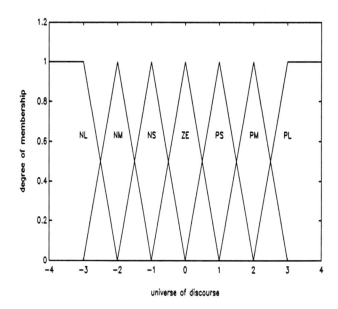

Figure 4. Fuzzy sets for force error and position error inputs.

The inference engine of the controller uses a seven by seven rule matrix. Each rule has the form [7]:

$$IF\ \mu_X(\alpha) \wedge \mu_Y(\beta)\ THEN\ Z\ at\ level\ \gamma$$

$$\gamma = MIN\{\ \mu_X(\alpha),\ \mu_Y(\beta)\ \}$$

where α and β are the inputs; $\mu_X(\alpha)$ and $\mu_Y(\beta)$ are the membership values for the linguistic variables **X** and **Y** respectively. **Z** is the rule consequent with activation level γ. The two, seven element fit vectors resulting from the fuzzification process trigger a total of four consequent control actions and their respective activation levels. Only four control actions are triggered since the input membership functions overlap by only 50%. The controller combines the control actions using correlation-product encoding [5, pg. 384]. If correlation-product encoding is used and the fuzzy sets are symmetric and unimodal, it can be shown that the calculation of the fuzzy centroid reduces to the following summation [5, pg. 388].

$$v = \frac{\sum_{i=1}^{7} m_o(y_i) y_i J_i}{\sum_{i=1}^{7} m_o(y_i) J_i}$$

Where $m_o(y_i)$ is the centroid of ith fuzzy output set, y_i is the membership value of the ith fuzzy set, J_i is area of the ith fuzzy set and v is the defuzzified output. The fuzzy centroid corresponds to the crisp or defuzzified control value which is the number of joint encoder counts to move. The output fuzzy sets are shown in figure 5.

The universe of discourse for the input fuzzy sets was chosen to span the entire range of inputs making data scaling unnecessary. The universe of discourse for the output fuzzy set was selected similarly. A 50% overlap in the membership functions of all the fuzzy sets was used so the controller would produce smooth control actions. Note that in the output set the zero membership function has been enlarged. Since the fuzzy centroid is weighted by the area of the membership function, increasing the size of the zero function causes a bias toward smaller control values. This was used to dampen the system and reduce oscillation.

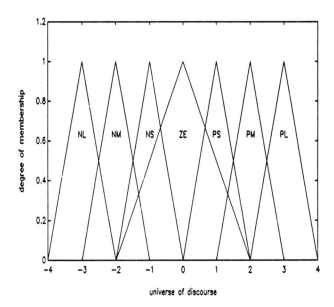

Figure 5. Fuzzy sets for force controller output in elbow encoder counts.

The first controller used was an empirically generated nonlinear controller. The controller was tuned so the robot was compliant for small force errors and increasingly stiff for large force errors. The experience with this controller gave a starting point for intuitively selecting the rule matrix for the fuzzy logic controller. Initially, the controller ignored the position error input and had a rule matrix that mimicked proportional control with a dead band for small force errors. Then, noting that the positional error dominated the open loop response of the system (figure 3), the magnitude of the position error was used to adjust the control value. If the position error was great, the control action was reduce to allow the error in position to settle. The final rule matrix is shown in figure 6. Figure 7 shows the resulting control surface for the controller. It graphs crisp control values versus position and force error. Note the influence of proportional control along the force error axis. One can also see that as the position error reaches its greatest magnitude the amount of the control action is reduced. The periodic ripples seen in the surface are caused by the method used to calculate the fuzzy centroid.

Experimental Results

The result of a step input to the fuzzy logic force controller is shown in figure 8. The initial transient response damped out within 2 seconds and there was a sustained oscillation of ± 0.4 pounds of force error. For comparison purposes, the step response of a proportional controller is shown in figure 9. A steady state force error of 0.3 pounds can be observed while the initial transient response requires approximately 3 seconds to damp out. The steady state error is due to the fact that elbow increment commands of less than one encoder count are discarded by the axis controller board. Attempts to raise the proportional gain to correct the steady state error resulted in an unstable system. Figure 10 shows the force response when a 0.5 Hz sinusoid was applied to the fuzzy logic force controller. A 2.5 pound overshoot can be observed.

Position Error

	NL	NM	NS	ZE	PS	PM	PL
NL	NM	NL	NL	NL	NL	NL	NM
NM	NS	NM	NM	NM	NM	NM	NS
NS	ZE	NS	NS	NS	NS	NS	ZE
ZE	ZE	ZE	ZE	ZE	ZE	ZE	ZE
PS	ZE	PS	PS	PS	PS	PS	ZE
PM	PS	PM	PM	PM	PM	PM	PS
PL	PM	PL	PL	PL	PL	PL	PM

(Force Error, rows)

Figure 6. Inference engine rule matrix.

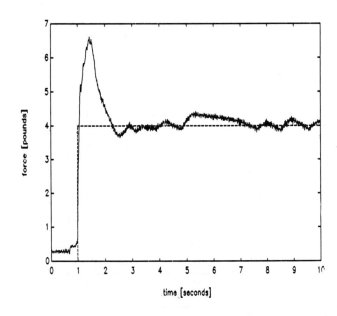

Figure 8. Response of fuzzy logic force controller due to step change in desired force.

Figure 7. Control surface of fuzzy logic controller.

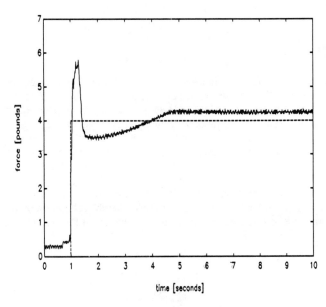

Figure 9. Response of proportional force controller due to step change in desired force.

The robustness of the fuzzy logic force controller was tested by commanding the MERLIN robot to follow a spiral trajectory on the table top while maintaining a 4 pound contact force. This task was programmed by continuously executing the force controller and periodically updating the six axes to reflect the inverse kinematic solution to the spiral trajectory. Figure 11 shows the position of the end-effector on the table top. The resulting contact forces, shown in figure 12, fluctuated erratically about the 4 pound set point but the robot never lost contact with the surface.

Conclusions

A fuzzy logic force controller has been implemented on a MERLIN 6540 Industrial Robot constructed with stepper motors. A fuzzy controller was selected because it is a non-model based controller capable of performing nonlinear control. These attributes make it well suited to solve the hard contact problem on this type of manipulator. The fuzzy controller was shown to be stable for force control experiments conducted on a hard surface and outperformed a proportional force controller. Robustness experiments showed the fuzzy controller was able to maintain contact while following a

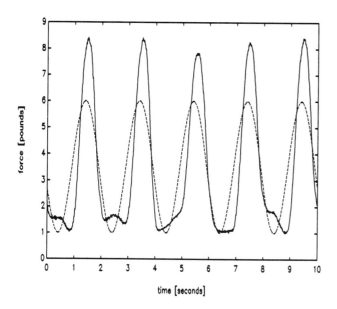

Figure 10. Response of fuzzy logic force controller due to sinusoidal change in desired force.

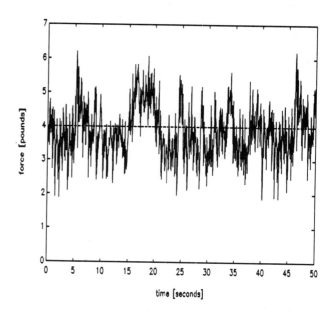

Figure 12. Force response of fuzzy logic force controller while tracing a spiral trajectory on a table top.

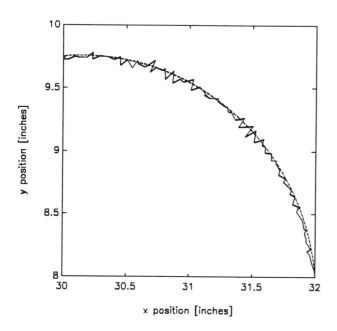

Figure 11. Trajectory of end-effector while tracing a spiral on a table top.

trajectory tangent to a surface. However, the resulting contact force fluctuated erratically about the set point. The force response could be improved best by adding compliance to the robot. This solution would preclude precise measurement of the robot's end-effector position unless a sensor was also included to measure the deflection of the compliance. A LVDT (linear variable displacement transducer) or an IRCC (instrumented remote center compliance) could be used for this task [10].

Acknowledgment

This work was supported by the Office of Naval Research under grant no. N00014-91-J-1621.

References

[1] El-Itaoui, A. H., and A. Eltimsahy, "Modeling and control of the MERLIN 6200 industrial robot," *IEEE International Conference on Robotics and Automation*, Philadelphia, PA, pp. 1670-1675, April 1988.

[2] Elosegui, P., R. W. Daniel, and P. M. Sharkey, "Joint servoing for robust manipulator force control," *Proceedings of the 1990 International Symposium on Intelligent Control*, 1990, pp. 246-251.

[3] Fu, K. S., R. C. Gonzalez, C. S. G. Lee, Robotics: Control, Sensing, Vision, and Intelligence, New York, NY: McGraw-Hill Book Company, 1987.

[4] Kazerooni, H., B. J. Waibel, and S. Kim, "On the stability of robot compliant motion control: theory and experiments," *ASME Journal of Dynamic Systems, Measurement, and Control*, vol. 112, pp. 417-426, September 1990.

[5] Kosko, B., Neural Networks and Fuzzy Systems, Englewood Cliffs, NJ: Prentice-Hall, 1992.

[6] Leahy, M. B., "Experimental analysis of robot control: a performance standard for the PUMA-560," *Proceedings of the IEEE Symposium on Intelligent Control*, Albany, NY, pp. 257-264, September 1989.

[7] Lee, C. C., "Fuzzy logic in control systems: fuzzy logic controller - Part I," *IEEE Transactions on Systems, Man, and Cybernetics*, vol. 20, no. 2, pp. 404-418, March/April 1990.

[8] Lee, C. C., "Fuzzy logic in control systems: fuzzy logic controller - Part II," *IEEE Transactions on Systems, Man, and Cybernetics*, vol. 20, no. 2, pp. 419-435, March/April 1990.

[9] Lim, C. M., and T. Hiyama, "Application of fuzzy logic control to a manipulator," *IEEE Transactions on Robotics and Automation*, vol. 7, no. 5, pp. 688-691, October 1991.

[10] Machida, K., Y. Toda, T. Iwata, and T. Komatsu, "Smart end effector for dexterous manipulation in space," *AIAA Journal of Guidance, Control, and Dynamics*, vol. 15, no. 1, pp. 10-16, January/February 1992.

[11] Shih, C.-L., W. A. Gruver, and Y. Zhu, "Fuzzy logic force control for a biped robot," *Proceedings of the 1991 Symposium on Intelligent Control*, Arlington, VA, pp. 269-274, August 1991.

[12] Veignat, N., "Microstep operation of stepper motors," *Motion Control*, vol. 2, no. 4, pp. 48-50, March 1991.

[13] Wen, J. T., "Position and force control of robot arms," *Proceedings of the IEEE Symposium on Intelligent Control*, Albany, NY, pp. 251-256, September 1989.

HIERARCHICAL INTELLIGENT CONTROL FOR ROBOTIC MOTION BY USING FUZZY, ARTIFICIAL INTELLIGENCE, AND NEURAL NETWORK

Toshio FUKUDA and Takanori SHIBATA

Dept. of Mechanical Engineering, Nagoya University,
1 Furo-cho, Chikusa-ku, Nagoya, 464-01, JAPAN
Phone +81-52-781-5111 ext. 4478
Fax +81-52-781-9243
e-mail: d43131a@nucc.cc.nagoya-u.ac.jp

ABSTRACT We present a new structure of intelligent control for robotic motion. This system is analogous to the human cerebral control structure for intelligent control. Therefore, the system has a hierarchical structure as an integrated approach of Neuromorphic and Symbolic control, including an applied neural network for servo control, a knowledge based approximation, and a fuzzy set theory for a human interface. The neural network in the servo control level is numerical manipulation, while the knowledge based part is symbolic manipulation. In the Neuromorphic control, the neural network compensates for the nonlinearity of the system and uncertainty in its environment. The knowledge base part develops control strategies symbolically for the servo level with a-priori knowledge. The fuzzy logic combined with the neural network is used between the servo control level and the knowledge based part to link numerals to symbols and express human skills through learning.

1. INTRODUCTION

In robotic fields, autonomous robots are required for industrial applications. The autonomous robots have to be intelligent and should have hierarchical control structures while mimicking human's cerebral control structure in order to achieve some tasks without human operators [1, 2] (Fig. 1). Here, we think that the intelligent robot should be able to act for various purposes in various environments by itself like human beings. Therefore, it is necessary for the autonomous robots to have reasoning mechanisms and adaptive servo controllers. The reasoning mechanism produces control strategies symbolically for complex and composite tasks by using knowledge and data base systems at a high level after recognition of the environment. Moreover, at a servo level, the controllers have to adapt to the environment having uncertainty. While the reasoning mechanism deals with symbols, the servo controller deals with numerals (Fig. 2).

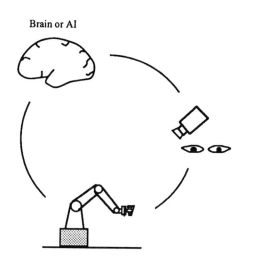

Fig. 1 Image of intelligent control

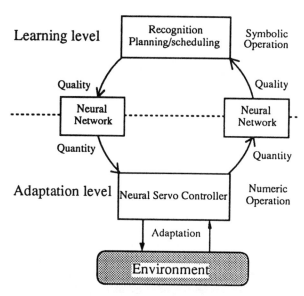

Fig. 2 Concept of Hierarchical Intelligent Control

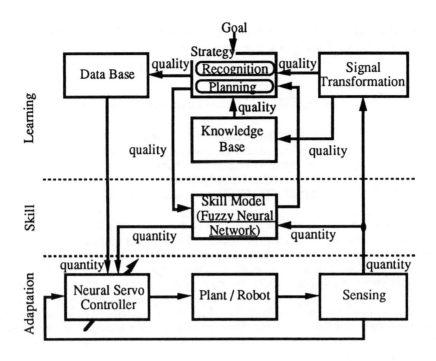

Fig. 3 Hierarchical Intelligent Control with skill based control by using Fuzzy Neural Network

The Artificial Intelligence (AI) techniques have been used to synthesize knowledge-based systems as expert systems. For intelligent control, there were some examples of *symbolic control* which uses the symbolic reasoning mechanism for the higher-level control [2 - 4]. However, it is difficult to classify sensed data in order to map the numerical data into the symbolic data for understanding of the process state. On the other hand, at the servo level, adaptive control like the Model Reference Adaptive Control and the Self Tuning Regulator has become available to complicated and composite systems having much uncertainty [5]. However, the adaptive control has drawbacks such as exponentially complicated calculation for a lot of unknown parameters and limitation on applicability to nonlinear systems [1]. In order to solve these problems, many groups have used the neural network for adaptive control [6 - 10]. We have demonstrated applicabilities of the neural network-based controllers to force control, stabbing control, and hybrid position/force control of robotic manipulators [7, 8]. Then, these research have shown that the neural network in the servo controller can compensate for the nonlinearities and uncertainty of the system and its environment. We call the neural network-based control *neuromorphic control*. The difficulty of the neuromorphic control is lack of generalization to various tasks [11]. That is, it is necessary for the neuromorphic control to expand task range so as to be intelligent control.

For an intelligent autonomous robotic system, we have proposed Hierarchical Intelligent Control, which is a hybrid control of the neuromorphic control and the symbolic control [11, 12] (Fig. 2). In this system, the control strategy is developed by symbolic reasoning with knowledge base system, and a neuromorphic controller is used at the servo level in order to adapt to the environment having uncertainty. The system used neural networks to link numerical real world and symbolical AI world. The system comprised two levels of feedback loops: a learning level of control process by the symbolic control and an adaptation level of dynamic process by the neuromorphic control. At the high level, the system planed a control strategy approximately and symbolically, and at the servo level, the neural network compensated for vagueness of the symbolic control strategy and for uncertainty of the environment. However, the system was not always effective with respect to a cost of time for planning or scheduling since the reasoning mechanism took much time. When there are changes in the characteristic of the environment at the same task, human beings change the way of task by using *skills* without symbolical reasoning. For autonomous robots, it is not always necessary to reason symbolically in the same case so as to save the cost of time. In this case, it is enough only to change the way of task without the planning. That is, it is enough only to tune the control references for the servo controller by using skills when the environment changes. In the past few years, several methods have developed for modeling and acquiring human skills and transferring them to robots and telerobots [13, 14]. For this purpose, Asada showed the neural network can acquire human skills by using teaching data [15].

However, the neural network can not cluster or classify the input space for the higher level to understand extracted human skills.

In this paper, we present an architecture of the Hierarchical Intelligent Control for intelligent robot motion. The system use the AI technique, the neural network, and the fuzzy logic [16]. We describe how to combine them for intelligent control of robot motion.

2. CONCEPT OF HIERARCHICAL INTELLIGENT CONTROL

The Hierarchical Intelligent Control system is a hybrid system of the neural network, the fuzzy logic, and the AI, and comprises three levels: a *learning* level, a *skill* level, and an *adaptation* level (Fig. 3). Therefore, there are three feed-back loops. The learning level is based on the expert system of AI technique for a reasoning mechanism and has a hierarchical structure: recognition and planning to develop a control strategy. The recognition level uses neural networks (NN) and fuzzy logic combined with the neural network (FNN) as nodes of decision tree [12, 17, 18]. In the case of the NN, input is numeric quantity sensed by some sensors and output is symbolic quality which indicates a process state (Fig. 4). In the case of the FNN, input and output are numeric quantity and the FNN clusters input signals by using membership functions (Figs. 5 and 6). That is, the FNN can transform numerical quantity into symbolic quality by using membership functions. Both the NN and the FNN were trained with the training data of a-priori knowledge obtained from human experts in order to transform various sensed data from numerical quantities to symbolic qualities, and to make *sensor fusion* and *meta-knowledge* at the learning level. The important information is sensed actively on using the knowledge base. The sensors of vision, weight, force, touch, acoustic, and others can be used as nodes of decision tree for recognition of the environment. Then, the planning level reasons symbolically for a strategic plan/schedule of robotic manipulation, such as task, trajectory, force, and other plannings in conjunction with the knowledge base. The system can include another *common sense* for robotic manipulation. Thus, the learning level reasons an unknown fact from a-priori knowledge and sensed information. Then, the learning level produces a control strategy for the adaptation level. Following the control strategy, the learning level selects initial data set for a servo controller at the adaptation level from a data base which maintains some gains and initial values of interconnection weights of a NN in the servo controller. Moreover, the recent sensed information from the skill level and the adaptation level updates the learning level through long-term learning process.

In the same task and different environments, it is necessary to change control references depending on the environment for the servo controller at the adaptation level. At the skill level as a middle level, we use the FNN for specific task following the control strategy produced at the learning level in order to generate appropriate control references. Input signals into the FNN are numerical values sensed by some specific sensors and some symbols which indicate the control strategy produced at the learning level. Output of the FNN is the control reference for the servo controller at the adaptation level. This output is based on the skill extracted from a human expert through learning training sets obtained from human experts. At the same moment, the FNN clusters the input signals in the shape of membership functions. We can use these membership functions as the symbolic information for the learning level.

At the adaptation level, the NN in the servo controller adjusts the control law to the current status of the dynamic process [1]. Particularly, compensation for nonlinearity of the system and uncertainty included in the environment must be dealt with by the NN. Thus, the NN in the adaptation process has to work more rapidly than that in the learning process. We have show that a NN-based controller, Neural Servo Controller, is useful to the nonlinear dynamic control with uncertainty, like force control of a robotic manipulator [8]. Eventually, the NNs and the FNN connect the neuromorphic control with the symbolic control for Hierarchical Intelligent Control while combining human skills.

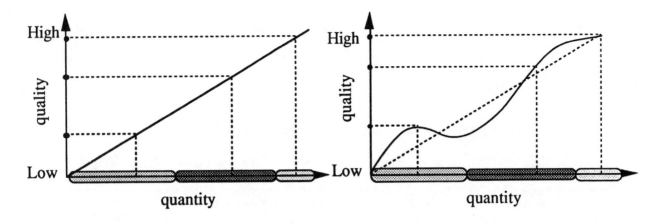

(a) 'If .. , then ..' rules: straight line function

(b) neural network: sigmoid-like curve function acquired through learning

Fig. 4 Signal transformation from quantity to quality

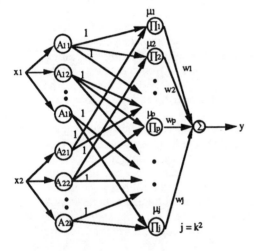

Fig. 5 Model of Fuzzy Neural Network

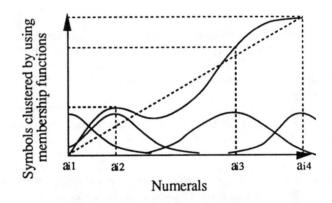

Fig. 6 Signal transformation from quantity to quality by using the membership functions in the FNN

3. Symbolic Control by Strategy

Symbolic control uses reasoning mechanism to produce a control strategy while using knowledge. The knowledge is acquired from a human expert and results of control at the adaptation level through long-term learning. Here, we assume that the manipulator grasps and carries objects whose characteristics are unknown. At first, the recognition level identifies the object approximately by visual sensing. There, we use the trained NN for an associative memory. The recognition level reasons symbolically from the knowledge concerned with the object, such as what the states and characteristics of the object are like, and then important information should be obtained by sensing. The recognition level produces the strategy to achieve the goal according to the states and the characteristics of the object based on the knowledge and the skill. For example, we consider the case of having the robot carry a container which has some water in it. By vision, the recognition level recognizes the object as a container at first. Several data about the material are also included in the knowledge. It is necessary to examine what the container is made of, such as paper, plastic, or glass. Therefore, the robot examines the stiffness of the glass. According to the knowledge, it is also necessary to investigate how much the container contains liquid in it. Therefore, the robot examines the weight of the container. In this way, the recognition level can recognize more precisely so that the planning level can execute task planning successfully by using skill, for example, to grasp the container without braking it and to carry it without spilling the liquid in the container. Thus, the learning level produces the control strategy for the adaptation level, reasoning with sensed data and knowledge. The control strategy select data of approximate control references and initial weights of the NN in the controller at the adaptation level. Therefore, the control strategy includes vagueness in itself because of symbolic expression. The neuromorphic control compensates for the vagueness by its adaptive capability.

4. CONFIGURATION OF FUZZY NEURAL NETWORK

The NN has capabilities of nonlinear mapping, parallel processing, and learning. On the other hand, the fuzzy logic is characterized as extension of binary Boolean logic [16]. The fuzzy logic is a class in which transition from membership to non-membership is gradual rather than abrupt. Both the NN and the fuzzy logic have some drawbacks. The NN can produce mapping rules from empirical training-data sets through learning, but the mapping rules in the network is not visible and is difficult to understand. On the other hand, since the fuzzy logic does not have learning capability, it is difficult to tune the rules. In order to solve these difficulties, recently, much research has been trying to synthesize the fuzzy logic and the NN [17, 18], and it is often called Fuzzy Neural Network (FNN). We use the FNN at the skill level in order to express human skills through learning. The FNN clusters the input parameters into fuzzy sub-space and identifies input-output relationships by using a weighted network. The FNN in Fig. 5 was proposed by Ichihashi [18] This FNN is based on the simplified fuzzy inference. A_{ik} is the membership function for i-th input variable x_i (i = 1 ... n) in k-th rule. w_k is the consequence of the k-th rule. Individual rules' results are given by μ_k. The membership functions and weights are modified through learning. After learning, we can acquire mapping rules as human skills from the FNN.

5. NEUROMORPHIC CONTROL

This section describes learning model for a neural network in a servo controller at the adaptation level. We call the NN-based control neuromorphic control. There are much research of adaptive control by using the NN. Figure 6 shows a structure of indirect inverse control of robotic manipulator, which uses a neural network as a controller combined with local regulators with fixed gain [8]. We train the neural network by using a error between actual output of the system and control reference value like feedback error learning [6]. The neural network has time delay elements so as to be dynamic configuration. In indirect inverse control, the neural network is used to compensate for nonlinearity and uncertainty. We have shown the usefulness of the neural servo controller [8]. The controller can adapt to the environment in short-term when the desirable initial state of the NN is given [11]. Therefore, if the data base system in the Hierarchical Intelligent Control system maintains weights obtained through learning, the controller adapts to its environment easily by using the control strategy [11, 12].

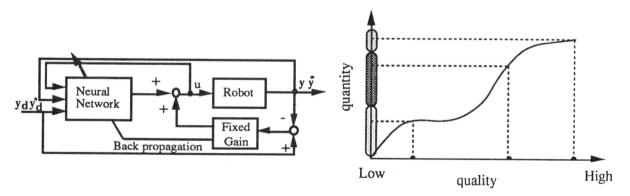

Fig. 7 Indirect inverse control: nonlinear regulator type

Fig. 8 Signal transformation from quality to quantity: the NN based controller compensate for vagueness in the control strategy by its adaptability.

6. CONCLUDING REMARKS

In this paper, we present a structure of hierarchical intelligent control for robot motion. The system comprises of neuromorphic control and symbolic control. The system uses the AI technique for reasoning mechanism and the NN for adaptive control. We showed some methods for connecting the neuromorphic control with the symbolic control by using the NN and fuzzy logic. The system would be useful for autonomous robot to achieve complicated tasks in various environments.

REFERENCES

[1] T. Fukuda and T. Shibata, Research Trends in Neuromorphic Control, Journal of Robotics and Mechatronics, Vol. 2 No. 4, pp. 4-18 (1991)
[2] J. Rasmussen, Skills, Rules, and Knowledges; Signals, Signs, and Symbols, and Other Distinctions in Human Performance Models, IEEE Trans. on Systems, Man, and Cybernetics, Vol. SMC-13, pp. 257-266, (1983)
[3] G. Saridis, Intelligent Robotic Control, IEEE Trans. on Automatic Control, Vol. 28, No. 5, pp. 311-321 (1983)
[4] H. Asada and S. Hirai, Towards A Symbolic-Level Force Feedback: Recognition of Assembly Process States, Proc. of Int'l. Symp. on Robotic Research, pp. 341-346 (1990)
[5] J. E. Slotine and W. Li, On the adaptative control of robot manipulators, Proc. of R&A (1986)
[6] M. Kawato, Y. Uno, M. Isobe, and R. Suzuki, A Hierarchical Neural-Network Model for Voluntary Movement with Application to Robots, IEEE Control Systems Magazine, 8(2), pp. 8-16 (1988)
[7] M. Tokita, T. Mitsuoka, T. Fukuda, T. Kurihara, Force Control for Robotic Manipulator by the Neural Network, JRSJ, Vol. 7 No. 2, pp. 47-51 (1989)
[8] T. Fukuda, T. Shibata, M. Tokita and T. Mitsuoka, Neural Network Applications for Robotic Motion Control; Adaptation and Learning, Proc. of IJCNN '90 - San Diego, pp. 447-451 (1990)
[9] K. S. Narendra and K. Parthasarathy, Identification and Control of Dynamical Systems Using Neural Networks, IEEE Trans. on Neural Networks, vol. 1, pp. 4-27 (1990)
[10] P. J. Werbos, Neurocontrol and Related techniques, Hand Book of Neural Computing Applications, Academic Press, Inc., pp. 345-380 (1990)
[11] T. Fukuda, T. Shibata et al., Adaptation and Learning for Hierarchical Intelligent Control, Proc. of IJCNN '91-Singapore, Vol. 2, pp. 1033-1038 (1991)
[12] T. Shibata, T. Fukuda et al., New Strategy for Hierarchical Intelligent Control of Robotic manipulator: - Hybrid Neuromorphic and Symbolic Control -, Proc. of IJCNN '91-Singapore, Vol. 1, PP. 107-112 (1991)
[13] G. Hirzinger and K. Landzettel, Sensory Feedback Structures for Robots with Supervised Learning, Proc. of Int'l Conf. on R&A, pp. 627-635, (1987)
[14] H. Asada and H. Izumi, Automatic Program Generation from Teaching Data for the Hybrid Control of Robots, IEEE Trans. on R&A, Vol. 5, No. 2, pp. 163-173 (1989)
[15] H. Asada and S. Liu, Transfer of Human Skills to Neural Net Robot Controllers, Proc. of R&A, pp. 2442-2448 (1991)
[16] L. A. Zadeh, Fuzzy Sets, Information and Control, Vol. 8, pp. 228, (1965)
[17] M. M. Gupta and J. Qi, On Fuzzy Neuron Models, Proc. of IJCNN'91-Seattle, Vol. 2, pp. 431-436, (1991)
[18] H. Ichihashi, Learning in Hierarchical Fuzzy Models by Conjugate Gradient Method using Backpropagation Errors, Proc. of Intelligent System Symp., pp. 235-240 (1991)

Hierarchical Control for Autonomous Mobile Robots with Behavior-Decision Fuzzy Algorithm

Yoichiro Maeda, Minoru Tanabe, Morikazu Yuta, Tomohiro Takagi
Laboratory for International Fuzzy Engineering Research (LIFE)
89-1, Yamashita-cho, Naka-ku, Yokohama, 231, JAPAN
Phone +81-45-212-8223, FAX +81-45-212-8255

Abstract

We proposed a construction method of the behavior-decision system using fuzzy algorithms capable of expressing sequence flow which includes a mixture of both crisp and fuzzy processing. We also propose in this paper a method of tuning algorithms for giving robots the autonomous ability to judge purposes of actions like human. In this method, we try to express ambiguous situation which a robot will encounter and behavior-decision algorithm flow by using the modified fuzzy algorithm with fuzzy branch controlled threshold, which we call it the behavior-decision fuzzy algorithm. Finally, we report some results of computer simulations and experiments concerning an evaluation of this method supposed simple in-door environment.

1: Introduction

The use of fuzzy logic in control applications in recent years has been remarkable. It is no exaggeration to say that fuzzy control has become recognized as the best model available for representing the knowledge of an experienced or skilled professional. In any attempt to produce a high-order intelligent robot with autonomous behavior-decision capabilities, it is important to develop some new methodology, based on intelligent processing and model building methods, which is at the same time capable of describing complex behavioral sequences both flexibly and using macros. Unfortunately, even though sequence flow of this kind can be described easily using a representation method based on conventional AI production rules, such a method cannot be used to handle ambiguous states. On the other hand, use of a representation method based on conventional fuzzy inference can handle ambiguities, but cannot be used where sequence flow is required.

On our research[1], we therefore try to express the macro behavior-decision algorithm close to the one which humans use every day by utilizing fuzzy algorithms. Fuzzy algorithms can both describe ambiguous situations and represent sequence flows which mix fuzzy as well as non-fuzzy (crisp) processing. Furthermore, we also propose in this paper a method of tuning algorithms for giving robots the autonomous ability to judge purposes of actions like humans, where behavior is self-determined based on the situation in which they find themselves.

2: Modified Fuzzy Algorithms

Fuzzy algorithms[2][3] are a concept first introduced by L. A. Zadeh as one possible tool for approximation analysis in decision making. According to Zadeh, fuzzy algorithms are "ordered sets of fuzzy instructions" and has the four following types of expressional capacity. These types are fuzzy definitional, generational, relational and decisional algorithms. In original fuzzy algorithms, processing after fuzzy branches, which a sequence of flow branches according to fuzzy processing, is supposed to be done in parallel and the consequent part (THEN or ELSE part) executing after the branch are necessary to inherit the weight (fuzziness) of grades in the previous part (IF part).

Features of fuzzy algorithms are shown as follows.

[Merits]
· Coding algorithms with both crisp and fuzzy quantities using a very natural representation
· Expressing the fuzzy processing in all types of algorithms including sequence flow

[Demerit]
· Large amounts of parallel processing with possibility of the algorithm growing out of control

Murofushi et al.[4] demonstrated the effectiveness of fuzzy algorithm through a simple experiment for the fuzzy control of model cars. For the fuzzy algorithm discussed in this paper, a constant threshold value has been placed on the grade values of the previous part in the fuzzy branch. Although this method is very effective in control applications, it results in crisp branching that loses some of the characteristics of the original fuzzy algorithm.

Fig.1 shows the general idea about control method of fuzzy branch in this paper. We call this method "the modified fuzzy algorithm" hereafter. If a previous part has been represented by only crisp processing, the consequent part (THEN part meaning "yes" branch or ELSE part "no" branch) is executed with perfect matching when the previous part is satisfied. If a previous part has included fuzzy processing, the THEN part is executed only when a

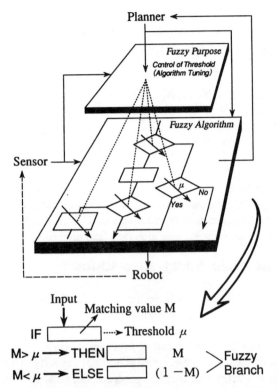

Fig.1 Method of Modified Fuzzy Algorithms

matching grade value in the previous part is greater than a given threshold (μ in the figure) and the ELSE part in the other case. In the modified fuzzy algorithms we propose here, this threshold value itself is tuned in accordance with the control purpose determined autonomously (as described later). And the processing of a branch in the previous part can be described with fuzzy labels in this method, but rule-executions in the consequent part are not performed with parallel firing processing as original fuzzy algorithm. When a matching value in a previous part is M, the THEN part is executed with the matching value M and the ELSE part with the complement of matching value (1-M).

3: Fuzzy Concepts and Fuzzy States

Before discussing how to code the main flow with a fuzzy algorithm, we first tried to define ambiguous condition which a robot will encounter during motion. We call it fuzzy concepts and fuzzy states, using the modified fuzzy algorithm as previously described.

First, as an example of fuzzy concepts which will be used in coding fuzzy states, attributes concerning obstacles (OB), walls (WA), forward free space (FS) and the robot itself (RO) are defined in the frame format as shown in Fig.2. We call it fuzzy-based frame, whose slot value can be defined with fuzzy labels. The attributes with these slot values can be expressed by using both fuzzy variables or crisp variables. For abbreviated labels with parentheses in fuzzy concepts, italic labels represent fuzzy values and all others represent crisp values.

The biggest problem when the robot is in locomotion, is the recognition by the robot of relatively ambiguous states such as whether it can move forward (passable) or avoid obstacles (avoidable). Since the explicit coding of ambiguous states of this kind is fairly difficult, we try to use an approximate representation of the fuzzy states as expressed with the modified fuzzy algorithm. The expressions in right column of Fig.2 show the fuzzy states of PASSABLE and AVOIDABLE represented by using above-mentioned fuzzy concepts.

Here, PASSABLE, IMPASSABLE, AVOIDABLE, UNAVOIDABLE are fuzzy labels. To determine if a state is passable in the example being presented, the minimum grade value about matching rules would be found for each in order of top to bottom, and as soon as any exceeded threshold ζ, the fuzzy state IMPASSABLE would result. If none of the rules exceeded the threshold, the fuzzy state PASSABLE would result. Since resulting states such as IMPASSABLE and others have also been represented by fuzzy labels, it is possible to get the degree of satisfaction about the states. It is possible to make fine adjustments to fuzzy state determinations by controlling the threshold value described above. Further, fuzzy state determinations can also be controlled indirectly by actually changing the slot value of fuzzy concepts which are used in fuzzy state definitional rules.

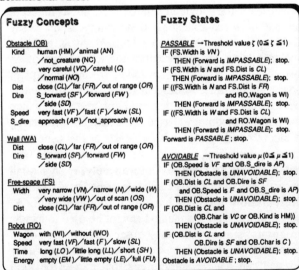

Fig.2 Fuzzy Concepts and Fuzzy States

4: Supervised Control based on a Fuzzy Purpose

Since this part of the algorithm is for fine tuning the entire algorithm as required by a control purpose in that time, it must be of a higher order than the fuzzy algorithm. By configuring a hierarchical structure in this way, we consider constructing a supervised control method of the algorithm based on macro-judgments with an evaluation function of fuzzy purposes. In the method being proposed, the configuration of the purpose judgement part includes two hierarchical determination parts of "strategy evaluation" and "tactic determination".

In the strategy evaluation part, modes of motion is evaluated by information currently being obtained from surroundings. We supposed that a robot has three types of motion modes SAFETY, ECONOMY and MIN-TIME. Expressive rules about these modes are described with the modified fuzzy algorithm. Degrees of satisfaction about each mode, Ms, Me and Mt, are set for each rule respectively. In this paper, only one mode whose this value is highest is selected as the purpose in that time.

Next, we explain about operations of tactics suited for carrying out the strategy selected in the strategy evaluation part. This part is capable of modifying the algorithm mainly according to the three following methods.

1) Boundary control on fuzzy branches
· modifying threshold values in the previous part of the rule (direct control)
· tuning membership parameters in fuzzy concepts (indirect control)

2) Modification of the control scheme
· modifying the degree of significance between path tracking and obstacle avoidance
· tuning the inference output value in danger evaluation

3) Changing control parameters
· modifying the universe of discourse about velocity and steering membership functions
· tuning the command output value of velocity and steering

Rules for these operations are described for each strategy. Tactics operations described above are determined based on the strategy selected every inference cycle. By applying such tactics for the behavior-decision fuzzy algorithm it is possible to perform online tuning of the fuzzy algorithm. In this method, since we use the modified fuzzy algorithm, it is possible to autonomously determine the control purpose on upper level of the algorithm and flexibly adjust the algorithm itself as compared with conventional fuzzy algorithms.

5: Overall Flow of Behavior-decision Fuzzy Algorithms

Fig. 3 shows a flow chart representing the behavior-decision fuzzy algorithm. In the figure, the upper left is the start of the algorithm and the lower center the finish. This processing is repeated for every sampling time. Meta

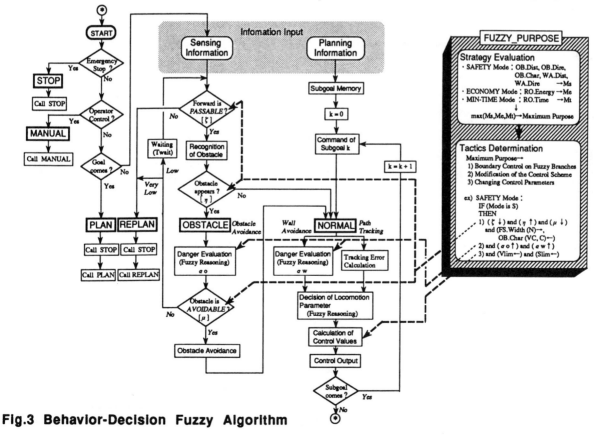

Fig.3 Behavior-Decision Fuzzy Algorithm

rules for supervised control of overall algorithm flow in the expression using the modified fuzzy algorithm style is described as shown as Fig.4 (Fuzzy-Algorithm META).

Here, IMPASSABLE represents a fuzzy state as defined in section 3. Gothic letters occurring within the algorithm represent the fuzzy subalgorithms. In this simulation,we used stop motions in place of PLAN and back-track motions in place of REPLAN since the simulator in this paper has not path planning routine. The ability of using "call" and "goto" statements and handling both fuzzy and crisp values is remarkable feature of fuzzy algorithms. Subgoal information given to the robot each time are assumed here to be input to memory as a target value while the robot is in offline.

The right column in Fig.4 shows a sample expression for the fuzzy subalgorithm OBSTACLE as it might be written for the sequence shown in Fig.3. It is assumed for this expression that the control purpose SAFETY has already been selected by the purpose judgement part. Here, italic letters represent fuzzy labels and all others crisp data. The phrase: "Obstacle is AVOIDABLE with Very Low", means that an obstacle is avoidable with very low grade in the possibility of avoiding it. In the step 5), after the recognition that an obstacle is avoidable with low grade, the system is waiting for a while to look at the obstacle motion. After this waiting period ends, step 6) makes a return to the main algorithm META so as to attempt a retry.

Fuzzy-Algorithm META

1) Input Subgoals (k=1,...,m).
2) k = 0.
3) IF Emergent Stop happens
   THEN call STOP;
4) ELSE IF Operator interrupts
   THEN call MANUAL;
5) ELSE IF Goal comes
   THEN call STOP; call PLAN;
6) ELSE call FUZZY_PURPOSE;
   IF Forward is *IMPASSABLE*
   THEN call STOP; call REPLAN;
7) ELSE IF Obstacle appears
   THEN call OBSTACLE.
8) call NORMAL.
9) Output Control commands.
10) IF Subgoal comes THEN k = k + 1.
11) go to 3).

Fuzzy-Algorithm OBSTACLE

1) Decide α o [Fuzzy Reasoning] .
2) IF Mode is *SF* THEN α o ↑ and μ ↓.
3) IF Obstacle is *AVOIDABLE* with *Very Low*
   THEN call STOP; call REPLAN;
4) ELSE IF Obstacle is *AVOIDABLE* with *Low*
5) THEN Waiting;
   IF (Waiting time ≤ Twait)
   THEN go to 5);
6) go to META 6);
7) ELSE Call Obstacle_Avoidance.
   RETURN.

Fig.4 Expressive Examples of Algorithm

6: Simulations

6.1: Mobile Robot Simulator

We constructed the robot simulator to check the decision and the motion of a robot. Fig.5 shows the locomotive model of a mobile robot. In this simulator, it suppose that a robot inputs following informations.

1) Planning information (Offline data):

(Psx, Psy, δ)k : Position of subgoals and permitted limit

2) Sensing information (Online data):

(Prx, Pry, θ r) : Present position and direction of a robot [from PSD position detector]

di (i=1,...,8) : Distance between a robot and an object [from 8 sonar sensors covered 120° area]

θ o : Direction for an obstacle in the robot coordinates [from image tracker by CCD camera]

More simplified rules than that one described before has been used in this simulation. Rules used in this simulation are shown in Fig.6.

The first rule about the fuzzy state PASSABLE is described based on fuzzy labels representing the width and distance of the forward space of the robot. Output of this rule is the degree of impassability (0<=IM<=1) of the forward space. If IM becomes higher than the threshold TH_imp, the forward space is decided as IMPASSABLE and a robot performs back-track in this simulation. In the second rule about the fuzzy state AVOIDABLE, it gives the degree of unavoidability (0<=UN<=1) for an obstacle. If UN exceeds the threshold TH_una, the obstacle is decided as UNAVOIDABLE and a robot performs back-track as well as IMPASSABLE. And if UN is more than half of TH_una and less than TH_una, a robot performs waiting motion (stopping for a while). The third rule about the fuzzy purpose SAFETY will get the degree of danger for control purpose evaluation. The robot is made to recognize the degree of danger (0<=DA<=1) according to this evaluation rule.

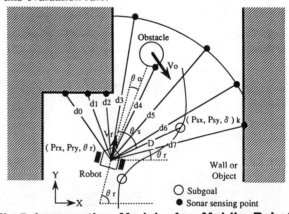

Fig.5 Locomotive Model of a Mobile Robot

Fuzzy state *PASSABLE*
IF P = [(FS.Width is *VN*) or (FS.Width is *N* and FS.Dist is *CL*)] is high
THEN (Forward is *IMPASSABLE*) → TH_imp <Back Track>
ELSE (Forward is *PASSABLE*).

Fuzzy state *AVOIDABLE*
IF A = [(OB.Speed is *VF* and OB.S_dire is *AP*) or
(OB.Dist is *CL* and OB.Dire is *SF* and OB.Speed is *F* and OB.S_dire is *AP*)] is high
THEN (Obstacle is *UNAVOIDABLE*) → TH_una <Back Track>
ELSE IF A is little high
THEN (Obstacle is *LITTLE UNVOIDABLE*) → TH_una/2 <Waiting>
ELSE (Obstacle is *AVOIDABLE*).

Fuzzy purpose *SAFETY*
IF S = [(OB.Dist is *CL* and OB.Dire is *SF*) or (WA.Dist is *CL* and WA.Dire is *SF*)] is high
THEN (Mode is *DANGER*).

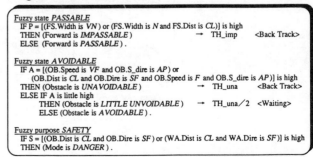

Fig.6 Fuzzy Algorithms In This Simulation

In these rules, fuzzy branches for THEN or ELSE are controlled by thresholds of IMPASSABLE and UNAVOIDABLE. For example, the threshold of IMPASSABLE is decided every sampling time as follows.

TH_imp=TH_imp+G_imp(TH_imp0(1-DA^2)-TH_imp)

Here, TH_imp and TH_imp0 shows the present and default threshold of IMPASSABLE, respectively. G_imp shows the control gain of the threshold. And the velocity of a robot is controlled by output value IM of the first rule as follows.

$$Vr = (1-IM^2)Vr0$$

In this equation, Vr0 means initial value of the robot velocity. Actual velocity Vr of a robot always becomes less than Vr0. These control operations correspond to the first and the third tuning method of the tactic determination rule in section 4.

6.2: Simulation Results

Fig.7 to 9 shows the trails of a robot and a moving obstacle in this simulation. The still objects within the environment (shown as black polygons) and the moving obstacle (shown as a large white circle going down at lower right area in the map) are detected using virtual ultrasonic sensors for a rough measurement of distance. During the simulation, the robot clears two subgoals (filled circles #3 and #11 in the figure) moving from the starting point #0 in the lower left, to the final goal #11 in the upper right.

The right side of a simulation window indicates each condition of fuzzy states, threshold values and fuzzy purpose. Two bar graphs on left side shows degrees of IMPASSABLE(IM) and UNAVOIDABLE(UN). In these graphs, black arrows means threshold values of TH_imp and TH_una, and a white arrow in the graph of UN indicates the threshold TH_una/2 for determining the waiting mode. And a black triangle shows the point when the value of fuzzy states exceeded against the threshold value. The right bar indicates the evaluating value DANGER(DA) on fuzzy purpose SAFETY. The threshold values of IM and UN are controlled by this value DA.

Fig.7 shows the simulation result which indicated the robot motion with the threshold values TH_imp and TH_una controlled by the value DA. In this simulation, a robot recognized impassability of his front space and decided back-track motion to the node #3. And, a robot entered the next little wide passage, but encountered an obstacle in right front. Also in this case, a robot goes back to the node #4 because the value UN exceeded the threshold value TH_una dropped by high degree of danger DA. From #6 to #7, the threshold values are fairly dropped for a robot encountered an moving obstacle, but the value UN indicates between black and white arrows because the forward passage is very wide. Therefore, a robot goes on waiting for the time when an obstacle passes by unless the value UN exceeds the TH_una.

On the other hand, the robot motion without threshold control is shown in Fig.8. In this case, each threshold does not change at any time and is kept the default value. Therefore, the value IM can not exceed the threshold value TH_imp and a robot passes the narrow passage from node #3 to #8 by force. Fig.9 shows the motion of a robot controlled only the threshold value TH_imp. In this case, a robot performed the back-track motion between node #3 and #8, but he passed by a standing obstacle near the node #7 because the value UN can't exceed the constant threshold value TH_una without control.

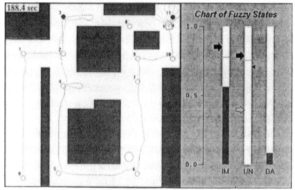

Fig.7 Simulation Results (1)

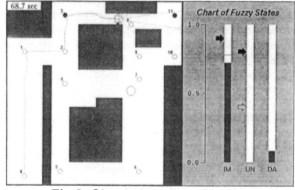

Fig.8 Simulation Results (2)

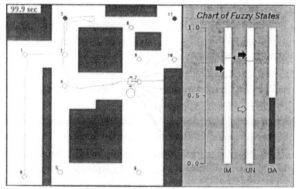

Fig.9 Simulation Results (3)

As shown in these figures, a robot clearly recognizes passage as being difficult whenever it enters a narrow region or approaches a surrounding objects or an obstacle. Especially, changing of thresholds in the graph shows that IMPASSABLE and UNAVOIDABLE recognition are affected by the fuzzy purpose SAFETY. Furthermore, the motion of a robot will become still more similar to that of human by switching several control purposes of SAFETY, ECONOMY, MIN_TIME and so on.

7: Experiments

The outward appearance of our autonomous mobile robot is shown in Fig.10. The total experimental system is mainly constructed with three parts : robot, controller and supervisor. The robot (HERO2000) is controlled by the controller (PC98RX) which is commanded by the supervisor (SUN4 workstation). And every control signal and sensor data are sent through RS232C wireless communication. In all systems, fuzzy reasoning and fuzzy algorithms are described with our original fuzzy shell FRASH (Fuzzy Real-time Auto-tuning SHell)[5] which has the inference engine and the online rule/frame editor in the library format, and the offline rule/frame editor and the real-time inference display process using multiwindows.

Fig.11 shows the appearance of this experiment. We performed experiments about autonomous locomotive function of our robot. A white curved line shows the tracking route of the robot. In the experiment whose environment is equal to the simulation, our robot exhibited almost similar motion to the simulation results in a previous section. Therefore, the utility of this method was proved by this practical experiment with an autonomous mobile robot.

8: Conclusion

We proposed a construction method of behavior-decision system using fuzzy algorithms capable of expressing sequence flow which includes a mixture of both crisp and fuzzy processing. The main features of this method are as follows.
1) Allows flexible and macro knowledge representation including the behavior-decision sequence through the utilization of fuzzy algorithms.
2) Allows the definition of fuzzy concepts expressed by using fuzzy-based frames and fuzzy states expressed by modified fuzzy algorithms.
3) Possesses algorithm tuning capability based on the threshold value controlled by fuzzy purpose.

Some results of computer simulations supposed a simple in-door environment and experiments using a real autonomous mobile robot were also shown to prove the effectiveness of this method proposed in this paper.

References

[1] Y.Maeda, M.Tanabe, M.Yuta and T.Takagi, "Control Purpose Oriented Behavior-Decision Fuzzy Algorithm with Tuning Function of Fuzzy Branch", *International Fuzzy Engineering Symposium '91*, pp.694-705 (1991)

[2] E.S.Santos, "Fuzzy Algorithm", *INFORMATION AND CONTROL 17*, pp.326-339 (1970)

[3] L.A.Zadeh, "Outline of a New Approach to the Analysis of Complex Systems and Decision Process", *IEEE Trans. on SMC*, Vol. SMC-3, No.1, pp.28-44 (1973)

[4] M.Sugeno, T.Murobushi and et. al, "Fuzzy Algorithmic Control of a Model Car by Oral Instructions", *Fuzzy Sets and Systems 32*, pp.207-219 (1989)

[5] Y.Maeda, J.Murakami and T.Takagi, "FRASH - A Fuzzy Real-Time Auto-tuning SHell with Expressive Function of Fuzzy Algorithms", *SICICI'92* (in Preparation)

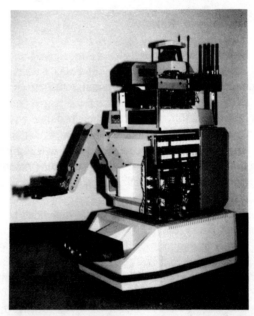

Fig.10 Configuration of the Mobile Robot

Fig.11 Robot Trajectory in the Experiment

Fuzzy Navigation of a Mobile Robot

Kai-Tai Song and Jen-Chau Tai
Institute of Control Engineering
National Chiao Tung University
1001 Ta Hsueh Road
Hsinchu, TAIWAN 300, R.O.C.

Abstract—A navigation system based on fuzzy logic controllers is developed for a mobile robot in an unknown environment. The structure of this fuzzy navigation system features the combination of sensor system, fuzzy controllers for motion planning and the motion control system for real-time execution. Six ultrasonic sensors on-board the mobile robot are used for distance measurement to the immediate obstacles. Sensor data are fuzzified to be the inputs of the fuzzy controller. Three states, each with five quantized levels are used to define the fuzzy set. Two fuzzy controllers are designed to handle the navigation problem. Each fuzzy controller, which corresponds to the turn right or turn left condition, has four inputs, two outputs and 81 rules. The outputs are the command velocities to the left and right wheels, which drive the mobile robot. These command velocities are sent to the lower level motion control system. The performance of this navigation system is tested by computer simulation. Satisfactory results have been obtained and are presented in this paper.

1 INTRODUCTION

The analysis of fuzzy systems is based on the theory of fuzzy sets [1]. Mamdani and others used it to develop controllers [2,3,4]. It is noticed that fuzzy logic controllers (FLCs) have been designed in many fields, such as water quality control, automatic train operation systems, elevator control, nuclear reactor control, automobile transmission control, automatic washing machine, vacuum cleaner, camera and many others [1,5]. In many cases fuzzy logic controllers perform better than conventional controllers. This is mainly because FLC can handle ill-defined processes, which in many cases can be controlled by a skillful human operator who normally never knows the dynamics of the plant.

Applying to the mobile robot, the capability of fuzzy navigation is concerned. It is a difficult task for a mobile robot to navigate through an unknown or unstructured environment. In a local unknown region, there exist some obstacles for a mobile robot. Traveling in such a region, the robot must not collide with any obstacle to reach the target. Therefore the mobile robot must be equipped with sensors to detect obstacles. But on the one hand, the sensors or sensor systems are still far from perfect for understanding the environment taking into account the present day technologies. On the other hand the sensory information is too complicated to explain the conditions to be deal with by the mobile robot. Therefore, special algorithms have to be provided to fuse sensor data for decision making. In this paper, fuzzy logic is applied to solve this problem.

Our experimental mobile robot has two independent driving wheels. It does not have the steering mechanism. The wheels are driven by DC servo motors. If the velocity of right wheel is greater than that of the left wheel, it turns to the left. If it wants to turn right, it must make the velocity of left wheel greater than that of the right wheel. To move along a straight line, the left and right velocities are kept the same. The motion control system of this vehicle is described in [6]. The mobile robot is equipped with six ultrasonic sensors to detect the environment. They are installed in pairs in the front side, left side, and right side of the vehicle. A PC-AT compatible is put on-board the robot to control the wheel velocities according to the sensory information. The mechanical data of the mobile robot is shown in Fig. 1. The original idea was to design an experimental vehicle to work in a factory environment. Its dimension is 74cm by 100cm. The distance between two driving wheels is 67.5cm. The radius of the driving wheels is 11cm. Two special chips from HP, HCTL-1100 motion controller are used to servo the wheel motors. If we send command 1 to the HCTL-1100 then the velocity of wheel is 2.318 cm/sec. If we send n to HCTL-1100

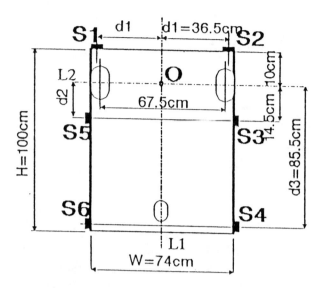

■: Ultrasonic transducer

◯: Driving wheel

○: Free wheel

Fig. 1. Mechanical dimensions of the mobile robot

then the velocity of wheel is $n*2.318$ cm/sec. Point O is the center of the mobile robot, it is in the center of two driving wheels. S1, S2, S3, S4, S5 and S6 represent sensors. S1 and S2 are installed on the front side to *see* the front environment. The distances detected are H1 and H2 respectively. S3 and S4 are installed on the right side of the mobile robot to *see* the right environment. The distances detected are R1 and R2 respectively. S5 and S6 are installed on the left side to *see* the left environment. The distances detected are L1 and L2 respectively. Fig. 2 shows the arrangement of this ultrasonic sensor system.

The characteristics of ultrasonic sensors are summarized as follows. When a sensor starts to measure the distance to an obstacle, it sends an ultrasonic wave, and counts the time elapsed until it detects a reflected echo wave. The distance is calculated according to the time counted. The beam opening angle of the ultrasonic wave is 22 degrees, so it can detect a cone-shaped region. The sensor needs some time for excitation and signal processing, so the distance measurement has some limitations. One set of measuring the environment is completed every 180ms and the minimum distance detectable is 30cm. If the distance is less 30cm, the sensor *see* it as 30cm.

Because the environment is ill defined, the mobile robot navigates practically in an unknown region. In

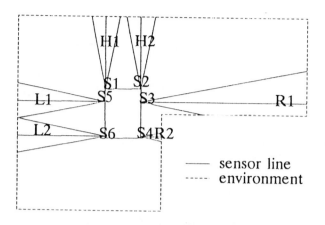

Fig. 2. Arrangement of ultrasonic sensors

other words, there exists unknown obstacles. FLC is used here to interpret the environment and control the navigation of the mobile robot. In the following, section 2 introduces the design of the Fuzzy Logic Controller. Section 3 describes the architecture of the fuzzy navigation system. Section 4 shows several simulations to demonstrate the algorithms. Section 5 gives general conclusion and discussions.

2 Fuzzy Logic Controller Design

The environment for a mobile robot to navigate is ill defined and too complex to analyze. This can be solved, however, by applying the fuzzy logic control theory. The FLC controls the velocities of the left and right wheels to avoid obstacles. But there are six sensors, which are too many to formulate the fuzzy rules. We break them into two group, namely the (S1, S2, S3, S4) and (S1, S2, S5, S6) respectively, to build two FLCs. When the vehicle turns to the right side, it uses (S1, S2, S3, S4) or RFLC (right fuzzy logic controller) to *see* the front and right environment and make desired command outputs for avoiding to collide with the obstacles. When the vehicle turns to the left side, it use (S1, S2, S5, S6) or LFLC (left fuzzy logic controller) to *see* the front and left environment and make desired command outputs for avoiding the obstacles.

Unfortunately, the number of sensors are still too many and make it very complex to formulate the rules, so we use only three linguistic states (*SMALL, MIDDLE, BIG*) to reduce the complexity of the FLCs. Every linguistic state has five quantized level (0,1,2,3,4), their membership function is shown in Fig. 3.

The inputs of FLC are the detected distances to the obstacles, and the outputs are the command velocities of HCTL-1100 to the wheel motor control. The command velocity to the left wheel is V_{dl} and to the right wheel

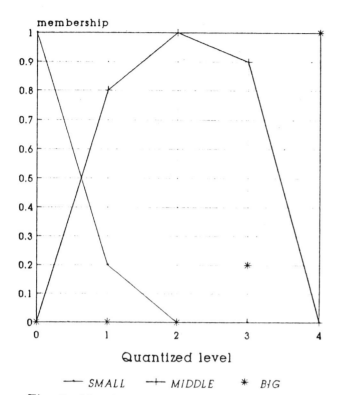

Fig. 3. Membership functions for the fuzzy set

Group 1: (H1,H2)

H1 and H2	Quantized level of H1 and H2
0 – 69cm	0
70 – 129cm	1
130 – 139cm	2
140 – 169cm	3
169 – 1000cm	4

Group 2: (L1,L2,R1,R2)

R1,R2,L1,L2	Quantized level of R1,R2,L1,L2
0 – 49cm	0
50 – 67cm	1
68 – 71cm	2
72 – 92cm	3
93 – 1000cm	4

The Height Method is used to create a decision table according to the rules. Then the quantized levels of output are found from the quantized level of inputs according to decision table. The rules used are of the following type:

IF LH1 is BIG, LH2 is BIG, LR1 is BIG, LR2 is BIG
THEN IL is BIG AND IR is SMALL.
IF LH1 is BIG, LH2 is BIG, LR1 is BIG, LR2 is MIDDLE
THEN IL is BIG AND IR is SMALL.
IF LH1 is BIG, LH2 is BIG, LR1 is BIG, LR2 is SMALL
THEN IL is BIG AND IR is SMALL.
...
...

And the decision table are of the following style:

inputs				outputs	
LH1	LH2	LR1	LR2	IL	IR
4	4	4	0	4	0
4	4	4	1	4	0
4	4	4	2	4	0
...	...	...	...	...	...

is V_{dr}. The commands are numbers of 1 to 5, which are chosen by taking consideration of odometry and the turning features of the vehicle. The IL and IR represent the quantized level of command velocities.

Sensor group (S1,S2,S3,S4) detect the distances (H1, H2, R1, R2). (H1, H2, R1, R2) are quantized to (LH2,LH1,LR1,LR2) by the quantization interface. The above mentioned fuzzy sets (*SMALL, MIDDLE, BIG*) and four variables are used to build 81 rules to make the mobile robot turn to the right and not collide with the obstacles. Then the decision table from 81 rules (by Height method) is constructed. This controller is called Right Fuzzy Logic Controller (RFLC). The decision table contains 625 combinations.

On the other hand, sensor group (S1,S2,S5,S6) detect the distances of (H1, H2, L1, L2). Similarly, (H1, H2, L1, L2) are quantized to (LH2,LH1,LL1,LL2) by the quantization interface. And the fuzzy sets (*SMALL, MIDDLE, BIG*) and four variables are used to build 81 rules to make the vehicle to turn to left and not collide with the obstacles. This controller is called Left Fuzzy Logic Controller (LFLC). The decision table contains also 625 combinations.

Because the distance which the sensors detect is from 30cm to 1000cm, it must be quantized firstly. The quantization length is divided into two groups–(H1, H2) and (R1,R2,L1,L2).

The purpose of the algorithm is to drive the mobile robot to aim at the target. According to the location of the target, the mobile robot decides to turn left, right or move straight forward. Normally, the vehicle turns to left if the target is on the left side of its heading. It turns to the right if the target is on the right side of its heading. It moves straight forward if the target is in the front. Therefore LFLC or RFLC is used according to

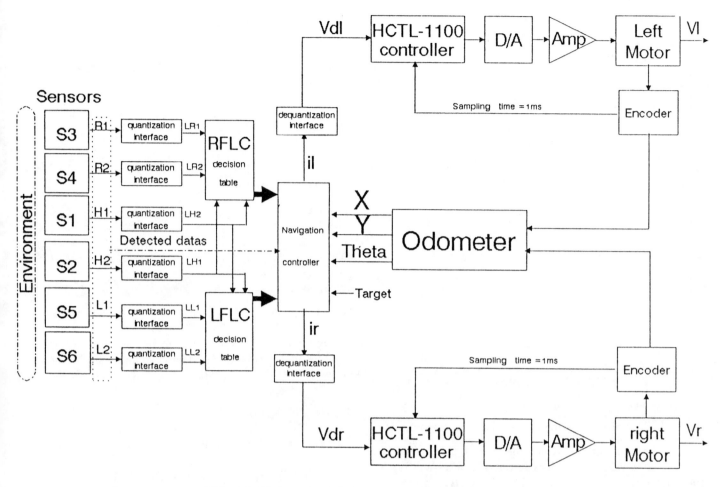

Fig. 4. Structure of the fuzzy navigation system

the location of the target is at the left or right side of it. If the target is in front of it, either LFLC or RFLC is used to navigate, depending on the condition that the left obstacle is *nearer* or the right obstacles is *nearer*. The mobile robot will move along its path if there is no near obstacle.

3 System Architecture

The fuzzy navigation system is shown in Fig. 4. The mobile robot is equipped with six ultrasonic sensors S1,S2,...,S6 to detect the environment. The detected data are quantified by the fuzzy quantifying interface. The FLC decision table is created in advance by off-line calculation. The navigation controller will check the current incoming information, which includes sensor data, target location and the location of the mobile robot, to decide which method should be used. Here an FLC, non-FLC or some virtual concept described latter can be selected. The outputs are the quantified command to the HCTL-1100. An odometer is provided to estimate the position and direction of the vehicle according the shaft encoder pulses.

3.1 Fuzzy v.s. Non-Fuzzy Controllers

If the mobile robot uses FLCs to travel through a corridor to attain a target somewhere else, it may give poor performance. This is mainly because only three linguistic terms are used, which gives rather poor interpretation of the immediate environment of the mobile robot. For instance, when travelling in a narrow corridor or following a wall, the navigation is like a snake-shaped path. It also collides with obstacles in some difficult situations. To cope with this, other non-fuzzy controllers are added to improve or compensate the weakness of the FLCs. These non-fuzzy controllers are put into function under special conditions; such as, if FLC decides to turn left, but at the right side of the vehicle there doesn't exist enough space for it to turn, then the non-fuzzy controller will take place of the FLCs to improve the performance. The switching between the fuzzy and non-fuzzy controllers occurs automatically and the mobile robot works well by the combination of FLC and non-FLC controllers.

3.2 Use of Virtual Wall

On the other hand, there exist some conditions that the mobile robot will collide with obstacles in the front side. This happens, for instance, when an obstacle and the target are both in the front of the mobile robot but with the target behind the obstacle. The first action of the vehicle will turn to the left, and next time it will find the target is on the right side, it will turn to right again, and mobile robot would not escape from this condition easily and travels like a swing. Fig. 5 shows this situation. The concept of introducing a virtual wall can solve this problem. The virtual wall will guide the vehicle to turn to the desired direction and won't turn back again.

Fig. 6 shows the concept of the virtual wall. The virtual wall is created by the navigation controller. It is a line located at the left or right side of the mobile robot. At the moment of creation, it is parallel with the heading of the mobile robot. The sensor-line at the position of the transducer is introduced for the convenience of calculating the virtual distance. The navigation controller finds the intersection of the virtual wall and the sensor-line. The virtual distance is the distance between the transducer and the intersection. If the virtual distance is smaller than the actual measured distance, the detected distance is replaced by the virtual distance and will be used by the FLCs. The mobile robot will turn to the desired direction by avoiding collision with the virtual wall.

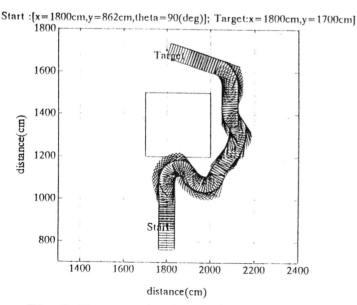

Fig. 5. Fuzzy navigation without virtual wall

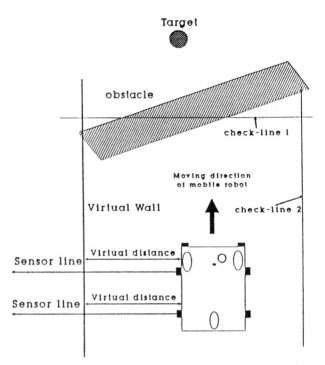

Fig. 6. Concept of the virtual wall

While the virtual wall guides the mobile robot to turn to the desired direction, it also restrict the normal behaviour of the mobile robot. Check-lines are added to facilitate the disabling the virtual wall. Some conditions are set to disable the functioning of the virtual wall. Firstly, as the real distance is smaller than the virtual distance, the virtual wall is disabled. This is because under this situation the real measured distance is small enough to make the the vehicle turn. Secondly, when the center point of the mobile robot is near the check-lines, the virtual wall will be disabled. This is because the target is no longer in the front side of the mobile robot. Fig. 7 shows the navigation result with the virtual wall.

3.3 Use of Virtual Target

A collision may happen when the mobile robot is passing the corner of an obstacle. Fig. 8 shows this situation. When the difference of H1 and H2 is large (e.g. bigger than 200cm), the mobile robot is in the corner side of an obstacle. If it attends to turn left, it will find H1 small; so it will turn to the right, then it finds H1 big enough to turn to left; the mobile robot will turn to left again and finally will collide with the obstacle. The method of avoiding to collide with corner is to create a virtual target to let the robot escape from the range of corner. The virtual target will guide the vehicle to pass the corner region. If it attains the virtual target, the controller will

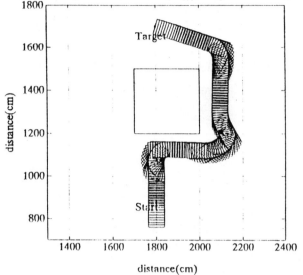

Fig. 7. Fuzzy navigation with virtual wall

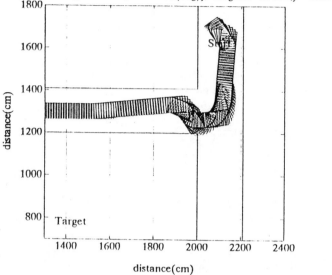

Fig. 8. Fuzzy navigation without virtual target

disable the effect of virtual target, because the environment is safe to navigate and virtual target will influence the controller to make wrong decision. Some conditions are defined to disable the effect of virtual target. the effects of adding virtual targets can be shown in Fig. 9.

4 Computer Simulation and Simulation Results

A simulation environment has been created to check the performance of the fuzzy navigation controller. The environment is composed of lines of fixed length. Sensors are stimulated by three lines as shown in Fig. 2 to represent the ultrasonic beam opening angle. The intersection of these sensor lines and environment can be calculated, and the distance between the vehicle position and the intersection can be determined. Sensor features such as the real sensor detect the minimum distance of 30 cm are put into this simulator. The sensor data are updated every 180 ms. Even this never represents the characteristics of real sensor completely, the simulation can still very useful for future implementation the fuzzy controllers to the real mobile robot. However, the dynamics of the vehicle is assumed to be perfect. The inertia of the actual wheels is neglected, that is, if we make the wheel move in some velocity there is no transition state to change from its original velocity.

Several typical environments are created to test the controller. Firstly in Fig. 10 is an aisle with a 135 degree corner. The vehicle must rotate 45 degrees to pass this

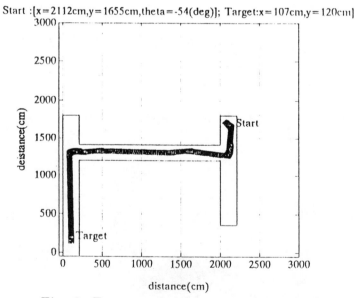

Fig. 9. Fuzzy navigation with virtual target

corner. In this 135 degree corner it doesn't have much space to rotate, but the mobile robot overcomes this problem. It passes the corner and attain the target. For comparison of the capability of obstacle avoidance, a special environment is created to run a test. From Fig. 11 and Fig. 12, it is clear that the fuzzy navigation controller gives acceptable results.

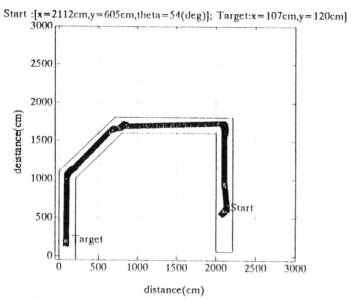

Fig. 10. Simulation result of passing corners

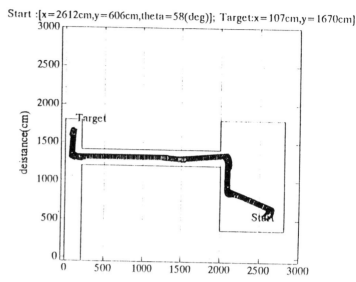

Fig. 11. Fuzzy navigation without an unexpected obstacle

5 Conclusion and Discussion

From the above simulations, it can be conclude that with a combination of the fuzzy and non-fuzzy controllers, a fuzzy navigation can be achieved based on the sensory information. Although the environment is too complex to interpret by the sensory information, FLC can handle this complexity. Even with poor sensors the fuzzy controller can still work satisfactorily. On the other hand, if more input variables, more linguistic states and more quantized level are used to interpret the navigation conditions, it is believed the non-FLC can be eliminated in the algorithm. However, it will be difficult to build such a huge system. This navigation system is to be implemented on our experimental mobile robot, the real-time performance such as obstacle avoidance in an unknown environment is expected.

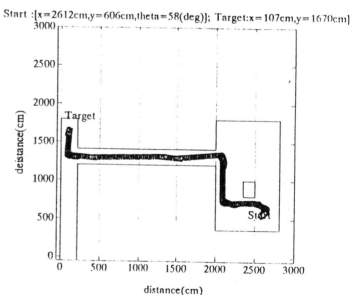

Fig. 12. Fuzzy navigation with an unexpected obstacle

References

[1] L. A. Zadeh, "Making Computers Think Like People", IEEE Spectrum, P.26-32, Aug. 1984.

[2] E. H. Mandani, "Application of Fuzzy Algorithms for Simple Dynamic Plant", Proc. IEEE, Vol. 121, no. 12, P. 1585-1588, 1974.

[3] E. H. Mamdani, "Development in Fuzzy Logic Control", Proceedings of 23rd Conference on Decision and Control, P. 888-893, 1984.

[4] Y. F. Li and C. C. Lau, "Development of Fuzzy Algorithms for Servo Systems", IEEE Control System Magazine, P. 65-71, 1989.

[5] J. A. Bernard, "Use of Rule-based System for Process Control", IEEE Control System Magazine, Vol.8, no.5, pp.3-13, 1988.

[6] C. E. Li, "Motion Control of Mobile Robot", Master Thesis, NCTU, Hsinchu, Taiwan, 1991.

Blending Reactivity and Goal-Directedness in a Fuzzy Controller

Alessandro Saffiotti*   Enrique H. Ruspini   Kurt Konolige

Artificial Intelligence Center, SRI International
Menlo Park, CA 94025, U.S.A.
saffiotti@ai.sri.com

Abstract— **Controlling the movement of an autonomous mobile robot requires the ability to pursue strategic goals in a highly reactive way. We describe a fuzzy controller for such a mobile robot that can take abstract goals into consideration. Through the use of fuzzy logic, reactive behavior (e.g., avoiding obstacles on the way) and goal-oriented behavior (e.g., trying to reach a given location) are smoothly blended into one sequence of control actions. The fuzzy controller has been implemented on the SRI robot Flakey.**

I. Introduction

Autonomous operation of a mobile robot in a real environment poses a series of problems. In the general case, knowledge of the environment is partial and approximate; sensing is noisy; the dynamics of the environment can only be partially predicted; and robot's hardware execution is not completely reliable. Though, the robot must take decisions and execute actions at the time-scale of the environment. Classical planning approaches have been criticized for not being able to adequately cope with this situation, and a number of reactive approaches to robot control have been proposed (e.g., [Firby, 1987; Kaelbling, 1987; Gat, 1991]), including the use of fuzzy control techniques (e.g., [Sugeno and Nishida, 1985; Yen and Pfluger, 1992]). Reactivity provides immediate response to unpredicted environmental situations by giving up the idea of reasoning about future consequences of actions. Reasoning about future consequences (sometimes called "strategic planning"), however, is still needed in order to intelligently solve complex tasks (e.g., by deciding not to carry an oil lantern downstairs to look for a gas leak [Firby, 1987].)

One solution to the dual need for strategic planning and reactivity is to adopt a two-level model: at the upper level, a planner decides a sequence of abstract goals to be achieved, based on the available knowledge; at the lower level, a reactive controller achieves these goals while dealing with the environmental contingencies. This solution requires that the reactive controller be able to simultaneously satisfy strategic goals coming from the planner (e.g., going to the end of the corridor), and low-level "innate" goals (e.g., avoiding obstacles on the way). A major problem in the design of such a controller is how to resolve conflicts between simultaneous goals.

In this paper, we describe a reactive controller for an autonomous mobile robot that uses fuzzy logic for trading off conflicting goals. This controller has been implemented on the SRI robot Flakey, and its performance demonstrated at the first AAAI robot competition, where Flakey finished second [Congdon et al., 1993]. The formal bases for the proposed controller have been set forth by Ruspini [Ruspini, 1990; Ruspini and Ruspini, 1991; Ruspini, 1991a] after the seminal works by Zadeh (e.g., [Zadeh, 1978]). In a nutshell, each goal is associated with a function that maps each perceived situation to a measure of desirability of possible actions from the point of view of that goal. The notion of a *control structure* is used for introducing high-level goals into the fuzzy controller. Intuitively, a control structure is an object in the robot's workspace, together with a desirability relation: typical control structures are locations to reach, walls to follow, doors to enter, and so on. Each desirability function induces a particular behavior — one obtained by executing the actions with higher desirability. Behaviors induced by many simultaneous goals can be smoothly blended by using the mechanisms of fuzzy logic. In particular, reactive and goal-oriented behaviors are blended in this way into one sequence of control actions.

The next section gives a brief overview of Flakey. Section III sketches the architecture of the controller, and describes the way behaviors are implemented, and how they are blended together. Section IV deals with the introduction of high-level goals into the reactive controller. Section V discusses the results, and concludes.

---

*On leave from Iridia, Université Libre de Bruxelles, Brussels, Belgium.

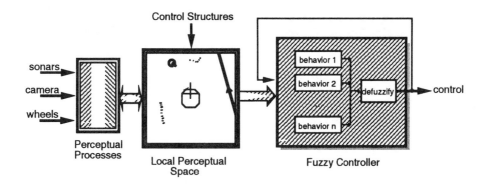

Figure 1: Architecture of Flakey (partial)

II. THE MOBILE ROBOT TEST-BED

Flakey is a custom-built mobile robot platform approximately 1 meter high and .6 meter in diameter for use in an indoor environment. There are two independently-driven wheels, one on each side, giving a maximum linear velocity of about .5 meters/sec. Flakey sensors include a ring of 12 sonars, giving information about distances of objects up to about 2 meters; wheel encoders, providing information about current linear and rotational velocity; and a video camera, currently used in combination with a laser to provide dense depth information over a small area in front of Flakey. On-board computers are dedicated to low-level sensor interpretation, motor control, and radio communication with an off-board Sparc station. Though it is possible to run the high level interpretation and control processes on board, they are normally run remotely for programming convenience.

Figure 1 illustrates the part of Flakey's architecture that is relevant to the controller. The sensorial input is processed by a number of interpretation processes at different levels of abstraction and complexity, and the results of interpretation are stored in the *local perceptual space* (LPS). The LPS represents a Cartesian plane centered on Flakey where all the information about the vicinity of Flakey is registered. In Figure 1, points corresponding to surfaces identified by the sonars and the camera are visible in the LPS — Flakey is the the octagon in the middle of the LPS, in top-view. The other objects in the LPS are "artifacts" associated to *control structures*, and are discussed in Section V. The content of the LPS constitutes the input to the controller: this checks its input and generates a control action every 100 milliseconds.

III. REACTIVE FUZZY CONTROLLER

The fuzzy controller is centered on the notion of *behavior*. Intuitively, a behavior is one particular control regime that focuses on achieving one specific, predetermined goal (e.g., avoiding obstacles). Hence, we can think of a behavior as a mapping from configurations in the LPS to actions to perform. More precisely, and following [Ruspini, 1991b], we say that each behavior B is associated with a desirability function

$$Des_B : \text{LPS} \times \text{Control} \to [0, 1]$$

that measures, for each configuration s of the LPS and value c of a control variable, the desirability $Des_B(s, c)$ of applying control values c in the situation s *from the point of view of B*. Equivalently, we can say that Des_B associates each situation s with the fuzzy set $\widetilde{C}$ of control values characterized by the membership function $\mu_{\widetilde{C}}(c) = Des_B(s, c)$. Notice that in general, c is a n-dimensional vector of values for all the control variables; in the case of Flakey, the control variables include linear acceleration and turning angle.

In practice, each behavior is implemented by a fuzzy machine structured as shown in Figure 2. The *fuzzy state* is a vector of fuzzy variables (each having a value in $[0, 1]$) representing the truth values of a set of fuzzy propositions of interest (e.g., "obstacle-close-on-left"). At every

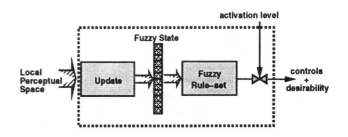

Figure 2: Implementation of a behavior.

cycle, the **Update** module look at the (partially) interpreted perceptual input stored in LPS, and produces a new fuzzy state. The **Fuzzy Rule-Set** module contains a set of fuzzy rules of the form "If A then c" where A is a fuzzy expression composed by predicates in the fuzzy states plus the fuzzy connectives AND, OR and NOT; and c is a vector of values for the control variables. Max, min, and complement to 1 are used to compute the truth value of disjunction, conjunction and negation, respectively. An example of a control rule is:

```
IF obstacle-close-in-front
AND NOT obstacle-close-on-left
THEN turn -6 degrees
```

Each "If A then c" rule computes the degree of desirability of applying control value c as a function of the degree at which the current state happens to be similar to A. The outputs of all the rules in a rule-set are unioned using the max T-conorm: the function computed in this way is meant to provide an approximation of the Des_B function above.[1] This desirability function is fed to the **Defuzzify** module for computing one single control value. We presently do defuzzification according to the centroid approach: the resulting control value is given by

$$\frac{\int c\, Des_B(c)\, dc}{\int Des_B(c)\, dc}.$$

As shown in Figure 1 above, many behaviors can be simultaneously active in the controller, each aiming at one particular goal — e.g., one for avoiding obstacles; one for keeping a constant speed; one for heading toward a beacon; etc. Correspondingly, many instances of the fuzzy machine depicted in Figure 2 simultaneously run in the controller, each one implementing one behavior's desirability function. All these desirability functions are merged into a composite one by the max T-norm; the defuzzification module converts the resulting tradeoff desirabilities into one crisp control decision. Care must be taken, however, of possible conflicts among behaviors aiming at different, incompatible goals. These conflicts would result in desirability functions that assign high values to opposite actions: simple T-norm composition should not be applied in these cases. The key observation here is that each behavior has in general its own *context* of applicability. Correspondingly, we would like that the impact of the control actions suggested by each behavior be weighted according to that behavior's degree of applicability to the current situation. For instance, the actions proposed by the obstacle avoidance behavior should receive higher priority when there is a danger of collision, at the expense of the other, concurrent behaviors. In order to do this, the output of each rule-set is discarded by the value of the corresponding *activation level*: typically, the activation level is represented by some variable in the fuzzy state. This corresponds to arbitrate the relative dominance of different behaviors by a set of meta-rules of the form

$$\text{IF } A' \text{ THEN activate\_behavior } B \qquad (1)$$

where A' is a LPS configuration. Notice that this solution is formally equivalent to transforming each rule "If A then c" in B into a rule "If A' and A then c" (see [Berenji et al., 1990] for a similar approach to conflict resolution.)

As an example, consider the way Flakey "wanders" around. In the wandering mode, three behaviors coexist in the controller: AVOID-OBSTACLES, AVOID-COLLISIONS and GO-FORWARD. GO-FORWARD just keeps Flakey going at a fixed velocity, given as a parameter. AVOID-OBSTACLES looks at the last 5 seconds' sonar readings in the LPS, and guides Flakey away form occupied areas. AVOID-COLLISIONS looks at the nearest sonar readings and proposes drastic actions (immediate stop and turn) when a serious risk of collision is detected. The activation levels of AVOID-OBSTACLES and AVOID-COLLISIONS are given by the fuzzy state variable "approaching-obstacle"; the complement of this value gives the activation level for GO-FORWARD. The visual result for an external observer is that Flakey "follows its nose", while smoothly turning away from obstacles as it approaches them.

IV. BEYOND PURE REACTIVITY

The behaviors discussed in the previous section are purely reactive: at each cycle, Flakey selects an action solely on the basis of the current state of the world as perceived by its sensors and represented in the local perceptual space. Engaging into more purposeful activities than just wandering around requires more than pure reactivity: we need to take explicit goals into consideration. For example, we may want Flakey to reach a given position at a given velocity, and still (reactively) avoid the obstacles on the way.[2]

In our approach, a goal is represented by a *control structure*. Intuitively, a control structure is virtual object (an *artifact*) that we put in the LPS, associated with a behavior that encodes the way to react to the presence of this object. For example, a "control-point" is a marker for a (x, y) location, together with a heading and a velocity: the associated behavior GO-TO-CP reacts to the presence of a control point in the LPS by generating the commands to reach that position, heading and velocity. In Figure 1 there are two artifacts: a control point to reach (left), and a wall to follow (right).

---

[1] See [Ruspini, 1991a; Ruspini, 1991b] for an account of fuzzy logic and fuzzy control in terms of similarity and desirability measures, and the use of T-norms and T-conorms in this context.

[2] Reactive behaviors are also associated with (innate) goals, hard-wired in the definition of the behavior. We are now interested in dynamically assigning specific strategic goals to Flakey.

Figure 3: Path families generated by actions with increasing values of Des_{FS}.

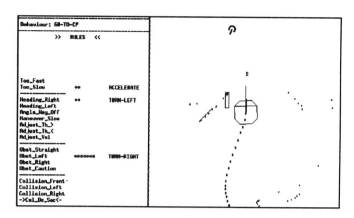

Figure 4: A snapshot of Flakey's control window while achieving a control point.

More precisely, a control structure is a pair

$$S = \langle Q_S, R_S \rangle,$$

where Q_S is an artifact, and R_S is a fuzzy relation between the position of Flakey and that of the artifact.[3] Such a control structure implicitly defines a goal: the goal to achieve, and maintain, the given relation between Flakey and the artifact Q_S. Intuitively, if Q_S is at position q, $R_S(q, p)$ says how much a position p of Flakey satisfies this goal. If the position of Flakey is such that $R_S(q, p) = \alpha$, we say that the control structure S is satisfied to the degree α.

The R_S relation induces a desirability function Des_S in the following way. Given the set P of possible positions of Flakey, and the set C of possible control values, let $\text{Exec}(p, c)$ denote the new position reached by applying control c from position p. Then, the desirability *from the viewpoint of the control structure S* of executing c when Flakey is at position p (and Q_S is at q) is given by

$$Des_S(q, p, c) = R_S(q, \text{Exec}(p, c))$$

However, not all positions are equally reachable by Flakey: moving to certain positions will require more effort (changes in velocity and/or direction, time, etc.) than moving to others. To account for this, we consider a second desirability function Des_F: $Des_F(p, c)$ measures the desirability *from the viewpoint of Flakey's motion capabilities* of executing control action c when Flakey is at position p. The desirability of control actions from the joint viewpoints of feasibility for Flakey, and effectiveness with respect to the control structure S, is measured by the combination $Des_S \otimes Des_F$ (where $\otimes$ is a T-norm). Figure 3 illustrates one such combined desirability function:

here, S is a control point, represented by the semi-circle near the top (the "tail" indicates the desired entry orientation.) The fading from black to white illustrates the increase in the value of $Des_S \otimes Des_F$ for some families of possible paths.

We have already seen how Des_S can induce, for each LPS configuration, a fuzzy set of possible controls. As we did in the case of reactive behaviors, we approximate this fuzzy set using rules on the form "If A then c". The only difference is that A now refers to artifacts rather than to sensorial input.[4] We have designed sets of rules for many "purposeful" behaviors, including going to a x, y position; achieving a control point; following a wall; crossing a door; and so forth. Each ruleset consists of a small number (four to eight) of rules. Purposeful behaviors can coexist with other behaviors, either purposeful and reactive: the context-dependent blending of behaviors explained above provides arbitration and guarantees the smooth integration of directed activities and reactivity.

Figure 4 exemplifies the performance of the integration. The picture shows Flakey's control window during an actual run: on the right is Flakey's local perceptual space. Flakey sits in the middle of the window, pointing upwards; the small points all around mark sonar readings, indicating the possible presence of some object; the rectangle on the left of Flakey highlight a dangerously close object. The window on the left lists all the currently active rules, grouped into rule-sets: topmost, the rules for the GO-FORWARD behavior; below, those for GO-TO-CP, for AVOID-OBSTACLES, and for AVOID-COLLISIONS. In the shown situation, Flakey is going too slow and heading right of the CP: hence, some desirability is given to the

---

[3]Positions are actually points in a (x, y, θ, v) 4-D space.

[4]Alternatively, these rules can be thought as responding to input from a "virtual sensor" that senses the position of an artifact.

accelerate and the *turn-left* actions. However, the close obstacle on the left causes the activation level of the AVOID-OBSTACLES behavior to be high, at the expenses of the other behaviors; hence, the *turn-right* action suggested by AVOID-OBSTACLES receives high total desirability (as indicated by the 7 stars). The small box in front of Flakey indicates the resulting turning control — some degrees on the right. The overall result of the blending is that Flakey makes its way among obstacles while *en route* to achieving the position and bearing of the given control point. The smoothness of the movement in evident in the wake of small boxes that Flakey left behind it (one box per second). Flakey's speed was between 200 and 300 mm/sec.

One word is worth spent on the problem of local minima, ubiquitous in approaches to robot navigation based on local combination of behaviors [Latombe, 1991]. The problem is illustrated in Figure 5 (top): the robot needs to mediate the tendency to move toward the goal, and the tendency to stay away from the obstacle. A straightforward combination of these two opposite tendencies (whether they are described by desirability measures, potential fields [Khatib, 1986], motor schemas [Arkin, 1990], or other) may result in the production of a zone of local equilibrium (local minimum): when coming from the left edge, the robot would be first attracted and then trapped into this zone. By using meta-rules like the 1 above to reason about the relative importance of goals, our context-dependent blending of behaviors provides a way around this problem. Figure 5 shows the path followed by Flakey in a simulated run (top), and the corresponding activation levels of the KEEP-OFF and REACH behaviors (bottom). In (a), Flakey has perceived the obstacle; as the obstacle becomes nearer, the KEEP-OFF behavior becomes more active, at the expenses of the REACH behavior. In this way, the "attractive power" of the goal is gradually shaded away by the obstacle, and Flakey responds more and more to the obstacle-avoidance suggestions alone. The REACH behavior re-gains importance, however, as soon as Flakey is out of danger (b).

V. CONCLUSIONS

We have defined a mechanism based on fuzzy logic for blending multiple behaviors aimed at achieving different, possibly conflicting goals. Goals are either built-in, as in most fuzzy controllers, or dynamically set from outside the controller. Typically, the built-in goals correspond to reactive behaviors (like avoiding collisions), while the dynamic ones are strategic goals communicated by a planner. Context-dependent blending of behaviors ensures that strategic goals be achieved as much as possible, while maintaining a high reactivity.

Our behavior blending mechanism has been originally inspired to the technique proposed by Berenji et al.

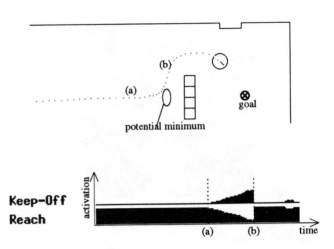

Figure 5: How context-dependent blending of behaviors avoids potential local minima.

[Berenji *et al.*, 1990] for dealing with multiple goals in fuzzy control. There are however two important differences: first, our context mechanism dynamically modifies the degrees of importance of each goal; second, we allow the introduction of high-level, situation-specific goals in the controller.

From another perspective, the work presented here fits in the tradition of the "two level" approaches to robot control, where a strategic planner is used to generate guidelines to a reactive controller (e.g., [Arkin, 1990; Payton *et al.*, 1990; Gat, 1991]). In our case, a plan consists in a sequence of control structures. For example, a plan to exit building E could consist in three successive *corridors* to follow, one *control point* in the entrance hall close to the door, and the exit *door* itself. The context of applicability of each control structure is used to decide when each control structure becomes relevant. (see [Saffiotti *et al.*, 1993; Saffiotti, 1993] for more on this issue). We believe that having based our architecture on fuzzy logic results in improved robustness (e.g., more tolerance to sensor noise and knowledge imprecision), while granting a better understanding of the underlying mechanisms.

Finally, many current approaches to robot control deal with multiple goals using the so-called "potential fields" method [Khatib, 1986]: goals are represented by pseudo-forces, which may be thought of as representatives of most desirable behavior from that goal's viewpoint. These optimal forces are then combined, as physical vectors, to produce a resultant force that summarizes their joint effect. In our approach, by contrast, the goals' desirability functions, rather than a summary description, are combined into a joint desirability function, from which a most desired tradeoff control is extracted. Moreover, this combination takes behaviors' context of applicability into account; this provides a key to eliminate the local minima arising from the combination of conflicting goals.

The technique proposed in this paper has been implemented in the SRI mobile robot Flakey, resulting in extremely smooth and reliable movement. The performance of Flakey's controller has been demonstrated at the first AAAI robot competition in San Jose, CA [Congdon et al., 1993]. Flakey accomplished all the given tasks while smoothly getting around obstacles (whose positions were not known beforehand) and people, and placed second behind Michigan University's CARMEL. Flakey's reliable reactivity is best summarized in one judge's comment: "Only robot I felt I could sit or lie down in front of." (What he actually did!)

Acknowledgments John Lowrance, Daniela Musto, Karen Myers and Leonard Wesley contributed to the development of the ideas presented in this paper. Nicolas Helft implemented a first version of Flakey's controller.

The first author has been supported by a grant from the National Council of Research of Italy. Research performed by the second author leading to the conceptual structures used in the autonomous mobile vehicle controller was supported by the U.S. Air Force Office of Scientific Research under Contract No. F49620-91-C-0060. Support for Kurt Konolige came partially from ONR Contract No. N00014-89-C-0095. Additional support was provided by SRI International.

References

[Arkin, 1990] Arkin, Ronald C. 1990. The impact of cybernetics on the design of a mobile robot system: a case study. *IEEE Trans. on Systems, Man, and Cybernetics* 20(6):1245-1257.

[Berenji et al., 1990] Berenji, H.; Chen, Y-Y.; Lee, C-C.; Jang, J-S.; and Murugesan, S. 1990. A hierarchical approach to designing approximate reasoning-based controllers for dynamic physical systems. In *Procs. of the 6th Conf. on Uncertainty in Artificial Intelligence*, Cambridge, MA.

[Congdon et al., 1993] Congdon, C.; Huber, M.; Kortenkamp, D.; Konolige, K.; Myers, K.; and Saffiotti, A. 1993. CARMEL vs. Flakey: A comparison of two winners. *AI Magazine*. To appear.

[Firby, 1987] Firby, J. R. 1987. An investigation into reactive planning in complex domains. In *Procs. of the AAAI Conf.*

[Gat, 1991] Gat, E. 1991. *Reliable Goal-Directed Reactive Control for Real-World Autonomous Mobile Robots.* Ph.D. Dissertation, Virginia Polytechnic Institute and State University.

[Kaelbling, 1987] Kaelbling, L. P. 1987. An architecture for intelligent reactive systems. In Georgeff, M.P. and Lansky, A.L., editors 1987, *Reasoning about Actions and Plans*. Morgan Kaufmann.

[Khatib, 1986] Khatib, O. 1986. Real-time obstacle avoidance for manipulators and mobile robots. *The International Journal of Robotics Research* 5(1):90-98.

[Latombe, 1991] Latombe, J.C. 1991. *Robot Motion Planning*. Kluver Academic Publishers, Boston, MA.

[Payton et al., 1990] Payton, D. W.; Rosenblatt, J. K.; and Keirsey, D. M. 1990. Plan guided reaction. *IEEE Trans. on Systems, Man, and Cybernetics* 20(6).

[Ruspini and Ruspini, 1991] Ruspini, E. H. and Ruspini, D. 1991. Autonomous vehicle motion planning using fuzzy logic. In *Procs. of the IEEE Round Table on Fuzzy and Neural Sys. and Vehicle Appl.*, Tokyo, Japan.

[Ruspini, 1990] Ruspini, E. H. 1990. Fuzzy logic in the Flakey robot. In *Procs. of the Int. Conf. on Fuzzy Logic and Neural Networks (IIZUKA)*, Japan. 767-770.

[Ruspini, 1991a] Ruspini, E. H. 1991a. On the semantics of fuzzy logic. *Int. J. of Approximate Reasoning* 5.

[Ruspini, 1991b] Ruspini, E. H. 1991b. Truth as utility: A conceptual systesis. In *Procs. of the 7th Conf. on Uncertainty in Artificial Intelligence*, Los Angeles, CA.

[Saffiotti et al., 1993] Saffiotti, A.; Konolige, K.; and Ruspini, E. H. 1993. Now, do it! Technical report, SRI Artificial Intelligence Center, Menlo Park, California. Forthcoming.

[Saffiotti, 1993] Saffiotti, A. 1993. Some notes on the integration of planning and reactivity in autonomous mobile robots. In *Procs. of the AAAI Spring Symposium on Foundations of Automatic Planning*, Stanford, CA.

[Sugeno and Nishida, 1985] Sugeno, M. and Nishida, M. 1985. Fuzzy control of a model car. *Fuzzy Sets and Systems* 16:103-113.

[Yen and Pfluger, 1992] Yen, J. and Pfluger, N. 1992. A fuzzy logic based robot navigation system. In *Procs. of the AAAI Fall Symposium on Mobile Robot Navigation*, Boston, MA. 195-199.

[Zadeh, 1978] Zadeh, L.A. 1978. Fuzzy sets as a basis for a theory of possibility. *Fuzzy Sets and Systems* 1:3-28.

Fuzzy Logic Based Robotic Arm Control

Robert N. Lea, Ph.D.
NASA/Johnson Space Center
Houston, Texas 77058

Jeffrey Hoblit
LinCom Corporation
Houston, Texas 77058

Yashvant Jani, Ph.D.
Togai InfraLogic Inc.
Houston, Texas, 77058

Abstract - Fuzzy logic and neural networks have emerged as new methods for decision making and control within the last decade. Non-linear complex systems can be easily controlled utilizing fuzzy logic principles. The dynamical behavior of robotic systems is complex, especially in presence of loads. Robotic joints experience and exhibit friction, stiction and gear backlash effects. Due to lack of proper linearization of these effects, modern control theory based on state space methods can not provide adequate control for robotic systems. Furthermore, inversion of Jacobian matrices, particularly when they are near singular, provide another computational problem especially for redundant joint robotic systems.

At the Johnson Space Center (JSC), we are investigating the feasibility of applying fuzzy logic based control for robotic systems. Functions such as tracking, approach and grapple are typically performed by a robotic arm in a manual mode or semiautomatic mode where a point of resolution is driven to a desired point by kinematic inversion. We are developing fuzzy logic based algorithms for semiautomatic mode so that the computational problem of Jacobian inversion can be eliminated. The difference between the desired location and current location is an input vector to the controller that generates joint rate commands. A six degree-of-freedom robotic arm simulation is used to evaluate the performance of the fuzzy logic based controller. In this paper, development of the control algorithm, simulation testing, results and the performance of the controller are reported.

I. INTRODUCTION

Industrial applications of fuzzy logic and neural networks have increased significantly within the last five years in the U.S. as evidenced by the Industrial Conferences [1, 2] as well as the recent FUZZ-IEEE '92 held at San Diego by the Institute of Electrical and Electronics Engineers [3]. Many of the applications are related to the control of non-linear and complex systems. Of particular interest are the control applications for a robotic arm. The control of planner arms [4,5] with redundant degrees of freedom has been demonstrated in simulation as well as hardware implementations. The control rules used in maintaining the trajectory are simple and do not require the inversion of a Jacobian matrix as in the traditional approach [6]. Advantages in computational speed and accuracy are realized.

Research activities in the Software Technology Laboratory at the JSC include the development of robust algorithms for decision making using fuzzy logic, neural networks, genetic algorithms, Dempster-Shafer theory and fuzzy clustering. We have demonstrated a robust control of the space shuttle relative trajectory during proximity operations [7], tether length control during deployment, on-station and retrieval phases [8], and control of Mars Rover trajectories and collision avoidance [9]. Fuzzy control algorithms utilized in these applications are primarily based on reducing the error and error rate which are related through the systems dynamics and are perceived via proper sensor measurements. This technique of reducing error and error rate was first applied to the attitude control of the space shuttle [10] where angle error and its rate error were reduced to within an acceptable range. Later the same technique was applied to reducing length and length rate errors for the tethered satellite system. Application of the same algorithm to the space shuttle robotic arm is presented here. We first describe the configuration of the Remote Manipulator System (RMS) of the shuttle in section 2 and then the development of the control algorithm for the point of resolution (POR) mode in section 3. The kinematic simulation test cases are then described in section 4 and test results and preliminary conclusions are discussed in section 5. Finally our future work in this area is summarized in section 6. References are provided at the end.

II. ROBOTIC ARM CONFIGURATION

The shuttle RMS is a six rotational joint arm as shown conceptually in fig. 1. The first joint is known as a yaw joint as it rotates the arm around the z-axis of the base frame. The next two joints are called pitch joints because the rotation is through the y-axis of the base frame when the arm is in its rest position. The next three joints are pitch, yaw and roll and correspond to the rotational sequence used in defining the shuttle orientation in on-orbit operations.

The first joint is about 13 inches from the base. The lengths of the six segments between the joints are approximately 23, 251, 278, 18, 22, and 13 inches, where the last length corresponds to the distance between the joint and POR. The state of the robotic arm can be specified in two different ways: 1) six joint angles will accurately describe the orientation of the arm, or 2) position and orientation of the POR in the base frame. The RMS base frame is related to the shuttle vehicle frame through a fixed coordinate transformation matrix. The relationship between the POR frame and the base frame is determined through the six

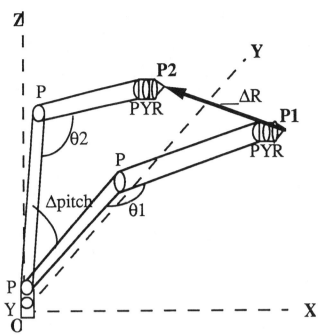

Fig. 1. The Shuttle RMS configuration and motion from point P1 to point P2.

rotations of the joints as well as the distance between the joints. Each joint angle providing the rotation is described in the coordinate frame at the joint. A kinematic forward transformation can be performed on the first state to obtain the position and orientation of the POR in the base frame. The position and orientation of the POR are related to joint angles via a Jacobian as follows.

$$[x, \theta]^T = J(\gamma) [\gamma]^T \quad (1)$$

where, x = position of the POR in base frame, θ = orientation of the POR, γ = joint angles, and J(γ) = Jacobian which is dependent on the joint angles only. Typically, sensors measure the joint angles and rates to provide proper feedback to the controller. Once the measurements are received, the motion is easily specified in terms of position and orientation of the POR in the base frame using forward kinematics. A camera mounted at the end of the arm can provide some inputs for the POR position in the base frame.

The objective of this study is to specify the desired position and orientation of the POR in the base frame, and let the controller manipulate the joint angles and rates to achieve that state. A path planner can easily work in the base frame, while the controller can work on the joint angle and rate. For a robotic arm, the Cartesian velocities of the POR are related to the joint velocities via the Jacobian matrix which is a non-linear trigonometric function of joint angles and lengths of arm segments.

$$[v, \omega]^T = J(\gamma, \gamma\text{-dot}) [\gamma\text{-dot}]^T \quad (2)$$

where, γ, γ-dot = joint angles and rates, v = POR velocity, ω = angular velocity and J (γ, γ-dot) is approximately equal to the Jacobian J (γ) when γ-dot is very small. Note that this is an instantaneous relationship and the Jacobian changes as the angle changes, even if the rates remain constant. If the joint rates are not very small, then, the Jacobian becomes a much more complex function of rates also.

III. DEVELOPMENT OF CONTROL ALGORITHMS

A typical control strategy is to difference the actual state vector from the desired state vector and use the deltas as the desired velocity and angular velocity in equation (2). Desired (or commanded) joint rates can then be derived by inverting the Jacobian if it is non-singular. Otherwise the inversion problems will prohibit the solution and control can not be achieved.

Our strategy is to use the fuzzy logic based attitude controller and avoid the Jacobian inversion. We noticed that if the deltas are transformed from the Cartesian coordinate frame [Δx, Δy, Δz] to spherical coordinate frame [ΔR, Δelev, Δazim], equivalently, range error, elevation error and azimuth error, then, each delta can be one to one related to each joint angle. For example, the azimuth error can be related to the first RMS joint which is a yaw joint. Thus, by moving the first joint, any azimuth error Δazim can be corrected. As the first joint moves, the rate of azimuth correction can be computed for the controller's use. Now, there are two parameters, azimuth error and its rate error that can be used in the fuzzy logic based attitude controller. The output of the controller is interpreted as the commanded joint rate instead of a jet firing command.

Similarly the elevation and range errors can be corrected by moving the second and third joint. The last three joints relate to the POR orientation angles directly because we use the same Shuttle rotational sequence to define POR orientation. In general, this in not true for any robotic arm, and one must clearly identify a correct correspondence. Thus, all six errors [Δazim, Δelev, ΔR, Δpitch, Δyaw, Δroll] can be corrected by moving the six joints in order. As soon as these errors are nulled, the state vector is at the desired value. The controller outputs the joint rate commands and the servo motors provide these commanded rates for each corresponding joint. Algorithmic steps as described below are simple : 1) compute deltas, 2) transform them into the spherical frame, and 3) apply the attitude controller to correct each error.

Inputs to the fuzzy guidance-controller are current and desired state vectors whose first three components are the position x, y, and z, and last three components are roll, pitch and yaw angles. The current state vector is derived by applying forward kinematics (equation. 1) to the current joint angles. The next step is to compute the deltas between the desired state and the current state by subtracting the two. These deltas [DEL[1] through DEL[6]] are then transformed into delta range, delta elevation, delta azimuth and delta angles as follows.

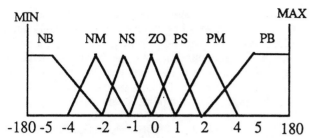

Fig. 2a - Fuzzy membership functions for input angle and angle rate.

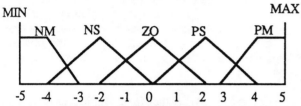

Fig. 2b - Fuzzy membership functions for output commanded joint rates

ΔR = sqrt (DEL[1]*DEL[1] + DEL[2]*DEL[2] + DEL[3]*DEL[3]);
Δelev = ATAN2 (DEL[3],sqrt(DEL[2]*DEL[2] + DEL[1]*DEL[1]));
Δazim = ATAN2 (DEL[1],DEL[2]);
Δpitch = DEL[5];
Δyaw = DEL[6];
Δroll = DEL[4];

Also the rates of these parameters are computed by using previous values.

$\Delta\Delta R = \Delta R - \Delta R\_previous$;
$\Delta\Delta$elev = Δelev - Δelev_previous ;
$\Delta\Delta$azim = Δazim - Δazim_previous ;
$\Delta\Delta$pitch = Δpitch - Δpitch_previous ;
$\Delta\Delta$yaw = Δyaw - Δyaw_previous ;
$\Delta\Delta$roll = Δroll - Δroll_previous ;

An attitude controller implementation [10] based on the phase plane is used six times to generate the commanded joint rates for the robotic arm. The membership functions for angle and angle rate are shown in fig. 2a and the membership functions for commanded (desired) joint rates are shown in fig. 2b. Derivation of the membership functions is based on the deadband values used in the phase plane, and the commanded values required to null these errors. Originally, these membership functions were derived for the Shuttle attitude control and no change has been made. The rulebase used to compute the commanded joint rate is shown in Table I.

This attitude controller has been tested for the shuttle operations in a very high fidelity simulation and has shown excellent results in comparison with the existing conventional controller [10]. This controller has also been integrated with the translational controller for relative trajectory and attitude control and has shown robust control as well as fuel savings [7]. A tether length controller has been derived based on this concept using length error and length rate error and has been tested in a high fidelity simulation with bead model to show advantages [8]. For the robotic arm control, the commanded joint rates are achieved by the servo motors to move the POR to the desired position and orientation.

For joint_1, Phase_plane (Δazim, $\Delta\Delta$azim, Gamma_1);
For joint_2, Phase_plane (Δelev, $\Delta\Delta$elev, Gamma_2);
For joint_3, Phase_plane (ΔR, $\Delta\Delta R$, Gamma_3);
For joint_4, Phase_plane (Δpitch, $\Delta\Delta$pitch, Gamma_4);
For joint_5, Phase_plane (Δyaw, $\Delta\Delta$yaw, Gamma_5);
For joint_6, Phase_plane (Δroll, $\Delta\Delta$roll, Gamma_6);

IV. SIMULATION AND TESTING

The RMS forward kinematics implementation based on equation (1) in the orbital operations simulation is used to test our approach. The simulation is initialized with an initial arm position and orientation and the POR is commanded to achieve a desired position and orientation. Based on the starting and final POR states, the algorithm generates the rate commands for each joint. Perfect servo motors with first order lag are used to achieve the commanded rate for each joint. The cycle time is 80 ms which is consistent with the shuttle cycle time. The simulation flow is shown in fig. 3 with our algorithm in the loop. Four tests are performed to verify the operation of our controller. Initial and final states of all four test cases are given in Table II. In all test cases, the new POR

TABLE I
Fuzzy Rulebase For The Robotic Joint Rate Control

		\multicolumn{7}{c}{Angle Error}
		NB	NM	NS	ZO	PS	PM	PB
Rate Error	NB	PM	PM	PS	PM			
	NM	PM	PM	PS	PM			
	NS	PS	PS	PS	PS			
	ZO	PS	PS	PS	ZO	NS	NS	NS
	PS				NS	NS	NS	NS
	PM				NM	NS	NM	NM
	PB				NM	NS	NM	NM

KEY:
NB - Negative Big
NM - Negative Medium
NS - Negative Small
ZO - Zero
PS - Positive Small
PM - Positive Medium
PB - Positive Big

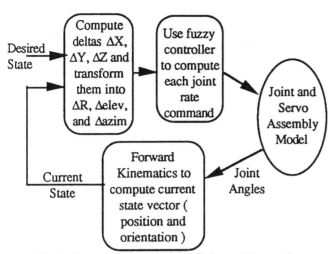

Fig. 3. Simulation flow with RMS forward kinematics and fuzzy controller

state was commanded 10 seconds into the run. Total run time of the simulation is 300 seconds.

The first test is a simple POR translation. The arm end point is commanded to go from its current position and attitude to a new position with no change in attitude. Motion was commanded on all three axes, with a large displacement on the X axis. The second test case commanded motion on the X axis as well as an attitude change. For the third test case, the arm is commanded to perform a three axes translation and three axes rotation. The fourth test case is a combination of maneuvers, simulating the "handling" of an object. The arm begins straight out and moves with a combined translation and rotation to line up on a location. It then moves straight in on the Z axis to grab the object, backs straight up on Z to lift the object, and then proceeds to a new location to drop the object. Thus, it has four desired points to reach in a given delta time.

V. RESULTS

In all cases the hybrid controller successfully achieved and held the commanded POR states. Approximately 60 seconds of run time was used in the maneuvering. The remaining run time demonstrated the ability of the control system to hold the desired position and orientation. Plots showing the POR translations and rotations as well as the joint angles and rates were generated for analysis. Observations from these plots show a few interesting characteristics of the controller.

First, the commanded POR states (x, y, z, pitch, yaw, roll) are not all reached at precisely the same time, and the path followed by each joint is not precisely smooth. The implication is that the POR did not traverse a straight line in the work space when going from its initial to commanded state. As the controller is currently implemented, each joint takes out its error as quickly as possible. Due to the configuration of the arm, some joints are able to eliminate error faster than other joints resulting in the arm tip moving "up and over" instead of straight to the desired position.

Second, there is some over shoot of the desired position along the Z axis in the first test case. The control of the arm is actually done in a spherical coordinate system. The work space, and the plots, are assumed to be represented by a Cartesian system. These different representations cause this over shoot. The control system is not controlling the shoulder pitch joint to drive the arm to a specific Z position, but to a specific elevation angle from the base of the arm to the arm tip. When the length (radial) of the arm is correct, as controlled by the elbow pitch joint, then the correct elevation angle will directly correspond to the desired Cartesian Z position. However, if the length of the arm is too large then the correct elevation angle will correspond to a larger than desired Cartesian Z position. So, when the over all length of the arm is commanded to decrease, there is a risk of over shooting the desired Z position when the elevation angle control (shoulder pitch joint) obtains its desired angle before the arm length control (elbow pitch joint) obtains its desired angle.

Finally, some error may occur in end tip orientation during large maneuvers, even when no orientation movement is commanded. The pitch plot of the first test case in fig. 4 shows this clearly around 15 seconds into the run. This is

Table II
Initial and final states of the end position (POR) for the test cases for RMS motion.

TC #	Starting POR						Final (Cmd) POR					
	X (Ft)	Y (Ft)	Z (Ft)	Pitch (Deg)	Yaw (Deg)	Roll (Deg)	X (Ft)	Y (Ft)	Z (Ft)	Pitch (Deg)	Yaw (Deg)	Roll (Deg)
1	50	0	1.2	0	0	0	20	5	5	0	0	0
2	50	0	1.2	0	0	0	30	0	0	30	40	50
3	50	0	1.2	0	0	0	30	10	10	30	40	50
4	50	0	1.2	0	0	0	30	0	-20	90	0	0
							30	0	-23	90	0	0
							30	0	-20	90	0	0
							30	0	0	0	0	0

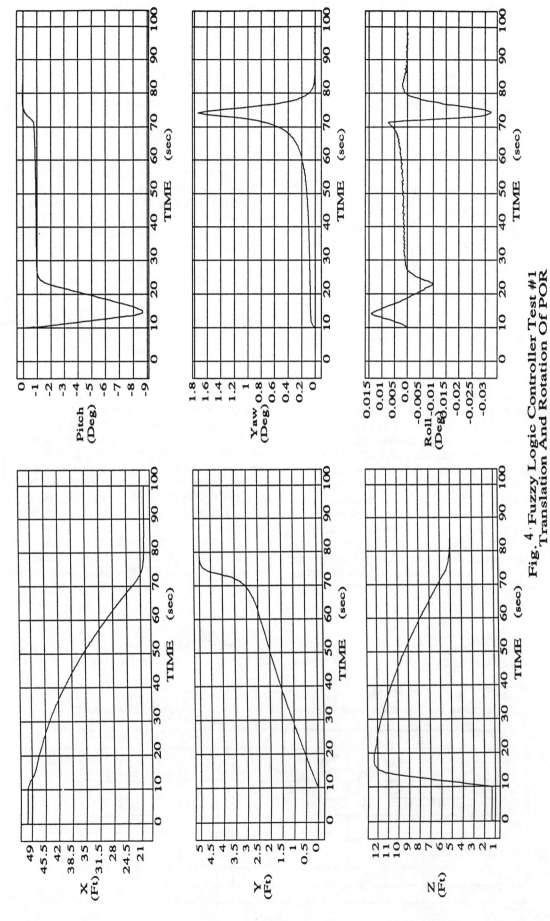

Fig. 4 Fuzzy Logic Controller Test #1
Translation And Rotation Of POR

caused by a brief period of time when both the shoulder pitch and the elbow pitch joints are driving negative. The controller commands are limited to 2 deg/sec in magnitude. Since two joints are pitching down at almost 2 deg/sec the combined change in pitch orientation is almost 4 deg/sec in the downward direction. With the controller limited to 2 deg/sec in command authority it is not possible for the wrist pitch joint to keep up with the accumulating error.

All of the behavior of the controller commented on above is caused to some degree by each joint controlling its own portion of the system independent of the other joints. Although the over all goal is clearly reached and held in each test case the exact path taken to the goal is not under complete control. If it is desired to exactly control the path taken to the desired POR this could be accomplished in several ways. For example, intermediate points along the desired path could be specified or the rate at which each joint moves towards its specified goal could be controlled by some set of overseeing rules. In either event the benefits of rule based control are still maintained.

VI. SUMMARY

It has been shown that there is a generic fuzzy algorithm that can be applied to the control of manipulator joints for maintaining a POR trajectory. Future work in this project will involve additional testing for many different cases, and tuning of rulebase and membership functions if necessary. As noted earlier, the motion of the joints needs to be correlated, especially when it is known that a particular joint has a large effect on a certain error and an adverse effect on another error. To minimize these adverse effects, the commanded joint rates should be properly related to each other. The relations among joint rates and their ratios need to be derived based on simulation results and hardware information. The fuzzy membership functions can be tuned for each joint to achieve rate matching. However, developing additional rules may offer advantages especially in light of collision avoidance and redundant degrees of freedom.

After thorough testing of the algorithm using a six degree of freedom arm, it will be extended to handle more than six degrees of freedom. Handling redundant degrees of freedom is very important, especially in light of obstacle avoidance and path planning. The rulebase for collision avoidance will also be developed, particularly for the objects within the work space which need to be avoided by each joint.

REFERENCES

[1] Proceeding of the Industrial Conference on Fuzzy Systems ICFS-91, sponsored by The Microelectronics and Computer Technology Corporation, Austin, Texas, June 27-28, 1991.

[2] Proceeding of the 1st International Workshop on Industrial Applications of Fuzzy Control and Intelligent Systems, sponsored by Center for Fuzzy Logic and Intelligent Systems Research, Texas A&M University, College Station, Texas, Nov. 21-22, 1991.

[3] Proceeding of the IEEE International Conference on Fuzzy Systems 1992, (FUZZ-IEEE 92), held at San Diego, California, March 8-12, 1992.

[4] A. Nedungadi : "A Fuzzy Robot Controller - Hardware Implementation", Proceedings of FUZZ-IEEE 92 held at San Diego, California, pp. 1325, March 8-12, 1992.

[5] G.V.S. Raju, & J. Zhou : "Fuzzy Rule Based Approach for Robot Motion Control", Proceedings of FUZZ-IEEE 92 held at San Diego, California, pp. 1349, March 8-12, 1992.

[6] J.J. Craig : Introduction to Robotics Mechanics & Control, Addison-Wesley Publishing Company, 1985. (Chapters 5 and 6)

[7] R.N. Lea, J. Hoblit and Y. Jani : "A Fuzzy Logic Based Spacecraft Controller for Six Degree of Freedom Control and Performance Results", Proceedings of AIAA Guidance, Navigation and Control Conference, New Orleans, August 12-14, 1991.

[8] R.N. Lea, J. Villarreal, Y. Jani, & C. Copeland : "Tether Operations Using Fuzzy Logic Based Length Control", Proceedings of FUZZ-IEEE 92 held at San Diego, California, pp. 1335, 1992.

[9] R.N. Lea, Y. Jani, M.G. Murphy & M. Togai : "Design and Performance of A Fuzzy Logic Based Vehicle Controller for Autonomous Collision avoidance", Proceedings of Fuzzy Neural Systems : Applications to Vehicle Control, Tokyo, Japan, November 1991.

[10] R.N. Lea, J. Hoblit & Y. Jani : "Performance Comparison of A Fuzzy Logic Based Attitude Controller with the Shuttle On-orbit Digital Auto Pilot", Proceedings of North American Fuzzy Information Processing Society (NAFIPS - '91) Workshop, Columbia, May 14-17, 1991.

Robotic Deburring based on Fuzzy Force Control

M.-H. Liu

Department of Control Engineering
Control System Theory & Robotic Group
Technical University of Darmstadt
Schlossgraben 1, 6100 Darmstadt, F.R. Germany

Abstract—The static and dynamic behaviour of the deburring process is analysed and with the help of fuzzy logic a linguistic description of the process is established. A fuzzy force control strategy for robotic deburring based on a corrective position/force control structure is put forward. This deburring strategy solves the problems of compensating positional inaccuracies and adjusting the feedrate according to unknown burr sizes in a systematic way. Experimental results have verified the effectiveness of the proposed deburring strategy.

I. Introduction

From the viewpoint of control three main problems are involved in the automatic deburring with a conventional industrial robot. The first problem is the programming of robot motion for complex workpieces. Because of the great number of teaching points and the required high teaching accuracy programming is much more time comsuming and tedious for deburring than for other tasks such as painting and welding. The second problem arises from the inaccuracies of the robot system introduced by workpiece tolerances, teaching errors, positioning errors, and the progressive tool wear, which must be taken into account in the design of a controller for robotic deburring. The third problem is the unpredictability of burr size and location. The robot motion should be controlled in such a way that the amount of material removed at different locations corresponds to different burr sizes. For the first problem we developed a contour following strategy, which was described in [3]. In this paper we will investigate the second and third problem. We assume here that the workpiece contour has already been teached in.

In recent years robotic deburring has been studied by many researchers. Much work has been done to solve these problems, such as development of sensor-aided programming systems [1, 2, 3], design of active end effectors [4, 5], and modelling and feedback control of the deburring process [6-15]. It has been shown that by controlling the tangential force the feedrate could be automatically adjusted according to burr sizes [6, 7, 8] and by controlling the normal force positional inaccuracies could be successfully compensated [9, 10, 11]. Because the tangential and normal forces are strongly related with each other an efficient deburring strategy has to be developed to solve the problems of both compensating positional inaccuracies and adjusting the feedrate according to burr variations. Puls and Barash [14] proposed a deburring strategy to adjust both the feedrate and the trajectory of the deburring tool by monitoring the normal force. When it got out of the range either the feedrate or the goal position was adjusted. Unfortunately that strategy was only simulated in practice and no systematic method was provided to adjust the feedrate and goal position.

In this paper a new deburring strategy is proposed based on a corrective position/force control structure and a fuzzy force controller. Under this strategy the cutting force (not only its components) is constantly controlled and the force control is realized by correcting the robot motion in the normal and tangential directions of the workpiece contour. A systematic method to analyse the deburring process and to design the motion corrections is provided. Taking the advantages of experiences, heuristics and intuitions of human experts the fuzzy control algorithm presents an easy way to design the deburring controller. The remainder of this paper is organized as follows. Section 2 presents the static, dynamic and fuzzy models of the deburring process. Section 3 develops the deburring strategy based on the fuzzy force control. Section 4 describes the implementation of the deburring strategy and shows some experimental results. Section 5 summarizes the results. For an overview about fuzzy logic control the reader is refered to [16].

II. Modelling of the Deburring Process

In this section we present a static model, a dynamic model and a fuzzy model of the deburring process.

A. Conventional Modelling

The static model of the deburring process relates a steady state change of the cutting force to changes of the process parameters. Although there are many parameters that influence the cutting force, the most important ones are the workpiece material, the feedrate (or tool travelling speed) and the cutting depth and width [17, 12]. In case of deburring the burr shape varies and therefore a better way than that used in [17, 12] to describe the burr size is to use the cross-sectional area instead of the height and width (Fig. 1). The cutting force can be expressed by the following nonlinear function

$$f = c\,\sigma^{n_1} v^{n_2} \stackrel{\text{def}}{=} \phi_{db}(\sigma, v) \qquad (1)$$

where f, σ and v are respectively the cutting force, the cutting cross-sectional area and the feedrate, shown in Fig. 1; n_1 and n_2 are positive constant exponents, and c is a cutting constant.

The linearized version of (1) can be expressed as

$$\delta f = \left[\frac{\partial \phi_{db}}{\partial \sigma}\right]_o \delta\sigma + \left[\frac{\partial \phi_{db}}{\partial v}\right]_o \delta v \stackrel{\text{def}}{=} c_\sigma \delta\sigma + c_v \delta v. \qquad (2)$$

The cutting cross-sectional area σ and feedrate v are nonlinear functions of the robot position given by

$$\sigma = \phi_\sigma(\mathbf{x}), \qquad (3)$$
$$v = \phi_v(\mathbf{x}, \dot{\mathbf{x}}) \qquad (4)$$

where $\mathbf{x}$ denotes the robot position vector in a cartesian coordinate system, given by $\mathbf{x} = [p_x\ p_y\ p_z\ \theta_x\ \theta_y\ \theta_z]^T$, and $\dot{\mathbf{x}}$ is the robot velocity vector. Then for a motion correction $\Delta\mathbf{x} = [\delta p_x\ \delta p_y\ \delta p_z\ \delta\theta_x\ \delta\theta_y\ \delta\theta_z]^T$ the static force change ($\dot{\mathbf{x}}$=const.) is given by

$$\delta f = \left[\frac{\partial \phi_{db}}{\partial \sigma}\frac{\partial \phi_\sigma}{\partial \mathbf{x}} + \frac{\partial \phi_{db}}{\partial v}\frac{\partial \phi_v}{\partial \mathbf{x}}\right]_o \Delta\mathbf{x}. \qquad (5)$$

The dynamic model of the deburring process relates the transient change of the cutting force to the changes of the process parameters. For a desired motion correction $\Delta\mathbf{x}_d(s)$ the transient change of the cutting force $\delta f(s)$ is determined by the dynamics of the robot system and of the cutting process. The robot control system is usually a positional one, the joint controller can be described as

$$\Delta\mathbf{q}(s) = \mathbf{G}_q(s)\Delta\mathbf{q}_d(s) \qquad (6)$$

where $\mathbf{q}$, $\mathbf{q} = [q_1\ q_2\ q_3\ q_4\ q_5\ q_6]^T$ for a 6-DOF robot manipulator, is the robot joint coordinate vector, $\Delta\mathbf{q}(s)$ and $\Delta\mathbf{q}_d(s)$ are respectively the Laplace form of the actual and desired joint motion correction vectors. Then the actual cartesian motion correction $\Delta\mathbf{x}(s)$ caused by $\Delta\mathbf{x}_d(s)$ is given by

$$\Delta\mathbf{x}(s) = \mathbf{J}(q)\,\Delta\mathbf{q}(s) = \mathbf{J}(q)\,\mathbf{G}_q(s)\,\mathbf{J}^{-1}(q)\,\Delta\mathbf{x}_d(s) \qquad (7)$$

where $\mathbf{J}(q)$ is the robot Jacobian matrix. The dynamics of the cutting process are of the form

$$\delta f(s) = g_\sigma(s)\,\delta\sigma(s) + g_v(s)\,\delta v(s) \qquad (8)$$
with $\lim_{s \to 0} g_\sigma(s) = c_\sigma$ and $\lim_{s \to 0} g_v(s) = c_v$.

The overall dynamic model can then be expressed as

$$\begin{aligned}
\delta f(s) &= g_\sigma(s)\,\frac{\partial \phi_\sigma}{\partial \mathbf{x}}\,\Delta\mathbf{x}(s) + \\
&\quad g_v(s)\left[\frac{\partial \phi_v}{\partial \mathbf{x}}\,\Delta\mathbf{x}(s) + \frac{\partial \phi_v}{\partial \dot{\mathbf{x}}}\,\Delta\dot{\mathbf{x}}(s)\right] \\
&= \left[g_\sigma(s)\,\frac{\partial \phi_\sigma}{\partial \mathbf{x}} + g_v(s)\left(\frac{\partial \phi_v}{\partial \mathbf{x}} + \frac{\partial \phi_v}{\partial \dot{\mathbf{x}}}s\right)\right] \cdot \\
&\quad \mathbf{J}(q)\,\mathbf{G}_q(s)\,\mathbf{J}^{-1}(q)\,\Delta\mathbf{x}_d(s). \qquad (9)
\end{aligned}$$

From (9) it can be seen that the overall system dynamics are rather complex and are dependent on the cutting conditions and the actual robot configuration, which makes it difficult to design an efficient deburring control strategy by using conventional model-based control techniques.

B. Modelling with Fuzzy Logic

To get a rough understanding of the deburring process in a qualitative way and to express the knowledge of an experienced human worker about the deburring process in a mathematical way we present a fuzzy modelling approach in this subsection.

A fuzzy model is generally represented by a set of fuzzy implication statements in the form

$$\text{IF } \mathcal{A} \text{ AND } \mathcal{B} \text{ THEN } \mathcal{C} \qquad (10)$$

where $\mathcal{A}$ and $\mathcal{B}$ are the antecedents (inputs) and $\mathcal{C}$ the consequent (output). Specifically the fuzzy model for the deburring process can be expressed by

$$\text{IF } \sigma \text{ is } \mathcal{A}_{\lambda_\sigma} \text{ AND } v \text{ is } \mathcal{B}_{\lambda_v} \text{ THEN } f \text{ is } \mathcal{C}_{\lambda_f}; \qquad (11)$$
$$\lambda_\sigma \in \{1, \cdots, n_\sigma\},\ \lambda_v \in \{1, \cdots, n_v\},\ \lambda_f \in \{1, \cdots, n_f\}$$

where $\mathcal{A}_{\lambda_\sigma}$, $\lambda_\sigma = 1, \cdots, n_\sigma$, are n_σ fuzzy labels defined on $\mathcal{U}_\sigma$, the universe of discourse of σ; $\mathcal{B}_{\lambda_v}$, $\lambda_v = 1, \cdots, n_v$, are n_v fuzzy labels defined on $\mathcal{U}_v$, the universe of discourse of v; and $\mathcal{C}_{\lambda_f}$, $\lambda_f = 1, \cdots, n_f$, are n_f fuzzy labels defined on $\mathcal{U}_f$, the universe of discourse of f. Equation (11) gives in fact $n_\sigma \times n_v$ implication statements.

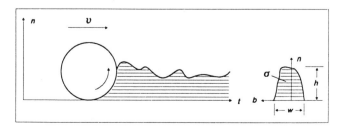

Fig. 1. Deburring Process

To get these implication statements the relation matrix $\mathcal{R}$ between the inputs and output may be first identified. The fuzzy model can be described with the help of the relation matrix by

$$\sigma \times v \xrightarrow{\mathcal{R}} f \qquad (12)$$

which maps the cutting cross-sectional area σ and the feedrate v onto the cutting force f.

Let $[\sigma_i, v_i, f_i]$, $i = 1, \cdots, N$, be N measured data pairs, and $\mathcal{R}(\lambda_\sigma, \lambda_v, \lambda_f)$ be an element of $\mathcal{R}$, representing the relation among $\mathcal{A}_{\lambda_\sigma}, \mathcal{B}_{\lambda_v}$ and $\mathcal{C}_{\lambda_f}$; $\lambda_\sigma \in \{1, \cdots, n_\sigma\}, \lambda_v \in \{1, \cdots, n_v\}, \lambda_f \in \{1, \cdots, n_f\}$. Then we have

$$\mathcal{R}(\lambda_\sigma, \lambda_v, \lambda_f) = \max_{i=1,\cdots,N} \left\{ \min \left[\mu_{\mathcal{A}_{\lambda_\sigma}}(\sigma_i), \mu_{\mathcal{B}_{\lambda_v}}(v_i), \mu_{\mathcal{C}_{\lambda_f}}(f_i) \right] \right\} \qquad (13)$$

where $\mu_{\mathcal{A}_{\lambda_\sigma}}(\sigma_i), \mu_{\mathcal{B}_{\lambda_v}}(v_i)$ and $\mu_{\mathcal{C}_{\lambda_f}}(f_i)$ denote respectively the grade of membership of σ_i, v_i and f_i in $\mathcal{A}_{\lambda_\sigma}, \mathcal{B}_{\lambda_v}$ and $\mathcal{C}_{\lambda_f}$. Having identified the relation matrix $\mathcal{R}$, the consequent of the statement (11) can then be obtained:

$$\mathcal{C}_{\lambda_f} = \left\{ \mathcal{C}_{\lambda_f} \mid \mathcal{R}(\lambda_\sigma, \lambda_v, \lambda_f) = \max_j \mathcal{R}(\lambda_\sigma, \lambda_v, j) \right\}. \qquad (14)$$

By suitably defining the membership function of the fuzzy subsets (14) may give very sharp results, that is, $\mathcal{R}(\lambda_\sigma, \lambda_v, \lambda_f)$ is much greater than any of the $\mathcal{R}(\lambda_\sigma, \lambda_v, j)$, $j = 1, \cdots, n_f$, $j \neq \lambda_f$. As an example, at first a fuzzy model of the deburring process was built by measuring 25 data pairs (with 5 different cutting cross-sectional areas and 5 different feedrates) from a set of experiments with a plastic workpiece. We got the relationships of Table one, in which S(Small), M(Medium) and B(Big) are fuzzy labels. One can read from the table for example the following statement: "IF σ is S AND v is B THEN f is M", which is in good accordance with the model (1). For a more detailed description of the process more fuzzy labels should be defined and more data pairs evenly distributed in their universes of discourse should be measured.

TABLE I. Fuzzy Deburring Model

f		v		
		S	M	B
	S	S	S	M
σ	M	S	M	B
	B	M	B	B

III. Deburring Strategy

Based on the analysis about the deburring process in the previous section a new deburring strategy is developed in this section.

A. Corrective Position/Force Control

Most of today's industrial robots are purely position-controlled devices with no means of directly controlling the contact force between the robot end effector and the workpiece. On the other hand force control is strongly necessary for many operations, such as deburring. During the past decade robot force control has been the subject of much attention in the robotics control literature. The proposed force control approaches can be categorized in three groups, that is, hybrid position/force control, compliance impedance control and joint torque control. A modified version of the hybrid position/force control [18], called "corrective position/force control" [3], is used to implement the force control for our robotic deburring and is explained here briefly.

Just as in the case of the hybrid position/force control force and position controls are performed in the constraint coordinate system for the corrective position/force control. By taking into account that most of the existing industrial robot controllers are positional ones and hence force control cannot be directly realized, the robot motion under the corrective position/force control is composed of two parts: a prespecified trajectory from the robot controller and motion corrections from the external corrective position/force controller. Because the prespecified trajectory cannot satisfy the control requirements completely due to positional inaccuracies the corrective position/force controller provides motion corrections to the prespecified trajectory to fulfil these requirements based on online sensor informations. In the directions in which force requirements exist force-controlled loops are introduced, while in the other directions position-controlled ones may be introduced. The motion corrections are given in the constraint coordinate system by:

$$^c\Delta \mathbf{x}_d = \sum_{i=1}^6 \mathbf{s}_i \delta_i \qquad (15)$$

where $\mathbf{x}$ is the position vector, given by $\mathbf{x} = [p_x\ p_y\ p_z\ \theta_x\ \theta_y\ \theta_z]^T$; $\mathbf{s}_i$ is the 6-dimensional unit specification vector, with the ith element equal to 1 and other elements equal to 0 (for example $\mathbf{s}_1 = [1\ 0\ 0\ 0\ 0\ 0]^T$), specifying that a motion correction is performed in the ith direction of the constraint coordinate system; and δ_i is the motion correction in that direction, given by

$$\delta_i = \begin{cases} \psi_i(\delta x_i) & \text{if the } i\text{th direction is position-controlled} \\ \varphi_i(\delta f_i) & \text{if the } i\text{th direction is force-controlled} \\ 0 & \text{otherwise} \end{cases}$$

where ψ_i and φ_i are position and force compensation functions.

There are two main differences between the "hybrid" and "corrective" position/force control approaches. The first difference is that the "hybrid" approach assumes that torque-servoed control loops are available in the robot control system and therefore force control can be directly implemented, while the "corrective" approach is in principle a position-control-based approach and the force control is realized by correcting the prespecified robot motion. In the "hybrid" approach positions and forces are controlled directly and separately while in the "corrective" approach only positions can be directly controlled and force corrections should be converted to position ones, for example with the help of a stiffness matrix. The second difference is that in the "hybrid" approach every direction of the constraint coordinate system is utilized for either position or force control: force control along those directions in which the robot is constrained by its environment, and position control along those directions in which the robot is unconstrained and is free to move, while in the "corrective" approach some of the directions of the constraint coordinate system may not have to be additionally controlled ($\delta_i = 0$), either because some of the position constraints have been already satisfied through the prespecified trajectory on the robot controller or because in some of the directions no constraints are required in a specific application. As will be explained in the following subsections, for the deburring task motion corrections are only required in the normal and tangential directions of the workpiece contour, that is, in the x_c- and y_c-directions. Then we have $\delta_i = 0$, $i = 3, 4, 5, 6$.

B. Deburring Control Algorithm

Based on (2) we explain at first how the problems of compensating positional inaccuracies and adjusting the robot motion according to unknown burr sizes can be separately solved by controlling the cutting force. (a) To compensate positional inaccuracies the robot motion should be adjusted in the normal direction of the workpiece contour ($\delta\sigma$ motion correction). This can be done by controlling the cutting force constantly, with the deburring tool travelling at a constant speed. A zero or too small cutting force means that the deburring tool is too far away from the workpiece contour because of the inaccuracies and thus the force controller should provide motion correction in the normal direction in such a way that the tool cuts the workpiece more deeply and the cutting cross-sectional area becomes greater and consequently the cutting force becomes greater. On the other hand, when the cutting force is too great it means that the tool cuts too deeply into the workpiece and therefore the force controller should generate motion correction in such a way that the tool moves slightly away from the workpiece. (b) To adjust the robot motion according to unknown burr sizes motion correction should be performed in the tangential direction of the workpiece contour so that the feedrate varies according to unknown burr sizes (δv motion correction). From (1) it can be seen that a big burr (big σ) generates a great cutting force when the feedrate stays constant. The bigger the burr ist, the greater the cutting force will be. In the case of unknown burr sizes we may control the cutting force constantly by adjusting the feedrate, so that a big burr is removed at a low feedrate, and the bigger the burr is, the lower the feedrate should be. To solve the problems of compensating positional inaccuracies and adjusting the robot motion according to unknown burr sizes at the same time we propose the following deburring control algorithm.

The first and primary goal of the proposed strategy is to apply the $\delta\sigma$ motion correction to control the cutting force to compensate positional inaccuracies because this is the prerequisite for a successful deburring. After the cutting force has come into the desired region, $|e(k-m)|, |e(k-m+1)|, \cdots, |e(k-1)|, |e(k)| \leq \eta_1$, in which $e(k) = f_d - f(k)$, where f_d and $f(k)$ are respectively the desired and actual force, k the actual sampling time, and m an integer, then the positional inaccuracies have been compensated and the value of $\delta\sigma$ at that moment indicates the amount of the inaccuracies, which should also be compensated hereafter. After that the $\delta\sigma$ motion correction is switched to the δv motion correction to adjust the feedrate according to burr sizes. The feedrate is increased when a small burr is encountered and decreased when a big burr is encountered, because the cutting force is controlled to be constant. In this way an unknown burr can be completely removed with a corresponding feedrate. In many cases, however, the positional inaccuracies may vary along the workpiece contour and they must be accordingly recompensated. Fortunately these inaccuracies can be observed and detected from the variation of the feedrate and the cutting force. If the slowest or biggest prespecified feedrate is reached or exceeded ($v \leq V_{min}$, or $v \geq V_{max}$, where V_{min} and V_{max} corresponds respectively to the feedrate at which the greatest and smallest possible burr is exactly removed under the desired cutting force), the possible causes are: (a) the workpiece is not properly positioned, (b) the workpiece tolerances are not evenly distributed, or imperfection exists in the workpiece, (c) teaching errors exist near that point, and/or (d) the deburring tool is worn out. The same causes may also bring the cutting force far away from the desired value ($|e(k)| \geq \eta_2$). In these cases the δv motion correction should be switched again to the $\delta\sigma$ motion correction to compensate the new positional inaccuracies. The proposed deburring control algorithm is summarized as follows:

REPEAT
{ IF $|e(k-m)|, |e(k-m+1)|, \cdots, |e(k-1)|,$
$|e(k)| \leq \eta_1$ during the $\delta\sigma$ motion correction
THEN switch to the δv motion correction

ELSE take the $\delta\sigma$ motion correction
IF $e(k) \geq \eta_2$ OR $v \leq V_{min}$ OR $v \geq V_{max}$
during the δv motion correction
THEN switch to the $\delta\sigma$ motion correction
ELSE take the δv motion correction }

By defining a reference coordinate system $\{B\}$, whose origin is located at the base of the robot, and a constraint coordinate system $\{C\}$, whose x-, y- and z-axis coincide respectively with the normal, tangential and binormal direction of the workpiece contour at the deburring point ($\mathbf{n}$, $\mathbf{t}$ and $\mathbf{b}$ shown in Fig. 1), the $\delta\sigma$ motion correction for the force control can be implemented through a motion correction in the x_c-direction and the δv motion correction through a motion correction in the y_c-direction. The $\delta\sigma$ and δv motion corrections are established on the robot controller based on the corrective position/force control structure through specification vectors $\mathbf{s}_1$ and $\mathbf{s}_2$, and are given as follows:
for the $\delta\sigma$ motion correction

$$^c\Delta\mathbf{x}_d(k) = \mathbf{s}_1\,\delta_1(k) = -\mathbf{s}_1\,[\delta\sigma_d(k)/\hat{w}] \quad (16)$$

and for the δv motion correction

$$\begin{aligned}^c\Delta\mathbf{x}_d(k) &= \mathbf{s}_2\,\delta_2(k) \\ &= \mathbf{s}_2\,[\delta_2(k-1) + \delta v_d(k)\,T_0]\end{aligned} \quad (17)$$

where $^c\Delta\mathbf{x}_d(k)$ is the motion correction vector expressed in the constraint coordinate system; $\delta_1(k)$ and $\delta_2(k)$ are motion corrections respectively in x_c- and y_c-direction, that is, respectively in the normal and tangential direction of the contour; T_0 is the sampling period; $\hat{w}$ is an estimate of the cutting width as shown in Fig. 1, and $\delta\sigma_d(k)$ and $\delta v_d(k)$ are desired motion corrections generated from separate force controllers, which are explained in the next subsection. To compensate the positional inaccuracies during the δv motion correction δ_1 should also be applied after the switch, taking the value of δ_1 at the switch moment. Therefore the control action during the δv motion correction should be:

$$^c\Delta\mathbf{x}_d(k) = \mathbf{s}_1\,\delta_1(k-1) + \mathbf{s}_2\,[\delta_2(k-1) + T_0\,\delta v_d(k)]. \quad (18)$$

The motion correction vector $^c\Delta\mathbf{x}_d$ is transformed to be expressed in the reference base coordinate system and then in joint coordinate system:

$$\Delta\mathbf{q}_d = \mathbf{J}^{-1}(q)\,^b\Delta\mathbf{x}_d = \mathbf{J}^{-1}(q)\,^b\mathbf{MP}_c\,^c\Delta\mathbf{x}_d \quad (19)$$

where $\mathbf{J}(q)$ is the Jacobian matrix of the robot manipulator; $^b\mathbf{MP}_c$ denotes the differential transformation matrix from $\{C\}$ to $\{B\}$. The whole control strategy is shown in Fig. 2.

C. Fuzzy Control Algorithm

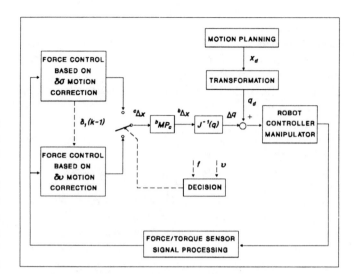

Fig. 2. Control Strategy for Robotic Deburring

To get the desired motion corrections $\delta\sigma_d$ and δv_d we present in this subsection a controller design approach based on fuzzy logic. The use of fuzzy logic control comes from three considerations. Firstly, the force data provided by the force/torque sensor are disturbed by noises. In spite of the use of filters within the sensor the delivered forces have an uncertainty of more than one Newton. An additional digital filter in the force controller would result in an additional time delay in the control loop. Secondly, the dynamic behaviour of the deburring process is nonlinear and depends on the working conditions. A detailed mathematical model is very complex and is difficult to obtain. Thirdly, the burr size is generally irregular and is expensive to measure. Fuzzy logic control provides a powerful design tool for control systems that involve uncertainties. For the design of a fuzzy logic controller no detailed mathematical model of a process is needed. It is realized as a set of heuristic decision rules, which are derived based upon operator's intuition and experiences. We try to solve or avoid the problems mentioned above by using the fuzzy logic control. The desired motion corrections $\delta\sigma_d$ and δv_d are generated based on the force error and the feedrate. They can be expressed in the following form:

$$\delta\sigma_d(k) = F_\sigma[e(k), v(k)], \quad (20)$$
$$\delta v_d(k) = F_v[e(k), v(k)] \quad (21)$$

where F_σ and F_v are nonlinear functions. For a simple implementation $\delta\sigma_d$ and δv_d are designed to be functions of the force error, and parameters in the functions are adjusted according to the feedrate.

Define $\tilde{e}(k) = \mathcal{K}_e e(k)$, $\widetilde{\delta\sigma_d}(k) = \mathcal{K}_{\delta\sigma}(k)\delta\sigma_d(k)$ and $\widetilde{\delta v_d}(k) = \mathcal{K}_{\delta v}(k)\delta v_d(k)$, where $\mathcal{K}_e$, $\mathcal{K}_{\delta\sigma}(k)$ and $\mathcal{K}_{\delta v}(k)$ are appropriate scaling factors, such that the universes of discourse of $e(k)$, $\delta\sigma_d(k)$ and $\delta v_d(k)$ are normalized to a special universe of discourse: $\tilde{e}(k)$, $\widetilde{\delta\sigma_d}(k)$, $\widetilde{\delta v_d}(k) \in [-\mathcal{D}, \mathcal{D}]$. Then the force controllers are given as:

$$\widetilde{\delta\sigma_d}(k) = \mathcal{G}_1[\tilde{e}(k)], \quad (22)$$
$$\widetilde{\delta v_d}(k) = \mathcal{G}_2[\tilde{e}(k)] \quad (23)$$

where $\mathcal{G}_1$ and $\mathcal{G}_2$ are fuzzy relations to be designed. The scaling factors in the force controllers are adjusted according to the feedrate based on another set of fuzzy relations, given by:

$$\mathcal{K}_{\delta\sigma}(k) = \mathcal{G}_3[v(k)], \quad (24)$$
$$\mathcal{K}_{\delta v}(k) = \mathcal{G}_4[v(k)]. \quad (25)$$

On the normalized universe of discourse seven fuzzy labels with triangular membership functions are defined: $\mathcal{A}_1 = NB$ (Negative Big), $\mathcal{A}_2 = NM$ (Negative Medium), $\mathcal{A}_3 = NS$ (Negative Small), $\mathcal{A}_4 = ZE$ (ZEro), $\mathcal{A}_5 = PS$ (Positive Small), $\mathcal{A}_6 = PM$ (Positive Medium), $\mathcal{A}_7 = PB$ (Positive Big), as shown in Fig. 3(a).

The fuzzy control algorithm is expressed by a set of implication statements or rules, and is given for the $\delta\sigma$ and δv motion corrections respectively by (26) and (27):

$$\mathcal{R}_i: \text{ IF } \tilde{e}(k) \text{ is } \mathcal{A}_{\lambda_{ei}} \text{ THEN } \widetilde{\delta\sigma_d}(k) \text{ is } \mathcal{A}_{\lambda_{\sigma i}}, \quad (26)$$
$$\text{IF } \tilde{e}(k) \text{ is } \mathcal{A}_{\lambda_{ei}} \text{ THEN } \widetilde{\delta v_d}(k) \text{ is } \mathcal{A}_{\lambda_{vi}}; \quad (27)$$
$$\lambda_{ei}, \lambda_{\sigma i}, \lambda_{vi} \in \{1, \cdots, 7\}; \quad i = 1, \cdots, 7.$$

Table 2 illustrates the utilized implication statements. Every rule of (26) or (27) provides a control action, expressed by a fuzzy label. The degree of fulfillment of rule $\mathcal{R}_i$ is given by $\mu_{\mathcal{A}_{\lambda_{ei}}}[\tilde{e}(k)]$. Therefore the defuzzified control value can then be computed based on the weighted average method by

$$\widetilde{\delta\sigma_d}(k) = \mathcal{K}_{\delta\sigma}(k)\delta\sigma_d(k) = \frac{\sum_{i=1}^{7} \mu_{\mathcal{A}_{\lambda_{ei}}}[\tilde{e}(k)]\, \gamma_{\sigma i}}{\sum_{i=1}^{7} \mu_{\mathcal{A}_{\lambda_{ei}}}[\tilde{e}(k)]}, \quad (28)$$

$$\widetilde{\delta v_d}(k) = \mathcal{K}_{\delta v}(k)\delta v_d(k) = \frac{\sum_{i=1}^{7} \mu_{\mathcal{A}_{\lambda_{ei}}}[\tilde{e}(k)]\, \gamma_{vi}}{\sum_{i=1}^{7} \mu_{\mathcal{A}_{\lambda_{ei}}}[\tilde{e}(k)]} \quad (29)$$

where $\gamma_{\sigma i}$ and γ_{vi} are respectively the central value of $\mathcal{A}_{\lambda_{\sigma i}}$ and $\mathcal{A}_{\lambda_{vi}}$, given by rule $\mathcal{R}_i$; $\lambda_{\sigma i}, \lambda_{vi} \in \{1, \cdots, 7\}$.

TABLE II. RULES FOR FUZZY FORCE

$\tilde{e}(k)$	NB	NM	NS	ZE	PS	PM	PB
$\widetilde{\delta\sigma_d}(k)$	NB	NM	NS	ZE	PS	PM	PB
$\tilde{e}(k)$	NB	NM	NS	ZE	PS	PM	PB
$\widetilde{\delta v_d}(k)$	NB	NM	NS	ZE	PS	PM	PB
$\tilde{v}(k)$	VS	S	SM	M	MB	B	VB
$\widetilde{\mathcal{K}_{\delta\sigma}}(k)$	VS	S	SM	M	MB	B	VB
$\tilde{v}(k)$	VS	S	SM	M	MB	B	VB
$\widetilde{\mathcal{K}_{\delta v}}(k)$	VB	B	MB	M	SM	S	VS

The scaling factors $\mathcal{K}_{\delta\sigma}(k)$ and $\mathcal{K}_{\delta v}(k)$ should be designed as functions of the feedrate because the feedrate is one of the most important parameters reflecting the working conditions. For example when the feedrate is very big a small change of the cutting cross-sectional area can produce a big change of the cutting force; on the contrary when the feedrate is very small a very big change of the cutting cross-sectional area is required to produce the same force change. To get the scaling factors we define $\tilde{v}(k) = a_1 v(k) + b_1$, $\widetilde{\mathcal{K}_{\delta\sigma}}(k) = a_2 \mathcal{K}_{\delta\sigma}(k) + b_2$ and $\widetilde{\mathcal{K}_{\delta v}}(k) = a_3 \mathcal{K}_{\delta v}(k) + b_3$, where a_j and b_j, $j = 1, 2, 3$, are constant scaling factors, such that the universes of discourse of $v(k)$, $\mathcal{K}_{\delta\sigma}(k)$ and $\mathcal{K}_{\delta v}(k)$ are normalized to the following universe of discourse: $\tilde{v}(k)$, $\widetilde{\mathcal{K}_{\delta\sigma}}(k)$, $\widetilde{\mathcal{K}_{\delta v}}(k) \in [0, 2\mathcal{D}]$. On this normalized universe of discourse the following seven fuzzy labels with triangular membership functions are defined: $\mathcal{B}_1 = VS$ (Very Small), $\mathcal{B}_2 = S$ (Small), $\mathcal{B}_3 = SM$ (between Small and Medium), $\mathcal{B}_4 = M$ (Medium), $\mathcal{B}_5 = MB$ (between Medium and Big), $\mathcal{B}_6 = B$ (Big), $\mathcal{B}_7 = VB$ (Very Big), as shown in Fig. 3(b). The fuzzy rules are defined similarly to (26) or (27) and are listed in Table 2. For example one can read from the Table the following rule: "IF $\tilde{v}(k) = \mathcal{B}_1 = VS$ THEN $\widetilde{\mathcal{K}_{\delta v}}(k) = \mathcal{B}_7 = VB$". The defuzzified outputs of the fuzzy relations, $\widetilde{\mathcal{K}_{\delta\sigma}}(k)$ [or $\mathcal{K}_{\delta\sigma}(k)$] and $\widetilde{\mathcal{K}_{\delta v}}(k)$ [or $\mathcal{K}_{\delta v}(k)$], are computed based on the weighted average method, similarly to (28) or (29). Then the desired motion corrections $\delta\sigma_d(k)$ and $\delta v_d(k)$ are given by (28) and (29), and the motion corrections to the robot control unit are given by (16), (18) and (19).

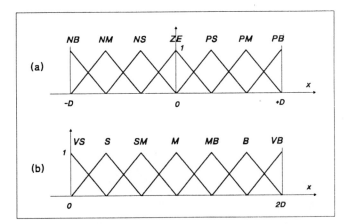

Fig. 3. Membership Functions

IV. IMPLEMENTATION AND EXPERIMENTS

The control system used for deburring is composed of a robot manipulator Manutec r3 with an air-motor-driven deburring tool of cylindric type, a DFVLR 6-DOF force/torque sensor, a robot controller RCM 3 and a control computer. The Manutec r3 is a free programmable 6-DOF joint manipulator, which is controlled by a Siemens RCM 3 robot control unit. Through ISRA-PRCI (PC Robot Control Interface) the RCM 3 is connected to a control computer. It receives motion corrections from the control computer via the PRCI interface and coordinates the motion corrections with its planned path. The 6-DOF force/torque sensor is mounted between the robot hand and the deburring tool. It works at the principle of strain gauge. Its internal electronics resolve the strain gauge outputs into forces and torques. The transmission of the force/torque data to the control computer is conducted through a dual-port memory interface. The uncertainty of the filtered force data is about $1N$ when the deburring tool is off and about $1.8N$ when it is turned on. The control computer is an IBM AT with processors 80386/80387. It acquires the forces and torques from the force/torque sensor and robot positions from the RCM control unit via the PRCI interface, performs the fuzzy force control, and transmits the motion corrections to the RCM control unit via the PRCI interface. On the control computer the PRCI interface is implemented through a dual-port memory. The control algorithm is realized at a sampling period of 8 ms. The programm is written in Microsoft C version 5.1.

Experiments were conducted on plastic and aluminium workpieces with step burrs. For simplicity the experimental results conducted on a plastic workpiece with a linear edge and a step burr are presented. The workpiece was mounted in such a way that the linear edge was in the y-direction of the base coordinate system and the step burr was in the z-direction, as shown in Fig. 4. The step burr had a cross-sectional area of $2\ mm^2$ at the two ends, and $5\ mm^2$ in the middle. To demonstrate the ability to compensate positional inaccuracies the workpiece was positioned intentionally with an error in the z-direction. The force controller provided first $\delta\sigma$ motion correction to compensate the positioning error and then δv motion correction to remove the burr completely. The cutting force, the feedrate and the z-coordinate of the tool are shown respectively in Fig. 5, Fig. 6 and Fig. 7. It can be seen that the positional error was successfully compensated and the feedrate was adjusted according to the burr size. It should be pointed out that the control behaviour was worse in the middle than in other parts of the burr, as can be read from the time history of the cutting force. In the case of too large a burr the burr could not be distinguished from positional inaccuracies under the deburring algorithm and therefore the burr could not be removed completely. In this case the deburring procedure must be repeated. To solve this problem we are working for a man-machine communication system with the help of fuzzy logic. The burr size will be teached into the force controller with linguistic variables, such as "too big", "very big", etc. Furthermore we are going to improve our experiments by using a high-frequency electric tool in place of the air-motor-driven tool to get a relatively stable tool rotational speed.

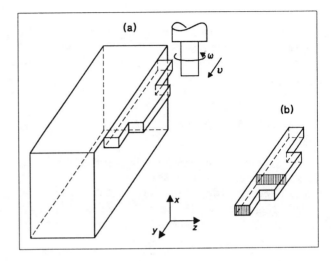

Fig. 4. Experimental Setup

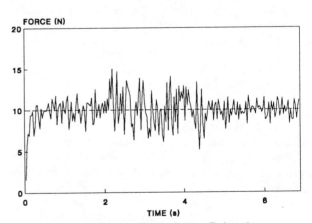

Fig. 5. Cutting Force during Deburring

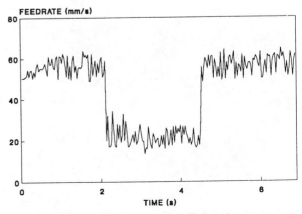

Fig. 6. Feedrate during Deburring

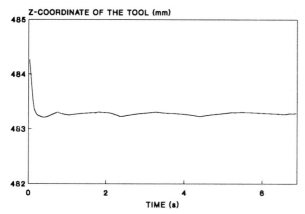

Fig. 7. Z-Coordinate of the Tool

V. Conclusions

In this paper the problems involved in robotic deburring are investigated and solutions based on force feedback control are proposed. A fuzzy control algorithm is employed to provide motion corrections in normal and tangential directions of the workpiece contour in a systematic way, which compensate positional inaccuracies and adjust the feedrate according to unknown burr sizes. Based on a corrective position/force control structure the proposed deburring strategy is easy to be implemented on a commercially available positional robot controllers. To get a better understanding of the deburring process static, dynamic and fuzzy models are established and analysed. Experimental results have shown that the proposed deburring strategy is able to deburr successfully in case of positional inaccuracies and unknown burr sizes. Our present work is going in the following two directions. (1) A three-step deburring strategy will be developed and investigated, in which the first step is the contour following of a sample workpiece to get the desired trajectory of the robot manipulator, the second step is the contour following of the workpiece to be deburred to get the variation of the burr size, and the third step is the deburring procedure under fuzzy force control. (2) A knowledge-based system will be developed to support the design of the fuzzy force control, for example to get the desired cutting force and other controller parameters, e.g., $\mathcal{K}_e$ and $\mathcal{K}_v$. This knowledge-based system will be supplemented by the man-machine interface equipped with fuzzy logic.

Acknowledgment

We are very grateful to Dipl.-Ing. E. Ersü, the President of the company ISRA-Systemtechnik GmbH, Darmstadt, where our experiments were performed. We would also like to thank Dipl.-Ing. St. Wienand and Dipl.-Ing. A. Seyffer for their support during the implementation.

References

[1] E.Abele, W.Sturz, and D.Boley, "Sensor and tool development for robotic deburring," *Technical Paper SME*, MR85-840, 1985.

[2] F.M.Proctor, R.J.Norcross, and K.N.Murphy, "Automating robot Programming in the cleaning and deburring," *Technical Paper SME*, Nr. 89-138, 1989.

[3] M.-H.Liu, "Theoretical and experimental preparations for automated deburring of plastics with an industrial robot," *Internal Report*, Dept. of Control Engineering, Techn. Uni. of Darmstadt, 1990 (in German), unpublished.

[4] J.J.Bausch, B.M.Kramer, and H.Kazerooni, "Development of compliant tool holders for robotic deburring," *Proc. of the Winter Annual Meeting of the ASME, Robotics: Theory and Applications*, pp.79-89, 1986.

[5] H.Asada and N.Goldfine, "End effector design for robotic grinding," *Proc. of the 10th IFAC World Congress*, pp.198-203, 1987.

[6] G.Plank and G.Hirzinger, "Controlling a robots motion speed by a force-torque-sensor for deburring problems," *Proc. of the 4th IFAC/IFIP Symp. on Infor. Contr. Problems in Manufactoring Technology*, pp.97-102, 1982.

[7] H.Kazerooni, "Automated robotic deburring using impedance control," *IEEE Control System Magazine*, pp.21-25, 1987.

[8] H.Kazerooni, "Robotic deburring of two-dimensional parts with unknown geometry," *J. of Manufacturing Systems*, vol.7, pp.329-338, 1988.

[9] T.M.Stepien, L.M.Sweet, M.C.Good, and M.Tomizuka, "Control of tool/workpiece contact force with application to robotic deburring," *IEEE J. of Robotics and Automation*, vol.3, pp.7-18, 1987.

[10] K.B.Haefner, P.K.Houpt, T.E.Baker, and M.E.Dausch, "Real time robotic position/force control for deburring," *Proc. of the Winter Annual Meeting of the ASME, Robotics: Theory and Applications*, pp.73-78, 1986.

[11] R.Hollowell and R.Guile, "Analysis of robotic chamfering and deburring," *Proc. of the Winter Annual Meeting of the ASME, Modeling and Control of Robotic Manipulator and Manufacturing Process*, pp.73-79, 1987.

[12] G.M.Bone and M.A.Elbestawi, "Robotic force control for deburring using an active end effector," *Robotica*, vol.7, pp.303-308, 1989.

[13] D.E.Whitney, A.C.Edsall, A.B.Todtenkopf, T.R.Kurfess, and A.R.Tate, "Development and control of an automated robotic weld bead grinding system," *Trans. ASME: J. DSMC*, vol.112, pp.166-176, 1990.

[14] F.M.Puls and M.M.Barash, "An adaptive control algorithm for robotic deburring," *Journal of Manufacturing Systems*, pp.169-178, 1985.

[15] D.A.Dornfield, and E.Erickson, "Robotic deburring with real time acoustic feedback control," *Proc. of the Winter Annual Meeting of the ASME, Mechanics of Deburring and Surface Finishing Processes*, pp.13-26, 1989.

[16] C.C.Lee, "Fuzzy logic in control systems: fuzzy logic controller," *IEEE Tran. on Systems, Man and Cybern.* vol.20, pp.404-435, 1990.

[17] M.Schulz, "Contribution to applying industrial robots to deburr aluminum castings," VDI-Verlag, 1988 (in German).

[18] M.H.Raibert and J.J.Craig, "Hybrid position/force control of manipulators," *Trans. ASME: J. DSMC*, vol.102, pp.126-133, 1981.

Manipulator for Man-Robot Cooperation
(Control Method of Manipulator/Vehicle
System with Fuzzy Inference)

Yoshio Fujisawa*, Toshio Fukuda*,
Kazuhiro Kosuge*, Fumihito Arai*,
Eiji Muro**, Haruo Hoshino**, Takashi Miyazaki**,
Kazuhiko Ohtsubo*** and Kazuo Uehara***

*Department of Mechano-Informatics and Systems, Nagoya University
Furou-cho, Chikusa-ku, Nagoya 464-01, JAPAN
**Takenaka Corporation Technical Research Laboratory
2-chome, Minamisuna, Koutou-ku, Tokyo 136, JAPAN
***Komatsu Ltd. Technical Research Center
1200 Manda, Hiratsuka-shi, Kanagawa 210, JAPAN

Abstract

In this paper, we propose a control method for the manipulator/vehicle System for Man-Robot Cooperation. Generally, the manipulator/vehicle system has the redundant degrees of freedom. In order to control the manipulator/vehicle system, we have to decompose the motion of the endeffector in the inertial coordinate system into the vehicle's motion and the manipulator's motion. Because the motion of the endeffector with respect to the inertial coordinate system is realized by both the manipulator's motion and the vehicle's motion. How to decompose the endeffector's motion is one of the key issues for the manipulator/vehicle system. If one wants to move the system to another place, the motion of the endeffector should be realized by the vehicle's motion. While if one wants to manipulate an object, the motion of the endeffector should be done by the manipulator's motion. That is, the decomposition of the endeffector's motion in the inertial coordinate system should be done based on the human intention. We use the fuzzy inference in order to model the human intention. The experimental results illustrates the effectiveness of the proposed control method for the manipulator/vehicle system for man-robot cooperation.

1. Introduction

We can predict that the labor shortage will be serious social problem in many fields. In order to settle this problem, many robots will be needed more and more in the near future. However, most of robots which are used in practice now is controlled by off-line way and they can not be applied to the fields which have structurally changeable tasks. Now let us consider the conventional industrial robots as an example. The reason the industrial robots are practically used is that the environments are structurally determined in advance so as to be suitable for the tasks carried out by robots. And all the robots have to do is repeating the tasks according to the task sequences programmed in advance. We can not apply these conventional industrial robots to the changeable tasks, because we can not consider the every case of the environments. To carry out the structurally changeable tasks, we need robots which have high intelligence and execute tasks by using their intelligence. But it is very difficult for us to make the high intelligent robotic system which can deals with changeable tasks from the current view of robotic technology. The robotic manipulator system which can execute tasks in cooperation with a human operator may be one of the solutions for this problem.

Many researches have been done on the robotic system which can carry out the tasks in cooperation with a human operator. G. Hirzinger has proposed the direct teaching method for the manipulator with force/torque sensor attached to it [1]. H. Kazerooni has proposed the extender. The human operator wears it and extends his power [2][3]. We have proposed a manipulator/vehicle system for man-robot cooperation, which is designed for handling heavy objects in cooperation with the human operator [4][5][6]. The manipulator/vehicle system for man-robot cooperation is the robot manipulator system for man-robot cooperation with mobile mechanism. Figure 1 shows the concept of the manipulator/vehicle system for man-robot cooperation proposed in this paper. Unlike the conventional manipulators, a human operator exists in the same working space with the robotic system and the human operator commands the motion of the manipulator directly. The manipulator for man-robot

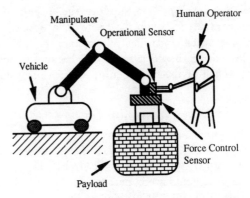

Fig. 1 Concept of Man-Robot Cooperation

cooperation can be used in various kinds of unknown environments/tasks.

In this paper, we propose a control method for manipulator/vehicle system for man-robot cooperation. We should design the motion of the manipulator considering the operational force and interaction between the manipulator and the environment to control the manipulator for man-robot cooperation proposed. In order to realize this motion of the manipulator, we use two impedance controllers. One is the human impedance controller and the other is the environmental impedance controller.

The manipulator/vehicle system is the robot manipulator system mounted on the mobile mechanism (vehicle). Compared with the manipulator which is fixed to the floor/ground, the manipulator/vehicle system has many merits; we can can move the system anywhere we want to execute tasks, and the system realizes a large working space without designing a large manipulator. The manipulator/vehicle system has redundant degrees of freedom and the motion has to be decomposed into the motion of the manipulator and the motion of the vehicle, when the desired motion of the manipulator's endeffector in the inertial coordinate system is given. The motion of the end effector on the ground surface should be realized by the vehicle if we want to move the system to another location, while the motion of the end effector should be realized by the motion of the manipulator if we want to manipulate an object. The decomposition of the motion of the end effector should be done based on the operator's intention. How to decompose the motion of the end effector is one of the key issues of the manipulator/vehicle system for man-robot cooperation. Fuzzy inference system is used to decompose the motion based on the operator's intention.

In the sequel, we first introduce the control method of the manipulator for man-robot cooperation using the human impedance controller and the environmental impedance controller. Second we explain the decomposition of the motion of the end effector. Third we discuss how to decompose the motion of the end effector using the fuzzy logic which models human intention. Finally we carried out experiments to illustrate the effectiveness of the proposed control method for the manipulator/vehicle system for man-robot cooperation.

2. Motion of Endeffector for Man-Robot Cooperation

Figure 2 shows the control system for the manipulator/vehicle system for man-robot cooperation proposed in this paper. We have to design the control system of the manipulator for man-robot cooperation considering the operational force by the human operator and the contact force between the manipulator and the. We assume that the operational force can be measured by the operational force sensor which is attached to the final link of the manipulator and the contact force can be measured by the environmental force sensor which is located between the final link of the manipulator and the end effector of the manipulator. The human impedance controller relates the motion of the manipulator and the operational force applied by the human operator, while the environmental impedance controller relates the contact force and the motion of the manipulator. The final motion of the manipulator is determined by both the motion calculated by the human impedance controller and the motion calculated by the environmental impedance controller as follows:

$$X_h = H^{-1} \cdot F_h \qquad (1)$$

$$X_e = E^{-1} \cdot F_e \qquad (2)$$

$$X = X_h + X_e \qquad (3)$$

where H^{-1} is the human impedance controller, E^{-1} is the environmental impedance controller, X_h is the motion of the manipulator determined by the human impedance controller and X_e is the motion of the manipulator calculated by the environmental impedance controller. X is the final motion of the manipulator. We determine this final motion of the manipulator by both X_h and X_e.

Using these two impedance controllers, we can control both the motion of the end effector of the manipulator based on the operator's intentional force and the interaction between the end effector and the environment.

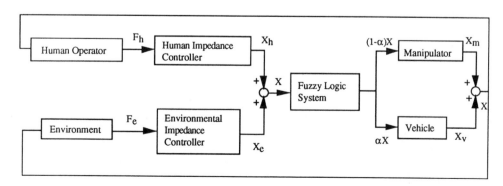

Fig. 2 Control System

3. How to Decompose Motion of End Effector

To model the manipulator/vehicle system, we introduce two coordinate systems as shown in fig. 3.

(1) **The inertial coordinate system** (O_0-$x_0 y_0 z_0$). The position of the end effector of the manipulator and the position of the vehicle are given in this coordinate system.

(2) **The vehicle coordinate system** (O_v-$x_v y_v z_v$). This coordinate system is fixed to the vehicle. The manipulator/vehicle system for man-robot cooperation is the robot manipulator system mounted on the mobile mechanism. The manipulator/vehicle system has redundant degrees of freedom. When the desired motion of the manipulator's end effector in the inertial coordinate system is given, the motion has to be decomposed into the motion of the manipulator and the motion of the vehicle. That is, the desired position of the manipulator's end point is decomposed as follows:

$$X = X_m + X_v \qquad (4)$$

$$X_m = (1-\alpha) \cdot X \qquad (5)$$

$$X_v = \alpha \cdot X \qquad (6)$$

where X is the desired motion of the manipulator's end point in the inertial coordinate system and the motion is determined by equation (3). Xm is the motion of the manipulator in the vehicle coordinate system and Xv is the motion of the vehicle in the inertial coordinate system. a is the decomposition ratio of the motion of the end effector in the inertial coordinate system; if the decomposition ratio a is set nearly equal to zero, the motion of the end effector is realized by the motion of the manipulator and if the ratio a is set nearly equal to unity, the motion of the end effector is realized by the motion of the vehicle.

4. Control of Manipulator/Vehicle System Based on Human Intention

The decomposition of the motion of the end effector should be done based on the operator's intention. How to decompose the motion of the end effector is one of the key issues of the manipulator/vehicle system for man-robot cooperation. In this section, we propose a control system of manipulator/vehicle system for man-robot cooperation based on the human intention. Fuzzy logic is used to decompose the motion based on the operator's intention, that is, we determine the decomposition ratio of the motion of the system using fuzzy logic. In this fuzzy logic system, we use two kinds of the fuzzy rules; the rules based on the operational force detected by the operational force sensor and the rules based on the manipulability of the manipulator. The first one is used to take the operator's intention into account. One may

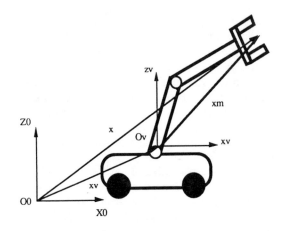

Fig. 3 Coordinate System

apply the large operational force when he wants to move the vehicle, while one may apply the small operational force when he wants to manipulate an object. The second one is to avoid the singular point of the manipulator and to realize the large working area.

The manipulability ω of the manipulator proposed by T. Yoshikawa [11] is calculated as follows.

$$\omega = \sqrt{\det\{J(\theta)J(\theta)^T\}} \qquad (7)$$

where $J(\theta) \in R^{m \times n}$ is the Jacobian matrix of the manipulator and q is the joint coordinate vector. In the case of m=n, we can calculate the maneuverability of the manipulator as follows:

$$\omega = |\det\{J(\theta)\}| \qquad (8)$$

The manipulability of the manipulator has the relation with the distance between the singular point of the manipulator and the end point of the manipulator. If the manipulability is small, the end point of the manipulator is close to the singular point of the manipulator, that is, the manipulator is hard to be manipulated. On the other hand, if the manipulability is large, the end point of the manipulator is far from the singular point of the manipulator, that is, the manipulator is easy to be manipulated.

We should consider the singular point of the manipulator when we control the manipulator/vehicle system cooperatively. If we do not take the singular point of the manipulator into account, the system can not move anymore when the manipulator reach the singular point. For these reasons, we decide the fuzzy rules for the singular point of the manipulator with respect to the manipulability of the manipulator system as follows.

If the manipulability of the manipulator is small then the vehicle should move. On the other hand, if the manipulability of the manipulator is large then the manipulator should move.

5. Experimental System

Figure 4 shows the photograph of the experimental system and fig. 5 shows the structure of the manipulator vehicle system, which we have developed. The manipulator has a parallel link mechanism with four degrees of freedom and each joint is driven by a DC motor through reducers (Harmonicdrives). It has a vacuum sucker attached to the end of the arm to manipulate the heavy object. The manipulator has two force/torque sensors. One is "the operational force sensor", which is attached to the end of the final link of the manipulator. The other is "the environmental force sensor, which is attached to the position between the end effector and the final link of the manipulator. We can measure the operational force and the contact force by these two sensors.

The vehicle is driven by two DC motors through reducers. The vehicle can do the rotary motion and the straight motion with these two DC motors.

6. Experiments

We used the experimental system as two degrees of freedom manipulator and one degree of vehicle. Figure 6 shows the fuzzy rules used for the manipulator/vehicle. Fh means the operational force applied by a human operator and the force can be measured by the operational force sensor. ω means the manipulability of the manipulator. B represents big, M represents medium and S means small, respectively.

Figure 7 shows the membership function used in this fuzzy logic. The manipulability of the manipulator in this experimental system w is calculated as follows;

$$\omega = |\det\{J(\theta)\}| = |l_1 \cdot l_2 \cdot \sin(\theta_2)| \quad (9)$$

$$J = \begin{bmatrix} -l_1 \cdot \sin(\theta_1) - l_2 \cdot \sin(\theta_1 + q_2) & -l_2 \cdot \sin(\theta_1 + \theta_2) \\ l_1 \cdot \cos(\theta_1) + l_2 \cdot \cos(\theta_1 + q_2) & l_2 \cdot \cos(\theta_1 + \theta_2) \end{bmatrix} \quad (10)$$

The link parameters of this experimental manipulator system is as follows;

$l_1 = 0.32$ [m], $l_2 = 0.42$ [m]

The largest manipulability of the experimental manipulator system ω_{max} is equal to $0.1344 [m^2]$ when the joint angle θ_2 equals ±90 deg., while the smallest manipulability of the experimental manipulator ω_{min} is equals to zero when the joint angle θ_2 equals ±180 deg.

We applied the proposed fuzzy logic system to the manipulator/vehicle system for man-robot cooperation, which consists of a manipulator with two degrees of freedom and vehicle with one degree of freedom.

Figure 8 illustrates the experimental results without fuzzy logic system. It is an example of the experimental results when the decomposition ratio a is equal to 0.5.

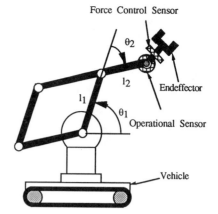

Fig. 4 Experimental System

Fig. 5 Photograph of System

F_h \ ω	S	M	B
S	V	V	V
M	M	MV	V
B	M	M	M

Fig. 6 Fuzzy Rules

Figure 9 also illustrates the experimental results with fuzzy logic system proposed in this paper. X is the desired position of the end point of the manipulator, xm is the position of the manipulator's end point in the x-axis direction with respect to the vehicle coordinate system, ym is the position of the manipulator's end point in the y-axis direction in the vehicle coordinate system and xv is the position of the vehicle in the inertial coordinate system.

From fig.8(x), the manipulator/vehicle system can not move anymore when the system reach 0.25 [m], because both the manipulator and the vehicle try to move in spite of the manipulator's end point reaches the singular point. From fig. 9(x), the system can move over

171

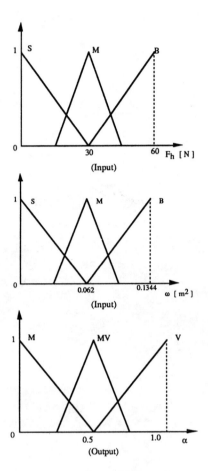

Fig. 7 Membership Functions

the position 0.25 [m], because the decomposition ratio a is determined with respect to the manipulability of the manipulator ω. As discussed in section 4, when the manipulability of the manipulator is small then the decomposition ratio is determined close to unity and the manipulability of the manipulator is large then the decomposition ratio is determined close to zero. From these experimental results, we can understand that the manipulator/vehicle system for man-robot cooperation can extend the working area using the proposed fuzzy logic system.

7. Conclusions

Generally, the manipulator/vehicle system has redundant motion degrees of freedom. How to decompose the motion of the system on the ground into the motion of the vehicle and the motion of the manipulator has discussed, and a control algorithm using fuzzy inference which expersses the human operator's intention was proposed in this paper. We designed the motion of the end effector using two impedance controllers, so that the end effector is controlled based on the operator's intentional force and the interaction with the environment. The proposed control algorithm extends the working space of the manipulator/vehicle system, which keeping the manipulability of the system. Finally experimental results illustrated the effectiveness of the proposed control system.

References

[1] G. Hirzinger and K. Landzettel, "Sensory Feedback Structures for Robotics with Supervised Learning", IEEE International Conference of Robotics and Automation, pp.627-635, (1985).
[2] H. Kazerooni and S. L. Mahoney, "Dynamically and Control of Robotic Systems Worn by Humans", IEEE International Conference of Robotics and Automation, pp.23992405, (1991).
[3] H. Kazerooni, "Human Machine Interaction via the Transfer of Power and Information Signals", IEEE International Conference on Robotics and Automation, pp.1632-1642, (1989).
[4] T. Fukuda, Y. Fujisawa, et. al., "A New Robotic Manipulator in Construction Based on Man-Robot Cooperation Work", Proc. of the 8th International Symposium on Automation and Robotics in Construction, pp.239-245, (1991).
[5] T. Fukuda, Y. Fujisawa, K. Kosuge, et. al., "Manipulator for Man-Robot Cooperation", 1991 International Conference on Industrial Electronics, Control and Instrumentation, Vol. 2, pp.996-1001.
[6] T. Fukuda, Y. Fujisawa, F. Arai et. al., "Study on Man-Robot Cooperation Work-Type of Manipulator, 1st Report, Mechanism and Control of Man-Robot Cooperation Manipulator", Trans. of the JSME, pp.160-168 (in Japanese).
[7] N. Hogan, " Impedance Control part 1-3", ASME Journal of Dynamic System Measurement, and Control, pp.1-24, (1985).
[8] K. Furuta, K. Kosuge, Y. Shiote, and H. Hatano, "Master-Slave Manipulatory Based on Virtual Internal Model Following Control Concept", IEEE International Conference on Robotics and Automation, pp.567-572, (1987).
[9] D. E. Whitney, "Historical Perspective and State of the Art in Robotic Force Control", The International Journal of Robotics Research, pp.3-13, (1987).
[10] D. E. Whitney, "Resolved Motion Rate Control of Manipulators and human protheses", IEEE Trans. on Man-Machine System, pp.47-53, (1969).
[11] T. Yoshikawa, "Manipulability of Robotic Mechanisms", The International Journal of Robotics Research, Vol. 4, No. 2, MIT Press, (1985).

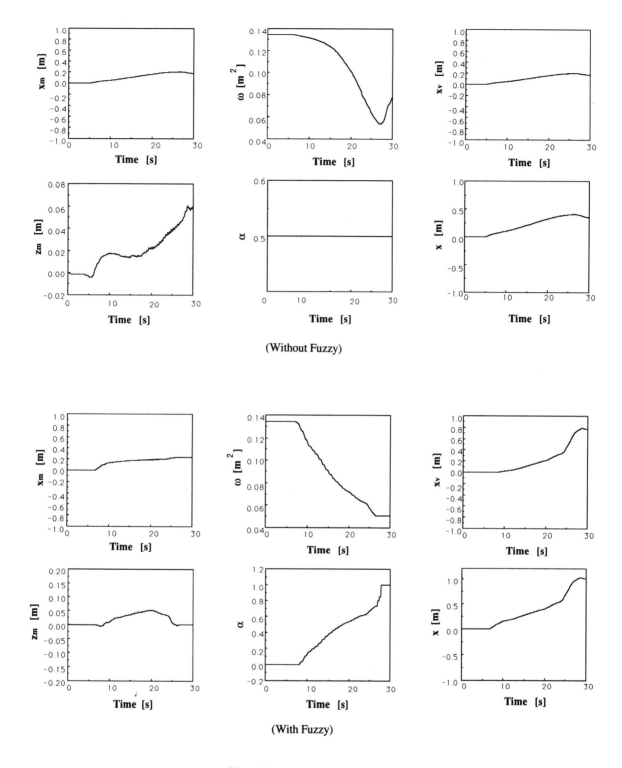

Fig. 8 Experimental Results

Chapter 4: Motors, Servos and Drives

ADAPTIVE FUZZY CONTROL OF HIGH PERFORMANCE MOTION SYSTEMS

E. Cerruto, A. Consoli, A. Raciti, A. Testa

Dipartimento Elettrico Elettronico e Sistemistico
Università di Catania
viale Andrea Doria, 6
95125 - Catania, Italy

Abstract: The paper describes a speed and position control system based on a fuzzy logic approach, integrated with a simple and effective adaptive algorithm. The proposed control scheme can be usefully applied to any electrical motor drive with decoupled torque and flux control. An experimental prototype, based on a PM brushless drive, has been realized using a mixed structure composed of recently introduced fuzzy logic chips and conventional digital circuits.

Introduction

Recently the fuzzy logic approach has been the object of an increasing interest due to its inner capabilities to adequately control ill-known or complex systems. Although the first studies about fuzzy logic were developed almost twenty years ago [1,2], only very few practical applications have been proposed. This is due to the high costs and difficulties of implementing fuzzy logic based algorithms with conventional computing systems. Nevertheless, the recent availability of hardware implemented fuzzy systems [3] has opened new possibilities, making reasonable in short times a fast increase of new industrial applications. In fact hardware based fuzzy logic devices are able to perform fuzzy logic algorithms hundred times faster and at a much lower cost than programmed digital systems.

The speed and position control of electrical drives using the fuzzy logic approach was recently proposed, mostly to improve the performances of conventional regulators [4, 5, 6]. Such controllers are based on a general scheme where the speed error and its derivative are evaluated and fuzzified exploiting suitable membership functions. The fuzzified signals are then sent to the inference block where control signals are deduced evaluating linguistic rules. Finally, the variation of the control signal is obtained with a defuzzification, achieved for example by means of the centroid algorithm. No experimental results have been presented in the above mentioned papers.

Although it has been shown by simulations that such controllers work better than conventional PI regulators, it has been realized that their performances can be even improved by adding an adaptive mechanism. Adaptive fuzzy systems proposed in literature [7] realize the adaption by on line tuning the rule base in the inference block, using a fuzzy logic based performance evaluator. This system examines the actual state of the controlled process and modifies the rules according to the differences between the actual performances and a set of reference performances. Such a control has only been proposed but never realized until now due to its inner complexity and difficulties in defining the adaptive rules.

In the present work a different approach is presented for the adaptive algorithm. In fact, by exploiting our knowledge of the physical structure of the drive, we can use the fuzzy logic to compensate unknown variations in the system parameters. The proposed method is based on the mentioned previous scheme, but differs as the output of the fuzzy regulator is the input of an adaptive block rather than be directly connected to the drive torque control system. By adopting the new method a speed and position control for high performance motion systems is implemented in this paper. A Permanent Magnet Synchronous (PMS) motor drive is considered as the basis of an experimental prototype that has been realized to practically confirm the validity of the proposed procedure.

Background of the Fuzzy Sets Theory

Traditional techniques of control are in some extent limited by the need to make a compromise between the best correspondence of the mathematical model to the actual plant and the complexity of the related algorithm. Sometime it could happen that inadequate control actions arise from a poor knowledge of the physical phenomena and/or from an inaccurate modeling. It is easy to verify that more complex and of high order is the system, much easier can fail the control objectives. Fortunately, in case quantitative relations are missed, qualitative informations are in general available that can result very helpful in performing the control actions.

Starting from such background, the "Fuzzy Sets" theory [1] has been conceived as a technique able to correctly manipulate, through a rigorous mathematical formalism, qualitative informations with an inductive action similar to the human approach. In fact, the specialist operator who controls the process in a plant, takes decisions on the basis of experience and/or intuition, performing a series of actions after an euristic reasoning. It follows that such decisions are not the consequence of a single valued measurement but, instead, the consequence of a number of informations from different sources, often unmeasurable and including some personal estimation.

A policy of euristic control in such a way is not easy to implement in terms of quantitative statements in order to automate the plant. The starting point is the "Fuzzy Set", a collection of objects of a certain universe having some common properties, characterized by a function of membership μ_i, that is variable between zero value (unpertaining) and one (full pertaining). The elementary operations on the fuzzy sets are: *union*, *intersection*, and *complement*, that are correspondent to the boolean operators OR, AND, NOT. The *union* of two fuzzy sets A and B of a universe of discourse X, characterized by the membership functions μ_A and μ_B, is a new fuzzy set with a membership function defined by:

$$\mu_{AB}(\alpha) = \text{Max}\,[\mu_A(\alpha);\ \mu_B(\alpha)] \qquad (1)$$

The *intersection* has a membership function defined by:

$$\mu_{A \cap B}(\alpha) = \text{Min} [\mu_A(\alpha); \mu_B(\alpha)] \quad (2)$$

Finally, the *complement* of a fuzzy set has a membership function:

$$\mu_{\neg B}(\alpha) = 1 - \mu_A(\alpha) \quad (3)$$

To establish a relation between two fuzzy sets A and B belonging to two different universes of discourse U and V, the definition of fuzzy *conditional expression* or *linguistic implication* has been established:

If X is A then Y is B (4)

where the two fuzzy sets A and B are called *antecedent* and *consequent*. To the relation (4), defined in U and having values in V, and expressed in terms of a cartesian product by R=AxB, the following membership function is associated:

$$\mu_R(u,v) = \mu_{A \times B}(u,v) = \text{Min}[\mu_A(u); \mu_B(v)],$$

$$u \in U, \quad v \in V \quad (5)$$

If a subset A' of A, and the relation R=AxB are assigned, the correspondent values B' can be deduced by the inferential rule of *composition*:

$$B' = A' \circ R = A' \circ (A \times B) \quad (6)$$

with a membership function:

$$\mu_{B'}(v) = \text{Max Min} [\mu_{A'}(u); \mu_R(u, v)] \quad (7)$$

A fuzzy controller is defined by a certain number M of conditional expressions as (4), where the antecedent carries informations related to the system variables to be controlled, and the consequent initializes the control variables based on a principle that may be dictated by the inferential rules of the composition. Attention must be focused on the continuous nature of the input and output variables of the controlled plant. In fact, since the Fuzzy regulator is able to manipulate only linguistic variables, the operations of "fuzzification" for the input variables, and defuzzifications for the output variables are needed, thus allowing the fuzzy regulator to operate in the control chain (Figure 1).

The fuzzification is made by establishing both the variation domains and the membership functions of the input variables to the controller. Moreover, the membership functions determine the conditional instructions to be initialized so allowing to choose the most effective control actions. The system performances will be affected both by the shapes and the number of the membership functions, that are written based on the knowledge of well experienced people, according to the euristic criterion. The most popular membership functions used are the triangular, trapezoidal, and exponential shapes.

The defuzzification is accomplished with preference by the Centroid or the Height methods. The former method allows to determine the output in correspondence of the gravity center of the overall membership function given by the controller:

$$u_0 = \frac{\int \mu(u) \, u \, du}{\int \mu(u) \, du} \quad (8)$$

while the Height method requires to evaluate the centroid of every output and then their average value, weighted by the proper degrees of pertinence:

$$u_0 = \frac{\sum_{i=1}^{M} \mu(u_i) \, u_i}{\sum_{i=1}^{M} \mu(u_i)} \quad (9)$$

Synthesis of a Fuzzy regulator

The speed control of a Salient Permanent Magnet Synchronous motor drive system is now analyzed. According to the two axis theory, written in the rotor reference frame, two equations respectively on the q- and d-axis can model the electric behaviour of a PMS motor:

$$v_q = r_s i_q + \frac{1}{\omega_b} \frac{d\psi_q}{dt} + \frac{\omega_r}{\omega_b} \psi_d \quad (10)$$

$$v_d = r_s i_d + \frac{1}{\omega_b} \frac{d\psi_d}{dt} - \frac{\omega_r}{\omega_b} \psi_q \quad (11)$$

The following expressions for the flux linkages of the stator circuits can be used [8], due to the anisotropic structure of the rotor:

$$\psi_q = X_{\ell q} i_q + X_{aq} i_q \quad (12)$$

$$\psi_d = X_{\ell d} i_d + m (i_d + I_0) \quad (13)$$

where I_0 is the constant value of a current source supplying a fictitious excitation wound circuit that gives a constant flux equivalent to the magnet action; $X_{\ell d}$ and $X_{\ell q}$ are the leakage reactances of the stator windings, here considered as unsaturable parameters; X_{aq} and m are the mutual reactances, accounting for the common flux linkages respectively between q- and d-axis quantities, and experimentally calculated since they deeply depend on the saturation level.

Under the hypothesis that the stator current vector is kept orthogonal to the rotor magnet flux vector, a simplified model of the drive can be considered, since the electromagnetic torque T_e depends only on the q-axis stator current and is given simply by:

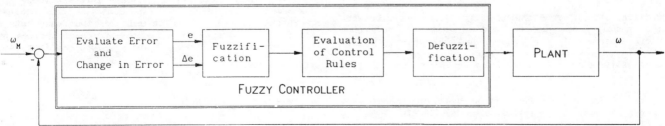

Figure 1 Basic structure of a fuzzy controlled system.

$$T_e = K_T i_q \tag{14}$$

where K_T is the torque constant and $i_q = I_s$ as $i_d = 0$.

To complete the system modeling it is necessary to write the relation that accounts for the mechanical equilibrium on the motor shaft:

$$\frac{d\omega_m}{dt} = \frac{1}{J}(T_e - T_\ell) \tag{15}$$

where J is the inertia of the overall rotating masses; T_ℓ is the load torque including all the frictions of the system.

The most significative variables entering the fuzzy regulator have been selected as the speed error and its time variation. The regulator output is the variation of the command current, established through an euristic logic according to the following items:
1) if the error and its derivative are zero, the previous value of the reference current is to be held;
2) if the error and its derivative are not zero, and tend to correct their-self in a very short time, the previous value of the reference current is to be held too;
3) in all other cases, a variation of the reference current has to be established depending on the error and its derivative.

To proceed on the way of input fuzzification, the previous items need to be transduced from the level of assertions to the level of conditional instructions typical of the fuzzy logic. First, an error band of the speed reference has been defined where the fuzzy control take place. In fact, in order to improve the performances of the fuzzy controller its parameters have been optimized for a fine control, close to the null error. Outside such band a simple proportional regulator has been adopted. The error inside the defined band as well as the range of the error derivative have been normalized between -128 and 128, according to the features of the used fuzzy chip, and split into seven different levels. A linguistic value has been attributed to every level in the following way:

NB   Negative Big
NM   Negative Medium
NS   Negative Small
Z   Approximately Zero
PS   Positive Small
PM   Positive Medium
PB   Positive Big

The same concepts have been applied to the variation of the reference current in the regulator output, and due to the strong control that can be required, two more linguistic values have been added:

NVB   Negative Very Big
PVB   Positive Very Big

In order to trace the euristic membership functions we need to define some quantities:

$e(k) = \omega_r^*(k) - \omega_r(k)$ speed error
$\Delta e(k) = e(k) - e(k-1)$ error variation
$\Delta i_q(k)$ variation of the output q-axis current
k sampling time

Membership functions have been associated respectively to the three variables (two inputs, error speed and its derivative, Figures 2, and 3; the output, the current command, Figure 4) that for the sake of simplicity have been chosen with triangular shapes.

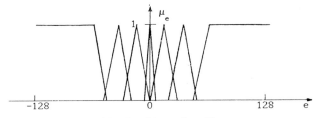

Figure 2 Membership functions for the error e.

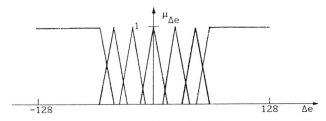

Figure 3 Membership functions for the error variation Δe.

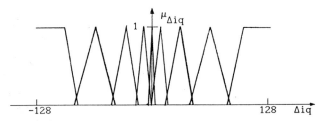

Figure 4 Membership functions for the Δi_q variation of the current.

The thresholds defining the different levels as well as the variation of the output current, have been optimized by successive iterations. Moreover, in order to get a good sensitivity, a higher density of levels has been built near to the zero value of the control variables. The fuzzy sets are seven for the error as well as for its derivative, thus a total of 7x7=49 combinations take place. Every combination is associated to a conditional instruction as follows:

$$\text{If } e \text{ is } \ldots \text{ and } \Delta e \text{ is } \ldots \text{ then } \Delta i_q \text{ is } \ldots \tag{16}$$

The statements have been established being associated to a PD controller and are collected in Table I.

Table I
Statements for speed control

Δe \ e	NB	NM	NS	Z	PS	PM	PB
NB	NVB	NVB	NVB	NB	NM	NS	Z
NM	NVB	NVB	NB	NM	NS	Z	PS
NS	NVB	NB	NM	NS	Z	PS	PM
Z	NB	NM	NS	Z	PS	PM	PB
PS	NM	NS	Z	PS	PM	PB	PVB
PM	NS	Z	PS	PM	PB	PVB	PVB
PB	Z	PS	PM	PB	PVB	PVB	PVB

The control law is obtained by connecting the 49 statements such as:

If e is NB and Δe is NB then Δi_q is NVB or
If e is NB and Δe is NM then Δi_q is NVB or
IF .... or
If e is PB and Δe is PB then Δi_q is PVB or

Every relation is evaluated by previous determination of the membership degree as the minimum between different memberships of fuzzy sets concurrent to their definition. The actual output is finally deduced by the fuzzification. The step by step control procedure is calculated at every sampling interval by:
1) evaluation and normalization of the error e(k) and its variation Δe(k);
2) determination of the membership degree for all the fuzzy sets $\mu_{ei}(e)$, and $\mu_{\Delta ei}(\Delta e)$;
3) determination of the membership degree μ_{Rij} associated to every relation using the Min operator: $\mu_{Rij} = Min(\mu_{ei}, \mu_{\Delta ej}(\Delta e))$;
4) defuzzification and denormalization of $\Delta i_q(k)$;
5) calculation of the command current by the following iterative procedure:

$$i_q(k) = i_q(k-1) + \Delta i_q(k) \qquad (17)$$

According to the deffuzification algorithm implemented in the used fuzzy microcontroller chip, the obtained output is simply the consequent of the most true antecedent. Then, in order to improve the resolution of the control, the output of the fuzzy controller is multiplied, after the denormalization, times the absolute value of the error.

Improved Fuzzy Control Scheme

Firstly, the outlined procedure has been applied to the speed control of the PMS drive. A trapezoidal trajectory for the speed reference has been selected as in Figure 5, characterized by an initial line with constant acceleration, a medium line with constant speed, and finally a constant deceleration. The corresponding trajectory of the acceleration reference is shown in Figure 6.

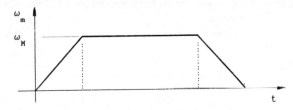

Figure 5   Trajectory of the speed reference.

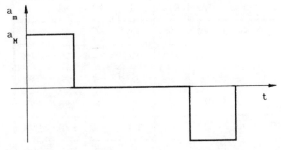

Figure 6   Trajectory of the acceleration reference.

It should be notice that the maximum speed ω_M and acceleration a_M have been established according to the mechanical ratings of the motor which impose superior limits to both such quantities. The shown speed trajectory, that is widely used in the field of robotic arms since it minimizes the time to move from different positions, has been used to analyze the system behaviour by computer simulation. In order to properly explore the system capabilities in detail, a control scheme with a conventional fuzzy logic based speed controller and a conventional MRAC scheme developed from the hyperstability theory have been compared. During the first acceleration the inertia has been suddenly increased eleven times from the rated value while a step load, equal to the rated stall torque, has been imposed during the first constant speed segment. The reported results in Figure 7 show as the proposed scheme is faster than a conventional MRAC in compensating the parameters variation while compared to a standard fuzzy controller it is able to nullify any error due to variations in the process dynamic.

As it is known, the system to be controlled and the reference speed trajectory are modeled by the following equations:

$$\frac{d\omega_r}{dt} = \frac{K_T}{J} i_q - \frac{T_\ell}{J} \qquad (18)$$

$$\frac{d\omega_m}{dt} = e_m \qquad (19)$$

where e_m is the acceleration reference. The two models have the same dynamic behaviour if the i_q current component is given by:

$$i_q = \frac{J}{K_T} e_m + \frac{T_\ell}{K_T} \qquad (20)$$

In the hypothesis that the inertia momentum, load torque, and torque constant are known, the relation (20) allows to obtain the desired dynamics. Unfortunately, although the ranging of such quantities is known a priori, the actual values depend on the operating conditions, so it is useful to arrange the control law in a different way:

$$i_q = k_p a_m + i_{qL} \qquad (21)$$

k_p and i_{qL} are suitable control parameters equal respectively to (J/k_T) and (T_ℓ/k_T) at rated conditions. As such relations are satisfied and the error at the output of the regulator is zero, equations (21) and (20) coincide and the speed control is realized as required. In other cases, an error will appear due to the difference between the actual speed and the reference speed. Due to such error, a corrective action will originate from the fuzzy regulator through a variation Δi_q in the command current in order to correct the speed. It may be useful to use such variation also to correct the values of the parameters k_p and i_{qL}. By differentiating the relation (21) we can account the effects of the parameter variations on i_q:

$$\Delta i_q = \Delta k_p a_m + \Delta i_{qL} \qquad (22)$$

If Δi_q is regarded as the output of the fuzzy regulator, that means as the particular variation of the current necessary to compensate the error, then according to relations (17), (21), (22) we can write:

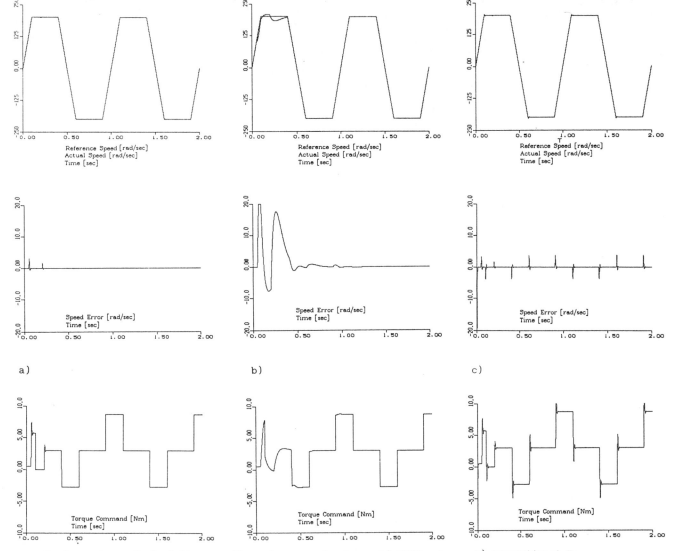

Figure 7 Comparison of simulation results; a) proposed scheme; b) MRAC scheme; c) conventional Fuzzy.

$$i_q(k) = i_q(k-1) + \Delta i_q(k) =$$
$$= \left[k_p(k-1) + \Delta k_p(k)\right] a_m(k) + i_{qL}(k-1) + \Delta i_{qL}(k) \quad (23)$$

Imposing that the current variation command in the fuzzy regulator is exactly equal to the value expressed by the relation (22), (22) can be used to determine the increments Δk_p and Δi_{qL}. However, the relation (22) is not able to give a single value for the unknowns Δk_p and Δi_{qL}, and a possible solution is given by the Moore-Penrose algorithm:

$$\Delta k_p(k) = \frac{h_p a_m(k)}{1 + a_m^2(k)} \Delta i_q(k) \quad (24)$$

$$\Delta i_{qL}(k) = \frac{h_{qL}}{1 + a_m^2(k)} \Delta i_q(k) \quad (25)$$

with h_p, $h_{qL} > 0$.

Equations (24)-(25) allow a physical understanding of the actions to be applied to the system under control. In fact, when the speed reference is constant, the error may only originate from a load torque variation, therefore the corrective action can be applied to the Δi_{qL} term only according to equation (25). Instead, as the speed is changing with strong acceleration the contribution to the error is essentially arising from the inertia torque which is much higher than the load torque. Therefore the corrective action is better achieved by acting on Δk_p, according to equation (24). A particular case happens if the inertia as well as the torque load are simultaneously changing, since the former is not considered separately from the latter, and as a consequence a wrong estimate of the ratio (J/k_T) would be accounted. Nevertheless, since the command current is in any case fixed by the fuzzy regulator, the system will always maintain the performances of robustness typical of such kind of control. At constant speed the current component i_{qL}, that is the ratio (T_ℓ/k_T), is correctly identified so that the value of the ratio (J/k_T) is rearranged in the successive phase of constant acceleration. In absence of other variations, the coefficients k_p and i_{qL} are correctly established and the exact value of current, needed to obtain the desired performances in the following transients, is fixed by equation (21), without any error variation in presence of slope changes.

The proposed control scheme can be seen as a particular case of Model Reference Adaptive Control (MRAC), in which the output of a fuzzy system is used to update the control parameters. In fact it can be observed that the proposed scheme still maintains the characteristic structure of a MRAC, including a speed reference and a regulator whose gains are updated to minimize the error between reference and system dynamics. In our case the reference dynamic is assigned by means of suitable speed profiles but more generally it can be also assigned by means of a reference transfer function. Compared to a conventional MRAC scheme a novel adaptive algorithm is used instead of a conventional algorithm based on gradient approach and others derived from the Lyapunov stability theory and the hyperstability theory of non linear systems [9, 10]. All these methods require a large amount of computations that make complex any practical application while giving often unsatisfying results especially when the order of the controlled process increases. Regarding to this aspect the proposed scheme seems to be an interesting alternative to conventional MRAC schemes, since it is in addition characterized by a simple and low cost implementation.

Simulation Results

Simulation tests, dealing with position transients, have been carried out using the ACSL language and here reported in order to confirm the features of the proposed control scheme. The simulated system is composed of a digital controller with 1 ms sampling time, a fully modeled 12 bit encoder, a current controlled PWM inverter and a PMS motor.

In Figure 8 the actual position, speed, load inertia and torque diagrams are reported for the considered control system when a square wave position reference signal is supplied and step variations of inertia and load torque take place. As it can be seen, using the fuzzy adaptive control scheme the dynamical error is kept small while the system is self retuning. Moreover, it is shown that due to the sudden load variation negative inertia coefficients are synthesized during the transient behaviour. Such negatives values are incorrect under a physical point of view, but they appear since the proposed control scheme try to compensate a positive increase of the load, during a negative acceleration, acting only on the inertia coefficient. Such characteristic shows as the fuzzy control works properly during a critical phase for the adaptive algorithm.

Experimental Results

A prototype system has been assembled and tested in order to practically evaluate the real performances and the implementation problems of the proposed control scheme. The realized system (Figure 9)

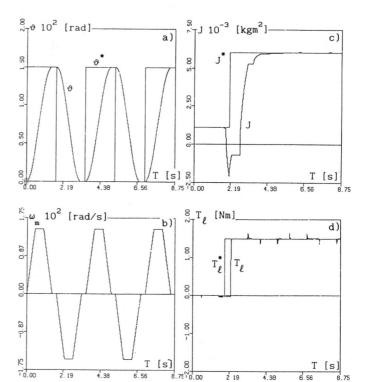

Figure 8 Simulation results: a) actual and reference positions; b) motor and reference speeds; c) inertia; d) reference and load torques.

is composed of a Fuzzy microcontroller board, a digital control system based on a 80486 PC system, a 12 bit encoder, a 1.5 kW six poles Permanent Magnet motor mechanically coupled to a DC PM generator trough a 1/64 ratio reduction gear, and an Hysteresis current controlled PWM inverter.

The used microcontroller, the NLX 320 Fuzzy chip from American Neuralogix, applies in parallel a set of fuzzy rules to the vector of inputs. This makes such device more fast and efficient in performing Fuzzy operations than conventional digital microprocessors that execute sequential algorithms. Unfortunately such first generation of Fuzzy device only supports linear symmetrical membership functions and the simplest "min/max inference method". Up to 64 rules can be stored and used with a processing rate of 30 million rules/second. The fuzzy microcontroller can process a vector of up to eight inputs of 8-bit, producing an immediate or accumulative 8-bit output.

As large on-line changes of the system inertia are very difficult to be experimentally realized, in order to test the features of the proposed control scheme, an initial incorrect value for the

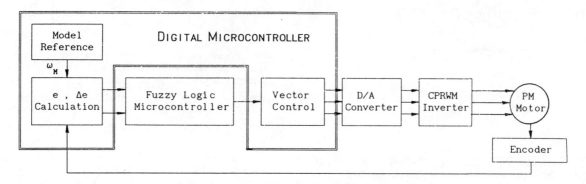

Figure 9 Block diagram of the experimental control system.

inertia coefficient in the control law has been imposed to obtain the equivalent actions exerted by a ten times increase of the system inertia. According to the above approach, the experimental responses of the proposed regulator to a square wave position reference signal and a ten times inertia equivalent increase are reported in Figure 10. A disturbance torque, with value depending on the sign of the speed, is present in the experimental system due to frictions.

As it can be seen in Figure 10, the expected position is unaffected by the inertia variation, moreover the error between the system and the reference speed trajectory, caused by the inertia change, is kept very small while the parameters of the control law are retuning. After the retuning procedure is completed, a periodical oscillation around the steady state value is present on the trace of the estimated inertia. This is due to the disturbance friction torques that are not considered in the adaption law. Such torques are compensated by the control system acting on the inertia coefficient of the control law.

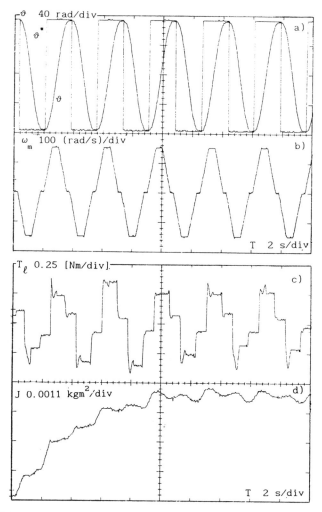

Figure 10 Experimental results: a) actual and reference positions; b) motor and reference speed; c) load torque; d) inertia.

Conclusions

The speed control of a permanent magnet synchronous motor drive by means of unconventional regulators has been investigated. In particular, the fuzzy logic approach has been applied due to its peculiarity to well meet the needs of plants with ill-known dynamic or multi-variable systems. Such an approach has been applied to compensate the variations of both load torque and inertia momentum that are subject to change during the normal behaviour of the drives. Under a reference speed trajectory with trapezoidal shape, the proposed regulator have shown good performances in simulation as well as in experimental tests. The system response has been optimized in correspondence of the changing slope of the speed trajectory, since error peaks arise in the output of the regulator, due to the intrinsic nature of the fuzzy algorithm. An experimental prototype has been realized for a practical evaluation of the proposed scheme using a recently introduced fuzzy microcontroller chip.

Acknowledgements

This work has been supported by the Italian National Council of Research (CNR) under the contract CNR 91.01911.PF67 "Progetto Finalizzato Robotica".

References

[1] L. A. Zadeh, "Outline of a new approach to the analysis of complex systems and decision processes", *IEEE Trans. on Systems Man and Cybernetics*, vol.3, No.1, pp. 28-44, January 1973.

[2] P. J. King and E. H. Mamdani, "The application of fuzzy control Systems to industrial processes", *Automatica*, Vol. 13, pp. 235-242, May 1977.

[3] Fuzzy Microcontroller Development System ADS230, Reference Manual, American NeuraLogix Inc, 1991.

[4] K. L. Tang and R. J. Mulholland, "Comparing fuzzy logic with classical controller designs", *IEEE Trans. on Systems Man, and Cybernetics*, vol. SMC-17, No. 6, pp. 1085-1087 November/December 1987.

[5] Ichiro Michi, N. Nagai, S. Nishiyama, and T. Yamada, "Vector control of induction motor with fuzzy PI controller", presented at the *IEEE Conference on Industry Application Society*, Dearborn, Michigan, September 28-October 4, 1991, pp. 341-346.

[6] G. C. D. Sousa, B. K. Bose, "A fuzzy set theory based control of a phase-controlled converter DC machine drive", presented at the *IEEE Conference on Industry Application Society*, Dearborn, Michigan, September 28 - October 4, 1991, pp. 854-861.

[7] J. C. Johnson and J. J Feeley, "Adaptive linguistic control of a two link robot arm", Proceedings of the *Conference on Applied Motion Control*, 1987, pp. 43-46.

[8] A. Consoli, and A. Testa, "A DSP sliding mode field oriented control of an interior permanent magnet motor drive", Proceedings of the *International Power Electronic Conference*, Tokyo, Japan, April 1990, pp. 296-303.

[9] K. J. Aström, B. Wittenmark, *Adaptive Control*. Boston: Addison Wesley Publishing Company, 1989.

[10] Y .D. Landau, *Adaptive Control - The Model Reference Approach*. New York: Marcel Dekker, 1979.

Fuzzy Algorithm for Commutation of Permanent Magnet AC Servo Motors without Absolute Rotor Position Sensors

Dong-Il Kim, Jin-Won Lee, and Sungkwun Kim

Control R/D Team, Production Engineering Division, Samsung Electronics,
416, Maetan-3-Dong, Kwonsun-Ku, Suwon City, Kyungki-Do 441-742, Korea

Abstract: A control method which drives the permanent magnet AC servo motor without the detection of the rotor position by the absolute position transducer such as an absolute encoder or a resolver is described. An incremental encoder is only coupled to the motor shaft in order to obtain the information for electrical commutation, motor speed, and motor position. A fuzzy algorithm based on the min-max compositional rule of inference is developed for the estimation of the absolute rotor position which is essential to electrical commutation. The nonfuzzy control input is generated from defuzzification by the center of gravity method. The motor speed and the motor position are measured by the method based on pulse interval measurement, which guarantees the same measurement resolution from a standstill to the rated speed. A digital control method which controls motor currents directly is proposed, whose main principle is based on high gain controller. The proposed control algorithm is implemented on the microprocessor control system. Experimental results show that the proposed control system is very effective in controlling the permanent magnet AC servo motor with high dynamic performance.

I. INTRODUCTION

Electrical servo systems are being used more and more owing to the ever-increasing automation of technological processes in industry. The field of application is wide and ranges from CNC (computer numerical controller) machine tools, manipulators, industrial robots, and precise positioning devices to specialized applications such as steppers for manufacturing semiconductors. The characteristics of the electrical servo systems are summarized as high servo response, robust noise suppression, high overload capability, wide speed range, and a good smoothness of movement at very low speeds. To date, DC servo motors have been widely used in obtaining such characteristics, which, however, have some disadvantages such as the mechanical commutation, the regular maintenance due to brush wear, and the losses in the rotor.

Permanent magnet AC servo motors have the inherent advantages such as rugged construction, easy maintenance, high efficiency, high power factor, and the precise synchronous operation compared with DC servo motors. The majority of the disadvantages of DC servo motors can be overcome by using permanent magnet AC servo motors. Recent researches have shown that permanent magnet AC servo motors could become serious competitors to DC servo motors in electrical servo systems [1]-[3].

However, permanent magnet AC servo motors require rotor position measuring devices which make electrical commutation possible. Resolvers, pole sensors, and absolute encoders are generally used as the measuring devices of the absolute rotor position for electrical commutation in permanent magnet AC servo motors [4], [5]. These measuring devices are generally expensive because auxiliary devices such as resolver to digital converters or incremental encoders must be incorporated for the precise detection of the motor speed and the motor position. On the other hand, incremental encoders, although appropriate to the detection of the motor speed and the motor position and relatively cheap from the viewpoint of the same measurement resolution, have a problem in detecting the absolute rotor position for electrical commutation. Both problems can be solved by coupling an incremental encoder to the permanent magnet AC servo motor and estimating the absolute rotor position based on the pulse train from it. Actually, the solution stems from the analysis of the movement of the rotor for the current command whose magnitude and phase are given arbitrarily [6]. Because the analysis can not be performed on dynamical modelling of the permanent magnet AC servo motor, most of the solutions have limitation in compensating nonlinearities and the variation of the moment of inertia completely. The fuzzy control technique can overcome this limitation because of the characteristic of inference mechanism which allows to infer the future values from current values of the qualitative variables.

Conventionally, current control processing is executed by custom analog circuits in the control systems of AC motors including permanent magnet AC servo motors because high execution speed is required in order to obtain impressed current control. However, there are some inherent disadvantages in the analog hardware circuit as follows: The analog circuit requires a lot of adjustments, and it shows system drift and parameter variation according to temperature and supply voltage. Full digital current regulation methods to solve these demerits have been proposed with the help of the recent high speed microprocessors such as digital signal processors (DSPs).

This paper discusses the improved fuzzy rule which converges much faster than the method in [7]. It has been motivated by the fact that the actual rotor position is equal to the arbitrarily selected value of the rotor position at the point where the motor speed is local maximum or minimum. This fuzzy rule for the estimation of the absolute rotor position is implemented with the help of look up tables. It is demonstrated that if combined with the divider circuit of the encoder pulse train, the speed measurement method based on pulse interval measurement becomes available for the accurate detection of the motor speed from a standstill to the rated speed with the same resolution [7]. In addition, the digital control method which controls motor currents directly is proposed, the main principle of which is based on high gain controller. The proposed control algorithm is implemented on the control system whose CPU is DSP ADSP2101. The experimental results demonstrate that the proposed fuzzy algorithm effectively estimates the absolute rotor position, which makes the electrical commutation possible, in spite of large load inertia coupled to the motor shaft and the disturbance torque. It is also shown that the high dynamic performance of the permanent magnet AC servo motor is obtained through the proposed speed detection, and speed and current control method.

II. ROTOR POSITION ESTIMATION

Before discussing the basic estimation algorithm of the rotor

position of the permanent magnet AC servo motor with an incremental encoder, let us consider the following assumptions:
1) Saturation is neglected although it can be taken into by parameter changes,
2) The back emf is sinusoidal,
3) Eddy currents and hysteresis losses are negligible.

Under this assumption, if the permanent magnet AC servo motor is being fed by impressed currents [8],[9], where motor currents are essentially sinusoidal and are regulated within fixed bands, the dynamic behavior of the permanent magnet AC servo motor can be simplified to

$$d\omega_r/dt = -B\omega_r/J + (T_e - T_L)/J, \quad (1)$$

where

$$T_e = K_T\Phi_m(u_1\sin\theta_r + u_2\sin(\theta_r + 2\pi/3) + u_3\sin(\theta_r + 4\pi/3)), \quad (2)$$

ω_r, θ_r are the motor speed and the rotor position, respectively, B is the damping coefficient, J is the moment of inertia, T_L is the disturbance torque, K_T is the torque constant, Φ_m is the magnetic flux of the permanent magnet rotor, and u_1, u_2, u_3 are stator current commands for phase A, B, C, respectively. If u_1, u_2, and u_3 are given by

$$[u_1\ u_2\ u_3]^T = [I_m\sin\theta_r'\ \ I_m\sin(\theta_r' + 2\pi/3) \ \ I_m\sin(\theta_r' + 4\pi/3)]^T, \quad (3)$$

where I_m is the amplitude of stator current commands and θ_r' is the arbitrarily assumed value of the absolute rotor position θ_r, then the resultant generated torque is expressed as

$$T_e = 1.5K_T\Phi_m I_m\cos(\theta_r - \theta_r'). \quad (4)$$

Equation (4) shows that θ_r' should be equal to θ_r in order to obtain maximum steady torque without pulsation. This means that the information of the absolute rotor position should be obtained.

The absolute rotor position can be easily obtained by coupling a resolver or an absolute encoder to the motor shaft, which outputs the absolute position value with respect to the reference point. However, such position measuring devices have demerits in cost or complexity in signal processing compared with incremental encoders with the same resolution. Therefore, it is required to develop the detection method of the absolute rotor position by an incremental encoder. In [7], the fuzzy rule was shown to be effectively applied to the estimation of the absolute rotor position.

Now, the basic estimation method of the absolute rotor position by using fuzzy logics is described [7]. First, consider the distance where the rotor moved and the number of passed sampling periods to be the situations (preconditions), and I_m and the distance where the rotor will move to be the action to be performed (postcondition). Then, the postconditions, which will be transformed into the actual inputs to the estimation algorithm of absolute rotor position, are specified by the production rules that satisfy preconditions at every sampling period. The fuzzy rules for the estimation of the absolute rotor position are expressed as "If situation, then action". In the proposed fuzzy rule, the following linguistic values are defined:

PB: Positive Big
PM: Positive Medium
PS: Positive Small
ZO: Zero
NS: Negative Small
NM: Negative Medium
NB: Negative Big

A fuzzy knowledge base and an inference mechanism determine the fuzzy rules in Table I for the estimation of the absolute rotor position on the basis of the aforementioned basic idea [10].

The access to the membership function must be performed in order to make the linguistic values in Table I be fuzzy. Here, the fuzzy rule will be implemented by look-up tables in the actual implementation, so the discrete membership function by Mamdani is introduced [11]. This process requires the quantization of the input and output variables. The quantized variables are shown in Table II, from which the discrete membership function given by Table III is derived.

TABLE I
The fuzzy rules.
(a) The fuzzy rule for the distance where the rotor will move.

N\D	NB	NM	NS	ZO	PS	PM	PB
ZO	ZO	PM	PS	ZO	NS	NM	ZO
PS	ZO	PM	PS	ZO	NS	NM	ZO
PM	ZO	PS	PS	ZO	NS	NS	ZO
PB	ZO	PS	ZO	ZO	ZO	NS	ZO

(b) The fuzzy rules for I_m.

N\D	NB	NM	NS	ZO	PS	PM	PB
ZO	ZO	PS	NS	ZO	NS	PS	ZO
PS	ZO	PM	PS	PS	PS	PM	ZO
PM	ZO	PB	PM	PM	PM	PB	ZO
PB	ZO	PB	PB	PB	PB	PB	ZO

N: Number of passed sampling periods
D: Distance where the rotor moved.

TABLE II
The quantization of input and output variables.

D(°)	N	Dis.(°)	I_m(%)	Q.V
-180		-90		-6
-80		-40		-5
-40		-20		-4
-20		-10		-3
-10		-5		-2
-5		-2		-1
-5 ~ +5	0	-2 ~ +2	0	0
5	20	2	20	1
10	40	5	40	2
20	60	10	60	3
40	75	20	75	4
80	90	40	90	5
180	100	90	100	6

Dis.: Distance where the rotor will move.
Q.V : Quantized variable

Then, the fuzzy rule, which is based on the min-max compositional rule of inference, is obtained from the membership functions defined by the linguistic rules in Table I. This control rule is defuzzified to generate the nonfuzzy control inputs, that is the amplitude of the current command and the assumed value of the absolute rotor position, by the center of gravity method.

$$I_m = \sum_{i=1}^{n}(u_{1i} * U_{1i}) / \sum_{i=1}^{n} u_{1i},$$

$$\theta_r' = \sum_{i=1}^{n}(u_{2i} * U_{2i}) / \sum_{i=1}^{n} u_{2i}, \quad (5)$$

where n is the number of rules, u_{1i} and u_{2i} are the membership functions for I_m and θ_r', respectively, and U_{1i} and U_{2i} are the representative values for I_m and θ_r' as the control inputs, respectively.

TABLE III
Discrete membership function.

	-6	-5	-4	-3	-2	-1	0	1	2	3	4	5	6
PB	0	0	0	0	0	0	0	0	0	0	0	0.3	0.7	1
PM	0	0	0	0	0	0	0	0	0.3	0.7	1	0.7	0.3
PS	0	0	0	0	0	0	0.3	0.7	1	0.7	0.3	0	0
ZO	0	0	0	0	0.3	0.7	1	0.7	0.3	0	0	0	0
NS	0	0	0.3	0.7	1	0.7	0.3	0	0	0	0	0	0
NM	0.3	0.7	1	0.7	0.3	0	0	0	0	0	0	0	0
NB	1	0.7	0.3	0	0	0	0	0	0	0	0	0	0

TABLE IV
Look up table for control input.
(a) Look up table for Table I (a).

N\D	-6	-5	-4	-3	-2	-1	0	1	2	3	4	5	6
0	0	2	4	3	2	1	0	-1	-2	-3	-4	-2	0
1	0	2	4	3	2	1	0	-1	-2	-3	-4	-2	0
2	0	2	4	3	2	1	0	-1	-2	-3	-4	-2	0
3	0	2	3	2	2	1	0	-1	-2	-2	-3	-2	0
4	0	1	2	2	2	1	0	-1	-2	-2	-2	-1	0
5	0	1	2	1	1	0	0	0	-1	1	-2	-1	0
6	0	1	2	1	0	0	0	0	0	-1	-2	-1	0

(b) Look up table for Table I (b).

N\D	-6	-5	-4	-3	-2	-1	0	1	2	3	4	5	6
0	2	2	2	1	0	0	0	0	0	1	2	2	2
1	3	2	2	1	1	1	1	1	1	1	2	2	3
2	4	3	2	2	2	2	2	2	2	2	2	3	4
3	5	4	3	3	3	3	3	3	3	3	3	4	5
4	6	5	4	4	4	4	4	4	4	4	4	5	6
5	6	5	5	5	5	5	5	5	5	5	5	5	6
6	6	6	6	6	6	6	6	6	6	6	6	6	6

As a result, the look up tables in Table IV for linguistic rule of I_m and θ_r' are obtained, which will be used in the actual microprocessor control system for the estimation of the absolute rotor position. At the instant when the execution of fuzzy estimation algorithm ends, the value $\pi/2 + \theta_r'$ becomes the absolute rotor position, as discussed in [7].

The proposed fuzzy algorithm is very effective in estimating the rotor position, but if it converges faster, the better control performance will be obtained at the transient state. The slight modification of the proposed estimation algorithm can make the convergence time much faster. Practically, if the algorithm is executed after the rotor position is roughly known, then the convergence time can be reduced greatly.

If the permanent magnet AC servo motor is initially fed by impressed currents commanded on the basis of (5) where I_m and θ_r' are given arbitrarily, then magnetic flux is produced at the position perpendicular to the arbitrarily selected position θ_r' in the stator. The permanent magnet rotor moves toward that position, and it oscillates around that position for a while until stop due to the inertia of the rotor. The generated torque at the position where the rotor has stopped is zero, in other words, the difference between θ_r' and θ_r is $\pi/2$, so θ_r' must be replaced by the previous θ_r' plus $\pi/2$ in order to always maximize the generated torque without pulsation. This is the basic idea of the rotor position estimation algorithm. The waveforms of the rotor position and the motor speed during this process is shown in Fig.1. From the scrutiny of Fig.1, one can obtain the idea which makes the aforementioned fuzzy estimation algorithm more efficient. The actual rotor position is shown to be equal to the arbitrarily selected θ_r' at the point t_1 or t_2 where the motor speed is local maximum or minimum in Fig.1. This apparently indicates that if $\pi/2$ is added to θ_r' at that point, then the permanent magnet AC servo motor with an incremental encoder can be ready to be driven within shorter time than by the above basic algorithm. Even though this method is very effective, the motor can not stop immediately because the rotor will move and pass θ_r' due to the inertia of the rotor, so not be in the complete ready state. Because the difference

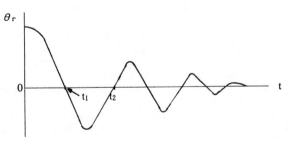

(a) The waveform of θ_r in case of $\theta_r' = 0$.

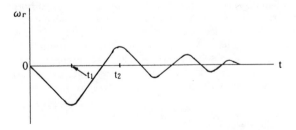

(b) The waveforms of ω_r.

Fig.1 The waveforms of the rotor position and the motor speed in case of arbitrarily selected I_m and θ_r'.

between the estimated and actual rotor positions around t_1 or t_2 is small, the magnitude of the following oscillation will be greatly reduced if the aforementioned fuzzy estimation algorithm is executed at t_1 or t_2. As a consequence, it becomes possible to estimate the actual rotor position just after the execution of the fuzzy estimation algorithm. This modified fuzzy estimation algorithm using the information of the rotor position can reduce the convergence time greatly than the fuzzy estimation algorithm in [7].

III. DIGITAL CURRENT CONTROL

The output of velocity control loop and the commutation information generate the stator current commands of desired magnitude and frequency for A, B, and C phases. They are compared with the actual phase currents by high gain controllers in the digital current control loop in order to make the permanent magnet AC servo motor fed by impressed currents. If the actual current exceeds the current command, the PWM signals are calculated to turn off the upper power switching element and turn on the lower power switching element in the MOSFET half-bridge inverter. On the contrary, if the current command exceeds the actual current, the PWM signals are calculated to turn on the upper power switching element and turn off the lower power swiching element in the half-bridge. These PWM signals calculated in the microprocessor are transmitted to the gate drive circuits of the MOSFET inverter through the digital PWM signal generators which consist of timer/counters. The block diagram of the digital PWM signal generator is shown in Fig.2. By this digital current control method, the dynamic behavior of the permanent magnet AC servo motor can be simplified to (1) which is similar to that of a DC servo motor.

IV. DIGITAL SPEED MEASUREMENT

When the power is on, the fuzzy estimation algorithm of the absolute rotor position is executed within very short time, then the ready signal is generated which indicates that the permanent magnet AC servo motor is ready to be driven in the speed or position control mode. The precise detection of the motor speed and the motor position is essential to the high control performance of the permanent magnet AC servo motor in the microprocessor control system.

In this paper, the speed measurement method based on pulse interval measurement shown in Fig.3 is proposed, which makes it possible to detect the motor speed precisely with the same resolution in the whole speed range [7]. From Fig. 3, it is apparent that the motor speed is given by

$$\omega_r = 6*10^4 f_c/(Mn), \qquad (6)$$

where f_c is the clock frequency in MHz, M the number of the pulses from the incremental encoder per revolution, and n the number of the clock pulses for interval T_w. In this method, the higher f_c is required in order to enhance the accuracy of the speed measurement because fewer clock pulses are obtained over T_w as the motor speed increases. However, increasing f_c both can not guarantee the same measurement resolution in the whole speed range and is limited by the physical limitation in hardware. Such a problem can be eliminated by utilizing the pulse train whose frequency is 1/N (N=1,...,127) times as high as that from the incremental encoder. The function of that pulse train is to provide the pulse interval for speed measurement, which is N times as long as that of T_w. The value of N is determined according to the motor speed so as to guarantee the detection of the motor speed with the same measurement resolution in the whole speed range. Consequently, the motor speed from a standstill to the rated speed is measured with the same resolution with the help of new pulse trains. The motor speed is calculated from (6).

Table V shows the relationship between the motor speed and the value of N which was used in experiments. In experiments, the whole speed range of the tested motor was divided into 9 steps for the simplicity of implementation. The block diagram of the hardware structure, which realizes the proposed speed measurement method, is shown in Fig. 4, where INA, INB are new encoder pulse trains.

Fig.3. Digital speed measurement principles from incremental encoders.

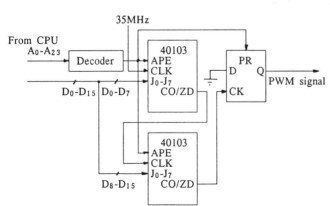

Fig.2. The block diagram of the digital PWM generator.

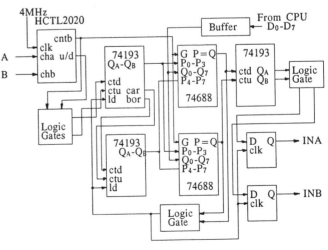

Fig. 4. The hardware construction for the proposed speed measurement method.

TABLE V
The relationship between 1/N and the motor speed.

Speed(rpm)	1/N	Speed(rpm)	1/N
0 - 200	4	1600 - 2000	1/5
200 - 400	2	2000 - 2400	1/6
400 - 800	1/2	2400 - 2800	1/7
800 - 1200	1/3	1600 - 3200	1/8
1200 - 1600	1/4	3200 -	1/8

over the pulse interval T_w of the pulse train from the incremental encoder. 100 and 300% load discs are coupled to the permanent magnet AC servo motor to investigate the performance of the proposed fuzzy estimation method of the rotor position under load inertias.

In the first experiments, the performances of the fuzzy estimation algorithm under 100% and 300% load inertias are investigated. Fig.7 shows the experimental results of estimating the absolute rotor position through the proposed fuzzy algorithm in case of 100% load inertia coupled to the motor shaft. On the other hand, The experimental results in Fig.8 demonstrates the procedure estimating the absolute rotor position in case of 300% load inertia. From both results, one can see that the convergence time and the magnitude of oscillation at the transient state are greatly reduced by the proposed fuzzy estimation algorithm compared with the method in [7]. The same quantization values as Table II and the

V. EXPERIMENTAL RESULTS

The performance of the proposed control algorithm is investigated in experiments. An acceleration feedback controller and a velocity feedforward controller are incorporated into the velocity PI controller in order to obtain both fast transient response, that is high dynamic performance, and robustness to the disturbance torque. A 4 pole permanent magnet AC servo motor is chosen for the experimental work, whose nominal data are listed in Table VI. The speed drive system implemented for the experimental work consists of a DSP ADSP2101, a MOSFET PWM inverter, and the pre-described permanent magnet AC servo motor. The block diagram representation of the speed drive system implemented for the experimental work is shown in Fig.5 while the actual speed drive systems are shown in Fig.6. Signals between the processor and the permanent magnet AC servo motor are processed through 12 bit A/D converters, D/A converters, and latches. For the detection of the motor speed and position, an incremental encoder whose resolution is 4000 pulses/rev. is coupled to the motor shaft and the pulses whose frequency is 4 MHz are integrated

Fig.6. The picture of the actual speed drive systems.

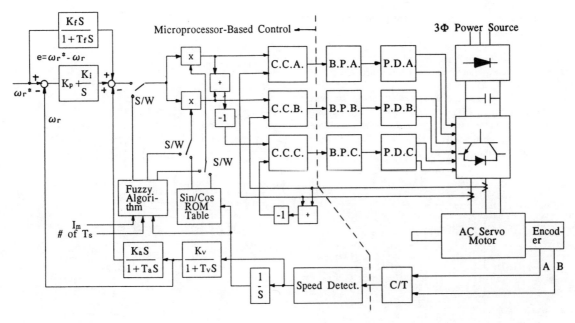

C.C.(A.,B.,C.): current controller for phase A(B,C)
P.D.(A.,B.,C.): PWM signal distributor for phase A(B,C)
$K_p(K_i)$: proportional (integral) gain
K_a, (K_f, K_v): Acceleration feedback (velocity feedforward, velocity feedback) gain
B.P.(A.,B.,C.): basic PWM signal of phase A(B,C)
C/T: counter/timer   S/W: switch
T_f, T_a, T_v: time constants T_s: sampling period

Fig.5. The speed drive system implemented for the experimental work.

same initial values of the rotor position were used in experiments to compare the performances of both algorithms.

The performance of speed measurement is demonstrated in the second experiment, where the speed command was given by a rectangular waveform with the amplitude of the rated speed. Fig.9 shows that the motor speed is measured with good performance by the proposed method based on pulse interval measurement.

From the experimental results, one can see that the proposed control algorithm consisting of the fuzzy estimation algorithm of the rotor position, the digital current algorithm, and the speed measurement algorithm can drive the permanent magnet AC servo motor with an incremental encoder with high dynamic performance.

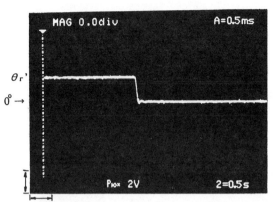

Fig.7. Experimental results in case of 100% load inertia.
(y-axis: 100°/div)

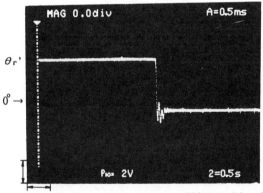

Fig.8. Experimental results in case of 300% load inertia.
(y-axis: 100°/div)

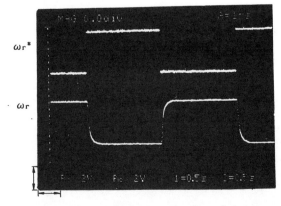

Fig.9. The experimental result of speed measurement.
(y-axis: 1600rpm/div)

TABLE VI
The nominal parameters of the tested permanent magnet AC servo motor.

number of pole pairs	4	rated speed	3000rpm
rated output	300W	rotor inertia	0.69gcms²
rated torque	9.7kgcm	torque constant	2.14kgcm/A
instant. max torque	29.1kgcm	mechanical time constant	1.9ms
power rate	8.01kW/s	electrical time constant	2.7ms
armature resistance	1.22Ω	armature inductance	3.3mH

VI. CONCLUSION

A fuzzy algorithm has been proposed which makes it possible to estimate the absolute rotor position essential to the electrical commutation of the permanent magnet AC servo motor with an incremental encoder coupled to the shaft. The improved fuzzy algorithm based on the method of [7] estimates the absolute rotor position with the desirable convergence time and little oscillation at the transient state. By this algorithm, the permanent magnet AC servo motor with an incremental encoder can be always controlled with maximum torque per ampere of stator current without pulsation except the transient state of the estimation.

The speed measurement method, which is based on pulse interval measurement and utilizes pulse train whose frequency is 1/N (N=1,...,127) times as high as that of the pulse train from the incremental encoder, has been shown to detect the motor speed precisely with the same measurement resolution in the whole speed range.

Experimental results show that the proposed control scheme including the fuzzy estimation algorithm of the rotor position, the digital current algorithm, and the speed measurement algorithm is very effective in controlling the permanent magnet AC servo motor with high dynamic performance.

Further researches should be directed toward the analysis of the overall control system mathematically.

REFERENCES

[1] Alfio, C., and Alfonso, A.,"Transient Performance of Permanent Magnet AC Servo Motors," IEEE Trans. Ind. Appl., vol. 22, pp.32-41, 1986.

[2] Pillay, P. and Krishnan, R., "Modelling of Permanent Magnet Motor Drives," IEEE Trans. Ind. Elec., vol.35, pp. 537-541, 1988.

[3] Gerhard, P., Alois, W, and Albert F. W., "Design and Experimental Results of a Brushless AC servo Drive, IEEE Ind. Appl., vol.20, pp.814-821, 1984.

[4] Erland K. P. and Saeid M., "Brushless Servo System with Expanded Torque-Speed Operating Range," MOTOR-CON'85, pp. 96-106,1985.

[5] Samsung Electronics, AC Servo Drives-SSV Series : User's Manual, 1990.

[6] Kim, D. I., Lee, J. W., and Kim, S. "Control of Permanent Magnet AC Servo Motors without Absolute Position Transducers," IEEE PESC'92, Boston, U.S.A., pp.578-585, 1991.

[7] Kim, D. I., Lee, J. W., and Kim, S. "Control of Permanent Magnet AC Servo Motors via Fuzzy Reasoning," will appear to IEEE '92 IAS Annual Meeting.

[8] Brod, D. M. and Novotny, D. W., "Current Control of VSI-PWM Inverter," IEEE Trans. Ind. Appl., vol.21,pp.562-570, 1985.

[9] Kim, D. I.," Control of Induction Motors via Singular Perturbation Technique and Nonlinear Feedback Control," IEEE IECON'90, Asilomar, U.S.A., pp.909-914, 1990.

[10]Zadeh, L. A.,"Fuzzy Sets," Inform. Contr., vol.8, pp.338-353, 1965.

[11]Mamdani, E. H., "Application of Fuzzy Algorithms for Control of a Simple Dynamic Plant," Proc. of IEEE, vol.121, pp.1585-1588, 1974.

FUZZY LOGIC-BASED CONTROL OF FLUX AND TORQUE IN AC-DRIVES

HOFMANN, WILFRIED and KRAUSE, MICHAEL

Technische Universität Dresden, Elektrotechnisches Institut, Mommsenstraße 13,
DO-8027 Dresden, Germany

ABSTRACT: A new control method based on fuzzy logic is proposed for the regulation of torque and flux in ac-drives fed by PWM-inverters. The input signals of the controller are transformed by fuzzy sets in six categories valued by membership functions. The interaction between the input signals and the outputs is realized by fuzzy logic consisting of special production rules corresponding with the well-known principle of Direct Self Control. Two strategies are derived to obtain the optimal voltage phasor and the pulse-time of the inverter switches. The experimental results show the advantages of the new control regime by keeping constant the pulse frequency using a nonlinear controller in connection with a pulse-width modulator.

1. INTRODUCTION

In general the regulation of ac-drives has been based on the precise measured values, exactly calculated model signal and an accurate preparation of output signals. It requires a high technical expense and has control properties depending on some parameters in the most of the cases and does not operate without problems in the entire range of speed. Besides a constant sampling rate combined with a nonlinear controller leads to exceeding of the given limits of tolerance. Some improvements are possible using a fuzzy logic-based control concept. This controller is combined with the principle of the Direct Self Control (DSC) created by DEPENBROCK [1] for the regulation of flux and torque in ac-drives.

2. FLUX AND TORQUE CONTROL
2.1. PHYSICAL BASIS

For using a regulating element, see fig.1, it is well-grounded to impress the stator flux with help of the directly influenced stator voltage by the inverter.
The stator flux is known as a mathematical space phasor in the form

$$\Psi_s = \int (u_s - R_s i_s) \, dt \qquad (1)$$

and shows that the stator flux can be directly controlled by the stator voltage disregarded the load depended voltage drop of the stator resistance.
The phasor diagram in fig.2 demonstrates the influence of the switched inverter voltage for building and rotating the stator flux space phasor.
The voltage induced in the squirrel-cage windings drives the rotor current and generates the rotor flux with:

$$\dot{\Psi}_r = (-\delta_r + j\omega)\Psi_r + \delta_r \Psi_s \qquad (2)$$

The rotor flux reacts on the changed stator flux with a slow reaction characterized by the time-constant $T_r = 1/\delta_r = L_r/R_r$.

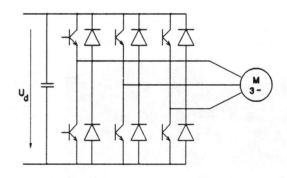

Fig.1.: Drive system

The contribute of the both fluxes follows to the torque impression:

$$m = \frac{3}{2} z_p \frac{1}{L_s} \Phi_s \Phi_r \sin\delta \qquad (3)$$

It follows that the stator flux amplitude and the angle between the both flux space phasors change the torque mainly.
According to the degree of freedom for controlling of the used inverter it results the seven possibilities of switching the voltage phasors. These have diverse effects on the amplitude and the phase angle of the stator flux phasor. In this operation those voltage phasors change the torque most effectively setting the stator flux phasor rotating.

2.2. METHODS OF DIRECT SELF CONTROL

The control methods known from [1] and [2] guiding the stator flux on a hexagonal or circular trajectory use the control action of the stator voltage in a rather different way. It turns out that a circular trajectory of flux has the advantage in the lower range and a hexagonal trajectory provides to the best results in the upper speed range. The combination of both methods is not possible without switching on the control structures.

Another problem is the turning up the switching or pulse frequency on the working point in the range of speed and load. It would be as well working in a great control range with a constant switching frequency on the standpoint of the power electronics for using optimally the installed power rectifier.

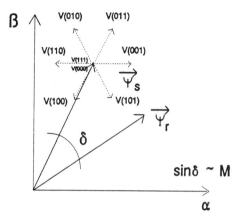

Fig.2.: Phasor diagram of the DSC

One possibility in this way is proposed in [3] with help of the pulse-width modulation for a circular flux trajectory. In that case the controller of flux and torque are linear by nature. Therefore a general method for both trajectories of flux is proposed with the advantage using a common regulation of torque and flux on condition that the pulse frequency is held on a constant value.

3. FUZZY CONTROLLER

3.1. BASIS PRINCIPLE

The first problem of a general controller with a nonlinear characteristic and the conversion of the generated action signals in quasi-continuous (pulse-width modulated) control signals consists in the fact that the direct analytical contribution is not making between the sharp evaluation limits of a two- or three-step controller and a pulse-width modulated output signal. In that reason it is necessary to evaluate the control errors with help of intermediate stages.

The elimination of threshold values leads to a certain graduated range.

A mathematical foundation is provided by the fuzzy theory [4],[5], whose logic is put on the instead of the conventional controllers. The new controller has to accomplish three essential functions:
- to give a relation between the control errors and their gradients,
- to support the implementation of diverse control rules,
- to generate a pulse-width modulated signal from the output signals of the nonlinear controller.

The fuzzy controller accomplishes the three elementary functions: fuzzification, fuzzy logic and defuzzification or decomposition.

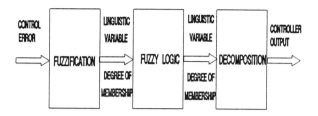

Fig.3: Fuzzy controller

The principle is demonstrated in the fig.5 . The Self controller with the sector selection is even realized by a conventional device and will should be removed in fuzzy sets later.

3.2. FUZZIFICATION

At first the evaluation limits of the control errors fixed for a two-step-controller are dissolved in intermediate values. The control errors of the stator flux and the torque, according to fig.4, is divided in 6 categories from large$^-$ to large$^+$, in which the continuous set membership values h replace the " true"-or-"false"-operations ranging from 0 to 1 with a positive gradient and than from 1 to 0 with a negative gradient.

Following it could be overlapped the membership functions of two adjoining categories.

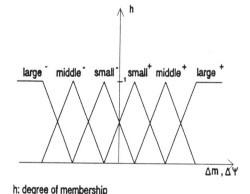

Fig.4: Fuzzy set

The termination of the categories in function of the control errors for the category large$^+$ with the membership value 1 can be compared with the hysteresis value of a conventional two-step-controller and determines the norm of the membership function's argument. Therefore it can be varied the gain of the control system.

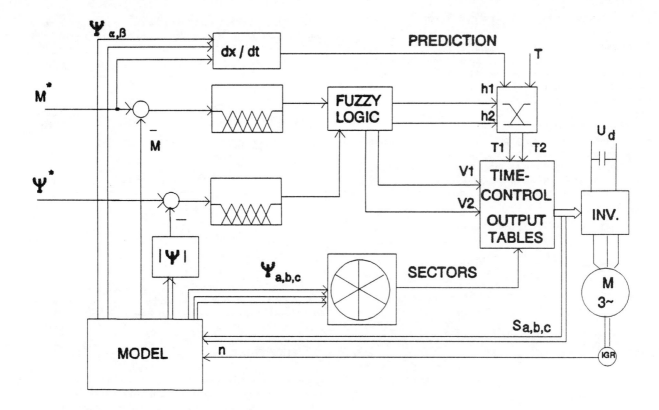

Fig.5: Direct flux and torque regulation with a fuzzy controller

3.3. FUZZY RULES

The fuzzy theory has a predictable quantity of rules with help of them it should be connected the linguistic variables taking along the membership functions. The combination of the composed variables flux- and torque error are prepared by the production rule. A new linguistic variable is produced applying to the change of the rotating direction and the amplitude of flux. Hence the variables are defined:
- Flux forward increase / decrease
- Flux backward increase / decrease
- Flux standstill increase / constant / decrease

The relation between the fuzzy sets of the control errors and the new linguistic variables can be described by the production rules in the form:

IF("*condition A* "AND" *condition B*") THEN "*conclusion C*"

The relation between the fuzzy sets of the control errors and the new linguistic variables, see fig.6, are determined by the control matrix shown in fig.7 . The contents of the control matrix is depended on the given flux-trajectory.

If more production rules lead to the same consequence they are string together by the OR-(fuzzy)-operator.

In the appendix the operations of the fuzzy controller are demonstrated on the hand of a calculated example.

The following table shows a choice of often used fuzzy operators in logical and arithmetical forms:

Logic: AND-operators:

minimum operator:
$m_R(A \wedge B) = \min\{m_R(A), m_R(B)\}$

product operator:
$m_R(A \wedge B) = m_R(A) * m_R(B)$

bounded difference:
$m_R(A \wedge B) = \max\{0, (m_R(A) + m_R(B) - 1)\}$

OR-operators:

maximum operator:
$m_R(A \vee B) = \max\{m_R(A), m_R(B)\}$

sum operator:
$m_R(A \vee B) = m_R(A) + m_R(B) - m_R(A) * m_R(B)$

bounded sum:
$m_R(A \vee B) = \min\{1, (m_R(A) + m_R(B))\}$

Useful rules for the fuzzy-logic are the rules of inference, these are the MAX-inference and the MAX-PROD-inference. If the MAX-PROD-inference is used in this example, the final membership function originates from multiplying the several membership functions h_m and h_ψ of the control errors applied parallel to the production rule. On the other side if the conclusion are differently the selection of the voltage phasor is determined by the greatest degree of membership.

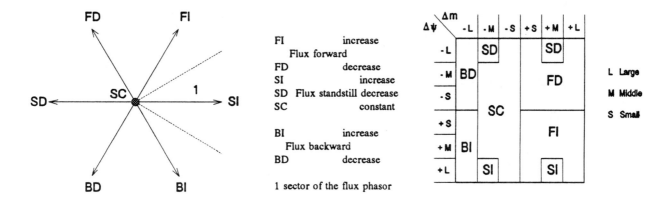

Fig.6: Definition of the new linguistic variable

Fig.7: Control matrix for a circular flux trajectory

3.4. DECOMPOSITION

For adapting the fuzzy values on the required controller output signal it is necessary to convert the determined linguistic variables in interpretable discrete controller output signals.

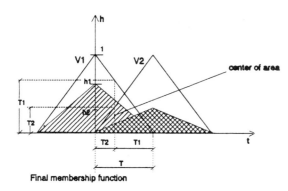

Fig. 8: Defuzzification and determination of pulse-times

The intention consists in the producing a pulse-width modulated voltage with the determination of the right on-times of the voltage phasors. The pulse-times are again calculated in every pulse-period for holding approximately constant the switching frequency. After determining the center of area with:

$$x_c = \frac{\int f(x)\,x\,dx}{\int f(x)\,dx} = \frac{h_2\,T}{2(h_1 + h_2)} \quad (4)$$

the x-coordinate can be written for ordering the on-times. Two methods of interpretation of the on-times are demonstrated in fig.8. These divides the pulse-period in two sections with the on-times T_1 and T_2 as the distance of the x-points of the both membership functions with the value 1 for the voltage vectors V_1 and V_2. Next the linguistic variables from the production rules are related to the voltage phasors V_0 to V_7 for every sector in the stator coordinate system. This can be done by a table of order.

4. FUZZY CONTROLLER WITH PREDICTION

The mode of operation represented above is going out from the instantaneous control errors of flux and torque without consideration of their gradients resulting from the selection of the optimal voltage vectors. From the linearized equations (1)-(3) the gradients of flux and torque can be written in per unit values:

$$\dot{\psi}_s = \hat{U}_s \cos\left(\alpha - (\nu - 1)\frac{\pi}{3}\right) + \frac{\Psi_r}{T_h}\cos\delta - \frac{\Psi_s}{T_\sigma} \quad (5)$$

$$\dot{m} = \frac{2\Psi_r}{L_\sigma}\left(\hat{U}_s \sin\left(\alpha - \delta - (\nu - 1)\frac{\pi}{3}\right) + \frac{\Psi_r}{T_h}\sin 2\delta - \frac{\Psi_s}{T_h}\sin\delta\right) \quad (6)$$

as trigonometrical functions with $T_\sigma = L_{m\sigma}/R_s$, $T_h = L_\sigma/R_s$, $\nu = 1...7$. The evaluation of the torque gradients can be taken into consideration for the structure of control by factorization of the on-times going out from determining the quotient dx_1/dt to dx_2/dt. With

$$T_1 + T_2 = T_1^* + T_2^* = T \quad (7)$$

it can be written the new pulse-times

$$T_1^* = T\,T_1 / \left(T_1 + \frac{\dot{x}_1}{\dot{x}_2}T_2\right) \quad (8)$$

$$T_2^* = T\,T_2 / \left(T_2 + \frac{\dot{x}_2}{\dot{x}_1}T_1\right) \quad (9)$$

5. EXPERIMENTAL RESULTS

The proposed control structure with help of fuzzy logic was tested for a 10 kVA inverter and a 1.5 kW ac-machine. The following motor parameters were taken as a basis:

$P_N = 1.5$ kW, $U_N = 220$ V, $I_N = 3,7$ A, $n_N = 1450$ min^{-1},
$R_s = 5\Omega$, $R_r = 4.8\Omega$, $L_m = 248$mH, $L_s = 260$ mH, $L_r = 257$ mH,

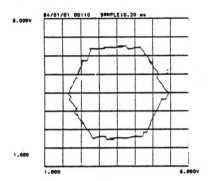

Fig.9: Control on a hexagonal flux trajectory

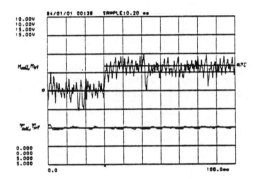

Fig.10: Torque step-response from 25% to 100% M_N

The limits of the control errors Δm and $\Delta \psi$ for the categories large$^+$ with the membership value $h = 1$ were fixed that the pulse frequency is about 0.7 kHz. By introducing of a common control structure with a variable production rules it could be controlled the flux not only on a hexagonal trajectory but also on a circular, see fig.9 and 11. The responses to a torque step from 25% to 100% of the nominal value are comparable with another, see fig.10 and 12, the rise time is 0.5 ms.

The described drive system is testing at present in the laboratory. The signal processing is realized by the microcontroller of the type SAB 80C166.

6. SUMMARY

In the survey was proposed a new control structure for a drive system with a fast flux- and torque controller producing a pulse-width modulated stator voltage from a graduated evaluation of the control errors with help of a fuzzy controller. The described method is common as far as the possibility is given to combine a original nonlinear controller with a linear pulse-width modulator. The new structure can generate different flux trajectories by modification of the production rules without change-over the two controllers. In future it should be important to use special devices for fuzzy control with components for programmable logic and conversion of values in fuzzy world and back.

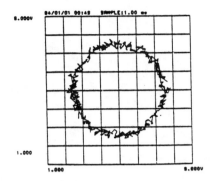

Fig.11: Control on a circular flux trajectory

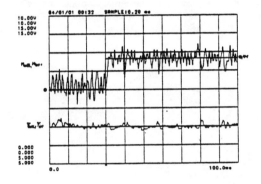

Fig.12: Torque step response from 25% to 100% M_N

7. REFERENCES

[1] Depenbrock,M.: "Direkte Selbstregelung (DSR) für hochdynamische Drehfeldantriebe mit Stromrichterspeisung" in etz-Archiv 7 (1985), No 7, pp. 211-218

[2] Takahashi,I.: "A new quick response and high efficiency control strategy of an induction motor" in IEEE Transactions on Industry Applications, Vol.22, 1986, No.5, pp. 820-827

[3] Baader,H.: "Hochdynamische Drehmomentregelung einer Asynchronmaschine im ständerflußbezogenen Koordinatensystem" in etz-Archiv 11,1989, No.1, pp.11-16

[4] Zadeh,L.A.: "Fuzzy Sets" in Information and Control, 1965 pp.338

[5] Bandemer,L.: "Einführung in die Fuzzy-Methoden" Akademie Verlag Berlin 1989

Appendix:

Fuzzification of torque error:

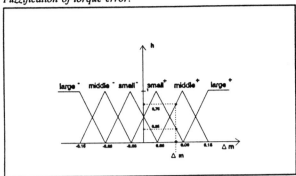

degrees of membership:

$h(middle^+) = 0.75 \quad h(small^+) = 0.25$

Fuzzification of flux error:

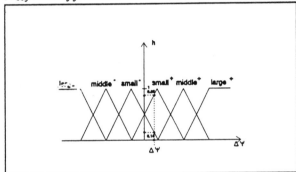

degrees of membership:

$h(small^+) = 0.85 \quad h(small^-) = 0.15$

↓

Production rules:

IF (M_middle$^+$ AND F_small$^+$) THEN (flux forw. increase) →
$h_p = h(M\_middle^+) * h(F\_small^+) = 0.63$

IF (M_middle$^+$ AND F_small$^-$) THEN (flux forw. decrease) →
$h_p = h(M\_middle^+) * h(F\_small^-) = 0.112$

IF (M_small$^+$ AND F_small$^+$) THEN (flux forw. increase) →
$h_p = h(M\_small^+) * h(F\_small^+) = 0.212$

IF (M_small$^+$ AND F_small$^-$) THEN (flux forw. decrease) →
$h_p = h(M\_small^+) * h(F\_small^-) = 0.038$

↓

table of order:

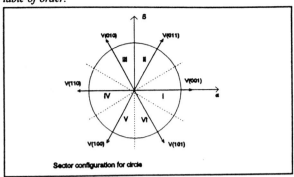

SECTOR II
flux forw. increase → V(010) → $h_p = 0.63$
flux forw. decrease → V(110) → $h_p = 0.112$
↓

Decomposition:

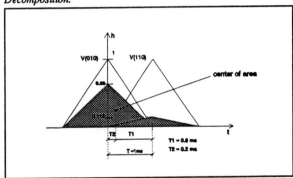

Response of flux phasors and torque:

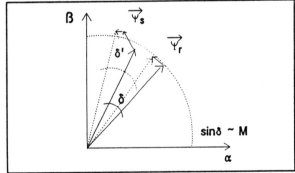

ADAPTIVE FUZZY TECHNIQUES FOR SLIP-RECOVERY DRIVE CONTROL

L.E.Borges da Silva[1]   G.Lambert-Torres[1]   V.Ferreira da Silva[1]

K.Nakashima[1]   G.E.April[2]   G.Olivier[2]

1- Escola Federal de Engenharia de Itajuba - Brazil
2- Ecole Polytechnique de Montreal - Canada

ABSTRACT

Wound-rotor induction motors have been used where large mechanical power and adjustable speed are required such as water pumping stations, cement plants, and so on. Speed adjustments may be achieved by slip power control, but power factor problems limit the usefulness of the scheme. In the modified Scherbius system described in this paper, a generalized hybrid inverter (composed by a six thyristor bridge, supplemented by GTO across the dc terminals and another one in series with the inverter bridge), is utilized in order to regulate the slip-energy recovery and the power factor of the system, for all regimes. A fuzzy adaptive control, based in a three levels control structure manipulates the system variables, improving the drive performance as the system evolves.

INTRODUCTION

Scherbius scheme have been largely used for speep control of wound-rotor induction motors, where large mechanical power are required. Therefore, power factor problems limit the usefulness of the scheme. This paper describes a modified Scherbius scheme (Fig.1), with a generalized hybrid inverter [1] (composed by a six thyristor bridge, supplemented by GTO across the dc terminals and another one in series with the inverter bridge). This systeme regulates the slip-energy recovery and the power factor of the system, for all regimes.

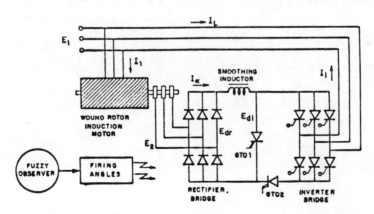

Fig.1 Modified Slip-Recovery Drive

A fuzzy adaptive control based in a three levels control structure manipulates the system variables, improving the drive performance. The control system must generates the inverter firing angles and the GTO's firing pulses according to a previously implemented strategy in order to maintain the power factor of the drive close to unity. After extensive analysis and simulations of the system, the complexity of the control have shown one of the most important restriction for this implementation. Unstable areas and multi-variables control structure are some of the aditional complications.

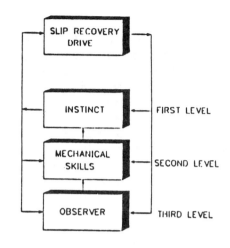

Fig.2 Control Structure

To solve these problems a three level control structure is proposed (Fig.2). Each level, described by a set of linguistic fuzzy rules, is responsable for a specific task. Together, the three levels constitute a flexible adaptive control structure, able to cope with slip power recovery, speed control, dc link current control, protections, desired working behaviour criterion, improving continualy performance, etc.

The control system, implemented using a rule based production system named FUZZY-FORTH [2], is composed basically by a data base, containing production rules, and a inference engine. The information contained in a set of rules produces meaningful inferences by using a loop type of inference engine. The FUZZY-FORTH, compiler accepts rules and place them in a "passive rule set" in such a way that the control system need only evaluate those rules that are relevant to the task at hand. It does so by placing the relevant rules in the "active rule set", which will be tested repeatedly by the inference engine. This allows the creation of multiple levels of control which may be likened to instinct, reflexes or even intelligence, if we may be so bold.

MODIFIED SLIP-RECOVERY DRIVE

Any power converter can be considered as a more or less complex set of switches, which by appropriate control, allow the adjustment of the electrical variables involved in the system. This implementation uses a hybrid converter due its capacity of generate or consume reactive power.

The schematic diagram of Fig.3 shows the power balance for this drive.

It is apparent that the speed control depends on the firing angle of the thyristors and the duty-cycle of the GTO's. It is well known in the literature that the machine torque is essentially proportional to the dc link current, so subject to some restrictions, a full range of speed control is possible.

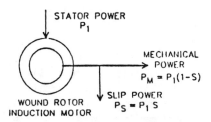

Fig.3 Power Balance

The theoretical behaviour of the drive may be understood by the analysis of the phase diagram shown in fig.4a. This diagram shows the loci of the stator, inverter and line currents as a function of the inverter firing angle and the GTO's duty-cycle. In this diagram I_1 represents the stator current with a phase angle of $\emptyset_1$ (lagging), I_1 represents the inverter current returned to the line, with a phase angle of $\emptyset_1$, and I_{L1} is the total line current. The curve $N_1 - N_3$ represents the line current (fundamental) locus, as would be obtained with a classical subsynchronous cascade.

The segment $U_1 - U_3$ represents the locus of the line current I_{L1} for a unit displacement factor, the main idea behind this paper is to keep the line current I_{L1} along this segment over all operating conditions.

The cross-hatched area (delimited by F_1, C, N_1, B, F_1) represents the region where the dc link current goes to zero. This means that the rectifier voltage E_{dr} is smaller than the inverter voltage E_{d1}.

Assuming that the stator current remain constant, (both in modulus and in phase), the area delimited by F_1, F_3, N_3, N_1, F_1, represents the attainable region, or the locus of the total line current I_{L1} for all possible firing angle and duty-cycle variations. The inverter firing angle controls the angular position of I_{L1} and the duty-cycle controls the amplitude.

Circle F_3, F_1, C, N_1, N_3, represents inverter current locus for duty-cycle iqual to zero.

The operating point shown in fig.4a, represents a condition where the inverter firing angle is around 120 degrees and the GTO's duty-cycle is equal to zero, for a given inductive power factor of the induction motor. In this condition, is visible that line current I_{L1} is larger than the stator current.

By manipulating the firing angle and duty-cycle it is possible to achieve the situation presented in fig.4b, where the displacement factor is unity and the line current is actually smaller than the stator current. This means that the inverter is working with a capacitive power factor, in other words, the inverter is working as a reactive power source.

This theoretical analysis shows that is possible to maintain a unity displacement factor over its full range of operation.

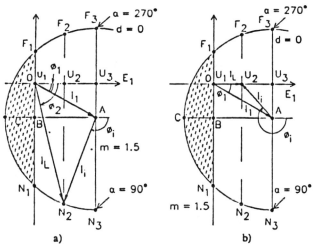

Fig.4 Theoretical stydies

SIMULATION OF THE DRIVE SYSTEM

The purpose of the simulation was to determine if it is possible, by judicious adjustments of the inverter firing angle and GTO's duty-cycle, to maintain unity displacement factor under all conditions of load and speed. In theory, through the analysis of the presented diagram, for a certain mechanical power required by the load, there is a corresponding real component of the line current, represented by the dashed line N_2-F_2.

The operational point N_2 has only to remain along this line and the desired objective will be reached. The results show that there are some unstable conditions as shown in Fig.5, but the disired behaviour can be reached.

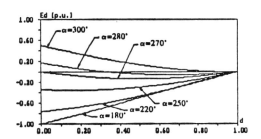

Fig.5 Operating Conditions

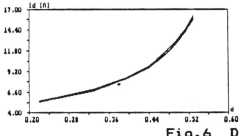
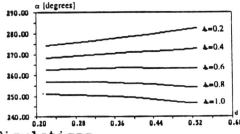

Fig.6 Drive Simulations

Fig.6 a,b presents the characteristics of: dc link current versus duty-cycle and inverter firing angle versus duty-cycle, taking in account first a unity power factor for constant speed (or rotor slip) as parameter. It is immediately apparent, that there is, for each desired load torque and rotor speed, a certain combination of inverter firing angle and duty-cycle that guarantees unity displacement factor. Using these informations it is possible to conceive the control structure.

CONTROL STRUCTURE

The basic idea behind this controller, is to implement a set of linguistic rules that describe the strategy of a hypothetical human operator, based in his experience about the behaviour of the drive. Naturally the response time will be many times faster than that of a real human operator. This approach has some advantages, for example: possibility to cope with unstable regions in the control space, as well as to perform on line parameters adaptation, response optimization, etc. The controller proposed here is based on the three level hierarchical structure shown in Fig.2. Each level is responsible for a specific task, and a certain degree of intelligence can be assigned to each. The first, which we could liken to the "<u>Instinct</u>" of the system, copes just with dc link current control, limiting and protection. This level is a standard Fuzzy Controller in the sense of Mandani [3]. The inputs to this controller are the speed error (Ω) and the speed error variation ($\delta\Omega$), and the output is the increment (δd) of duty-cycle. the rules are written in the form:

IF Ω IS X_1 AND $\delta\Omega$ IS Y_1 THEN δd IS Z_1
...   .
IF Ω IS X_n AND $\delta\Omega$ IS Y_n THEN δd IS Z_n

The desired output will be inferred from a set of "Fuzzy Conditional Statements" or "Linguistic Rules" by the "Compositional Rule of Inference" [4] and defuzzyfied using a center of gravity criterium to produce a numerical output value. This kind of controller is well known in the technical literature and the results are very good. This level also introduces some protection rules, against converter over-current or avoid dangerous operating regions like shown in Fig.5. It is immediately visible that, for certain values of the firing angle (around 270 degrees), there are two solutions for the duty-cycle, this is indicative of an unstable region.

For example some of these protection rules might be:

IF DC_Link_Current IS Very high
   THEN Decrease_Duty_cycle Very_much END-RULE

IF Firing_Angle IS Around_270 THEN Limit_Duty-cycle END-RULE

Such rules can, easily be placed alongside the actual control rules, in the active rule set.

The second level, responsible for the reference value that the lower level needs to controls correctly the motor, manipulates knowledge about: start conditions, acceleration rate, working parameters, and so on. That means, this level is responsible for the "Mechanical skills" of the drive. The input used for this level are: speed reference, actual speed and duty-cycle. The output is the inverter firing angle. It also evaluates the speed error which the lower level needs to work properly. The fuzzy implications describing this level have the form [5]:

IF x_1 IS A_1 OR B_1 OR ... AND
   ...
   x_k IS A_k OR B_k OR ...
THEN $y = p_0 + p_1 x_1 + ... + p_k x_k$

where:
$x_1 - x_k$ — variables of premise that also appear in the consequence
$A_1 - A_k$ — fuzzy sets with linear membership functions
$p_0 - p_k$ — parameters in consequence
y — variable in consequence whose value is inferred

The final desired output y, inferred from the set of implications, is given by:

$$y = \frac{\Sigma \mu_i \cdot y_i}{\Sigma \mu_i}$$

where μ_i is the membership function of the premise given by the MIN operator. This structure is helppful to create a certain desired transient behaviour.

The third level receives, as input, all system variables plus the information about the actual displacement factor $\emptyset_1$. This level named "Observer", in the sense that it looks at the overall behaviour of the system to decide what has to be done, i.e, determine the required adjustement of the system parameters. The structure of the "Observer", (Fig.7), is responsible for progressive improvement of the behaviour of the system and can be written with simple rules like:

IF System is in a steady state condition AND
   the operating point does not match the model
 THEN call the Investigator END-RULE

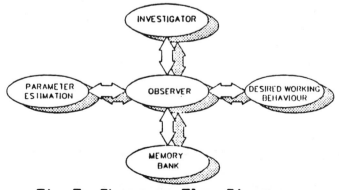

Fig.7 Observer Flow Diagram

The "<u>Investigator</u>" is a set of fuzzy conditional statements wich, based in the knowledge obtained from the simulations, adjust the operating point until the desired workink criterion is attained. After the working point have been attained, the "<u>Estimator</u>" have to find the parameter vector θ so that the performance index τ is minimized. And the performance index is defined as root mean square of the output errors, i.e. of the differences between the inferred values of the consequence and the output data values. An alternative on-line algorithm proposed in [6], that cope with fuzzy conditional statements as shown in the description of the second level, was used.

The variables used in the rules must be calculated from the actual process, before fed into the rules. Therefore, scale the calculated values by a suitable factor and quantize the result to the closest element of the universe of reference is mandatory. So a special rule is introduced to run before all others, in the active rule set. This rule is in the form:

RULE: Rule0 fixpoint calcul END-RULE

Since it has no conditional part, so it is always executed by the Inference Engine. The routine "<u>fixpoint</u>" freezes the converter variables, and the routine "<u>calcul</u>" calculates the variables to be fed into the rules. These routines may be written in assembler language, for speed.

FUZZY-FORTH RULE BASED PRODUCTION SYSTEM

The FUZZY-FORTH rule based production system used in this implementation, have extended the ordinary meaning of a Boolean conditional statement to cope with the concept of membership function. This allows for managing fuzzy variables, in order to assume labels of fuzzy sets such as "<u>big</u>", "<u>small</u>" or even "<u>neibourhood of zero</u>" instead of numerical values.

The information contained in a set of rules, produces meaningful inferences by using a loop type of "Inference Engine".

The FUZZY-FORTH rule editor is contained in a set of FORTH defining words to help construct the rules. It produces rules and places them in the FORTH dictionary adding them to the available knowledge base (the "<u>Passive Rule Set</u>"). Only those rules placed in the "<u>Active Rule Set</u>" are tested by the Inference Engine.

The rules are defined in the form:

RULE: name_of_rule
          IF (variable#1) IS (fuzzy set) OR ...AND
                    ...
             (variable#n) IS (fuzzy set) OR ...
          THEN (actions) .... END-RULE

In order to simplify the explanation, the action part of the rule is assumed to be deterministic, although, it can accept all kinds of structures: deterministic, fuzzy, or even conditional (IF THEN).

Another very important function implemented into FUZZY-FORTH is the "modifier", which allows to modify the defined fuzzy sets by words like "very", "not", "extremely", etc. The modifications can be programmed by:

MOD: (name) (operation) ;MOD

Defined in this way, they may then be introduced in a rule like:

..(variable)IS(modifier)(fuzzy set)OR..

The MOD: ;MOD structure is designed in such way that several modifiers can be placed before the fuzzy label (for instance IF variable IS not very very low THEN increase the voltage).

The action part of a rule can even add rules to the Active Rule Set, or, using "EXCLUDE" function, remove some. In this way, the rule-base can effectively modify itself as the system evolves.

BIBLIOGRAPHY

[1] G.Olivier, G.E.April, S.Manias, C.I.El Hadjri, "Generalized Analysis of Line Commuted Converters an Close Relatives" PESC-87 Power Electronics Specialists Conference, pp.642,649, 1987.

[2] L.E.Borges da Silva, G.E.April, G.Olivier, "FUZZY-FORTH Rule Based Production System for Real Time Control Systems" The Seventh Rochester Forth Conference, Rochester, pp.79-82, 1987.

[3] E.H.Mandani, S.Assilian, "An Experiment in Linguistic Synthesis with a Fuzzy-Logic Controller" Int. J. Man Machine Studies-7, pp.1-13, 1975.

[4] L.A.Zadeh "Outline of New Approach to the Analysis of Systems and Decision Processes" IEEE Trans. SMC-3-1, pp.28-44, 1975.

[5] T.Takagi, M.Sugeno, "Fuzzy Identification of System and Its Application to Modeling and Control IEEE Trans. SMC-15-1 pp. 116-132, 1985.

[6] G.Lambert Torres, D.Mukhedkar, "Writing a Fuzzy Knowledge Base" Int. Conf. on System Man and Cybernetics, 1990.

Fuzzy Controller For Inverter Fed Induction Machines

Sayeed A. Mir   Donald S. Zinger   Malik E. Elbuluk

Department of Electrical engineering
University of Akron
Akron, OH 44325-3904
Phone (216)972-7649
Fax (216)972-6487

Abstract—— An induction machine operated with a direct self controller (DSC) shows a sluggish response during startup and under changes of torque command. Fuzzy logic is used in conjunction with direct self control to minimize these problems. A fuzzy logic controller chooses the switching states based on a set of fuzzy variables. Flux position, error in flux magnitude and error in torque are used as fuzzy state variables. Fuzzy rules are determined by observing the vector diagram of flux and currents.

The operation of the direct self controller becomes difficult at low speeds due to the effect of change in stator resistance on the flux measurements. To improve the system performance at low speeds a fuzzy resistance estimator is proposed to eliminate the error due to the the change in stator resistance. At constant flux and torque commands any change in stator resistance of the induction machines causes an error in stator current. This error is utilized by the fuzzy resistance estimator to correct the stator resistance used by the controller to match the machine resistance.

Both fuzzy controller and fuzzy resistance estimator are simulated for a 3hp induction motor. The simulation results demonstrate a good performance.

I. INTRODUCTION

The vector control of induction motor drives has made it possible to use induction motors in applications requiring fast torque control such as traction. In a perfect field oriented control, the decoupling characteristics of the field oriented induction machines are affected highly by the parameter changes in the motor [1]. Stator direct self control (DSC), a variation of field oriented control, uses only the stator resistance in its calculations making the controller less sensitive to parameter changes [2,3]. Such a control is shown in fig.1. A detailed block diagram of the stator flux and the torque estimator is shown in fig.2. In stator direct self control, the errors in the torque and flux along with the flux position are directly used to choose the switching state. In conventional implementation, there is no method of distinguishing between very large errors and relatively small errors. Therefore the switching states chosen for the large error that occurs during the startup or during a step change in command torque or command flux, are the same as the switching states chosen for fine control during normal operation. This causes a sluggish system response during startup and during a change in command flux or command torque. If the states used by the system are chosen in accordance with the range of torque error, flux error, and flux position, the response of the system at the startup and during a change in command torque or command flux can be

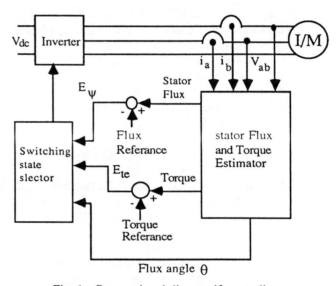

Fig. 1. Conventional direct self controller.

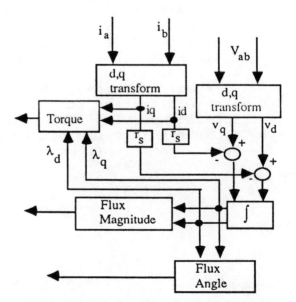

Fig.2. Stator flux and electric torque estimator

improved. Also, in a conventional controller each switching state is defined for a flux position in a band width of $\pi/6$ radians. The response of the controller can be improved by selecting the switching state in accordance with the position of the flux within this band width. Another limitation of the stator direct self control is that at low speeds, the operation becomes difficult due to the error in stator resistance used to estimate the stator flux [2,3].

The DSC uses errors in torque and flux, whether they are too big or too small, to determine switching states. Too big and too small are relative terms that contain a certain amount of fuzziness. This seems to be a natural situation for use of a fuzzy controller. A fuzzy controller converts a set of linguistic rules, based on expert knowledge, into an automatic control strategy [4,5]. Such controllers are often been found to be superior to conventional controllers especially when information being processed is inexact or uncertain. A typical block diagram of a fuzzy logic controller is shown in fig.3. In general, it consists of a fuzzification interface, a knowledge base, decision making logic and a defuzzification interface.

This paper presents a fuzzy controller for stator direct self control of an induction machine. Fig.4 shows a block diagram of proposed fuzzy controller which modifies the direct self controller by incorporating fuzzy logic into it. Also, a fuzzy resistance estimator is discussed to estimate the change in the stator resistance. The change in the steady state value of stator current for a constant torque and flux command is used to change the value of stator resistance used by the controller to match the machine resistance.

II. THE FUZZY CONTROLLER

A. *Fuzzy State and Control Variables*

The fuzzy controller is designed to have three fuzzy state variables and one control variable for achieving constant torque and flux control. Each variable is divided into fuzzy segments. The number of fuzzy segments in each variable is chosen to have maximum control with minimum number of rules. The first variable is the difference between the command stator flux ψ_s^* and the estimated stator flux magnitude ψ_s (error in stator flux E_ψ) given by:

$$E_\psi = \psi_s^* - |\psi_s|$$

The actual stator flux can be calculated from the voltage and current information in stationary reference frame as [6]:

$$\psi_{qs}/\omega_b = \int (V_{qs} - i_{qs} r_s) \, dt$$
$$\psi_{ds}/\omega_b = \int (V_{ds} - i_{ds} r_s) \, dt$$
$$|\psi_s| = \sqrt{(\psi_{qs}^2 - \psi_{qs}^2)}$$

The universe of discourse of the flux error fuzzy variable is divided into three overlapping fuzzy sets: Positive flux error (PE_ψ), zero flux error (ZE_ψ), and negative flux error (NE_ψ). The grade of membership distribution is given in fig.5(a) which uses a triangular distribution.

The second fuzzy state variable is the difference between command electric torque te^* and the the estimated electric torque te (error in torque E_{te}) given by:

$$E_{te} = te^* - te$$

The electric torque is estimated from the flux and current information as [6]:

$$te = (i_{qs}\psi_{ds} - i_{ds}\psi_{qs}) p/4\omega_b$$

To make the torque variations smaller, the universe of discourse of torque error is divided into five overlapping fuzzy sets: Positive large error (PLE_{te}), positive small error (PSE_{te}), Zero error (ZE_{te}), negative small error (NSE_{te}), and negative large error (NLE_{te}). The grade of membership distribution is shown in fig.5(b).

The third fuzzy state variable is the angle between stator flux and reference axis (stator flux angle θ) given by:

$$\theta = -\tan^{-1}(\psi_{ds}/\psi_{qs})$$

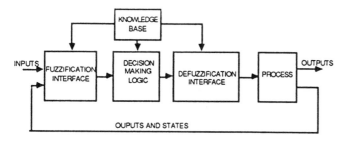

Fig.3 Basic fuzzy controller

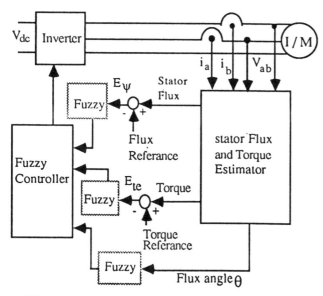

Fig.4 Fuzzy controller for direct self control of induction machine

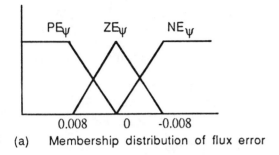
(a) Membership distribution of flux error

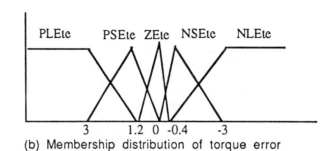
(b) Membership distribution of torque error

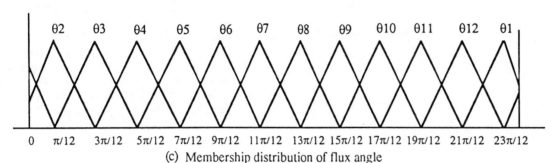
(c) Membership distribution of flux angle

Fig.5. Membership distribution of fuzzy variables for fuzzy controller

The universe of discourse of this fuzzy variable is divided into 12 fuzzy sets (θ_1 to θ_{12}). The membership distribution of is shown in fig.5(c).

The control variable is the inverter switching state (n). In a six step inverter, seven distinct switches states are possible [7]. The switching states are crisp thus do not need a fuzzy membership distribution.

B. Fuzzy Rules for self control

Each control rule, can be described using the state variables E_ψ, E_{te}, and θ and the control variable n. The ith rule R_i can be written as:

R_i : if E_ψ is A_i, E_{te} is B_i and θ is C_i then n is N_i

Where A_i, B_i, C_i and N_i represent the fuzzy segments.

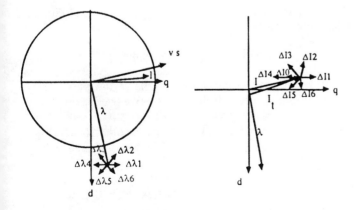

Fig. 6 Vector diagram used for knowledge base

The control rules are formulated, using the vector diagram for direct self control as shown in fig.6 [2]. Looking at the position of the flux in fig. 6, state 5, 6 and 1 will increase the flux while states 2, 3 and 4 will decrease it. Similarly states 6, 1 and 2 will increase the torque while states 3, 4 and 5 will decrease it. For a large increase in flux and a small increase in torque, state 6 is selected. For a small increase in flux and a large increase in torque, state 1 is selected. For a small decrease in flux and a small increase in torque, state 2 is selected. For a large decrease in flux and a small decrease in torque, state 3 is selected. For a a small decrease in flux and a large decrease in torque, state 4 is selected. For a small increase in flux and a large decrease in torque, state 5 is selected. For a small decrease in torque and constant flux, state 0 is selected. This selection changes as the position of flux vector changes. The total number of rules is 180 as shown in fig 7. Each cell in this diagram shows the best switching state for the given angle.

C. Fuzzy Interface

The interface method used is basic and simple and is developed from the minimum operation rule as a fuzzy implementation function [5]. The membership functions of A, B, C and N are given by μ_A, μ_B, μ_C and μ_N respectively. The firing strength of ith rule α_i can be expressed as:

$$\alpha_i = \min(\mu_{Ai}(E_\psi), \mu_{Bi}(Ete), \mu_{Ci}(\theta))$$

By fuzzy reasoning using Mamdani's minimum operation rule as a fuzzy implication function [5], the ith rule leads to the control decision

$$\mu_{Ni}'(n) = \min(\alpha_i, \mu_{Ni}(n))$$

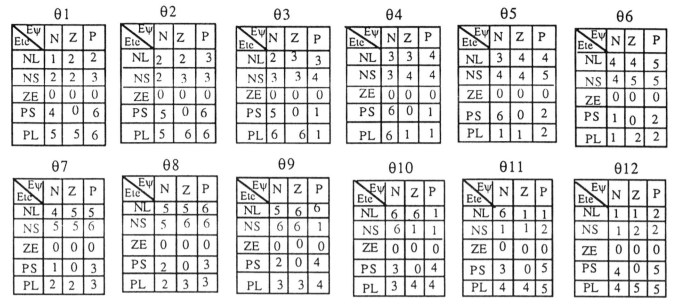

Fig.7. Set of fuzzy rules for fuzzy control of induction machine

Thus the membership function μ_N of the output n is pointwise given by:

$$\mu_N(n) = \max_{i=1}^{180} (\mu_{N_i'}(n))$$

Since the output is crisp, the maximum criterion method is used for defuzzification [5]. By this method, the value of fuzzy output which has the maximum possibility distribution, is used as control output.

III. SIMULATION RESULTS

A 3 hp., 220V 4 pole induction motor with the controller has been simulated. The parameters of the motor are given in table 1. A switching time of 50 μsecs is used in simulation. A constant dc voltage of 225 volts is used as the input to the inverter. The command torque and command flux used are 11.9 N-m (rated torque) and 0.234 volt-sec respectively. Figs. 8(a) and (b) show the torque and the stator flux responses of the system during startup for the conventional DSC and the fuzzy controller. The response of the fuzzy controller is faster than the conventional DSC. In fuzzy controller initial stator flux error is very large. Thus controller chooses the states giving higher increase in the flux. The change in torque during this time is small. Once the flux error becomes small, controller chooses the states giving faster increase in torque. Fig. 9 shows the response of the system for a step change in torque from 11.9 N-m to

Rs	Rr	Xs	Xr	Xm
.435	.816	26.884	26.884	26.13

Table 1. Parameters of 3hp induction machine

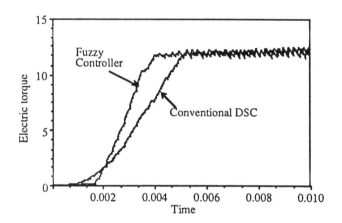

Fig.8(a). Torque response of fuzzy controller and conventional DSC during startup.

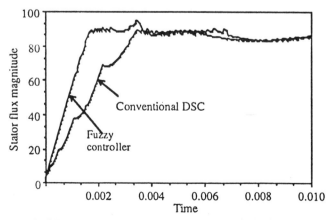

Fig.8(b). Stator flux response of fuzzy controller and conventional DSC during startup.

207

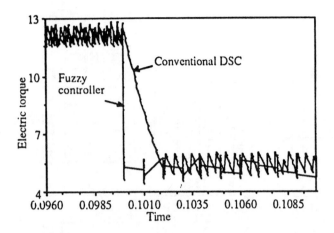

Fig.9. Torque response of fuzzy controller and conventional controller for step change in command torque.

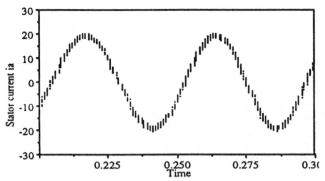

Fig.10. steady state stator line current with the fuzzy controller.

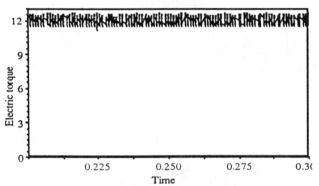

Fig.11. Steady state electromagnetic torque with the fuzzy controller

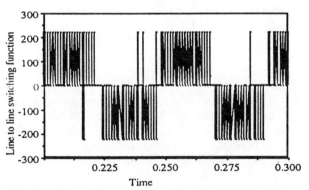

Fig.12. Line to line switching function with the fuzzy controller

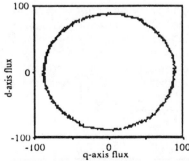

Fig.13. Steady state stator flux vector with the fuzzy controller

5 N-m keeping the flux command constant. The response of the fuzzy controller is faster than the conventional controller.

The steady state behavior of fuzzy controller is seen in Figs. 10-13. The line current in fig.10 shows a sinusoidal response. The electric torque in fig.11 has about the same ripple as that for the conventional DSC. The average torque is close to the commanded value while in the conventional controller it is slightly higher. The switching function of fig. 12 shows an occasional switching into the states opposite to the current half cycle. This is not desirable for optimum current regulation but for optimum torque and flux behavior it is not considered a problem. The steady state flux vector in fig. 13 shows nearly a circular path indicating a good flux regulation.

IV. CORRECTION OF ERROR CAUSED BY CHANGE IN STATOR RESISTANCE

In direct self control schemes estimation of stator flux is based upon the knowledge of stator resistance. This is especially true at low speeds where the resistive drop ($I_s r_s$) is the major portion of measured terminal voltage. The stator resistance error would cause improper flux estimation making the controller perform poorly. The magnitude of stator current vector can be used, to correct the stator resistance used by the controller, during any change in stator resistance of the machine. The magnitude of stator current vector in direct self control is a function of torque and flux. It is not affected by any change in the input dc voltage or a change in load. Also, the model used in the direct self controller is independent of all machine parameters other than stator resistance. Change of any parameter other than stator resistance does not change the magnitude of stator current vector. For any change in current vector, during a change in input voltage or the motor parameters other than stator resistance, the controller chooses the switching states so that the stator current changes back to original value to have constant flux and torque. During a change in stator resistance the actual and the estimated stator flux are different.

Therefore, the switching states selected by the controller for constant flux and torque do not change the current to its constant value. Thus, for a constant value of torque and stator flux, any change in the magnitude of the stator current vector is due to the change in stator resistance. With knowledge of the magnitude of current vector, for the given values of stator flux and torque, a fuzzy resistance estimator can be developed for the correction of changes in stator resistance. The fuzzy resistance estimator suggested is shown in fig.14. The estimator requires the magnitude of the stator current vector to obtain the change in stator resistance. This magnitude is obtained by measuring the stator currents and calculating the current vector. This is filtered (to get rid of the high frequency ripple) and send to the fuzzy resistance estimator.

A. *Fuzzy Resistance Estimator*

To estimate the error in stator resistance the stator current vector error and the change in the current vector error are employed. Current vector error and change of current vector error are defined as:

$$e(k) = Is(k)^* - Is(k)$$
$$\Delta e(k) = e(k) - e(k-1)$$

Where Is^* is the current vector corresponding to the flux and torque commands and Is is the measured stator current vector given by:

$$Is(k) = \sqrt{i_{qs}^2 + i_{ds}^2}$$

The universe of discourse of the two fuzzy input variables and the output variable, which is the change in stator resistance, Δr_S are divided into five fuzzy sets each as shown in figs. 15 (a), (b) and (c). The fuzzy rule applied can be written as:

If e is A_i, Δe is B_i then Δr_S is C_i

The error in stator current vector for a linear change in stator resistance (as shown in fig.16(b)) is shown in fig.16(a). Fig.16(c) shows the relation between the error in stator current vector and the error in stator resistance. Using this response we can formulate fuzzy rules to change the stator resistance used by the controller. There are twenty five rules as shown in fig.17.

Mamdani's minimum operation rule is used as interface method and, finally, the value of resistance error (Δr_S) can be obtained by center of gravity method used for defuzzification [5]. The value of stator resistance used by the controller is then given by

$$r_S(k) = r_S(k-1) + \Delta r_S(k)$$

C. *Simulation Results*

The resistance of 3 hp motor is changed from 0.435Ω to 0.82Ω linearly in 4 secs. This value is retained

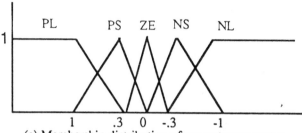
(a) Membership distribution of current error vector

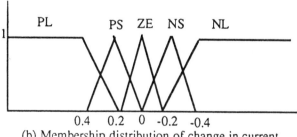
(b) Membership distribution of change in current vector error

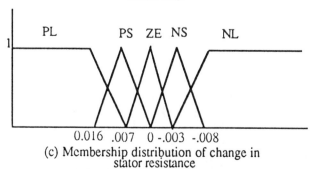
(c) Membership distribution of change in stator resistance

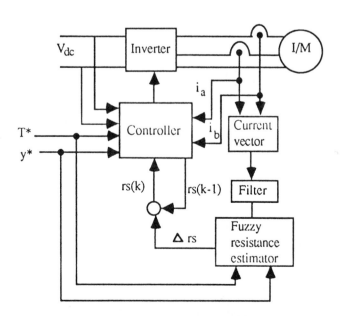

Fig.14. Controller with fuzzy error estimator.

Fig.15 Membership distribution of fuzzy input Variables

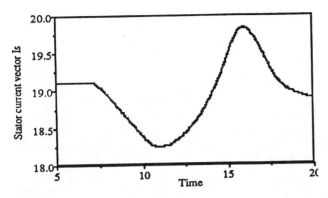

Fig.16(a). Error in stator resistance for linear change in stator resistance

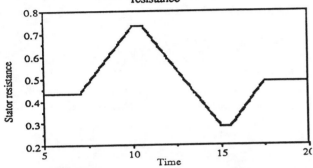

Fig.16(b). Change in stator resistance

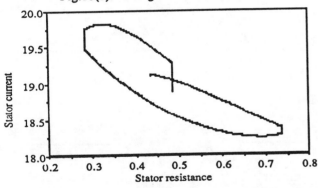

Fig.16(c). Magnitude of stator current vector vs. stator resistance

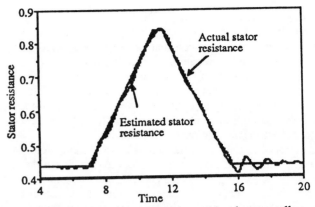

Fig.18. Estimated stator resistance used by the controller and actual stator resistance with fuzzy resistance estimator

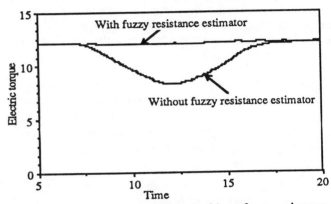

Fig.18(a). Electric torque with and without fuzzy resistance estimator

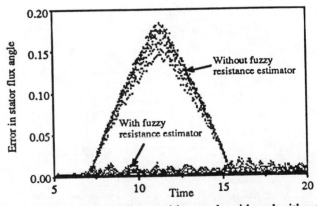

Fig.19(b). Error in flux position angle with and without fuzzy resistance estimator

Δe \ e	PL	PS	ZE	NS	NL
PL	PL	PL	PL	PS	ZE
PS	PL	PL	PS	ZE	NS
ZE	PL	PS	ZE	NS	NS
NS	PS	ZE	NS	NL	NL
NL	ZE	NS	NL	NL	NL

Fig.17. Fuzzy Rules for fuzzy resistance estimator

for .5 sec and than decreased back to 0.435Ω at same rate. The switching frequency of the inverter is 20 khz and the resistance in in the controller is updated every 50 switching cycles using the fuzzy resistance estimator. The command torque and the command flux are kept constant at 11.9 N-m and 0.234 volt-sec. The magnitude of stator current vector under these conditions is 19.1 amps. Fig 18 shows the actual stator resistance and stator resistance used by the controller with the fuzzy resistance estimator. The stator resistance used by controller follows very close to the actual stator resistance of the motor. Fig 19 (a) and (b) show the filtered electric torque and error in the flux position angle with and without the fuzzy error estimator. There is very small error in actual electric torque of the machine during change in resistance using fuzzy error estimator. Without the correction the actual machine torque drops down from 11.9 to 8.2. The error in stator flux position, which is more important as it may cause the controller to choose wrong switching states, is negligible, when the fuzzy error estimator is used. With out the correction the value of error reaches up to 0.2 radians.

C. *Failure of Fuzzy Resistance Estimator:*

The current implementation of stator resistance estimator has been found to give trouble at low torque levels. Under low torque levels, resistance estimates were found to vary irradically causing the controller to produce improper torque and flux. It is necessary to turn off the estimator at low torque levels to ensure proper operation. Further investigation is being made to improve the control behavior.

V. CONCLUSION

A fuzzy logic controller for a direct self control of an induction machine has been presented in this paper. A response faster the than conventional DSC during startup and during a step change in torque is achieved. The performance has been tested by simulations. Also, a fuzzy resistance estimator has been proposed to correct the stator resistance used by the controller to match the actual stator resistance of the machine. The stator resistance used by the controller follows the actual stator resistance of the machine. The algorithms are constructed upon experience with the motor drive system. The performance of fuzzy resistance estimator has been tested by simulation.

REFERENCES:

[1] K. H. Nordin, D. W. Novotny and D. S. Zinger, "The influence of motor parameter deviation in feedforward field oriented control.," *IEEE Transaction on Ind. Appl.*, vol. IA-21, No. 4, Aug. /Sep., 1985, pp. 1009-1015.

[2] T. Habetler and D. Divan, "Control strategies for direct control of induction machine using space vector modulation," *IEEE Transaction on Ind. Appl.*, Sept./oct. 1991, pp 893-901.

[3] Isao Takahashi and Toshihiko Noguchi, " A new quick-response and high Efficiency control strategy of an induction machine," *IEEE Transactions on Ind. Appl.*, vol. IA-22, No. 5. Sep./ Oct., 1986, pp 820-827.

[4] C. C. Lee, "Fuzzy logic in control systems: Fuzzy logic control - part 1," *IEEE Transaction on Systems, Man and Cybernetics*, vol 20, No. 2, March/April, 1990, pp 404-418.

[5] C. C. Lee, "Fuzzy logic in control systems: Fuzzy logic control - part 2," *IEEE Transaction on Systems, Man and Cybernetics*, vol 20, No. 2, March/April, 1990, pp 419-435.

[6] Paul C. Krause, *Analysis of Electric Machinery*, MaGraw-Hill, 1986.

[7] D. M. Brod and D. W. Novotny, "Current control of VSI-PWM inverter," *IEEE Transactions on Ind. App.*, May/June, 1985, pp 562-570.

A FUZZY CURRENT CONTROLLER FOR FIELD-ORIENTED CONTROLLED INDUCTION MACHINE BY FUZZY RULE

*Seong-Sik Min, Kyu-Chan Lee, Jhong-Whan Song and Prof. Kyu-Bock Cho

R&D Institute, Hyosung Industries Co., Ltd.
4, 5-Ka, Dangsan-dong, Yeongdeungpo-ku, Seoul, Korea. ZIP 150-045

Abstract: *A current controlled pulse width modulation(PWM) for voltage source inverter(VSI) is one of the control scheme which controls output current directly. Because current controlled PWM scheme reduces the orders of differential equations of the drive system, high performance a.c. motor drives can be implemented. This paper proposes a Fuzzy current controlled PWM which minimizes a current ripple using Fuzzy theory in a constant switching frequency. This technique is applied to an electrical drive system with an induction motor(IM) and studied by simulation. In comparison with the known classical methods such as ramp comparison and hysteresis band control method, simulation results show the better performances.*

INTRODUCTION

Because the operational performance of an induction motor(IM) is robust normally and free of maintenance comparing with d.c. machines, it has been one of the targets to realize high performance machine control with IM in the electrical drive technology community. The model equation of IM for speed regulation has the 5th-order nonlinear differential equation. Now if the input current of IM is controlled directly, the model equation can be reduced to the 3rd-order nonlinear differential equation and so it is much simpler to control IM by means of the field-oriented methods employing the current controller.

Ever since the current controlled voltage source inverter(CC-VSI) was introduced by A.B.Plunkett[1], this method has been rapidly expanded in AC motor drive systems because it can reduce the modeling order of differential equations of the drive system as above mentioned. With the wide use of CC-VSI, the control methods have also been developed[2][3]. The goals of CC-VSI are first to yield the control system high performance, secondly to reduce the harmonic losses of the load by minimizing the current ripple, thirdly to reduce the switching losses at the power component device, and to make an easier filter design at the load. A curent controller with hysteresis band[4] is simple in its structure and has the capability to limit peak current by itself. But the switching frequency to retain the current within hysteresis band can not be bounded at will, and it is varied with load and speed, and so may contain excess harmonics. To maintain the switching frequency constantly, one can use the ramp comparison controller[5][6]. But this method generates a phase delay which deteriorates the system performance, and if the load time constant is smaller than that of the ramp, then multiple modulation waves would be generated in a period of reference ramp signal.

There are also suggested two types of predictive current controllers on the basis of space vector. One type considers a constant switching frequency method[7], and the other employs a minimum switching frequency method[8]. The constant switching frequency method does not have an inherent current limiting function and a specified current ripple as in a hysteresis band current controller. While the minimum switching frequency method requires a calculation time to get the next switching state, which makes a control system delay.

Adaptive hysteresis band current controller maintains switching frequency constant[9][10], and it shows the simple structure of the hysteresis band current controller, and has the ability to limit current by itself. But when it is applied to an AC motor which has an open neutral point, the periods of switching on and off called a respective applied voltage coefficient can not be determined identically and is varied with load and speed conditions. So it was difficult to determine respective applied voltage coefficient accurately. Our institute has published in large numbers for the purpose of working out above entioned problems[11-13].

This paper presents a adaptive fuzzy current controller, which controls the switching pattern of inverter using fuzzy logic. This is one of the AI based emerging and highly advanced technologies, which can improve and resolve the problem remaining nowadays in other current control methods. Especially when this method is used, it is not necessary an exact modelling and analyzing the system mathematically. Furthermore because this method makes effective use of experience, optimal control whithin the expert's knowledge can be implemented maintaining a constant inverter switching frequency. Therefore it is easy to design a power filter reducing harmonic components at the load.

In this paper, the described voltage-fed current controlled PWM inverter adapted fuzzy logic is applied to IM drive system which has complexity coupled parameters, and it is generally difficult to apply any known analytic method. An IM drive system with a Fuzzy current controller is developed and simulated to show the availability and performance of proposed method. An on-line method is introduced to apply to inverter and shows much improved performances.

THE ARCHITECTURE OF A GENERAL PURPOSED FUZZY SIMULATOR

A fuzzy control has the several advantages ; it is not needed for the modeling and analysis of the controlled system, it can be applied to highly non-linear system, expert's experiences can be used. If it has big amount of parameters and variables, its calculation time is not so hardly consumed, because of parallel processing characteristics[14]. This fuzzy control part is seperately created in the form of a general purposed fuzzy simulator, and used for the simulation.

This simulator is developed in three different parts of whole software package ; first fuzzy rule editor which assists it's experience saved in the form of fuzzy rule and membership function, second a simulator which shows the fuzzy outputs produced by the assumed inputs, and third a fuzzy controler(emulation part) which produces the control outputs by the real input values(Fig. 1).

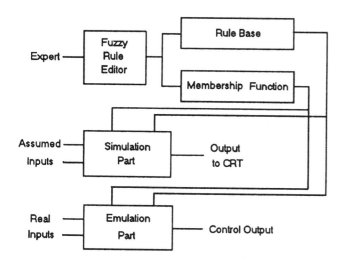

Fig. 1 A general purposed fuzzy simulator

This fuzzy controller system is consisted of the three parts, a fuzzification which translates a crisp input to a fuzzy function using the membership function, a fuzzy inference engine which induces the fuzzy outputs from the fuzzy inputs, and a defuzzification which translates a fuzzy output to a crisp value to be able to control the system(Fig. 2).

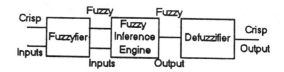

Fig. 2 The architecture of fuzzy controller

Fig. 3 and Fig. 4 demonstrate the general purposed fuzzy simulator. In a this simulator all inputs and output values have range from 0 to 255(8 bit). Fig. 3(a) and Fig. 3(b) show the membership function of the difference between the reference current and the actual current(error), and the corresponding output of the fuzzy controller in order to control a PWM inverter switching. Fig. 3(c) indicates the fuzzy output in case that input 1 and 2 each is 201 and 134, and the corresponding crisp output. Fig. 4(a) and Fig. 4(b) display the membership functions of the difference between an existing error and a last one, and the corresponding output membership function of the fuzzy controller. Fig. 4(c) exhibits the fuzzy output and the corresponding crisp value in case that each input 1 and input 2 is 120 and 100.

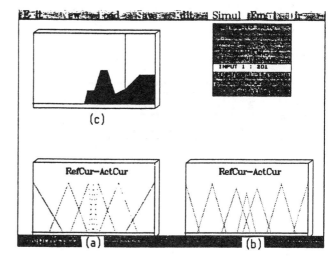

Fig. 3 The general purposed fuzzy simulator
  (a) Input membership function of error
  (b) Output membership function of error
  (c) The simulation results in case that error and error difference sre 201 and 134.

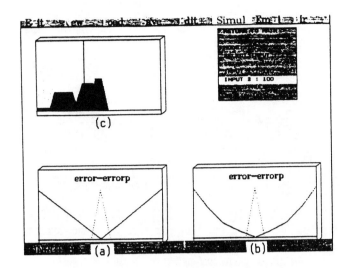

Fig. 4 The general purposed fuzzy simulator
(a) Input membership function of error difference
(b) Output membership function of error difference
(c) The simulation results in case that each error and error difference are 120 and 100.

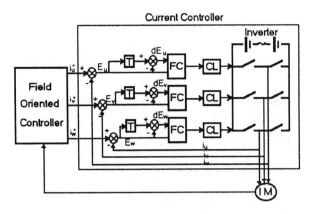

T : Time delay circuit
FC : Fuzzy Controller
CL : Calculation Logic

Fig. 6 The architecture of a proposed fuzzy current controller

A FUZZY CURRENT CONTROLLER

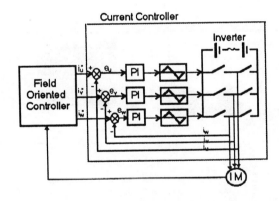

Fig. 5 The architecture of a ramp comparison current controller

Fig. 5 is the architecture of a ramp-comparison current controller used in the many commercial product. Fig. 6 is the architecture of a proposed fuzzy current controller. In this scheme, the inputs of a fuzzy controller are the differences(error) of the output of field oriented controller, i.e. reference current, and actual current, and time variance of this error. The output is the variance of the inverter switching time. The calculation logic produces the control signal of each phase in order to implement the optimal inverter switching. The three fuzzy controller with the same functions and characteristics are imported for the three phase control.

In this scheme, a fuzzy controller first produces the control signal to fit the controlled inverter and increases on-time according to the actual current less than the reference or viceversa.

THE DESIGN OF MEMBERSHIP FUNCTION

In the fuzzy control scheme, the fuzzy knowledge representation for controling the induction motor is kept in the form of membership function, and it has a triangular shape form shown in Fig. 3 and Fig. 4. The important fuzzy rules for inverter control is as follows.

E : error
O : variance of inverter on-time
Ė : time variance of error
P : positive
PB : positive big   PS : positive small
N : negative
NB : nagative big   NS : nagative small

If E is P and Ė is P, then O is PB.
If E is P and Ė is zero, then O is PS.
If E is P and Ė is N, then O is PS.
If E is zero and Ė is P, then O is PS.
If E is zero and Ė is zero, then O is zero.
If E is zero and Ė is N, then O is NS.
If E is N and Ė is P, then O is NS.
If E is N and Ė is zero, then O is NS.
If E is N and Ė is N, then O is NB.

A scale factor is used so that the scale of error and error variance locates in the active region of fuzzy logic.

THE SIMULATION RESULTS

In order to demonstrate the improved performance of the proposed scheme, it is selected to compare with other conventional method such as a hysteresis band and a ramp comparison current controller. which is used currently in many commercially avalible industrial drive systems. Table 1 illustrates the specification data of a selected 5HP three-phase induction motor for simulation test.

Table 1. Specification data of three-phase induction motor

Parameters	Values
Rating Power	5 HP
Rating Voltage	220 V
Number of Poles	4
Input Frequency	60 Hz
R_s	1.57 Ω
R_r	1.31 Ω
L_s	6.31 mH
L_r	8.19 mH
M	221.3 mH

```
switching frequency of inverter : 4 kHz
bandwidth of hysteresis        : 1 Ampere
A.C. input voltage             : 3Ph. 220V
velocity of rotor              : 100 rad/sec
sampling time                  : 1 msec
```

Fig.7, Fig.8 and Fig.9 show the responses of inverter output current waveform for a fuzzy, a hysteresis and a ramp-comparison current controller respectively at no-load condition. Fig.10, Fig.11 and Fig.12 are the results of each system with load(10 [N.m]). From these figures, it can be known that a new controller show the better current tracking performance than other system. Also it can be found out that the switching frequency of a hysteresis band scheme is very unstable and a ramp-comparison method with load exposes a delayed response, while a proposed scheme has stable current tracking capability in a constant switching frequency.

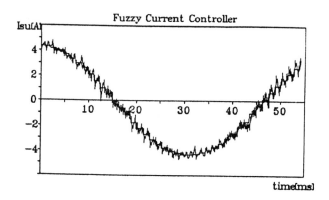

Fig. 7 Current tracking ability of a proposed current controller at no-load condition

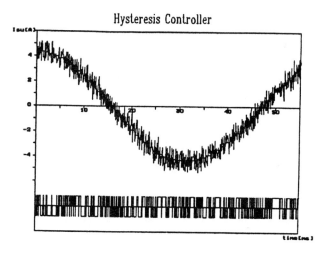

Fig. 8 Current tracking of a hysterisis band current controller at no-load condition

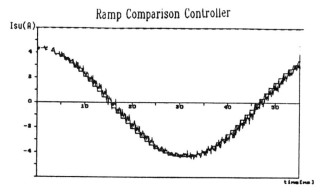

Fig. 9 Current tracking capability of a ramp comparison current controller at no-load condition

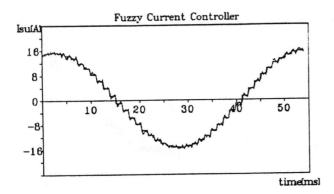

Fig. 10 Current tracking ability of a proposed current controller with load(10 [N.m])

CONCLUSION

In this paper, a fuzzy current controller using the general purposed fuzzy simulator is proposed and investigated the adaptability to a inverter-fed vector controlled ac drive system. From the simulation results, it is verified that the proposed controller has a better current control capability than other conventional controller.

For our next exploration regarding this study, an experimentation is made and published.

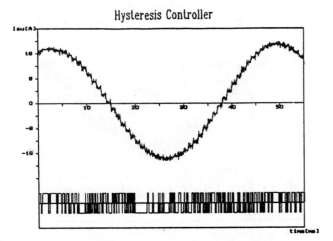

Fig. 11 Current tracking of a hysterisis band current controller with load(10 [N.m])

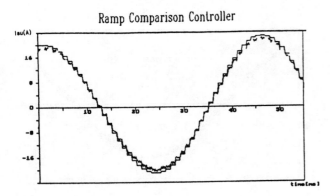

Fig. 12 Current tracking capability of a ramp comparison current controller with load(10 [N.m])

REFERENCES

[1] A.B.Plunkett,"A current-controlled PWM transistor inverter drive," in Proc. Conf. Rec. 14th Annual Meeting. IEEE/IA 1979, pp.785-792.

[2] D.M.Brod and D.W.Novotny,"Current control of VSI-PWM inverters," IEEE Trans. on IA Vol.21, No.4, May/June 1985, pp.562-569.

[3] P.Freere and P.Pillay,"Design and evaluation of current controllers for PMSM drives," in Proc. Conf. Rec. IEEE/IAS 1990, pp.1193-1198.

[4] A.W.Green and J.T.Boys,"Hysteresis current-forced three-phase voltage sourced reversible rectifier," IEE Proc., Vol.136, Pt.B, No.3, MAY 1989, PP.113-120.

[5] A.Schonung and H.Stemmler,"Static frequency changers with 'subharmonic' control in conjucation with reversible variable speed a.c. drives," Brown Boveri Review, Aug/Sept 1964, pp.555-577.

[6] Y.Itoh et.al.,"Stability analysis of a digital current controller for a PWM inverter using a neural network," in Proc. Conf. Rec. IEEE/IAS 1990, pp.1129-1134.

[7] G.Pfaff, A.Weschta, and A.Wick,"Design and experimintal risults of a brushless ac servo-drive," in Conf. Rec. 17th Annual Meet. IEEE/IAS, 1982, pp.692-697.

[8] J.Holtz and S.Stadtfeld,"A predictive controller for the stator current vector of ac machines fed from a switched voltage source," in Conf. Rec. Annual Meet. IEEE/IPEC, 1983, pp.1665-1675.

[9] B.K.Bose,"An adaptive hysteresis band current control technique of a voltage-fed PWM inverter for machine drive system," IEEE Trans. on IA Vol.37, No.5, Oct. 1990, pp.402-408.

[10] J.M.Ruiz et.al.,"Minimal UPS structure with sliding mode control and adaptive hysteresis band," in Proc. Conf. Rec. IEEE/IAS 1990, pp.1063-1067.

[11] J.W. Song, K.C. Lee, K.B. Cho and J.S. Won, "An Adaptive Learing Current Controller for Field-Oriented Controlled Induction Motor by Neural Network," IECON'91, pp.469-474.

[12] S.S. Min, K.C. Lee, J.W. Song, and K.B. Cho, "A Fuzzy Current Controller using A Gerneral Purposed Fuzzy Simulator," Conf. of Korea Institute of Electrical Engineers, Nov. 22, 1991, pp.341-344.

[13] S.S. Min, K.C. Lee, J.W. Song and K.B. Cho, "A Neuro-Fuzzy Current Controller for Field-Oriented Induction Machine," Accepted, International Symposium on Power Electronics, Seoul, Korea, April 9-11, 1992.

[14] H.J. Zimmermann, "Fuzzy Sets Theory and It's Applications," Kluwer-Hijhoff Publishing, 363p., 1985.

Chapter 5: Power Systems

A FUZZY KNOWLEDGE-BASED SYSTEM
FOR BUS LOAD FORECASTING

G. Lambert-Torres[1]   L.E.Borges da Silva[1]   B.Valiquette[2]

H.Greiss[3]   D.Mukhedkar[2]

1. Escola Federal de Engenharia de Itajubá - Brazil
2. École Polytechnique de Montréal - Canada
3. Harris Computer Systems - Canada

ABSTRACT

This paper describes an alternative approach to short-term load forecasting. The approach merges traditional mathematical techniques and fuzzy concepts in a knowledge base. The rules of this knowledge base are devised using the historical data of the bus and are represented by fuzzy conditional statements. An example using real data from an actual power system is presented.

INTRODUCTION

In Operation Control Centers, many programs and computational routines [1] are available to operators to provide the best possible solutions for power system operation problems. Today many advisor programs using IA techniques are available to aid in operators' decidion-making [2]. Short-term load forecasting is one such program available to operators.

There are many reasons to study short-term load forecasting and they can be classified according to their focus on either objectives or time. The economical operation of power plants, and solution of the distribution systems, are examples of load forecasting focusing on objectives [3]. Day-to-day operation, scheduling of fuel supplies, and planning operations, are examples which focus on time.

This paper describes an alternative approach to short-term load forecasting to assist in the operation and planning of distribution systems. The proposed alternative approach is a knowledge based system composed of three knowledge bases. The first knowledge base includes historical data of the load, the second includes data of recent performance of the load. These knowledge bases are composed of fuzzy conditional statements which are obtained using a technique described in [4]. The third knowledge base enables the inclusion of operator knowledge of the load from the next hour to the next 24 hours. This base includes fuzzy values in the description of the rules. The operator knowledge inclusion is very important for special loads or for special characteristics due to external factors. The load forecasted value is calculated by an inference engine using fuzzy logic.

Following an outline of load forecasting classification, the proposed alternative approach is described and demonstrated using Hydro-Quebec Power System data (Canada).

CLASSIFICATION OF LOAD FORECASTING

This section presents a classification of load forecasting focusing on either objectives or time. A brief description of each focus is presented below.

Load Forecasting According to Objectives

There are two different objectives for studying load forecasting economical operation of power plants, and economical solution of the distribution systems. The first objective is to study the global performance of the load without taking into account the individual performance of each feeder. The second objective takes into account the characteristics of each load. The first objective is applied to power system's operation and planning, while the second is applied to the distribution system's operation and planning.

Load Forecasting According to Time

Currently, there are three possible classifications for forecasting focusing on time: short-term (the next half-hour to twenty four hour ahead), medium-term (the next day to the next year), and long-term (beyond the next year). Criteria's for selecting classification are different. As an example, the most important factors for short-term load forecasting include: the day of the week, temperature, humidity, seasonal effects and so on. Long-term load forecasting factors include: economic aspects, political aspects, industrial development degree of a region etc. Thus, the objectives for each classification range from: short-term forecasting which concerns itself with the day-to-day operation and scheduling of the power system, medium-term forecasting whith deals with the scheduling of fuel supplies and maintenance operations and long-term forecasting which involves planning operations.

DESIGN OF THE ALTERNATIVE APPROACH

There are many methods to calculate load forecasting. In all cases, a complete database is suitable as it allows the major degree of accuracy.

Database

The database must contain historical information about load and weather parameters. Dry bulb temperature, wet bulb temperature, relative humidity, wind direction and wind speed are the most common weather parameters. For special load, additional factors such as sun inclination and cloudiness may also bear an influence.

The data structure discriminates all data: its value (v),

hour(h), day(d), year(y) and season(s). The general forms for the data are presented below.

Real Power:	P(v,h,d,w,y,s)
Dry Bulb Temperature:	D(v,h,d,w,y,s)
Wet Bulb Temperature:	M(v,h,d,w,y,s)
Relative Humidity:	H(v,h,d,w,y,s)
Wind Direction:	WD(v,h,d,w,y,s)
Wind Speed:	WS(v,h,d,w,y,s)

Real power values, temperature values, relative humidity and wind speed are stored in [MW], [^{o}C], [%] and [km/h], respectively. For the wind direction values, the following convention is employed: Nort is 1, and each additional 22.5o in a clock-wise sense is a further 1. For a calm day 0 is used. So for example, the direction WSW is 12 and NE is 3.

Twenty four hour notation is used for hourly values. The day values are numbered as 1 for Monday through 7 for Sunday. Holidays receive the number 6 (Saturday). The week values are numbered by season, the most recent receiving the number 1. This database is built to manipulate the current year through three years previous. The yearly values range from 1 for the current year to 4 for three years ago. The seasonal values are as follows: winter - 1, spring - 2, summer - 3, and fall - 4.

For a maximum of two missing values (v) of one data type, a single average between the previous and next values is calculated. When more than two values are missing, the data is forgotten.

Special days (such as: snow storm) are stored in the database with a flag. These values are not used for current calculations, except when this particular weather condition is forecasted.

Alternative Approach

The proposed alternative approach is based on a knowledge base formed by three bases. The first knowledge base contains information about the standard load B(t) where the historical data set is represented. The second knowledge base contains information about the deviation load R(t) where the recent load performance is represented. The third knowledge base contains rules-of-thumb about special characteristics of the load. The first two knowledge bases expresses information by fuzzy conditional statements, whilst the third base is composed of information given by operator experience. A fuzzy mechanism is able to make inferences using the actual data, to provide the load forecast.

First Knowledge Base: Standard Load

The standard load represents the historical data set. An hourly load shape must be represented using weather conditions. The relationship between hourly load and weather conditions must be established directly from a mathematical function:

$$B(t) = f(P,K,M,H,WD,WS)$$

or through a standard shape from each of the elements.

The first konwledge base is composed of a standard shape for load and weather conditions. The standard shape is created using a weighted average with a decreasing degree of membership for each year of the database. If the load evolution has not been large over the last years, the decrement is small. If the load evolution has been large, however, the decrement is consequential. Thus, the membership $\mu_1(year)$, and $\mu_2(year)$ represent a large and a not so large evolution of the load, respectively.

$$\mu_1(year) = \{(1|1),(2|0.5),(3|0.1),(4|0)\}$$

$$\mu_2(year) = \{(1|1),(2|0.8),(3|0.6)(4|0.4)\}$$

Equation (1) shows a standard shape value calculation for time t, of the season s.

$$B_\phi(t) = \frac{\sum_{y-1}^{4} \sum_{t_0=1}^{k_3} \sum_{d=k_1}^{k_2} \mu_i(y) \cdot \phi(t,d,w,y,s)}{\sum_{y-1}^{4} k_3 \cdot (k_2-k_1) \cdot \mu_i(y)} \quad (1)$$

Where $\phi(...)$ represents each one of the database elements namely, real power, dry bulb temperature, wet bulb temperature, etc. The values k_1 and k_2 depend on the day (working days or weekend) of the specific week, year, and season. The value k_3 depends on the specific year and season.

After calculations of all the times-of-day t, an equation of the shape must be calculated. An algorithm to express a shape by fuzzy conditional statements has been proposed in [4]. These statements present a linear membership function as a primise and a straight line as a consequence.

Second Konwledge Base: Deviation Load

The second knowledge base represents the deviation load, and contains information from the most recent four weeks, and represents recent load performance. This data is merged in a multiple linear model through the method of least squares. Equation (2) shows the general form for this model, using differences between actual, and standard weather information.

$$R_1(t) = a_0 + a_1 \cdot (D_a(t) - D_s(t)) + a_2 \cdot (D_a(t) - D_s(t))^2 + a_3 \cdot (D_a(t) - D_s(t))^3 + a_4 \cdot (M_a(t) - M_s(t)) + a_5 \cdot (M_a(t) - M_s(t))^2$$

$$+ a_6 \cdot (H_a(t) - H_s(t)) + a_7 \cdot (H_a(t) - H_s(t))^2$$

$$+ a_8 \cdot (D_a(t) - D_a(t-1)) + a_9 \cdot (M_a(t) - M_a(t-1))$$

$$+ a_{10} \cdot (H_a(t) - H_a(t-1)) + a_{11} \cdot (W_a(t) - W_a(t-1)) \quad (2)$$

Where $\phi_a(t)$ and $\phi_2(t)$ represent actual, and standard values of the elements and of time t, respectively.

In equation(2), the square of the difference between the dry bulb temperatures (due to the strong correlation between the load this temperature) was considered. The increase of parcel number is possible (for example, other square differences), but it is not recommended since problems of multivariables increase rapidly in complexity, in this case. Without creating a better representation of the deviation load.

Third Knowledge Base: Operator Experience

In special cases, load forecast can be manipulated by a set of rules according to certain characteristics. Two sets of rules are presented, namely those of daily routine load and thermic inertia. A daily routine load has regular attributes for specific parts of the day. For example, is there is a load point at 9:00 a.m. and the actual load at 8:00 a.m. is greater than the forecasted load calculated for the same weather conditions, then part of the load point has already been consumed. Likewise, if the load at 8:00 a.m. is less than the forecasted load, then the load point may possibly increase. This special characteristic is applicable in load forecasting for predicting a maximum of 2 hours ahead.

Rule: If the percentage error between the actual load and the forecasted load is big then add 50% of this error, times the membership value for one hour ahead forecasting, and add 25% of this error, times the membership value for two hours ahead forecasting.

'Big' is a fuzzy value for a deviation. The values for 'big' 50% and 25% were obtained from many forecasts and were considered acceptable for all simulated residential buses. When the error is negative, the extra values can be subtracted.

Thermic inertia is the difficulty in changing from a better to a a worse temperature. For example, if the average winter temperature for the previous day was $-50^{o}C$, and today is $-20^{o}C$, the load will increase due to the temperature difference plus a certain thermic inertia. If the next day, the temperature remains at $-20^{o}C$, this inertial component will decrease. In Canada for example, this factor is very important due to the excessive temperature ranges. This special characteristics is mainly applicable for load forecasting 24 hours in advance.

Rule: If the 24 hour temperature variation is big then add 10% to forecasted load. The values used in this rule must be calculated for each bus of the system using the historical data. In our case, these values ('big temperature variation' and 10%) reference the example in the next section.

Mechanism of Inference

Following the calculation of the knowledge bases, the inference engine is activated to determine the load forecast value. The mechanisms of the inference engine are described below.

The inference is made in several parts, namely, the initial estimate, the second estimate and the final estimate. The initial estimate uses only the first knowledge base and the actual values of load and weather conditions. It updates the standard load values. Equation (3) presents the initial estimate.

$$L'(t+\Delta) = B(t+\Delta) \cdot \frac{P_a(t)}{B(t)} \quad \text{for } \Delta = 1, 2, \ldots \quad (3)$$

The second estimate is a rated average between the initial estimate and the second knowledge base (deviation load). The ratios of this average are calculated for each bus of the system and can be modified according to the kind of forecasting and the time-of-day. Equation (4) presents this expression.

$$L''(t+\Delta) = \alpha_1 \cdot L'(t+\Delta) + \alpha_2 \cdot R(t+\Delta) \quad (4)$$

The final estimate is obtained after a treatment of the value $L''(t+\Delta)$ by the third knowledge base. If any rule of this knowledge base is applicable then the final estimate is like a second estimate.

ILUSTRATIVE EXAMPLE

This example uses data from the Hydro-Quebec Power System, Canada. The numerical data extends from April 1984 to July 1988. It contains information related to real power flows in lines 1288 and 1289, as well as actual weather conditions. The weather conditions include parameters such as dry bulb temperature, wet bulb temperature, relative humidity, wind speed and direction, all of which were measured hourly. Since the two lines worked in parallel to supply the Plouffe bus, load performance is a function of the addition of these lines. The example forecasts load for the winter month of January 1988.

Application of the algorithm to compute the model of the shape proposed in [4], for the real power, results in:

R_1: If t is $\mu_1(t)$ then $P_1(t) = 0.85t + 116.25$

R_2: If t is $\mu_2(t)$ then $P_2(t) = -1.70t + 165.33$

R_3: If t is $\mu_3(t)$ then $P_3(t) = 4.556 + 71.60$

R_4: If t is $\mu_4(t)$ then $P_4(t) = -1.30t + 153.87$

Where:

$\mu_1(t) = -0.0419t + 0.3881$ for $3 \leq t < 13$

$\mu_2(t) = 0.1079t - 0.6190$ for $5 < t \leq 15$

$\mu_3(t) = -0.0184t + 0.4495$ for $15 \leq t < 25$

$\mu_4(t) = 0.0254t - 0.4725$ for $16 < 6 \leq 27$

Note: 1. Outside the limits for t, the membership values are zero.
2. 24 hour notation is used.
3. 27 is 24 + 4, i.e., 3 a.m. of the next day.

For the recent load performance, it is possible to calculate the following rules using the same algorithm.

R_1: If t is $\mu_1(t)$ then $P_1(t) = 138.5613 - 1.7476(D_C(t) - D_S(t)) + 2.0460(M_C(t) - M_S(t)) - 0.6361(H_C(t) - H_S(t))$

R_2: If t is $\mu_2(t)$ then $P_2(t) = 172.5990 + 6.6206(D_c(t) - D_s(t))$
$- 5.4491(D_c(t) - D_s(t))^2 - 5.5781(M_c(t) - M_s(t))$
$+ 0.00968(H_c(t) - H_s(t))$

R_3: If t is $\mu_3(t)$ then $P_3(t) = 175.8391 - 9.7388(D_c(t) - D_s(t))$
$+ 1.9356(D_c(t) - D_s(t)^2 + 1.7729(M_c(t) - M_s(t))$

R_4: If t is $\mu_4(t)$ then $P_4(t) = 181.1597 - 6.4085(D_c(t) - D_s(t))$
$+ 1.8975(D_a(t) - D_s(t))^2$

Where:
$\mu_1(t) = -0.3380t + 3.0280$ for $1 \leq t < 8$
$\mu_2(t) = 0.7908t - 4.5357$ for $6 < t \leq 15$
$\mu_3(t) = -1.2000t + 22.6000$ for $15 \leq t < 20$
$\mu_4(t) = 1.8033t - 33.2623$ for $18 < t \leq 25$

Load forecasts for January 22, 1988 were obtained for three different periods, an hour ahead, two ahead and three hours ahead.

The initial estimate was made using equation (1). The values produced are displayed in figure 1. The second estimate was subsequently made. The average ratio of 0.4 was used for the initial estimate, and 0.6 for the deviation load. Table 1 presents the actual and forecasted loads. The average error for the forecast of one hour ahead was 0.79%, for two hours ahead was 1.02%, and for three hours ahead was 1.13%.

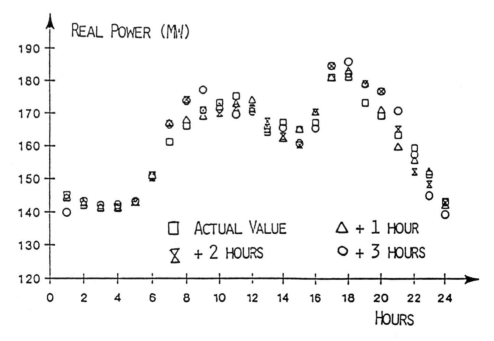

Figure 1 - Initial Estimate Values.

Table 1: Actual and Forecasted Loads

Hour	Real Load [MW]	Forecast-ed Load +1h [MW]	Error [%]	Forecast-ed Load +2h [MW]	Error [%]	Forecast-ed Load +3h [MW]	Error [%]
1h	145,00	144,06	-0,65	143,94	-0,73	142,32	-1,85
2h	142,00	142,99	0,70	143,19	0,84	143,27	0,89
3h	141,00	141,06	0,40	141,30	0,21	141,39	0,28
4h	141,00	141,09	0,06	141,08	0,06	141,45	0,32
5h	143,00	145,31	1,61	145,43	1,70	145,55	1,78
6h	151,00	148,41	-1,71	148,13	-1,90	148,42	-1,71
7h	161,00	163,09	1,30	162,95	1,21	162,82	1,13
8h	166,00	169,65	2,20	172,14	3,70	172,00	3,61
9h	171,00	170,61	-0,23	171,35	0,20	173,91	1,70
10h	173,00	172,43	-0,33	171,60	-0,81	172,37	-0,36
11h	175,00	173,52	-0,85	173,08	-1,10	172,26	-0,56
12h	171,00	172,60	0,93	171,66	0,38	171,23	0,13
13h	164,00	164,16	0,10	165,24	0,76	164,33	0,20
14h	167,00	164,82	-1,31	165,09	-1,14	166,16	-0,50
15h	165,00	164,94	-0,04	162,98	-1,22	163,25	-1,06
16h	167,00	168,81	1,08	168,81	1,08	166,79	-0,12
17h	181,00	180,62	-0,21	182,09	0,60	182,09	0,60
18h	181,00	181,55	0,30	181,23	0,13	182,71	0,95
19h	173,00	174,95	1,33	174,95	1,13	174,87	1,08
20h	169,00	170,60	0,95	172,93	2,32	172,93	2,32
21h	163,00	159,64	-2,06	161,84	-0,71	164,10	0,67
22h	159,00	159,31	0,19	157,93	-0,67	160,03	0,65
23h	151,00	150,79	-0,14	149,34	-1,10	148,02	-1,97
24h	143,00	142,54	-0,32	142,82	-0,13	141,45	-1,08

CONCLUSIONS

The alternative approach to load forecasting discussed here, uses a knowledge based approach divided into three knowledge bases containing information about standard load and standard weather values, recent performance of the load, and rules-of-thumb from operator experience.

REFERENCES

[1] Wood,A.J.;and Woolenberg,B.F.- "Power Generation, Operation and Control" John Wiley & Sons, 1984.
[2] Valiquette,B;Lambert-Torres,G.;and Mukhedkar,D.-"An Expert System Based Diagnosis and Advisor Tool for Teaching Power System Operation Emergency Control Strategies",IEEE Trans. on Power Systems, Vol.6, No.3, p.1315-1322, August 1991.
[3] Handschin,E.;and Dornemann,C.- "Bus Load Modelling and Forecasting", IEEE Trans. on Power Systems Vol.3, No.2, p.627-633, 1988.
[4] G.Lambert Torres, L.E.Borges da Silva, and D. Mukhedkar- "Fuzzy Conditional Statements to Modelling and Control", Proc. of Third Int Fuzzy Systems Association (IFSA),Seatle, p.606-609, 1989.

A Symptom-Driven Fuzzy System for Isolating Faults

Jiann-Liang Chen, Ronlon Tsai

Advanced Technology Center
Computer & Communication Research Lab.
Industrial Technology Research Institute

Huan-Wen Tzeng

Department of Industrial Education
National Taiwan Normal University
Taipei, Taiwan, R.O.C.

Abstract The article discusses recent work toward the simplification of fault isolation within distribution networks using a symptom-driven fuzzy system. Based on a symptom description and situation-specific background data, the system draws upon the prior experiences of distribution networks to determine which component is the most likely one to be at fault. The developed fuzzy system narrows the search to a relatively few suspicious components by means of the symptom description and supporting data. At the same time, the system draws upon the prior experiences of networks to determine the membership-grade (i.e. the degree that each suspicious component is indeed faulty) and executes an investigation that focuses on the most likely fault component. The system periodically seeks feedback until the fault location is investigated. The system is running under Sun workstation and applied to the fault isolation of a distribution network in the west of Taipei which contains four substations and thirty-eight feeders. The results from several trials indicate that the fuzzy system performs well in minimizing the amount of time taken to isolate faults and will be a valuable tool to assist distribution troubleshooters in handling the task of isolating faults.

Key Words: Fuzzy system, Fault isolation, Membership-grade

1 INTRODUCTION

The concept central to the operation of a distribution system in Taiwan Power Company is to supply power to customers and to meet their load demands. In order to get through the work, there are two main mechanisms of a distribution system. One is to increase the capacity of the distribution system, and the other is to reduce the effect of emergency faults. Usually the latter, whether the customers are supplied with power or not, deserves more attention than the former. Hence, research on the restoration procedure is of great importance, especially the reduction of the restoration time. To handle and scrutinize the restorative state, fault diagnosis is difficult and time-consuming. Therefore, the fault diagnosis, especially on isolating faults, is the first step in restoration procedure.

Manuscript received August 1, 1992. This paper is a partial result of the project no. 37H1200 conducted by ITRI under sponsorship of Minister of Economic Affairs, R.O.C.

The task of isolating faults within large and complex distribution systems may be extremely demanded. Several approaches based on switching operations and logic programming have been reported in the literatures for isolating faults of distribution systems [1-3]. Moreover, considerable progress in the applications of expert systems [4-8] and artificial neural networks [9-11] to fault management have been achieved in recent years. However, a large system may consist of hundreds of components and a blind search would be costly and extremely time-consuming.

The article discusses recent work toward the simplification of fault isolation within distribution systems using a symptom-driven fuzzy system. Much like any expert diagnostician, based on a symptom description and situation-specific background data supplied by the customer and the trouble database, the system draws upon the prior experiences of the distribution system to determine which component is the most likely one to be at fault. Then, the task of isolating faults will be proceeded. To demonstrate the effectiveness of the proposed fuzzy system, the isolating fault on a distribution system which consists of four substations and thirty-eight feeders is examined.

The remaining parts of this paper are organized as follows. Section 2 introduces the initial task domain. The subsequent section describes the symptom-driven fuzzy system for isolating faults. Trial results on a distribution system are given in section 4. Section 5 discusses our works. Finally, some conclusions are described in section 6.

2 INITIAL TASK DOMAIN

The system under study is a distribution system in the west of Taipei which contains four substations and thirty-eight feeders [12,13]. For clarity, Figure 1 shows the system structure with omitted feeders. At the same time, the four substations in Figure 1 are labeled as YC, YD, YE, and YL and the feeders are labeled as YC21, YC22, ..., YC28, YD21, YD22, ..., YD30, YE21, YE22, ..., YE30, and YL21, YL22, ..., YL30. In addition, the voltage level of each feeder is 11.4 KV and all kinds of components contain 55 types such as transformer, PC-pole, Wood-pole, CT, PT, ..., LT.-cable [14].

As any complex system, components occasionally go wrong. For example, a customer may be unable to establish a connection with the distribution network such that the outage is generated. Facing the mentioned problem, most distribution system installations maintain a hot-line(telephone) where system customers can report problems. Typically after a complaint reaches the service center, a trouble ticket is recorded including a symptom description and the name and location of the customer who reports the trouble and other details. The fault associated with these symptoms may be invoked within the system facilities or within the transformer, cable, ..., or any number of external sources. Thus, the task of isolating faults may be perceived as excessive.

The process of judging isolating faults is the most difficult aspect of fault management. Hence, it is the main objective of the paper to develop a symptom-driven fuzzy system to isolate faults on the distribution system shown in Figure 1.

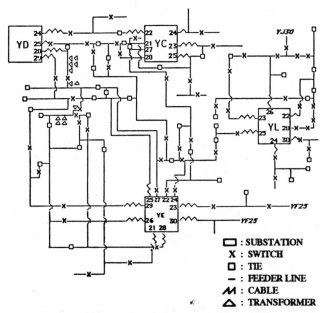

Fig. 1 The system structure

3 SYSTEM OPERATION

The symptom-driven fuzzy system has several technologies of operation that are intended to accommodate different skill levels and task demands. Figure 2 depicts the investigation of trouble reported by customers and will be discussed in detail as follows.

- **STEP 1: Encoding the Problem**

  A few symptom descriptions are automatically incorporated into the fuzzy system and associated with a fault condition. Then, the troubleshooters begin an investigation by gathering the same sort of information that would be included within a trouble ticket.

- **STEP 2: Identifying the Candidate Set**

  A large and sophisticated distribution network may consist of hundreds of components and a blind search would be costly and extremely time-consuming. Starting the process of fault isolation, the symptom-driven fuzzy system restricts the search range to a candidate set. The set is made up of all the related components which may turn out the symptom of specific fault in distribution networks.

- **STEP 3: Building the Routing Table**

  The troubleshooters set a routing table that specifies the order in which each component within the candidate set will be investigated. At the same time, the routine table is arranged such that the components deemed most likely to be at fault are investigated first.

  Much like any expert diagnostician, the system draws upon the prior experiences of distribution networks to determine the membership-grade of each component at fault. A number of theories had been devised to deal with the membership-grade, but the research described here focuses on the Zadeh's fuzzy theory [15-17]. In the fuzzy theory, a component may belong partially to a fuzzy set. The degree of membership in a fuzzy set is measured by a membership function which maps the experiences of distribution networks into the codomain of real numbers defined in the interval from 0 to 1. By the prior experiences of distribution networks stored in the trouble database, the membership function is referred to as S-function which approaches the half bathtub hazard function. The bathtub hazard function is often classified according to their tendency to increase, decrease, and remain constant in time for most of components. Figure 3 illustrates the S-function and its definition. In this definition, α, β, and γ are parameters which may be adjusted to fit the desired membership data by the given prior experiences of components of candidate set.

- **STEP 4: Detecting Fault**

  Once a component is selected for investigation, the system takes the component to determine whether its operation is normal. If the component is found to be normal, the next component specified on the routing table is selected. Consequently, the fault will be detected. Then, the actions of isolating faults will be suggested by a switching operation expert system which is a consultant system for fault management.

- **STEP 5: Testing**

  If step 4 is successful, the troubleshooters or customers will be asked to indicate whether the initial symptom has disappeared. If the symptom persists, the symptom-driven fuzzy system will infer that there are multiple faults and feedback to step 4 to continue testing at the next most likely fault component. The investigation terminates only when the fault underlying the reported symptom is isolated. Finally, the system will update the trouble database to record the symptom and component that is found to be at fault.

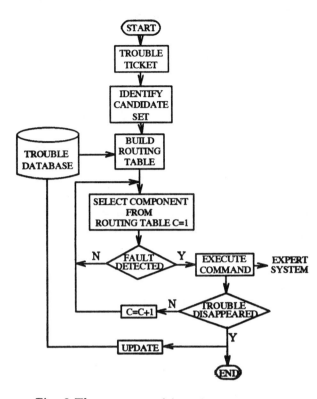

Fig. 2 The symptom-driven fuzzy system

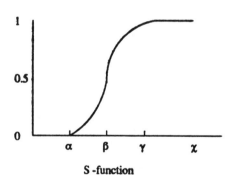

$$S(X; \alpha, \beta, \gamma) = \begin{cases} 0 & \chi \leq \alpha \\ 2\left(\dfrac{\chi - \alpha}{\gamma - \alpha}\right)^2 & \alpha \leq \chi \leq \beta \\ 1 - 2\left(\dfrac{\chi - \gamma}{\gamma - \alpha}\right)^2 & \beta \leq \chi \leq \gamma \\ 1 & \gamma \geq \chi \end{cases}$$

Fig. 3 The S-function and its definition

4 RESULTS

Based on the fuzzy theory and algorithm described in last section, a symptom-driven fuzzy system is implemented. The trial system is a distribution system located within the Taipei West District of Taiwan Power Company and shown in Figure 1. The routing table shown in Table 1 is derived from the trouble report given by the customer labeled as C12-2 of feeder YD28.

Table 2 The routing table

COMPONENT	MEMBERSHIP GRADE	ROUTING ORDER
o.c.b.	0.0	26
ug 3C500xp1	0.2	17
oh 3A477	0.5	12
oh 3A477	0.0	27
ftie-sw.-D.S.	---	
ftrx.-1	---	
ftrx.-2	---	
oh 3A477	0.53	11
ftrx.-3	---	
ftrx.-4	---	
oh 3A477	0.12	23
oh A#2	0.0	28
ffuse(T)	---	
fload point-1	---	
oh 3A477	0.18	20
oh A#2	0.2	18
ffuse(T)	---	
fload point-2	---	
oh 3A477	0.24	15
oh A#2	0.0	29
ffuse(T)	---	
fload point-3	---	
oh 3A477	0.55	10
oh A#2	0.15	21
ffuse(T)	---	
fload poin	---	
oh 3A477	0.62	8
ftrx.(H)-5	---	
oh 3A477	0.62	9
oh A#4/0	0.0	30
tfuse(N)	---	
fload point-5	---	
oh 3A477	0.70	6
fjumper	1.00	1
oh 3A477	0.68	7
fswitch-D.S.	0.0	31
oh 3A477	0.21	16
ftie-sw.-D.S.	---	

Some components are omitted

Legends: 1. "---" stands for the unrelative component to the trouble report
2. the trouble report is given by the customer labeled as C12-2 (YD28)
3. some relative components are omitted due to limited space

5 DISCUSSIONS

From the results presented so far, several observations are in order:
1. From Table 1, a huge number of connected components are observed. To reduce the running time for building the routing table, the components of distribution system within the same zone section-

alized by protective or switching devices and connected in series logically may be merged into a single equivalent component.

2. It is enhanced that traditional approaches to task simplification have focused on properties of the user interface. The use of colors, graphics, and zooms and techniques of many coding and display can provide distribution system troubleshooters with critical information required to make decisions.

3. The diversity of components has the same weight before building the routing table. For increasing the reliability, reducing the damage degree, and saving the cost investment, every component may be given a critical point by the objective function.

6 CONCLUSIONS

The symptom-driven fuzzy system has been developed for isolating faults of a distribution system in Taiwan. The system parallels the troubleshooters procedure in the fault isolation and is implemented on a Sun workstation. To simplify fault isolation, the routing table is built by the membership-grade given by the fuzzy theory. In view of the several trials, it is authors' belief that the symptom-driven fuzzy system will be a valuable tool to assist distribution system troubleshooters in handling the task of isolating faults.

ACKNOWLEDGMENTS

The authors would like to express their gratitude to the Advanced Technology Center of Computer and Communication Research Laboratories of Industrial Technology Research Institute for computer facility support.

REFERENCES

[1] C.H. Castro, J.B. Bunch, and T.M. Topka, "Generalized algorithms for distribution feeder deployment and sectionalizing," IEEE Trans., PAS-99, pp.549-557, 1980.

[2] B.M. Aucoin and B.D. Russell, "Distribution high impedance fault detection utilizing high frequency current components," IEEE Trans., PAS-101, pp.1596-1606, 1982.

[3] K.P. Wong and C.P. Tsang, "A logic programming approach to fault diagnosis in distribution ring networks," Electr. Power Syst. Res., pp.77-87, 1988.

[4] S.N. Talukdar, E. Cardozo, and T. Perry, "The operator's assistant-an intelligent, expandable program for power system trouble analysis," IEEE Trans., PWRS-1, pp.182-187, 1986.

[5] C.C. Liu, S.J. Lee, and S.S. Venkata, "An expert system operational aid for restoration and loss reduction of distribution systems," IEEE Trans. on Power Systems, Vol.3, No.2, pp.619-626, 1988.

[6] J.J. Keronen, "An expert system prototype for event diagnosis and real-time operation planning in power system control," IEEE Trans. on Power Systems, Vol.4, No.2, pp.544-550, 1989.

[7] D.S. Kirschen and T.L. Volkmann, "Guiding a power system restoration with an expert system," IEEE Trans. on Power Systems, Vol.6, No.2, pp.558-564, 1991.

[8] F. Eickhoff, E. Handschin, and W. Hoffmann, "Knowledge based alarm handling and fault location in distribution networks," IEEE Trans. on Power Systems, Vol.7, No.2, pp.770-776, 1992.

[9] S. Ebron, D.L. Lubkeman, and M. White, "A neural network approach to the detection of incipient faults on power distribution feeder," IEEE Trans. on Power Delivery, Vol.5, No.2, pp.905-914, 1990.

[10] N. Kandil, V.K. Sood, K. Khorasani, and R.V. Patel, "Fault identification in an AC-DC transmission system using neural network, IEEE Trans. on Power Systems, Vol.7, No.2, pp.812-819, 1992.

[11] J.L. Chen and C.L. Chen," A connectionist expert system for fault diagnosis," Journal of Electr. Power Systems Research, Vol.24, No.2, pp.99-103, 1992.

[12] J.L. Chen and Y.Y. Hsu, "An expert system for load allocation in distribution expansion planning," IEEE Trans. on Power Delivery, Vol.4, No.3, pp.1910-1918, 1989.

[13] Y.Y. Hsu, L.M. Chen, and J.L. Chen, "Application of a microcomputer-based database management system to distribution system reliability evaluation," IEEE Trans. on Power Delivery, Vol.5, No.1, pp.343-350, 1990.

[14] Y.Y. Hsu, J.L. Chen, and L.M. Chen, Development of a software program for evaluation of distribution system reliability, Research Report of Taiwan Power Company, 1988.

[15] L.A. Zadeh, Fuzzy logic, IEEE Computer, pp.83-93, 1988.

[16] K.S. Leung and W. Lam, "Fuzzy concepts in expert system," IEEE Computer, pp.43-56, 1988

[17] J. Giarratano and G. Riley, Expert systems principles and programming, PWS-KENT Publishing Company, Boston, pp.185-341, 1989.

COMPARISON OF FUZZY LOGIC BASED AND RULE BASED POWER SYSTEM STABILIZER

JUAN SHI   L.H.HERRON   A. KALAM

Department of Electrical and Electronic Engineering
Victoria University of Technology
Ballarat Road, P.O.Box 64, Footscray
Australia, 3011

Abstract: This paper presents a comparison result of applying fuzzy logic based and rule based power system stabilizer for a synchronous machine. To achieve good damping characteristics over a wide range of operating conditions, speed deviation and acceleration of a synchronous machine are chosen as the input signal to the stabilizers. The stabilizing signal is determined from certain rules for rule-based power system stabilizer. For fuzzy logic based power system stabilizer, the supplementary stabilizing signal is determined according to the fuzzy membership function depending on the speed and acceleration states of the generator. The simulation result shows that the proposed fuzzy logic based power system stabilizer is superior to rule-based stabilizer due to its lower computation burden and robust performance.

1. INTRODUCTION

The success of excitation control in improving power system dynamic performance in certain situations has led to greater expectations as to the capacity of such control. Because of the small effective time constants in the excitation system control loop, it was assumed that a large control effort could be expended through excitation control with a relatively small input of control energy. But the excitation system introduces a large phase lag at low system frequencies just above the natural frequency of the excitation system. Thus it can often be assumed that the voltage regulator in the excitation system introduces negative damping. The application of a power system stabilizer (PSS) is to generate a supplementary stabilizing signal, which is applied to the excitation control loop of a generating unit, to introduce a positive damping torque [1]. The most widely used conventional power system stabilizer is the lead-lag PSS where the gain settings are fixed at certain values which are determined under particular operating conditions. The design of the conventional power system stabilizer is based on linear approximation of nonlinear power plant. Since the operating point of a power system drifts as a result of continuous load changes or unpredictable major disturbances such as a three-phase fault, the fixed gain conventional PSS can not adapt the stabilizer parameters in real time based on on-line measurements. A self-tuning power system stabilizer has been employed to adapt the stabilizer to maintain good dynamic performance over a wide range of operating conditions.

Although the self-tuning PSS has offered better dynamic performance than the fixed-gain PSS, it suffers from a major drawback of requiring model identification in real-time which is very time consuming, especially for a microcomputer with limited computational capacity. To overcome this problem, a rule-based power system stabilizer (RBPSS)[6] and a fuzzy logic based power system stabilizer (FLPSS)[7,10] are developed without real-time model identification. Both RBPSS and FLPSS could be easily constructed using a simple microcomputer associated with A/D and D/A converters. The operating conditions of the synchronous machine are expressed by the quantities of speed deviation and acceleration in the phase plane.

In this paper, the previous fuzzy logic based power system stabilizer used by T.Hiyama in reference [10] has been modified by using a nonlinear membership function. The proposed fuzzy logic based power system stabilizer has been compared with the rule-based and fuzzy logic based power system stabilizer[10]. The simulation results indicate that the proposed fuzzy logic based power system stabilizer is superior to the rule-based stabilizer and the previous fuzzy logic based power system stabilizer.

2. RULE-BASED POWER SYSTEM STABILIZER (RBPSS)

A single-machine infinite-busbar power system shown in Figure 1 is considered in this study. The supplementary stabilizing signal u_s is added to the excitation control loop as shown in Figure 1. The accelerating control of the study unit is achieved by applying a negative stabilizing signal to the excitation loop, because the electrical output of the study unit can be decreased by the negative stabilizing signal. Correspondingly, decelerating control is achieved by applying a positive stabilizing signal to the excitation loop with the increased electrical output through the positive stabilizing signal.

According to T.Hiyama [6], the synchronous generator condition can be expressed with the quantities of speed deviation and acceleration in the phase plane. The phase plane is divided into six sectors as shown in figure 2, with simple control rules for each one of them. From these control rules, together with the generator condition at each sampling time, the supplementary stabilizing signal is determined and applied to the excitation control loop.

The stabilizing signal $u_s(t)$ is given by:

$$u_s(t) = U_s(k), \quad \text{for } k\Delta T \leq t < (k+1)\Delta T \quad (2.1)$$

in a discrete form, where k indicates the time $k\Delta T$, and ΔT represents the sample interval. The generator condition at the time $t = k\Delta T$ is given by the point p(k) in the phase plane as shown in figure 2.

$$p(k) = [\ \Delta\omega(k),\ \{\Delta\omega(k)-\Delta\omega(k-1)\}/\Delta T\ ] \quad (2.2)$$

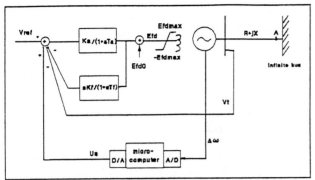

Figure 1. A single machine infinite-busbar power system

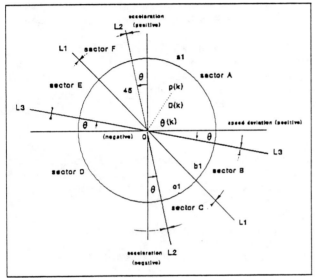

Figure 2 Six sectors in phase plane

The origin in the phase plane is the desired equilibrium point in figure 2. All the control efforts should be directed to moving the current condition p(k) towards the origin as soon as possible.

The control rules are as follows:

Rule1: If $p(k) \in$ sector A, then $U_s(k) = G(k)U_{max}$ (2.3)
Rule2: If $p(k) \in$ sector B, then $U_s(k) = G(k)U_{min}$ (2.4)
Rule3: If $p(k) \in$ sector C, then $U_s(k) = -G(k)U_{min}$ (2.5)
Rule4: If $p(k) \in$ sector D, then $U_s(k) = -G(k)U_{max}$ (2.6)
Rule5: If $p(k) \in$ sector E, then $U_s(k) = -G(k)U_{min}$ (2.7)
Rule6: If $p(k) \in$ sector F, then $U_s(k) = G(k)U_{min}$ (2.8)

where $G(k) = D(k)/D_r$, for $D(k) < D_r$ (2.9)

$G(k) = 1.0$, for $D(k) \geq D_r$ (2.10)

$D(k) = |p(k)|$ (2.11)

The term G(k) indicates the gain factor at the time $t = k\Delta T$, and G(k) is given by a nonlinear function as shown in figure 3.
The maximum and minimum value of the stabilizing signal U_{max} and U_{min} depends on the generating unit. Distance parameter D_r and angle θ can be adjusted to their optimal values according to a performance index as defined in the following:

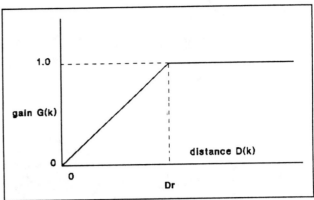

Figure 3 Gain factor G(k) related to distance D(k)

$$J = \sum_{K=1}^{M} [t_k \Delta\omega(k)]^2 \quad (2.12)$$

3. FUZZY LOGIC BASED POWER SYSTEM STABILIZER (FLPSS)

3.1. Introduction to fuzzy set theory

Fuzzy set

A fuzzy set F in a universe of discourse U is characterised by a membership function μ_F which takes values in the interval [0,1] namely, $\mu_F: U \rightarrow [0,1]$. A fuzzy set may be viewed as a generalisation of the concept of an ordinary set whose membership function only takes two values {0,1}. Thus a fuzzy set F in U may be represented as a set of ordered pairs of a generic element u and its grade of membership function: $F = \{(u, \mu_F(u)) | u \in U\}$. When U is continuous, a fuzzy set F can be written concisely as $F = \int_U \mu_F(u)/u$. When U is discrete, a fuzzy set F is represented as

$$F = \sum_{i=1}^{n} \frac{\mu_F(u_i)}{u_i} \quad (3.1)$$

Support, Crossover Point, and Fuzzy Singleton:

The support of a fuzzy set F is the crisp set of all points u in U such that $\mu_F(u) > 0$. In particular, the element u in U at which $\mu_F = 0.5$, is called the crossover point and a fuzzy set whose support is a single point in U with $\mu_F = 1.0$ is referred to as fuzzy singleton.

3.2 Fuzzy logic based power system stabilizer

The phase plane is divided into two sectors in [10] as shown in figure 4. The required control strategies are the same as in rule-based stabilizer.

Two membership functions, $N\{\theta_i(k)\}$ and $P\{\theta_i(k)\}$, are defined as shown in figure 5 to represent both the sector A and B respectively. The term $\theta_i(k)$ indicates the phase angle of the point $p_i(k)$ as shown in Figure 4. By using these membership functions, the stabilizing signal is computed as follows:

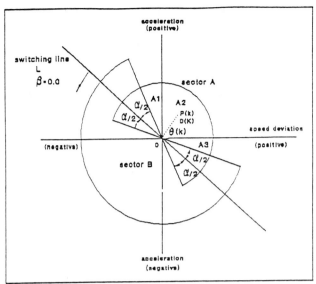

Figure 4 Two sectors in phase plane

$$U_s(k) = G(k)[N\{\theta_i(k)\}U_{max} - P\{\theta_i(k)\}U_{max}]/[N\{\theta_i(k)\} + P\{\theta_i(k)\}]$$
$$= G(k)[2N\{\theta_i(k)\}-1]U_{max} \quad (3.2)$$

Where $P\{\theta_i(k)\} = 1 - N\{\theta_i(k)\}$, (3.3)

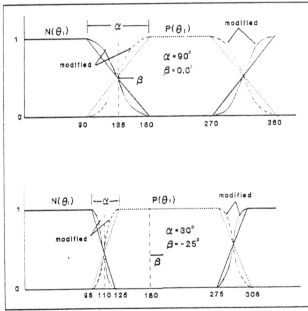

Figure 5 Membership functions

$G(k) = D(k)/Dr$, for $D(k) < Dr$ (3.4)

$G(k) = 1.0$, for $D(k) \geq Dr$, (3.5)

$D(k) = |p(k)|$. (3.6)

3.3 The proposed fuzzy logic based power system stabilizer

In order to improve the performance of the above fuzzy logic based power system stabilizer, the function $N\{\theta_i(k)\}$ and $P\{\theta_i(k)\}$ have been modified as shown in Figure 5. The function $P\{\theta_i(k)\}$ exists for $\theta > \theta_0$. The function $P\{\theta_i(k)\}$ can be expressed as follows:

0	for $0 \leq \theta_i \leq \theta_0$
$2[(\theta_i - \theta_0)/\alpha]^2$	for $\theta_0 \leq \theta_i \leq \theta_0 + \alpha/2$
$1 - 2\{[\theta_i - (\theta_0 + \alpha)]/\alpha\}^2$	for $\theta_0 + \alpha/2 \leq \theta_i \leq \theta_0 + \alpha$
1	for $\theta_0 + \alpha \leq \theta_i \leq \theta_0 + 180$ (3.7)
$1 - 2\{[\theta_i - (\theta_0 + 180)]/\alpha\}^2$	for $\theta_0 + 180 \leq \theta_i \leq \theta_0 + 180 + \alpha/2$
$2\{[\theta_i - (\theta_0 + 180 + \alpha)]/\alpha\}^2$	for $\theta_0 + 180 + \alpha/2 \leq \theta_i \leq \theta_0 + 180 + \alpha$
0	for $\theta_0 + 180 + \alpha \leq \theta_i \leq 360$

where all the angles in (3.7) are in degrees. $\theta_0 + \alpha/2$ and $\theta_0 + 180 + \alpha/2$ are crossover points.

4. SIMULATION STUDY

Consider a single machine connected by a transmission line of $0.02 + j0.4$ pu impedance to a large system as shown in Figure 1. The machine is represented by a 7th order flux linkage state-space model with saturation neglected.

The following disturbances were considered in the simulation study:

a) step change in mechanical torque Tm for 0.3 sec.
b) step change in reference voltage Vref for 0.3 sec.
c) A 0.1 sec three-phase to ground fault at point A in Figure 1.

The simulation has also been done under different operating conditions as shown in Table I.

Table I

Operating Conditions			
Operating condition	P1	P2	P3
Real Power P	1.0	1.0	0.4
Reactive Power Q	0.62	0.4	0.2

4.1 Optimal setting of stabilizer parameters

There are several parameters to be adjusted to their optimal values. The location of the switching line L between the positive and negative stabilizing signals, the size of the cross sections between the sector A and B for FLPSS and the distance parameter Dr. The maximum and minimum value of the stabilizing signal U_{max} and U_{min} depends on the generating unit to be studied.

For optimal setting, a discrete-type quadratic performance index is defined as in equation (2.12).

The index J is specified to investigate the time optimality of the study unit. According to the values of the performance index, the optimal settings of the adjustable parameters can be determined. Namely, the adjustable parameters are set to the values which minimise the above index.

The values of the performance index corresponding to the change of Dr values for different power system stabilizers have been studied. Figure 6 shows the performance index under operating condition P2 for different Dr values. The optimal Dr is 0.014.

The values of performance index corresponding to the change of switching line L have also been studied. For rule based power system stabilizer, the optimal value of θ is 0. For fuzzy controller [10] and the proposed fuzzy logic based power system stabilizer, the optimal setting of α and β are: $\alpha=30$, $\beta=-25$ and $\theta_0=95$ degree. Figure 7 shows the performance index under operating condition P2 for different setting of α and β values with proposed fuzzy logic based power system stabilizer.

4.2 Step change in mechanical torque Tm

Performance of the proposed fuzzy logic based power system stabilizer compared with the rule based [6], the fuzzy [10] and conventional power system stabilizer (CPSS) for 10% step increase in mechanical torque Tm for 0.3 sec under three different operating conditions is shown in Figure 8. These results show that the proposed fuzzy logic based power system stabilizer improves in performance. Fuzzy[10] gives poor result under operating condition P2.

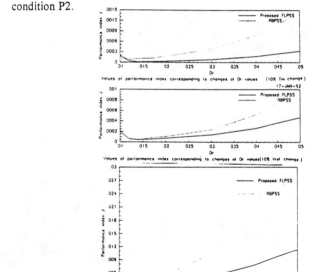

Figure 6

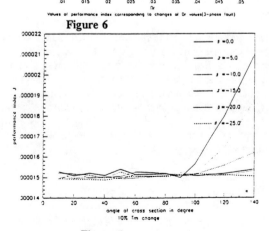

Figure 7.

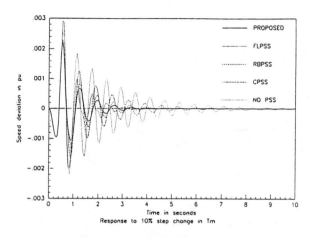

Figure 8 (a) operating condition P1

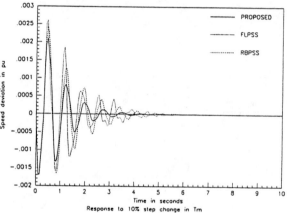

Figure 8 (b) operating condition P2

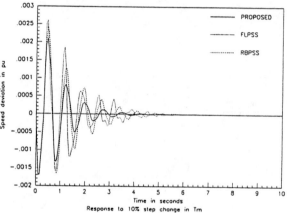

Figure 8 (c) operating condition P3

4.3 Step changes in reference voltage Vref

Results of the proposed fuzzy PSS for 10% step increase in Vref for
0.3 sec for different operating points compared with the rule-based [6] and fuzzy [10] are shown in Figure 9. It can be seen that the proposed fuzzy PSS gives considerably better result.

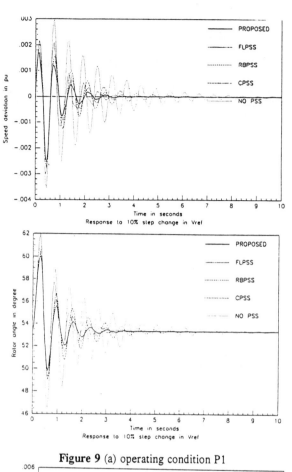

Figure 9 (a) operating condition P1

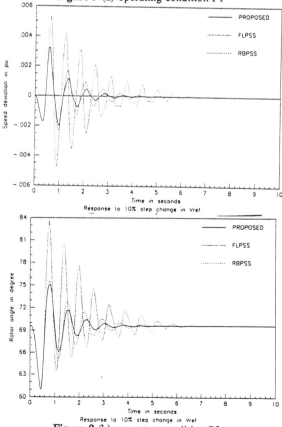

Figure 9 (b) operating condition P2

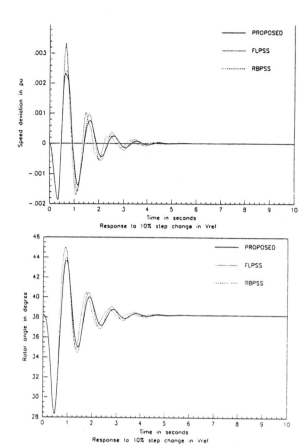

Figure 9 (c) operating condition P3

4.4 Three-phase to ground fault at point A

Figure 10 shows the performance of the proposed fuzzy PSS for three phase to ground fault at point A as shown in Figure 1. System response with the proposed fuzzy PSS is compared with the rule-based [6] and fuzzy [10]. For operating point P1, the results with different PSS are very close. For operating point P2 and P3, the proposed fuzzy PSS gives better performance. For operating point P2, fuzzy [10] gives a poor result. Figure 11 shows the stabilizing signal for Three-phase to ground fault at different operating points with the proposed fuzzy logic based PSS.

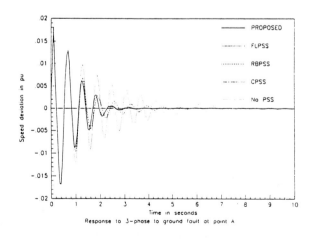

Figure 10 (a) operating condition P1

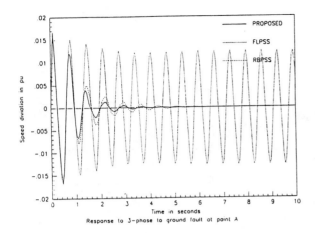

Figure 10 (b) operating condition P2

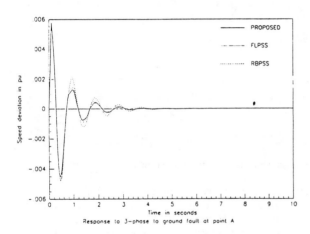

Figure 10 (c) operating condition P3

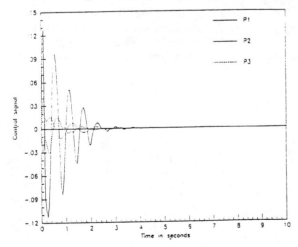

Figure 11

5. CONCLUSION

From the simulation results, we can see that the proposed fuzzy logic based power system stabilizer is superior to the rule-based PSS and fuzzy logic PSS [10] due to its robust performance. The proposed fuzzy logic and rule based power system stabilizer provide better dynamic performance under disturbance conditions than conventional PSS. They are also superior to a self-tuning stabilizer because they do not require real-time model identification as the self-tuning PSS does, therefore, they require less computational capacity on a microcomputer with its ease and simplicity of practical implementation.

6. REFERENCES

[1]. LARSEN,E.V., and SWANN,D.A.:"Applying Power System Stabilisers", IEEE Trans.,1981,PAS-100,pp 3017-3046.

[2]. ANDERSON,P.M., and FOUAD,A.A.:"Power System Control and Stability", (Iowa State University Press, Ames,Iowa,1977')

[3]. DEMELLO, F.P., and CONCORDIA, C.: "Concepts of Synchronous Machine Stability as Affected by Excitation Control", IEEE trans., 1969, PAS-88, pp316-329.

[4]. ZIMMERMANN,H.J.:"Fuzzy Set Theory and Its Applications" (Kluwer-Nijhoff Publishing Company, 1985)

[5]. HSU,Y.Y., and LIOU,K.L.: " Design of Self-tuning PID Power System Stabiliser for Synchronous Generators", IEEE Trans., 1987,EC-2,pp.343-348.

[6]. T.HIYAMA.:"Application of Rule-Based Stabilising Controller to Electrical Power System". IEE Proc. C, Vol.136,No.3,1989, pp.175-181.

[7]. HSU,Y.Y. and CHENG, C.-H.: " Design of Fuzzy Power System Stabilisers for Multimachine Power Systems" IEE Proceedings, Vol.137,Pt.C,No.3,pp233-238.

[8]. LEE,CHUEN CHIEN." Fuzzy Logic in Control Systems: Fuzzy Logic Controller-Part I". IEEE Trans. Syst. Man Cybern., Vol.SMC-20,No.2,pp404-418, 1990.

[9]. LEE,CHUEN CHIEN."Fuzzy Logic in Control Systems: Fuzzy Logic Controller-Part II". IEEE Trans. Syst.Man Cybern., Vol.SMC-20,No.2,pp419-435,1990.

[10]. T.HIYAMA. and T.SAMESHIMA. 'Fuzzy logic control scheme for on-line stabilization of multimachine power system'. Fuzzy Sets and Systems 39 (1991).pp181-194.

7. APPENDIX

System constants

Generator constants
Inertia constant	$M=4.74s$
d-axis synchronous reactance	$Xd=1.70$, p.u.
d-axis transient reactance	$Xd'=0.245$, p.u.
q-axis synchronous reactance	$Xq=1.64$, p.u.
Time constant of d-axis circuit	$Tdo'=5.9$, s

AVR constants
AVR gain	$Ka=400$
AVR time constant	$Ta=0.02$,s
compensation loop gain	$Kf=0.03$
compensation loop time constant	$Tf=1$, s

ANALYSIS OF POWER SYSTEM DYNAMIC STABILITY VIA FUZZY CONCEPTS

Pei-Hwa Huang

Department of Electrical Engineering
National Taiwan Ocean University
Keelung 20224, Taiwan, R.O.C.

Abstract — This paper deals with the study of power system dynamic stability problem using fuzzy concepts. From the eigenstructure of the electromechanical oscillation modes, fuzzy sets which associate each system generation unit with the relative effect the respective modes impose on it can be constructed. Analysis of these fuzzy sets provides deeper insight into system dynamic stability problems. The proposed fuzzy-based analysis approach is illustrated through an application to a four-machine study system.

Keywords: power systems, dynamic stability, fuzzy set theory.

I. INTRODUCTION

Since 1960s, considerable efforts have been placed on the study of low frequency spontaneous oscillations in power systems [1]. These oscillations may be sustained for minutes and grow to cause system separation. The low frequency oscillations are attributed to the oscillations of the electromechanical modes of the system, *i.e.* oscillations associated with system generator rotors. Such stability problem is commonly referred to as power system dynamic stability [2-5].

The analysis of power system low frequency oscillating phenomena requires a good understanding of the characteristics of the oscillatory modes, and eigenstructure analysis technique, *i.e.* investigation of both system eigenvalues and eigenvectors, has proven to be a powerful tool in studying power system dynamic stability problems [4-8]. This approach requires first the representation of the system in the state space form and then the determination of its eigenstructure, including eigenvalues and the corresponding eigenvectors. The eigenstructure provides detailed information about dynamic characteristics of the system and hence provides a better insight into the nature of the dynamic stability problems.

The main purpose of this paper is to apply fuzzy concepts [9-11] to the analysis of power system dynamic stability problem. The basic concepts of eigenstructure analysis and fuzzy set theory are discussed and characterization of relations between system eigenstructure and fuzzy sets is addressed. From the eigenstructure of the electromechanical oscillation modes, fuzzy sets can be constructed with the view to associating each system generator with the relative effect which the respective modes impose on it. These fuzzy sets provide great insight into system dynamic stability problem and analysis can be performed on them via fuzzy concepts such as α-cut, cardinality, and ranking. Such fuzzy-based approach is more systematic and informative than the traditional empirical analysis on system eigenstructure. The described approach is illustrated through an application to a sample power system.

II. ANALYSIS APPROACH

A. Eigenstructure Analysis

The time response of a linear system can be expressed in terms of the system eigenstructure, including eigenvalues and the corresponding eigenvectors [12]. An eigenvalue and its associated eigenvector is referred to a mode. The stability of an oscillation mode is determined by its eigenvalue of which the real part gives the damping and the imaginary part indicates the frequency of oscillation. The eigenvector depicts the mode shape of an oscillation mode. The relative activities of state variables in a certain oscillation mode are given by the corresponding

---

Manuscript received July 30, 1992. This work was supported in part by the National Science Council of the Republic of China under Grant No. NSC 81-0404-E-019-525.

elements in the eigenvector. Combining a certain state variable with the corresponding eigenvector element in a certain mode gives clearly how much the state variable is affected by the oscillation mode and thus provides deeper insight into the dynamic stability problem. Since the study of power system dynamic stability concerns mainly the system electromechanical oscillation modes, eigenstructure of all the system electromechanical oscillation modes is to be figured out and analyzed in detail.

B. Fuzzy Concepts

The fuzzy set [9-11], which is a generalization of the classical crisp set, extends the values of set membership from two values $\{0,1\}$ to the unit interval $[0,1]$. A fuzzy set can be defined mathematically by assigning to each possible element of the set a value representing its grade of membership in the set. A fuzzy set F can written as:

$$F = \mu_1/x_1 + \mu_2/x_2 + \cdots + \mu_n/x_n \quad (1)$$

where the slash is employed to link the i-th element x_i with its membership grade μ_i and the plus sign indicates that the listed pairs of elements and membership grades collectively form the definition of the fuzzy set F. The membership grades of the elements in a fuzzy set are of key importance in any theoretical or practical application of fuzzy set theory.

Another definitions useful in this paper are the α-cut and the cardinality. An α-cut of a fuzzy set F is a crisp (nonfuzzy) set F_α that contains all the elements with the membership grade in F greater than or equal to the specified value of α. This definition can be written as:

$$F_\alpha = \{ x \mid \mu_F(x) \geq \alpha \} \quad (2)$$

The cardinality of a fuzzy set F, denoted by $|F|$, is defined to be the summation of the membership grades of all the elements of F. Thus,

$$|F| = \sum_{x \in F} \mu_F(x) \quad (3)$$

Note that the cardinality of a fuzzy set is a scalar and can be interpreted as the actual number of elements in the set. For the crisp (nonfuzzy) set, cardinality is defined to be the number of elements.

C. Eigenstructure and Fuzzy Concepts

For the study of power system dynamic stability using eigenstructure analysis, the low frequency electromechanical oscillation modes can be interpreted as, from the viewpoint of fuzzy concepts, some kind of fuzzy sets. These fuzzy sets have system generators as their elements. One certain oscillation mode constitutes a fuzzy set in which each system generator takes the magnitude of the corresponding electromechanical eigenvector element (rotor speed deviation $\Delta\omega$) as its membership grade value in this particular mode. A fuzzy set constructed in this way represents the "severity" of the associated electromechanical mode since it links the system generators with the effect resulted from the electromechanical mode. These fuzzy sets form the core of this study and can be analyzed via fuzzy concepts.

D. Analysis Procedure

Based on the technique of eigenstructure analysis and the concepts of fuzzy sets, the procedure proposed for analyzing power system dynamic stability can be summarized as the following steps:

(1) Find the linearized state equation of the power system under study.
(2) From the linearized state equation in *(1)*, find the system eigenstructure.
(3) From the eigenstructure in *(2)*, construct the severity fuzzy set for each electromechanical mode.
(4) Analyze the severity fuzzy sets constructed in *(3)* using fuzzy concepts.

As an illustration example, the presented analysis procedure will be applied to the study system in the next section with emphasis placed on the interpretation of constructed severity fuzzy sets.

III. EXAMPLE

In this section, dynamic stability study on a sample power system is reported. The system considered in this study is a four-machine, including one infinite bus, power system as shown in Fig. 1. The detailed system data can be found in [13].

According to the analysis procedure described in last section, the linearized system state equation is first derived:

$$\dot{x} = Ax \quad (4)$$

where x and A are the state vector and system matrix and are of dimensions 12×1 and 12×12, respectively.

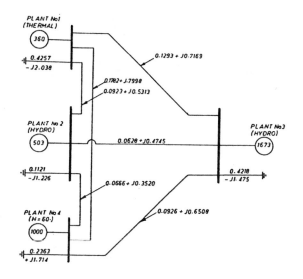

Fig.1 One-line diagram of the four-machine system.

The state vector x has the following structure:

$$x = [\Delta\omega_1 \; \Delta\omega_2 \; \Delta\omega_3 ... (\text{others})...]^T \quad (5)$$

where $\Delta\omega_i$ denotes the rotor speed deviation of the i-th machine. The system is of order 12 and the 12 eigenvalues are figured out and listed in TABLE I. Since there are four machines in the study system, there will be three electromechanical oscillation modes. The eigenstructure of the three modes can be tabulated as TABLE II in which the magnitudes of electromechanical eigenvector elements associated with the rotor speed deviation state ($\Delta\omega_i$) are shown.

TABLE I
EIGENVALUES OF THE FOUR-MACHINE SYSTEM

System Eigenvalues
$0.264 \pm j\,4.091$*
$-0.063 \pm j\,7.370$*
$0.095 \pm j\,7.836$*
-18.871
-17.052
-15.190
-5.891
-3.431
-1.511

* electromechanical modes

Now we can construct the fuzzy set describing the severity of i-th mode, denoted by S^i, from the information listed in TABLE II. The three severity fuzzy sets S^1, S^2, and S^3 are given in equations (6)–(8):

$$S^1 = 0.59/m_1 + 0.76/m_2 + 1.00/m_3 \quad (6)$$

$$S^2 = 1.00/m_1 + 0.19/m_2 + 0.05/m_3 \quad (7)$$

$$S^3 = 0.83/m_1 + 1.00/m_2 + 0.14/m_3 \quad (8)$$

where m_i stands for machine i.

TABLE II
EIGENSTRUCTURE OF THE FOUR-MACHINE SYSTEM

e.v. \ e.$\vec{v}$.	Mode 1 ($0.264\pm j4.091$)	Mode 2 ($-0.063\pm j7.370$)	Mode 3 ($0.095\pm j7.836$)
$\Delta\omega_1$	0.59	1.00	0.83
$\Delta\omega_2$	0.76	0.19	1.00
$\Delta\omega_3$	1.00	0.05	0.14

e.v.: eigenvalue   e.$\vec{v}$.: eigenvector

The cardinality for each severity fuzzy set can be calculated as following:

$$|S^1| = 0.59 + 0.76 + 1.00 = 2.35 \quad (9)$$

$$|S^2| = 1.00 + 0.19 + 0.05 = 1.24 \quad (10)$$

$$|S^3| = 0.83 + 1.00 + 0.14 = 1.97 \quad (11)$$

Before proceeding to further analyze the severity fuzzy sets, different levels of α-cuts for S^1, S^2, and S^3 can be obtained as shown in TABLE III.

From the above results, we can make the following observations:

(1) It is obvious from TABLE I and TABLE II that the study system is dynamically unstable. Modes 1 and 3 have negative damping and the damping of mode 2 is not sufficient.

(2) From the α-cuts in TABLE III, we can see that the effect of mode 1 on the system is the most far reaching (system mode) while those of mode 2 and mode 3 are localized (local modes). Note that mode 1 is of the worst damping and with the lowest oscillation frequency. Many works have reported that modes with widespread influence often have poor dampings and low oscillation frequencies [1]. By taking α-cuts of several different levels, relative effects of electromechanical modes on the system can be examined in detail. Such α-cut approach will exhibit its power in studying the dynamic stability problems of systems with large number of machines.

(3) Cardinality of a fuzzy set can be interpreted as the "actual" number of elements of the set since the membership grades of the elements are equal to or less than 1.0. The cardinality of a severity fuzzy set is the summation of the extent of being affected for each machine and can be realized as the "severity" or "power" of a certain electromechanical oscillation mode. From equations (9)–(11), we have

$$|S^1| > |S^3| > |S^2| \quad (12)$$

Such relation is consistent with the order of mode dampings if ranked from the poorest to the finest. Hence the cardinality can be considered as a severity index which can be used for the problem of fuzzy sets ranking [14]. Therefore, we obtain the following severity ranking:

$$S^1 > S^3 > S^2 \quad (13)$$

Eq.(13) states that mode 1 is the most critical one among the three electromechanical oscillation modes.

TABLE III
α-CUTS OF DIFFERENT LEVELS

α	S^1_α	S^2_α	S^3_α
1.00	$\{m_3\}$	$\{m_1\}$	$\{m_2\}$
0.83	$\{m_3\}$	$\{m_1\}$	$\{m_1, m_2\}$
0.76	$\{m_2, m_3\}$	$\{m_1\}$	$\{m_1, m_2\}$
0.59	$\{m_1, m_2, m_3\}$	$\{m_1\}$	$\{m_1, m_2\}$
0.19	$\{m_1, m_2, m_3\}$	$\{m_1, m_2\}$	$\{m_1, m_2\}$

(4) Another concern in this study is the installation site of power system stabilizers (PSS) [4-8] which have been widely adopted to improve system damping. Elements with membership grades of value 1.00 are the most representative members in a fuzzy set. In the severity fuzzy set, the unit with membership grade value 1.00 plays the most important role and is the suitable installation site for the PSS designed for improving the damping of the corresponding electromechanical oscillation mode. Therefore, to improve the damping of mode 1 which is the most critical one, machine 3 will be the optimum location for the PSS. As for mode 2 and mode 3, suitable locations will be machine 1 and machine 2, respectively.

IV. CONCLUSION

This paper presents the application of fuzzy concepts to the analysis of power system dynamic stability. The analogy and relations between system eigenstructure and fuzzy sets are discussed and employed in the analysis of electromechanical oscillation modes. From the eigenstructure of the electromechanical oscillation modes, we can construct fuzzy sets to associate each system generator with the relative effect which the respective modes impose on it. These fuzzy sets provide deeper insight into system dynamic stability problem and analysis can be performed on such fuzzy sets via fuzzy concepts. Analysis of these fuzzy sets provides deeper insight into system dynamic stability problem.

The proposed approach is illustrated through an application to a sample power system. It is found that the proposed fuzzy-based approach is more systematic and informative than the traditional empirical analysis on system eigenstructure.

REFERENCES

[1] C. Barbier, E. Farrari and K.E. Johansson, "Questionnaire on electromechanical oscillation damping in power systems: Report on answers," *ELECTRA*, vol. 64, pp. 59–90, 1979.

[2] CIGRE, "Definitions of general terms relating to the stability of interconnected synchronous machines," Supplement to paper no. 334, 1966.

[3] IEEE Task Force on Stability Terms and Definitions, "Proposed terms and definitions for power system stability," *IEEE Trans. on Power Apparatus and Systems*, vol. PAS-101, pp. 1894–1898, 1982.

[4] P.M. Anderson and A.A. Fouad, *Power System Control and Stability*, Iowa State University Press, Ames, Iowa, 1977.

[5] Y.N. Yu, *Electric Power System Dynamics*, New York: Academic Press, 1983.

[6] Y.Y. Hsu, P.H. Huang, C.J. Lin and C.T. Huang, "Oscillatory Stability Considerations in Transmission Expansion Planning," *IEEE Trans. on Power Systems*, vol. PWRS-4, pp. 1110–1114, 1989.

[7] P.H. Huang and Y.Y. Hsu, "Eigenstructure Assignment in a Longitudinal Power System Via Excitation Control," *IEEE Trans. on Power Systems*, vol. PWRS-5, pp. 96–102, 1990.

[8] P.H. Huang, *Power System Dynamic Stability Study Via Eigenstructure Analysis*, Ph.D. Dissertation, National Taiwan University, June 1989.

[9] L.A. Zadeh, "Fuzzy sets," *Information and Control*, vol.8, pp. 338–353, 1965.

[10] H.J. Zimmermann, *Fuzzy Set Theory—and Its Applications*, 2nd ed., Boston: Kluwer Academic Publishers, 1991.

[11] G.J. Klir and T.A. Folger, *Fuzzy Sets, Uncertainty, and Information*, Englewood Cliffs, N.J.: Prentice-Hall, 1988.

[12] T. Kailath, *Linear Systems*, Englewood Cliffs, N.J.: Prentice-Hall, 1980.

[13] Y.N. Yu and C. Siggers, "Stabilization and Optimal control Signals for a Power System," *IEEE Trans. on Power Apparatus and Systems*, vol. PAS-90, pp. 1469–1481, 1971.

[14] G. Bortolan and R. Degani, "A Review of Some Methods for Ranking Fuzzy Subsets," *Fuzzy Sets and Systems*, vol. 15, pp. 1–19, 1985.

BIOGRAPHY

Pei-Hwa Huang received his B.Sc. and Ph.D. degrees, both in electrical engineering, from National Taiwan University in 1985 and 1989, respectively. Currently, he is with the Department of Electrical Engineering, National Taiwan Ocean University. His present research interests are in power systems analysis, linear control systems, and applications of fuzzy set theory. He is a member of IEEE.

Chapter 6: Industry Applications

Application of Neuro-Fuzzy Hybrid Control System to Tank Level Control

Tetsuji Tani
Maintenance & System Development Section
Manufacturing Department
Idemitsu Kosan Co., Ltd., Japan

Shunji Murakoshi and Tsutomu Sato
Institute of Information Technology
Information Systems Department
Idemitsu Kosan Co., Ltd., Japan

Motohide Umano
Department of
Precision Engineering
Osaka University, Japan

Kazuo Tanaka
Department of
Mechanical Systems Engineering
Kanazawa University, Japan

Abstract : This paper proposes a practical control method using neural networks and fuzzy control techniques, where neural networks estimate the target of fuzzy control. Neural networks estimate the transient state of the plant which has non-linear process such as refrigerating and filtering. Based on the estimation, the suitable control target pattern for fuzzy control is selected.

This method is applied to the tank level control of the solvent dewaxing plant. And it is shown that this proposed system can control the tank level effectively not only in steady state but also in transient state.

1. Introduction

Since phenomena in the real plant are too complicated to build a theoretical model, it is very difficult to design a control system of such a plant. The operator, however, can control such a plant using his experience. Recently, fuzzy logic, neural network, or both are applied to the real process rather than mathematical models[1, 2, 3]. Fuzzy logic deals with the linguistic and imprecise rules by expert's knowledge. Neural network is also applied to control plants.

This paper deals with a tank level control including non-linear process such as refrigerating and filtering.

A real process often has more than one purpose of control. Our purposes of control are,

(1) to change the flow rate from the tank smoothly,
(2) to keep the tank level stable.

These are contrary to each other.

To overcome these problems, we observe an experienced operator's procedure. He can estimate the suitable target of the tank level and keep the tank level stable. The aim of this paper is to design a neuro-fuzzy hybrid control system which replaces expert's operation. Neural networks estimate the transient state of the plant and based on this estimation the control target pattern of fuzzy control for the smooth change of the flow rate is selected.

This neuro-fuzzy hybrid method is applied to the real tank level control of the solvent dewaxing plant, and good results are obtained not only in steady state but also in transient state.

2. Description of Process

In the process of the vacuum distillation for producing lubricant oil, distillate oil and reduced oil are produced, including wax. Such oils including wax generally have a high

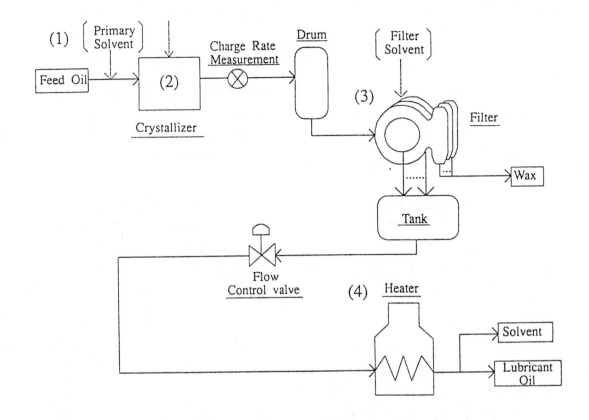

Fig.1. Process flow of the solvent dewaxing plant

solidifying point, so we have to remove the wax not to freeze at low temperature.

The solvent dewaxing plant for removing the wax is outlined in Fig.1. This plant uses solvent to remove the freezed wax easily. We have several steps as follows:

(1) The primary solvent is added to the feed oil.
(2) The feed oil is refrigerated in crystallizer on adding secondary solvent.
(3) The congealed wax is removed by the filter on spraying the filter solvent. The filter is composed of a vacuumed rotating drum which separates the congealed wax from the feed oil. The dewaxing oil, which is a mixture of lubricant oil and solvent, is sent to the tank.
(4) The heater makes solvent evaporate. As a result, low fluid point lubricant oil is produced.

Our control purpose is to keep the tank level constant in all conditions. However, we have the following difficulties for keeping the tank level constant.

(1) The inflow rate to the tank varies with the filter plugging. Since the response of the filter plugging has a long delay time when the feed oil is switched (the delay time also depends on kinds of feed oil), it is difficult to keep the tank level constant using a feed-forward controller.
(2) The heater has a limit in the changing of the flow rate.
(3) We have the feed oil switching frequently (every three or four days).

These factors are combined in a complicated fashion, where an experienced operator used to control the flow rate manually.

3. Operator's Procedure

We have two operation states. One is steady state for everyday operation and the other is transient state of the feed oil switching.

(1) Steady State

Several filters are in operation and one is stopped periodically for washing. The tank level, therefore, goes down and up periodically. An experienced operator,

(a) estimates the flow rate roughly by observing the tank level over several hours,

(b) compensates the flow rate by observing the tank level and the time of filter washing.

(2) Transient State

When the feed oil will be switched, an operator controls the tank level beforehand to make the change of the flow rate more smoothly. For example, if the inflow rate of the tank is expected to be lower after the feed oil switching, the flow rate is beforehand decreased to keep the tank level constant. This prevents the tank from becoming empty and the flow rate from changing rapidly.

We model such experienced operator's procedures to control the tank level as follows:

Step1 : To obtain long-time tendencies of the flow rate, we calculate the average from operation data. This is equivalent to the expert's estimation of rough flow rate.

Step2 : To compensate the flow rate, we use a fuzzy control system based on expert's knowledge. The input variables of this fuzzy control system are the tank level and the time of filter washing.

Step3 : To control the tank level beforehand, we find transient state by using neural networks. This is equivalent to the expert's predictions. The expert predicts the transient state and changes the control target of the tank level.

4. Structure of Neuro-Fuzzy Hybrid Controller

We design neuro-fuzzy hybrid control system which

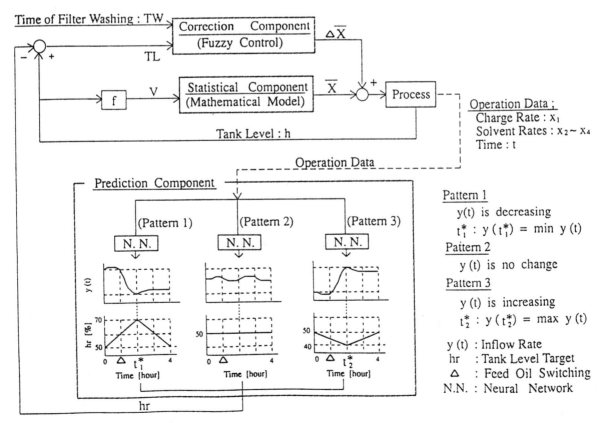

Fig.2. Outline of neuro-fuzzy control system

replaces expert's operation[4, 5, 6]. The controller consists of three components, (1) a statistical component, (2) a correction component (fuzzy controller) and (3) a prediction component (neural networks) in Fig.2.

4.1 Statistical Component

The statistical component is a statistical model for calculating long-time tendencies of the flow rate from operation data. An experienced operator sets the flow rate based on long-time tendencies for the tank level. For example, if the level has a tendency to increase, the flow rate gradually increases.

The difference of the tank amount ΔV is defined as
$$\Delta V(t) = f(h(t)) - f(h(t-1)) \quad (1)$$
where h(t) and h(t-1) are the tank level and f converts its level to the corresponding amount. And the average flow rate $\overline{X}$ is defined as
$$\overline{X}(t) = \overline{X}(t-1) + \alpha \Delta V(t) \quad (2)$$
where α is a real number ($0 < \alpha \leq 1$) and determined by experience. And $\alpha = \alpha_1$ in the steady state and $\alpha = \alpha_2$ is in the transient state, where $\alpha_1 < \alpha_2$.

4.2 Correction Component

The correction component is a fuzzy controller for compensating the flow rate from the statistical component to stabilize the tank level. We use a simplified method of fuzzy reasoning[7].

The control rules of experienced operators to stabilize the tank level are shown in Table 1. These rules mean that,
- when the tank level is near the target, operators focus on the rate of level changing,
- when the tank level is far from the target, operators focus on the time until the next washing.

As a example : If TL is PS and ΔTL is PS then $\Delta \overline{X}$ is PS, where $\Delta \overline{X}$ is compensation of $\overline{X}$.

The tank level target for fuzzy control is 50% of the tank capacity in a steady state. In a transient state, it is set by neural networks which will be described in the next section.

4.3 Prediction Component

When the feed oil will be switched, we have to predict the inflow rate of the tank. But it is too complicated process to build a mathematical model. We use a neural network approach to predict the inflow rate.

The prediction component is neural networks for predicting the inflow rate to estimate the target of fuzzy controller. We use a three layers model whose learning method is back propagation algorithm[8]. Our neural network is shown in Fig.3. The input layer has five units, the hidden layer has ten and output layer has one. We had an interview with experienced operator to decide the input variables. Input are charge rate of

Table 1. Control rule table

		ΔTL					TW		
		PB	PS	ZE	NS	NB	PB	PS	ZE
TL	PB	—	—	—	—	—	PB	PB	PS
	PS	PB	PS	—	ZE	NB	—	—	—
	ZE	ZE	ZE	ZE	ZE	ZE	ZE	ZE	ZE
	NS	PB	ZE	—	NS	NB	—	—	—
	NB	—	—	—	—	—	NS	NB	NB

(Input of fuzzy control)

```
  TL  : h - hr , where h is tank level and
                 hr is tank level target
  ΔTL : rate of the changing level
  TW  : time until next filter washing
```

(Compensation of the average flow rate)

```
  PB : Positive Big
  PS : Positive Small
  ZE : Zero
  NS : Negative Small
  NB : Negative Big
```

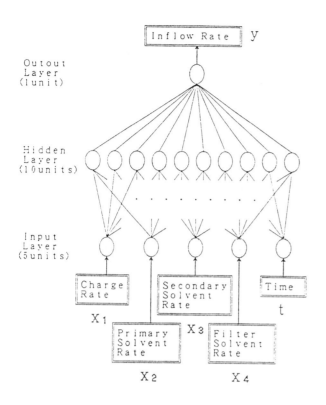

Fig.3. Outline of neural network

the feed oil, primary solvent rate, secondary solvent rate, filter solvent rate and time. Output is the inflow rate. We use several neural networks for the different feed oil switching patterns, e.g., oil A to oil B and oil B to oil C.

From the prediction of the inflow rate by the trained neural networks, we can find the followings:

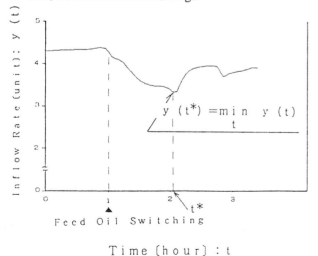

Fig.4. Example of inflow rate pattern

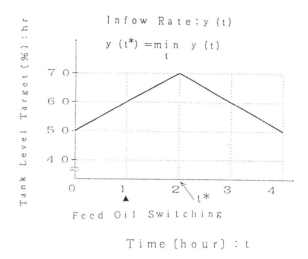

Fig.5. Example of tank level target pattern

(1) The inflow rate changing pattern after the feed oil switching: increasing, decreasing and no change.
(2) The time of the highest stage or the lowest stage for the inflow rate.

For example, Fig.4 shows a prediction of a inflow rate pattern by using the trained neural network when the oil is switched oil A to oil B. We can find the followings:

(1) The inflow rate of the tank is decreasing.
(2) The inflow rate of the tank becomes the lowest stage about an hour after the feed oil switching.

Fig.5 shows the target pattern when the oil is switched from oil A to oil B. This means that it takes 2 hours that the tank level must be increased for compensation to the lowest stage of the inflow rate, and then it is decreased to 50% of the tank capacity. This is equivalent to the expert's action. He makes the tank level the highest at the lowest stage of the inflow rate.

5. Results

We applied the proposed method to the real plant of the solvent dewaxing plant at Idemitsu Chiba refinery. As results of on-line test, we got the stability for the tank level, and

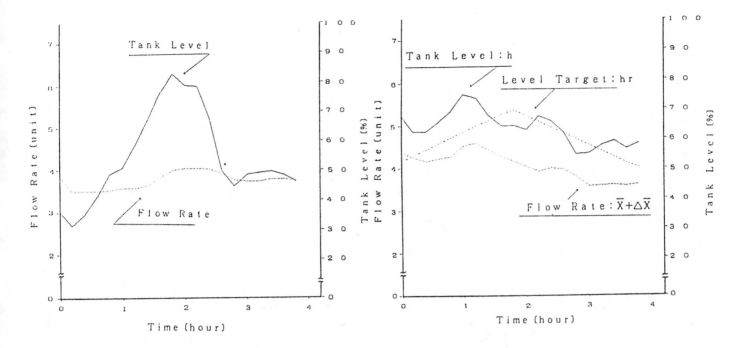

Fig.6. Example of manual operation

Fig.7. Example of neuro-fuzzy control system

smoothness for the flow rate not only in the steady state but also in the transient state.

Fig.6 shows a result of the manual control when oil A is switched to oil B. The experienced operator has raised the tank level before oil switching. On the other hand, Fig.7 shows a result of the proposed method for the same oil switching. The tank level ranges from 35% to 75% by the operation of proposed method, although it raises from 30% to 80% by manual operation. And the flow rate by the proposed method is as smooth as that by the well-experienced operator.

6. Conclusion

A practical control method of neuro-fuzzy hybrid control is proposed. Neural networks are used for estimating the target of fuzzy control.

This method is applied to the tank level control of the real process. The process has two purposes that are contrary each other. The hybrid controller shows its usefulness not only in steady state but also in transient state.

Reference

[1] M. Sugeno, Ed. : Industrial Applications of Fuzzy Control, North-Holland(1985).

[2] A.Guez, J.L.Elibert and M.Kam : Neural Network Architecture for Control, IEEE Control Systems Magazine, Vol.8, No.2, pp.22-25(1988).

[3] K. Suzuki and Y. Nakamori : Model Predictive Based on Fuzzy Dynamic Models, 4th Inter. Conf. on Process Sys. Engg, Vol.II, pp.18.1-18.15(1991).

[4] T. Tani and K. Tanaka : A Design of Fuzzy-PID Combination Control System and Application to Heater Output Temperature Control, Trans.Society of Instrument and Control Engineers, Vol.27,No.11, pp.1274-1280(1991) (in Japanese).

[5] T. Takagi and M. Sugeno : Fuzzy Identification of Systems and Its Application to Modeling and Control, IEEE Trans. on Sys. Man and Cybernetics, Vol.SMC-15, No.1, pp.116-132(1985).

[6] G. T. Kang and M. Sugeno : Fuzzy Modeling and Control of Multilayer Incinerator, Fuzzy Sets and Systems, Vol.18, pp.329-346(1986).

[7] M. Mizumoto : Fuzzy Controls by Product-Sum-Gravity Method, Advancement of Fuzzy Theory and Systems in China and Japan (ed. by X.H.Liu and M.Mizumoto), International Academic Publishers, pp.c1.1-c1.4(1990)

[8] D. E. Rumelhart, G. E. Hinton, and R. J. Williams : Learning internal representations by error propagation, Parallel Distributed Processing, Vol.1, MIT press, Cambridge, pp.318-362(1986)

Minimization of Combined Sewer Overflows Using Fuzzy Logic Control

Sheng-Lu Hou and N. Lawrence Ricker
Department of Chemical Engineering, BF-10
University of Washington
Seattle, WA 98195

ABSTRACT

Combined sewer overflows (CSOs) have caused serious environmental pollution in many U. S. cities. Combined sewers carry both sanitary sewage and storm runoff. During heavy storms the runoff can exceed the capacity of the collection system, at which point the mixture of municipal sewage and storm runoff must be released to the environment untreated. These combined sewer systems are typical large-scale systems and have proven to be hard to control using conventional control technologies. This paper presents an application of fuzzy logic control in minimizing the CSOs for a small but representative section of the Seattle Metro collection system. The results are encouraging and it is intended that this approach be expanded to encompass the full-scale system.

KEYWORDS

fuzzy logic control, combined sewer overflows (CSOs), optimization, real time control

INTRODUCTION

Many urban drainage systems in larger cities are "combined" sewer systems in which both sanitary sewage and storm runoff are collected. Overflows from these combined sewer systems during storms can cause serious pollution problems, since the mixture of municipal sewage and storm runoff is released to the environment untreated. Along with population growth and industrial development, the pollution load due to CSOs has been increasing.

One solution is construction of storage facilities or separated sewers to handle peak loads, but the cost is often prohibitive. Since the sewers are dimensioned to convey the runoff of rare heavy storms, which occur not more than once in every ten years, idle transport and storage capacity exists most of the time. Furthermore, the real loading of such a system is so variable in space and time that in almost all situations only parts of the systems are at capacity. Thus, an attractive and low-cost alternative to construction is real-time process control. In a real-time control system, gates and pumps within the system can be manipulated from a central control room to maximize storage of peak loads, thereby minimizing CSOs. Many researchers have proposed real-time control methods for such reservoir systems, such as using dynamic programming (Harboe, 1983; Georgakakos and Marks, 1987; Foufoula-Georgiou and Kitanidis, 1988; Soliman and Christensen, 1988), predictive control formulation with a linear programming solution (Patry, 1983), predictive control with a quadratic programming solution (Papageorgiou, 1983), model predictive control (Ricker, 1989) and optimization (Neugebauer, Schilling and Weiss, 1991).

The specific problem considered here is the real-time control of the CSOs in the Seattle Metro collection system, which consists of about 160 km of interconnected pipes ranging from 0.3 m to 3.6 m in diameter. There are 13 pump stations and 19 regulator stations that can be used to control the flows within the whole systems. Besides, there are outfall gates at most regulator stations that can be opened to release untreated wastewater to receiving waters in order to avoid backups in low connections. Overflow weirs are usually located upstream of the gates in the trunks to prevent backups and flooding in low sewer connections. Available measurements include about 40 liquid levels and rainfall intensity at 17 locations, which reports once every minute. Figure 1 shows the schematic of a pipe section in the Metro network.

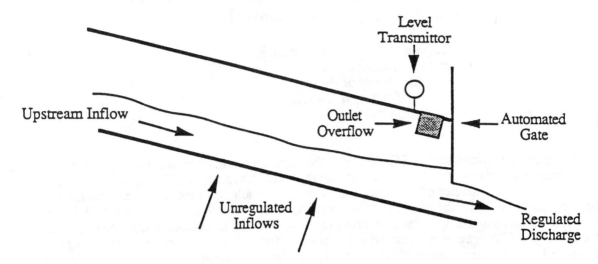

Figure 1: Schematic of a section of a sewer line in the Metro network

What makes this application a challenging task for fuzzy logic control, as well as for other control techniques, is that this plant-wide control problem involves: the optimization of the dynamic response of a plant in which there are many (over 30) manipulated variables and output variables; forecasting of unmeasured plant disturbances (storm loads); inequality constraints rather than setpoints on most output variables; nonlinear plant dynamics with significant lags and time delays. This problem is also of considerable practical importance, not only in Seattle but in other urban areas as well.

Metro's combined sewer system has been controlled by operators for many years. The operators base their control actions on heuristic rules, which are gained from their experience. Therefore, a rule-based controller is attractive for this problem because it makes good use of the available control strategies developed by experts and skilled operators and it avoids the difficulty of building mathematical models for such a large and complex system. As a start, we present here the results of fuzzy logic control on a three-reservoir model junction of the combined sewer system.

THE CONTROL OF THE THREE-RESERVOIR SYSTEM:

A three-reservoir system, shown in Figure 2, was used in this study. Such a system represents the most common physical structure in the combined sewer system – a junction of two sewer branches. The reservoirs, shown as triangles, model variations in liquid holdup within the sewer pipes. They have a specified maximum capacity. Flows leaving each can be regulated (within specified physical limitations).

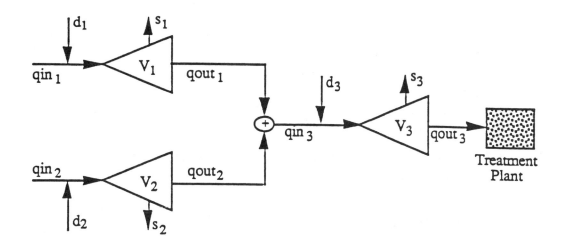

Figure 2: Diagram of the three-reservoir system
(representing a junction in the combined sewer network)

The variables used in Fig. 2 are as follows:

qin_i is the flowrate entering reservoir i from an upstream transport element (m³/s)
d_i is the average total disturbance inflow entering reservoir i, i.e., the sum of all the lateral inflows (m³/s)
s_i is the average overflow rate for reservoir i (m³/s)
$qout_i$ is the average regulated flowrate leaving reservoir i (m³/s)
V_i is the liquid holdup in reservoir i (m³).

Here, the $qout_i$ variables are the decisions to be made by the controller.

There are four principal components in every fuzzy logic controller: a fuzzification interface, a knowledge base, a decision-making logic and a defuzzification interface. The first step in the controller design is the selection of controller's inputs. The proper choice of process state variables is essential to the characterization of the operation of a fuzzy system. In many applications, the final control goal is to make the system's outputs track specified setpoints. Similar to the information required by a discrete PID controller, inputs to a fuzzy logic controller often employ current error (current error = setpoint - current output) and change in error (change in error = current error - error in the last sampling period) and sometimes sum of errors (sum of error = current error + all previous errors). However, there is no setpoint for each time step during the operation in a combined sewer system, that is, there is no specification of how much combined sewage should be in each reservoir at the end of each time step. More important is the relative fullness of the reservoirs. It is preferred that the storm runoff is equally distributed in all areas, avoiding the situation where there is an overflow occurring at a particular location while the rest of the system still is under-utilized. Therefore, df_{1-2} and $df_{1,2-3}$, which represent the relative fullness among the three reservoirs, are chosen as the controller inputs in our work and are defined as follows:

$$df_{1-2} = \frac{V_1}{C_1} - \frac{V_2}{C_2}$$

$$df_{1,2-3} = \frac{V_3}{C_3} - \max\left(\frac{V_1}{C_1}, \frac{V_2}{C_2}\right)$$

where V_i is the liquid holdup in reservoir i (m³)
        C_i is the maximum capacity of reservoir i (m³)
        df_{1-2} is the difference between the fullness of reservoir 1 and that of reservoir 2
        $df_{1,2-3}$ is the difference between the fullness of reservoir 3 and that of the higher occupied one in reservoirs 1 and 2.

We then define the following linguistic variables and membership functions, shown in Figure 3, for both df_{1-2} and $df_{1,2-3}$ in the input space:

    Negative Large (NL);   Negative Small (NS);   Zero (ZE);
    Positive Large (PL);   Positive Small (PS).

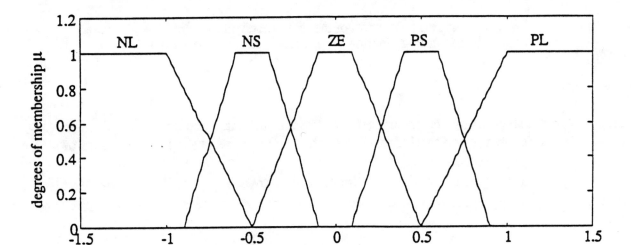

Figure 3: The membership functions for the input space

The output space of the controller includes the two manipulated variables qout$_1$ and qout$_2$, both of which have the saturation limits of 25 m³/s. This output space is classified by six primary fuzzy sets, which are as follows:

    Small Minus (S-);   Small Plus (S+);   Medium Minus (M-);
    Medium Plus (M+);   Large Minus (L-);   Large Plus (L+).

The membership functions for these linguistic variables are also defined as trapezoidal shapes shown in Figure 4.

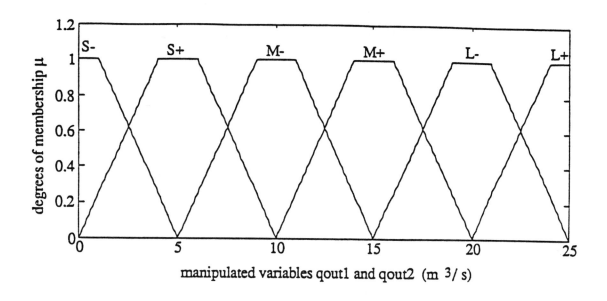

Figure 4: The membership functions for the output space

The fuzzy control rules for the three-reservoir system are governed by the general control goals of such CSO problem. These include: (1) minimizing the overall overflows rather than some particular local overflows (unless exceptions have been stipulated for sensitive locations), and (2) making full use of the available storage facilities, in other words, avoiding overflows until all the reservoirs are full. Guided by these goals, the fuzzy control strategy is generated, which is to keep all the reservoirs equally filled during each sampling period. For example, in the case that

$\frac{V_3}{C_3} \ll \frac{V_2}{C_2} < \frac{V_1}{C_1}$, what an experienced operator would decide to is to open the outflow control valve from both upstream reservoirs wide, but to different degrees, *i.e.*, $qout_1$ should be larger than $qout_2$. This yields the following rule:

Rule i: if (df_{1-2} is PS and $df_{1,2-3}$ is NL), then ($qout_1$ is L+ and $qout_2$ is M+).

All the other control rules in the rule base are of the same form as this one. The variables and fuzzy sets used in this rule have the definitions given above. Overall, the fuzzy logic controller for the three-reservoir system is a two-input-two-output controller.

One reason why a fuzzy logic controller may be regarded as a means of emulating a skilled human operator is because it simulates inexact or approximate reasoning. The approximate reasoning in this fuzzy logic controller employs the Mamdani's minimum operation rule as its fuzzy implication function. Before the inferred fuzzy control actions are executed on the system, the center of area method is used within the defuzzification interface in order to produce a crisp control action.

RESULTS AND DISCUSSION

Figures 5 and 6 show the performances of the system under two different conditions:
- with the same initial capacity for all three reservoirs (Fig.5); and
- with different starting points for each reservoir (Fig.6).

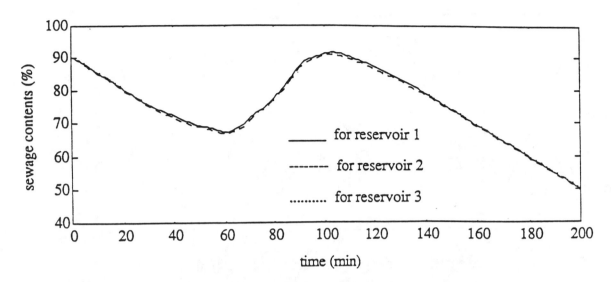

Figure 5: The simulation results for the three-reservoir system
(with the same initial sewage contents for both reservoirs)

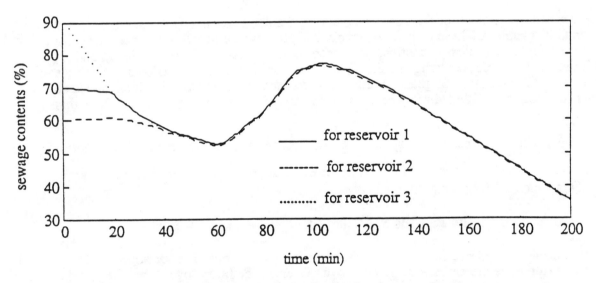

Figure 6: The simulation results for the three-reservoir system
(with a different initial sewage content for each reservoir)

As mentioned in the previous section, the control goals for the combined sewer system are to make full use of the available storage facilities and at the same time minimize the sum of all overflows than particular instantaneous overflows. $qout_i$ ($i = 1, 2, 3$), the average regulated flowrate leaving reservoir i , are used as decision variables for the fuzzy controller and are controlled simultaneously in order to achieve the above goals. The saturation limits for both $qout_1$ and $qout_2$ are 25 m^3/s, whereas $qout_3$ has a limit of 35 m^3/s and is kept at this value throughout the entire operation (unless reservoir 3 is nearly empty). This maximizes the amount of sewage entering the treatment plant. Since there are five fuzzy subsets for both df_{1-2} and $df_{1,2-3}$ in the controller's input space, 25 linguistic rules are generated based on our knowledge and operators' experience with the CSO systems. Each rule has form of the example in the previous section.

As shown in both figures, the sewage content curve in each reservoir, which is defined as
$$\frac{\text{sewage volume in the reservoir}}{\text{total volume of the reservoir}} \times 100\%,$$
has a similar shape. The capacity curves first decrease, typical of dry-weather operation of the system - emptying the reservoirs as much as possible in anticipation of a storm. Once the storm begins (after ~1 hour), the curves increase to a peak that represents the dynamic equilibrium between the loading and release rates of sewage in the reservoirs. Finally, operation reverts to the dry-weather pattern. It is clear from these two figures that the sewage contents in all three reservoirs are kept almost the same throughout the storm, even when the initial sewage holdups differ. This reflects our control strategy and is believed to be an efficient way to reach the control goal for this system. Since all reservoirs were kept below 100% at all times, no overflows were generated.

Figure 7 shows the performance of the system during such a heavy storm that overflows were inevitable. The two upstream reservoirs fill and overflow about 8 minutes before the downstream one. This is not ideal but understandable since our current fuzzy controller only includes instantaneous feedback – there is no explicit planning for future events. We hope to improve performance by adding a predictive capability. With predictive fuzzy control, the control actions at each time step can compensate for the possible impact of future disturbances on the system. This should help to smooth the control action and may also allow an increase in the sampling period.

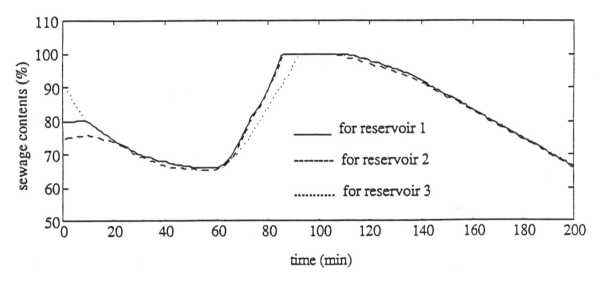

Figure 7: The simulation results for the three-reservoir system
(the situation when overflows must occur in this system)

CONCLUSIONS

Combined sewer system is a typical large scale and complex system. Most such systems are still manually controlled. A fuzzy logic controller has been designed for a three-reservoir subsystem representing the most common connection in the combined sewer systems. This application of fuzzy logic control on the three-reservoir system suggests that it is quite promising for on-line control and minimization of CSOs. Research is proceeding to expand the scope of the controller to the full scale system and to improve performance by including a predictive mode.

REFERENCES

Bare, W.H., Mulholland, R.J., and Sofer, S.S., "Design of a self-tuning rule based controller for a gasoline refinery catalytic reformer", IEEE transactions on automatic control, Vol. 35, No. 2, p156-164, (1990).

Batur, C., Kasparian, V., "Predictive fuzzy expert controllers", Computer and industrial engineering, Vol. 20, No. 2, p199-209, (1991).

Bernard, J.A., "Use of a rule-based system for process control", IEEE control systems magazine, Vol. 8, No. 5, p3-13, (1988).

Dubois, D., and Prade, H., "Fuzzy sets and systems: theory and applications", Vol. 144 in "Mathmatics in science and engineering", Academic Press, (1980).

Gupta, M.M., Ragade, R.K. and Yager, R.R., "Advances in fuzzy set theory and applications", North-Holland publishing company, New York, (1979).

Kandel, A., "Fuzzy mathematical techniques with applications", Addision-Wesley publishing company, Inc., (1986).

Kaufmann, A., "Fuzzy mathematical models in engineering and management science", Elsevier science publishers B.V., (1988).

Kaufmann, A. and Gupta, M.M., "Introduction to fuzzy arithmetic: theory and applications", Van Nostrand reinhold company, New York, (1984).

Kickert, W.J.M., and Lemke, J.R.V.N., "Application of a fuzzy controller in a warm water plant", Automatica, Vol. 12, p301-308, (1976).

King, P.J., and Mamdani, E.H., "The application of fuzzy control systems to industrial processes", Automatica, Vol. 13, p235-242, (1977).

Lee, C.C., "Fuzzy logic in control systems: Fuzzy logic controller", IEEE transactions on system, management and cybernetics, Vol.20, No.2, p404-435, March/April, (1990).

Mamdani, E.H. and Assilian, S., "An experiment in linguistic synthesis with a fuzzy logic controller", International journal of man-machine studies, Vol. 7, No. 1, p1-13, (1975).

Negoita, C.V., "Fuzzy systems", Abacus Press, (1981).

Neugebauer, K., Schilling W., and Weiss, J., "A network algorithm for the optimum operation of urban drainage systems", Water science technology, Vol. 24, No.6, p209-216, (1991).

Novak, V., "Fuzzy sets and their applications", Adam Hilger publishing Inc., (1989).

Raju, G.V.S., Zhou, J. and Kisner, R.A., "Fuzzy logic controller to a steam generator feedwater flow", American Control Conference, p1491-1492, (1990).

Ricker, N.L., Vitasovic, Z., Swarner, R., and Speer, E., "Modeling stratedgies for minimization of combined sewer overflows", (1990).

Sugeno, M., (editor), "Industrial applications of fuzzy control", Amsterdam: North-Holland, (1985).

Zadeh, L.A., "Outline of a new approach to the analysis of complex systems and decision processes", IEEE transactions on systems, management and cybernetics, Vol. SMC-3, No. 1, p28-44, (1973).

Zadeh, L.A., "The birth and evolution of fuzzy logic", International journal of general systems, Vol. 17, No. 2-3, p95-105, (1990).

Identification and Analysis of Fuzzy Model for Air Pollution
- An Approach to Self-learning Control of CO Concentration -

Kazuo TANAKA, Manabu SANO
Department of Mechanical Systems Engineering
Kanazawa University
2-40-20 Kodatsuno Kanazawa 920 Japan

Hiroyuki WATANABE
Department of Computer Science
The University of North Carolina
CB#3175, Sitterson Hall, Chapel Hill, N.C. 27599 U.S.A.

Abstract

This paper presents identification and control for a fuzzy prediction model of CO (carbon monoxide) concentration which is one of important factors in air pollution problems. We have many uncertainty (imprecise) factors for predicting CO concentration. Our basic approach is to handle this imprecision by fuzzy logic based techniques. The fuzzy modelling technique, proposed by Kang and Sugeno, is used for identifying a fuzzy prediction model. It is shown that the identified fuzzy model is very useful for predicting CO concentration. Furthermore we attempt to simulate a self-learning control of CO concentration by Widrow-Hoff learning rule which is a basic learning method in neural networks. Simulation results show that this self-learning controller is useful for CO concentration control.

1. Introduction

Many kinds of fuzzy models in control processes have been developed since Mamdani's paper [1] was published. Most of these models are expressed by a set of fuzzy linguistic propositions which are derived from the experience of skilled operators and knowledge of manual control. However, we have the following difficulties for modeling a complex system such as air pollution models:
 (1)non-linearity, and
 (2)interference of many predictor variables.

Fuzzy modeling for such a complex system is very difficult. One of the possible approaches to overcome the difficulties is to develop a new type of fuzzy model. One of the authors has reported in previous paper [2] that the modeling method using a new type of fuzzy model, Takagi and Sugeno's model, is useful in identification of a complex system. In this paper, we attempt to identify CO concentration model in the air at a traffic intersection point of a large city of Japan.

Furthermore, we attempt to simulate self-learning control of CO concentration using an identified fuzzy model. The purpose of this control is to keep CO concentration at a constant level. In environment control, in this case, control of air pollution,
(1)there are some state variables such as wind velocity, temperature and amount of sunshine which can not be manipulated,
(2)it is, in practice, impossible to perfectly manipulate input variables such as the volume of traffic,
(3)it is difficult to identify a perfect prediction model for air pollution.

For these reasons, it is not easy to analytically solve optimal parameters of controller. We successively adjust parameters of controller by a self-learning method [7, 8, 9]. This self-learning method is based on Widrow-Hoff learning rule which is a basic learning method in neural networks.

2. Fuzzy modeling

2.1 Takagi and Sugeno's fuzzy model

A new type of fuzzy model, proposed by Takagi and Sugeno [3], is described by fuzzy IF-THEN rules which locally represent linear input-output relations of a system. This fuzzy model is of the following form:

Rule i : IF x_1 is A_{i1} and $\cdots$ and x_n is A_{in}
   THEN $y_i = c_{i0}+c_{i1}x_1+\cdots+c_{in}x_n$,   (1)

where $i=1, 2, \cdots, r$, r is the number of IF-THEN rules, y_i is the output from the i-th IF-THEN rule, and A_{ij} is a fuzzy set.

Given an input $(x_1, x_2, \cdots, x_n)$, the final output of the fuzzy model is inferred by as follows [8]:

$$y = \sum_{i=1}^{r} w_i y_i \quad (2)$$

where y_i is calculated for the input by the consequent equation of the i-th implication, and the weight w_i implies the overall truth value of the premise of the i-th implication for the input calculated as

$$w_i = \prod_{k=1}^{n} A_{ik}(x_k) , \quad (3)$$

where $A_{ik}(x_k) = \exp(-(x_k-d_{ik})^2/b_{ik})$, d_{ik} and b_{ik} are parameters of the membership functions.

2.2 Outline of identification algorithm

We explain outline of identification algorithm of a fuzzy model proposed by Sugeno and Kang [4,5]. Fig.1 shows the identification algorithm. The identification procedure is classified into three steps:

(Step 1) choice of the premise structure and the consequent structure;
(Step 2) identification of the parameters of the structure determined in (Step 1);
(Step 3) verification of the premise structure and the consequent structure.

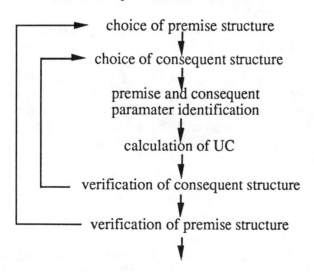

Fig.1 Identification algorithm

(Part 1)
The structure identification of a fuzzy model consists of two parts. The first part is related to the premise structure. This has two problems. One is that we have to find which variables are necessary in the premise. The other is that we have to find an optimal fuzzy partition of the input space, which is a problem peculiar to fuzzy modeling. For example, suppose that a fuzzy model of a three inputs-one output process is represented by three fuzzy IF-THEN rules such as

Rule 1: IF x_1 is $Small_1$ and x_3 is $Small_2$
   THEN $y_1 = 0.5+1.2x_2-0.6x_3$,
Rule 2: IF x_1 is $Small_1$ and x_3 is Big_2
   THEN $y_2 = 1.2+0.2x_1+x_2-2.3x_3$,
Rule 3: IF x_1 is Big_1
   THEN $y_3 = 1.5+2.3x_1-0.9x_3$.

The variable x_2 does not appear in the all premise parts and Rule 3 has the only one variable x_1 in its premise part. For example, Rule 2 represents the input-output relation in the fuzzy subspace defined by $x_1=Small_1$ and $x_3=Big_2$. Therefore, the premise structure identification is the same as the problem to find out the fuzzy partition of the input space, where the number of the fuzzy subspaces corresponds to that of the fuzzy IF-THEN rules. The premise parameters are those of the fuzzy variables of premise parts.

The second part of the structure identification is related to the consequent structure identification. We have to find which variables are necessary in the consequent parts of IF-THEN rules. For example, Rule 1 has two variables x_2 and x_3 in its consequent part, while Rule 3 has x_1 and x_3.

In addition, we need to find a criterion for the verification of an assumed structure. The problem will be discussed in Step 3.

(Step 2)
We apply the least squares method to the identification of the consequent parameters since each of the consequent parts is described by a linear equation. In the identification of the premise parameters, the well known complex method is used since the problem of finding the optimum premise parameters minimizing a performance index is reduced to a non-linear programming.

(Step 3)
As a proper criterion for the verification of the structure of the fuzzy model, we utilize the unbiasedness criterion (UC) used in GMDH [6].

To calculate UC, we must divided the observed data into two sets N_A and N_B. UC is calculated as

$$UC = \sum_{i=1}^{n_A} \{(y_{iAA}-y_{iAB})^2 + (y_{iBA}-y_{iBB})^2\}^{0.5} \quad (4)$$

where n_A is the number of the data set N_A, y_{iAA} the estimated output for the data set N_A from the model identified using the data set N_A. y_{iAB} the estimated output for the data set N_A from the model identified using the data set N_B. This criterion is based on the fact that the parameters of the fuzzy model with the true structure are the least sensitive to the changes of the observed data which are used for identifying the parameters.

The identification of the premise structure can be regarded as the problem to partition the inputs space into the fewest fuzzy subspaces so that the identified fuzzy model represents the object system adequately.

When identifying the premise structure, we use the following idea. As the number of fuzzy subspaces, i.e., the number of fuzzy IF-THEN rules, increases, the UC of the fuzzy model decreases. But, if the number of fuzzy subspaces exceeds that of the optimal premise structure, the parameters of the model become sensitive to the changes of the data used for identifying the parameters, and the UC of the model increases.

Here we choose the premise structure which minimizes the unbiasedness criterion UC, and use the following algorithm resembling the forward selection of variables which is a method for finding variables in a linear model. That is, we start the process from the identification of a model with one IF-THEN rule, i.e. a linear model, and increase the number of fuzzy IF-THEN rules until the UC of the fuzzy model begins to increase. In the process of the premise structure identification, the premise parameters and the consequence of the model are also identified as shown in Fig.1.

3. CO concentration model

We attempt to predict CO concentration in the air at a traffic intersection point of a large city of Japan. Inputs and an output of a fuzzy model are shown in Fig.2. x_1 is wind velocity, x_2 is volume of traffic, x_3 is temperature, x_4 is amount of sunshine, and y is CO concentration. For each variable, we perform normalization so that the mean and the variance of normalized variable equal 0 and 1, respectively. In other words, we transform the distribution to $N(0,1)$.

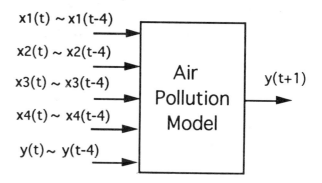

Fig.2 Inputs and a output of air pollution model

The data used for identification and prediction of a fuzzy model is collected at the most busy traffic intersection of a large city of Japan. The number of data used for identification and prediction are 480 input-output data pairs and 253 input-output pairs, respectively. The sampling interval is 15 minutes. Of course, the prediction data is not used for identification of a fuzzy model. It is used only for checking the validity of a fuzzy model identified by using the identification data.

Eq.(5) shows the performance index of the model.

$$J = \frac{1}{m} \sum_{t=0}^{m-1} \left|\frac{y(t+1)-y^*(t+1)}{y(t+1)}\right| \times 100, \quad (5)$$

where m is the number of input-output data ($m=253$). $y^*(t+1)$ and $y(t+1)$ are the outputs of a fuzzy model and the real system at time instant $t+1$, respectively. $y^*(t+1)$ and $y(t+1)$ are raw data and are not normalized.

Table 1 Performances of models

	Linear model	Fuzzy model
J_1	5.7	4.8
J_2	11.8	5.9

Table 1 shows the values of performance index for a linear model and the fuzzy model. The fuzzy model have two IF-THEN rules. J_1 and J_2 are the values of performance index for the identification data and the prediction data, respectively. The performance index of the fuzzy model is superior to that of linear model. This means that the air pollution model is essentially non-linear. The result shows the validity of the identified fuzzy model.

Fig.3 shows the identification result.

Rule 1 : IF x2(t-1) is $\exp(-(x2(t-1)+1.90)^2/5.80)$
  THEN y1(t+1) =
   -0.008x1(t-2)+0.026x1(t-1)-0.032x1(t)
       -0.205x2(t-1)+0.249x2(t)
   -0.138x4(t-4)-0.181x4(t-3)+0.335x4(t-2)
       +0.088x4(t-1)-0.112x4(t)
   +0.090y(t-4)-0.106y(t-3)+0.011y(t-2)
   -0.460y(t-1)+1.356y(t)-0.018

Rule 2 : IF x2(t-1) is $\exp(-(x2(t-1)-2.06)^2/5.72)$
  THEN y2(t+1) =
   -0.013x1(t-2)+0.059x1(t-1)-0.032x1(t)
       -0.497x2(t-1)+0.827x2(t)
   +0.006x4(t-4)+0.066x4(t-3)-0.130x4(t-2)
       +0.095x4(t-1)-0.036x4(t)
   -0.009y(t-4)+0.007y(t-3)-0.037y(t-2)
   -0.103y(t-1)+0.748y(t)-0.002

Fig.3 Identification result

4. Self-learning control of CO concentration

We attempt to simulate a self-learning control of keeping CO concentration at a constant level using Widrow-Hoff learning rule. Fig.5 shows self-learning control system, where x2(t) is a manipulated variable, y(t) is a controlled variable and r is a setpoint of CO concentration.

In CO concentration control,
(1) there are some state variables such as wind velocity x1, temperature x3 and amount of sunshine x4 which can not be manipulated,
(2) it is, in practice, impossible to perfectly manipulate the volume of traffic which is an input variable,
(3) it is difficult to identify a perfect CO concentration model.

It is, however, assumed in this simulation that
(1) with respect to wind velocity x1, temperature x2 and amount of sunshine x4, we use real values of the prediction data,
(2) it is possible to perfectly manipulate the volume of traffic,
(3) the identified fuzzy model perfectly represents real system.

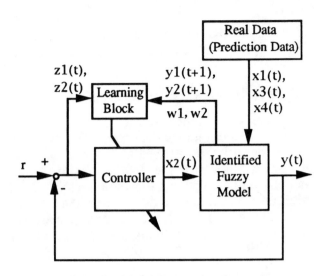

Fig.5 Self-learning control system

We use the following linear controller.
$$\Delta x2(t) = a_1 z1(t) + a_2 z2(t),$$
$$x2(t) = x2(t-1) + \Delta x2(t),$$
where z1(t)=r-y(t) and z2(t)=z1(t)-z1(t-1) and a_i (i=1, 2) is a parameter of the controller. For the above reasons, it is not easy to analytically solve optimal parameters of controller. So we apply Widrow-Hoff learning rule to controller design. The idea which automatically adjusts parameters of fuzzy controller using Widrow-Hoff learning rule was first introduced by Ichihashi [8].

Let us consider the following performance function.
$$J = \frac{1}{2}(r - y(t+1))^2 \qquad (6)$$
By partially differentiating J with respect to each controller parameter a_i, we obtain
$$\frac{\partial J}{\partial a_i} = -\left(r - \sum_{j=1}^{2} w_j y_j(t+1)\right) z_i(t) \sum_{j=1}^{2} w_j p_j , \qquad (7)$$
where w_j is a membership value of j-th rule of fuzzy model at time instant t and p_j is a consequent parameter of x2(t), that is, p_1 = 0.249 and p_2 = 0.827. We can successively modify controller parameters using Eq.(8).
$$a_i^{NEW} = a_i^{OLD} + \varepsilon_i \left(r - \sum_{j=1}^{2} w_j y_j(t+1)\right) z_i(t) \sum_{j=1}^{2} w_j p_j \qquad (8)$$
where ε_i is a learning factor.

Fig.6 ~ Fig.8 show simulation results of self-learning control system, where ε_i = 0.003 (i=1, 2). The setpoint r is set as follows. If t < 100 then r = 32 else r = 16. Fig.9 shows a relation between the number of learning and summation of squared error, that is,

$$SE = \sum_{t=1}^{m-1} (r - y(t))^2 \qquad (9)$$

where m=253. It is found from these figures that this self-learning controller is useful for CO concentration control.

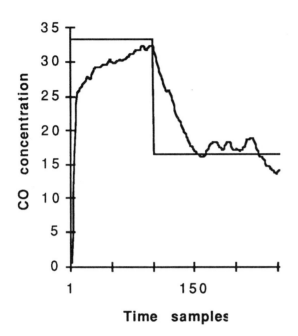

Fig.6 Control result (number of learning : 1)

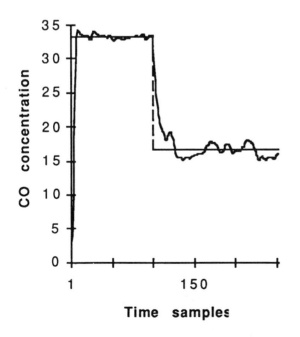

Fig.7 Control result (number of learning : 5)

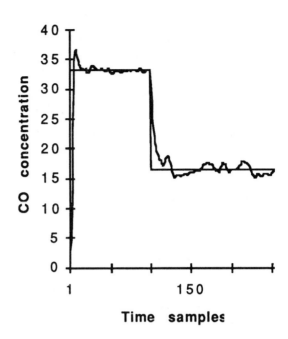

Fig.8 Control result (number of learning : 10)

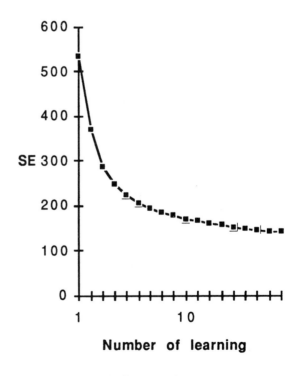

Fig.9 Learning process

5. Conclusion

We have identified CO concentration model in the air at a traffic intersection point using fuzzy modeling method. Identification results show the validity of this method.

Furthermore, we have attempted to simulate a self-learning control of CO concentration. The purpose of this control is to keep CO concentration at a constant level using Widrow-Hoff learning rule. It has been assumed in this simulation that the identified fuzzy model perfectly represents real system. However, this is not generally the case. Therefore, we should investigate robustness of this control system. This is a subject for future study.

References

[1] E.H.Mamdani:Applications of fuzzy algorithms for control of a simple dynamic plant, Proc.IEE 121 (12), pp.1585-1588 (1974).

[2] M.Sugeno and K.Tanaka:Successive identification of a fuzzy model and its applications to prediction of a complex system, FUZZY SETS AND SYSTEMS 42, no.3, pp.315-334 (1981).

[3] T.Takagi and M.Sugeno:Fuzzy identification of systems and its applications to modeling and control, IEEE Trans. SMC 15, pp.116-132 (1985).

[4] M.Sugeno and G.T.Kang:Fuzzy modeling and control of multilayer incinerator, FUZZY SETS AND SYSTEMS 18, pp.329-346 (1986).

[5] M.Sugeno and G.T.Kang:Structure identification of fuzzy model, FUZZY SETS AND SYSTEMS 28, pp.15-33 (1988).

[6] A.G.Ivakhnenko at el.:Principle versions of the minimum bias criterion for a model and an investigation of their noise immunity, Soviet Automat. Control 11, pp.27-45 (1978).

[7] K.Tanaka and M.Sano : Design of Fuzzy Controllers Based on Frequency Characteristics and Its Self-tuning, proceedings of 8th fuzzy system symposium, pp.525 - 528 (1992). (in Japanese)

[8] H.Ichihashi and T.Watanabe:Learning Control by Fuzzy Models Using a Simplified Fuzzy Reasoning, vol.2, no.3, pp.429-437 (1990). (in Japanese)

[9] H.Ichihashi : Iteraive Fuzzy Modeling and a Hierarchical Network, proceedings of IFSA'91, pp.49-52 (1991).

Self-Learning Fuzzy Modeling of Semiconductor Processing Equipment

Raymond L. Chen and Costas J. Spanos
Department of Electrical Engineering and Computer Sciences
University of California, Berkeley CA 94720

Abstract

A qualitative equipment model for a Low Pressure Chemical Vapor Deposition (LPCVD) process is presented. The model is based on fuzzy representation of input-output relationships and utilizes self-tuning membership functions. To demonstrate this concept we have built a fuzzy inference system for polysilicon grain size prediction based on deposition and annealing temperatures. After we train the system with experimental data, it automatically tunes its membership functions to accommodate additional experimental data.

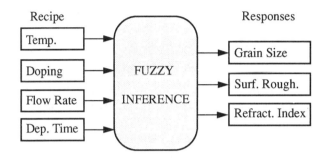

Figure 1 A fuzzy inference system for LPCVD modeling

1.0 Introduction

The Berkeley Computer-Aided Manufacturing (BCAM) System is a framework being built to facilitate the experimentation with various CAM applications [1][2]. One objective of the BCAM system is to support recipe generation for certain processes. Despite recent advances in equipment modeling, recipe generation is still extremely difficult to automate and remains as one of the most costly, time-consuming and error prone operations in today's semiconductor manufacturing. One of the reasons is the inability of most equipment models to describe important qualitative aspects of the process. For example, although it is possible to accurately predict the thickness of a deposited polysilicon film given the equipment settings, it is extremely difficult to predict other aspects of the process, such as grain size and orientation, surface roughness, step coverage properties, etc.

Our goal is to simplify this task through the development and application of an artificial intelligence system that uses fuzzy models for the inference of qualitative aspects of a process (Figure 1). Within this system we are developing models that describe process "attributes", which are usually understood by an experienced operator, but ignored by most automated approaches. A process for polysilicon low pressure chemical vapor deposition (LPCVD) has been selected as a test vehicle for this investigation [3]. The long-term objective is to develop prototype models that can be used to predict process responses such as grain size, surface roughness, grain orientation and step coverage of a deposited film. This prediction should be based on process settings, such as deposition and annealing temperatures, doping profile, silane flow rate, pressure, deposition time, etc.

Once these models are in place, we will install them within the BCAM equipment model library. A number of quantitative models have already been incorporated into this library. The combined quantitative-qualitative models will then be used for equipment and process simulation, recipe generation, diagnosis, and control. Additional qualitative models will also be developed for other equipment and processes in the future.

In this paper, we will describe a simple qualitative model that predicts the average grain size of polysilicon deposition films through a fuzzy inference system. We will first review some background knowledge of the LPCVD process (Chapter 2), then we will introduce the concept of a fuzzy inference system (Chapter 3). In Chapter 4, we will discuss a self-learning fuzzy inference model for predicting the LPCVD grain size. Future goals will be discussed in Chapter 5.

2.0 Background Knowledge

2.1 LPCVD process

Figure 2 depicts the Tylan horizontal glass tube reactor used for polysilicon LPCVD [4]. The wafers are stacked perpendicularly along the tube, and are placed 1.2 cm apart. Aluminum cantilever rods support the quartz boats, which house the wafers. Silane (SiH_4), mixed with an inert carrier gas (Nitrogen), is injected at the lower front end of the tube, and is pumped out from the back end. Three main heating coils around the tube maintain a proper operating temperature. A heat baffle in the front portion of the furnace reduces radiative heat loss and also serves to stabilize the temperature and to smooth the flow in the middle and back sections of the tube.

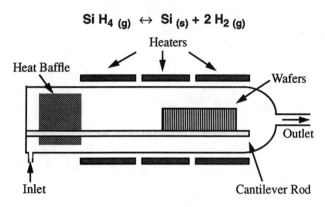

Figure 2 Deposition furnace

There are many qualitative aspects of this process which cannot be modeled numerically, even though extensive research has been done and much knowledge has been collected. These aspects of the polysilicon film include, among other things, its grain size, and its step coverage properties.

2.2 Grain size prediction

The average grain size ($\bar{s}$) of a polysilicon film is an important quantity that will affect the performance of semiconductor devices [5]. This is of great interest in static RAMs that use polysilicon p-channel transistors, since the minority-carrier lifetime increases with increasing grain size. On the other hand, many factors in the deposition process affect the outcome grain structure, including the average grain size. Among those factors are the deposition temperature (T_d) and annealing temperature (T_a). While many other factors may also be decisive, we will concentrate on the dependence of grain size $\bar{s}$ to the T_d and T_a in this paper.

At very low T_d the deposited polysilicon film is amorphous. When T_d is higher than some transition temperature, a film with polycrystalline structure will form, and its grain size will increase when T_d increases [5]. The high-temperature (T_a) annealing process enlarges the grain size, however, this annealing enlargement has stronger effect on film deposited with lower T_d. Hence, after high temperature (T_a) annealing, the films deposited with lower T_d have larger average grain size than those deposited with higher T_d [6]. To date, there are no widely accepted theories to explain the complicated relations between deposition/annealing temperatures and grain size. The expert knowledge in this field is rather qualitative and sometimes even contradictory. Our model (Chapter 4) manages to capture the qualitative knowledge into a computer software system.

3.0 An Introduction to Fuzzy Logic

In this section, we will briefly define the basics of fuzzy logic and describe the concept of a fuzzy inference decision-making system [7].

3.1 Fuzzy sets and membership function

The fuzzy set [8] theory is in many ways a generalization of the classical set theory. A *classical* (crisp) set A is defined as a collection of elements of a superset X ($x \in X$) which satisfy certain conditions specifying A. Each element $x \in X$ can either belong to or not belong to the crisp set A, where $A \subseteq X$. To generalize this definition, we can introduce a membership function μ for each element of X in order to specify its degree of belongness to a set. If we allow this membership function to be continuous, we can define a *fuzzy* set B as a collection of elements $x \in X$ with membership function $\mu(x)$, where $\mu(x)$ can be any real number between 0 and 1:

$$B = \{(x, \mu_B(x)) \mid x \in X\} \quad (3\text{-}1)$$

Fuzzy logic can be derived as a generalization of the Boolean logic by allowing the "membership function" $\mu(x)$ to be any number between 0 and 1. Equivalently, Boolean logic can be seen as a special case of fuzzy logic, when $\mu(x)$ can only be 0 or 1.

3.2 Boolean rules and fuzzy logic rules

In Boolean logic, the most basic logic operations we need to consider are "PASS", "COMPLEMENTARY", "AND" and "OR". These Boolean rules can be generalized in a fuzzy context listed in Table 1:

Table 1: Boolean and fuzzy logic rules

Name	Boolean Logic Rule	Fuzzy Logic Rule
PASS	IF $x \in A$, THEN $z \in Z$	IF $(x, \mu_A(x)) \in A$, THEN $(z, \mu_Z(z)) \in Z$ [$\mu_Z(z) \equiv \mu_A(x)$]
COMPL.	IF $x \notin A$, THEN $z \in Z$	IF $(x, \mu_A(x)) \in A$, THEN $(z, \mu_Z(z)) \in Z$ [$\mu_Z(z) \equiv 1 - \mu_A(x)$]
OR	IF $x \in A$ OR $y \in B$, THEN $z \in Z$	IF $(x, \mu_A(x)) \in A$ OR $(x, \mu_B(x)) \in B$, THEN $(z, \mu_Z(z)) \in Z$ [$\mu_Z(z) \equiv \max(\mu_A(x), \mu_B(x))$]
AND	IF $x \in A$ AND $y \in B$, THEN $z \in Z$	IF $(x, \mu_A(x)) \in A$ AND $(x, \mu_B(x)) \in B$, THEN $(z, \mu_Z(z)) \in Z$ [$\mu_Z(z) \equiv \min(\mu_A(x), \mu_B(x))$]

3.3 The concept of the linguistic variable

A linguistic variable takes any of several *linguistic* values [8]. Each of these linguistic values is associated with a value of a membership function. For example: in LPCVD of polysilicon, the average *grain size* can be a linguistic variable taking the linguistic values as *"small"*, *"medium"*, *"large"*, etc. Thus a grain size of 180*nm* might be expressed with the help of the appropriate membership functions as shown in Figure 3,

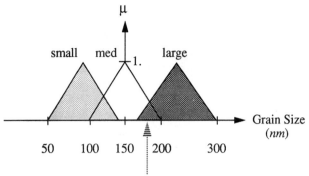

Grain Size = {(small, 0.0), (medium, 0.3), (large, 0.2)}

Figure 3   Grain Size

which means that the linguistic variable *grain size* in this example has the value *"medium"* with a membership of 0.3 and the value *"large"* with a membership of 0.2, and *"small"* with a membership value of 0.0. Such a transformation of a numerical value (grain size in *nm*) to a fuzzy linguistic variable is often called *"fuzzification"*.

3.4   Fuzzy inference rules

We can now illustrate the concept of fuzzy inference, which is an approximate reasoning technique, based on fuzzy logic rules, linguistic variables, and their membership functions. We will use a simplified model of polysilicon LPCVD process to illustrate the concept. This model determines the grain size $\bar{s}$ from two inputs T_d and T_a, i.e., the deposition and annealing temperatures of a process. We fuzzify these three parameters (T_d, T_a, and $\bar{s}$) to linguistic variables by representing them through fuzzy sets and membership functions.

Subsequently, the expert knowledge can be summarized by the following fuzzy inference rules for annealed polysilicon films:

- IF the deposition temperature is low, THEN the grain size is large.
- IF the annealing temperature is high, THEN the grain size is larger.

These rules are depicted in Table 2:

Table 2:   Fuzzy rule table for simplified model

Grain Size	T_a low	T_a med	T_a high
T_d low			large
T_d med		med	
T_d high	small		

Now let us consider a pair of input values (T_d, T_a). After fuzzifying the inputs into linguistic values through their membership setting, we have:

T_d = {(low, 0.8), (med, 0.25), (high, 0)};
T_a = {(low, 0), (med, 0.45), (high, 0.5)};

By applying the fuzzy rules of Table 3, we obtain the following information:

- IF (T_d, T_a) = (high, low), THEN $\bar{s}$ = small,
  $\mu_{G.S.1}$ = min(0.0, 0.0) = 0.0.
- IF (T_d, T_a) = (med, med), THEN $\bar{s}$ = med,
  $\mu_{G.S.2}$ = min(0.25, 0.45) = 0.25;
- IF (T_d, T_a) = (low, high), THEN $\bar{s}$ = large,
  $\mu_{G.S.3}$ = min(0.8, 0.5) = 0.5;

Thus our output grain size $\bar{s}$ is:

$\bar{s}$ = {(small, 0), (med, 0.25), (large, 0.5)}.

We need to point out that there are many ways other than the above "minimum" operation to obtain the weights *w1* and *w2* from the "AND" rules for multiple inputs. An alternative is to use the product of membership values of each individual input (see Chapter 4).

Once the linguistic value of the output is determined, a "defuzzification" step may be applied to obtain a numerical value for that output. There are many ways to defuzzify an output. A simple way is to do a "weighted average":

$$\bar{s} = \frac{\mu_{G.S.2} \cdot A + \mu_{G.S.3} \cdot B}{\mu_{G.S.2} + \mu_{G.S.3}} \quad (3\text{-}2)$$

where *A* and *B* are corresponding numerical values for $\bar{s}$ = {(med,1)} and $\bar{s}$ = {(large,1)}. This algorithm is illustrated in Figure 4.

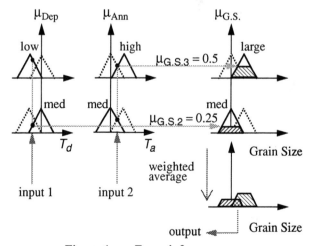

Figure 4   Fuzzy inference

In the next chapter, we will present a more generic grain size prediction model based on values of (T_d, T_a). The rules for the inference system will be derived directly from the existing experimental "training" data.

4.0 Self-Learning LPCVD Models

4.1 A general fuzzy model

Without losing generality, let us start with a fuzzy inference system with 2 independent inputs $\{x_1, x_2\}$ and one output y. All results shown here can be directly generalized for a k-input system [9].

4.1.1 Representation and inference

Assume that x_1 is fuzzified into a linguistic variable with a series of membership functions as shown in Figure 5. x_2 is

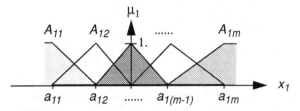

Figure 5   Membership function for x_1

fuzzified in the same way. The inference rules of such a 2-input fuzzy system take the following generic form:

- Rule-$\{i,j\}$:
  IF x_1 is A_{1i} and x_2 is A_{2j} THEN y is w_{ij}.

where $\{i,j\}$ ($i=0,1,...,m$, $j=0,1,...,n$) are rule numbers, A_{1i} and A_{2j} are the respective linguistic values (represented by the membership functions) of x_1 and x_2, with μ_{1i} and μ_{2j} as their corresponding membership values. w_{ij} is initially the actual value of the output y for the respective input combination. w_{ij} will be subsequently tuned along with $\{a_{1i}, a_{2j}\}$ to accommodate new input data.

The membership value μ_{1i} is expressed in Eq.(4-1):

$$\mu_{1i} = \begin{cases} \dfrac{x_1 - a_{1(i-1)}}{a_{1i} - a_{1(i-1)}}, & \text{when } a_{1(i-1)} < x_1 \leq a_{1i}; \\ \dfrac{a_{1(i+1)} - x_1}{a_{1(i+1)} - a_{1i}}, & \text{when } a_{1i} < x_1 \leq a_{1(i+1)}; \\ 0, & \text{otherwise} \end{cases} \quad (4\text{-}1)$$

A similar expression applies for μ_{2j}.

When applying the Rule-$\{i,j\}$, however, we will employ a different approach to the "AND" rule from those in Chapter 3. Namely, we will use the product of μ_{1i} and μ_{2j} as the corresponding output membership μ_{yij}, instead of their minimum value. The output y thus can be derived by the weighted average:

$$y = \sum_{i,j} \mu_{yij} \cdot w_{ij} = \sum_{i,j} (\mu_{1i} \cdot \mu_{2j}) \cdot w_{ij} \quad (4\text{-}2)$$

Note that the sum of weights $\sum \mu_{yij} = \sum \mu_{1i} \cdot \mu_{2j} = 1$ from the above definition of membership functions.

4.1.2 An algorithm for self-tuning membership functions

A self-tuning algorithm was presented in [9] and it was further developed for this work. Let us assume a set of experimental data points y_k^r, corresponding to process settings $\{x_{1k}, x_{2k}\}$ ($k=1,...,q$). The original data set is chosen so that $q=m \times n$. Our goal is to teach the system from these data by adjusting the parameters $\{a_{1i}, a_{2j}, w_{ij}, i=1,...,m, j=1,...,n\}$. In other words, we want to decide the values of $\{a_{1i}, a_{2j}, w_{ij}\}$ based on data set $\{x_{1k}, x_{2k}, y_k^r\}$. To do this, we must define a "cost function" for the system-predicted output values y_k as:

$$E = \frac{1}{2} \sum_{k=1}^{q} (y_k - y_k^r)^2$$
$$= \frac{1}{2} \sum_{k=1}^{q} \left(\sum_{i,j} \mu_{1i} \cdot \mu_{2j} \cdot w_{ij} - y_k^r \right)^2 \quad (4\text{-}3)$$

where μ_{1i} and μ_{2j} are also functions of x_{1k} and x_{2k} as shown in their definitions above. To solve the problem, we need to calculate the derivatives of E, which are:

$$\frac{\partial E}{\partial w_{ij}} = \sum_{k=1}^{q} (y_k - y_k^r) \cdot (\mu_{1i} \cdot \mu_{2j}) \quad (4\text{-}4)$$

and,

$$\frac{\partial E}{\partial a_{1i}} = \sum_{j=1}^{n} \sum_{k=1}^{q} (y_k - y_k^r) \left(\sum_{l=i-1}^{i+1} u_l \cdot w_{ij} \right) \quad (4\text{-}5)$$

where,

$$u_i = \begin{cases} \dfrac{-\mu_{1i} \cdot \mu_{2j}}{a_{1i} - a_{1(i-1)}}, & \text{when } a_{1(i-1)} < x_1 \leq a_{1i}; \\ \dfrac{\mu_{1i} \cdot \mu_{2j}}{a_{1(i+1)} - a_{1i}}, & \text{when } a_{1i} < x_1 \leq a_{1(i+1)}; \\ 0, & \text{otherwise} \end{cases} \quad (4\text{-}6a)$$

$$u_{i-1} = \begin{cases} \dfrac{(1 - \mu_{1(i-1)}) \cdot \mu_{2j}}{a_{1i} - a_{1(i-1)}}, & \text{when } a_{1(i-1)} < x_1 \leq a_{1i}; \\ 0, & \text{otherwise} \end{cases} \quad (4\text{-}6b)$$

$$u_{i+1} = \begin{cases} \dfrac{(\mu_{1(i+1)} - 1) \cdot \mu_{2j}}{a_{1(i+1)} - a_{1i}}, & \text{when } a_{1i} < x_1 \leq a_{1(i+1)}; \\ 0, & \text{otherwise} \end{cases} \quad (4\text{-}6c)$$

A similar expression applies for $\frac{\partial E}{\partial a_{2j}}$.

We also specify the constraints for those parameters:

$$\begin{cases} a_{11} \leq a_{12} \leq \ldots \leq a_{1m} \\ a_{21} \leq a_{22} \leq \ldots \leq a_{2n} \end{cases} \quad (4\text{-}7)$$

There are many optimization techniques available. After testing with different optimization methods, however, we chose a non-linear variable-metric constraint optimization algorithm to find the values of the $m+n+m\times n$ parameters $\{a_{1i}, a_{2j}, w_{ij}\}$.

4.2 Simulation for grain size prediction

The 12 "training" data points in Figure 6 are the experimental results of X-ray-measured average grain size of phosphorus-doped LPCVD polysilicon films [6]. We first choose 9 data points as our training data to derive 9 (=3×3) rules for our inference system (i.e. $m=n=3$, see previous section). By applying the fuzzy inference algorithm illustrated in Section 4.1, the system interpolates all the data points for the 2-dimensional region of input space (T_d, T_a), as shown by the curves in Figure 6(a).

For instance, for recipe (T_d, T_a) = (620, 900), (620, 950), and (620, 1000), our system predicts predicted average grain size, $\bar{s}$ = 80nm, 120nm, and 148nm, respectively.

However, suppose we have completed experiments with these three different recipe settings and measured the resulting average grain size, $\bar{s}$ = 75nm, 95nm, and 100nm [6], which are different from our model prediction, even though the system does tell the qualitative characteristics. We then introduce these 3 new experimental data points to our system and let the system to adjust itself to fit all the 12 data points by minimizing the cost function as defined in Eq. (4-3). Note that we will not change the number of rules. We just shift (tune) the membership function parameters $\{a_{1i}, a_{2j}\}$ and the corresponding output parameters w_{ij} ($i,j=1,2,3$ in this case). The total number of optimizing points equals 15 (=$m+n+m\times n$). The result is shown in Figure 6(b), where we see the curves are now fitting all of the 12 points. It took 28 iterations and the final cost function value was reduced from 1.44×10^5 to 1.18×10^4, as shown in Figure 7.

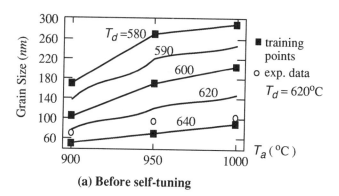

(a) Before self-tuning

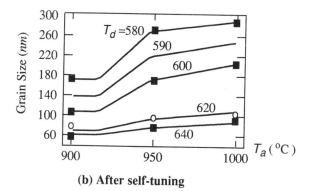

(b) After self-tuning

Figure 6 X-ray grain size versus annealing temperature and deposition temperature for phosphorus-doped LPCVD.

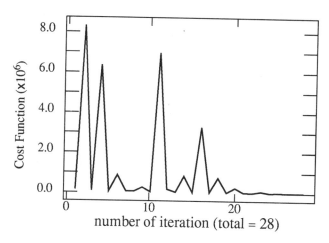

Figure 7 Cost function versus the number of iteration.

Finally, the membership functions before and after the optimization process are shown in Figure 8. The numerical data are summarized as follows:

♦ 9 initial training data points:

$$a_{1i} = \begin{bmatrix} 580 \\ 600 \\ 640 \end{bmatrix}, \quad a_{2j} = \begin{bmatrix} 900 \\ 950 \\ 1000 \end{bmatrix},$$

$$w_{ij} = \begin{bmatrix} 170 & 270 & 290 \\ 103 & 170 & 205 \\ 50 & 70 & 90 \end{bmatrix}. \quad (4\text{-}8)$$

- 3 additional data points:

$$T_d[k] = 620,$$
$$T_a[k] = \begin{bmatrix} 900, & 950, & 1000 \end{bmatrix},$$
$$s^r_k = \begin{bmatrix} 75, & 95, & 100 \end{bmatrix}.$$
(4-9)

where the unit for temperature parameters a_{1i}, a_{2j} and T_d, T_a is °C, and the unit for the grain size parameter w_{ij} and the desirable output data points s^r_k is *nm*. After optimization, the 15 new membership function parameters are:

$$a_{1i} = \begin{bmatrix} 579.8 \\ 598.0 \\ 624.5 \end{bmatrix}, \quad a_{2j} = \begin{bmatrix} 915.4 \\ 947.4 \\ 999.2 \end{bmatrix},$$

$$w_{ij} = \begin{bmatrix} 170.8 & 270.0 & 291.0 \\ 108.5 & 176.8 & 213.5 \\ 57.4 & 72.5 & 84.6 \end{bmatrix}.$$
(4-10)

as plotted in Figure 8(a) and (b):

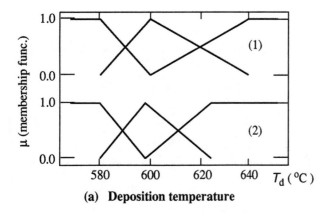

(a) **Deposition temperature**

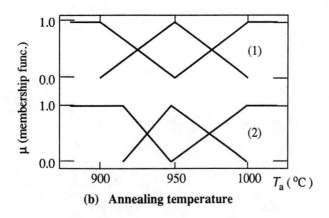

(b) **Annealing temperature**

(1) Before iteration (2) After iteration

Figure 8 Membership functions before and after iterations

The programs are written in C. It usually takes a few minutes run-time to obtain the above results on a SPARC-1 workstation. Our results show that the fuzzy logic based inference system can represent qualitative knowledge and adaptively adjust the system parameters to fit new data. In a real-world problem, however, the input variables are likely to be interdependent. In such a case, a so called "Neural-Network Fuzzy Reasoning" approach [12] might work better.

5.0 Future Plans

The combined quantitative and qualitative models can be used for LPCVD recipe generation and process control. The knowledge base may be further modified and enhanced after taking into account actual process responses. (See Figure 9 below).

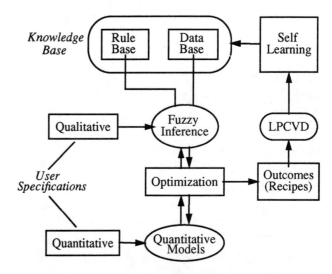

Figure 9 Future architecture of a LPCVD equipment model

We will start from the simple model we have shown in Chapter 4, in order to develop and refine complete qualitative models to describe grain size, film texture, surface roughness and refractive index with respect to the recipe. Those qualitative LPCVD models can be incorporated with BCAM's dynamic equipment model library. Also, the entire BCAM system will be exported to SIMPL-IPX of TCAD (Technology Computer-Aided Design) [13] process simulation system at U.C. Berkeley. Our goal is to collect experimental data dynamically from the actual process so that our fuzzy inference system can eventually perform both process simulation and real-time process control.

6.0 Conclusion

Due to the large number of process variables and their complex interactions in a semiconductor manufacturing environment, the pertinent expert knowledge is mostly qualitative and incomplete. A computer-aided integrated system for a future automated fabrication factory can be facilitated by fuzzy logic based qualitative models which can capture the qualitative human knowledge into a computer software simulation system. This fuzzy inference system can be self-learning to accommodate the updated expert knowledge.

According to the fuzzy methodology, the process parameters are first fuzzified into linguistic variables before the inference rules are applied. These linguistic values of the process parameters are then mapped to appropriate linguistic values of the output parameters. The mapping is implemented as a series of fuzzy set operations. The proper input-output relationships are captured by means of designing the appropriate membership functions and inference rules through the existing training data. The output linguistic values obtained from the fuzzy inference system are finally defuzzified into conventional numerical values. When the new process data is obtained, the system automatically adjusts the membership functions and the corresponding fuzzy rule parameters to best fit the new coming data (and old training data as well, depends on the situation). This self-learning algorithm has great potential in applications on real-time process control in a computer-aided manufacturing environment.

7.0 Acknowledgment

One of the authors (RLC) benefited from discussions with Dr. Hideyuki Takagi, Dr. Lixin Wang and others who are members of Professor Zadeh's research group at U.C. Berkeley. This work is being funded by the SRC (contract # 92-DC-008).

8.0 References

[1] D.A. Hodges, L.A. Rowe and C.J. Spanos, "Computer Integrated Manufacturing", 1st *IEEE/SEMI International Semiconductor Manufacturing Science Symposium,* San Francisco, California (September 1989).

[2] Costas J. Spanos, "The Berkeley Computer-Aided Manufacturing System", *SRC/DARPA CIM-IC Workshop,* North Carolina State University (August 1991).

[3] Kuang-Kuo Lin and Costas J. Spanos, "Statistical Equipment Modeling for VLSI Manufacturing: An Application to LPCVD Reactors", *IEEE Transactions on Semiconductor Manufacturing,* Vol.3, No.4, pp. 216-229 (November 1990).

[4] Sherry F. Lee, "A Three-Dimensional Physically-Based LPCVD Model", *M.S.E.E. thesis,* EECS Dept., U.C.Berkeley (1992).

[5] Ted Kamins, "Polycrystalline Silicon for Integrated Circuit Applications", *Kluwer Academic Publishers* (1988).

[6] G. Herbage, L. Krausbauer, E.F. Steigmeier, A.E. Widmer, H.F. Kappert, and G. Neugebauer, "LPCVD polycrystalline silicon: Growth and physical properties of *in situ* phosphorus doped and undoped films", *RCA Review,* Vol.44, 287-312 (June 1983).

[7] Lotfi A. Zadeh, "Fuzzy Sets", *Information and Control,* Vol.8, 338-353 (1965).

[8] H. -J. Zimmermann, "Fuzzy Set Theory - and its Applications", *Kluwer Academic Publishers* (1991).

[9] Hiroyoshi Nomura, Isao Hayashi and Noboru Wakami, "A Learning Method of Fuzzy Inference Rules by Descent Method", proceedings of *IEEE International Conference on Fuzzy Systems,* pp.203-210, San Diego (March 8-12, 1992).

[10] Costas J. Spanos, "Statistical Parameter Extraction for IC Process Characterization", *Ph.D dissertation,* Dept. of Electrical and Computer Engineering, Carnegie-Mellon University (May 1985).

[11] M.J.D. Powell, "A Fast Algorithm for Nonlinearly Constrained Optimization Calculation", *Proceedings of 1977 Dundee Conference on Numerical Analysis* (June 1977).

[12] Hideyuki Takagi, "Design of Fuzzy System by NNs and Realization Adaptability", preprint (1992).

[13] Tom L. Luan, "Manufacturing-Based IC Process and Device Simulation", *M.S.E.E. thesis,* Memorandum No. UCB/ERL M91/55, U.C.Berkeley (May 1991).

Range Tests Made Fuzzy:
An Alternate Perspective on the Built—in—Test of Real Time Embedded Systems

Jennifer D. Brown
General Electric Aircraft Controls
600 Main St.
Johnson City, NY 13790

George J. Klir
Department of Systems Science
State University of New York at Binghamton
School of Engineering and Applied Science
Binghamton, NY 13902–6000

Abstract – In this paper, we use fuzzy logic to formulate a general strategy for constructing a range test, a common form of built–in–test employed by real time embedded systems. The logic of an actual test used in industry is redesigned using the strategy. The resulting test is shown to be superior to the original, both theoretically and experimentally.

I. INTRODUCTION

Since 1975, when Zadeh formulated the extension principle [1], it has been known that any domain of mathematics based on set theory can be fuzzified. Although the principle prescribes how to fuzzify, it does not explain why to fuzzify. "Until you demonstrate that fuzzification is useful," argued many skeptics, "any efforts to fuzzify must be viewed solely as a mathematical game, sort of *l'art pour l'art*." Demonstrations of great utility of fuzzification are now plentiful, thanks, primarily, to effective cooperation of many theoreticians and practitioners in Japan [2–4].

In this paper, we explore the use of fuzzy logic in the area of built–in–tests. We show that, similarly as in many other application areas, a design based on fuzzy logic outperforms the traditional design based on crisp logic. We assume that the reader is familiar with basic ideas of fuzzy set theory [5].

II. FROM BUILT-IN-TEST TO THE RANGE TEST

Built–in–Test is an industry term which is loosely defined as *a system's ability to evaluate its own state of repair, and take appropriate action in the event of an anomaly*. The term may be unfamiliar to most, but the concept itself should not be. Consider, for example, an automobile with a temperature warning light. If the engine coolant temperature rises above a certain threshold, the warning light is illuminated. In this case, the automobile could be said to be performing built–in–test; its state of repair is evaluated using coolant temperature and, if necessary, appropriate action is taken by illuminating the warning light rather than, say, turning the engine off!

Classically, we think of warning lights such as the one in the example above as being actuated by mechanical systems. Increasingly, however, tasks such as this are the responsibility of a computer, resident in an embedded control or monitoring system. In such systems, the above example represents an instance of a class of built–in–tests known as *range tests*. In the general case, the computer determines if a quantity, for example coolant temperature, is within the range of values deemed "acceptable", i.e. representative of a non–failure condition. If it is within the range, no further action is taken; if it not, it is said to be "unacceptable"; a failure is recognized, and an appropriate action is taken.

Typically, range tests are implemented in a somewhat *ad hoc* fashion using crisp sets to represent the range of acceptable and unacceptable values. It seems, however, that there must exist conditions where the distinction is not so obvious.

III. DISCUSSION

In simple terms, the range test must determine if a quantity is acceptable; typically, the quantity is a function of the values of one or more signals monitored by the system, for example the signal alone, the average of several signals, or the difference between two signals. If the quantity is determined to be unacceptable, an appropriate action must be taken quickly enough to allow the prevention of adverse effects (permanent damage to the automobile engine, for example).

Before a general strategy can be developed, the concepts of *acceptable, unacceptable,* and *quickly enough* must be examined and understood. In the ideal case, we can crisply define the range or set of values for the given quantity which are acceptable in terms of system operation. If it is further assumed that the system is noiseless, and that the computer has complete, error–free knowledge of the quantity's behavior at all instants in time, the definition of unacceptable, i.e. representative of a failure condition, from the computer's viewpoint is very straightforward:

A quantity is unacceptable if it violates the range of acceptable values.

In reality, however, the amount of noise introduced due to the physical characteristics of the electronics and/or the harshness of the operation environment is not always insignificant. Here, it must be recognized that an occasional violation of the range of acceptable values is not necessarily indicative of an unacceptable condition. In order to avoid declaring failures due to noise or some other inexplicable fluke of nature, the idea of consistency must be introduced, and the definition of unacceptable revised as follows:

A quantity is unacceptable if it <u>consistently</u> violates the range of acceptable values.

Interpreting the use of "consistently" in this context to mean "more often than not", the distinction between the acceptable and unacceptable conditions becomes fairly clear in all but the most borderline cases, where a judgement call must be made. It must also be recognized, however, that the computer does not have access to complete, error free information about the behavior of the quantity being evaluated.

Using the terminology of the General Systems Problem Solver [6], the excitation signal can be considered a *source system*; the computer obtains data about this system through an *observation channel*, which typically consists of some conditioning circuitry and a digital sampler as shown in Fig. 1. (Figures are included at the end of the paper.) The observation channel introduces both information loss and error into the data. Behavior is known only for the instants in time at which samples are taken; the accuracy of the behavior information is limited by the resolution of the sampler and the precision of the additional electronics required to fully implement the observation channel. The computer, then, is really making an educated guess about the goodness or badness of the evaluated quantity based on the sampled data.

Again, our definition of "unacceptable" requires modification. In this case, the concept of belief must be introduced to qualify judgements made about the behavior of the actual quantity based entirely on the behavior of the sample values obtained; rather than declare failures based on the knowledge that the quantity conforms to the revised definition of "unacceptable", the computer is forced to declare failures based on the belief that the quantity conforms to this definition. From this point of view, an unacceptable quantity is defined in terms of its sample values as follows:

A quantity is unacceptable if its sample values violate the range of acceptable values often enough to believe that the quantity itself consistently violates this range.

Next, we must challenge the crisp representation of the range of acceptable values. This range may be influenced by many different factors, including:

- The values, if taken by the quantity, which will cause adverse effects must be prevented, e.g., "coolant temperatures of $\geq 250°F$ will cause automobile engine damage."

- The values of the quantity which the system was designed to produce under worst case conditions, e.g., "the system was designed such that if it is working properly, automobile coolant temperature will never exceed the worst case condition of 225°F".

- Requirements dictated by specifications governing system design, e.g., "the procurement specification requires that the temperature light be illuminated for automobile coolant temperatures exceeding 220°F."

In the third case, the use of a crisp range is justifiable. In the first two, however, it may not be. In general, while there are a range of values for which it may be said that the system is definitely performing acceptably and another range of values for which it may be said that the system is definitely performing unacceptably, there is also a region of values for which the performance of the system is marginal in either direction; the exact point at which the system stops performing acceptably and starts performing unacceptably is generally not known. This is impossible to represent using crisp sets, but easily represented by a fuzzy interval comprised of two overlapping fuzzy sets, such as that shown in Fig. 2. Again, our definition of an unacceptable quantity requires modification:

A quantity is unacceptable if its sample values violate the range of acceptable values to a sufficiently high degree often enough to believe that the quantity itself consistently violates this range.

Finally, we must examine the concept of *quickly enough*. While it is evident from the above discussion that more than one sample value should be collected before declaring a failure, there is a limit to the amount of time which may be used to make the decision. In the boundary condition where all sample values of the quantity are definitely unacceptable, this limit is governed by the shortest amount of time required by the system to incur adverse effects. Thus, we need to make sure that in this case, the number of samples required to record a failure is large enough to provide sufficient noise intolerance and overall belief that the actual signal is indeed bad, yet small enough that the associated execution time is safely less than the worst case reaction time of the system. At the other boundary condition, as long as the sample values of the quantity are definitely acceptable, a failure should never be declared. In between boundary conditions, the general rule of thumb is as follows:

The larger the deviation of the quantity value from the required range, the greater the chances are of incurring adverse effects, and the faster a failure must be declared.

IV. FORMALIZATION OF GENERAL STRATEGY

Using the results of the above discussion and assuming an iterative implementation, a general strategy for constructing a range test is formalized as follows:

1. Define a universal set S consisting of all possible sample values for the given quantity.

2. Define fuzzy sets A and U on S, where:
   A = the set of possible sample values which represent acceptable system performance, and
   U = the set of possible sample values which represent unacceptable system performance.

3. Define an overall level of belief that the actual behavior of the quantity is unacceptable, b, as a function of the accumulation over time of the membership of the individual sample values in the sets A and U:
   $$b(s_n) = f(b(s_{n-1}), \mu_U(s_n), \mu_A(s_n))$$
   where n represents the nth iteration of the built-in-test logic and s_n represents the nth sample value of the quantity being evaluated.

4. Define a threshold value, t, such that $b(s_n) \geq t$ is a necessary and sufficient condition to declare a failure, and such that the time required for $b(s_n) \geq t$ to occur is consistent with the needs of the boundary conditions as discussed above.

V. APPLICATION

Consider the following example, drawn from industry (the physical numbers used have, however, been changed for proprietary reasons):

A digital control monitors turbine blade temperature for use in its control strategy. Since the accuracy of the temperature signal provided by the sensor is dependent upon the accuracy of an excitation signal provided to the sensor by the engine control, the excitation signal is also monitored and subjected to a range test. In this case, the quantity being evaluated is the excitation signal in engineering units (dc volts); ideally, it should be constant at 15.0 volts dc. The range of acceptable values is a function of the worst-case-error analysis performed for the excitation function; the action taken in the event of a failure is the recording of a failure in non-volatile memory for later retrieval by maintenance personnel, and the adjusting of the control strategy to compensate for the loss of the temperature signal. The built-in-test logic is iterated every 50 msec; a failure must be declared in time to avoid overheating the turbine blades.

The logic currently implemented by the engine control for this test is as follows:

> IF 10 consecutive excitation signal samples violate the range [14.9, 15.1] THEN record a failure against the excitation signal.

Applying the terminology of the general procedure, μ_U, μ_A, b, and t are as shown in Fig. 3. When evaluated at the boundary conditions, this approach performs as would intuitively be expected:

- Given $\mu_A(s_n) = 1$, $\mu_U(s_n) = 0$ for all n, no failures will be recorded;
- Given $\mu_A(s_n) = 0$, $\mu_U(s_n) = 1$ for all n, a failure will be recorded in the 10th iteration; as the logic is iterated every 0.05 seconds, the corresponding elapsed time is 0.50 seconds, which intuitively seems quite fast enough to avoid overheating the turbine blades.

Exploring its performance between boundaries, however, the following shortfalls are discovered:

- By defining A and U as crisp sets, no allowance is made for the inherent fuzziness in the distinction between acceptable and unacceptable values.
- $b(s_n)$ is defined such that a single acceptable sample outweighs all the contributions of up to nine unacceptable samples.

In addition to being counterintuitive, its performance due to the second shortfall can be disturbingly poor from a mathematical standpoint. Consider, for example, a case in which $b(s_{n-1}) = 9$: if the value of $s_n = 15.1 - \varepsilon$, $0.2 \geq \varepsilon \geq 0$, then $b(s_n) = 0$, and at least 10 more samples must be taken before a failure is recorded; if however, $s_n = 15.1 + \varepsilon$, $\varepsilon \geq 0$, then $b(s_n) = 10$, and a failure is recorded immediately. As ε is allowed to become infinitesimally small, the difference between $15.1 - \varepsilon$ and $15.1 + \varepsilon$ approaches 0, yet the difference in the action taken is drastic. As a result, such logic will not declare a failure against the excitation signal shown in Fig. 4, though it is obviously unacceptable from an intuitive standpoint.

As an alternative, consider the fuzzified definitions of μ_U, μ_A, and b given in Fig. 5, which were derived using the general strategy presented in Section IV. The equivalent block diagram is shown in Fig. 6.

By representing U and A as fuzzy sets, we are recognizing and preserving information about the inherent fuzziness of the range of acceptable values. Combining this with a more intuitive definition of the function b, the adjustment made to $b(s_n)$ as a result of each new sample is a continuous function of the sample value, and the possibility of drastically different actions based on an infinitesimally small difference in sample values is eliminated.

VI. Empirical Comparison

Both the original built–in–test logic and the proposed alternative were modeled using the STELLA software package and an Apple MacIntosh II personal computer, and their responses evaluated for various specific cases of the following general scenarios:

1. Excitation signal is always unacceptable: all signal samples s_n are such that $\mu_U(s_n)=1$ and $\mu_A(s_n)=0$.
2. Excitation signal is always acceptable: all signal samples s_n are such that $\mu_U(s_n)=0$ and $\mu_A(s_n)=1$.
3. Input signal is marginally unacceptable: some signal samples are within the range [14.9, 15.1], but on the whole the samples violate the range more often than not.
4. Input signal is marginally acceptable: some signal samples violate the range [14.9, 15.1], but on the whole the samples are within the range more often than not.

As expected, the two approaches performed identically in all test cases from the two boundary condition scenarios (1 and 2). Neither accumulated any belief that an "always acceptable" signal was unacceptable, while both belief measures reached the threshold t in the minimum number of iterations (ten) when faced with an "always unacceptable" excitation signal.

In an effort to approximate typical situations, the signal samples s_n in the test cases for the marginal scenarios (3 and 4) were generated from a simulated random normal distribution. The distribution mean represented the value of the signal given ideal (noiseless) conditions, while the standard deviation controlled the level of noise present. For each scenario, a variety of mean / standard deviation combinations were tested; the computer simulation for each combination was replicated several times using different initial seeds for the random number generator.

The test cases performed for scenarios 3 and 4 provided very interesting results. While both approaches showed a tendency to take progressively longer to declare a failure as the unacceptability of the excitation signal was made more and more marginal (i.e. as the distribution of signal samples became closer and closer to being 50% within the range [14.9, 15.1] and 50% outside it), the original logic was much more affected by changes in the random number generator seed than the proposed alternative, and therefore much less consistent in the number of iterations required to declare a failure for a given set of sample distribution parameters.

For example, one test case performed used a distribution with mean (theoretical) 15.1255, standard deviation 0.03. Theoretically, 80% of the signal samples generated should have had values outside the range [14.9, 15.1], while 20% should have had values within it. Ten replications were run, each using the same distribution parameters with a different seed. The results are tabulated in Table 1.

TABLE 1:
Summary of Results for Sample Test Case

Replication	# iterations to declare failure	
	Original	Proposed
1	30	23
2	17	28
3	19	22
4	10	23
5	28	25
6	127	18
7	12	22
8	55	31
9	13	18
10	25	21

While in the end both approaches correctly declared a failure in every replication, the difference in variation of the number of iterations (and hence time) required to do so was somewhat startling. For the ten replications, the traditional logic declared a failure in an average of 33.6 iterations, with a standard deviation of 11.8 iterations; the proposed logic, however, declared a failure in an average of 23.1 iterations, with a standard deviation of only 1.4 iterations. This difference is also illustrated in Fig. 7, which shows the simulation data collected for two of the replications.

VII. Conclusion

We used fuzzy logic to formulate a general strategy for constructing a range test, a common form of built–in–test employed by real time embedded systems. We then used the strategy to redesign the logic of a test developed by traditional methods, and showed that the result was superior to the original both theoretically and empirically.

While we have made substantial progress since first addressing this subject, efforts to date have been focused on the formulation of the general strategy and providing evidence that the area is, in fact, worthy of future study. Having done this, the logical next steps include further study of how to optimally choose the membership functions $\mu_U(s_n)$ and $\mu_A(s_n)$, the exploration of the performance of different belief functions $b(s_n)$, and performing further case studies.

REFERENCES

[1] Zadeh, L. A, *The Concept of a linguistic variable and its application to approximate reasoning.* Information Sciences, 8, 1975, pp. 199–249, 301–357; 9, pp. 43–80.

[2] Lewis, H., *Letter to the Editor*. International Journal of General Systems, 19, No. 2, 1991, pp. 171–175.

[3] Bellon, C., P. Bosc, and H. Prade, *Fuzzy Boom in Japan.* International Journal of Intelligent Systems, 7, No. 4, 1992, pp. 293–316.

[4] Schwartz, D. G. and G. J. Klir, *Fuzzy logic flowers in Japan.* IEEE Spectrum, 29, No. 7, 1992, pp. 32–35.

[5] Klir, G. J. and T. A. Folger, Fuzzy Sets, Uncertainty, and Information. Prentice Hall, Englewood Cliffs, N.J., 1988.

[6] Klir, G.J., Architecture of Systems Problem Solving. Plenum Press, New York, 1985.

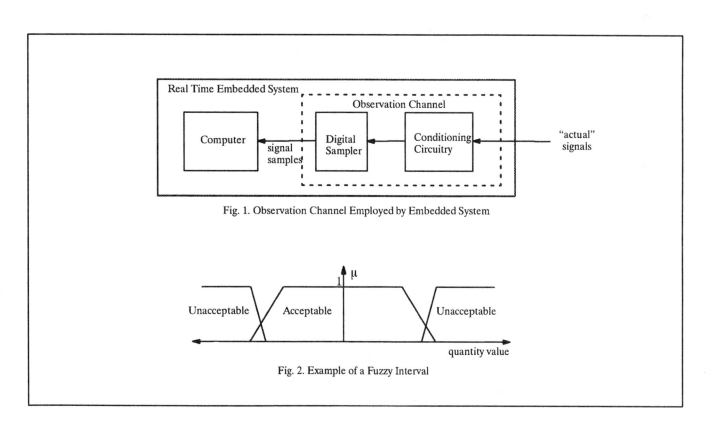

Fig. 1. Observation Channel Employed by Embedded System

Fig. 2. Example of a Fuzzy Interval

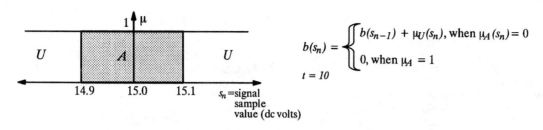

Fig. 3. General representation of original logic

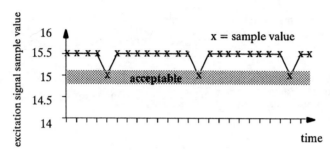

Fig. 4. Obviously unacceptable excitation signal

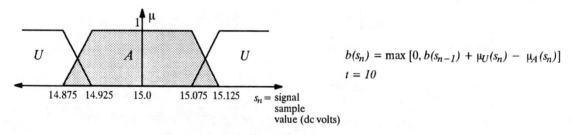

Fig. 5. Definition of alternative logic

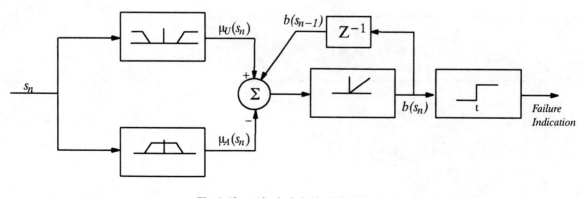

Fig. 6. Alternative logic in block diagram form

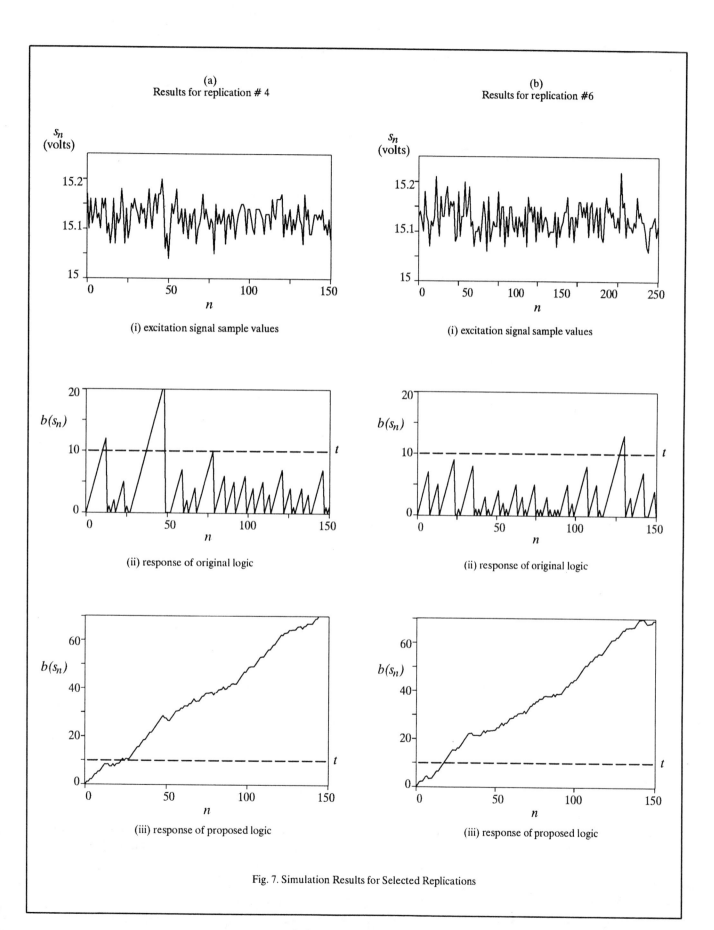

Fig. 7. Simulation Results for Selected Replications

FUZZY SEAM-TRACKING CONTROLLER

Yoshito Sameda

Heavy Apparatus Engineering Laboratory
Toshiba Corporation
1, Toshiba-cho, Fuchu-shi, Tokyo 182, Japan

Abstract

The welding spot of an automatic laser welding system is very small, thus requiring real-time accurate control of the welding position. We have developed a seam-tracking controller that uses fuzzy logic. This controller incorporates an operator's seam recognition ability and uses a method for auto-tuning the fuzzy rules. The major difficulty of an optical-sensing seam-tracking controller is to estimate the seam location from an image which may be deteriorated by changes in the workpiece's surface condition. The inference rule for the seam detection using fuzzy logic can incorporate ambiguous descriptions. Hence, it can infer the appropriate seam location from a deteriorated image which does not exactly match the rules.

1. Introduction

The CO_2 laser welding system is very useful for speedy, non-distortional welding. As its welding spot is very small, the laser beam must be kept exactly on the seam. An automatic laser welding system must provide real-time welding position control and robust seam detection. It must be able to estimate a seam location for a variety of joint types and workpiece surface conditions.

A visual system can achieve high resolution of the seam's image with a lens of high magnifying power. Several visual sensors for arc-welding are utilized[1]. However, the image may be deteriorated by several causes, such as dust, stains and flaws on the workpiece surface. The conventional image processing techniques such as the binary image processing and template matching don't work well in these conditions. However, a man can recognize a seam from a deteriorated image. Therefore, we incorporated an operator's knowledge into the fuzzy rules. In fuzzy logic, the inference rule for seam detection can include ambiguous descriptions. Hence, it can infer the appropriate seam location from a deteriorated image which does not exactly match the rules. We applied fuzzy logic to the seam detecting process.

It takes a long time to tune fuzzy rules by trial and error. Hence, we need an automatic method of tuning fuzzy rules. Several auto-tuning systems are utilized, such as those using fuzzy neural controllers[2] and self-organizing fuzzy controllers[3]. We have developed an auto-tuning system which tunes the rules based on data provided by the operator.

2. Laser welding system

This system welds a straight seam of a moving workpiece, as shown in figure 1. The seam may be a butt, a lap, or a T joint.

The controller observes the seam to be welded. It employs a pattern projection and optical-triangulation technique to obtain the cross-section of the workpiece. The two laser diodes project slit rays and a CCD camera produces two corresponding images. Figure 2 shows a pipe workpiece. There are some differences in brightness distribution between images A and B. These differences depend on the angle of the workpiece surface.

The image processor extracts certain features

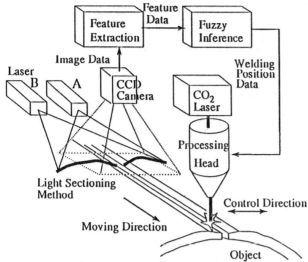

Fig. 1 CO$_2$ Laser Welding System

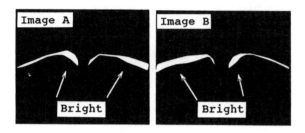

Fig. 2 Input Image

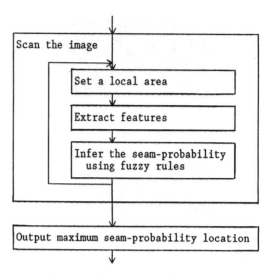

Fig. 3 Detecting Algorithm

from the image. It then estimates the seam location from the brightness distribution and the profile of the image using fuzzy rules. The processing head is moved to keep the CO$_2$ laser beam on the seam within $\pm$ 0.1 mm because the laser beam is 0.2 mm in diameter.

There is a time delay from when the seam position is detected to when the processing head is moved. This time depends on the welding speed and the gap between welding spot and observed point.

3. Seam detection with fuzzy inference

Figure 3 shows the seam detecting algorithm using fuzzy inference. The image processor sets a local area on the input image and extracts features from it. It then infers the probability of the seam appearing in the local area using fuzzy rules. If the probability value for the area is high, there is a high probability that it contains the seam. The image processor scans the local areas over the image until it finds the local area with the highest probability. The local area size is variable for differences in seam gap width.

We formulated the feature extraction method and the fuzzy rules to simulate an an operator's ability to identify a seam.

For example, the knowledge for brightness distribution of a pipe seam is described as follows:

Knowledge 1: If the center of the local area is dark and both sides of the local area are bright, then the local area may include the seam. The seam does not reflect light.

Knowledge 2: If the center of the local area is dark and the left side of the local area is bright as in image A, and the right side of the local area is bright as in image B, then the local area must include the seam. This is the normal pipe seam brightness distribution.

Knowledge 3: If the center of the local area is dark and both sides of the local area are a little bright, then the local area may contain dust or a defect.

Figure 4 shows the local area and its brightness features: x_1, x_2 (the center of the area), x_3, x_4 (the left of the area), and x_5, x_6 (the right of the area).

Figure 5 shows the membership functions defined for each feature. The shapes of the membership functions are triangles and trapezoids to simplify the fuzzy inference calculation.

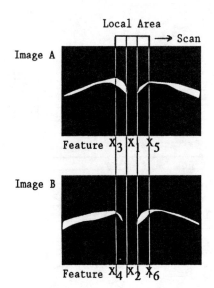

Fig. 4 Local Area & Feature

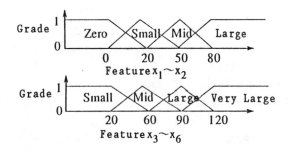

Fig. 5 Membership Function

The knowledge stated are described by fuzzy rules:

Rule 1: If x_1 is Large and x_2 is Large, then y = 0.

Rule 2: If x_1 is Zero and x_2 is Zero, and x_3 is Mid and x_4 is Mid, and x_5 is Mid and x_6 is Mid, then y = 40.

Rule 3: If x_1 is Zero and x_2 is Zero, and x_3 is Large and x_4 is Mid, and x_5 is Mid and x_5 is Large, then y = 80.

"Large", "Mid" and "Zero" are fuzzy labels, y indicates seam-probability and is not a fuzzy value. These are sample of the rules for a pipe seam. The fuzzy inference operation is a simplified Mamdani method for speedy processing. The fuzzy rules are represented by a table.

4. Auto-tuning of fuzzy rules

To simplify descriptions, the fuzzy rules are described for 2-dimensional inputs as follows:

Rule$_{11}$: If x_1 is A_{11} and x_2 is A_{21}, then y = b_{11}
...
Rule$_{ij}$: If x_1 is A_{1i} and x_2 is A_{2j}, then y = b_{ij}
...

The inputs x_1, x_2 are the feature data, A_{1i}, A_{2j} are fuzzy labels based on membership functions, and b_{ij} is the seam-probability.

We should tune the membership functions of A and and the seam-probability b_{ij}. Figure 6 is a diagram of the auto-tuner. The auto-tuner obtains several training data that are features of the images and the correct seam locations as taught by an operator. It then infers the seam-probability from the training data and estimates the seam location.

The auto-tuner defines the membership functions with the training data value distributions, and optimizes seam-probability b using the descent method of nonlinear programming.

The seam-probability y of features x_1, x_2 is calculated by

$$y = \frac{\sum_{ij} w_{ij} b_{ij}}{\sum_{ij} w_{ij}} \quad (1)$$

$$w_{ij} = \min(\mu A_{1i}(x_1), \mu A_{1j}(x_2))$$

μA is the membership function of the fuzzy label A.

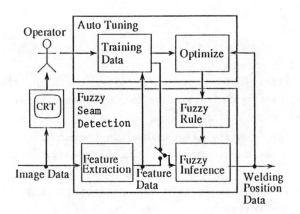

Fig. 6 Auto Tuning of Fuzzy Rule

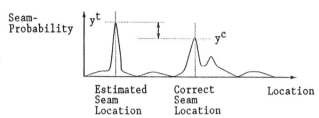

Fig. 7 Wrong Estimation

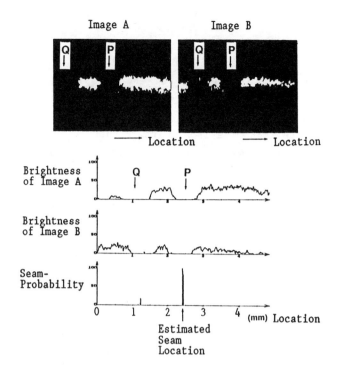

Fig. 8 Experimental Result

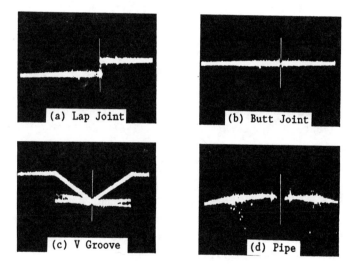

Fig. 9 Experimental Result

If the estimate of seam location equals the taught seam location, the rules are right. If the estimated seam location differs from the taught seam location, the fuzzy rules must be tuned.

We describe the training data of the seam features as x_1^c and x_2^c, and of the non-seam features as x_1^t and x_2^t. The seam-probability y^c and y^t as shown in figure 7 are calculated by x^c and x^t using equation (1).

The auto-tuner evaluates the fuzzy rules using a function in which the differences between the taught location's seam-probability and the inferred wrong location's seam-probability are summed in a training image. We defined the objective function for the descent method as follows:

$$E(b) = \sum_t g(a + y^t - y^c) \qquad (2)$$

$$g(y) = \text{if } y > 0 \text{ then } y \text{ else } 0$$

The auto-tuner optimizes the value of b to minimize the objective function E(b) value. We can obtain the direction vector to optimize b from E(b) differentiated with respect to b_{ij}.

$$\frac{d\,E(b)}{d\,b_{ij}} = \frac{\sum_t w_{ij}^t}{\sum_{trs} w_{rs}^t} - \frac{\sum_c w_{ij}^c}{\sum_{crs} w_{rs}^c} \qquad (3)$$

5. Experimental result

We experimented on a pipe workpiece which couldn't be tracked with a conventional system. Figure 8 shows images of the workpiece, and graphs indicating the brightness, and seam-probability.

Location P is the seam and location Q is non-reflective dust. You can see that the maximum seam-probability indicates the correct seam location.

Figure 9 shows a lap joint, a butt joint, a V groove and a pipe, and their estimated seam locations. The controller could detect the correct seam location for any type of joint.

The control accuracy was ±0.1mm, and the processing time was 66 ms. The controller could track a seam for 10 m/minute welding speed.

6. Conclusion

We have developed a fuzzy rule-based visual system for detecting a seam from a deteriorated image and an auto-tuning system for a seam-tracking controller. The fuzzy seam-tracking controller can be applied to laser welding of any joint type.

Reference

[1] A.Kabe, "On-line Visual Sensor for Arc-welding Robots," in Denshi Tokyo, 25, 1986, pp 61-64

[2] H.Takagi, "Fusion Technology of Fuzzy Theory and Neural Network," in Proceeding of the International Conference on Fuzzy Logic & Neural Networks IIZUKA '90, 1990, pp. 13-16

[3] T.J.Procyk and E.H.Mamdani, "A Linguistic Self-Organizing Process Controller," Automatica, 15, 1979, pp. 15-30

Neural Network Based Decision Model Used for Design of Rural Natural Gas Systems

W. Pedrycz[1], J. Davidson[2], and I. Goulter[3]

[1] Dept. of Electrical & Computer Engineering
University of Manitoba
Winnipeg, Manitoba, Canada R3T 2N2

[2] Dept. of Civil Engineering
University of Manitoba
Winnipeg, Manitoba, Canada R3T 2N2

[3] Dept. of Civil Engineering and Building
University College of Central Queensland
Rockhampton, Queensland, Australia

Abstract

The architecture of a neural network for the design of rural natural gas systems is developed based on logical constructs arising from the theory of fuzzy sets. Multi-layer architecture is implemented to capture "local" and heterogeneous features of the decision process. The neural networks ability to model the decision making process is illustrated with an example set of alternative layouts of rural natural gas systems.

1. Introduction

Decision-makers are often faced with several decision objectives (criteria) usually conflicting and fuzzy in their nature. From the beginning fuzzy sets have been studied as a flexible conceptual framework for decision-making [1][3][5][8]. A key issue that has been identified is how to combine (aggregate) objectives existing in the decision problem. This, in fact, calls for the solution of several sub-problems, e.g., modelling importance of decision objectives, developing suitable models of logical operations (connections) and modelling varying influence of objectives [8]. These problems have been addressed in many ways.

The aspect of aggregation essentially involves finding appropriate logical operations. A number of papers have been written on this topic. Most of them refer to generalized AND and OR operations modelled as t (s) norms, compensative or averaged operations [3][8]. It is worth noting that despite their diversity the operators are highly structured as implied by simple well-defined expressions defined over all range of grades of membership.

This paper examines the design and application of neural networks in the context of a specific decision-making problem, the selection of an appropriate layout for rural natural gas distribution system. The neural networks considered exploit one-stage classes of single-layer and multi-layer architectures driven by logical constructs arising from the theory of fuzzy sets. This type of neural network has been introduced and analyzed in previous work [2].

The discussed application is a multicriteria optimization problem. The problem is explained in more detail later in the paper. It can be viewed in an abstract sense as consisting of a finite set of "N" alternatives evaluated in terms of a collection of criteria (objectives) $o_1, o_2, ..., o_n$. For a given alternative the terms $x_1, x_2, ..., x_n$ will denote degrees to which the alternative satisfies the individual criteria. As such these numbers are viewed as grades of membership existing within the unit interval [0,1]. Neural networks will be utilized to model relationships (mapping) between the space of objectives and the space of preference of alternatives.

2. Neural Network Model of Decision-Making

This section develops the basic architecture of the neural network starting from a description of a single node and proceeding with the overall network. The single node [6] can serve as an appropriate model of simple and homogeneous decision-making while the network [4] can capture all "local" properties of decision-making yielding its heterogeneous aspects.

2.1 Single Node Neural Network Structure

The basic logical neuron represents the mapping between a space of degrees of suitability (preference) of a given alternative and the values of the degrees of satisfaction of its criteria (objective). The simplest transformation realized is described as:

$$d = x_1 \text{ AND } x_2 \text{ AND } ... \text{ AND } x_n$$

where $x_1, x_2, ..., x_n$ stand for the degrees of satisfaction of objectives of the alternative and "d" describes the degree of its preference. Note that in the above expression all the objectives are treated uniformly. The importance (relevance) of the objectives in the decision problem is incorporated by affecting x_i's by adding weights via the OR operation. This modifies the previous formula:

$$d = (x_1 \text{ OR } w_1) \text{ AND } (x_2 \text{ OR } w_2) \text{ AND } ... \text{ AND } (x_n \text{ OR } w_n) \qquad (1)$$

The weight w_i influencing x_i can be regarded as a threshold value. The higher the value of w_i the lower the influence of the i-th objective on a final decision. For $w_i = 1$, (x_i OR w_i) equals 1 and the influence of the objective is completely eliminated. Recalling that OR and AND operations are realized with the aid of s and t-norms respectively [3], (1) reads as:

$$d = \mathop{T}_{i=1}^{n} (w_i \text{ s } x_i) \qquad (2)$$

We will refer to (2) as a generic t-s computational node. The weights are also named connections of the node. The vectors $\mathbf{x} = [x_1 \, x_2 \, ... \, x_n]$ $\mathbf{w} = [w_1 \, w_2 \, ... \, w_n]$ summarize the evaluation criteria and the respective connections.

The learning process involves adjusting connections so that an overall performance realized in terms of a sum of squared errors

$$Q = \sum_{k=1}^{N} (t_k - d_k)^2$$

is minimized.

Where t_k and d_k stand for a target and the output of the network occurring at the k-th instance in the learning set, say $(\mathbf{x}_k, t_k)$. Plugging (2) into the computations yields an expression describing changes in the connections:

$$w_\ell = w_\ell - \frac{\alpha}{2} \frac{\partial Q}{\partial w_\ell}$$

$\ell = 1,2,...,n$, $\alpha \in (0,1)$ namely

$$\frac{\partial Q}{\partial w_\ell} = -2(t_k - d_k) \frac{\partial d_k}{\partial w_\ell}$$

Since the computation of the last derivative involves triangular norms all calculations should be postponed until the form the triangular norms is specified.

$$\frac{\partial d_k}{\partial w_\ell} = \frac{\partial}{\partial w_\ell} (\mathbf{x}_k \text{ AND\_OR } \mathbf{w}) = \frac{\partial}{\partial w_\ell} \left(\underset{i=1}{\overset{n}{T}} x_i^k \text{ s } w_i \right) =$$

$$= \frac{\partial}{\partial w_\ell} \left(A \text{ t} \left(x_\ell^k \text{ s } w_\ell \right) \right)$$

where

$$A = \underset{i \neq \ell}{\overset{n}{T}} \left(x_i^k \text{ s } w_i \right)$$

This generic model of decision-making can be augmented by including complemented (negated) values of objectives **x**. The purpose of this extension is that one can properly model the contribution of the objectives having a "reverse" influence on the final decision (i.e., the lower the satisfaction of the objective the more visible its influence on the suitability of the studied alternative). Denote by **x'** the following extended vector of objectives including complements of x_i, $\bar{x}_i = 1 - x_i$.

$$\mathbf{x'} = [x_1, x_2, ..., x_n \mid \bar{x}_1, \bar{x}_2, ..., \bar{x}_n]$$

Then (2) is replaced by

$$d = \mathbf{x'} \text{ AND-OR } \mathbf{w'} \quad (2')$$

From this point no distinction is made between (2) and (2') since it is assumed that the difference can be deduced from the context.

Two scenarios which serve as examples of the decision model discussed to this point are summarized as DATA1 and DATA2, respectively. The decision (d) involves several alternatives described by two objectives x_1 and x_2.

	DATA1 x_1	x_2	t	DATA2 x_1	x_2	t
1	1.0	0.6	0.7	0.8	0.5	0.15
2	0.2	0.7	0.3	0.1	0.8	0.65
3	0.7	0.2	0.25	0.7	0.1	0.0
4	0.0	0.1	0.15	0.4	0.5	0.25
5	0.6	0.7	0.5	0.4	0.5	0.2
6	1.0	0.2	0.35	0.2	0.9	0.9

The triangular norms utilized by the neuron are specified as a product (t-norm) and probabilistic sum.[1] The results of learning (for α = 0.25) are shown in Fig. 1.

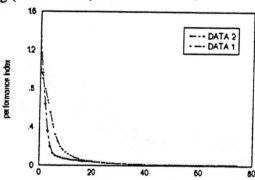

Fig. 1. *Learning: performance index vs. learning epochs*

Both direct and complemented objectives are used in the neuron. The computed connections are equal to:

$$w = \begin{matrix} x_1 & x_2 & \bar{x}_1 & \bar{x}_2 \\ [0.264 & 0.196 & 1.0 & 0.98] \end{matrix}$$

The final grade of preference d is expressed with the aid of x_1 and x_2 while $\bar{x}_1$ and $\bar{x}_2$ are practically excluded (the connections equal to 1.0). Thus

$$d = (0.264 \text{ OR } x_1) \text{ AND } (0.196 \text{ or } x_2)$$

The mapping that is achieved shows a strong coincidence between the results of the network and the data, see Fig. 2.

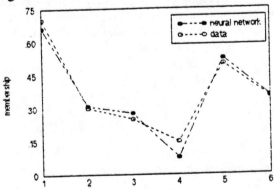

Fig. 2. *Data and results produced by the neural network*

For the second data set the network is described as:

$$d = x_2 \text{ AND}(0.21 \text{ OR } \bar{x}_1)$$

So the complemented form of x_1 has a clear impact on d. Again the network follows the data, Fig.3.

---

[1] The probabilistic sum reads as asb = a + b - ab,

a, b ∈ [0,1]

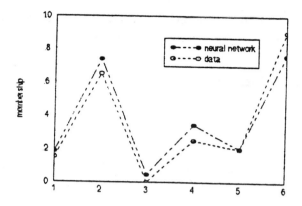

Fig. 3 Data and results produced by the neural networks

2.2 Logical Three-Layer Neural Networks in Representation of Decision-Making Processes

Single node (neuron) models straight forward and homogeneous decision procedures (i.e., the cases in which objectives have a uniform although unknown influence on a final decision). It could happen that the objectives and decisions are not associated in the same way across all numerical values of their grades of membership. This phenomenon is captured by a series of single AND-OR computational nodes. Each of them describes a region of homogeneous aggregation of objectives. The regions are merged with the use of the OR operation. Thus the resulting architecture consists of three layers as shown in Fig. 4.

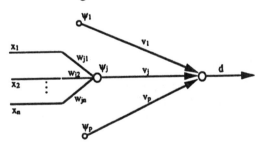

Fig. 4 Three layer neural network

The hidden layer comprising of "p" AND-OR nodes builds "local" homogeneous models of decision-making. The output signals $\psi_2, \psi_2, \ldots \psi_p$ characterize a varying contribution of decision objectives. These relationships are formally expressed as:

- for the hidden layer:

$$\psi_j = \mathop{T}_{i=1}^{n} (w_{ji} \, s \, x_i) \qquad (3)$$

$j = 1,2,\ldots,p.$

- the output layer uses a single AND-OR neuron.

$$d = \mathop{S}_{j=1}^{p} (v_j \, t \, \psi_j) \qquad (4)$$

The signals of the hidden layer ψ_j are summarized using the OR operator (s-norm) while their contribution to "d" is modulated by weights (connections) v_j. In virtue of the AND operation applied there, higher values of connections v_j emphasize a stronger impact of the ψ_j on the resulting decision d. Of course the vector **x** could include complemented variables. If this is the case the previous remarks remain valid.

The learning in the network described by (3) and (4) is carried out by adjusting its connections.

$$\mathbf{w} = [w_{ji}] \quad \text{and} \quad \mathbf{v} = [v_j]$$

$$j = 1, 2,..., p, \quad i = 1, 2,..., n$$

This facet of learning refers to the parametric component of the network. All relevant formulas for the number of nodes of the hidden layer can be found in [4]. Again, the general scheme reads as:

$$r = r - \alpha \frac{\partial Q}{\partial r} \tag{5}$$

$$v = v - \alpha \frac{\partial Q}{\partial v} \tag{6}$$

with α denoting a learning rate, $\alpha \in (0,1)$.

The choice of the size of the hidden layer "p" pertains to the non-parametric aspect of learning. Due to its character one should follow a heuristic rule: Start from the smallest number of nodes, say p = 1. Complete the parametric learning (5) - (6). If the results obtained are not satisfactory increase the number of elements in the hidden layer (which leads to enhanced representation capabilities of the network).

3. Application of Neural Networks to Rural Gas Network Design

In previous work [2] a method was developed to select the most preferred layout for a rural natural gas distribution system from a set of alternatives. The layout was selected in accordance with the conflicting objectives of minimizing cost while maintaining "generally rectilinear" geometry. The previous work modelled the decision-making process as the intersection of two membership functions corresponding to the two objectives. Quantitative measures for rectilinearity and cost savings (both normalized on a unit interval) were developed to determine the membership functions for the two objectives. Saaty's method of priorities was used to derive the intersection of the two objectives based on pairwise comparison of the alternatives in accordance with engineering judgment. The intention was to calibrate the original membership functions based on the derived intersection.

In the present work the same decision is modelled though the use of the described neural network. The experimental data set includes 21 gas networks. Examples of these are shown in Figure 5.

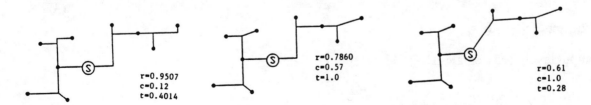

Fig. 5 *Examples of gas networks*

The same two decision criteria (objectives) are used as in the previous work [2], rectilinearity (r) and cost savings (c), and these criteria are calculated by the same method. In addition, the resulting decisions (t) are derived by Saaty's priority method. The evaluation of all the cases is shown in Fig. 6

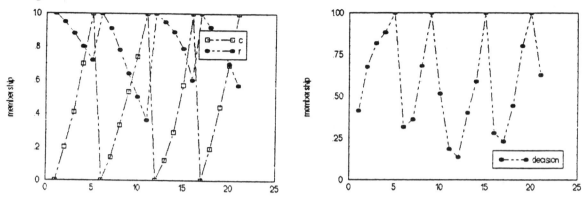

Fig. 6 Membership functions of objectives and decisions

The series of neural networks used to form the model of decision-making involves changes in the hidden layer starting from two nodes and proceeding up to 8. The first 50 learning epochs carried out for $\alpha = 0.1$ are reported in Fig. 7(a) while the final values of the performance index (after 400 learning epochs) are shown in Fig. 7(b).

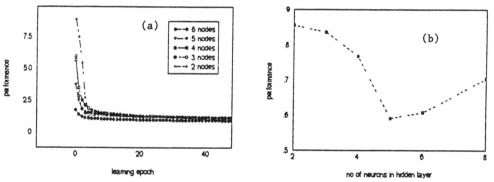

Fig. 7. Results of learning (a) and performance index vs. size of the hidden layer (b)

It is obvious that 5 nodes of the hidden layer led to the lowest value of the performance index. The results of the network vs data are given in Fig. 8

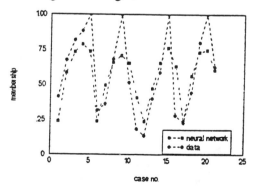

Fig. 8 Decisions vs. results of neural network

In most cases the values of the membership coincide. The evident exception is for the grades of membership equal to 1; the neural network produced significantly lower values, however higher than in the remaining situations.

The neural network has the following connections:

output-hidden layer (s-t neurons):
    1.0   1.0   0.98   0.22   0.3

hidden-input layer:

x_1	x_2	$\bar{x}_1$	$\bar{x}_2$
1.0	0.0	1.0	0.0
1.0	0.0	1.0	0.0
0.0	0.14	0.99	0.98
0.0	0.54	1.0	1.0
1.0	0.0	1.0	0.0

After pruning weak connections of the network (see Section 2) the resulting decision model is described by the following formula:

$$d = (c \text{ AND } \bar{c}) \text{ OR } (r \text{ AND } (c \text{ OR } 0.14))$$

4. Conclusions

The neural network approach to the layout design problem has the following desireable features:
- the neural network captures the relative degree of preference for layouts to the extent that the neural network produces the same solution to the problem as engineering judgement.
- once the connections have been established the neural network provides solutions to the problem efficiently without the effort required by other methods such as the Saaty procedure.

Acknowledgements

Financial support for this work came from the Natural Sciences and Engineering Research Council of Canada.

References

1. R.E. Bellman, L.A. Zadeh, Decision making in a fuzzy environment, Management Sci., B17, 1970, 141-164.
2. J. Davidson, W. Pedrycz, I. Goulter, Formulation and validation of fuzzy decision-making models: An application to the design of rural natural gas networks, Proc. of 10th NAFIPS Workshop, May 14-17, 1991, pp. 343-347.
3. D. Dubois, H. Prade, Possibility Theory: An Approach to Computerized Processing of Uncertainty, Plenum Press, New York, 1988.
4. K. Hirota, W. Pedrycz, Logic-based neural networks, IEEE Trans. on Neural Networks, submitted.
5. W.J. M. Kickert, Fuzzy Theory on Decision-Making, Martinus Nijhoff, Leiden, 1978.
6. W. Pedrycz, Neurocomputations in relational systems, IEEE Trans. on Pattern Analysis and Machine Intelligence, vol. 13, 1991, p. 289-296.
7. U. Thole, H.J. Zimmermann, P. Zysno, On the suitability of minimum and product operations for the intersection of fuzzy sets, Fuzzy Sets and Systems, 2, 1979, 167-180.
8. H.J. Zimmermann, Fuzzy Set Theory and Its Applications, Kluwer/Nijhoff, Dordrecht, 1985.

APPLICATION OF FUZZY CONTROL SYSTEM TO HOT STRIP MILL

Naoki Sato, Noriyuki Kamada, Shuji Naito, Takashi Fukushima, Makoto Fujino

Electrical & Instrumentation Engineering Department
Yawata Works, Nippon Steel Corporation
1-1 Tobihatacho, Tobata-ku, Kitakyushu City 804, JAPAN

Abstract: The application of conventional fuzzy control to the iron & steel making processes has brought significant results in various fields by building up fuzzy models for the processes which had previously been difficult. The currently developed system, unlike the conventional sytem, includes a hybrid control system. In the hybrid control system, strict mathematical expression models are applied to a process requiring very high accuracy to compensate errors using the fuzzy model. This has accomplished fruitful results.

Introduction

The requirements for product dimensional accuracy, particularly on a strip thickness in a hot strip mill in iron & steelmaking processes, have come to be stricter year after year. For improvement of thickness accuracy on the head of the strip, it is mandatory to increase the setup control accuracy by applying the rolling theoretical model. However, the rolling theoretical model includes indeterminate parameters, and therefore non-linearity is not negligible.
For these reasons, no more improvement of thickness accuracy seemed to be expected.
Then, the accuracy improvement of theoretical models was attempted by reviewing the model and the learning-only control.
As a result, the "hybrid fuzzy control system" has been developed. The theoretical model is set up after the error correction by applying the fuzzy rules to the theoretical model errors and to the actual rolling load when a sheet bar is grasped by the upstream stands of the finishing mill. Through this system the accuracy can be improved by 10%, when compared to the conventional control where the theoretical model alone is applied.

Conventional Gauge Control System and Its Problems in the Hot strip Mill

A strip thickness control system in the hot strip mill is shown in Fig. 1. The finishing mill consists of 6 stands and measures the temperature and thickness (gauge) of a sheet bar at the finishing stand entry and sets feed-forwardly an initial gap based on the rolling models in order to obtain the specified thickness. The sheet bar is then grasped by the first finishing stand. The set-up control range is up to this point. After the sheet bar has been grasped by the finishing stand, the feedback control operates to dynamically control the thickness using the actual load resulting from the load cells and the measured thickess by X-ray.

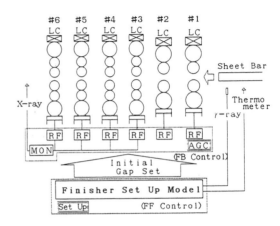

Fig. 1 Gauge Control System

The strip thickness, except for its head end, can be highly accurately controlled by the feedback control in this manner using sensors. On the head of a strip, the thickness accuracy is greatly dependent on the model accuracy.
A rolling mill, once the material to be rolled has been grasped, undergoes elastic deformation due to the load of material, which may cause a thicker strip thickness than a set-up gap. For this reason, the rolling load must be accurately measured for setting the proper gap in order to realize exacting thickness control. As the parameters of the model formula for prediction of rolling load and for estimation for rolling mill elongation include errors, the accuracy of the parameters has been tried to be improved by learning. However, in the Hot Strip Mill, Yawata Works of Nippon Steel, we have faced the difficulty that the improvement of accuracy for the mathematical expression model by learning only is limited to a certain extent. Because, the changing frequency of set points has increased along with the increase of operational flexibility.

Dynamic Setup System by Using Fuzzy Theory

To deal with the above-mentioned problem, we have developed the system as shown in Fig. 2. The system comprises the fuzzy controller and its learning model. The fuzzy controller predicts the thickness error at the finishing stand exit by application of fuzzy reasoning based on an error between the predicted load and the loading results on the intermediate stands (hereinafter called "load error"), immediately after the start of the rolling with an initial gap set by means of the conventional setup model. The necessary gap correction is then figured.
The learning model provides consecutive learning for parameters of the gap model corrected in accordance with the resulting thickness error, after the head of a strip has passed through the X-ray on the finishing stand exit. In application of fuzzy theory of the

objective thickness control at the finishing stage in hot rolling, not only high-speed response but also a powerful learning function is needed to secure high accuracy in the operational flexibility. In the currently developed system, by regarding the antecedent as trapezoidal membership function and the consequent as linear equation, the reasoning with high speed has been materialized, and thus excellent learning ability has been successfully obtained.

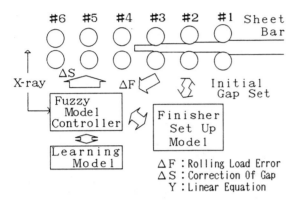

Fig. 2 Fuzzy Dynamic Setup System

Fuzzy Controller

The present fuzzy model has been built up taking the following matters into consideration.

1) Structural identification between the antecedent and the consequent (determination of variables and number of domains for fuzzy division)

2) Parameter identification of membership function

The membership function will be later discussed respectively for the antecedent and the consequent.
Discussion now is on the structural identification. The present system, as mentioned previously, features the prediction of a strip thickness on the finishing stand exit from the load error in the upstream stands. The system uses load error as variables for the antecedent and the consequent. A stand from which the load error is picked up, and the number of fuzzy divisions have been determined on the following basis.

1) As intense as possible correlation between the objective output variables (or, for this system, thickness error on the finishing stand exit).

2) Diversified correlation tendency for each divided domain by input variables

In other words, through an analysis of the correlation between the load error and the thickness error at each stand an identification error analysis on the estimated thickness error at the finishing stand exit, the load errors of the two intermediate stands can be adopted for optimum variables. In addition, the load errors of the two stands are divided into three (positive, zero, and negative) for optimum fuzzy divisions.
Letting the fuzzy variables of the load errors be ΔF_i and ΔF_{i+1}, the fuzzy model is composed of 9 rules in the combination of ΔF_i and ΔF_{i+1} as shown in Table 1.

Table 1. Fuzzy rule

R1 IF ΔF_i is PO and ΔF_{i+1} is PO then ΔS_i is Y_1
R2 IF ΔF_i is PO and ΔF_{i+1} is ZE then ΔS_i is Y_2
R3 IF ΔF_i is PO and ΔF_{i+1} is NE then ΔS_i is Y_3
R4 IF ΔF_i is ZE and ΔF_{i+1} is PO then ΔS_i is Y_4
R5 IF ΔF_i is ZE and ΔF_{i+1} is ZE then ΔS_i is Y_5
R6 IF ΔF_i is ZE and ΔF_{i+1} is NE then ΔS_i is Y_6
R7 IF ΔF_i is NE and ΔF_{i+1} is PO then ΔS_i is Y_7
R8 IF ΔF_i is NE and ΔF_{i+1} is ZE then ΔS_i is Y_8
R9 IF ΔF_i is NE and ΔF_{i+1} is NE then ΔS_i is Y_9

<Antecedent>

The antecedent membership functions are configured in the following statistical processing from a relationship between the load error and the thickness error. We developed a system in which a distribution range is made up based on the data of some 500 coils. The results are plotted in the coordinate as shown in Fig. 3, with a load error put to the abscissa and a thickness error to the ordinate. Then, the center line of distribution is drawn by applying the least square method. The intersects between the abscissa and the two straight lines σ (variable in a range from 0.3σ to 3.0σ) away from the center line, are regarded as intersects of trapezoidal membership function intersects.
According to the configuration of the membership function, with an increase of error distribution inclination or a decrease of variance σ, the ZE domain and fuzziness section become smaller. This makes the ambiguity smaller and brings a large gap correction rate in either a positive or negative direction. Adversely, with a small inclination of error distribution and a large variance σ, the ZE domain and fuzziness section becomes larger, increasing the ambiguity and issuing no instruction for gap correction. This results in suppression of excessive control and erroneous control. Further, the identifying configuration allows the later describing membership function to be learned quite easily.

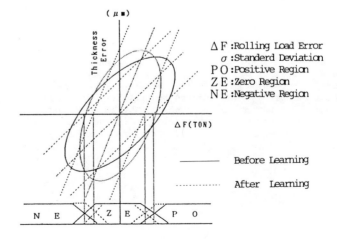

Fig. 3 Membership function

<Consequent>

The consequent membership function calculates the gap correction rate from the linear expression of Equation (1) which has been given by the least square method on the basis of the load errors of Stand i and Stand i+1.

$$y_i = a^j_0 + a^j_1 \Delta F_i + a^j_2 \Delta F_{i+1} \quad \ldots (1)$$

$(j = 1 \text{ to } 9)$

This has been accomplished from both sides on the high-speed response of the fuzzy controller and the simplified learning algorithm for each slab.
As stated above, approximating the membership fucntions of the antecedent and the consequent has realized a high-speed operation by software processing to enable all processings within 200ms to be completed. Thus, the dynamic control system in a high-speed line of 1000mpm has been established.

Correction Rule

The gap correction rules for fuzzy control are expressed by the fuzzy model as shown in Table 1. Nine fuzzy rules are used to configure nine domains: Y1 to Y9. With respect to the input spaces divided into the ambiguous domain portions (combination of load errors), the entire non-linear relationship is expressed by the linear relationship formed in each domain portion in accordance with the input variables (load error).
For example, as shown in Fig.4, if the gap correction domain is assumed to be Point A based on the difference between the load error of stand i and that of stand i+1, the four domains of Y1, Y2, Y4 and Y5 are related simultaneously at Point A. As a result, the certainty showing establishment grade for each domain (g_1, g_2, g_4 and g_5) is, as shown in Table 2, calculated from the establishment grades for each membership function; α_1, α_2, β_1 and β_2. From this certainty and the aforementioned linear pattern of the consequent, the final gap correction rate is given on the average load[1] of the correction rate for each domain by applying Equation (2).

Table 2. Correction Quantity

Section	Certainty	Linear Equation
Y_1	$g_1 = \alpha_1 \times \beta_1$	y_1
Y_2	$g_2 = \alpha_1 \times \beta_2$	y_2
Y_4	$g_4 = \alpha_2 \times \beta_1$	y_4
Y_5	$g_5 = \alpha_2 \times \beta_2$	y_5

$$\Delta S = \frac{\sum_{i=1}^{\ell} g_i \times y_i}{\sum_{i=1}^{\ell} g_i} \quad \ldots (2)$$

Learning Model

<Antecedent>

Figure 5 represents the learning of the membership function for the antecedent. A distribution range is calculated based on the data of some 500 coils in terms of the load and thickness errors acquired for every coil. The disribution range is updated at every exchange of the finishing mill rolls to change the membership function.
The solid lines represent the membership function before learning, and the dotted lines the membership function after learning. As shown in the figure, the variance is nearly equal, but the inclination of distribution is different between before and after learning. Because of the steep inclination in distribution for the function after learning ($a_2 < a_1$), the relationship of intersects between the membership function domains and the abscissa becomes $b_2 < b_1$.
This indicates that the ZE domain after learning is narrower than that before learning. Consequently, the control rule can be so modified as to meet an increased gap correction rate, while keeping the load error constant.
In other words, with a steeper inclination in distribution, the ambigrous portion of the antecedent will be divided again toward promoting the correcting action. Likewise, the rearrangement of the fuzzy divisions, depending on the situation, will keep the linear equation accurate for each domain of the consequent, as well as increase the learning effect of the consequent. This should result in the optimization of the gap correction rate.

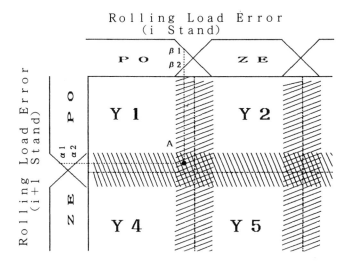

Fig. 4 Fuzzy Reasoning Image

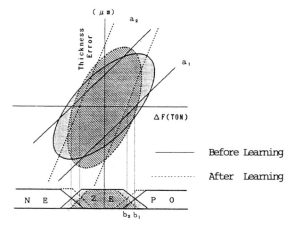

Fig. 5 Learning Model for Membership Function

<Consequent>

Because of approximation by the linear expression as mentioned previously for the consequent, the factor "a" in the linear equation is learned for every coil to prevent the degradation of controllability. The learning uses the weighted successive least squaring method, as shown in Equations (3) & (4), to follow the change of operational conditions. From the forgetting factor λ, the learning grade has been determined.

$$\hat{a}_{(t)} = \hat{a}_{(t-1)} + P_{(t)} X_{(t)} \{y_{(t)} - \hat{y}_{(t)}\} \quad \ldots (3)$$

$$P_{(t)} = \frac{1}{\lambda_{(t)}} P_{(t-1)} - \frac{1}{\lambda_{(t)}} \left\{ \frac{P_{(t-1)} X_{(t)} X^T_{(t)} P_{(t-1)}}{1 + X^T_{(t)} P_{(t-1)} X_{(t)}} \right\} \ldots (4)$$

In this method, making the λ small, to keep the extent of grain matrix P for Equation (4), will increase a rate of learning correction for the parameters given by Equation (3).

We have extended the above-mentioned idea and made the λ variable, rather than fixed, in accordance with the ambiguity grade of the fuzzy model[2], achieving adequate learning on the ambiguity portion of the nearby boundaries between domains. For example, if the four rules hold for Point A as mentioned previously, the learning grade is changed in accordance with the certainty for each domain, assuming the forgetting factor λ as functions of certainty for each domain: g_1 through g_5. Consequently, when the certainty factor in the current-state recognition is larger, the λ is decreased to place an emphasis on the current value. On the contrary, when the ambiguity is larger, λ is increased to emphasize the past information. The learning proceeds automatically in this way.

Results of Development

Figure 6 represents the effects of the application of the dynamic setup system using currently-developed fuzzy logic. The graph ordinates indicate the thickness error on the finishing stand exit, and the abscissas indicate the load error on the same. The dotted lines show the control range of thickness error. The results of steel grade A show that the distribution of thickness error to load error has been flattened in comparison with that obtained before control, resulting in small-error dispersion. The thickness accuracy shows an improvement of approx. 8% in hit ratio (on-gauge ratio) inside the control range. On the other hand, the results of steel grade B show a large change of distribution inclination, particularly in the enclosed and large error-indicating portions in Figure 8, indicating dramatic improvement of some 12% in on-gauge ratio.

Figure 7 shows the transition of the improvement grade in on-gauge ratio. For steel grade A, an average of approx. 10% of the on-gauge ratio improvement has been achieved since the beginning of the application of this system, and significant results exceeding the target level have brought in since 100% of the system has been put into service.

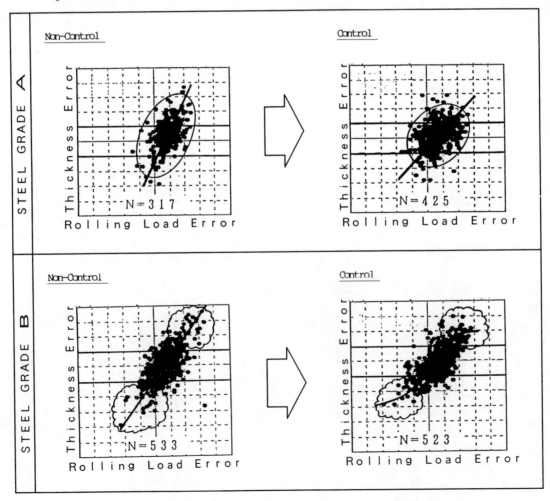

Fig. 6 Application Results of the Dynamic Setup System

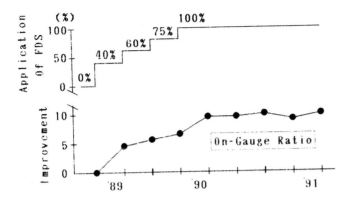

Fig. 7 Transition of On-gauge Ratio

Conclusion

In the currently developed system, while placing the base onto the rolling theoretical model, the model has been built up in the use of the fuzzy model for the error mergence comprising the theoretical model balance and disturbance. This leads to a so-called hybrid system in which new strict model is combined with the fuzzy model. As a result, great accuracy improvement has been achieved.
We believe that this approach is applicable for a control method that can cope with increasingly diversified production styles.

References

[1] Sugano: "Fuzzy control" Nikkan Kogyo Shinbunsha, 1988

[2] K. Ueyama, K. Ezaki, S. Hirayama, T. Niidome: "Application of Fuzzy Model to Preset of Tandem Mill", IFSA, 1986

A Fuzzy Classification Technique for Predictive Assessment of Chip Breakability for Use in Intelligent Machining Systems

J. Fei and I.S. Jawahir
Center for Robotics and Manufacturing Systems
University of Kentucky
Lexington, KY 40506

Abstract—This paper presents a new methodology developed for predicting the chip breakability levels achievable in machining operations. This method is based on a fuzzy classification technique involving neighboring cluster groups. The newly developed chip control diagrams with classified regions are suitable for use in computer-aided process planning systems facilitating intelligent machining.

1. INTRODUCTION

Developing efficient and effective means of information processing for intelligent manufacturing has now become an important focal point in research. In particular, an intelligent machining system requires reliable information relating to the operating parameters of the actual machining process. Many innovative techniques and methodologies have been developed and successfully implemented for on-line monitoring of the various machining performance parameters[1], [2]. With the growing need for innovative designing and planning of the process of machining, a greater emphasis is being placed on developing intelligent process planning systems for use in off-line modes. The success of such a computer-aided intelligent machining process planning system very heavily depends on the use of comprehensively predictive cutting models along with extensive knowledge databases, and on the implementation of new techniques for knowledge representation and inferencing to enable the most effective and highly intelligent decisions. Intelligent process planning systems require realistic cutting process models which could provide predictions of machining performance for a given set of cutting conditions, tool geometries and work/tool material combinations. Predictive models needed for such a system also would be required to provide optimum solutions, based on the best choice of cutting parameters such as feeds and depths of cut. Most recent works on knowledge-based predictive modeling and optimum machining performance assessment indicate the feasibility of introducing such capabilities in the computer-aided process planning systems through appropriate databases and rules [3], [4], [5].

The traditionally known computer-aided planning systems suffer from the lack of adequate and realistic information about the actual process of machining due to the vary complex nature of the machining process. Among other machinability factors, the need for including a predictive assessment about the levels of achievable "chip breakability" in the computer-aided process planning systems has long been felt among the manufacturing community although no significant attempt has been made in this regard[6], [7]. The cutting tool manufacturers have given only partial solutions to this problem by providing "chip control diagrams" (also known as the "chip charts") for selected work material/tool insert combinations and cutting conditions. Also, the users of cutting tools for various machining operations are at a disadvantage due to the lack of knowledge on the application range, levels of chip breakability achievable, etc., for the most common combinations of work material/tool insert. Indeed, with the alarmingly increasing rate of cutting tool design features being introduced in the market at present, it seems almost impossible to develop any guidelines for the effective use of these tool inserts for achieving "optimum performance levels" [8]. The CIRP (International Institution for Production Engineering Research), which sponsored an authoritative survey on chip control systems in 1979 [9], formulated a working group on chip control in 1990, and this working group consisting of over 30 internationally known researchers has met 5 times since then and had over 25 formal presentations on the subject of chip control in advanced machining systems [10].

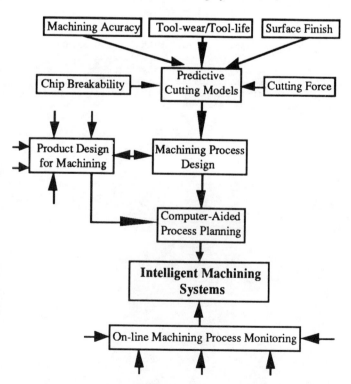

Fig. 1 The Role of Predictive Assessment of Chip Breakability in Intelligent Machining Systems

Unfortunately, the well known deterministic theories of machining do not address the problem of chip breakability and hence none of the existing predictive models are suitable for use in computer-aided process planning systems for assessing the chip breakability. The inherent "fuzziness" in the term "chip breakability" has called for the need for quantifying chip breakability and this now has lead to the development of a fuzzy rating system for chip breakability [11], [12]. This work has now been extended in the work presented in this paper to include a systematic grouping of the various levels of chip breakability using a fuzzy classification technique, which could be integrated into the computer-aided process planning system for providing a predictive assessment of chip breakability at the planning stage, prior to the actual machining. Fig. 1 illustrates the integral role of this methodology in the process of intelligent machining.

2. EXPERIMENTAL WORK

Machining experiments were conducted to set up the database for chip breakability and to establish the fuzzy classification technique.

2.1 Experimental Set-up

Lathe: Model: Leblond with spindle RPM 10,000 (max.) and a feed range of 0.056-3.87 mm/rev. and Motor HP: 150

2.2 Cutting Conditions

Cutting Speed: 230 m/min
Feed: 0.056 - 0.312 mm/rev
Depth of Cut: 0.25 - 2.54 mm

The range of feeds and depths of cut were extended to include light to medium cutting conditions, mainly to see the continuing trends of the process variables that were measurable. However, the analysis of the experimental work was restricted only to the finish turning low feed - low depth of cut range. No cutting fluids were used during the cutting tests.

2.3 Tool Inserts

TNMG type cermet tool inserts with four different chip groove geometries (labeled as FCB1, FCB2, etc.) manufactured by four major cutting tool manufactures were used in the experimental work. A total of eight types (four sets, each with two nose radii (subscripts 1 and 2 representing nose radius r_ϵ = 0.4 mm and 0.8 mm respectively) were used.

2.4 Work Material

Medium Carbon Steel (AISI 1045) was used as work material. The hardness values of this work material was BHN = 187.

3. FUZZY CLASSIFICATION OF CHIP BREAKABILITY IN THE CUTTING STATE SPACE

3.1 Fuzzification for Chip Breakability

The term "chip breakability" is defined as a fuzzy variable due to the inherent "fuzziness" in the understanding of the acceptable levels of chip forms and shapes. In the present work a new methodology for describing chip breakability has been derived based on the fundamentals of fuzzy reasoning. First, it is assumed that the following three factors are used to determine the levels of chip breakability with the corresponding contribution weights:

(a) size of the chip   (0.4);
(b) shape of the chip   (0.3); and
(c) chip producibility   (0.3).

The dimensional features, such as length and other geometric parameters (e.g. diameter or curl radius), of the chip or the chip coil are described by (a), above, while the geometric configuration of the chip such as helical form, spiral form, comma shape, etc. are addressed by (b) and the difficulty or easiness to produce the chip (concerning the material property) is covered by (c).

It is quite obvious that all three features given above are fuzzy in nature due to the "vagueness" involved in the definitions. Therefore, in this paper, the term chip breakability is defined as a fuzzy variable which takes the following five possible linguistic values: excellent (E), good (G), fair (F), poor (P), and very poor (VP). The chips produced in our experiments are assigned appropriate linguistic values based on the evaluation made using the above contributing factors. Tables 1 and 2 show the linguistic values assigned to the corresponding chip control diagrams shown in Figs. 2 and 3.

Representative chip breakability levels for 20 different combinations of feeds and depths of cut in the cutting condition state space were estimated for each of the eight chip control diagrams. In order to assess chip breakability for <u>any combination</u> of feed and depth of cut in the cutting condition state space, an interpolation and classification algorithm was developed.

3.2 Linear Interpolation

In machining with tool inserts having complex chip groove geometries, the chip breakability is greatly influenced by small variations in cutting conditions. It is however reasonable to assume that the chip breakability changes gradually and consistently as cutting conditions change. If two neighboring representative states associated with two non-neighboring linguistic values of chip breakability, a new representative state can be interpolated at the midpoint between them and this state could be associated with a linguistic chip breakability value between the two non-neighboring ones. If the two non-neighboring linguistic values are separated by two other values,

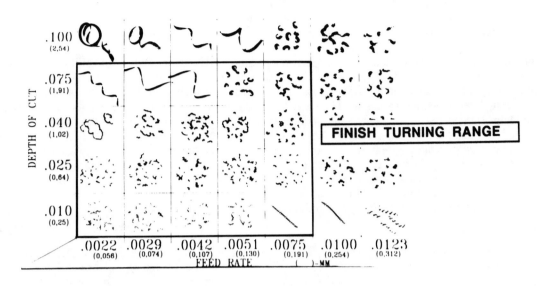

Work Material: AISI 1045   Cutting Speed: 230 m/min   Nose Radius: 0.4 mm

Fig. 2 Chip Control Diagram for Tool Insert FCB2$_1$

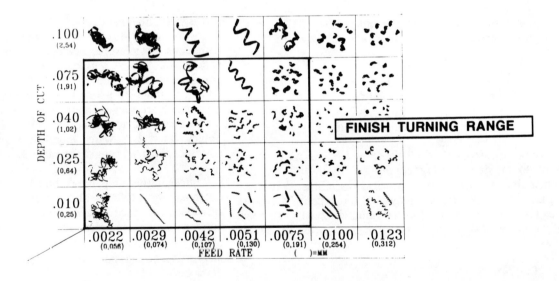

Work Material: AISI 1045   Cutting Speed: 230 m/min   Nose Radius: 0.8 mm

Fig. 3 Chip Control Diagram for Tool Insert FCB3$_2$

two representative states can be linearly interpolated between the two original sample states, and they could be associated with the two linguistic values separating the two original linguistic values. If two different linguistic values are assigned to the same interpolated point, these two values will be assigned to two different points nearby. These two points are selected in such a way that they are closer to the original sample states associated with the same linguistic values.

3.3 The Fuzzy Classification Methodology

After the above interpolation, the original representative states and the newly interpolated representative states are used as cluster centers associated with corresponding linguistic chip breakability values. Any point in the cutting condition state space (combination of feed and depth of cut) can be classified into one cluster or two neighboring clusters representing

TABLE 1
Linguistic Values of Chip Breakability for Tool Insert FCB2₁
(Nose Radius: 0.4 mm)

d \ s	Feed (mm/rev)				
Depth of Cut (mm)	0.056	0.074	0.107	0.130	0.191
1.91	P	P	P	F	F
1.02	P	F	F	F	E
0.64	E	E	E	E	E
0.25	E	E	E	E	F

Work Material: AISI 1045   Cutting Speed: 230 m/min

TABLE 2
Linguistic Values of Chip Breakability for Tool Insert FCB3₂
(Nose Radius: 0.8 mm)

d \ s	Feed (mm/rev)				
Depth of Cut (mm)	0.056	0.074	0.107	0.130	0.191
1.91	VP	VP	VP	P	F
1.02	VP	VP	F	G	G
0.64	VP	VP	G	G	G
0.25	VP	P	F	F	F

Work Material: AISI 1045   Cutting Speed: 230 m/min

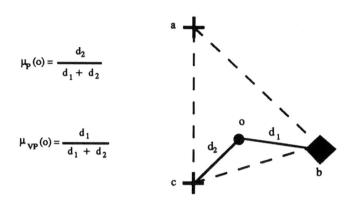

$$\mu_P(o) = \frac{d_2}{d_1 + d_2}$$

$$\mu_{VP}(o) = \frac{d_1}{d_1 + d_2}$$

Fig. 4 The Fuzzy Classification Method

In this figure d_1 is the minimum distance from o to P cluster center and d_2 is the minimum distance to a VP cluster center. Fig. 5 shows the membership grade of "very poor" chip breakability for tool insert FCB3₂ corresponding to the points in the cutting condition state space as a result of this fuzzy classification method. In this figure, the membership grade of a point represents the certainty level of very poor chip breakability when using a given combination of feed and depth of cut. It can be easily seen that there is a very close correlation between Fig. 5 and the chip control diagram shown in Fig. 3. When using small feeds at various depths of cut, it is most likely that the chips will look like the ones in the first two columns in Fig. 3. The irregular (at various certainty levels) transformation of "very poor" chip breakability to "poor" level at lighter feeds is seen quite clearly in Fig. 5. Similar figures can be drawn for other levels of chip breakability as well.

different chip breakability levels with corresponding membership grades using the minimum-distance fuzzy pattern classification method proposed in this work. The fuzzy classification algorithm can be described as follows:

1. If any point in the cutting condition state space is surrounded by some cluster centers associated with the same linguistic values, this point is assigned with that linguistic value and a membership grade 1.0.

2. If any point in the cutting condition state space is surrounded by some cluster centers with two different linguistic values, only these two linguistic values are assigned to this point. Its membership grades are determined by the minimum Euclidean distance to each cluster.

Suppose a point 'o' in the cutting condition state space is surrounded by cluster centers a, b, c which belong to two different clusters (see Fig. 4). a and c are associated with the linguistic value VP, and b is associated with the linguistic value P. The point 'o' belongs to these two clusters at the same time. The membership grades are then determined as shown in Fig. 4.

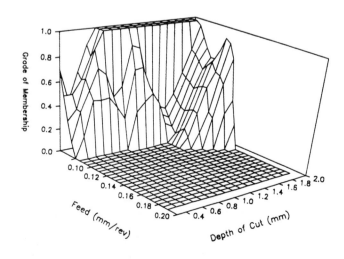

Fig. 5 The Membership Grade Distribution for "Very Poor" Chip Breakability with Varying Cutting Conditions for FCB3₂

By using this method, any point in the cutting condition state space can be classified into no more than two neighboring clusters with corresponding membership grades.

Figs. 6 and 7 show two α-level (α = 0.5) classifications for two different tool inserts. In these figures, the entire cutting condition state space is divided into a number of sub-regions

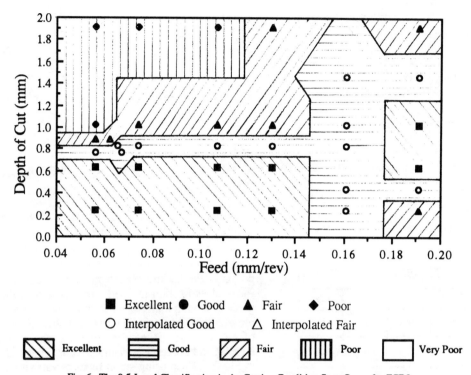

Fig. 6 The 0.5-Level Classification in the Cutting Condition State Space for FCB2$_1$

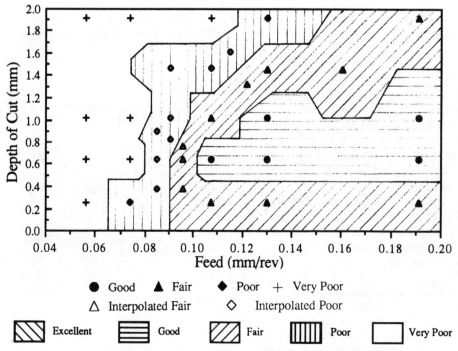

Fig. 7 The 0.5-Level Classification in the Cutting Condition State Space for FCB3$_2$

associated with different linguistic chip breakability values. Each point in a particular sub-region has a 0.5 or higher membership grade to be classified as a member of that sub-region. Every point on the boundaries has equal distance to the closest cluster centers on both sides. In other words, every point on the boundary has equal certainty level (0.5 membership grade) to be classified into chip breakability levels on both sides of the boundary.

This algorithm can be used for assessing chip breakability within the finish turning operation range. The basic advantage of using the method presented in this paper is that, based on a very small number of representative experimental work, it is possible to predict chip breakability for any combination of feed and depth of cut values. This prediction then can be used in process planning. The chip breakability assessment combined with other machining performance assessments, such as surface finish, cutting force, can also be used in a finish turning optimization system (under investigation by authors) which can help machine operators select the best cutting tools and the best cutting conditions to improve quality and productivity.

4. Conclusions

1. The traditionally known exhaustive method of producing chip control diagrams does not fully reveal the achievable levels of chip breakability as they are based on the chip forms/shapes produced at discrete levels of cutting conditions. The fuzzy classification method developed and presented in this paper shows a continuous approach which clearly identifies regions (islands) of chip breaking and their variation levels.

2. The fuzzy classification method requires very few carefully selected experimental cutting conditions (namely, feed and depth of cut) and thus, eliminates the need for time consuming experimental work.

3. The partitioned cutting condition state space and the corresponding classification patterns can easily be stored and implemented in a computer-aided process planning system to provide the predictive knowledge on chip breakability. This in turn could facilitate the planning, monitoring and control processes in intelligent machining systems.

References

[1] H. K. Tonshoff, J. P. Wulfsberg, H. J. J. Kals, W. Konig and C. A. van Luttervelt, "Developments and Trends in Monitoring and Control of Machining Processes", Annals of the CIRP, Vol. 37 (2), 1988.
[2] M. Shiraishi, "Scope of In-process Measurement, Monitoring and Control Techniques in Machining Process: Part1: In-Process Techniques for Tools", Precision Engineering, Vol. 4 (10), 1988, pp. 179-189.
[3] J. Fei and I. S. Jawahir, "A Fuzzy Knowledge-Based System for Predicting Surface Roughness in Finish Turning", Proc. FUZZ-IEEE, Mar. 8-12, 1992, San Diego, pp. 899-906.
[4] J. Fei, X. D. Fang and I. S. Jawahir, "Towards A Fuzzy Set-Based Optimization of Finish Turning Operations", Proc. IFSIC Conference, Mar. 16-18, 1992, Louisville, pp.28-36.
[5] M. Wang, J. Y. Zhu and Y. Z. Zhang, "Fuzzy Pattern Recognition of the Metal Cutting States", Annuls of the CIRP, Vol. 34 (1), 1985, pp. 133-136.
[6] I. S. Jawahir, "A Survey and Future Predictions for the Use of Chip Breaking in Unmanned Systems", Int. J. Adv. Mfg. Tech., 3 (4), 1988, pp. 87-104.
[7] I. S. Jawahir and P. L. B. Oxley, "New Developments in Chip Control Research; Moving Towards Chip Breakability Predictions for Unmanned Manufacture, Proc. Int. Conf. ASME (MI '88, Atlanta, USA), April 1988, Vol. 1, pp. 311-320.
[8] I. S. Jawahir, "The Tool Restricted Contact Effect as a Major Influencing Factor in Chip Breaking: An Experimental Analysis, Annals of the CIRP, 37 (1), 1988, pp. 121-126.
[9] W.Kluft, W.Konig, C.A.Van Luttervelt, K.Nakayama, A. J. Pekelharing, "Present Knowledge of Chip Control, Keynote Paper, Annals of the CIRP, 28 (2), 1979, pp. 441- 455.
[10] Selected Presentations, CIRP STC-C Working Group on Chip Control (Aug.1990, Berlin, Jan.1991, Paris, Aug. 1991, Stanford, Jan.,1992, Paris, Aug.1992, Aix-en-Provence).
[11] X. D. Fang and I. S. Jawahir, "An Expert System Based on a Fuzzy Mathematical Model for Chip Breakability Assessments in Automated Machining", Proc. 2nd Int. ASME Conf., (MI'90 Atlanta, Georgia, USA, March 1990), Vol. IV, pp. 31-37.
[12] X. D. Fang and I. S. Jawahir, "On Predicting Chip Breakability in Machining of Steels with Carbide Tool Inserts Having Complex Chip Groove Geometries", J. of Materials Processing Technology, Elsevier Publishers, Vol. 28, 1991, pp. 37-41.

A RULE-BASED FUZZY LOGIC CONTROLLER FOR A PWM INVERTER IN PHOTO-VOLTAIC ENERGY CONVERSION SCHEME.

ROHIN. M. HILLOOWALA
Student Member, IEEE

ADEL. M. SHARAF
Senior Member, IEEE

ELECTRICAL ENGINEERING DEPARTMENT
UNIVERSITY OF NEW BRUNSWICK
FREDERICTON NEW BRUNSWICK
CANADA E3B5A3

ABSTRACT:

The paper presents a Rule based controller based on fuzzy set theory to control the output power of a PWM inverter in Photo-Voltaic Energy conversion interface scheme. The objective is to track and extract maximum available solar power from the PV array under varying solar irradiation levels. To achieve this the power error $e = (P_{ref} - P_{pv})$ and the rate of change of this error $\dot{e}$ are used as the input signals to the rule-based fuzzy controller and the output signal is used to control the PWM inverter. The input signals are defined by a set of linguistic variables or labels characterised by their membership functions which are preassigned for each class. A fuzzy relation matrix relates the input signals (e, $\dot{e}$) to the fuzzy controller output U and using Fuzzy set theory and associated fuzzy logic operations, the desired fuzzy controller output is obtained. The fuzzy output (in terms of linguistic variable or label) is defuzzified to obtain the actual numerical (analog) output signal of the controller. This output (analog) signal is then fed to the PWM inverter to control only the output voltage (output frequency being fixed at 60 Hz.) and hence power drawn from the PV array. The proposed rule based controller is simulated and the results are experimentally verified on a PV energy conversion scheme consisting of an emulated PV array and a Pulse width modulated Inverter and is found to give a good power tracking performance.

I INTRODUCTION:

Solar energy conversion/interface schemes using a photo-voltaic solar array and line commutated/PWM inverter have been modelled, analyzed and implemented [1,2]. For a given solar insolation level and ambient temperature the voltage versus current and output power versus current characteristics are as shown in Fig:1. It is seen that there is a particular operating point (I_{opt}, V_{opt}) at which maximum output power P_{opt} is obtained. Most of these schemes use a PID controller to track and extract maximum power under varying solar insolation levels. The conventional PID controller requires quite a bit of tuning to obtain a fast and

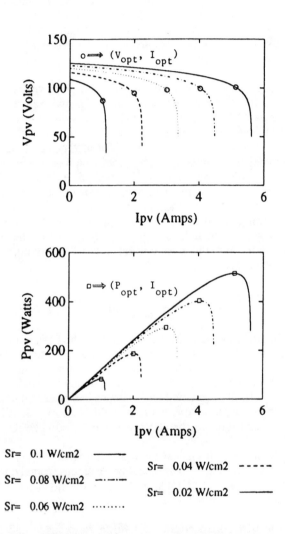

Figure:1 Characteristics of PV array at T=28°C and different solar insolation levels.

dynamically acceptable response and are usually implemented using analog circuits which have the tendency to drift with age and temperature. This causes degradation of the system performance. In this paper a new type of controller using Fuzzy set theory is proposed. The rule based fuzzy controller to track and extract maximum power P_{opt} from the solar array under varying conditions of solar irradiation, uses two real time measurements namely error $e=(P_{opt} - P_{pv})$, and the rate of change of error $\dot{e}$ as the control input signals. These input signals (e, $\dot{e}$) are first expressed in terms of linguistic variables or labels such as LP (large positive), MP

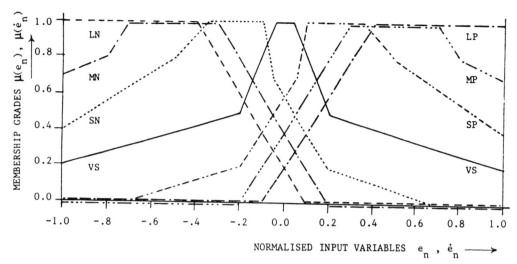

Figure:2 Reference fuzzy sets to represent input variables in lingustic labels characterised by membership grades.

(medium positive), SP (small positive), VS (very small), LN (large negative), MN (medium negative), SN (small negative) using fuzzy reference sets. The fuzzy reference sets are defined for each variable to cover the entire range of interest, with considerable amount of overlap between reference sets, as shown in Fig:2. This considerably simplifies the problem of fuzzy set definition. The input signals are nonfuzzy (crisp) values which must be fuzzified to be used as input signals to the fuzzy controller. The result of fuzzification will be a set of grades of membership of each of the linguistic label involved, as shown in Fig:2. Since the reference sets are overlapping, each nonfuzzy value of the variable will belong to at least two reference sets and the grade of membership to the other sets (labels) will be zero. Next, the relationship between the input signals expressed as linguistic labels and the fuzzy controller output is developed using fuzzy set theory and described as a fuzzy relation matrix using assigned membership functions. Finally using fuzzy logic operations, the fuzzy controller output is found, defuzzified (changed from linguistic label to numerical values) and used to control the pulse width of the PWM inverter. The proposed scheme with fuzzy rule-based controller is simulated and experimentally verified and found to give good power tracking performance.

II SYSTEM CONFIGURATION:

The proposed scheme consists of an emulated PV solar array, a DC link and a Pulse Width Modulated Inverter feeding some local load as shown in Fig:3. A brief description of each system subsection is as follows.

A. PHOTO VOLTAIC ARRAY:

The solar array is emulated using the characteristic equation relating the solar cell's voltage and current as is shown below;

$$V_s = (AKT/q)*\ln[(I_{ph} - \beta I_s - I_0)/I_0] - I_s R_s$$
$$I_{pv} = I_s * N_p \qquad (1)$$
$$V_{pv} = V_s * N_s$$

where, V_s is the cell voltage, I_s is the cell current, q is the charge of an electron, K is Boltzman's constant, A is completion factor and its value is in the range 1 to 3 [1], T is the absolute temp, I_{ph} is the photo current, I_0 is the reverse saturation current, R_s is the equivalent series resistance of the cell, I_{pv} is the array current, V_{pv} is the array voltage, N_p is the number of strings in parallel, N_s is the number of cells in series in a string. The photo current I_{ph} is a function of the solar insolation level S_r and its variation with S_r is given by

$$I_{ph} = K * S_r \qquad (2)$$

where K is a constant of value 0.56 Amps/W/cm^2 and S_r is the solar insolation in W/cm^2.

The characteristic equation relating the PV cell's voltage and current and associated parameters such as N_p, N_s, etc. are programmed into a digital computer and used to control a power amplifier whose output characteristics are made to match those of the PV array. Using a data translation board DT2821, the solar insolation level S_r and the array current I_{pv} are read into a digital computer and used to compute I_{ph} and I_s. Knowing the cell current I_s and using the above characteristic equation, the cell's output voltage V_s and hence the array's output voltage V_{pv} are computed and a voltage reference signal is fed to the power amplifier through D/A converter of the data translation board DT2821. By appropriately setting the gain of the power amplifier, the solar array's voltage-current characteristic can be obtained at it's output. This PV array simulator is used to experimentally verify the fuzzy rule-based controller in the lab.

B. INPUT FILTER AND DC LINK:

The input filter consists of a series reactor and a shunt capacitor as shown in Fig:3. The series reactor reduces the current ripple content in the array current and the shunt capacitor reduces the ripple content in the DC link voltage and provides a relatively stiff voltage source for the PWM inverter. The DC link current is governed by the following differential equation

$$p I_{pv} = 1/L_{DC} [V_{pv} - V_I - R_{DC} I_{pv}] \qquad (3)$$

where R_{DC} and L_{DC} are the DC link reactor's resistance and inductance respectively, $I_{pv}=I_{DC}$ is the DC link current, V_{pv} is the PV array's voltage and V_I is the DC voltage at the inverter input. The output power of the PV array is given by

$$P_{pv} = V_{pv} I_{pv} \qquad (4)$$

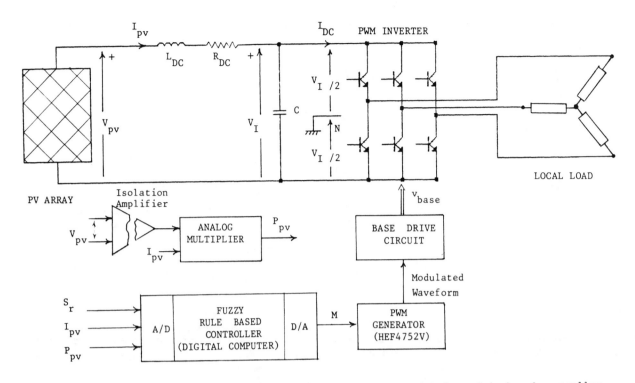

Figure:3 Block schematic of proposed solar energy conversion scheme with fuzzy Rule-based controller.

C. PWM INVERTER:

The DC power available at the output of the PV array is converted to AC power using a Pulse width modulated (PWM) inverter employing double edged modulation. The PWM signal used to switch the transistors in the inverter is generated using a purpose designed LSI circuit type HEF4752V. The IC provides three complementary pairs of output drive waveforms which when applied to a three phase six-element bridge inverter, produces a symmetrical three phase output. The output waveforms consists of sinusoidally modulated train of carrier pulses, both edges of which are modulated such that the average voltage difference between any two of the output three phases varies sinusoidally. This is illustrated in Fig:4 (courtesy Signetics application manual for HEF4752V [8]) for a carrier wave having 15 pulses for each cycle of the inverter output. Fig:4a shows the 15-fold carrier, Fig:4b the double edged modulated R phase, and Fig:4c and 4d show the double edged modulated Y and B phases respectively. The line to line voltage waveform obtained by subtracting Y-phase from R-phase is shown in Fig:4e. Each edge of the carrier wave is modulated by a variable angle δ_x as shown in Fig:5, and can be mathematically represented by

$$\delta_x = M \cdot \sin(\alpha_x) \cdot \delta_{max} \quad (x = 1, 2, \ldots 2r+1) \quad (5)$$

where, M is the modulation index, subscript x denotes the edge being considered, r is the ratio of carrier wave frequency to fundamental frequency at the inverter output, α_x is the angular displacement of the unmodulated edge and δ_{max} is the maximum displacement of the edge for the chosen frequency ratio r.

In the chosen PV energy conversion scheme, the inverter output frequency is held constant at 60 Hz. In this range of inverter output frequency, the PWM generator HEF4572V generates a carrier wave with frequency 15 times that of the fundamental frequency at inverter output. Such a choice, results in 15 pulses per half cycle in the line to line voltage waveform at inverter output. By modulating the carrier wave and hence the phase voltages, the fundamental and harmonic voltage content can be varied. There are 15 pulses and 15 slots of 12° each as shown in Fig:5. In each slot two edges are modulated. For 100 % modulation (i.e. M=1) the maximum amount by which the edge can be modulated is 6°. Any further displacement of the edge will cause the pulse in the modulated phase voltage waveform to merge, resulting in a reduction of the number of pulses in the line to line voltage waveform.

Fourier analysis of the modulated phase voltage waveform shown in Fig:5b shows that the amplitude (peak value) of the nth voltage harmonic component is given by

$$V_{n\,phase} = \frac{V_I}{n\pi} \left[\sum_{k=0}^{r} \Big(\cos(n\theta_1 - nM\delta_{max}\sin(\theta_1)) \right.$$
$$- \cos(n\theta_2 + nM\delta_{max}\sin(\theta_2)) \quad (6)$$
$$+ \cos(n\theta_3 - nM\delta_{max}\sin(\theta_3))$$
$$\left. - \cos(n\theta_2 + nM\delta_{max}\sin(\theta_3)) \Big) \right]$$

where, V_I is the input DC voltage, n is the harmonic number, r is the carrier to fundamental frequency ratio (r=15 in this case), δ_{max} is the maximum displacement of the edge ($\delta_{max}=6°$ in this case) and Θ_1, Θ_2 and Θ_3 are defined as follows

$$\Theta_1 = (2k-2)\cdot\pi/15; \quad \Theta_3 = (2k)\cdot\pi/15 \quad (7)$$
$$\Theta_2 = (2k-1)\cdot\pi/15;$$

and the peak value of the nth harmonic component of the line to line voltage waveform is

$$V_{n\,line} = \sqrt{3} \cdot V_{n\,phase} \quad (8)$$

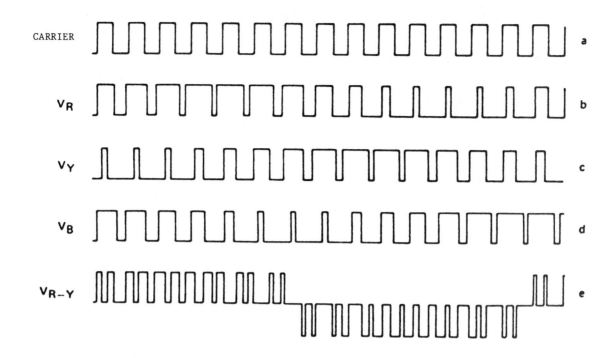

Figure:4 15 - pulse double edge sinusoidal pulse width modulated waveforms.

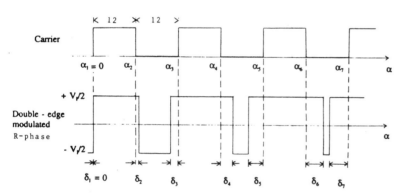

Figure:5 Carrier waveform and double edge modulated phase voltage waveform.

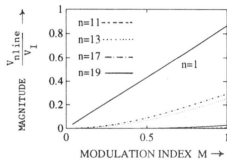

Figure:6 Variation of nth harmonic amplitude (peak) with modulation index M.

The variation of the nth harmonic component of the line to line voltage waveform (expressed as a per unit of the input DC voltage V_I) with modulation index M is shown in Fig:6. It is to be noted that triplen harmonics are present in the modulated phase voltage waveform. However, since the triplen harmonics in all the three phases have zero phase displacement, they will cancel out and not appear in the line to line voltage waveform.

Assuming the inverter to be lossless, and equating the input DC power to the output AC power, the following relation is obtained

$$V_I = \frac{2I_{DC}\sqrt{(R_L^2 + w_e^2 L_L^2)}}{f^2(1, M)\cos(\phi)} \quad (9)$$

where w_e is the electrical frequency at the inverter output, R_L and L_L are the resistance and inductance of the per phase load, ϕ is the power factor angle between fundamental components of voltage and current, $f(1, M)$ is a non-linear function of the modulation index M, relating the peak fundamental line to line voltage to the DC input voltage V_I of the inverter.

III FUZZY RULE BASED CONTROLLER:

The control objective is to track and extract maximum power from the PV array for a given solar insolation level. The maximum power corresponds to the optimum operating point (P_{opt}, I_{opt}) which is determined for different solar insolation levels S_r using off-line simulations. The data obtained is used to relate P_{ref} ($P_{ref}=P_{opt}$) to S_r using second order polynomial curve fit as shown below

$$P_{ref} = -11.575 + 4785.7*S_r + 4706.8*S_r^2 \quad (10)$$

Using a digital computer with data translation card DT2821, the solar insolation level and the power output of the PWM inverter are sampled at regular intervals ($\Delta T=300\mu s$). The reference power is computed using the above equation and compared to the actual measured power output of the PV array. The error in power and its rate of change are used to adjust the modulation index of the PWM inverter. This changes the output voltage of the PWM inverter and hence the power drawn from the PV array.

As shown in Fig:3, the input signals to the fuzzy rule-based controller are solar insolation level S_r and the power output of the PV array P_{pv}, and these are used to compute the error in power output e and its rate of change $\dot{e}$. The fuzzy controller's output is change in modulation index $U=\Delta M$ and is determined as follows:

Step :1 Calculate the normalised power error at the kth instant

$$e_n(k) = [P_{ref}(k) - P(k)]/P_{ref}(k) \tag{11}$$

Step :2 Calculate the normalised rate of change of error

$$\dot{e}_n(k) = [e_n(k) - e_n(k-1)]/(\Delta T * K_s) \tag{12}$$

where ΔT is the sampling interval selected as 0.1 ms and K_s is the scaling factor chosen such that $\Delta T * K_s = 1$ to allow normalisation in the range -1 to 1.

Step :3 Use assigned membership functions shown in Fig:2 to represent the normalised error e_n and rate of change of error $\dot{e}_n$ in fuzzy set notations using linguistic labels (LP, MP, SP, VS, SN, MN, LN).

Step :4 Use the generalized decision table as proposed by MacVicar-Whelan [3] and shown in Table I, to determine the fuzzy controller output for a given error and its rate of change.

Table:I MacVicar Whelam's Decision table [3]

		RATE OF CHANGE OF ERROR $\dot{e}_n$						
		LN	MN	SN	VS	SP	MP	LP
ERROR e_n	LP	VS	SP	MP	LP	LP	LP	LP
	MP	SN	VS	SP	MP	MP	LP	LP
	SP	MN	SN	VS	SP	SP	MP	LP
	VS	MN	SN	VS	SP	SP	MP	LP
	SN	LN	MN	SN	SN	VS	SP	MP
	MN	LN	LN	MN	MN	SN	VS	SP
	LN	LN	LN	LN	LN	MN	SN	VS

Step :5 Using the decision matrix the fuzzy controller output in linguistic variable such as LP, MP, SP, VS, SN, MN, LN is decided. It is seen that there are (7*7) = 49 combinations of error e_n and its rate of change $\dot{e}_n$. Each combination corresponds to a particular rule. Hence there are 49 rules on the basis of which the fuzzy controller's output is decided. A typical rule would be

Rule 7   if e is LP and $\dot{e}$ is LP
         then the controller output U should be LP

Step :6 Using Fuzzy set theory [4,5], the decision matrix is converted to the fuzzy relation matrix R shown in Table:II, which gives the relationship between the fuzzy set characterising controller inputs and fuzzy set characterising controller output (U). The controller output obtained by applying a particular rule is expressed in linguistic labels characterised by membership grades. For example, Rule 7 is now expressed as

Rule 7':   if e_n is LP and $\dot{e}_n$ is LP
           then the controller output U is described by the fuzzy set
           {(LN, 0), (MN, 0), (SN, 0), (VS, 0), (SP, 0), (MP, 0.5), (LP, 1.0)}

Step :7 The membership grade of the condition part is determined using fuzzy set theory. The condition part consists of two predicates 'e_n is LP' and '$\dot{e}_n$ is LP' combined by an 'AND' operator. Using law of intersection of two fuzzy sets the grade of membership of the condition part is determined.

$$\mu(x_7) = \mu('e_n \text{ is LP' AND } '\dot{e}_n \text{ is LP'}) \tag{13}$$
$$= \min (\mu('e_n \text{ is LP'}) \; \mu('\dot{e}_n \text{ is LP'}))$$

Step :8 Knowing the membership grade for the condition part and the fuzzy relation matrix, the membership grade for the controller output characterised by the linguistic labels LP, MP, SP, VS, SN, MN, LN can be obtained using the intersection rule of Fuzzy set theory. The membership grade for the linguistic label LP is computed as follows

$$\mu_{U,7}(LP) = \min(\mu_R(x_7, LP) \; \mu(x_7)) \tag{14}$$

Step :9 This procedure is repeated for all the 49 rules and the final grade of membership is determined using the composition rule of fuzzy set theory. For example, the controller output characterised by the linguistic label 'LP' can be evaluated as follows

$$\mu_U(LP) = \max_{x_i} (\min(\mu_R(x_i, LP) \; \mu(x_i))) \quad i=1,2,..49 \tag{15}$$

Step :10 Step 9 is repeated for the controller output characterised by the other linguistic labels (MP, SP, VS, SN, MN, LN)

Step :11 The final controller output can be decided using
(i) The Mean of Maxima (MOM) criteria
(ii) The Center of Area (COA) criteria
(iii) The Maximum algorithm

In this paper the maximum algorithm is used wherein, the linguistic label with the highest membership grade is chosen as the controller's output.

To demonstrate the fuzzy controller's action, let the controllers input signals be $e_n = 1.0$ and $\dot{e}_n = -0.2$. Using reference fuzzy sets defined in Fig:2, the controller inputs can be described by the following fuzzy sets.

e_n: {(LP, 1), (MP, 0.7), (SP, 0.4), (VS, 0.2), (SN, 0), (MN, 0), (LN, 0)}

$\dot{e}_n$: {(LP, 0), (MP, 0), (SP, 0.2), (VS, 0.5), (SN, 1.0), (MN, 0.8), (LN, 1.0)}

where the numbers correspond to the membership grade of the particular label. The membership grade of the condition part of rule 7 is given by

$$\mu(x_7) = \mu('e_n \text{ is LP' AND } '\dot{e}_n \text{ is LP'})$$
$$= \min (\mu('e_n \text{ is LP'}) \; \mu('\dot{e}_n \text{ is LP'}))$$
$$= \min (1, \; 0) = 0$$

The membership grade for the linguistic label LP can be computed as follows

$$\mu_{U,7}(LP) = \min(\mu_R(x_7, LP) \; \mu(x_7))$$
$$= \min(1, 0) = 0$$

Table: II Fuzzy relation matrix R

x_i	Controller inputs $(e_n, \dot{e}_n)$	Controller Output Membership grades						
		$\mu_R(x_i,LN)$	$\mu_R(x_i,MN)$	$\mu_R(x_i,SN)$	$\mu_R(x_i,VS)$	$\mu_R(x_i,SP)$	$\mu_R(x_i,MP)$	$\mu_R(x_i,LP)$
x_1	(LP, LN)	0	0	0.5	1.0	0.5	0	0
x_2	(LP, MN)	0	0	0	0.5	1.0	0.5	0
x_3	(LP, SN)	0	0	0	0	0.5	1.0	0.5
x_4	(LP, VS)	0	0	0	0	0	0.5	1.0
x_5	(LP, SP)	0	0	0	0	0	0.5	1.0
x_6	(LP, MP)	0	0	0	0	0	0.5	1.0
x_7	(LP, LP)	0	0	0	0	0	0.5	1.0
x_8	(MP, LN)	0	0.5	1.0	0.5	0	0	0
x_9	(MP, MN)	0	0	0.5	1.0	0.5	0	0
x_{10}	(MP, SN)	0	0	0	0.5	1.0	0.5	0
x_{11}	(MP, VS)	0	0	0	0	0.5	1.0	0.5
x_{12}	(MP, SP)	0	0	0	0	0.5	1.0	0.5
x_{13}	(MP, MP)	0	0	0	0	0	0.5	1.0
x_{14}	(MP, LP)	0	0	0	0	0	0.5	1.0
x_{15}	(SP, LN)	0.5	1	0.5	0	0	0	0
x_{16}	(SP, MN)	0	0.5	1.0	0.5	0	0	0
x_{17}	(SP, SN)	0	0	0.5	1.0	0.5	0	0
x_{18}	(SP, VS)	0	0	0	0.5	1.0	0.5	0
x_{19}	(SP, SP)	0	0	0	0.5	1.0	0.5	0
x_{20}	(SP, MP)	0	0	0	0	0.5	1.0	0.5
x_{21}	(SP, LP)	0	0	0	0	0	0.5	1.0
x_{22}	(VS, LN)	0.5	1.0	0.5	0	0	0	0
x_{23}	(VS, MN)	0.5	1.0	0.5	0	0	0	0
x_{24}	(VS, SN)	0	0.5	1.0	0.5	0	0	0
x_{25}	(VS, VS)	0	0	0.5	1.0	0.5	0	0
x_{26}	(VS, SP)	0	0	0	0.5	1.0	0.5	0
x_{27}	(VS, MP)	0	0	0	0	0.5	1.0	0.5
x_{28}	(VS, LP)	0	0	0	0	0.5	1.0	0.5
x_{29}	(SN, LN)	1	0.5	0	0	0	0	0
x_{30}	(SN, MN)	0.5	1.0	0.5	0	0	0	0
x_{31}	(SN, SN)	0	0.5	1.0	0.5	0	0	0
x_{32}	(SN, VS)	0	0.5	1.0	0.5	0	0	0
x_{33}	(SN, SP)	0	0	0.5	1.0	0.5	0	0
x_{34}	(SN, MP)	0	0	0	0.5	1.0	0.5	0
x_{35}	(SN, LP)	0	0	0	0	0.5	1.0	0.5
x_{36}	(MN, LN)	1.0	0.5	0	0	0	0	0
x_{37}	(MN, MN)	1.0	0.5	0	0	0	0	0
x_{38}	(MN, SN)	0.5	1.0	0.5	0	0	0	0
x_{39}	(MN, VS)	0.5	1.0	0.5	0	0	0	0
x_{40}	(MN, SP)	0	0.5	1.0	0.5	0	0	0
x_{41}	(MN, MP)	0	0	0.5	1.0	0.5	0	0
x_{42}	(MN, LP)	0	0	0	0.5	1.0	0.5	0
x_{43}	(LN, LN)	1.0	0.5	0	0	0	0	0
x_{44}	(LN, MN)	1.0	0.5	0	0	0	0	0
x_{45}	(LN, SN)	1.0	0.5	0	0	0	0	0
x_{46}	(LN, VS)	1.0	0.5	0	0	0	0	0
x_{47}	(LN, SP)	0.5	1.0	0.5	0	0	0	0
x_{48}	(LN, MP)	0	0.5	1.0	0.5	0	0	0
x_{49}	(LN, LP)	0	0	0.5	1.0	0.5	0	0

This is the membership grade of the controller output 'LP' for rule:7. Considering all the 49 rules, the final value of the membership grade for the linguistic label 'LP' is determined using the composition rule as follows

$$\mu_U(LP) = \max_{x_i} \left(\min(\mu_R(x_i,LP)\, \mu(x_i)) \right) \quad i=1,2,3,\ldots 49$$
$$= 0.5$$

The same procedure is repeated for the six other linguistic variables representing the probable controller output and their membership grades are found to be as follows

$$\mu_U(MP) = 1.0; \quad \mu_U(SP) = 0.8; \quad \mu_U(VS) = 0.7;$$
$$\mu_U(SN) = 0.6; \quad \mu_U(MN) = 0.5; \quad \mu_U(LN) = 0.4; \quad (16)$$

Using the 'maximum algorithm' the controller output in linguistic terms is "MP" since it has the highest membership grade. The reference signal representing the modulation index is an analog signal. Hence, the fuzzy linguistic label has to be defuzzified, that is converted to numerical value. Based on previous experience with power tracking controllers, defuzzification is done using the conversion table shown in Table:III.

Table III Conversion from linguistic labels to numerical values

	LP	MP	SP	VS	SN	MN	LN
U=ΔM p.u.	0.1	0.05	0.025	0	-0.025	-0.05	-0.1

Note: 1.0 p.u. corresponds to M=1.0 which is represented by 0 V
0.5 p.u. corresponds to M=0.5 which is represented by 10 V

The modulation index M at any instant of time is given by

$$M(k) = M(k-1) - U_d(k-1) \qquad (17)$$

where U_d is the defuzzified output of the controller representing the actual change in the modulation index. Taking the change introduced by the fuzzy controller into account, the modulation index M is computed at regular intervals ($\Delta T=300$ μs) and an analog signal of appropriate amplitude is sent to the PWM inverter to control it's output voltage and power and hence the power drawn from the PV array.

IV SIMULATION AND EXPERIMENTAL VERIFICATION:

The complete model of the PV energy conversion/interface scheme is represented by equations (1), (3), (4), (6) and (8). To study the system response using a fuzzy rule-based controller, under varying solar insolation levels, these equations have to be solved. This is achieved by using a simulation software package TUTSIM. The various equations are solved at regular intervals ($\Delta T=0.1$ ms). The PV array's output power P_{pv} is computed and compared to the P_{ref} at the given solar insolation level S_r. The error in power $e=(P_{ref}-P_{pv})$ is used to change the modulating index M by $U_d=\Delta M$. This changes the DC voltage V_I at the inverter input and alters the DC link current $I_{pv} = I_{DC}$, thereby affecting the PV array's output power. Simulation results depicting the variation of various variables for step changes in solar insolation level are shown in Fig:7. It is seen that as the solar insolation level increases, the modulation index M increases, which causes the PV array's output power to increase. This continues till the PV array's output power becomes equal to the P_{ref} at which state, the array is operating at it's optimum operating point.

The proposed rule based fuzzy controller is implemented using a digital computer and the data translation card DT2821, for the simple PV energy conversion scheme as shown in Fig:3. Using the data translation card, the solar insolation level S_r, the solar array voltage V_{pv} and the array current I_{pv} are sampled at regular intervals ($\Delta T=300$ μs). The array's maximum power output P_{ref} is computed for the insolation level S_r. The actual power output of the array is computed as the product of the array voltage and current ($P_{pv}=V_{pv}I_{pv}$). Knowing P_{ref} and P_{pv}, the error e and the rate of change of error ė are computed and used to determine the fuzzy controller's output which is then used to control the pulse width of the inverter's output voltage waveform. The inverter's output is fed to a three phase load which is either a resistive load bank or a 3-phase induction motor driving a DC generator. By changing the pulse width of the PWM inverter, within limits (0-6°), it is possible to change the power drawn from the PV solar array, thereby making maximum utilization of the available solar energy. The experimental results shown in Fig:8 indicate that the proposed controller is successful in tracking and extracting maximum solar power from the solar array under varying insolation levels by maintaining the output power as near as possible to the optimal (maximum) power P_{opt}.

V CONCLUSIONS:

An alternative rule-based controller based on Fuzzy set theory is proposed for a PV energy conversion scheme. The objective is to track and extract maximum available solar power from the PV array under varying solar insolation levels. To achieve this the power error $e=(P_{ref}-P_{pv})$ and the rate of change of this error ė are used as input signals to the fuzzy rule-based controller and it's output signal is used to control the PWM inverter. The input error signals are fuzzified and expressed as linguistic labels characterised by their membership grades. Using a fuzzy relation matrix (which relates the input error signals to the fuzzy output signal expressed as a linguistic label), a set of 49 rules and fuzzy logic operations, the controller output is obtained. The fuzzy controller output expressed in linguistic labels is defuzzified to obtain the actual analog signal to control the PWM inverter. The proposed fuzzy rule-based controller is simulated and experimentally verified and is found to give good power tracking performance.

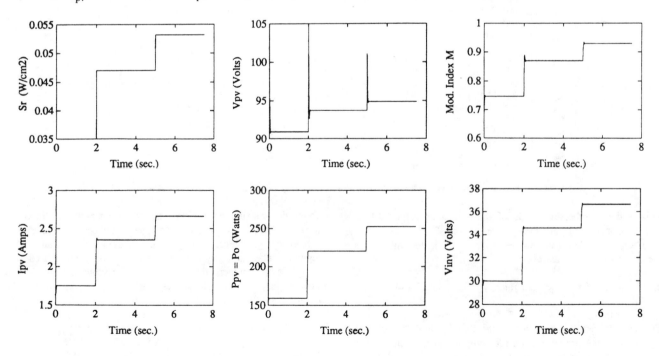

Figure:7 Simulation results depicting variation of various variables for step change in solar insolation level S_r.

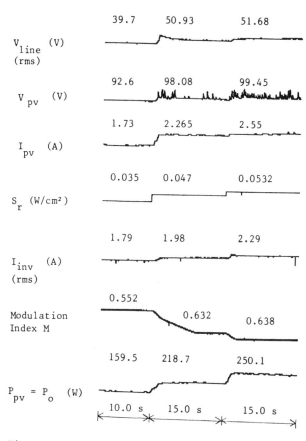

Figure:8 Experimental results depicting variation of various variables for step change in solar insolation level S_r.

References:

[1] H.S. Rauschenbach, 'Solar cell array design handbook: the principles and technology of photovoltaic energy conversion', Van Nostrand Reinhold Company, New York, 1980.

[2] J. Appelbaum, 'The operation of loads powered by separate sources or by a common source of solar cells', IEEE Trans. Energy Conversion, Vol. 4, No. 3, pp.351-357, Sept. 1989.

[3] P.J. MacVicar - Whelan, ' Fuzzy sets for man machine interactions', Int. Journal Man-Machine Studies, Vol. 8, pp. 687-697, Nov. 1976.

[4] L. Zadeh, ' Outline of a new approach to the analysis of complex systems and decision processes', IEEE Trans. System Man Cybernetics, Vol.28, pp. 28-44, 1978.

[5] Zimmermann H.J., ' Fuzzy set theory and its applications', Kluwer-Nijhoff Publishing Company, 1985.

[6] Hsu Y.Y. and Cheng C.H., 'Design of fuzzy power system stabilizer for multimachine power systems', IEE proceedings, Vol 137, pt.C, No.3, pp 233-238, May 1990.

[7] Hilloowala R.M. and Sharaf A.M., ' Single phase induction motor drive scheme for pump irrigation using photovoltaic source', Proceedings of 22nd Annual North American Power Symposium, pp 415-427, Auburn, Alabama, Oct 1990.

[8] Signetics reference manual 'HEF4752V application guidelines Advance information', April 1981.

FUZZY CONTROL OF WIRE FEED RATE IN ROBOT WELDING

by Satoshi Yamane[*], Guillermo Alzamora[**], and Takefumi Kubota[***]

[*]Maizuru College of Technology, Maizuru, Kyoto 625, JAPAN
[**]Saitama University, Urawa, Saitama 338, JAPAN
[***]Himeji Institute of Technology, Himeji, Hyougo 670, JAPAN

Abstract

This paper deals with the problem concerning the controlling of wire feed rate and of the weld pool shape in robot welding. First, the control method of wire feed rate is discussed to keep an arc length constant. Next, it is proposed how to take the clearer image of the weld pool with a CCD camera. The information of weld pool is obtained by processing the image. The fuzzy control method is introduced for controlling the wire feed rate and the weld penetration. At last, the validity of the proposed method is illustrated in the experiments.

Introduction

We can obtain a good quality of welding result when arc length and weld shape are kept constant regardless of the disturbance.

The arc length is related to the wire feed rate and the welding current. The authors use the welding current to control the weld pool shape. First, the authors discuss how to keep the arc length constant by adjusting the wire feed rate. In robot welding, the wire is supplied with wire feed motor. Even if the voltage across the wire feed motor is kept constant, the wire feed rate is unstable by a rotation of the wire reel. Since an arc length also becomes unstable, the wire feed rate has to be kept constant. The induce of state equations which describe the motion of the wire feed motor was tried. The state equations are complicated, as the torque changes by the rotation of wire reel, wire tension, and so on. To stabilize an arc length may be difficult by the method based on modern control theory. Therefore a new method based on the fuzzy control is developed for controlling of the wire feed rate. Next, authors discuss the method to control weld pool shape by adjusting the welding current. For this purpose, sensing of the weld pool shape is important. One of useful sensing method is to take the weld pool with a camera in a real time[1]. However, it has been difficult to watch the molten pool under welding arc with TV camera because of its after image feature. The authors propose the method for detecting the pool shape in pulsed MIG welding. In order to take the clearer image under the arc, the relation between the shutter operation time of a camera and the pulsation phase of the welding current is discussed. A number of useful information have been obtained from the image processing.

Since the welding phenomena are so complicated, we can not precisely describe it only with the differential equations. Furthermore, there are many disturbances in the system parameters. It is difficult to apply the modern control to the plant from above mentioned. Therefore, the fuzzy control system has been designed for controlling the pool shape.

The weld penetration is well corresponding to the pool width. Hence, it is necessary to control the pool shape to obtain the uniform back-bead.

It seems possible to make the visual and intelligent welding robot by utilizing the developed technique.

Control System of Wire Feed Rate

The fuzzy control system for controlling the wire feed rate is shown in Fig. 1. r is the reference value of the wire feed rate. The roller rotates while the power is supplied to the wire feed motor. The wire is coiled on the reel. The wire is sent from the reel to the base metal through the welding torch by the rotation of the roller. The tacho generator is the sensor of the wire feed rate, since the output V of the tacho generator is in proportion

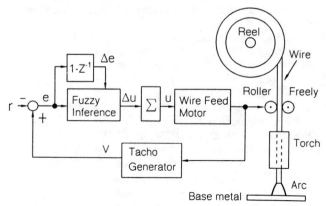

Figure 1 Fuzzy control system for the wire feed.

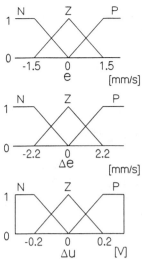

Figure 2 Fuzzy variables for wire feed control.

Table 1 Control Rules.

Δe \ e	N	Z	P
P	Z	N	N
Z	P	Z	N
N	P	P	Z

P=Positive
Z=Zero
N=Negative

↓ Δu

to the rotation speed of the wire feed motor. The difference between the wire feed rate and its reference is used as the deviation. The voltage across the wire feed motor is inferred from the deviation and its variation by the fuzzy inference.

Fuzzy Controller for Wire Feed Rate Control

The deviation e (=V-r) and the variation Δe of the deviation are adopted as the input variables of the fuzzy controller. The variation Δu of the voltage across the wire feed motor is manipulating variable and is inferred form both the control rules and the input variables with the fuzzy controller. The fuzzy variables are determined from fundamental experiments.

The control rules are considered from the various kinds of situations of the wire feed rate and of its variation. For example, if the wire feed rate is slower than the reference and the wire feed rate has decreased, then experts should increase the motor voltage to accelerate the wire feed rate. Let slow, decrease, and increase be represent with the symbol N (Negative), N (Negative), and P (Positive), respectively. The rule at this situation is described by following if-then form.

    if e is N and Δe is N then Δu is P   (1)

where N and P are the fuzzy variables.

The if-parts are determined from the control knowledge of experts in the situation. The control rule is described by using the ambiguousness P, Z, and N. The membership functions of P, Z, and N are shown in Fig. 2. Another control rules are constructed from the knowledge of experts and listed in Table 1, which is constructed form 9 kinds of rules.

Control Experiment Result of Wire Feed Rate

First, the experiment is performed without the fuzzy controller. The experiment result is shown in Fig. 3. The wire feed rate is unstable, since it vibrates accordance with the rotation of the reel. The resultant arc length becomes unstable.

Next, the experiment is performed with the fuzzy controller. The experiment result is shown in Fig. 4. The wire feed rate was controlled and kept constant regardless of the rotation of the reel. Therefore, the arc length is also kept constant.

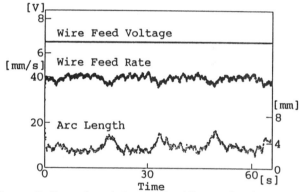

Figure 3 Experimental result without the fuzzy controller.

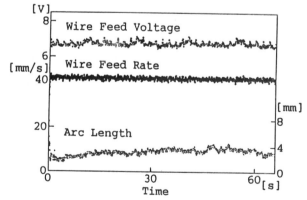

Figure 4 Experimental result with the fuzzy controller.

Control System for Weld Pool Shape Control

The block diagram for control system of weld pool using the CCD camera is shown in Fig. 5. CCD camera takes the weld pool image and sends out the video signal of the image to the image memory. The personal computer processes the taken image in the image memory and calculates the pool width by

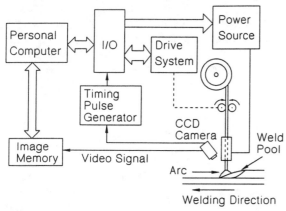

Figure 5 Control system for the weld pool.

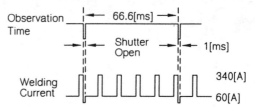

Figure 6 Timing chart to open shutter.

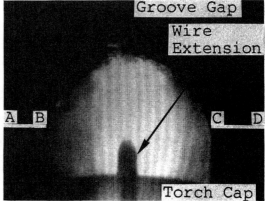

(a) Information of the pool image

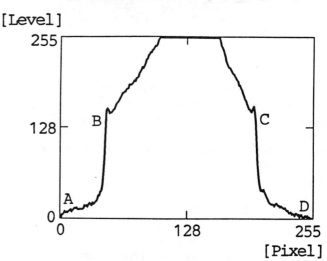

(b) Brightness distribution
Figure 7 Typical image of weld pool.

image processing. After that, the personal computer determines the welding current and the welding speed so as to keep the pool shape constant. The reference value of wire feed rate is determined from the welding current so as to keep an arc length short.

<u>Sensing Weld Pool with CCD Camera</u>

Some methods have been suggested to detect the weld pool image with CCD camera in pulsed MIG welding process. One of them is to reduce the arc current as small as possible to avoid the affection of an arc light while the shutter of the CCD camera has opened. Moreover, authors proposed to use the high speed shutter function of the CCD camera [2].

The corresponding welding current waveform are represented in Fig. 6. The typical image of the weld pool and its brightness distribution are shown in Fig. 7 while the CCD camera is located in front of the welding torch. In Fig. 7 (b), it can be seen that the brightness changes steeply near the pool boundary. Thus, the computer will calculate the pool width. The experiments yield that the pool width is in proportion to the bead penetration. Because, it is impossible to directly detect the depth of the weld pool in welding, we can select the pool width, for instance, as an input information for the penetration control system.

It is well known that when the groove gap changes, the different amount of filler metal and heat input into weld joint are required to maintain the uniform quality of the whole bead. The groove gap can be detected from the pool image as shown in Fig.7(a). This information can be utilized to resist the disturbance in groove gap.

In other hand, while the welding current is constant, the distance between the pool front edge and the arc center (Fig. 8(a)) is longer with the smaller gap than that with the wider gap (Fig. 8 (b)). This indicates that the change in welding condition can be checked out by investigating the position of the front pool edge relative to the arc center. This distance is also considered as is deeply involved to the back bead [3] that we can select it as another input variable of the penetration control system.

<u>Fuzzy Controller for Weld Pool Control</u>

The block diagram of the fuzzy controller for weld pool control is illustrated in Fig. 9. Let r and W be reference value and the weld pool width, respectively. Two inputs e and Δe of the fuzzy controller are the deviation (W-r) of the pool width and its variation, respectively. The variation Δu of the welding current is inferred from e and Δe.

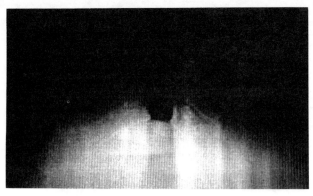
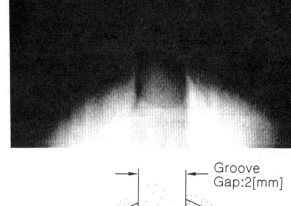
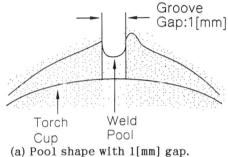
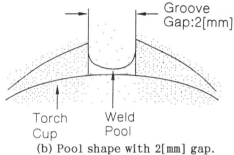

(a) Pool shape with 1[mm] gap.   (b) Pool shape with 2[mm] gap.

Figure 8 The relationship between groove gap and pool position.

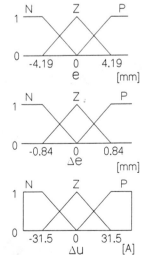

Figure 9 Fuzzy controller for weld pool control.

Generally, the control rules are based on the experience of experts. The control rule table is the same as the control table for wire feed control. The fuzzy inference acts as if e is P and Δe is Z, then Δu is N. i.e. the current should be decreased when the pool width is wider than the reference and the pool width has not changed. The fuzzy variables upon which the construction of the fuzzy sets depend must be decided carefully from experiments and is shown in Fig.10.

Figure 10 Fuzzy variables for weld pool control.

Experimental Result of Weld Pool Control

The practical experiment has been carried out in butt welding with the following conditions. The base metal is of 3.2[mm] thickness, 350[mm] length, and 100[mm] width. Two plates are assembled with 1[mm] groove gap at the beginning section and 0[mm] gap at the ending section.

Figure 11 (a) shows the transient behaviors of the welding current, the welding speed, the pool width, and the back bead without the fuzzy controller.

Figure 11 (b) shows the experiment result with the fuzzy controller. The cross sections of the penetration with the variation of the groove gap are shown in Fig. 12. (1), (2), (3), and (4) in Fig. 12 is correspond to the locations (1), (2), (3), and (4) in Fig. 11, respectively.

Based on the fact that the weld pool images taken at points (3) and (4) appear same in shape (seen in Fig. 13), it is necessary to control the weld pool shape to obtain the uniform back bead regardless of the changing in groove gap.

Conclusion

A good quality of welding result can be obtained when arc length and weld shape are kept constant regardless of the disturbance in robot welding. For this purpose, the control method of the arc length and of weld pool shape was developed with the fuzzy controller. The arc length is kept constant by adjusting the wire feed rate.

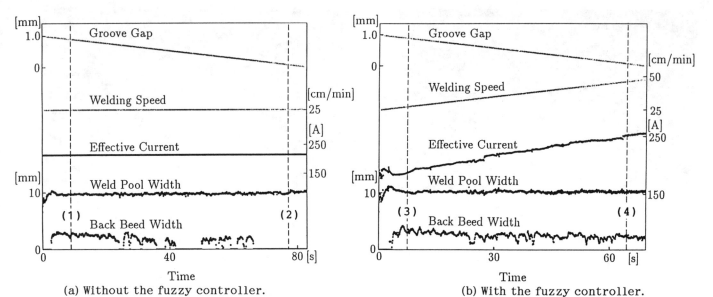

(a) Without the fuzzy controller.   (b) With the fuzzy controller.
Figure 11 Experiment results of the weld pool control.

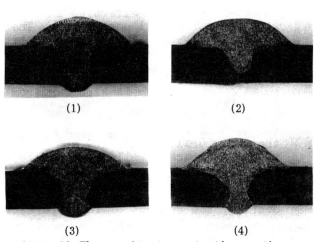

Figure 12 The resultant penetration sections.

In the practical welding, we should regulate the welding current to keep the pool width constant and adjust the welding speed to resist the disturbance in the groove gap, so that the uniform penetration can be obtained. With the fuzzy controller, the desired response of the pool width has been achieved from experiments. The uniform back bead can be obtained by the control of the weld pool shape.

Reference

[1] K.Ohshima, et al, "Observation and Digital Control of Weld Pool in Pulse MIG welding", Quarterly Journal of Japan Weld. Soc., vol. 5, No.3, pp.304-311.

[2] M.Yamamoto, et al, "Adaptive Control of Pulsed MIG Welding Using Image Processing System", Conf. Rec. of the IEEE IAS, 1988, pp.1381-1386.

[3] Y.Kitazawa, "Through-the-Arc Sensing Control of Welding Speed for One-Side Welding", Sensor and Control System in Arc Welding, Technical Commission on Welding Processes Japan Welding Society, 1991,pp.II-121-124.

[4] Welding Handbook, American Welding Society, vol. 1,Miami, Florida,1976.

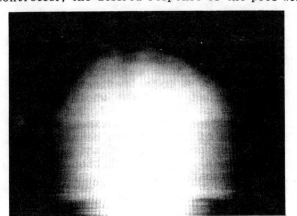 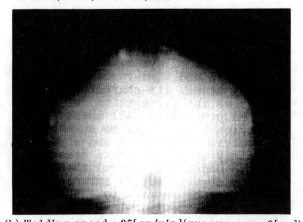

(a) Welding speed : 50[cm/min](groove gap : 0[mm])   (b) Welding speed : 25[cm/min](groove gap : 0[mm])
Figure 13 The image of controlled pool shape.

Chapter 7: Electronics

Autonomous Navigation of a Mobile Robot Using Custom-Designed Qualitative Reasoning VLSI Chips and Boards

François G. Pin, Hiroyuki Watanabe,* Jim Symon,* and Robert S. Pattay

Oak Ridge National Laboratory
Oak Ridge, TN 37831-6364, USA

Abstract

Two types of computer boards including custom-designed VLSI chips have been developed to add a qualitative reasoning capability to the real-time control of autonomous mobile robots. The design and operation of these boards are first described and an example of their use for the autonomous navigation of a mobile robot is presented. The development of qualitative reasoning schemes emulating human-like navigation in a-priori unknown environments is discussed. The efficiency of such schemes, which can consist of as little as a dozen qualitative rules, is illustrated in experiments involving an autonomous mobile robot navigating on the basis of very sparse and inaccurate sensor data.

1: Introduction

One of the greatest challenges in the motion planning and control of autonomous mobile robots in a-priori unknown or dynamic environments is to provide the reasoning modules with methods for handling and/or coping with the many imprecisions, inaccuracies, and uncertainties present in the system. These typically arise from three major sources: (1) errors in the sensor data (there are no perfect sensor systems) which lead to inaccuracies and uncertainties in the representation of the environment, the robot's estimated position, etc., (2) imprecisions or lack of knowledge in our understanding of the system, i.e., our inability to generate complete and exact (crisp) mathematical and/or numerical descriptions of all the phenomena contributing to the system's and environment's behavior, and (3) approximations and imprecisions in the information processing schemes (e.g., discretization, numerical truncation, convergence thresholds, etc.) that are used to generate decisions or control output signals.

Qualitative reasoning (also termed approximate reasoning) refers to a set of methodologies which have been developed to provide alternative solution methods for decision-making problems when the uncertainties can not be fully engineered away (e.g., there are limits on maximum sensor precision, predictability of the environment, etc.). The general approach underlying these methodologies consists in capturing some aspects of the reasoning methods typically exhibited by humans when controlling systems, i.e., by implicitly incorporating uncertainties in the information gathering and reasoning processes, rather than attempting to explicitly determine and propagate them through numerical calculations or representations. Several approximate reasoning theories and associated mathematical algebra have been developed over the past two decades,[1] the most commonly used today for applications to control systems being Zadeh's Theory of Fuzzy Sets.[2-5] This theory is at the basis of very successful implementations varying from control of subway cars, cement kilns, washing machines, still and video cameras, inverted pendulums, to painting processes and color image reconstruction, to even ping-pong playing robots.[6-12]

One of the important factors which have prevented the wide-spread utilization of approximate reasoning in real-time systems has been the unavailability of computer hardware allowing processing and inferencing directly in terms of approximate or linguistic, or "fuzzy" variables (e.g., far, fast, slow, left, faster, etc.) and approximate rules (e.g., if obstacle is close, then go slower; if temperature is high and pressure is increasing, then decrease power a lot, etc.). Prospective implementations thus had to rely on simulation of the approximate reasoning schemes on conventional hardware and computers based on "crisp" (numerical) processing, with a resulting significant penalty in speed of operation, prohibiting applications in most "hard real-time" systems.

In cooperation with MCNC, Inc., unique computer boards have recently been developed using custom-designed VLSI chips[13,14] which can be programmed to directly communicate and interface in terms of qualitative variables and rules. Additionally, the boards' architecture is reconfigurable on-line to allow several levels of reasoning (meta level, non monotonic, etc.) and to allow full inferences with up to 350 rules and 28 input channels to take place in 30 μ sec, i.e., at a rate of 30,000 Hz (at least two orders of magnitude faster than video frame rate). This paper provides an overview of the design and operation of these boards and discusses

---

* Department of Computer Science, University of North Carolina at Chapel Hill, Chapel Hill, NC 27599

their first implementation in the development of approximate reasoning methodologies and schemes for CESAR's series of HERMIES (Hostile Environment Robotic Machine Intelligence Experiment Series) testbed robots.

2: Qualitative reasoning on a VLSI chip

The qualitative reasoning methodology utilized for the VLSI implementation is inspired from the Theory of Fuzzy Sets, in which the functions $\mu_X(x)$ defining the membership of an element x to a subset X of a universe of discourse U can take any value in the interval $[0, 1]$, rather than only the discrete $\{0, 1\}$ values (0 for does not belong, 1 for belongs) used in conventional (crisp) Set Theory. The function $\mu_X(x)$ thus defines the *degree* of membership of the element x in X. Such a subset X of U is termed a qualitative (or approximate, conceptual, or fuzzy) variable for reasoning on the universe of discourse U.

For the current VLSI implementation, reasoning is embodied in programmable "production rules" operating on four sets of qualitative input variables and two sets of output qualitative variables, as in

IF (A is A_1 and B is B_1 and C is C_1 and D is D_1)
THEN (E is E_1 and F is F_1) , (1)

where $A_1, B_1, \ldots F_1$ are qualitative variables whose representative membership functions define the rule, and $A, B, C \ldots F$ are the time-varying qualitative input and output variables analogous to memory elements in conventional production systems.

With the above representation, the Fuzzy Set Theoretic Operations can be directly applied to the qualitative variables on their universe of discourse: given two subsets A and B of U,

$$\mu_{A \cap B}(x) = \min(\mu_A(x), \mu_B(x)) \quad (2)$$

$$\mu_{A \cup B}(x) = \max(\mu_A(x), \mu_B(x)) \quad (3)$$

The laws of logical inferences including modus ponens, cartesian product, projection and compositional inferences (e.g., see [3] and [4] for detailed description of these laws of inferencing) can also be applied to multivariable systems. In particular, the extension principle[3,4] is used in the mapping between a set A of the input universe of discourse U and its extension through F to the output universe of discourse V, as:

$$\mu_{F(A)}(v) = \underset{u}{Sup}\, \mu_A(u) \quad (4)$$

where $v = F(u)$, $u \epsilon U$, $v \epsilon V$.

For their VLSI implementation, each qualitative variable is represented by its membership function discretized over a (64×16) array of $(x, \mu(x))$ values. Equations (1), (2), (3), and (4) can thus be easily implemented using series of min. and max. gates as shown in Fig. 1 for one rule. Figure 2 schematically represents an inference with two rules of the form IF (A is A^1 and B is B^1) THEN (E is E^1) operating on two input A and B and producing a composite membership function for E.

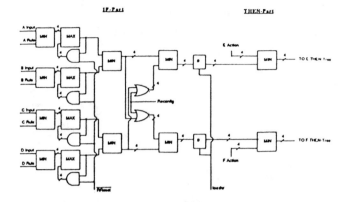

Fig. 1. Data path for rule execution.

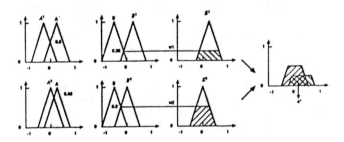

Fig. 2. Schematic of a qualitative inference using two rules operating on two input and one output channels.

Because conventional sensors typically provide data in "crisp" form, i.e., they provide a single number and the uncertainty on the measurement is ignored, it is desirable to add this uncertainty on the measurement, effectively mapping it to a qualitative variable, prior to processing through approximate reasoning. This step (which has been termed fuzzification) is of course not necessary if the data is already in the form of a qualitative variable, such as in interchip communications, and therefore has been implemented as a programmable optional data path on the VLSI chip. Similarly, an optional defuzzying step which calculates the "center of weight" of the output composite membership function (see Fig. 2), can be used to send "crisp" data to conventional actuators if these are used in the process control hardware as depicted on Fig. 3. To provide added flexibility, the chip architecture is reconfigurable, allowing either 50 rules operating on four input and two output channels or 100 rules operating on two input and one output channels. Since all rules are processed in parallel, the speed of operation of the chip is independent of the configuration or the number of rules involved in the inferencing, and reaches 30,000 FFIPS (Full Fuzzy Inferences Per Second). In other words, full qualitative data processing and inferencing schemes can take place at 30 KHz, (i.e., at least two orders of magnitude faster than the sampling rate of typical sensors) making feasible the control of very fast systems or motions, such as those involved in reflex behaviors based on very approximate or uncertain information.

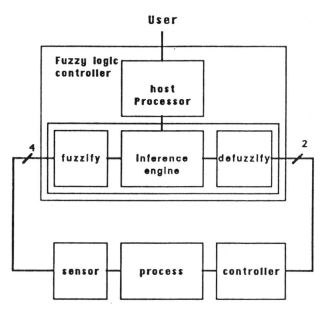

Fig. 3. Schematic of a typical qualitative control system for a real-time process.

Two types of VMEbus-compatible printed boards and associated software were developed to allow interfacing of the chips with sensors and actuator data channels for application to "intelligent machines" and in particular autonomous mobile robots. The first type of board includes one chip and is therefore limited to inferencing involving only 4 input and 2 output channels. The second type of board includes 7 chips and some multiplexer circuits which allow on-line reconfiguration of the input, output and interchip communication paths. This provides the capability to implement qualitative reasoning schemes with up to 350 rules and 28 input channels (with all chips in parallel), multi-level reasoning schemes (e.g., 4 chips in a first layer feeding into 2 chips in the second layer feeding into 1 chip in the third layer), or non-monotonic reasoning (e.g., with feedback of the output of some of the downstream chips into the input of some of the upstream chips, in a series or "cascade" of chips). The speed of operation of each layer of parallel chips remains the same than on the single chip board, with the multi-layer configurations reaching rates in the KHz order of magnitude.

3: Test implementation for mobile robots navigation

The problem of autonomous mobile robots navigating in a-priori unknown and unpredictable environments was selected for initial testing of the qualitative reasoning systems because its characteristics rank very high on the list of criteria that typically indicate suitability of a reasoning scheme for representation and implementation using qualitative logic: the input to the control system, particularly when provided by sonar range finders and odometry wheel encoders, is extremely inaccurate, sparse, uncertain and/or unreliable; there exist no complete mathematical and/or numerical representation of the behavior termed navigation, although, as demonstrated by humans, a logic for this behavior exists which can typically be represented and successfully processed in terms of linguistic variables; by its given nature the behavior of the environment is unpredictable, leading to large uncertainties in its representation; the approximations involved in the numerical representation of the system and its environment (e.g., geometric representations, map discretization in grid, etc.) are significant.

The single chip board and a recently developed omnidirectional mobile platform[15,16] were used for these initial experiments. Because of the limitations to 4 input channels using this board, data from only 3 frontal sonars were used for perception of the environment, while the fourth channel produced information on angular direction to the given navigation goal based on odometry sensor data. The output channels provided translational speed and steering velocity commands to be sent to the motor controls. No a-priori environmental data or maps were input to the system, nor were any generated during motion, and in this sense, the initial investigations focussed on reactive navigation. The development and testing of the first series of qualitative rule-bases led to empirical findings providing significant insights for efficient implementation of qualitative reasoning schemes in autonomous robots:

- Modularity and consistency of the rule-base can best be achieved through decomposition of the decision-making scheme into elemental and independent "behaviors."
- Independence of these elemental behaviors is assured if each can be formulated as a direct mapping between a subset of the input and a subset of the output, with no redundancy in the qualitative values spanned by the input variables of different behaviors.
- Independent behaviors can be singly developed and tested, and their independence experimentally verified prior to merging with other behaviors.
- Once developed, tested and verified, each behavior can be assigned a normalized "weight" (in [0,1]) corresponding to its relative importance with respect to other behaviors with which it is to be merged, (e.g., safety from obstacle vs. speed of operation, etc.). The weighting is implemented by a direct scaling of the membership functions of either the input or the output variables.
- Merging of the behaviors is handled directly and continuously through the laws of combinatorial inferencing, therefore providing a formal resolution to one of the major problems with which the "behaviorist" community (e.g., see [17] and [18] and references therein) has struggled: the real-time selection and/or conflict resolution in multi-behavior systems.

Building upon the experience and empirical results gained during the development of the first series of rule-bases, a new rule-base conforming with the observations listed above was developed for the single chip board and the CESAR's omnidirectional platform pictured in Fig. 4. The photograph in the figure shows the ring of acoustic range sensors at the edge of the platform deck (only 3 frontal sensors are used) and the disk drive unit, the battery pack (rear right) and the seven-slot VME-bus (rear left) which hosts the qualitative inferencing board. The control system of the platform (detailed in [15] and [16]) includes a velocity loop servoing at 100 Hz on the commanded translational and rotational velocities, which will be hereafter referred to as speed control and turn control, respectively. Thus, behaviors corresponding to speed control (S.C.) and turn control (T.C.) as functions of goal orientation (G.O.) and obstacle proximity (O.P.) where developed as follows:

$$\begin{array}{rcll}
\text{G.O.} & \longrightarrow & \text{S.C.} & (1\ \text{rule}) \\
\text{O.P.} & \longrightarrow & \text{S.C.} & (4\ \text{rules}) \\
\text{G.O.} & \longrightarrow & \text{T.C.} & (2\ \text{rules}) \\
\text{"far" O.P.} & \longrightarrow & \text{T.C.} & (2\ \text{rules}) \\
\text{"near" O.P.} & \longrightarrow & \text{T.C.} & (2\ \text{rules}) \\
\text{"very close" O.P.} & \longrightarrow & \text{T.C.} & (3\ \text{rules})
\end{array}$$

where the three latter behaviors embody the fact that different navigation behaviors are utilized depending on whether all obstacles are still "far," "near," or "very close," thus reflecting differences in safety concerns (i.e., priority of the behavior) implemented using different weights. For each sampling period and decision, several behaviors are typically triggered and merged through the Fuzzy Set Theoretic laws of Combinatorial Inferencing, resulting in a smooth and continuous sensor-based navigation control.

Fig. 4. The CESAR omnidirectional robotic platform prototype.

The rules for TC and SC as a function of GO express the very intuitive fact that if the goal is to the left (respectively right), then a small increment of turn to the left (respectively right) needs to be made during the loop rate cycle; and when the direction of the goal increases from 0° (front) to ± 180°, then the speed is correspondingly decreased. The rules for SC as a function of OP express that when the distances to any obstacles (i.e., the sensor returns in all three directions) are increasing, then the speed can be increased toward its maximum value (one rule), while when distance to an obstacle (sonar return) in any of the three sonar directions (thus, three rules) decreases toward a safety threshold (here selected as 30 cm), then the speed needs to be decreased, down to zero at or below that threshold.

Once these speed control and goal tracking behaviors were designed, they were merged and tested in environments with no obstacles. Since the chip used for this initial implementation allowed only four input channels, no information related to distance to the goal could be provided to the qualitative reasoning scheme in order to make the robot stop when reaching the goal. This was easily remedied in these tests by using the odometry data in the master program to stop both the reasoning scheme and the robot when it approached to within a given radius (2.5 cm) of the goal. In future implementations using the seven-chip board allowing up to 28 input channels, the distance to the goal could be input to the qualitative inference scheme and the stopping at the goal could be simply implemented as an additional behavior in the reasoning scheme.

Once these behaviors were tested, the rules for the TC as a function of OP behaviors were developed. When all sonar returns are "far" (further away than approximately 2 m), the turn should be away from the closest obstacle. However, the weight on that behavior must be less than that for the TC as a function of GO, to ensure that when it is far away from any obstacles, the robot's priority is still to move in the general direction of the goal. When at least one of the sonar returns is "near" (between about 30 cm and 2 m), the turn is away from the obstacle, increasing in magnitude with decreasing distance, such that at the lowest distance of 30 cm, this obstacle avoidance behavior has more weight on TC than the goal tracking behavior. Finally when any sonar return is less than 30 cm (the robot is stopped as required by the behavior on SC as a function of OP), the turn is always to the right. Note that setting the turn away from the closest obstacle in this latter behavior would often result in the dead-lock situations in which the robot reaches a limit cycle, and continuously oscillates between two orientations. This type of situation constitutes one of the very serious drawbacks of the reactive navigation methods using potential field techniques, and has been alleviated here using this behavior of TC as a function of "very close" OP. Also note that this behavior allows the robot to travel to the end of dead-end corridors, turn around, and backtrack to a more open area, a situation which would lead to a (local minimum) dead-end point in potential field techniques.

Figures 5 and 6 show plots of sample runs made with the robot to illustrate the overall reactive navigation using the qualitative inferencing scheme and, in particular, the two characteristics just discussed. In the figures, the lightly shaded areas represent the obstacles which were placed in the room, while the path of the robot is illustrated using the dark succession of circles. In Fig. 5, the robot initially starts toward the goal, encounters the wall, moves along the wall, passes the point directly opposite to the goal on the perpendicular to the wall (at which a dead-lock would be encountered using potential field techniques), and continues until it reaches the end of the wall where it can turn to reach the goal. In Fig. 6, the robot starts toward the goal and, when facing obstacle A head-on, moves in the opening on its right which is closest to the goal direction. When reaching the end of this blocked corridor, the robot turns around (using the TC as a function of "very close" OP behavior), exits the corridor, turns in a direction closest to the goal direction, avoids the small obstacles and then moves to the goal.

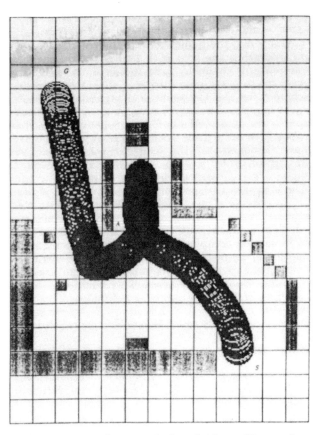

Fig. 6. Sample run of the platform illustrating obstacle avoidance in more complex environments, motion in corridors, and no "trapping" in the local minimum at the end of the blocked corridor.

4: Summary and concluding remarks

Autonomous robot control in a-priori unknown, unpredictable, and dynamic environments requires many calculational and reasoning schemes to operate on the basis of very imprecise, incomplete, sparse or unreliable data, knowledge or information. In such systems, for which engineering all the uncertainties away from the hardware is not currently fully feasible, approximate reasoning may provide an alternative to the complexity and computer requirements of conventional uncertainty analysis and propagation techniques.

Two types of computer boards including custom-designed VLSI chips have been developed to investigate the implementation and real-time use of approximate reasoning in autonomous robotic systems. The methodologies embodied on the VLSI hardware utilize the Fuzzy Set Theoretic operations to implement a production rule type of inferencing on input and output variables that can directly be specified as qualitative variables through membership functions. All rules on a chip are processed in parallel, allowing full inferences to take place in about 30 μsec. This speed of operation makes real-time reasoning feasible at rates much faster than sensor

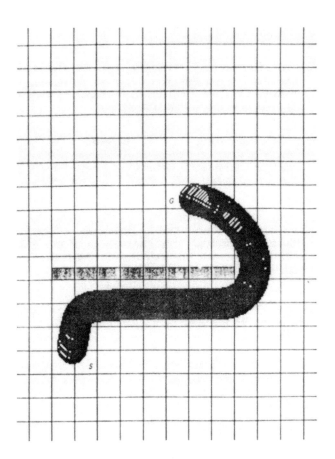

Fig. 5. Sample run of the platform illustrating basic obstacle avoidance, stable wall following, and no "trapping" in local minima. S and G denote the start and goal locations.

data acquisition, therefore, making control of "reflex-type" of motions envisionable.

One of the qualitative inferencing boards, incorporating one chip with four input channels and two output channels, was installed on a test-bed platform to investigate the use of qualitative reasoning schemes for the autonomous navigation of a mobile robot in a-priori unknown environments on the basis of sparse and imprecise data. Experiments in which the robot uses only three acoustic range (sonar) sensors have demonstrated the feasibility of basic reactive navigation with a scheme including six elemental behaviors represented in fourteen qualitative rules. The approach using superposition of behaviors allows to progressively merge additional behaviors into the scheme to resolve any specific additional situation which may be encountered in particular environments of increasing complexity. Our ongoing work focusses on this area, utilizing the recently completed multi-chip board (which allows up to 28 inputs and 14 outputs) to investigate schemes with additional input variables and greater numbers of behaviors, for which we were limited in this first series of experiments by the four-input-only restriction of the single-chip board.

References

[1] "Uncertainty in Artificial Intelligence," eds. L. N. Kanal and J. F. Lemmer, North-Holland, New York (1988).

[2] L. A. Zadeh, "Fuzzy Set," *Information and Control* 8, 338–353 (1965).

[3] L. A. Zadeh, "Outline of a New Approach to the Analysis of Complex Systems and Decision-Making Approach," *IEEE Transactions on Systems, Man, and Cybernetics* SME-3(1), 28–45 (January 1973).

[4] L. A. Zadeh, "Fuzzy Logic," *IEEE Computer* 21(4), 83–93 (April 1988).

[5] "Fuzzy Sets and Their Applications to Cognitive and Decision Processes," eds. L. A. Zadeh, K. S. Fu, K. Tanaka, and M. Shinmra, Academic Press, Inc., New York (1975).

[6] S. Yasunobu, S. Miyamoto, T. Takaoka, and H. Ohsihima, "Application of Predictive Fuzzy Control to Automatic Train Operation Controller," in *Proc. IECON '84* (1984), 657–662.

[7] L. P. Holmblad and J. J. Ostergaard, "Control of a Cement Kiln by Fuzzy Logic," *Fuzzy Information and Decision Processes*, eds. M. M. Gupta and E. Ssnchez, 389–399 (1982).

[8] L. I. Larkin, "A Fuzzy Logic Controller for Aircraft Flight Control," *Industrial Applications of Fuzzy Control*, ed. M. Sugeno, 87–103 (1985).

[9] H. Ono, T. Ohnishi, and Y. Terada, "Combustion Control of Refuse Incineration Plant by Fuzzy Logic," in *Proc. 2nd Inter. Fuzzy Systems Association Congress* (July 1987), 345–348.

[10] M. Sugeno, et al., "Fuzzy Algorithmic Control of Model Car by Oral Instructions," *Fuzzy Sets and Systems* 32, 207–219 (1989).

[11] T. Yamakawa, "Stabilization of an Inverted Pendulum by a High-Speed Fuzzy Logic Controller Hardware System," *Fuzzy Sets and Systems* 32, 161–180 (1989).

[12] K. A. Hirota and S. Hachisu, "Fuzzy Controlled Robot Arm Playing Two Dimensional Ping-Pong Game," *Fuzzy Sets and Systems* 32, 149–159 (1989).

[13] H. Watanabe, W. Dettloff, and E. Yount, "A VLSI Fuzzy Logic Inference Engine for Real-Time Process Control," *IEEE J. of Solid State Circuits* 5(2), 376–382 (1990).

[14] J. R. Symon and H. Watanabe, "Single Board System for Fuzzy Inference," in *Proc. Workshop on Software Tools for Distributed Intelligent Control Systems* (September 1990), 253–261.

[15] F. G. Pin and S. M. Killough, "A New Family of Omnidirectional and Holonomic Wheeled Platforms for Mobile Robots," submitted to *IEEE Trans. on Robotics and Automation* (in review).

[16] S. M. Killough and F. G. Pin, "Design of an Omnidirectional and Holonomic Wheeled Platform Prototype," in *Proceedings of the 1992 IEEE Conference on Robotics and Automation*, May 10–15, 1992, Nice, France.

[17] R. A. Brooks, "Elephants Don't Play Chess," *Robotics and Autonomous Systems* 6(1-2), 3–15 (1990).

[18] R. C. Arkins, "Integrating Behavioral, Perceptual and World Knowledge in Reactive Navigation" *Robotics and Autonomous Systems* 6(1-2), 105–122 (1990).

FUZZY CONTROL OF AN INDUSTRIAL ROBOT IN TRANSPUTER ENVIRONMENT

Jarmo Franssila and Heikki N. Koivo
Tampere University of Technology, Control Engineering Laboratory
PO Box 692, SF-33101 Tampere, Finland

Abstract

The dynamics of the robot is highly nonlinear, offering a demanding field for different control methods. The fuzzy logic controller (FLC), as nonlinear one, is one solution to the control problem of the robot. The application of the FLC to the control of the industrial robot and experiment results are described in this paper. The performance of this controller was investigated by feeding step inputs to several joints at the same time. The results are compared with those resulting from a conventional PD controller. One disadvantage of the FLC is the difficulty to design optimal rule base. To overcome this problem a self organising controller was implemented and tested.

Due to the demand of small sample time, multiprocessor environment was used in experiments. A calculation of the control of individual joints was shared by the network of transmitters. In this network processes can run effectively in parallel.

Introduction

The principles of fuzzy logic were first published by Zadeh [19]. Mamdani brought control engineering point of view to fuzzy logic [10]. There after a lot of research has been made concerning development and application of FLC's. The earliest subject to apply the FLC was found in process industry [5], where the conventional control methods had given unsatisfactory results. An apriori knowledge of an experienced operator was mapped into control rules of the FLC. In comparison to different processes, mechanical systems need faster and more accurate responses. In recent years the amount of research and applications has increased rapidly, aiming to improve the performance of the mechanical systems with fuzzy logic. The FLC has proven to be the best alternative in certain circumstances like in commercial container crane [18].

Robots are nowadays involved in many industrial manufacturing phases and therefore the improvement of their efficiency is in interest of manufacturers. The dynamic of robots include many unpleasant effects from control engineering point of view. Nonlinearities like friction, backlash and position dependent inertia are reasons why it's difficult to know the dynamical behaviour of the robot. Up to now robots are controlled mainly with linear fixed parameter controllers, adaptive controllers and model based controllers. Each has its own disadvantages- fixed parameter controller doesn't take into account nonlinearities and can't adapt to changes in environment, there is no guarantee that parameters of the adaptive controller will converge fast enough, the model based controller is cumbersome and time consuming to develop and in spite of this there is always unknown effects. The conventional FLC lies between fixed parameter and model based controller. Inputs to the FLC can be the same as to PD controller, however, apriori knowledge of behaviour of the system is necessary in form of control rules.

The FLC has been successfully applied to position control of the manipulator [2], [4], [9], [11], [12]. The authors have reported the FLC to have better performance than conventional control methods in their problems. In this paper the implementation of the FLC to position control of the industrial robot and results of experiments are described.

The quality of self organising was added to the FLC. The ideas behind self organising controller (SOC) were first introduced by Procyk and Mamdani [13], [14]. The SOC has an ability to adjust its control strategy as required by environment.

Again improvement of the position control of the robot has been one aim of the research. Scharf et al. [15] controlled one joint and Tanscheit and Scharf [17] brought tests further by controlling two joints at the same time with two independent SOC's. According to authors SOC had better performance in comparison to PID controller in their tests.

The sampling time in tests, mentioned above, was between 20 ms and 40 ms using the main processor of the robot. One way to improve the response, is decrease the sampling time. In our tests three joints were controlled using three independent SOC's with sampling interval 4 ms maximum. This was possible using the transputer network for calculation instead of the main processor of the robot. The ability of learning of these independent SOC's is also described in this paper.

FLC

The mathematical background of fuzzy logic is very complicated, but in spite of this only a minor part of mathematics is needed for control. Set operators min, max and complement are tools, with which the FLC is implemented.

FLC is composed of three main parts [6]: a fuzzification interface, a decision making logic and a defuzzification interface Fig. 1.

0-7803-0582-5/92$3.00©1992 IEEE

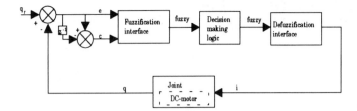

Fig. 1: The configuration of fuzzy logic controller.

The first task is to choose input variables. The most used input pair is error and change of error, where error is set value minus present value. Input could also be for example direct measured quantity like speed or height. In the case, where the dimension of input is larger than two, a further handling of inputs is more complicated - especially building up control rules. If the output of FLC is control signal direct, the PD like fuzzy controller is concerned. Because DC-motor has an integrator in it no integral term is necessarily needed in joint controller. Before fuzzification the scaling of "crisp" input signals transfers the input values into corresponding universes of discourses for fuzzification. If operator f describes the nonlinear function which is performed by FLC then the input output relation can be expressed by

$$u = GU \cdot f(GE \cdot e, GC \cdot c), \quad (1)$$

where e is error, c is change of error and u is control signal. GE, GC and GU are scaling factors respectively. These three scaling factors are the fine tuning parameters of the FLC for which it is sensitive. After scaling one should decide whether a continuous [4] or a discretized [8], [2] universe of discourse of input signals is used. We preferred the continuous one to discretized because of demand on position accuracy. In the case of discretized universe the whole function of FLC can be calculated beforehand and saved in the form of a look-up table. On the contrary in the continuous case each individual rule has to be saved and the effects of each rule have to be calculated at every control cycle. The next choice to make is the shape of the membership functions. The most used shape is triangle, which we have also used.

One membership function represents one fuzzy subset. In normal case fuzzy subset has to cover the whole universe of discourse of input and output signals. When sensitive behaviour of FLC is expected, then more subsets corresponding to universe of discourse are needed. The subsets have been labelled with ZE (zero), SN (small negative), SP (small positive) and so on, corresponding to the location of their support in universe of discourse. In our implementation universes of signals are covered with seven fuzzy subsets.

The next stage is decision making logic, which forms fuzzy output based on control rules. Control rules take into account the dynamical behaviour of the robot arm. Seven subsets against universe of discourse of both input signals results in 49 control rules. Due to the overlapping of membership functions, one input pair activates more than one rule at a time. Each rule is expressed as conditional "if - then " statement. The fuzzy control rule is implemented with fuzzy relation R_i and its membership function is defined as follows

$$\mu_{R_i}(e,c,u) = \mu_{E_i}(e) \wedge \mu_{C_i}(c) \wedge \mu_{U_i}(u), \quad (2)$$

where μ_E, μ_C and μ_U are the membership functions of input and output signals. From this implication the fuzzy output of FLC can be obtained with compositional rule of inference. Based on this inference fuzzy output set u_i corresponding rule i is

$$\mu_{u_i}(u) = \mu_{e'}(e) \wedge \mu_{c'}(c) \wedge \mu_{R_i}(e,c,u), \quad (3)$$

where $\mu_{e'}$ and $\mu_{c'}$ are membership functions of fuzzy input. Because inputs are "crisp" values input sets e' and c' are singletons. Previous equations simplify to

$$\mu_{u_i}(u) = \mu_{E_i}(e^0) \wedge \mu_{C_i}(c^0) \wedge \mu_{U_i}(u), \quad (4)$$

where e^0 and c^0 are "crisp" input values. The overall fuzzy output is formed by combining individual subsets u_i. The membership function of overall fuzzy output U is

$$\mu_U(u) = \bigvee_{i=1}^{n} \mu_{u_i}(u). \quad (5)$$

Operators $\vee$ and $\wedge$ are usually implemented with max and min, and therefore the previous operation is often called max-min -composition. However, this max-min-composition isn't the only possibility to make the inference. Instead max-min, generalised T-operators can be used. Max-min-pair is only one T-norm, T-conorm pair. Variety of T-norm and T-conorm operators can be chosen. In fact, according to Gupta et al. [1], they achieved better results in certain control tests using other norm-conorm-pair than max-min.

The last stage is defuzzification. Like before there are again several methods to perform defuzzification. The most widely used method is center of gravity, which gives the "crisp" output by,

$output = \Sigma$ *(height of the membership function • center of support)*/Σ *height of the membership function.*

To other methods belong max, mean of max and center of gravity with alpha cut methods. Mean of max method has been mentioned to have a good transient performance, while center of gravity method has better steady state performance [7].

As has been pointed out, there are several parameters, which a designer has to choose and which affect the quality of control. One way to overcome the difficulty to design the suitable amount of fuzzy subsets and the correct rule base, is self organising controller (SOC).

SOC

The SOC is a hierarchical controller, which can self organise its rule base [13]. The basic level is a conventional FLC, in the higher level the rule acquisition takes place. The higher level can improve the performance of FLC by modifying its rule base. The structure of SOC is depicted in Fig. 2.

One should note that the universes of discourse of input signals have been quantized with the block Q.

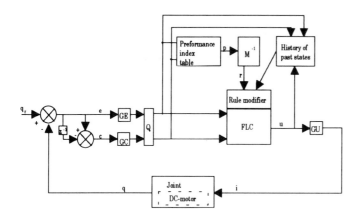

Fig. 2: The configuration of the SOC.

The function of the FLC is similar to the continuous FLC described in previous chapter, but taking into account quantization. In the FLC the fuzzy relation corresponding to rule k is expressed as a three dimensional matrix $\mathbf{R}_k$ by outer product

$$\mathbf{R}_k = \mathbf{E}_k \times \mathbf{C}_k \times \mathbf{U}_k, \qquad (6)$$

where $\mathbf{E}_k$, $\mathbf{C}_k$ and $\mathbf{U}_k$ are vectors representing fuzzy subsets of input and output signals. The overall relation matrix $\mathbf{R}$ is formed from individual relation matrices by union

$$\mathbf{R} = \mathbf{R}_1 \cup \mathbf{R}_2 \cup \ldots \cup \mathbf{R}_n, \qquad (7)$$

where n is the total number of rules.

Self organising level consists of two main parts: the performance index table and the rule modifier. If the system output wasn't the desired one, it is assumed that some earlier control action caused the bad response. The performance index table gives measure, how the rule modifier has to change the rule base of the FLC, so that the response would be better. The performance index table is a two dimensional matrix in the case of two input signals. The matrix elements represent distance between the present state and the desired state.

Let's assume that at time nt quantized error $e(nt)$ and change of error $c(nt)$ are fed into the FLC. The FLC gives output $u(nt)$ corresponding to its input. After some delay, say d, $u(nt)$ affects the output $y(nt + dt)$ of the system causing error $e(nt + dt)$ and change of error $c(nt + dt)$. The error and change of error are then fed to the performance index table, indexing some correction from matrix $\mathbf{PI}$ by $p(nt + dt) = \mathbf{PI}[e(nt + dt), c(nt + dt)]$. Thereafter the incremental inverse model M^{-1} of the system is used to find out, what is the desired correction to control $u(nt)$. This phase can be expressed by $r(nt + dt) = M^{-1}p(nt + dt)$. If the inverse model is unknown, then scalar coefficient can be used instead of M^{-1} in scalar case. So the corrected control is now $v(nt) = u(nt) + r(nt + dt)$. The rule modification is based on signals $e(nt)$, $c(nt)$ and $v(nt)$.

Now the modifier should replace the implication

$$\mathbf{E}(nt) \to \mathbf{C}(nt) \to \mathbf{U}(nt), \qquad (8)$$

with implication

$$\mathbf{E}(nt) \to \mathbf{C}(nt) \to \mathbf{V}(nt), \qquad (9)$$

where $\mathbf{E}(nt)$, $\mathbf{C}(nt)$, $\mathbf{U}(nt)$ and $\mathbf{V}(nt)$ are fuzzy subsets constructed around elements $e(nt)$, $c(nt)$, $u(nt)$ and $v(nt)$. These two implication form relation matrixes $\mathbf{R}'(nt)$ and $\mathbf{R}''(nt)$ given by the outer product

$$\begin{aligned} \mathbf{R}'(nt) &= \mathbf{E}(nt) \times \mathbf{C}(nt) \times \mathbf{U}(nt), \\ \mathbf{R}''(nt) &= \mathbf{E}(nt) \times \mathbf{C}(nt) \times \mathbf{V}(nt). \end{aligned} \qquad (10)$$

Let $\mathbf{R}(nt)$ represent the overall relation matrix and $\mathbf{R}(nt + t)$ the new modified one. Procyk and Mamdani [14] have used the statement

$$\mathbf{R}(nt+t) = \left\{ \mathbf{R}(nt) \text{ but not } \overline{\mathbf{R}'(nt)} \right\} \text{else } \mathbf{R}''(nt) \qquad (11)$$

in modification. The statement can be presented with set operations like

$$\mathbf{R}(nt+t) = \left\{ \mathbf{R}(nt) \wedge \overline{\mathbf{R}'(nt)} \right\} \vee \mathbf{R}''(nt). \qquad (12)$$

However, relation matrices are not used in implementation of the SOC, due to the fact, that the use of these matrixes is time and memory consuming. Assume, that universes of discourse of signals e, c and u are quantized to n_e, n_c and n_u level respectively. Then the size of relation matrices is $n_e \times n_c \times n_u$.

Instead of relation matrices both the controller and the modifier are implemented to use individual rules. Some approximation to modification algorithm itself has also been suggested [14]. Shao [16] and Ho [3] have skipped over the individual rules and modified the look-up table of the FLC directly.

Control test environment

The control experimentation system is composed of the host computer, the transputer network, the interface and the robot. The configuration of the system is depicted in Fig. 3.

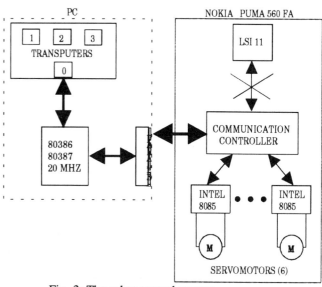

Fig. 3: The robot control system.

The central piece of equipment is the personal computer running MS-DOS. It connects to the robot using parallel interface card. In the robot, the parallel connection is mounted at arm interface board, replacing the main processor of the robot. This board can be used to communicate to the six joint microprocessors. There is one processor for each joint. Joint processors perform tasks such as servoing the joint to the desired position and reading the position of the joint. The processors know a command set, with which one can for instance control the joint specifying the motor current from an external computer. This makes it possible to close the control loops in the external computer and have complete freedom in experimenting with the servo control.

The processor in the PC is essentially an I/O processor. It transmits data between transmitters and the arm interface board of the robot. Position encoders of joints are read and control output is written to joints.

The transputer network consists of four INMOS T800 transmitters on MICRO WAY QUADPUTER 2 board, which is an add-on board in PC. Each transputer has 4 Kbytes of fast internal on-chip ram and up to 1 Mbyte external memory. Tranputers communicate with each other over serial links. Each transputer has four two directional links. The host has been linked to operate as a serial link. Only one transputer can be connected to each end of the link. Therefore the host is connected via link only to transputer number 0. The communication of all other transmitters must go through transputer number 0 for any host input-output.

Programs in PC and in transmitters are written in the C language (Microsoft C and Logical Systems C respectively). Logical systems C cross compiler has standard C routines and in addition transputer specific routines. The computational task for the first, second and third joints is assigned to transmitters 1, 2 and 3 respectively. Transputer 0 takes care of timing and data routing between host and other transmitters.

The robot is six degrees of freedom PUMA 560 FA industrial robot.

Control tests and results

The performance of the conventional FLC was tested first and compared to PD controller. The controlled joints were the first three, which cause a waist, shoulder and elbow rotation. The step inputs were fed at the same time to those three joints. The sampling time in tests was 2 ms.

Based on these tests, the performance of the FLC was quite similar to PD controller. There wasn't any difference between responses, but the control signal of the FLC was a bit smoother. As an example the responses and control signals concerning the second joint are depicted in Fig. 4 and 5.

Next the SOC was tested in the control of the joints with a series of input steps. In our application the error was discretized to 19 levels and the change of error to 13 levels. The sampling interval was 3 ms in all tests concerning SOC. The sampling time depends on the number of control rules - the more rules the longer time is needed for calculation.

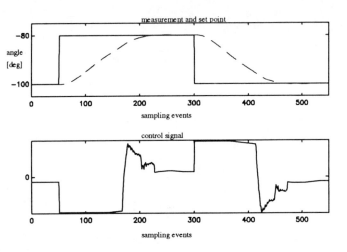

Fig. 4: The results of the second joint using FLC.

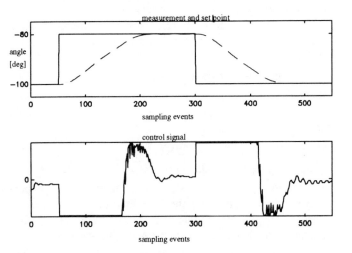

Fig. 5: The results of the second joint using PD controller.

It is difficult to estimate beforehand, how many rules will be generated. Instead of zero, there were three rules at the beginning. These three control rules were quite obvious, namely: "if *error* is zero and *change of error* is zero then *control* is zero", "if *error* is small negative and *change of error* is zero then *control* is small negative" and "if *error* is small positive and *change of error* is zero then *control* is small positive". This initial rule set improves the stability near the set point.

Only the first joint was controlled at first using SOC, while the other two joints were motionless. The ability of SOC to improve the performance by organising the rule base is depicted in Fig. 6 and 7.

As can be seen SOC improve its performance during the learning cycle and generates less than 45 rules, whereas the conventional FLC before had 49 control rules.

After the control of the individual joint, three SOC's operated in parallel controlling three joints at the same time. The responses and the control of each joint are depicted in Fig. 8, 10 and 12. Fig. 7, 9 and 11 indicate the convergence of the SOC's.

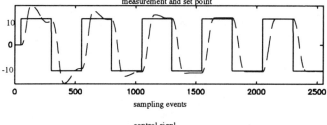

Fig. 6: The response and the control signal of the SOC for the first joint.

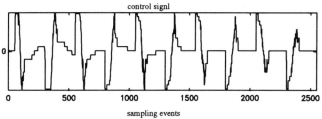

Fig. 7: The total number of the control rules for the first joint.

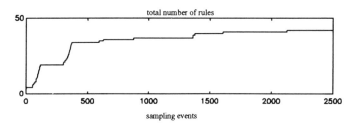

Fig. 8: The response and the control signal of the SOC for the first joint (three joints are moving).

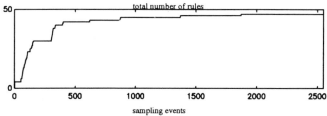

Fig. 9: The total number of the control rules for the first joint.

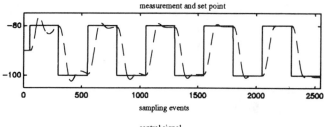

Fig. 10: The response and the control signal of the SOC for the second joint.

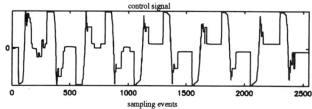

Fig. 11: The total number of the control rules for the second joint.

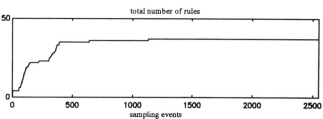

Fig. 12: The response and the control signal of the SOC for the third joint.

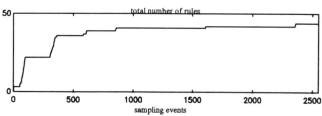

Fig. 13: The total number of the control rules for the third joint.

The fast movements of each joint effect to the other joints and disturb the convergence of their SOC's. The SOC results in better convergence and response when only one SOC at a time is used for the control of some joint, while the other joints are motionless. During the tests two problems were noticed. The first problem is quite obvious, because discretized input signals decrease the accuracy of the response. The second problem is a sensitiveness of the SOC to the scaling factors. In our implementation only two scaling factors was used for the fine tuning, because of control restrictions GU factor was fixed. A small change in these scaling factors didn't affect instability, but made the response worse.

Conclusion

The FLC and the SOC were implemented and tested in the position control of the industrial robot. The performances of those controllers were investigated by feeding step inputs to the first three joints of the robot at the same time. The control algorithm for each joint run in parallel in different transputer making the small sampling time possible.

For the design of the FLC deep knowledge about control engineering is not needed. The controller is implemented using simply linguistic statements, which are based on operator's knowledge. On the other hand, there are several factors in the basic structure of the FLC that affects its performance. A designer has to take into account all these factors and choose the best combination for the specific problem.

The scaling factors, the control rules and the shape and size of the membership function belong to the tuning parameters. The lack of the systematic tuning method makes the tuning difficult or at least time consuming.

The SOC is one possibility to overcome the difficulty to tune the control rules, due to the self organising quality. It uses fixed membership functions but, however, the scaling factors still need correct values.

In the robot control the SOC has shown to operate satisfactorily in the specific tests. The weaknesses of the SOC are the accuracy, the sensitiveness to scaling factors and their tuning. Before mentioned subjects need further investigation and improvements.

References

[1] M. M. Gupta, J. Qi, " Design of fuzzy logic controllers based on generalised T-operators ", Fuzzy Sets and Systems, vol 40, 1991.

[2] K. Hirota, Y. Arai, S. Hachisu, " Fuzzy Controlled Robot Arm Playing Two Dimensional Ping-Pong Game ", Fuzzy Sets and Systems, vol 32, pp. 149-159, 1989.

[3] J. M. Ho, S. R. Lin, " A Learning Algorthm for Fuzzy Self-Organising Controller ", presented at the IEEE International Workshop on Intelligent Motion Control, Istanbul, Turkey, August 20-22, 1990.

[4] L. J. Huang, M. Tomizuka, " A Self-Paced FuzzyTracking Controller for Two-Dimensional Motion Control ", IEEE Transactions on System, Man and Cybernetics, vol. 20, pp. 1115-1124, 1990.

[5] P. J. King, E. H. Mamdani, " The Application of Fuzzy Control Systems to Industrial Processes ", Automatica, vol. 13, pp. 235-242, 1977.

[6] C. C. Lee, " Fuzzy Logic in Control Systems: Fuzzy Logic Controller, Part 1", IEEE Transaction on Systems, Man and Cybernetics, vol. 20, pp. 404-418, 1990.

[7] C. C. Lee, " Fuzzy Logic in Control Systems: Fuzzy Logic Controller, Part 2" IEEE Transaction on Systems, Man and Cybernetics, vol. 20, pp. 419-435, 1990.

[8] Y. F. Li, C. C. Lau, " Development of Fuzzy Algorithms for Servo Systems ", presented at the IEEE International Conference on Robotics and Automation, Philadelphia, USA, April 24-29, 1988.

[9] C. M. Lim, T. Hiyama, " Application of Fuzzy Logic Control to a Manipulator ", IEEE Transactions on Robotics and Automation, vol. 7, pp. 688-691, 1991.

[10] E. M. Mamdani, " Application of Fuzzy Logic to Approximate Reasoning Using Linquistic Synthesis ", IEEE Transactions on Computers, vol. c-26, pp. 1182-1191, 1977.

[11] S. Murakami, F. Takemoto, H. Fujimura, E. Ide, " Weld-Line Tracking Control of ARC Welding Robot Using Fuzzy Logic Controller ", Fuzzy Sets and Systems, vol. 32, pp. 221-237, 1989.

[12] R. Palm, " Fuzzy Controller for a Sensor Guided Robot ", Fuzzy Sets and Systems ", vol. 31, pp. 133-149, 1989.

[13] T. J. Procyk, " A Self Organising Controller for a Dynamic Processes ", Ph.D. Thesis, Queen Mary College, University of London, 1977.

[14] T.J. Procyk, E. H. Mamdani, " A Linguistic Self-Organising Process Controller ", Automatica, vol. 15, pp. 15-30, 1979.

[15] E. M. Scharf, N. J. Mandic', E. H. Mamdani, " A Self-Organising Algorithm for the Control of a Robot Arm", International Journal of Robotics and Automation, vol. 1, pp. 33-41, 1986.

[16] S. Shao, " Fuzzy Self-Organising Controller and Its Application for Dynamic Processes ", Fuzzy Sets and Systems, vol. 26. pp. 151-164, 1988.

[17] R. Tanscheit, E. M. Scharf, " Experiments with the Use of a Rule-Based Self-Organising Controller for Robotics Applications ", Fuzzy Sets and Systems, vol. 26, pp. 195-214, 1988.

[18] S. Yasunobu, T. H. Hasegawa. " Evaluation of an Automatic Container Crane Operation System Based on Predictive Fuzzy Control ", Control Theory and Advanced Technology, vol. 2, pp. 419-432, 1986.

[19] L. A. Zadeh, " Outline of a New Approach to theAnalysis of Complex Systems and Decision Processes", IEEE Transactions on System, Man and Cybernetics, vol 3, pp. 28-44, 1973.

Fuzzy Logic Approach to Placement Problem

Rung-Bin Lin and Eugene Shragowitz

Computer Science Department
University of Minnesota, Minneapolis, MN 55455

Abstract

A contemporary definition of the placement problem is characterized by multiple objectives. These objectives are : minimal area, routability, timing and possibly some others. This paper contains a description of the placement system based on the fuzzy logic approach. Linguistic variables, their linguistic values and membership functions are defined. Fuzzy logic rules govern the placement process. Details of implementation and experimental results are provided.

1. Introduction

Placement is one of the important steps in VLSI design. Many placement algorithms [1-7] have been introduced in the past few years. All these algorithms do not explicitly address two important aspects of the placement problem : presence of multiple conflicting objectives and utilization of expert knowledge in decision making. These aspects of the placement problem can be addressed by the technique based on the fuzzy set theory [9-11]. One attempt to apply the fuzzy set theory to problems of physical design was made in [8]. In this work the connectivity matrix was modified and then the fuzzy set theory was used to construct clusters for placement. The aforementioned work does not address the most interesting multi-objective aspect of the design problem, which is in the center of this paper. In this paper, we apply fuzzy reasoning to the placement of sea-of-gate arrays. We compare results obtained with and without fuzzy logic and draw some conclusions from this comparison.

2. Basics of Fuzzy Set Theory

A fuzzy set A, as defined by L.A. Zadeh, is a class of objects with continuum of grades of membership. An element may partially belong to a fuzzy set. This is contrary to the ordinary set theory, where an element is either in a set or not in a set. Formally, a fuzzy set is defined as follows [9] :

Definition 1 : *A fuzzy subset A of a universe of discourse X is defined as $A = \{ (x, \mu_A(x)) \mid all\ x \in X \}$, where X is a space of points and $\mu_A(x)$ is a membership function of $x \in X$ being an element of A.*

In general, a membership function $\mu(x)$ is mapping from X to the interval [0,1]. If $\mu_A(x)=1$ or 0, the fuzzy set A becomes an ordinary set. One possible membership function for the fuzzy set A is illustrated by Fig. 1. On this figure point x_1 belongs to the fuzzy set A with degree of membership $\mu_A(x_1)$. An example of the fuzzy set A is *a person's age near 50*. In this case, the space X is defined as the interval $[0,t]$, where t is the maximum age humans have ever had. A person of 45 has a degree of membership in the fuzzy set *a person's age near 50*. We will return to this example later in the paper.

Set operations such as union, intersection, and complementation, etc., are naturally introduced into the theory of fuzzy sets. Due to the nature of the membership function, results of fuzzy set operations are also fuzzy sets. The membership function of the resulting fuzzy set for basic operations on two fuzzy sets A and B is usually defined in one of the following forms :

$$\mu_{A \cap B}(x) = min(\mu_A(x), \mu_B(x)). \quad (1)$$

$$\mu_{A \cup B}(x) = max(\mu_A(x), \mu_B(x)). \quad (2)$$

$$\mu_{\neg A}(x) = 1.0 - \mu_A(x). \quad (3)$$

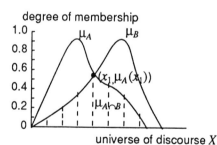

Fig. 1 Membership function for fuzzy set $A \cap B$ realized by function $min(\mu_A, \mu_B)$

Some other algebraic operators may also be used to define these fuzzy set operations. For instance, $\cup$ could be defined as addition and $\cap$ could be defined as multiplication. Selection of operators solely depends on the problems of applications. The shaded region in Fig. 1 shows the membership function of the fuzzy set $A \cap B$, which results from the intersection of fuzzy sets A and B. Here μ_A and μ_B are the membership functions for the fuzzy sets A and B respectively. For example, μ_A could be a membership function for the fuzzy set of *a person's age*

---

*This work is supported in part by the Mentor Graphics Corporation.

near 50 and μ_B could be a membership function for the fuzzy set of *a person's height around 6 feet*.

3. Approximate Reasoning by Linguistic Variables

In the Webster's dictionary logic is defined as a science of correct reasoning. In the domain of problem solving, imprecise information always prevents us from making precise reasoning. Reasoning by a human is seldom exact. Therefore, the traditional binary logic based on Boolean theory of ordinary sets is not adequate to model human reasoning. Fuzzy logic is defined as "the art of fuzzy reasoning". Whenever "art" is mentioned, it involves more or less subjective judges and empirical studies about underlying domains. Reasoning made on imprecise domain is called fuzzy reasoning. A concept of fuzzy logic can be better understood if a "degree of membership" in a fuzzy set is interpreted as a "degree of truth". For example, equation (1) can be interpreted as that an element $x \in X$ is in the fuzzy set A and in the fuzzy set B with the degree of truth $\mu_{A \cap B}(x)$. That is, A and $B = \{(x, \mu_{A \cap B}(x)) | all\ x \in X\}$.

As defined by L.A. Zadeh, a linguistic variable is a variable whose values are words or sentences in natural or artificial language. The formal definition is referred to [10]. Informally, the concept of a linguistic variable can be explained by an example.

Example 1: As it was mentioned earlier, the *age near 50* is a fuzzy set. Here, *age* is a linguistic variable. The linguistic variable *age* may have many linguistic values. For instance, the set of linguistic values of *age* may be defined as {very young, somewhat young, young, somewhat old, old, very old, near 50, not close to 60 and not too old, ... }. Each linguistic value defines a fuzzy set on X, where X is a possible range of *age* measured in years. For instance, X can be the interval [0,200].

By using hedges such as *some, more, very, quite, fair, extremely*, etc., and connectives such as *and* and *or*, we can make approximate reasoning based on linguistic variables and their values. For instance, suppose we have another linguistic variable called *health* defined for each person. Linguistic values of *health* are defined as *very good, good, bad,* and *very bad*. Then we can make some reasoning based on values of these variables. For example,

If a person's *age* is *young* and his *health* is *very good*, he can donate *some blood*.

Here *donate blood* is another linguistic variable with values *donate no blood, donate some blood, donate more blood, donate a lot of blood*. As it can be observed from above descriptions, fuzzy sets based on linguistic variables are used to model the imprecise information. Fuzzy logic is then used to make reasoning about the correct actions from the inputs. In the same manner, we will use this approach for the physical layout problems.

4. Multi-objective Optimization by Fuzzy Logic Approach

Consider a multi-objective optimization problem [11]

$$Maximize\ F(x), \qquad (4)$$

where $F(x)=(f_1(x), f_2(x), ... f_m(x))$ is a vector-function, X is a solution space, and $f_i(x)$ is mapping from X to the set of the real number for all $i \in I = \{1,2,3,....,m\}$. Unlike the single objective optimization problem, no concept of optimal solution is universally accepted for the multi-objective optimization. In practical cases, the rating of an individual objective reflects the preference of decision makers. At best, a compromise between competing objectives can be expected. Fuzzy approach has some properties useful for this situation.

Let $g_i(\bullet)$ be mapping from the domain Y_i to the interval [0, 1] for all $i \in I$, where $Y_i \subset R$ is the range of $f_i(\bullet)$. $g_i(\bullet)$ is a measurement of the objective $f_i(\bullet)$ according to the degree of satisfaction of decision makers. $g_i(y_i)=1$, $y_i \in Y_i$ means complete satisfaction of the i-th objective, while $g_i(y_i)=0$ could mean 100% dissatisfaction. Therefore, a good compromise for multiple objectives will be the solution, which is the most satisfactory to decision makers. Therefore, the multi-objective optimization problem can be presented as follows:

$$Maximize\ G(x), \qquad (5)$$

where $G(x)=G(g_1(f_1(x)), g_2(f_2(x)),....,g_m(f_m(x)))$ is a function of membership functions, $f_i(x) \in Y_i$ for all $i \in I$. The same definition can be applied to minimization. From now on, the function (5) is a target for optimization.

5. Fuzzy Logic Approach to Constructive Placement

When placement is performed manually by experts, they should select a cell and a position for it. Experts have to use their visual perceptions of partial placements to make the best selections. It is quite often that qualitative characteristics are available instead of quantitative measurements. Layout experts usually reason using qualitative characteristics and often are able to obtain good layouts. For example, one line of reasoning based on qualitative characteristics could be: *If a cell has strong connectivity to the partial placement and it has critical timing delays on its nets, it is a good candidate to be selected for placement*. The reasoning statement is called a "rule" in a fuzzy logic system. Three linguistic variables are employed here to make an approximate reasoning. They are *connectivity, timing delay*, and *candidate. The strong connectivity, the critical timing delay,* and *the good candidate* are respectively their linguistic values. Each linguistic value defines a fuzzy set. It is likely that more than one rule is necessary to make the best selections. For example, another rule could be: *If a cell has weak connectivity to the partial placement or it has small timing delays on its nets, it is a bad candidate*.

It seems, that the negation of the first rule is equivalent to the second rule. It is actually so in the binary logic reasoning. However, in fuzzy logic it is not. The reason is that *does not have strong connectivity* is not equivalent to *has weak connectivity* in terms of linguistic values. For the same reason the statements from the following pairs *not critical timing delay, good timing delay,* and *not bad candidate, good candidate* are not equivalent to each other. In case when rules are used to make a decision, we need to get a *crisp value* based on evaluations of these rules. A *crisp value* is a precise value in fuzzy logic system. In case of the placement problem, it is an evaluation of the degree of urgency for candidate cells and evaluation of the degree of quality for candidate positions. One can see that the given above rules of selection of candidate cell describe the multi-criteria decision making based on the fuzzy information (sets). In the following text, we will use the fuzzy approach to solve these problems.

5.1 Constructive Strategy of Placement

Two basic strategies of placement are known : constructive and iterative. Sometimes combinations of these two are used when the initial solution is obtained by the constructive placement and improved by the iterative technique.

The constructive placement consists of two major tasks (1) *select a cell,* and (2) *select a position for placing the selected cell*. These two tasks are repetitively executed until all cells are placed. The placement process can be defined as a quadruple $S(P,H,L,F)$, where

(1). P is a partial placement characterized by the set of already placed cells, area used, number of used and remaining routing tracks, amount of remaining space and other values;

(2). H is the set of remaining unplaced cells;

(3). $L = L^{\alpha} \cup L^{\beta}$ is a set of rules and functions that are used to generate a new state. Here L^{α} is a set of rules for decision making, and L^{β} is a set of objective functions involved in decision making;

(4). F is the set of final states.

Each instance of (P,H) defines a state of the placement process. A state is called the final state when $H = \emptyset$ or no space is available for placement. The placement process stops when it reaches the final state. A final state is called feasible, if all cells are successfully placed and no violations of timing constraints are reported. The purpose of the placement process is to reach a feasible final state in a reasonable amount of time.

Let a state of the constructive placement process at time t be described by (P_t, H_t), where P_t and H_t are respectively partial placement and set of remaining unplaced cells at time t. A state transition from (P_t, H_t) to (P_{t+1}, H_{t+1}) is the result of solving the two major tasks. Selection of a cell and selection of a position for placement are governed by the set L of rules and functions. Multi-objective optimization problems arise in the process of solving these tasks.

In general, timing performance and usage of silicon area are the most important objectives for optimization in design of VLSI circuits. In order to optimize these two goals, the constructive placement must consider all the factors influencing the course of the process.

5.2 Criteria Related to Selecting a Cell

Placement is constructed by adding one cell at a time to the partial placement starting from the left side of the chip and moving to the right. Incremental global routing is performed after each cell is placed. For a partial placement P_t, some nets are completely routed, some nets are partially routed and the remaining nets are not routed. To select a cell η from the set H_t of remaining unplaced cells, we consider several criteria :

(1). connectivity $f_1((P_t,H_t),\eta)$ of selected cell η to partial placement P_t,

(2). connectivity ratio $f_2((P_t,H_t),\eta) = f_1((P_t,H_t),\eta)/f_\eta$, where f_η is the total number of pins for cell η,

(3). approximate average timing delay $f_3((P_t,H_t),\eta)$ on nets computed for the cell η.

Timing delays in this work were computed according to the methodology described in [6]. As one can see multiple criteria are important for selecting a candidate cell. By simple reasoning we can explain that a good layout is hard to obtain if a single criterion is used for selection of cells. For example, suppose criterion (1) is used alone. The cells with the largest numbers of pins are going to be selected first. In general, such cells are larger. Therefore, the larger cells will very likely concentrate on the left side of the chip. This phenomenon was observed in the solutions described in [6]. If criterion (1) is used alone, cells with $f_2((P_t,H_t),\eta) = 1$ will be delayed from being selected, even though these cells could be perfect candidates to be placed next. As a result, some interconnections will become very long. But, if the criterion (2) is used alone, the larger cells will be delayed from being selected because of the large value of f_η. This explanation illustrates why multiple criteria should be taken into consideration.

5.3 Applications of Fuzzy Logic to Selection of Cells

As it was defined earlier, the placement process is characterized by a quadruple $S(P,H,L,F)$. For each new state, a cell is selected first and then a position is found for it. The state transition is accompanied by updating information on resource usage in partial placement P. Suppose, the current state of a placement process is (P_t, H_t), and the space of points for selecting a cell η is defined as $X_t = \{all\ \eta \in H_t\}$. To select a cell is to find a point $x_\eta \in X_t$ such that it is the most satisfactory to the layout experts.

A set of key parameters that influence the placement process must be identified. For selecting a cell, these parameters are identified in the previous section. Three objective functions $f_1((P_t,H_t),\eta)$, $f_2((P_t,H_t),\eta)$, and $f_3((P_t,H_t),\eta)$ were defined there. Three linguistic variables *connectivity, connectivity ratio,* and *timing delay* are defined for these functions. Then the problem of selecting a cell may be formulated as follows :

$$Maximize\ F_1(x), \qquad (6)$$

where $F_1(x)=(f_1(x_\eta),f_2(x_\eta),f_3(x_\eta))$ is a vector-function. A set of linguistic values is defined for each component of the objective function. In our case, only one linguistic value is defined for each variable. That is *strong connectivity, large ratio,* and *critical timing delay*, respectively.

As it was described in section 4, instead of finding maximum of (6), we are solving another problem :

$$Maximize\ G_1(x), \qquad (7)$$

where $G_1(x)=G_1(g_1(f_1(x_\eta)),g_2(f_1(x_\eta)),g_3(f_3(x_\eta)))$ is a function. $g_i(\bullet)$ is a degree of satisfaction of layout experts with objectives $f_i(\bullet)$ for $i \in \{1,2,3\}$. These degrees of satisfaction are described by membership functions on fuzzy sets of linguistic values. That is to say, $g_1(\bullet)$, $g_2(\bullet)$, and $g_3(\bullet)$ are respectively the membership functions for *strong connectivity, large ratio, and critical timing delay*.

Membership functions are very important for successful applications of fuzzy logic. They quantify the qualitative characteristics of parameters. Empirical data, if they are available, or general knowledge are used to build the membership functions. Otherwise, the subjective judgement of decision makers is employed to define them. In our case, membership functions are easy to build. They are assumed to be non-decreasing functions, because each function $g_i(\bullet)$ measures degree of satisfaction of objective $f_i(\bullet)$ for all $i \in \{1,2,3\}$. For instance, the larger is the connectivity $f_i(\bullet)$ to the partial placement, the higher is the degree of membership in the fuzzy set *strong connectivity*. Specific instances of these membership functions for linguistic values of the linguistic variables *connectivity, connectivity ratio*, and *timing delay*, are given in Fig. 3, Fig. 4 and Fig. 5. These functions reflect some level of knowledge about cell library available to an expert prior to layout. The membership functions may be adjusted according to the physical and electrical characteristics of the circuits. For example, the membership function for *strong connectivity* may be adjusted with respect to the average connectivity for different designs. To simplify possible changes in the shapes of membership functions they are constructed as piece-linear functions.

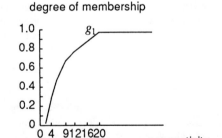

Fig. 3 Membership function for fuzzy set *strong connectivity*

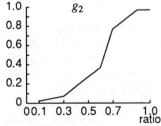

Fig. 4 Membership function for fuzzy set *large ratio*

In fuzzy logic maximization (or minimization) of the function $G_1(x)$ is formulated in terms of the values of linguistic variables as a rule-based algorithm. In our case, different rules represent alternative evaluations of the objectives. It is quite similar to the case when objectives are evaluated from different viewpoints. The results of evaluations of different rules are taken together to produce a value for making decisions. Two decision rules are formulated for this problem.

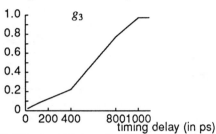

Fig. 5 Membership function for fuzzy set *critical timing delay*

Rule 1 : If a cell η has (*strong connectivity to the partial placement* or *large ratio*) and it produces *critical timing delay*, it is a *good candidate* to be placed next.

Rule 2 : If a cell η has *strong connectivity* and *large ratio* and it produces *critical timing delay*, it is a good candidate.

Here *a candidate* is a linguistic variable, and *a good candidate* is its linguistic value. The membership function of *a good candidate* is implicitly defined as the function $g(x_\eta)$ where $g(x_\eta)$ is a value resulting from an evaluation of a rule. At the first glance, it seems that Rule 2 is contained in Rule 1. But this is only true when these two rules are evaluated according to binary logic. It will be demonstrated below that in the fuzzy logic domain these two rules produce different results.

Finally, for each point $x_\eta \in X_t$, these rules are evaluated together to get a crisp value for decision making.

Evaluation of Rule 1 : It is easy to see that Rule 1 allows us to get a value of function $G_1(x)$ for any x_η in consideration in terms of the membership functions of linguistic values. Formally, it can be written according to laws of fuzzy logic in the following form :

$G'_1(x_\eta) = max\ [\ g_1(f_1(x_\eta)), g_2(f_2(x_\eta))] \times g_3(f_3(x_\eta))$.

In fuzzy logic "or" is realized by the function *max*. Fuzzy logic "and" is realized by algebraic *multiplication*. These two operators are mentioned in section 2. Notice, that *max* is not a compensatory operator. That is, $max\ (\ g_1(f_1(x_\alpha)),\ g_2(f_2(x_\alpha)))$ $>\ max\ (\ g_1(f_1(x_\beta)),\ g_2(f_2(x_\beta)))$ even if $min\ (\ g_1(f_1(x_\beta)),\ g_2(f_2(x_\beta))) >> min\ (\ g_1(f_1(x_\alpha)),\ g_2(f_2(x_\alpha)))$. For example, if cell β has a fairly large connectivity and a fairly large ratio, it is still ranked inferior to cell α, which has a very large ratio, but a very small connectivity. Obviously, it is not wise to make a decision solely based on Rule 1. The same argument can be applied to the situation when, instead of fuzzy "or", fuzzy "and" is used to combine the first two functions $g_1(\bullet), g_2(\bullet)$ and function *min* is used to realize Rule 1.

Evaluation of Rule 2 : Rule 2 proposes an another formulation of the function $G_1(x)$:

$$G''_1(x_\eta) = g_1(f_1(x_\eta)) \times g_2(f_2(x_\eta)) \times g_3(f_3(x_\eta)). \quad (8)$$

In fuzzy logic "and" operator is realized by algebraic *multiplication*, which is a compensatory operator. As a result, cell β may be ranked superior to cell α in the aforementioned case. However, if decision is made only based on Rule 2, it will not provide the best selection, either. Notice, that $g_3(\bullet)$ always participates in making decisions to account for timing factor. Evaluations for Rule 1 and Rule 2 are weighed according to following rule to produce a crisp value $G_1(x_\eta) = \lambda \times G'_1(x_\eta) + (1.0 - \lambda) \times G''_1(x_\eta)$, where $0 < \lambda < 1.0$ is the weighing factor. In our case, $\lambda = 0.5$, because it is assumed that these two rules are of equal importance. A function $G_1(x_\eta)$ gives a crisp value for the function $G_1(x)$ from formula (7). We compute for each cell η a value $G_1(x_\eta)$ and choose the cell with the largest value as the next cell for placement.

5.4 Numerical Example of Selecting a Cell

Let us assume, that *connectivity* $f_1=12$ for a candidate cell η, that corresponds to $g_1(12)=0.78$ degree of participation in the set *strong connectivity* from Fig. 3. Let *connectivity ratio* $f_2=0.7$, that corresponds to $g_2(0.7)=0.8$ degree of membership in the set *large ratio* according to Fig. 4. Let the average *timing delay* $f_3=800$ picoseconds, i.e., the degree of membership in the fuzzy set *critical timing delay* is $g_3(800)=0.75$ according to Fig. 5. Then, evaluation of the Rule 1 for these data gives $G'_1(x_\eta) = max(0.78, 0.8) \times 0.75 = 0.8 \times 0.75 = 0.6$. Evaluation of the Rule 2 gives $G''_1(x_\eta) = 0.78 \times 0.8 \times 0.75 = 0.468$. And, finally, the crisp value for the cell is $G_1(x_\eta) = 0.5 \times 0.6 + 0.5 \times 0.468 = 0.534$. This value is be used to compare the given cell with other cells.

5.5 Criteria for Selecting a Position

After a cell has been chosen, a position must be selected for this cell. The decision-making process involves analysis of the current status of the partial placement. The following criteria could be considered when a new position α is selected for a cell η added to the partial placement P_t:
(1). timing delay $f_4((P_t, H_t), \eta, \alpha)$,
(2). wasted area $f_5((P_t, H_t), \eta, \alpha)$ = *wasted area / (cell area + wasted area)*,
(3). effect on future utilization of chip area $f_6((P_t, H_t), \eta, \alpha)$.

A feasible position α for the selected cell η must first satisfy constraints on geometric fit and routability. It is clear that the set of feasible positions varies for different cells due to these two constraints. As the partial placement grows, its right boundary will be indented. Effect of this phenomenon is registered by criterion (3). The unevenness of the border in y-dimension may result in wasting a substantial chip area if large cells are placed next to the indented boundary of the partial placement. As one can see, selection of a position is a multi-criteria decision problem. Optimization based on any single objective seldom creates good layouts.

5.6 Applications of Fuzzy Logic to Selection of Positions

The approach similar to the one described in the section on cell selection can be used to find positions for cells. A position should be considered as the most satisfactory one by the layout experts.

Three objective functions $f_4((P_t, H_t), \eta, \alpha)$, $f_5((P_t, H_t), \eta, \alpha)$, and $f_6((P_t, H_t), \eta, \alpha)$ were defined for this purpose in the previous section. The problem of deciding a position α for the selected cell η becomes

$$\text{Minimize } F_2(x), \quad (9)$$

where $F_2(x) = (f_4(x_\alpha), f_5(x_\alpha), f_6(x_\alpha))$ is a vector-function. There is no unique way to introduce a norm for a vector-function. In classical norms such as Eucleadian norm and others are counterintuitive in such situations and produce bad results in applications. Fuzzy logic provides solutions for such situations.

Three linguistic variables *timing delay, wasted area,* and *border evenness* are respectively defined for the above three functions. A set of linguistic values is defined for each linguistic variable. In our case, two linguistic values *small timing delay* and *large timing delay* are defined for *timing delay*. Two linguistic values *large wasted area* and *small wasted area* are defined for *wasted area*. Two linguistic values *good border evenness* and *bad border evenness* are also defined for *border evenness*.

Instead of finding an optimum of (9), we are solving the following problem:

$$\text{Maximize } G_2(x), \quad (10)$$

where $G_2(x) = G_2(g_4(f_4(x_\alpha)), g_5(f_5(x_\alpha)), g_6(f_6(x_\alpha)))$ is a function of membership functions. $g_i(\bullet)$ is a degree of satisfaction of layout experts with value of objective $f_i(\bullet)$ for $i \in \{4,5,6\}$. Simultaneously, we minimize a function $H_2(x) = H_2(h_4(f_4(x)), h_5(f_5(x)), h_6(f_6(x)))$ which characterizes a degree of dissatisfaction of layout expert with values of objectives $f_i(\bullet)$ for $i \in \{4,5,6\}$.

Membership functions for these linguistic values are omitted here due to the lack of space. These membership functions are piece-linear and could be easily changed during tuning. They are constructed based on analysis of available data from the underlying domain. A clock period is an important element to shape membership functions for the *critical timing delay*. In our case, empirical data about average timing delay and average wiring length are available. They become great assets in building membership functions. If these data are not immediately available, the subjective judgement of decision makers will play a major role. In this case, after some trial runs, the shapes of membership functions can be easily adjusted.

Two rules have been introduced for selection of positions:
Rule 1: If a position produces *small timing delay* and *small wasted area* and *good border evenness*, it is a *good position*.
Rule 2: If a position does *not* produce ((*large timing delay* or *large wasted area*) and *bad border evenness*), it is a *good*

position.

Here *position* is a linguistic variable, and *good position* is its linguistic value. These two rules are then evaluated to get a crisp value for each position α. Our approach works as follows :

Evaluation of Rule 1 : Rule 1 is based on the evaluation of the function $G_2(x)$. According to the laws of fuzzy logic, Rule 1 can be realized in the following form :

$$G'_2(x_\alpha) = min[g_4(f_4(x_\alpha)), g_5(f_5(x_\alpha)), g_6(f_6(x_\alpha))]$$

Evaluation of Rule 2 : Rule 2 is based on the evaluation of the function $H_2(x)$. Respectively, Rule 2 has the following formulation :

$$G''_2(x_\alpha) = 1.0 - H''_2(x_\alpha), \quad \text{where}$$
$$H''_2(x_\alpha) = max[h_4(f_4(x_\alpha)), h_5(f_5(x_\alpha))] \times h_6(f_6(x_\alpha)).$$

Values obtained from evaluation of Rule 1 and Rule 2 are used to obtain a crisp value for the position in consideration, i.e. $G'_2(x_\alpha)$ and $G''_2(x_\alpha)$ are combined to get a crisp value $G_2(x_\alpha) = \lambda \times G'_2(x_\alpha) + (1.0 - \lambda) \times G''_2(x_\alpha)$, where $0 < \lambda < 1.0$ is the weighing factor. In our case, we take $\lambda = 0.5$. Crisp values $G_2(x_\alpha)$ are compared for all possible positions α and a position with the largest value of $G_2(x_\alpha)$ is selected for placement. In the case of cell selection we used two different ways to characterize satisfaction with each possible solution. In this case the crisp value $G_2(x_\alpha)$ is obtained by maximizing the degree of satisfaction with the solution and by minimizing the degree of dissatisfaction with the same solution.

6. Experimental Results

The approach described above is tested on three industrial chips provided by Control Data Corporation. The implementation is based on the system developed in [6]. Only the decision-making procedures for selection of cells and positions are modified to implement the fuzzy logic approach. Table 1 lists the main characteristics of these chips. Table 2 compares our results with the results from [6]. It shows a dramatic improvement in timing performance even though the average connection length does not improve too much. It is worth to mention that we always obtained good timing for these three circuits for different shapes of membership functions. Moreover, no unrouted connections are produced in ETA test case. This is a very important characteristic of the solution because the unrouted connections could produce many long paths when they are routed. The increase in running time for all three chips is small, since the membership functions are relatively easy to evaluate.

	clock period	# of cells	# of nets	# of pins	utilization of cells
CDC-A	120 ns	1156	2515	7484	66%
CDC-B	120 ns	1705	2537	7330	61%
ETA	21 ns	5092	4983	13781	76%

Table 1. Three industrial sea-of-gates designs

	JUNE3	Fuzzy Approach
slack of the longest path (CDC-A)	+19 ns	+31 ns
# of long paths	0	0
average connection length	19	17
# of unrouted connections	0	0
slack of the longest path (CDC-B)	+8 ns	+65 ns
# of long paths	0	0
average connection length	17	21
# of unrouted connections	0	0
slack of the longest path ((ETA)	0 ns	+2 ns
# of long paths	0	0
average connection length	15	15
# of unrouted connections	3	0

Table 2. Placement results

7. Conclusion

We successfully applied fuzzy logic to optimize a process of decision making in physical design. Multiple objectives such as utilization of area, routability, and timing were considered simultaneously and balanced by fuzzy logic algorithms. Our experiments demonstrated that solutions obtained by fuzzy logic were of much better quality than those achieved by standard techniques.

8. References

[1] Michael Burstein and Mary N. Youssef, "Timing Influenced Layout Design", in Proc. of ACM/IEEE 22nd Design Automation Conference, pp. 124-130, 1985.

[2] A.E. Dunlop, V.D. Agrawal, and et. al., "Chip Layout Optimization Using Critical Path Weighting", Proc. 21st Design Automation Conf., pp. 142-146, 1987.

[3] M. Igusa, M. Beardslee, and A. Sangiovanni-Vincentelli, "ORCA : A Sea-of-Gates Place and Route System", Proc. 26th Design Automation Conf., pp. 122-127, 1989.

[4] M. Jackson, E.S. Kuh, "Performance-Driven Placement of Cell Based IC's", Proc. 26th Design Automation Conf., pp. 370-375, 1989.

[5] M. Marek-Sadowska, S.P. Lin, "Timing Driven Placement", in Proc. ICCAD'89, pp. 94-97.

[6] Suphachai Sutanthavibul and Eugene Shragowitz, "JUNE: An adaptive Timing-Driven Layout System", in Proc. of ACM/IEEE 27th Design Automation Conference, pp. 90-95, 1990.

[7] Suphachai Sutanthavibul and Eugene Shragowitz, "Dynamic Prediction of Critical Paths and Nets for Constructive Timing-Driven Placement", in Proc. of ACM/IEEE 28th Design Automation Conference, pp. 632-635, 1991

[8] M. Razaz and J. Gan, "Fuzzy Set Based Initial Placement for IC Layout", In Proc. European DAC, pp. 655-659, 1990.

[9] L.A. Zadeh, "Fuzzy Sets", Information and Control 8, 338-353 (1965).

[10] L.A. Zadeh, "The Concept of a Linguistic Variable and its Application to Approximate Reasoning-I", Information Science 8, pp. 199-249 (1975).

[11] Hans J. Zimmermann, "Fuzzy Sets , Decision Making, and Expert Systems", Kluwer Academic Publishers, Boston, 1987.

A Fuzzy Algorithm for Multiprocessor Bus Arbitration

Robert T. Tran, Timothy R. Slator and Anaikuppam R. Marudarajan
Department of Electrical and Computer Engineering
California State Polytechnic University, Pomona
Pomona, CA 91768-4065

Abstract—The performance of a multiprocessor system is largely contingent upon the efficiency of the elected bus arbitration algorithm. In many cases the bus arbitration task is not a precise one, rather, it is environment dependent and thus this situation makes this task a likely candidate for fuzzy implementation. In this paper a fuzzy algorithm for multiprocessor bus arbitration is proposed. Its hardware design strategies are also addressed.

Keywords: Arbitration Protocol, Multiprocessor System, Fairness, System Throughput, Fuzzy Arbitration Protocol.

I. INTRODUCTION

A multiprocessor architecture improves a system's throughput and reliability because it is a network of processor modules where each module usually contains its private memory and local I/O channels [2]. The performance of such a system would degrade very severely unless the contention for shared resources is arbitrated properly. Although the traditional operating system scheduling disciplines such as shortest job first (SJF), highest response ratio next (HRN) can, in principle, be adopted to handle multiprocessor arbitration, they are rarely used owing to their high implementation cost. Also, the performance of customarily used heuristic based multiprocessor arbitration protocols such as equal-priority, unequal-priority, rotating-priority (round-robin), random-delay and queuing protocols [1] is primarily limited by their rigid nature of operation. These results along with randomness associated with the actual operating environment of a multiprocessor system strongly motivate one to devise an adaptive arbitration protocol which would optimize the twin performance goals: average wait time and fairness by making decisions that vary with time. At this point, the use of a fuzzy protocol appears to be very attractive because it not only blends human expertise and heuristics into an adaptive framework but also has the ability to deal with imprecise information.

The rest of this paper is organized as follows. Section II of this paper describes the fuzzy arbitration protocol (FAP) and the structure of the required knowledge base. The actual simulation results are reported and analyzed in Section III. Section IV is devoted to address the hardware implementation of the proposed FAP and related issues. Finally, section V presents the conclusions.

II. FUZZY ARBITRATION PROTOCOL

Current arbitration protocols use a single parameter, the priority of a processor, to describe the system status and attempt to optimize the system performance around that parameter. The performance of current protocols depends on the probity of the input parameter. As a result, the protocols are insensitive to changes unrelated to the input parameter. A fuzzy arbitration protocol on the other hand uses many parameters to describe the system status. These parameters represent the actual work load and work distribution in a multiprocessor system: processor wait time before receiving the bus grant, processor job length or frequency of requests of a processor. FAP can adapt to a changing environment and can also be modified to weight certain parameters more than others.

FAP can be implemented using a typical fuzzy system architecture: fuzzifier, fuzzy knowledge base and defuzzifier. FAP fuzzifies various parameters using their membership functions. FAP then applies these fuzzified values to the rule base and performs a fuzzy inference process to obtain a defuzzified output.

Three parameters are chosen to represent the status of a processor: wait time (WAI), job length (LEN) and request frequency (FRE). The status parameters describe with as much detail as possible the resource needs of each processor. The output of the FAP is the arbitration index (AIX). The arbitration index represents the likelihood that a processor will get a bus grant. An arbitration index is calculated every arbitration cycle for each processor. The processor with the highest arbitration index is given the bus grant.

The status parameters and the output of the FAP are described in details as follows:

a. WAI: Wait time measures the time that a processor has waited since it made a resource request. A long wait time implies a higher likelihood due to the need to prevent a processor from being starved. Conversely, a short wait time implies a lower likelihood. The wait time is incremented at the end of an arbitration cycle if the request is not acknowledged. The wait time is reset when a request is granted. WAI is fuzzified to five levels with the following hedges: *very short, short, medium, long, very long*.

b. LEN: Job length indicates the number of bus cycles needed for a processor to access a resource and it is proportional to the data transfer between a processor and a resource. A long job length implies a lower likelihood than short job length because it has been shown that the shortest job first algorithm minimizes the average wait time of a multiprocessor system. The job length for memory access may be determined from the size of a DMA transfer from common memory to local memory. LEN is fuzzified to five levels with the following hedges: *very short, short, medium, long, very long*.

c. FRE: Request Frequency indicates the bus traffic intensity. Request Frequency measures the number of requests from a processor. Priority for a request frequency may be assigned in one of two ways: low priority for high request frequency or high priority for high request frequency. The former guards against starvation while the latter ensures that I/O intensive processors are not bogged down. In this case, the priority of a request frequency is assigned to low for high request frequency. System characteristics determine how this parameter is set. FRE is fuzzified to five levels with the following hedges: *very seldom, seldom, average, often, very often*.

d. AIX: The arbitration index indicates the likelihood that a processor will be given the bus grant. AIX is fuzzified to five levels with the following hedges: *very low, low, medium, high, very high*.

Unlike a control problem, arbitration cannot make use of output feedback. The current arbitration decision is independent of the previous arbitration decision, therefore a unique output must exist for every input combination. The consequence of this observation is that all rules must be specified. The number of rules in the rule base are given by f^i, where f is number of fuzzifiers or levels of an input and i is the number of inputs. In this case, 125 rules are needed to completely specify the arbitration system.

Each input is weighted according to human intuition. No single parameter appears to be more important. Wait time and request frequency govern fairness while job length minimizes the average wait time. However, in some instances, it is very difficult to achieve both goals of high system throughput and high degree of fairness. In specifying the rule base, job length is considered to be more important. Assigning an arbitration index in this manner is not completely arbitrary, but based on the idea that even if a processor has been waiting a long time it is best to give the bus grant to the processor with a shorter job.

Typical rules are as follows:

if   *WAI is very short and*
        LEN is very short and
        FRE is very seldom
then   *AIX is medium*

if   *WAI is medium and*
        LEN is medium and
        FRE is very often
then   *AIX is low*

if   *WAI is very long and*
        LEN is long and
        FRE is average
then   *AIX is high*

if   *WAI is very long and*
        LEN is very short and
        FRE is often
then   *AIX is very high*

III. SIMULATION RESULTS

To study the performance of FAP and compare its performance with other protocols, a computer program is written to simulate the bus arbiter in a multiprocessor system with a number of processors requesting access to a single resource. The following arbitration protocols are simulated: fixed-priority (unequal priority), rotating-priority, last-grant least-priority (LGLP), random priority, shortest job first (SJF) and FAP. This simulation program assumes the following:

a. No preemption is assumed for the processors. Once a processor is given the bus grant, it will have the bus until its job is finished.
b. When a request is rejected, the processor must resubmit the request on the next arbitration cycle.
c. When a request is pending for a processor, any subsequent requests from that processor are queued.
d. The universe of discourse is chosen to be [1,32].
e. In FAP, the membership functions for the inputs and outputs are triangular functions. The output defuzzification is calculated using the centroid method.
f. The number of processors may be from 3 to 20.
g. Bus traffic intensity is the probability of a processor

generating a request and is a settable simulation parameter.

The job request arrivals and the job lengths are simulated by uniformly distributed random number sequences.

The performance of the simulation program is assessed by calculating the following measures: the average wait time and the standard deviation of the wait time. When a processor receives the bus grant, the elapsed time between the request and the grant is recorded. When a processor does not get the bus grant, its wait time is incremented. The average wait time indicates the system throughput while the standard deviation of the wait time indicates the degree of fairness of the arbitration.

In this simulation, random sequences of requests of up to 1000 cycles are generated for 3, 5, 10 and 20 processors and fed into the arbitration simulator for each of the protocols.

Average wait time and fairness are normalized and plotted as a function of the number of processors to identify a protocol's success at meeting the performance goals. The ideal protocol minimizes the average wait time and maximizes fairness. A 45° line passing through the origin represents the ideal protocol performance. Figures 1, 2, and 3 show performance plots for traffic intensity of 0.2, 0.5, and 0.8, respectively. LGLP, rotating priority, and random priority protocols exhibit the highest degree of fairness; however, the average wait time increases dramatically as the number of processors increase. Fixed priority and SJF protocols minimize the average wait time but are the least fair. FAP follows the ideal performance line the closest indicating that FAP responds positively to changes in the system environment.

IV. HARDWARE DESIGN STRATEGIES

The intensive computations required by the fuzzy inference make traditional sequential processing impractical in real time applications. A useful fuzzy arbiter can be obtained when the inference process for each processor is done concurrently and with a hardware based centroid calculation. At present, fuzzy logic hardware can be built in three ways: using commercially available fuzzy logic controller chips [4], using Field Programmable Gate Array [3] and using custom VLSI fuzzy logic controller [5]. Due to the cost and design cycle, the authors favor the first two approaches.

In the following paragraph, a brief summary of the experiment using the first approach is described. Using the Neuralogix NLX230 chip, a fuzzy arbitration unit is built. It is capable of reading input parameters from two processors: wait time and job length and provides the arbitration index of the winner. This fuzzy arbitration unit can be easily expanded to handle systems with more than two processors. This experiment is consistent with the assumptions made in the previous section.

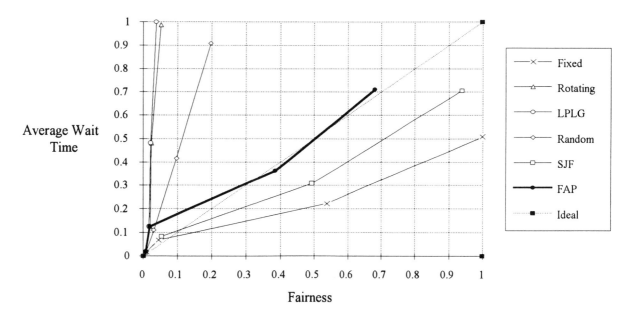

Fig. 1. Performance Plot with Traffic Intensity 0.2

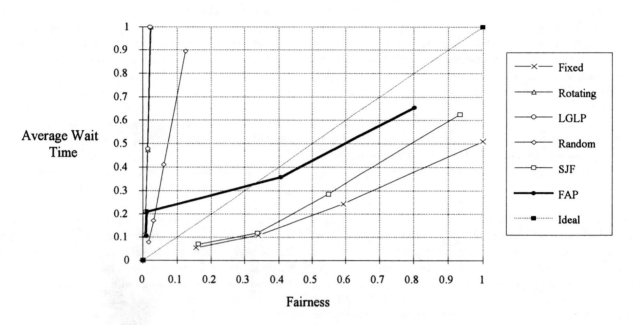

Fig. 2. Performance plot with Traffic Intensity 0.5

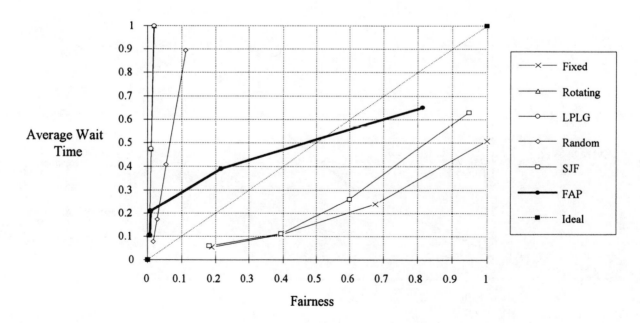

Fig. 3. Performance Plot with Traffic Intensity 0.8

V. Conclusions

The ideal arbitration protocol balances the system's throughput against fairness. The performance goals are balanced when the average wait time is minimized and fairness in arbitration is maximized. FAP clearly shows superior performance in optimizing the performance goals. Consequently, FAP appears to be a viable alternative method for handling multiprocessor bus arbitration. In addition, FAP demonstrates immunity to changes in the system environment such as additional processors and bus traffic intensity. Using the hardware approach, one can implement a practical fuzzy arbiter for multiprocessor systems. By modifying the rule base, a designer can easily tune or enhance the performance of the FAP for a specific system environment. Finally, the authors find that with the same rule base but with manipulated input parameter combinations, FAP can mimic any other protocol.

Acknowledgment

The authors wish to express their thanks to Dr. Y Cheng and Dr. C. Ford-Livene of California State Polytechnic University, Pomona, CA for their suggestions during the course of this research work.

References

[1] F. E. Guibaly, Design and Analysis of Arbitration Protocols, *IEEE Transactions on Computers*, vol. 38, pp.161-171, February 1989.

[2] K. Hwang and F. A. Briggs, *Computer Architecture and Parallel Processing*, New York, NY: McGrawHill 1984.

[3] M. A. Manzoul and D. Jayabharathi, Fuzzy Controller on FPGA Chip, *IEEE International Conference on Fuzzy Systems 1992*, pp. 1309-1316, March 1992.

[4] NeuraLogix Fuzzy MicroController NLX230 Data Sheet, Sanford, Florida, August 1991.

[5] H. Watanabe, W. D. Dettloff and K. E. Yount, A VLSI Fuzzy Logic Controller with Reconfigurable, Cascadable Architecture, *IEEE Journal of Solid-State Circuits*, vol. 25, pp. 376-382, April 1990.

Chapter 8: Sensors

IMPROVING DYNAMIC PERFORMANCE OF TEMPERATURE SENSORS WITH FUZZY CONTROL TECHNIQUE

Wang Lei, Volker Hans
Measurement and Control, University of Essen
Schützenbahn 70, D-4300 Essen 1

Summary

The dynamic and precise measurement of temperatures is limited by the transient reaction of the sensor. The self-heating process of the sensor can be accelerated by a feeding current which is controlled by a fuzzy-controller.

1. Introduction

In industrial processes temperatures with dynamic variations in time request a quick response of measuring and controlling devices. The speed of the measuring system is essentially specified by the time behaviour of the primary sensor. Therefore it is necessary to improve the dynamic behaviour of the temperature sensor.

In a first approximation the transient reaction of the sensor is assumed to be exponential. The step response can be described by a single of by a combination of several e-functions. The relatively long time constants are conditioned by the mechanical construction of the sensor. They lead to long times until the final value of the temperature is determined. There are only a few possibilities to accelerate the exponential characteristics: the changing of the construction of the sensor which mostly cannot be realized, and secondly the series connection of analog components which approximately show the inverse response character of the sensor /1/.

In the following a new method for improving the dynamic response of the slow temperature resistance sensor shall be presented.

2. Conception for improving the dynamic performance

2.1 Fundamental ideas

The problem is to force a fast response of the slow resistance sensor. This can be realized by a fast approximation of the eigen-temperature of the sensor to the temperature to be measured by feeding the sensor with matched current. The procedure is only qualified for a positive rise of temperature, naturally. But in most cases this is of greater importance than temperature drops. It is presupposed that the temperature rises exponentially:

$$(T - T_0) = (T_u - T_0)(1 - e^{-t/T_K}), \qquad (1)$$

where T_0 ist the initial temperature, T_u is the temperature to be measured and T_K is the time constant of the sensor. The eigen-temperature of the sensor which is fed by a current with rise on account of the heat exchange between the thermistor and the temperature to be measured. The temperature rise can be described by

$$(T - T_0) = \frac{R_{th}}{C} \cdot I^2 + (T_u - T_0)(1 - e^{-t/T_K}). \qquad (2)$$

The first term is the self heating of the sensor, the second term describes the heat exchange approximately.

2.2 Conception of electrical heating

There are two problems which must be taken into consideration:

1. How does the sensor know that the eigen-temperature is higher, lower or equal to the temperature to be measured?

2. Which current is necessary to feed the sensor particulary since the temperature to be measured is unknown and even the temperature difference to the eigen-temperature of the sensor is unknown, too?

To solve the problem a new method with intelligent comparison of temperatures has been developed. The principle is described by means of a temperature step function. At first the sensor is fed with nominal current for an exactly defined time interval and the change of the sensor temperature is observed. After this change which is proportional to the temperature difference between the current intrinsic temperature and the temperature to be measured. The current necessary for an accelerated intrinsic heating is calculated and applied to the sensor. After this the sensor is fed again for an exactly defined time interval with nominal current and the change of the sensor temperature is observed. After this change of temperature the sensor is fed again with a high fuzzy-controlled current for further self-heating.

As mentioned above the feeding current must be controlled. It is of greatest difficulty to control the current with conventional analog control devices. Several factors influence the current, and a mathematical solution has a very high complex. The feeding current must not lead to any overshoot of the temperature.

The fuzzy-theory is an alternating method to exact mathematics. It is well-suited where mathematical descriptions are not possible. Fuzzy-control basing on fuzzy-theory is a non-linear control /2, 3/. The advantages are high speed, no overshooting and robust characteristics.

3. Fuzzy-controller

The feeding current of the sensor is controlled as output-variable of a fuzzy-controller with the two input-variables temperature difference ΔT and measured temperature T. Figure 1 shows the principle

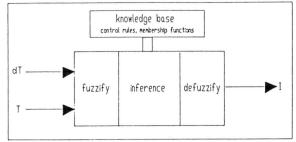

Figure 1: The applied fuzzy-controller

The fuzzy-controller is working in three steps: fuzzification, fuzzy-inference and defuzzification by application of fuzzy-variables, membership function and control rules which must be determined by experimental and subjective knowledge. The fuzzification means the linguistic interpretation of a technical quantity in general, the temperature in particular. The defuzzification correspondingly is the transformation of a linguistic term into a technical quantity, particularly the feeding current of the sensor. The influence draws the conclusions from the knowledge based rules, which fulfil the qualifications according to the membership functions.

The following look-up table with COG (Center of Gravity) defuzzification shows the relation between input and output variables. The values indicate a reference measurement for the heating current.

dT / T	-20	-10	0	10	20	30	40	50	60	70
-20	xxxx	xxxx	0.01	0.35	0.49	0.54	0.70	0.84	0.90	1.00
-10	xxxx	0.02	0.02	0.41	0.54	0.60	0.75	0.90	0.96	1.09
0	0.03	0.03	0.03	0.54	0.70	0.75	0.80	0.96	1.01	xxxx
+10	0.05	0.05	0.05	0.60	0.75	0.90	0.96	1.09	xxxx	xxxx
+20	0.08	0.08	0.08	0.75	0.80	0.96	1.01	xxxx	xxxx	xxxx
+30	0.12	0.12	0.12	0.90	0.96	1.09	xxxx	xxxx	xxxx	xxxx
+40	0.18	0.18	0.18	0.96	1.01	xxxx	xxxx	xxxx	xxxx	xxxx
+50	0.26	0.26	0.26	1.09	xxxx	xxxx	xxxx	xxxx	xxxx	xxxx

Figure 2: Look-up table with COG

4. Simulation and experimental results

At first the procedure presented has been simulated by software, where the heat compensation within the sensor is not taken into account. The result is shown in figure 3:

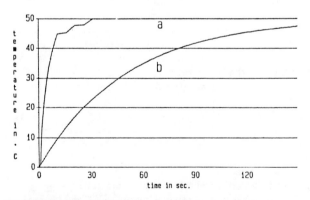

Figure 3: Simulation of fuzzy-controlled dynamic temperature measurement.
a. Characteristic with fuzzy controlled current
b. Natural characteristic of the sensor

The figure shows that the final temperature of the fuzzy controlled sensor is reached in comparison to the sensor without controller within considerable short time.

For an experimental test a relatively great and slow sensor of type YSI 44031 has been applied and has been fed with a fuzzy-controlled current. The experimental result of the sensor which is heated with fuzzy-controlled current is shown in curve a, figure 4, and the sensor which is fed with nominal current is shown in curve b, figure 4, for a temperature step of $41K$.

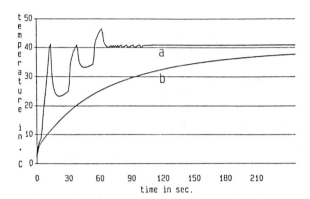

Figure 4: Experimental result of fuzzy-controlled dynamic temperature measurement.
a. Characteristic with fuzzy controlled current
b. Natural characteristic of the sensor

It is evident that the final temperature is already reached after the first heating procedure. After switching off the heating current the temperature decreases on account of heat compensation effects within the sensor. After a minimum the sensor is heated again with fuzzy-controlled current and the temperature rises. The process is repeated until the final temperature is reached. The measuring time is considerably reduced in comparison to the sensor which is fed with nominal current.

5. Conclusion

A fuzzy control conception has been realized for improving the dynamic performance of a slow temperature sensor. Simulations and experiments show that the sensor response can be forced by a fuzzy-controlled feeding current which accelerates the step response time evidently. Although the procedure described is only qualified for positive rise of temperatures the resulting dynamic performance is evidently improved.

6. References

1. Hofmann, D.: MSR.10 (1967), pp. 20-27.

2. Zadeh, L. A.: Information and control 8 (1965), pp. 338-353.

3. Zadeh, L. A.: Information and control 12 (1968), pp. 94-102.

Multi-Sensor Integration System with Fuzzy Inference and Neural Network

Toshio FUKUDA, Koji SHIMOJIMA, Fumihito ARAI, and Hideo MATSUURA

Dept. of Mech. Eng., Nagoya University,
Furo-cho, Chikusa-ku, Nagoya, 464-01, Japan

Abstract

In an intelligent robotic system, sensors are important for the recognition of the system state and environmental state. Consequently, the sensor integration system (SIS) has been studied to apply to more fields. In this paper, we show that SIS with multiple sensors can expand the measurable region with high accuracy and that operators can use the system as easily as a single high performance sensor system.

Previous systems reported so far were based on statistical techniques. Therefore, these systems did not have the flexibility to the change or replace of sensors. This paper presents an approach to the SIS using the knowledge data base of sensors; the proposed SIS allows for the changing/replacing of sensors. The system consists of four subsystems: 1) sensors performing as hardware sensing devices, 2) knowledge data base of sensors (KBS), 3) fuzzy inference, 4) neural network (NN).

Sensor's measurement value includes a measurement error. This system estimates the measurement error of each sensor using the fuzzy inference. Fuzzy rules are made from the KBS. Measurement values are integrated by the NN. All inferred measurement errors and measurement values are put into the NN. Then NN's output gives the integrated measurement value of multiple sensors. The proposed system is shown to be effective through a series of extensive experiments.

1. Introduction

We are using many kinds of sensors in many fields. Sensors will be applied to many other fields. The demands of sensors' specifications spread out higher accuracy and wider application. For example, a sensor for recognition environmental states of a moving robot system must have wide measurement range and stability against environmental change, and a sensor for measuring a super high precision product must have high accuracy and wide measurement range. It is too difficult for a single sensor to satisfy all these demands. Even if it could satisfy these requirements, it would be very expensive. In order to satisfy these demands, the sensor integration system (SIS)[Luo and Kay 1989] has been studied. The SIS uses a combination of multiple sensors and operators can use this system as easily as a single high performance sensor system.

Multi-sensor system has two major problems. One is how to recognize the most accurate sensor. The most accurate sensor depends on some environments of sensors. Various SIS methods have been reported so far: hypothesis testing by sensor models and Bayesian approach [Durrant-Whyte 1988], confidence distance matrix [Luo and Lin 1988], estimation by performance and cost criteria [Zheng 1989], estimation of cost by Bayesian [Richardson and Marsh 1988], and using Kalman filter [Nakamura and Xu 1989]. Most of these systems are based on statistical techniques, therefore they can use sensors effectively in stable environment. Without having the knowledge data bases of the sensors' specification, they do not have the flexibility against a changing/replacing of sensors or a change of sensor's environment.

Another problem is how to judge the most accurate sensor is the most suitable for using in the robotic system. The suitable sensor depends on the relation among sensors' position. The relation among sensors' position sometimes causes failure of measuring.

In this paper, we present the SIS which has the flexibility against the changing/replacing of sensors and the changing of sensor's environment. This system has the knowledge data base of sensors' specification (KBS) and estimates the measurement error of each sensor's measurement value by the fuzzy inference with the KBS. Then, the robotic system can recognize the most accurate sensor by the inferred measurement error. In addition, we use neural network (NN) for judgment the most suitable sensor. The estimated measurement errors and measurement values are put into the neural network (NN). Then the NN gives the integrated value of multiple sensors.

2. Sensor Integration System

Proposed system consists of four subsystems: 1) sensors as hardware sensing devices, 2) knowledge data base of sensors (KBS), 3) fuzzy inference, 4) neural network (NN). The total system is shown in fig. 1.

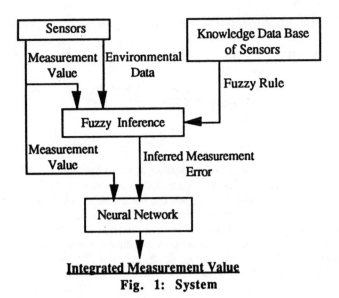

Fig. 1: System

2.1 Knowledge Data Base of Sensors

Sensors' specifications are written in the knowledge data base of sensors. They are: 1) measurement region, 2) measurement precision, 3) resolution of sensor, 4) environmental restrictions to use sensor such as condition of temperature, brightness and so on, and 5) error affected by environments. For 4) and 5), every factor affecting measurement values of each sensor is written in this data base. The followings are data examples of the non-contact gap sensor, which is used in the experiments:

    Measurement range: 0 - 2 mm
        error rate: 0.5 %
    Recommended temperature range: 0 - 100 Degree.
    Temperature drift.   : 0.06 %

2.2 Fuzzy Inference

Measurement value of sensors may contain some errors. Measurement error is caused by the principle of measurement method and by the affection of environments. If magnitude of measurement error can be estimated, we can obtain the measurement value which is the closest to the true value and the most accurate.

Possible cause of sensor measurement error is considered as followings: if the sensor uses semiconductors, the cause will be drifts of temperature, and in the case of optical techniques, the cause will be the brightness of the environment. These physical phenomena are vague and the influence on sensors is not clear either.

Therefore, using the fuzzy inference, this system estimates these values and infers considerable maximum errors. This system uses the simplified fuzzy inference, which has characteristics of less calculation, and the defuzzificated values are continuous. The followings are the case of two inputs (X,Y) and one output (Z). Equation (1) represents the i-th rule, where C_i is the real number, and the individual rules' results are given in eq. (2). The over-all result of fuzzy inference is given by eq. (3).

The Rule of Fuzzy Inference:
    R_i : If X is A_i Y is B_i then Z is C_i (i=1,···,n)   (1)
The Fitness of the i-th Rule:
    $F_i = F_{Ai}(X) \cdot F_{Bi}(Y)$ (2)

$$Z = \frac{\sum_{i=1}^{n} F_i \cdot C_i}{\sum_{i=1}^{n} F_i} \qquad (3)$$

Membership function and the deduced action clause of each rule are determined automatically based on the sensor measurement region, sensor precision, usable sensor region for each environment and effect from each environment to the measurement value. In this way, the system can estimate the maximum measurement error easily when the sensor is replaced. The membership function has the shape of trapezoid in this paper. Representation of the membership function for the i-th sensor measurement region is shown as fig. 2. Figure 3 shows the system outline of error estimation fuzzy inference.

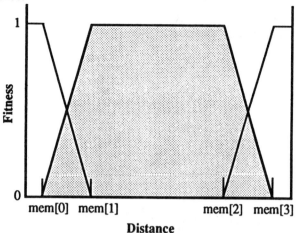

Fig. 2: Membership Function

mem[0]=r[0][i]
mem[1]=r[0][i]+p[i]•(r[1][i]-r[0][i])
mem[2]=r[1][i]-p[i]•(r[1][i]-r[0][i])
mem[3]=r[1][i]
r[0][i]: Lower limit of the i-th sensor measurement region.
r[1][i]: Upper limit of the i-th sensor measurement region.
p[i]: Measurement precision of the i-th sensor.

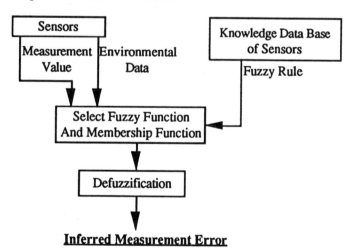

Fig.3: System outline of error estimation with fuzzy inference

2.3 Neural Networks

The neural network (NN) is expressed as an artificial mathematical model. Characteristics of NN are:
1) A large and varied quantity of inputs can be processed quickly by parallel distributed processing, 2) NN gives expected outputs based on learning, 3) NN gives almost the expected outputs against inputs without learning experience, if learning patterns are optimized. 4) NN can change the processing pattern not by changing the NN's program but by changing NN's learning pattern.

The NN's learning method adopted here is the back propagation method [Rumelhart et al., 1986]. The NN is used for the integration of measurement values from multi-sensors. According to the NN's 4th characteristics, by changing the learning data patterns, we can integrate a new multi-sensor system. The input/output relationship is represented as follows:

$$\text{Unit}_i = \sum_{j=1}^{J} W_{ij} \cdot O_j + B_i$$

$O_j = f(\text{Umit}_j)$

$f(X) = 1/(1+\exp(-X))$

Unit i: Input value of the i-th Unit
O_j: Output value of the j-th Unit in the previous layer $(j=1,...,J)$
W_{ij}: Weight connecting the j-th Unit with the i-th Unit
B_i: Bias of the i-th Unit
$f(X)$: Sigmoid function.

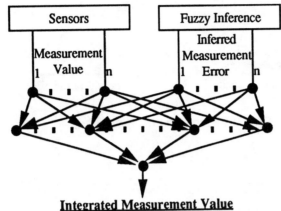

Fig. 4: Outline of Neural Networks

Learning is carried out by changing connection weights. When an input pattern is p, the output of the i-th unit is Opi, and the expected output of the i-th unit is Dpi. The error function E is shown by eq.(4). Connection weights are changed in order to reduce the value of E. Connection weights change is given by eq.(5), where η is the coefficient of learning:

Error function:
$$E = \sum_{i=1}^{I} \frac{(D_{pi} - O_{pi})^2}{2} \quad (4)$$

Value of changing connection weight:
$$\Delta w_{ij} = \eta \frac{\partial E}{\partial w_{ij}} \quad (5)$$

In this paper, we use 3-layered NN employing a hidden layer and one input/output layer.

Figure 4 shows the outline of NN used in this paper. NN's inputs are the normalized measurement values and the normalized inferred error values of each sensor. These values are given as follows:

$M_i^* = M_i / M_{max}$
$E_i^* = E_{min} / E_i \quad (i=1,2,...,n)$

M_i: Measurement value of the i-th sensor
E_i: Inferred error of the i-th sensor
M_{max}: The maximum of all measurement values
E_{min}: The minimum of all inferred errors.

3. Experimental Methods and Results
3.1 Experimental Methods

Learning pattern data of the NN are made by the sensor simulator. Simulated sensors are two different types of non-contact gap sensors which are used in experiments. Each sensor has the different measurement region and precision. Sensor 1's measurement region is 0 mm to 2 mm and precision is $1*10^{-6}$ m. Sensor 2's measurement region is 0 mm to 10 mm and precision is $1*10^{-5}$ m. The sensor simulator is designed considering the following matters: 1) measurement region, 2) resolution of the A/D board and 3) setting position of two sensors.

The experiments are carried out by changing distance between sensors and the measurement object. Each sensor is set at the tip of 5 axis industrial robot shown in fig. 4. As the robot moves, the distance between the sensors and the object, which is the surface of metal materials, changes. These changes are measured by all sensors at the same time. The measurement unit of sensors is voltage, and each data is gotten into the memory by the A/D board. Then it is transferred into the computer.

When we use multiple distance measurement sensors, the offset caused by the difference in each sensor's setting position is the most important. It is very difficult to cancel this offset because the offset value cannot be obtained. Based on the offset values, the selection of best sensor is changed. In this experiment, when sensor 1 is closer to the object than sensor 2, we define the offset as "plus," and when sensor 2 is closer to the object than sensor 1, the offset is "minus."

Experiments are carried out 3 times. At first, sensors are installed so that the offset is nearly equal to 0 m. Secondly, sensors are set so that the offset is to be "plus" by adjusting the sensor 1's position. Finally, sensors are set so that the offset becomes "minus" by adjusting the sensor 1's position.

In these experiments, three methods of SIS are used:
   1) If Clause: When measurement value of sensor 2 is within the measurement region of sensor 1, sensor 1's measurement value is selected.
   2) Fuzzy Inference: The sensor having the minimum inferred error which is estimated by the fuzzy inference of all sensors is selected.
   3) NN+Fuzzy: Each sensor's measurement value and inferred error by fuzzy inference are put into the NN. Then the NN's output is used as the integrated measurement value of all sensors. (See fig. 1)

3.2 Experimental Results

Experimental results are shown in fig. 5-7. Each fig (a) in fig. 5-7 shows the measurement values of sensor 1 and 2, each fig. (b) shows outputs of the "NN+Fuzzy's" and the "If Clause's" SIS and each fig. (c) shows the outputs of the "NN+Fuzzy's" and the "Fuzzy Inference's" SIS.

Figure 5 shows the experimental results, where the offset of sensor sets is nearly equal to 0 m, so that there is no difference among three SIS.

Figure 6 shows the offset is "plus." In fig. 6 (b), "If Clause's" SIS cannot use the sensor 1 efficiently, because sensor 1 is only selected when the measurement value of sensor 2 is in the measurement region of sensor 1. Other systems can use sensor 1 when sensor 1 is measurable.

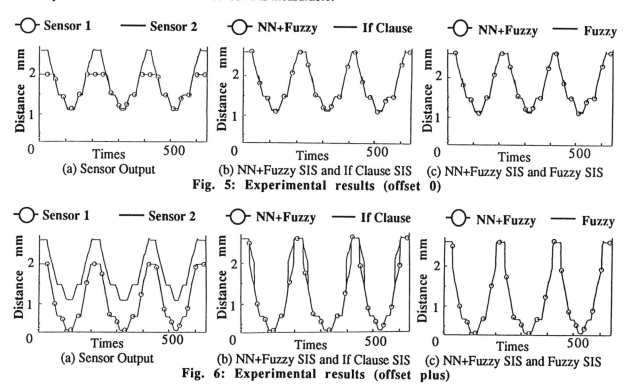

Fig. 5: Experimental results (offset 0)

Fig. 6: Experimental results (offset plus)

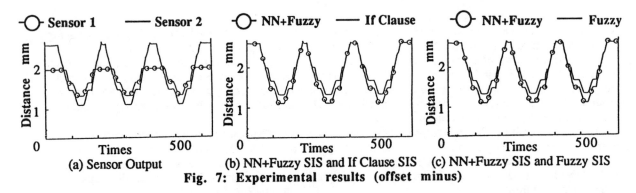

Fig. 7: Experimental results (offset minus)

Figure 7 shows the case of the offset sets "minus." When SIS selects sensor 1, both "If Clause's" and "Fuzzy Inference's" SIS have errors as large as the value of the offset. "If Clause's" SIS had the insensitive region, because SIS selects sensors not by consideration of the sensor 1's state, but by consideration of sensor 2's. As well, the "Fuzzy Inference's" SIS has discontinuous changes of value when SIS selects sensor 1 over sensor 2 or sensor 2 over sensor 1. "NN+Fuzzy's" SIS is the least error system of all, because this SIS is not affected by the sensor position offset. This is achieved by considering NN's learning data. When the offset of sensors is larger than 0 m, then the NN's expected value is based on sensor 1, while it is based on sensor 2 for the other case. Table 1 shows the quality of each SIS for each offset condition.

Table 1: The quality of each SIS for the offset

Offset	IF	Fuzzy	NN+Fuzzy
+	−	O	O
0	O	O	O
−	−	−	O

4. Conclusions

In this paper, a new SIS is proposed, so that SIS can be constructed easily by linking the sensor specification to the knowledge data base. The following results were shown by experiments using two different distance measurement sensors with the proposed SIS:

1) The "Fuzzy Inference's" SIS can use sensor 1 efficiently all the time. If the offset between sensor 1 and sensor 2 becomes "minus," SIS's output value has errors as large as the offset's value.

2) The "NN+Fuzzy's" SIS can use each sensor as efficiently as any states of sensors. The difference between the SIS's integrated value and the true value of the distance was minimal.

Using fuzzy inference, we can use sensors more efficiently, because we can get more information of sensors' state. Using NN, we can treat sensors more freely, because we can design outputs of NN for various states of sensors easily. The usefulness of the "NN+Fuzzy's" SIS has been shown through these experiments.

Reference

[1] H. F. Durrant-Whyte, "Sensor models and multi-sensor integration," Int'l. J. Robot. Res., Vol. 7, No.6, Dec., 1988, pp. 97/113.
[2] R. C. Luo, M. Lin, "Dynamic multi-sensor data fusion system for intelligent robots," IEEE J. Robot. Automat., Vol. 4, No. 4, 1988, pp. 386/396.
[3] R. C. Luo, M. G. Kay, "Multi-sensor integration and fusion in intelligent systems," IEEE Trans. on System, Man, and Cybernetics, Vol. 19, No. 5, 1989, pp. 901/931.
[4] Y. Nakamura, Y. Xu, "Geometrical fusion method for multi-sensor robotics system," Prof. IEEE Int'l. Conf. Robotics and Automat., 1989, pp. 668/673.
[5] J. M. Richardson, K. A. Marsh, "Fusion of multi-sensor data," Int'l. J. Robot. Res., Vol. 7, No. 6, Dec., 1988, pp 78/96.
[6] Rumelhart, McClelland, and The PDP Reserch Grop, "Parallel Distributed Processing," 1986, The MIT Press.
[7] Y. F. Zheng, "Integration of multiple sensors into a robotic system and its performance evaluation," IEEE Tran. Robotics and Automat., Vol. 5, No. 5, Oct., 1989, pp. 658/669.

A FUZZY LOGIC APPROACH FOR HANDLING IMPRECISE MEASUREMENTS IN ROBOTIC ASSEMBLY

H.B. Gurocak   A. de Sam Lazaro

Intelligent Systems Laboratory
Department of Mechanical and Materials Engineering
Washington State University
Pullman, WA 99164-2920

Abstract. The problem of part mating and assembly with close tolerances has been addressed in the past by active or passive compliance and by force/position control. With ambient sensor noise and imprecise measurements it is often difficult to attain high precision in manipulator control. A fuzzy logic approach to ascertaining the positional error during an unsuccessful assembly attempt is presented. The methodology employed here involves correlating sensory inputs, via a fuzzy algorithm, into an offset vector.

1. INTRODUCTION

Automated assembly is a geometric problem. In part mating, most of the problems arise either because the parts cannot be made perfectly identical or, more importantly, they cannot be perfectly positioned. Precision assembly with programmable automation is usually accomplished with the help of auxiliary hardware and software to control the manipulator at the terminal point. Typically, a vision system is used to determine system and random errors at some finite interval of time prior to assembly and re-positioning commands are issued to the manipulator. If such positioning errors are not corrected before the assembly operation continues, forces and associated moments will be exerted on the part being carried by the manipulator. The forces experienced by the part provide information which may be advantageously used to determine the relative positioning errors between the mating parts. Determining these forces is possible with relatively inexpensive devices (strain gages). Thus, the need for an expensive vision system may be eliminated. However, the measurement system used as a substitute for vision should be sensitive enough in order to determine small positioning errors (in the order of 0.1 mm).

Following a brief description of a position sensing compliant wrist developed for this research, the method of determining positional errors using this system will be introduced. A presentation of the experimental results obtained using this device and a fuzzy algorithm will conclude this paper.

2. POSITION SENSING COMPLIANT WRIST

In automated assembly of a tight tolerance rigid cylindrical peg in a hole (both chamferless), contact between the face of the peg and the rim of the hole may occur due to lateral positioning errors. Any further axial motion would cause the end effector carrying the peg to alter its path as it approaches the hole. In order to prevent this, the support for the peg must have a axial compliance.

When mating occurs, it may not be possible to determine the exact line of action of the force acting on the peg due to a plane-plane contact. However, tilting the peg slightly into the hole results in a two-point contact. If the location of the line connecting these contact points, relative to the center of the peg is obtained, the lateral error and its direction may be computed. This suggests that, in addition to the axial compliance, the support would require an angular compliance allowing the peg to rotate about its tip.

In order to achieve the above, a position sensing compliant wrist has been designed [1] (Figure 1). The wrist is composed of a 3D slider-crank mechanism with the crank rotation axis being the line connecting contact points. It has a measurement system that consists of two parallel triangular plates and three cantilever beams on which strain gages are attached.

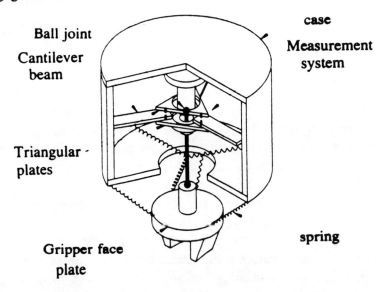

Figure 1. Position Sensing Compliant Wrist

When a positioning error takes place, the mechanism causes the beams to deflect resulting in strain on the beams. The offset r (Figure 2) is given by

$$r = \frac{l_5^2 - (l_5 + l_1 - \Delta)^2 + 2(l_5 + l_1 - \Delta)\cos\theta - l_1^2}{2 l_1 \sin\theta} \quad (1)$$

where Δ is the distance through which the mechanism is guided by the robot from the instant the peg touches the surface till the robot stops.

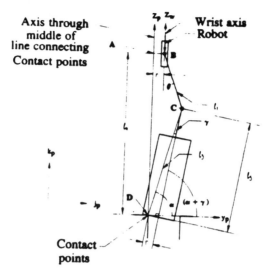

Figure 2. Side view of the mechanism

When the mechanism is displaced, the angle (θ) and the direction of the lateral error (β) can be obtained from the unit normal to the triangular plates as follows (Figure 3)

$$\theta = \cos^{-1}(n_z) \tag{2}$$

$$\beta = \cos^{-1}\left(\frac{-n_x}{\sqrt{n_x^2 + n_y^2}}\right) \tag{3}$$

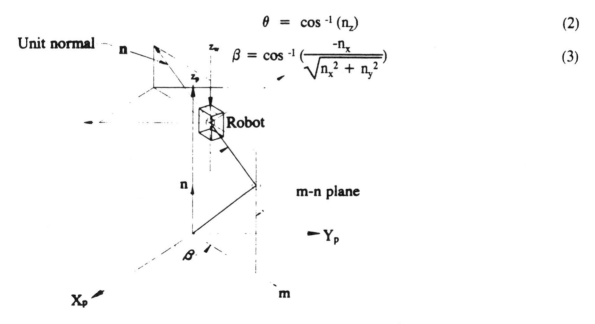

Figure 3. Direction of Lateral Error

where n_x, n_y and n_z are the components of the unit normal vector. Finally, the lateral error can be computed from

$$e_r = \sqrt{\left(\frac{d_h}{2}\right)^2 - \left(\frac{d_p}{2}\right)^2 + r^2} - r \qquad \begin{aligned} d_h &= \text{Hole diameter,} \\ d_p &= \text{Peg diameter.} \end{aligned} \tag{4}$$

Using θ, β, and e_r a positional error vector is obtained. The sensor readings (in this case strain gage measurements) are imprecise, mainly due to noise and the effect of temperature. This inevitably results in a reduction of the sensitivity of the wrist mechanism, especially in determining the direction of the lateral error. In order to handle this imprecision, a fuzzy logic approach has been adopted.

3. DETERMINING POSITIONAL ERROR USING FUZZY LOGIC

While a deterministic method has been presented above, a practical approach of dealing with noise is to adopt a fuzzy logic system. In this system, fuzzy logic technique involves the use of linguistic variables to define sets for sensor input values. These sets are manipulated by a rule base to obtain a single output set. The rule base consists of rules which are of the form :

$$\text{IF} <\text{condition}> \text{THEN} <\text{situation}> \qquad (5)$$

where
<condition> : is the combination of sensor readings and offset expressed in linguistic terms,
<situation> : is the possible direction of the lateral error in linguistic terms.

The fuzzy sets for these linguistic terms are defined for 'offset' (OFS), 'sensor readings' (S1, S2, S3) and the 'direction of lateral error' (DLE). The deterministic value of the direction of error is obtained in four steps :
    a. The sensor outputs are monitored directly and are used, after
       amplification, as inputs to the fuzzy system (fuzzification),
    b. These readings are then converted into fuzzy variables, each with an
       appropriate membership function,
    c. The rules are evaluated,
    d. Deterministic value of the direction of lateral error is computed from
       the resulting fuzzy set (defuzzification).

Rule Base. This application, 48 rules have been employed in the rule base. The inputs to the rule base are fuzzified sensor readings and the offset. Explicitly, each rule has the form

IF OFS is *SML* AND S1 is *MED* AND S2 is *BIG* AND S3 is *BIG* THEN
       DLE is *AROUND 30 degrees*
ALSO
  IF OFS is *MED* AND S1 is *BIG* AND . . . .   (6)

In order to formulate the rules, an experiment was conducted. First, the offset and the direction of lateral error were set to a known value and the corresponding sensor readings were recorded. Then, the sensor readings were assigned fuzzy labels (see Table. 1) by dividing the maximum possible range into nine equal sectors. The experiment was repeated at every 1 mm and for every 5 degrees.

Fuzzy Sets. Each rule requires definition of labels such as *SML* (small), *MED* (medium), *B120* (direction of lateral error β around 120 deg.). For the fuzzy variables OFS, S1,..,S3 and DLE. The fuzzy sets employed in this application are summarized in Table 1.

Fuzzy sets for offset	Fuzzy sets for sensor readings	Fuzzy sets for dir. of lat. error
VSM : Very small	PBG : Positive big	B000: Beta around 0 deg.
SML : Small	PMD: Positive medium	B030: Beta around 30 deg.
MED: Medium	PSM: Positive small	. . .
BIG : Big	PVS: Pos. very small	. . .
	ZER: Zero	. . .
	. .	. . .
	. .	B330: Beta around 330 deg.
	NBG: Negative big	

Table 1. Fuzzy sets

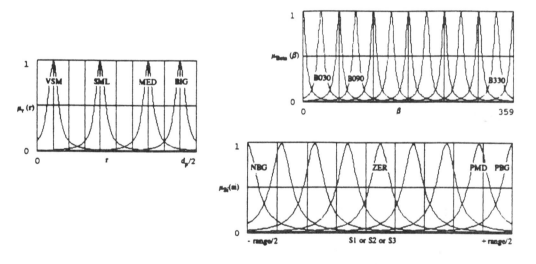

Figure 4. Membership functions

The universe of discourse for sensor readings is the maximum deflection range for the beams which occurs at minimum offset. In addition, the universe of discourse for offset is the radius of the peg, assuming that the offset can not be any bigger than the radius of the peg.

Each fuzzy set has been defined as a continuous set whose membership function is given by [2]:

$$\mu(x_i) = \frac{a}{a + b(x_i - x_0)^2} \quad (7)$$

where   a, b : are constants to adjust the spread of the set,

x_0 : is the point of minimum fuzziness,
x_i : is the element for which the membership value is to be determined.

Inference and Defuzzification. Based on sensor readings fuzzified as described above, an inference process is initiated. For a specific set of sensor readings, the rule base is executed and the most appropriate rule(s) fired [3]. The resulting fuzzy set has the membership function given by

$$\mu_{Beta}(\beta_i) = \underset{i}{MAX} \; MIN \; (\mu_{OFS_i}(r), \mu_{S1_i}(s1), \mu_{S2_i}(s2), \mu_{S3_i}(s3)) \qquad (8)$$
$$i=1...\# \text{ of rules}$$

The next step is defuzzification for finding the deterministic value of the output which is the direction of lateral error. For this purpose two approaches have been used. The first approach is center of area of the resulting fuzzy set

$$\beta = \frac{\int \beta_i \; \mu_{Beta}(\beta_i) \; d\beta_i}{\int \mu_{Beta}(\beta_i) \; d\beta_i} \qquad (9)$$

The second approach is to use the root-mean-square value [4]

$$\beta = \frac{\Sigma_i \beta_i}{n} \qquad (10)$$

where β_i is the value of the element in the universe of discourse of resulting fuzzy set at which the membership function reaches its maximum value and 'n' is the number of times it reaches this maximum value.

4. EXPERIMENTATION AND RESULTS

To assess the performance of the compliant wrist with fuzzy logic, an experiment was conducted. The experimental equipment consisted of the compliant wrist described earlier, an electronic circuit for sensor signals, an A/D card for an IBM-PC and a robot. The peg and the hole utilized in the experiment were 20.0 mm and 20.1 mm in diameter, respectively.

To examine the response in different directions of lateral error around the hole systematically, the experiment has been conducted by setting the direction of lateral error to values in the interval of 0 to 359 degrees, in increments of 30 degrees, for various values of offsets. The signals from the wrist were read into the computer. The direction of lateral error was first calculated by the deterministic model (3) and then by the rule base (6). When the rule base was utilized, defuzzification was accomplished first by center of area (9), then by root-mean-square value of the resulting fuzzy set (10).

The error in the output for different methods has been computed from:

$$e_{DM} = \beta_{DM} - \beta_{set}$$
$$e_{CA} = \beta_{CA} - \beta_{set}$$
$$e_{RMS} = \beta_{RMS} - \beta_{set} \qquad (11)$$

where β_{DM} is calculated deterministically by direct measurement (3),
β_{CA} is calculated by center of area method (9),
β_{RMS} is calculated by root-mean-square method (10),
β_{set} is the preset direction of lateral error.

The results are presented in Figure 5.

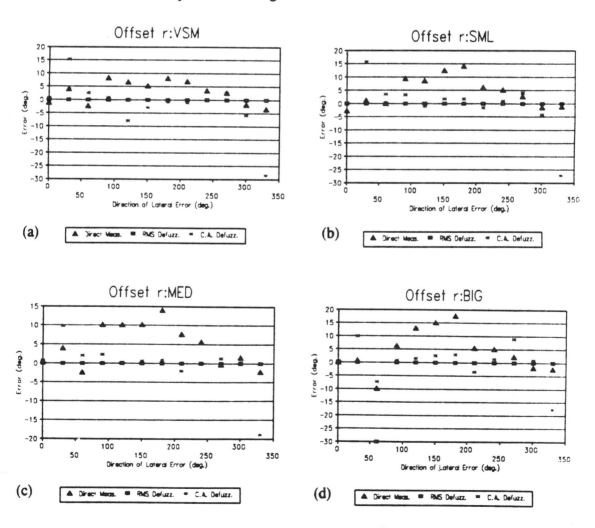

Figure 5a,b,c,d Results

With a few exceptions the fuzzy logic approach produced better results than the purely deterministic model. Furthermore, when the results of two methods of defuzzification were compared, it appeared that defuzzification of the inferred fuzzy set by root-mean-square gave superior results to those of defuzzification by center of area. This is true only if the direction of lateral error is a multiple of 30^0. Because, by its nature the method of root-mean-square converges the defuzzified answer to the peak value of the inferred fuzzy set. Further, if more than one peak occurs, the output is calculated to be the average of these peak values as in the case for 60 deg. setting in Figure 5.d.

Defuzzification by center of area also yielded results which were better than those of deterministic model. In case of an arbitrary direction of lateral error (not a multiple of 30^0), defuzzification of the inferred fuzzy set by center of area is more promising than both the deterministic model and the root-mean-square. It is concluded that in a generic case the center of area method would be most applicable.

5. CONCLUSION

A position sensitive compliant wrist has been used in conjunction with a fuzzy logic system for ascertaining and correcting positional error. Experimentation with this system was carried out with a peg-in-hole precision assembly scenario. It has been shown that this device along with the fuzzy logic approach gives better results than the purely deterministic model. Furthermore, it has been observed that defuzzification by center of area of the inferred fuzzy set was more promising for this particular application. In the second phase, the error vector obtained from the above system will be utilized to reposition the manipulator. These results will be reported separately.

6. REFERENCES

[1] Gurocak, H.B. and de Sam Lazaro, A. (1991). "A method of Position Sensing For Precision Assembly," to appear in the proceedings of the ASME Design Technical Conference, Sept. 22-25, 1991, Miami, FL.

[2] Zadeh, L.A., Fu, K-S., Tanaka, K. and Shimura, M. (1974). "Fuzzy Sets and Their Applications to Cognitive and Decision Processes. Acadamic Press, Inc.

[3] Zadeh, L.A. (1973) "Outline of an Approach to the Analysis of Complex Systems and Decision Processes," IEEE Transactions on Systens, Man and Cybernetics, SMC-3

[4] Kiszka, J.B., Kochanska, M.E., and Sliwinska, D.S. (1985). "The Influence of Some Parameters on the Accuracy of a Fuzzy Model ".Industrial Applications of Fuzzy Control. North-Holland, 1985.

Chapter 9: Aerospace

Space Shuttle Attitude Control by Reinforcement Learning and Fuzzy Logic

Hamid R. Berenji*
AI Research Branch
NASA Ames Research Center
Moffett Field, CA 94035

Robert N. Lea
Software Technology Branch
NASA/Johnson Space Center
Houston, TX 77058

Yashvant Jani
Togai InfraLogic Inc.
17000 El Camino Real
Houston, TX 77058

Pratap Khedkar
CS Division, Dept. of EECS
University of California
Berkeley, CA 94720

Anil Malkani*
AI Research Branch
NASA Ames Research Center
Moffett Field, CA 94035

Jeffrey Hoblit
LinCom Corporation
NASA/Johnson Space Center
Houston, TX 77058

Abstract— In this paper, we discuss the results of applying the ARIC and GARIC architectures, which have been developed for reinforcement learning using fuzzy logic, to the attitude control of the Space Shuttle. This paper demonstrates that it is possible to control the pitch, roll, and yaw of the Space Shuttle within a specified deadband by using fuzzy control rules and automatically adapt to a reduced error tolerance. The performance of this controller is compared with a controller using conventional control theory and also a non-adaptive fuzzy controller. Our results, using the Orbital Operations Simulator (OOS) system, demonstrate that more difficult tasks can be learned by our controller while the fuel efficiency remains very high.

I. Introduction

The non-linear behavior of many space systems and unavailability of quantitative data regarding the input-output relations makes the analytical modeling of these systems very difficult. Fuzzy logic control can play an important role in development of intelligent systems for space applications [4, 10, 11, 12]. A difficulty in design of these systems relates to fine-tuning the membership functions of the labels used in the rules. A few approaches have been recently suggested which use neural networks to define and fine-tune the membership functions (e.g., [9]). However, these have been mostly off-line and supervised learning approaches. In [5, 6, 13] the idea of using reinforcement learning for developing fuzzy membership functions has been proposed. After successful application of the ARIC and GARIC architectures to cart-pole balancing and truck backing [7], in this paper, we study the performances of these algorithms in the attitude control of the Space Shuttle.

In the next section, fuzzy reinforcement learning is briefly reviewed followed by the description of the attitude control problem in the Space Shuttle. We then apply the ARIC and GARIC architectures to this problem and compare their performances with a conventional controller and a non-adaptive fuzzy controller.

II. Fuzzy Reinforcement Learning

In reinforcement learning, one assumes that there is no supervisor to critically judge the chosen control action at each time step. The learning system is told indirectly about the effect of its chosen control action. The study of reinforcement learning relates to *credit assignment* where, given the performance (results) of a process, one has to distribute reward or blame to the individual elements contributing to that performance. This may be further complicated if there is a sequence of actions, which is collectively awarded a delayed reinforcement. In rule-based systems, for example, this means assigning credit or blame to individual rules (or their parts) engaged in the problem solving process. Barto, Sutton, and Anderson [3] used two neuron-like elements to solve the learning problem in cart-pole balancing. In these approaches, the state-space is partitioned into non-overlapping smaller regions and then

*Sterling Software

the credit assignment is performed on a local basis. In fuzzy reinforcement learning, these partitions can overlap leading to the use of fuzzy partitions in the antecedents and consequents of fuzzy rules. The reinforcements from the environment are then used to refine the definition of the fuzzy labels in the rules.

A. The ARIC Architecture

The Approximate Reasoning-based Intelligent Control (ARIC) architecure has been proposed in [5]. This architecture extends Anderson's method [2] by including the prior control knowledge of expert operators in terms of fuzzy control rules. In ARIC, a neural network is used to perform action and state evaluations. Also, two coupled neural networks are used to select a control action at each time step where the first network uses fuzzy inference to recommend an action and the second network calculates a degree to which the action recommended by the first network should be modified. The ARIC architecture tunes its fuzzy controller through updating the weights on the links in these networks. As this learning proceeds, the action recommended by the fuzzy controller is followed more often. Only monotonic membership functions are used in ARIC and the fuzzy labels used in the control rules are adjusted locally within each rule. For further details on the learning algorithm in ARIC, see [5].

B. The GARIC Architecture

In GARIC, we provide an algorithm to tune the fuzzy labels globally in all the rules and allow any type of differentiable membership function to be used in the construction of a fuzzy logic controller. The system determines a control action by using a neural network which implements fuzzy inference. In this way, prior expert knowledge can be easily incorporated. Another neural net learns to become a good evaluator of the current state and serves as an internal critic. Both networks adapt their weights concurrently so as to improve performance.

The architecture of GARIC is schematically shown in Figure 1. It has three components:

- The Action Selection Network (ASN) maps a state vector into a recommended action F, using fuzzy inference.

- The Action Evaluation Network (AEN) maps a state vector and a failure signal into a scalar score which indicates state goodness. This is also used to produce internal reinforcement $\hat{r}$.

- The Stochastic Action Modifier (SAM) uses both F and $\hat{r}$ to produce an action F' which is applied to the plant.

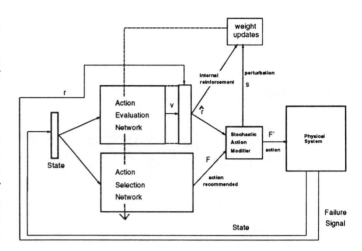

Figure 1: The Architecture of GARIC.

The ensuing state is fed back into the controller, along with a boolean failure signal. Learning occurs by fine-tuning of the free parameters in the two networks : in the AEN, the weights are adjusted; in the ASN, the parameters describing the fuzzy membership functions change. For further details on GARIC, see [6].

III. SPACE SHUTTLE ATTITUDE CONTROL

The Space Shuttle attitude controller is expected to perform four basic operations: 1) attitude hold or maintaining the desired attitude within a small region of the desired value, typically known as a deadband, 2) attitude maneuver or going from one attitude to another, 3) rate hold or maintaining a desired rate on a given axis, and 4) rate maneuver or going from one rate value to another rate value for a given axis. The Space Shuttle is the only operational vehicle which currently provides us with a reference rotational controller for comparison. Its on-orbit controller or Digital AutoPilot is based on modern digital control theory and is a highly optimized controller [1]. Its phase plane logic is complex and requires system specific parameters for a given mission and a given configuration. It uses two types of thrusters (two levels of jet thrusts), known as primary and vernier, and operates with two different sets of deadband values. It can perform rate maneuvers in pulse as well as discrete modes. Typical perturbations acting on the system include gravity gradient, aerodynamic torques, and translational burns.

Typical controllers based on the phase plane concept have angle errors and rate errors as input values. The output controller value is a command for generating a correcting torque. For the space shuttle, the rotational corrective torques are generated by thrusters.

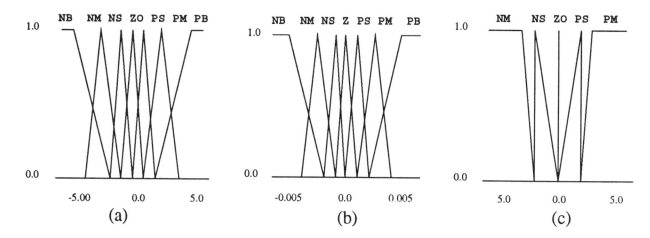

Figure 2: Membership functions used in ARIC for (a) angle error, (b) rate error, and (c) commanded acceleration

Table 1: Fuzzy control rules (jet firing commands) for attitude control

		\multicolumn{7}{c}{Angle Error}						
		NB	NM	NS	ZO	PS	PM	PB
Rate Error	NB	PM	PM	PS	PM			
	NM	PM	PM	PS	PM			
	NS	PS	PS	PS	PS			
	ZO	PS	PS	PS	ZO	NS	NS	NS
	PS				NS	NS	NS	NS
	PM				NM	NS	NM	NM
	PB				NM	NS	NM	NM

IV. SHUTTLE ATTITUDE CONTROL WITH ARIC

A fuzzy logic attitude controller has been developed where seven fuzzy labels: Negative Big (NB), Negative Medium (NM), Negative Small (NS), Zero (ZO),...., Positive Big (PB) have been defined for the input-output variables. The membership functions for angle error, rate error, and commanded acceleration are shown in Figure 2. Also, 31 control rules were used which are shown in Table 1. A typical fuzzy rule in this table is: If angle error is Negative Big and error rate is Zero, Then acceleration is Positive Small.

In the ARIC model for attitude control of the Space Shuttle, the input layer in each network includes 3 nodes which represent the angle error, angle error rate, and a bias node. Although ARIC does not require that the Action-state Evaluation Network and the Action Selection Network have equal number of hidden layer nodes, both have been modeled with 31 nodes in their hidden layers. Also, both networks have a single node at their output layer.

A. Results

We ran several experiments testing ARIC's performance using a high fidelity simulator called the Orbital Operations Simulator (OOS)[8]. Figure 3 illustrates the results of the ARIC model's performance in the attitude control problem.

The analysis of the defuzzification method and the definitions of PS and NS membership functions shows that a small hysteresis exists during rate reversal. We overlapped the NS and PS labels for several small values and repeated the tests to find out which overlap value could provide nearly the same fuel usage. Our results indicate that the overlap of 0.01 provides nearly the same fuel usage as the non-adaptive fuzzy controller. At this time ARIC does not have the desired capability of moving the end points of a membership function and it can only change the slopes. This capability is provided in GARIC. Furthermore, triangular functions are required for the output parameters to add inertia for no action. As currently implemented, the ZO output membership function does not provide any action to slow down the changes in series of actions. Again, GARIC provides this much desired capability and handles the actions appropriately.

V. SHUTTLE ATTITUDE CONTROL WITH GARIC

The structure of GARIC for space shuttle consists of the following:

- In ASN, there are two inputs, error and error rate, each using 7 labels, 31 rules with conclusions using five labels, and a single output. Hence the network has 2,14,31,5,1 neurons in its five layers.

- In AEN, there are two inputs, error and error rates and a biased unit, 31 hiden layer nodes, and a single

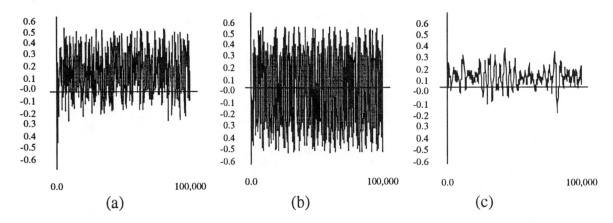

Figure 3: ARIC's performance on pitch angle control

output. Hence the network has 3,31,1 neurons in its three layers.

For each rule, seven labels (NB, NM, NS, ZE, PS, PM, PB) are used for angle error and angle error rate and five labels (NM, NS, ZE, PS, PM) are used for jet firing commands, as shown in Figure 4. In a learning experiment, a failure occurs when the value of a state variable goes beyond the allowed deadband. Everytime a failure occurs in a GARIC execution, we shift the control to a supervisory control routine to bring the state of the system back to within the deadband. This is needed to restart the the learning process after a failure. We selected our original fuzzy controller, with a small modification, to act as the supervisor. This modification was done on the definition of the Negative-Small (NS) label by shifting it to center on -3 instead of its original center of -2 in the definition of labels for commanded acceleration. This modification was sufficient in order to bring back the error to within a $\pm.4$ deadband after a failure, but was not sufficient to hold it there.

A. Results

In the following experiments, we start with the fuzzy controller developed in [11] with membership function definitions as shown in Figure 4. The fuzzy logic controller controls errors and error rates within $\pm.5$ as shown in Figure 5 but cannot control them within a $\pm.4$ deadband. However, with a small number of trials (less than 10), GARIC learns to do this task by revising the membership functions. A similar experiment was performed to train this newer controller to hold the error within a $\pm.3$ deadband. This time 5 trials were needed for GARIC to learn this new task. Figures 6, 7, and 8 show the controller's performance during the learning, its revised labels, and its performance after the learning, respectively.

Although an adaptive behavior has been added, the fuel consumption for the 100,000 time step simulation runs was about 222 lb which are in the same range as the original non-adaptive fuzzy controller and less than the conventional controller in OOS. However, the conventional controller in OOS has more constraints on its design (such as hardware concerns about jet life, which is improved by using as few firings as possible, but making each one longer). Therefore this comparison is done just to show that our results are within the acceptable range from the conventional controller's results.

VI. CONCLUSION

In this paper, we have applied two fuzzy reinforcement learning architectures to the difficult control problem of Space Shuttle attitude control. We have shown that the controller designed by this approach can automatically adapt to meet a new requirement by modifying its membership functions. For fuel consumption, we showed that the resulting adaptive fuzzy logic controller compares very well with the conventional controller and the original non-adaptive fuzzy controller, with the additional power of being able to automatically learn a new task.

REFERENCES

[1] Space shuttle orbiter operational level c functional subsystem software requirements. Technical Report Version STS 83-00009B, GN&C, Part C, Flight Control Orbit DAP, September 1987.

[2] C. W. Anderson. *Learning and Problem Solving with Multilayer Connectionist Systems*. PhD thesis, University of Massachusetts, 1986.

[3] A. G. Barto, R. S. Sutton, and C. W. Anderson. Neuronlike adaptive elements that can solve difficult learning control problems. *IEEE Transactions on Systems, Man, and Cybernetics*, 13:834–846, 1983.

[4] H. R. Berenji. On the integration of reinforcement learning and approximate reasoning for control. In *American*

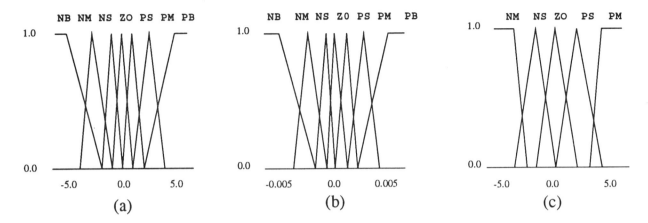

Figure 4: Original membership functions used in GARIC for (a) angle error, (b) rate error (c) commanded acceleration

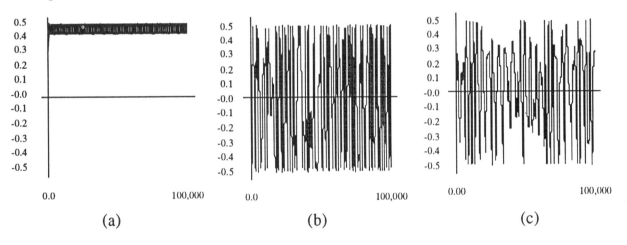

Figure 5: Fuzzy logic controller's performance (a) pitch, (b) roll, (c) yaw

Control Conference, pages 1900–1904, Brighton, England, 1991.

[5] H.R. Berenji. An architecture for designing fuzzy controllers using neural networks. *International Journal of Approximate Reasoning*, 6(2):267–292, February 1992.

[6] H.R. Berenji and P. Khedkar. Learning and tuning fuzzy logic controllers through reinforcements. *IEEE Transactions on Neural Networks*, 3(5), 1992.

[7] H.R. Berenji and P. Khedkar. Using fuzzy logic for performance evaluation in reinforcement learning. Technical Report FIA-92-28, AI Research Branch, July 1992.

[8] H. C. Edwards and R. Bailey. The orbital operations simulator user's guide. Technical Report ref. LM85-1001-01, LinCom Corporation, June 1987.

[9] J.S. Jang. Self-learning fuzzy controllers based on temporal back propagation. *IEEE Transactions on Neural Networks*, 3(5), 1992.

[10] R. Lea, R. H. Fritz, J. Giarratano, and Y. Jani. Fuzzy logic control for camera tracking system. In *8th International Congress of Cybernetics and Systems*, New York, NY, June 1990.

[11] R. Lea, J. Hoblit, and Y. Jani. Performance comparison of a fuzzy logic based attitude controller with the shuttle on-orbit digital auto pilot. In *North American Fuzzy Information Processing Society*, Columbia, Missouri, May 1991.

[12] R. Lea, J. Villarreal, Y. Jani, and C. Copeland. Learning characteristics of a space-time neural network as a tether skiprope observer. In *North American Fuzzy Information Processing Society*, pages 154–165, Puerto Vallarta, Mexico, December 1992.

[13] C.C. Lee and H.R. Berenji. An intelligent controller based on approximate reasoning and reinforcement learning. In *Proc. of IEEE Int. Symposium on Intelligent Control*, Albany, NY, 1989.

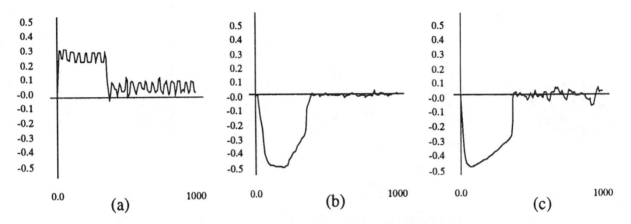

Figure 6: GARIC's performance during learning control of (a) pitch, (b) roll, (c) yaw

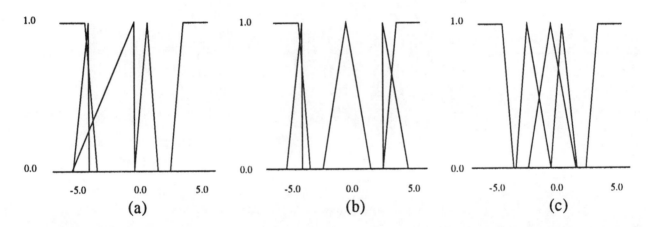

Figure 7: Revised membership functions (after learning a new error deadband of ±.3) for (a) pitch, (b) roll, (c) yaw

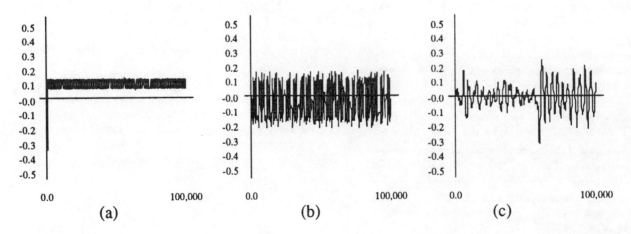

Figure 8: Controller's performance after learning the new deadband ±.3 (a) pitch, (b) roll, (c) yaw

INTELLIGENT CONTROL OF A FLYING VEHICLE USING FUZZY ASSOCIATIVE MEMORY SYSTEM

T. Yamaguchi*, K. Goto*, T. Takagi*,
K. Doya** and T. Mita***

*LIFE: Laboratory for International Fuzzy Engineering Research, Siber Hegner Building 3FL., 89-1 Yamashita-cho, Naka-ku, Yokohama-shi, 231, JAPAN. (Tel.)+81-45-212-8225, (Fax.) +81-45-212-8255.
**Faculty of Engineering, University of Tokyo, 7-3-1 Hongo, Bunkyo-ku, Tokyo, 113, JAPAN.
(Tel.)+81-3-3812-2111 ex. 6913, (Fax.) +81-3-3816-7805.
***Faculty of Engineering, Chiba University, 1-33 Yayoi-cho, Chiba-shi, 260, JAPAN.
(Tel.)+81-472-51-1111 ex. 2842, (Fax.) +81-472-51-7337.

ABSTRACT

We propose a flying vehicle intelligent control which simulates pilot's operation knowledge and its training steps. To simulate the operation knowledge and its training steps, we use a fuzzy associative memory system (called FAMOUS: Fuzzy Associative Memory Organizing Units System). FAMOUS use associative memory neural networks which represent fuzzy knowledge. There are two types of the operation knowledge; 1)DFK: Dynamic Fuzzy Knowledge, for example the circle flight operation pattern and 2)SFK: Static Fuzzy Knowledge, for example the hovering operation model corresponding to each flying condition (higher or lower flying). FAMOUS represents both dynamic and static fuzzy knowledge using its hierarchical knowledge representation. Pilot's fuzzy knowledge are initially put into FAMOUS by the pilot, and are refined by the pilot's ideal operations using the learning algorithm on FAMOUS. After the learning of the fuzzy knowledge, vehicle's flight ability is more excellent than before learning. We actually show the excellent flight of a small four-fans flying vehicle (like a helicopter) to show the usefulness of FAMOUS on these intelligent systems.

KEYWORDS: Associative memories, Fuzzy model, Learning control, Neural networks, Flying vehicle.

1. INTRODUCTION

VTOL vehicle (for example helicopter) easy flies in 3-dimensional space and has high flight ability. But the flight operation is not easy because the vehicle is unstable and its physical characteristics change corresponding to the flying altitude by the ground effect. The pilot acquires flying operation knowledge with some flight training. The pilot's knowledge consists of two knowledge types, DFK (Dynamic Fuzzy Knowledge) and SFK (Static Fuzzy Knowledge). At the circle flight, the pilot's knowledge is represented using "the operation pattern" with DFK. At the hovering flight, the pilot's knowledge is represented using SFK which shows "the operation model" corresponding to each "flying condition (higher or lower flying)" with fuzzy rule [1][2], and adapts hovering operations. We have realized both circle flight's DFK and hovering flight's SFK with FAMOUS (Fuzzy Associative Memory Organizing Units System) [3][4].

FAMOUS realizes SFK using the static relationships "A -> B" interpreted from the rules, and realize DFK using the dynamic relationships "A-up -> B-down". FAMOUS is an associative memory network system which has fuzzy concept nodes instead of normal input/output nodes. This fuzzy concept consists of a label and a truth value (or a linguistic truth value). The truth value is given by the activation value of the node. The label shows the meaning of its fuzzy concept. Fuzzy concept node has a function, which abstracts the label, under its nodes. FAMOUS is a hierarchical associative memory system. We use FAMOUS's hierarchical structure to realize both circle flight's DFK and hovering flight's SFK. As SFK adapts the operation models corresponding to the flying condition in the subordinate part, DFK operates circle flight under all conditions in the superordinate part. DFK is considered to be an abstraction knowledge. This knowledge structure simulates pilot's knowledge structure. The refinement of pilot's knowledge is also simulated by FAMOUS. As FAMOUS separately refines each part of acquired knowledge based on fuzzy concept nodes, its learning speed is fast. We show this learning ability with flight experiments.

In this paper, we first explain the experimental system of the VTOL vehicle. Nextly, we represent how to realize the intelligent control system using FAMOUS, and how to realize hovering flight and circle flight. The usefulness of this intelligent system is shown with these flight experiments.

2. THE FLYING VEHICLE EXPERIMENTAL SYSTEM

Fig.1 shows the experimental system construction, and shows the outline of a four-fan flying vehicle in the upper side. The experimental system consists of 32bit-computer for control, 16bit-AD/DA-interface, two PSD-cameras, filters for PSD-camera output signals, and a RC-transmitter which transmit the actuating signal from the computer. The computer controls the flying vehicle with sensing the position and velocity by PSD-camera signals. This system simulate the pilot operation. We use cameras instead of eyes, and put into the computer the skills of flight operations. **Fig.2** shows this computer processing diagram, and **Fig.3** shows photos of the experimental system.

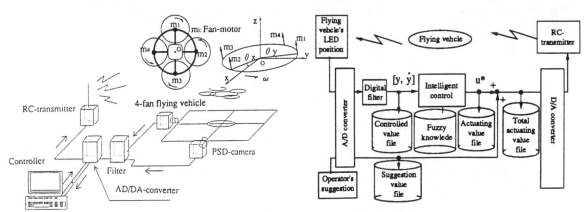

Fig.1 The vehicle experimental system.   **Fig.2** The computer processing diagram.

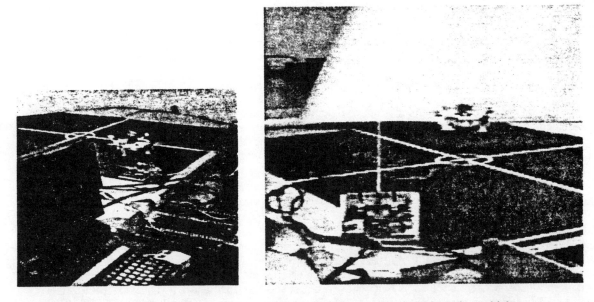

1) The complete view of the system.   2) RC-transmitter and hovering vehicle.
Fig.3 Photos of the experimental system.

The vehicle controls the pitching, rolling angle θ_x, θ_y and the rotating velocity ω, which are shown at the upper side in Fig.1, using gyrocompass sensors in vehicle's body. The control computer gets the controlled vector which represent the position x, y, the respective differential position, the rotating angle, the rotating velocity and the altitude. And the computer put the actuating vector which represents the x, y-axis actuating value, the rotating actuating value, and the altitude actuating value. In this paper, we mainly explain the x-axis control. The other controls are the same as the x-axis control. Each vector is saved in the computer to refine the pilot's knowledge.

3. INTELLIGENT CONTROL SYSTEM BY MEANS OF FAMOUS

Advantages on neural networks improve weak points on fuzzy logic, and advantages on fuzzy logic improve weak points on neural networks as shown in **Fig.4**. In order to improve each other, we combine neural networks and fuzzy logic as shown in **Fig.5**. We propose FAMOUS for the first step of fuzzy-neural network research.

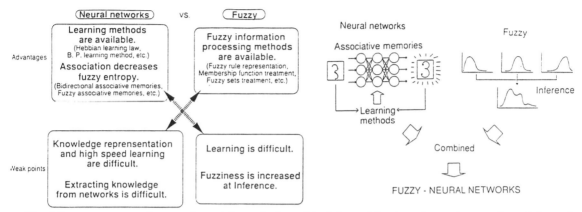

Fig.4 Features on neural networks and fuzzy logic.   **Fig.5** A combination of neural networks and fuzzy logic.

FAMOUS combines fuzzy knowledge representation and neural network's association and learning functions. We represent fuzzy knowledge using not only tree structure on fuzzy rule but also map structure based on fuzzy concept nodes [4]-[7]. Fuzzy concept node (or fuzzy proposition) has a function which abstracts its label. The abstraction function can be used the same structure of fuzzy concept node. FAMOUS is a hierarchical system. Fuzzy concept nodes are connected with the relationships interpreted from fuzzy knowledge. FAMOUS is an associative memory system which has fuzzy concept nodes instead of input/output nodes. FAMOUS refines each part of the fuzzy knowledge separately using neural network's learning functions. The learning speed on separated networks is faster than the speed on normal neural networks.

We use FAMOUS to realize flying vehicle intelligent control which simulates pilot's fight operation and its training steps. There are two types of operation knowledge which are DFK and SFK. Associative memory neural networks represent both dynamic pattern "A-up -> B-down" and static model "A -> B". (The equality in bidirectional connection weight is not need for dynamic associative memory networks but is need for static associative memory networks.) FAMOUS represents both DFK and SFK using fuzzy concept node networks. We use SFK to represent rules of the hovering flight with pairs of "the condition (higher flying or lower flying)" and "the operation model". We use DFK to represent the circle flight operation pattern. **Fig.6** shows the combination of DFK and SFK.

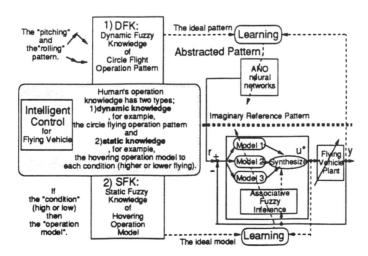

Fig.6 The vehicle intelligent control system.

SFK "If vehicle's condition (higher or lower flying), then its hovering operation model (controller)" is represented using BAM (Bidirectional Associative Memories) [8] and fuzzy model rule representation [9]. BAM is useful networks for representing pairs of if-then rule data [3]. Because the memorizing ability is given, and the learning algorithm is given. The DFK circle pattern "Pitching-front (x-axis operation is plus) -> Rolling-right (y-axis operation is minus) -> ···" is represented using FCM (Fuzzy Cognitive Map) [10] or ANO (Adaptive Neural Oscillator) [11] networks. FCM or ANO networks represent dynamic transition pattern. Both FCM and ANO networks have each learning algorithm. We use ANO networks to represent the circle flight operation pattern because ANO networks have high memorizing ability.

We propose a hierarchical intelligent system as shown in Fig.6. The physical dynamics of vehicle changes corresponding to the flying altitude by the ground effect. SFK adapts the operation model (controller) and let flying vehicle stable under all conditions. As the closed loop physical dynamics does not have large change with the effect of SFK, DFK circle flight operation pattern is used for all conditions. DFK in this intelligent system is an abstraction knowledge which represents circle operation pattern suitable for all conditions. We explain each part in detail at the following section.

4. HOVERING FLIGHT BY MEANS OF SFK

Fuzzy rule has a feature that expert knowledge is shown by a compact if-then rule set by means of membership functions and input-output functions. But it is not easy for experts to create all of the functions in fuzzy rule set. FAMOUS is useful for representing the outline of expert's fuzzy knowledge, and is useful for refining that knowledge with expert's training steps. FAMOUS recalls good inference results from the fuzzy knowledge using association. The pilot shows his control knowledge as the pair of "the condition (,for example higher flying or lower flying)" and "the operation model (controller)" using fuzzy rule. In fact, it is not easy that the pilot create the functions of each operation model in a fuzzy rule set. FAMOUS realizes the control knowledge using the relationships interpreted from the fuzzy rule set, and carries out fuzzy inference using its association even if there are uncertain operation models in the fuzzy rule set. Before learning the operation model, the pilot use the model which is not good fit for each condition, but it is robustly stabilizable for all of the conditions. At the learning algorithm, each uncertain operation model is individually trained under the condition, and the relationship between each condition and its operation model is reinforced. FAMOUS recalls well-trained operation models connection to the input condition, and controls a flying vehicle by synthesizing their operation models. We actually applied FAMOUS to controlling a small vehicle with four fans, and show its usefulness for simulating human training steps and human association at hovering control operations.

4.1 The knowledge of vehicle physical dynamics and its hovering operation skills

The pilot knows that the vehicle has two characteristic physical dynamics, lower flying and higher flying. The characteristic of lower flying is high sensitivity with much disturbance by the reaction from the floor. The characteristic of higher flying is low sensitivity. We show this knowledge for flying dynamics using fuzzy model [12].

The vehicle fuzzy model is given by Eq.1.

$$P_1 := \alpha_2 P_2 + \alpha_3 P_3 \quad (0 \leq \alpha_i \leq 1,\ \alpha_2 + \alpha_3 = 1). \tag{1}$$

In Eq.1, P_2 is lower flying model, P_3 is higher flying model, and P_1, which shows all the space of the parameter variance plant $P_{v(s)}$, is a parameter synthesizing model of P_2 and P_3. The α_i shows the membership grade which is shown in Fig.7. The membership function evaluate the distances between P_1 physical dynamics and P_2, P_3 physical dynamics. The approximate open loop transfer function $P_{v(s)}$ is given by Eq.2.

$$P_{v(s)} = \frac{ke_0}{s^4 + kd_3 s^3 + kd_2 s^2}. \tag{2}$$

And Eq.2 has a variance parameter k ($k_2 \geq k \geq k_3$). In Eq.2, k, e_i and d_i are plus parameters. When the condition is lower flying, a variance parameter k is equal to k_2. When the condition is higher flying, a variance parameter k is equal to k_3.

The initial hovering knowledge based on the fuzzy model is given by Eq.3.

Rule 1: If h is P_1, then $u_1=f_1(y)$ with C_{u1}.
Rule 2: If h is P_2, then $u_2=f_2(y)$ with C_{u2}. (3)
Rule 3: If h is P_3, then $u_3=f_3(y)$ with C_{u3}.

In Eq.3, the if-part represents the condition P_i and the then-part represents the operation model (controller) f_i. This fuzzy rule set uses Takagi and Sugeno fuzzy model representation [9], which uses a input-output function in the then-part, because the plant is approximately shown as Eq.1. We initially use a robust controller f_1, which is given by the pilot hovering operation model, for all controllers f_i. By the training of the flying operation, we refine these controllers. The connection rate C_{ui} represents the connection between the condition P_i and the controller f_i. Values of C_{u2} and C_{u3} are low before learning controllers f_2 and f_3 because both are initially equal to f_1. This fuzzy controller construction is shown in **Fig.8**. At fuzzy inference, the synthesized controller u^* is given by Eq.4,

$$u^* = \sum_{i=1}^{3} \gamma_i \cdot f_i(y) .$$ (4)

In Eq.4, the synthesizing rate α_i is given by normalizing the then-part's activation value γ_i of associative memory networks at fuzzy inference.

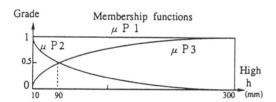

Fig.7 Membership functions of the fuzzy model.

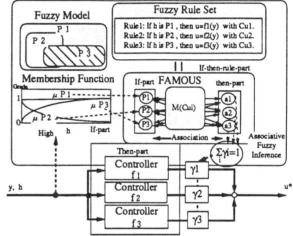

Fig.8 The SFK of hovering flight.

4.2 Fuzzy rule by means of FAMOUS

We use FAMOUS to realize fuzzy rules as shown in **Fig.9**. It consists of three elements: 1)if-part, 2)if-then-rule-part, and 3)then-part. The fuzzy rules represent expert knowledge describing the relationships between "conditions" and "operation models". The if-part has membership functions which abstract and characterize conditions of the rule set. The then-part has membership functions or input-output functions which represent operation models of the rule set. FAMOUS use associative memories to realize the relationships between if-part's conditions and then part's operation models using BAM. **Fig.10** shows BAM networks. BAM memorizes x_i-y_i pattern pairs in terms of correlation matrix M and its transpose matrix M^T. Assume that BAM memorizes x-y pair ($x,y \in [0,1]$) and x'' (which means a noisy x) is given in the x layer, BAM recalls the x-y pair on x, y layer. BAM recalls from memory use reverberation which is given by Eq.5.

$$Y_t = \phi(M \cdot X_t), \quad X_{t+1} = \phi(M^T \cdot Y_t) .$$ (5)

In Eq.5, $X_t = [ax_1, ax_2, ..., ax_m]^T$, $Y_t = [ay_1, ay_2, ..., ay_n]^T$ are each activation vector on the x, y layer at reverberation step t, and $\phi(\cdot)$ is a sigmoidal function of each neuron. The correlation matrix M and M^T are given by Eq.6.

$$M = \beta \sum_{i=1}^{N} y_i \cdot x_i^T, \quad M^T = \beta \sum_{i=1}^{N} x_i \cdot y_i^T .$$ (6)

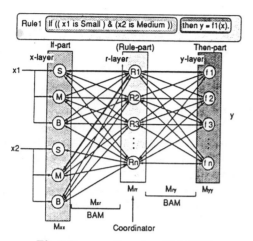
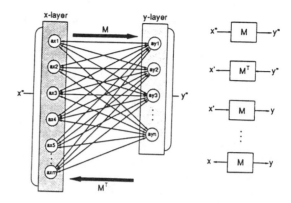

Fig.9 Fuzzy rule using FAMOUS.   **Fig.10** BAM: Bidirectional Associative Memories.

In Eq.2, x_i, y_i ($i=1{\sim}N$) are BAM memorizing pairs, β is an association parameter, and each element x_i, $y_i \in \{0,1\}$ is usually converted to bipolar element$\in \{-1,1\}$ in consideration of BAM energy function. BAM recalls by means of reverberation has the feature that the most suitable pair for the input conditions appear as a result of decreasing fuzzy entropy which shows fuzziness rate of inference output. We use this feature to control fuzziness at fuzzy inference. We realize fuzzy rules by means of associative memories such as BAM in consideration of BAM memorizing ability. Fig.9 shows the fuzzy rule realization by means of associative memories. Each neuron at x layer represents if-part membership function given by fuzzy rules as conditions. Each neuron at y layer represents then-part operation model $f_i(\cdot)$ given by fuzzy rules. Each neuron at r layer represents fuzzy rule i. We use BAM to connect the x layer to the r layer and the r layer to the y layer. The M_{xr} and M_{ry} are these correlation matrices. At the r layer, there are mutual minus weighted connections in order to recall the most suitable rule for the input conditions. And there is also a minor plus weighted recessive connections at the r layer in order to save each activation. The M_{rr} is r layer's correlation matrix called the coordinator. There are also minor plus weighted recessive connections at x and y layers in order to save each activation. The M_{xx} and M_{yy} are these x and y layers' correlation matrices.

At fuzzy inference, we carry out reverberation with those correlation matrices M_{xx}, M_{xr}, M_{rr}, M_{ry}, M_{yy} and each transpose correlation matrix. This associative fuzzy inference can improves one of the conventional fuzzy inference's weak points, which is increasing fuzziness of inference output, because this association can control increasing fuzziness. Since we use each activation value before value sufficiently converges, we can imitate some fuzzy inference methods without previous week point. We use some activated fuzzy rules to synthesize each control gain.

Assume that the γ_i is normalized value of activation a_i at the y layer, the total output of fuzzy rules is given by Eq.4.

We use "connection rates" C_{ui} in fuzzy rules of Eq.3. The C_{ui} represents the connection between the operation model $f_i(\cdot)$ and the optimum model $f_i(\cdot)^*$ using a value ($\in [0,1]$). Before learning the operation model, C_{ui} is low value because we use the model which is not good fit for each condition, but it is robustly stabilizable for all of the conditions. After learning the operation model, C_{ui} is set higher value because each uncertain operation model is individually trained under the condition and the connection between each optimum model and the operation model is reinforced. FAMOUS recalls well-trained operation models connection to the input condition and synthesizes their operation models because we multiple C_{ui} to the i-part of the correlation matrix M in Eq.6.

4.3 The acquisition of the initial hovering operation model

Initially, the pilot control the vehicle using a operation model suitable for all conditions. We get this robust operation model and set it to the controller f_1. The control knowledge consists of two elements: the feedback of a position vector y (whose gain is shown as h_{10}) and the feedback of the velocity $\dot{y}$ (whose gain is shown as h_{11}). The closed loop transfer function $P_{cv(s)}$ is given by Eq.7.

$$P_{cv(s)} = \frac{k e_0}{s^4 + kd_3 s^3 + kd_2 s^2 + k e_0 h_{11} s + k e_0 h_{10}}. \tag{7}$$

We check that the controller f_1 is a stabilizer for all conditions [13].

4.4 The refinement of each hovering operation model

By the training of the flying operations, we refine respective controllers. Before learning the operation model, we use the model which is not good fit for each condition, but it is robustly stabilizable for all conditions. At the learning algorithm, each uncertain operation model is individually trained under the condition, and the relationship between each condition and its operation model is reinforced. We use Widrow-Hoff learning algorithm [14] to refine each operation model under its condition. The operation error E is given by Eq.8.

$$E = \sum_{t=1}^{T} (\Delta u_t - \Delta h_{io} y_t - \Delta h_{i1} \dot{y}_t)^2 \rightarrow Min. \tag{8}$$

In Eq.8, Δu_t is the training signal of the pilot, Δh_{io} and Δh_{i1} are refine values of the controller f_i's parameters. We would like to minimize E. These refine values of the controller are given by Eq.9.

$$\Delta h_{io} = - \frac{\sum_{t=1}^{T} \Delta u_t y_t}{\sum_{t=1}^{T} y_t^2}, \quad \Delta h_{i1} = - \frac{\sum_{t=1}^{T} \Delta u_t \dot{y}_t}{\sum_{t=1}^{T} \dot{y}_t^2}. \tag{9}$$

Fig.11 shows synthesizing rates (g_is) of controllers (f_is) when higher flying condition is given. Before the learning of the controller f_3, the controller f_1 is activated. After the learning, the controller f_3 is activated. Because the value of C_{u3} (which is given in Eq.3, and represents the connection between the controller f_3 and the optimum controller f_3^*) is low before learning of the controller f_3, but the value of C_{u3} is higher after learning. We use the 4-th steps activation values for inference results in the dynamics shown in Fig.11.

Fig.12 shows the results (which represents controller synthesizing rates) of fuzzy inference for any conditions (,for example, higher flying or lower flying) at 1)no-trained and at 2)well-trained. In these cases, the connection rate value C_{ui} is changed from 1)$C_{u1}:C_{u2}:C_{u3}=6:1:1$ for no-trained to 2)$C_{u1}:C_{u2}:C_{u3}=1:9:9$ for well-trained. After the training of controllers, each controller is used for its condition, for example the controller f_2 is used for the lower flying and the controller f_3 is used for the higher flying.

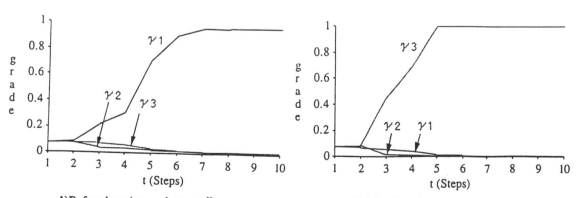

1)Before learning each controller.   2)After learning each controller.

Fig.11 The inference dynamics at association.

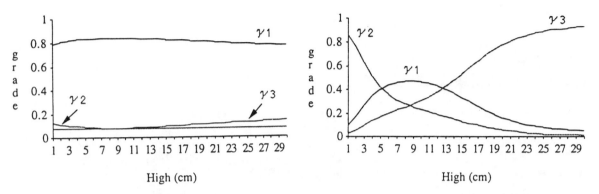

1)Before learning each controller. 2)After learning each controller.
Fig.12 The fuzzy inference results for all conditions.

4.5 Results of hovering flight experiments and discussions

Fig.13 shows results of hovering experiments 1)before learning and 2)after learning. Both success the hovering control. The vertical axis of Fig.13 shows the sensor output voltage which represents the position by multiplying 0.083. The horizontal axis shows the time (1step/200msec). We evaluate the hovering performance with the integration values for the position from the zero point. The integration value, which is given by the experiment before learning, is 8% larger than the integration value, which is given after learning. Therefore, the fuzzy controller after learning is more excellent than it before learning.

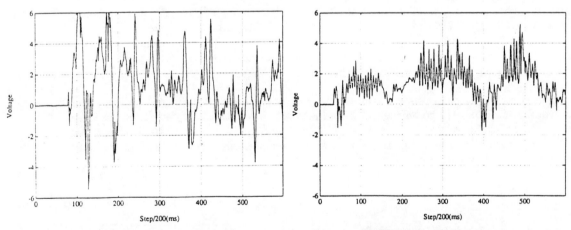

1)Before learning each controller. 2)After learning each controller.
Fig.13 Results of hovering experiments.

The refined controller parameters h_{ij} (where i shows controller f_i, and j shows parameter number of controller f_i) are given as $h_{2i} < h_{1i} < h_{3i}$. Each controller is a stabilizer for all flying conditions, and the fuzzy controller which synthesizing their controllers is also a stabilizer for all flying conditions. The lower flying controller f_2 after learning is more excellent than the controller f_1 (which is initially set for the controller f_2) for disturbance. Because the lower flying model P_2 has much disturbance by the reaction from the floor, the controller f_2 (after the learning) is excellent for disturbance. The higher flying controller f_3 after learning has higher gains than before learning because the higher flying model P_3 is low sensitivity. Each controller is refined to fit for each condition by means of the learning, and the fuzzy controller is more excellent than before learning.

5. CIRCLE FLIGHT BY MEANS OF DFK

We initially put pilot's circle fright operation pattern into ANO networks. The operation pattern is represented by the pair of "Pitching operation pattern (for example, Pitching-front-gradually -> Pitching-front-strongly -> ... -> Pitching-front-stop -> ...)" and "Rolling operation pattern" as DFK. And we refine ANO to fit the vehicle output y for the desired ideal pattern. This ANO system is shown in **Fig.14**. As SFK system adapts the controller F corresponding to each flying condition and keeps the vehicle stable in the subordinate part, ANO need not to memorize all operation pattern corresponding to each flying condition. ANO need to memorize the abstract pattern. This ANO networks are shown in **Fig.15**. There are three layers: input-layer, hidden-layer and output-layer in ANO. The ANO for memorizing circle flight pattern has eight nodes. These nodes are connected each other.

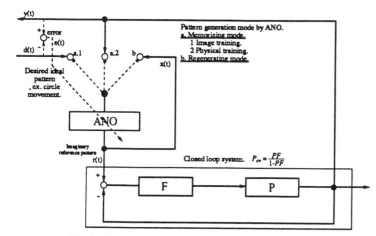

Fig.14 Dynamic fuzzy knowledge using ANO.   Fig.15 ANO: Adaptive Neural Oscillator.

5.1 The acquisition and the refinement of knowledge for circle flight operation pattern

Firstly, we initially set ANO the circle flight operation pattern which represents pilot's desired image pattern using ANO learning algorithm. This learning is called "the image training". At the image training, the node $a.1$ (shown in Fig.14) is connected to ANO input. And we assume that the plant P_{cv} is equal to 1 (or ANO output is equal to the system output y). We need only abstract image of circle flight operation pattern (which represent the outline of pilot's dynamic knowledge). We acquire the pair of a sine-wave and a cosine-wave as this abstract image. We set pitching operation pattern $r\sin(2\pi\omega t)$ and set rolling pattern $r\cos(2\pi\omega t)$ as the initial DFK. We also use this pair as the desired ideal pattern D_1, D_2.

Nextly, we refine the initial DFK in order to fit system output y for the desired ideal pattern. This learning is called "the physical training". At the physical training, the node $a.2$ (shown in Fig.14) is connected to ANO input. And SFK system P_{cv} is connected between ANO output and ANO input. ANO output is changed by the physical training to fit the system output y for the ideal pattern. So we call this ANO output "imaginary reference pattern".

We use the following learning algorithm for both "the image training" and "the physical training" [11]. As the dynamic model of ANO networks, we use Eq.10.

$$y_i(t) = g_i\left(h_i \sum_{j \in I \cup H \cup O} w_{ij} y_j(t)\right), \quad (i \in H \cup O). \tag{10}$$

And the output error function of total network is defined by Eq.11.

$$e(t) = \sum_{i \in O} e_i(t) = \sum_{i \in O} \frac{1}{2}(y_i(t) - d_i(t))^2. \tag{11}$$

In Eq.10 and Eq.11, y_i is the actual output, g_i is the differentiable non-linear function $g_i(net) = \frac{2}{1+e^{-net}} - 1$., d_i is the desired output of the i-th unit, h_i is a linear operator of time lag, and w_{ij} is the connecting weights from the j-th unit to the i-th unit. The external input y_i $(i \in I)$ and the desired output d_i $(i \in O)$ are externally given. I, H and O belong to the groups of the indices of input, hidden and output units respectively.

We use $h_i = \left(1 + \tau_i \frac{d}{dt}\right)^{-1}$ which the operator of first order time lag with the decay time constant τ_i. In this case, if we put the total input to the i-th unit $u_i(t) = \sum_{j \in I \cup H \cup O} w_{ij} y_j(t)$ and the internal state $x_i(t) = h_i u_i(t)$, the continuous-time neural network model is defined by Eq.12.

$$\tau_i \frac{d}{dt} x_i(t) = -x_i(t) + u_i(t), \quad (i \in H \cup O). \tag{12}$$

If we differentiate Eq.10 by $y_j(t)$ $(j \in H)$, we get Eq.13.

$$\frac{\partial y_i}{\partial y_j} = g_i'(x_i(t)) h_i w_{ij}, \quad (i \in O, j \in H). \tag{13}$$

Then the learning equation for the hidden units $(j \in H)$ are derived by Eq.14.

$$\frac{d}{dt} w_{jk} = -\frac{\partial e}{\partial w_{jk}} \varepsilon = -\sum_{i \in O} \frac{\partial e}{\partial y_i} \frac{\partial y_i}{\partial y_j} \frac{\partial y_j}{\partial w_{jk}} \varepsilon$$
$$= -\varepsilon \sum_{i \in O} (y_i(t) - d_i(t)) g_i'(x_i(t)) h_i w_{ij} g_j'(x_j(t)) h_j y_k(t), \quad (k \in I \cup H \cup O). \tag{14}$$

In Eq.14, ε is the learning rate. The approximate steepest descent method for the weight w_{jk} $(j \in H)$ from output units to hidden units is derived as above. We call this method "the direct back-propagation".

After the physical training, the node b (shown in Fig.14) is connected to ANO input. This structure is ANO regenerating mode.

5.2 Results of circle flight simulations

Fig.16 shows the top view of flying vehicle on circle flight operations before and after the physical training, and shows the transition of each node's activation value in ANO. In Fig.16, H_i, O_i and D_i represent each activation value $\in [-1,1]$. H_i, O_i and D_i show the hidden node, the output node and the input node (whose activation value is equal to the desired pattern's value) as shown in Fig.15. **Fig.17** shows the integration value of the errors from the desired ideal pattern on physical trainings. The horizontal axis shows the time (sec). In Fig.17, the horizontal axis shows the learning steps of physical training. DFK's circle flight operation after learning is more excellent than before learning.

6. CONCLUSION

We realized an intelligent system which controls flying vehicle using pilot's flight operation knowledge and simulating pilot's training steps. In order to realize this system, we proposed a fuzzy associative memory system FAMOUS which combine fuzzy knowledge representation and neural network's association and learning functions. We showed that FAMOUS is useful for representing pilot's SFK (: Static Fuzzy Knowledge, for example, hovering flight operation model) and DFK (: Dynamic Fuzzy Knowledge, for example, circle flight operation pattern) as the outline knowledge of flight operations. We also showed that FAMOUS is useful for refining both initial SFK and initial DFK. At representing SFK and DFK, we used FAMOUS's hierarchical structure. As SFK system adapted the controller corresponding to each flying condition and keep the vehicle stable in the subordinate part, DFK needed not to memorize all the operation patterns corresponding to flying conditions. DFK needed to memorize the abstract pattern. At refining SFK and DFK, FAMOUS separately refines each part of fuzzy knowledge based on fuzzy concept nodes using neural network's learning functions. The learning speed in separated networks, which based on fuzzy concept nodes, is faster than the speed in normal neural networks. We actually showed the excellent flight of a small four-fan flying vehicle and showed the usefulness of FAMOUS on the intelligent control system.

FAMOUS, which has the feature of memorizing dynamic knowledge based on fuzzy concept nodes, clears up some week points of fuzzy theory. Moreover, FAMOUS, which has the features; separately learning, good association and hierarchical structure, are useful for realizing other intelligent systems.

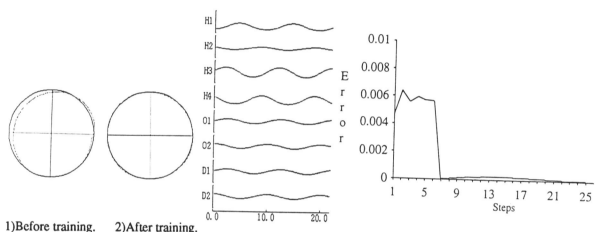

1) Before training.   2) After training.
Fig.16 Circle flight results at the physical training.

Fig.17 Error transition of the physical training.

REFERENCES

[1] L. A. Zadeh: Fuzzy Algorithms, *Information and Control*, Vol.12, 94/102 (1968).
[2] T. Terano: *Fuzzy System Theory and Its Applications*, Ohmu-sya (1987) (*in Japanese*).
[3] T. Yamaguchi, N. Imasaki and K. Haruki: Fuzzy Rule Realization on Associative Memory System, *IJCNN-90-WASH-DC Conf. Proc.*, Vol.2, 720/723 (1990).
[4] T. Yamaguchi, T. Takagi and K. Doya: Fuzzy Inference with An Associative Memory System, *6th Fuzzy System Symposium Conf. Papers*, 93/96 (1990) (*in Japanese*).
[5] T. Takagi, T. Yamaguchi and M. Sugeno: Conceptual Fuzzy Sets, *IFES'91 Conf. Proc.*, (1991).
[6] K. Goto and T. Yamaguchi: Fuzzy Associative Memory Application to A Plant Modeling, *Artificial Neural Networks*, 1245/1248 (1991).
[7] T. Yamaguchi, M. Tanabe, K. Kuriyama and T. Mita: Fuzzy Adaptive Control with An Associative Memory System, *IEE CONTROL91 Conf. Pub. No.332*, Vol. 2, 944/947 (1991).
[8] B. Kosko: Adaptive Bidirectional Associative Memories, *Applied Optics*, Vol.26, No.23, 4947/4960 (1987).
[9] T. Takagi and M. Sugeno: Fuzzy Identification of Systems and Its Application to Modeling and Control, *IEEE Trans.*, SMC-15-1, 116/132 (1988).
[10] B. Kosko: Fuzzy Cognitive Maps, *Int. Jour. of Man-Machine Studies*, Vol.24, 65/67 (1986).
[11] K. Doya and S. Yoshizawa: Adaptive Neural Oscillator Using Continuos-Time Back-Propagation Learning, *Neural Networks*, Vol. 2, 375/385 (1989).
[12] T. Yamaguchi, K. Goto, M. Yoshida, Y. Mizoguchi and T. Mita: Fuzzy Associative Memory System and Its Application to A Helicopter Control, *IFES'91 Conf. Proc.*, (1991).
[13] T. Yamaguchi, K. Kuriyama and T. Mita: Stabilizers Problem on Fuzzy Adaptive Control, *33th Automatic Control Joint Conf. Proc.*, 155/158 (1990) (*in Japanese*).
[14] B. Widrow and M. E. Hoff: Adaptive Switching Circuits, *WESCON conv. Rec.*, 4 (1960).

DEVELOPMENT AND SIMULATION OF AN F/A-18 FUZZY LOGIC AUTOMATIC CARRIER LANDING SYSTEM

Marc Steinberg
Naval Air Warfare Center, Aircraft Division, Warminster, PA 18974-0591

ABSTRACT - This paper describes the development and simulation of a Fuzzy logic Automatic Carrier Landing System(FACLS) for the F/A-18 aircraft. The FACLS controls three-dimensional flight path, speed, sink rate, and angular attitudes to allow a safe ship-board landing. The FACLS has limited control authority and is required to cope with carrier motion, air turbulence, sensor noise and delays, and the carrier's air wake. Fuzzy logic was used to create a system that combines elements of human pilot "intelligence" with more conventional automatic control laws. To do this, the FACLS has six fuzzy logic rule bases embedded in a classical control structure. The FACLS was tested in simulation and compared with the conventional F/A-18 Automatic Carrier Landing System. Results suggest significant benefits from fuzzy logic in both guidance and control tasks for aircraft outer loop control.

INTRODUCTION

The gulf war demonstrated the benefits of maintaining low observability, reducing pilot workload, and performing precision weapon delivery and single pass kills. These capabilities are important not only for lethality, but also for minimizing risk to Navy pilots and crews. One potential way of enhancing these capabilities is with advanced outer-loop flight control functions. For example, an advanced outer loop could be used to automate or assist the pilot in following a trajectory that yields maximum weapon launch opportunities while minimizing radar signature and enemy fire control solutions. Other examples of advanced outer loop control functions include automatic recovery and automated/assisted missile evasion. Performing these types of advanced outer loop functions will likely require more sophisticated techniques than those currently used in aircraft control. This is particularly true due to the desire to incorporate some degree of skilled pilot "intelligence" into the control technique and remain sensitive to complicated pilot concerns.

One technology that shows potential for this type of control problem is fuzzy logic[1]. Fuzzy logic may be effective for advanced outer loops because of its ability to incorporate intelligent rule-based behavior and improve dynamic response through highly nonlinear control strategies. To date, there have been many successful applications of fuzzy logic control[2-4]. Yet, aircraft are a unique and difficult application that is not very amenable to the majority of control techniques. While most advanced control techniques have been demonstrated with aircraft as the controlled plant, very few have been used in any way on production aircraft. There has been some work showing the potential use of fuzzy logic to flight control problems. Two examples using simplified aircraft models are aircraft longitudinal approach control[5] and aircraft roll control[6].

To examine fuzzy logic's capability on a complicated aircraft outer loop control problem, carrier landing was chosen[7]. The appeal of carrier landing as a demonstration problem is that it is the most difficult task routinely done by Naval aircraft and it includes both guidance and control aspects[8-19]. A carrier landing system must control three-dimensional flight path, speed, sink rate, and angular attitudes to allow a safe ship-board landing. The landing, itself, is essentially a precisely controlled crash onto a small moving target. The aircraft must clear the carrier ramp and capture one of four closely spaced wires with a hook on the back of the aircraft. The aircraft must also make the landing with the proper speed and sink rate to avoid damage to the aircraft. The small landing area combined with significant carrier motion makes this a particularly challenging problem. This is particularly true because most high performance aircraft have unforgiving dynamics at the speeds required for carrier approach. Further, a carrier landing system must operate with limited control authority and accommodate turbulence, the carrier's air wake, and significant sensor noise and delays.

This paper describes the development and simulation of a Fuzzy logic Automatic Carrier Landing System(FACLS) that replaces the ship-based elements of the conventional ACLS as well as the F/A-18 auto-pilot and auto-throttle. The FACLS specifies desired sink rates and roll angles, and then chooses stick and throttle commands to meet the sink rate and roll angle commands while also maintaining a nominal speed. Fuzzy logic was used to create a system that combines elements of human pilot "intelligence" with more conventional automatic control laws. To do this, the FACLS has six fuzzy logic rule bases embedded in a classical control structure. The FACLS also incorporates rules to remain sensitive to complicated pilot concerns that can improve overall pilot confidence in the system.

SIMULATION TESTBED

The simulation testbed was a 386-based F/A-18A six-degree-of-freedom non-linear carrier approach model. The simulation was required to have representative complexity and limitations of an actual F/A-18 on carrier approach. Thus, while this simulation is not high fidelity, it does contain:

- Nonlinear aerodynamics/engine model
- Nonlinear inner loop flight control system model
- Actuator models with rate/position saturations
- Atmospheric turbulence/Carrier Air Wake model
- Sensor noise, dynamics, and time delays
- Carrier motion model

For more details see ref. 7.

SYSTEM ARCHITECTURE

The FACLS was designed based on two main sources of knowledge. The first source was Automatic Carrier Landing System(ACLS) techniques and included the architecture of the conventional F/A-18 ACLS system, Navy ACLS requirements, and research reports on limitations and potential improvements for ACLS systems. The second source was pilot techniques and included discussions with Naval pilots, literature on piloting techniques, and Naval Flight Procedures. These two sources yielded very different control techniques that are often incompatible due to the differing capabilities and limitations of automatic systems and human pilots. Further, while automatic techniques for carrier landing are well understood, human pilot techniques are not, particularly close in to the carrier when dealing with carrier motion.

After completion of knowledge aquisition, it became apparent that a good system should have elements of both the automatic and human pilot control strategies. To implement this approach, a pure fuzzy logic controller would not be either practical or desirable. A pure fuzzy logic controller would require thousands of rules. Yet, the number of rules could be dramatically decreased by dividing the rules into separate rule-bases and implementing parts of the control strategy by conventional classical control elements such as scheduled gains. Another reason for avoiding a pure fuzzy logic controller was that the Fuzzy Logic Automatic Carrier Landing System(FACLS) requires dynamic elements such as lags and filters that cannot be easily implemented within a conventional fuzzy logic control system. Therefore, the fuzzy controller was designed as a classical control structure similar to the conventional F/A-18 ACLS with six blocks containing individual fuzzy rule bases. To provide a fair comparison with the conventional F/A-18 carrier landing system, the FACLS is constrained to have most of the same architecture limitations. This includes the same disengagement criterion and limits on allowable angular rates and attitudes. The FACLS also only uses sensor inputs that are available to the current ACLS system.

The final architecture of the FACLS is shown in Fig. 1. The close-in, glideslope, and lineup rule bases are the ones that contain most of the pilot ``intelligence''. All of the rule bases use max-dot encoding and centroid de-fuzzification. The glideslope and lineup rule bases set desired values of sink rate and roll angle respectively based on a stabilized glideslope(carrier motion is averaged out). The close-in rule base sets additional stick and throttle commands during the final part of the approach to deal with carrier deck motion and carrier air wake. The remaining three rule bases set stick and throttle commands

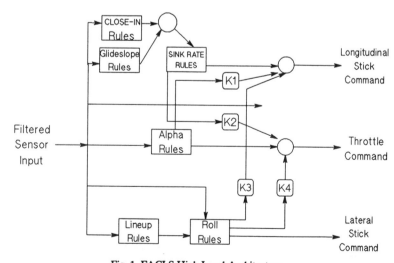

Fig. 1 FACLS High Level Architecture

to maintain nominal angle-of-attack and achieve the roll angle and sink rate commanded by the lineup and glideslope rule bases. These three rule bases are very similar in operation to the current F/A-18 auto-pilot/auto-throttle combination, except they use some nonlinear strategies to improve dynamic performance. The remaining elements in Fig. 1 are cross-feeds between the three lower level rule bases to compensate for the effect of each rule base. For example, a stick input to change sink rate would also implement a feedforward throttle change so that angle-of-attack does not significantly change during the sink rate correction.

MEMBERSHIP FUNCTIONS

The inputs to the FACLS are glideslope (vertical error), sink rate, sink acceleration, lineup(horizontal glideslope error), drift rate(rate of change of lineup), drift acceleration, roll rate, roll angle, angle of attack, normal acceleration, pitch rate, and carrier motion. The initial values of the membership functions were chosen based on what pilots consider in the linguistic ranges. The final values of the membership functions were determined through extensive tuning during simulation. Range to carrier was divided into four trapezoidal membership functions as shown in Fig. 2. The membership functions right to left are AT START, IN MIDDLE, IN CLOSE, and AT RAMP. Sink rate, sink acceleration, load factor, pitch rate, roll rate, drift rate, and drift acceleration had seven triangular membership functions which were symetrical about zero and had with small membership functions towards zero. This is shown in Fig. 3 for the positive roll rate membership functions. Angle of attack, lineup, carrier motion, and glideslope had membership functions with a trapezoid to represent near zero and half-trapezoids to represent positive and negative large membership functions. This is shown in Fig. 4 for the positive angle of attack membership functions. The membership functions right to left are The membership functions right to left are VERY SLOW, SLOW, LITTLE SLOW, and OK (pilots describe angle-of-attack by fast and slow due to its relation to speed on approach).

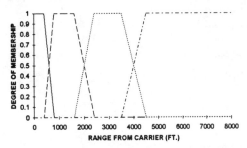

Fig. 2 Range to Carrier Membership Functions

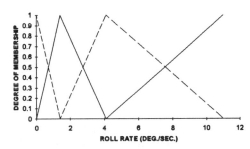

Fig. 3 Positive Roll Rate Membership Functions

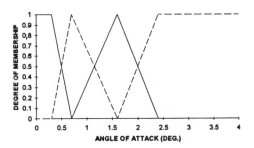

Fig. 4 Positive Angle of Attack Membership Functions

The outputs of the FACLS are throttle, longitudinal stick and lateral stick. Throttle command was an actual command to the Approach Power Compensator Actuator, but stick commands were electronic versions of the stick command that would be implemented within the flight control computers. Like most input membership functions, the initial output membership functions were triangular and symmetrical about zero. The only difference was the use of 19 membership functions for each output.

RULE BASE DEVELOPMENT

Of the six rule bases, the Roll, Sink Rate, and Alpha rule bases are the simplest and will be examined first. These three rule bases operate using largely similar strategies. To start with, each of these three rule bases has six rules that implement a high gain proportional control loop using roll, sink rate, and angle-of-attack error respectively. A typical proportional control rule, from the Roll rule base is

If ROLL ERROR *is* POSITIVE LARGE, *then* LATERAL STICK *is* POSITIVE LARGE

The Roll, Sink Rate, and Alpha rule bases also have rules that implement derivative action around zero error to null out rates. The derivative action is based on roll rate, vertical acceleration, and a mixture of load factor and pitch rate respectively. A typical derivative action rule, from the Sink Rate rule base is

If SINK RATE ERROR *is* NEAR ZERO *and* SINK ACCELERATION *is* POSITIVE LARGE, *then* LONGITUDINAL STICK *is* NEGATIVE MEDIUM LARGE

The Alpha rule base, due to the low bandwidth of throttle changes, also has some derivative based rules to increase control actions in the presence of adverse rates. An example is

If ANGLE-OF-ATTACK ERROR *is* POSITIVE SMALL *and* MIX *is* NEGATIVE LARGE, *then* THROTTLE *is* POSITIVE MEDIUM LARGE

Finally, the Roll, Sink Rate, and Alpha rule bases each have rules to phase out control actions close to the carrier. An example from the Roll rule base is

If RANGE *is* AT RAMP, *then* LATERAL STICK *is* ZERO

The next rule base that will be examined is Lineup. The lineup rules, which output desired roll angle, are much more complex than the three rule bases described above. This is because they are designed to be a combination between conventional human pilot strategy and automatic control strategy. The automatic strategy, used in the current ACLS system, requires double lead equalization because of the number of integrators between lineup error and aileron position in aircraft dynamics. In contrast, pilots use a discreet strategy in which they input a lateral stick doublet to change drift rate and then later put in an opposite doublet to nullify this drift rate. This doublet-wait-reverse doublet strategy taken to its optimal extreme is a bang-bang controller. Fuzzy logic is used to form a compromise between the pilot strategy and the continuous linear controller. This compromise has some improved performance of the nonlinear pilot control technique, but has more moderate commands so it is acceptable for an automatic system. For example, in the AT START range, the roll pulse would be started by a rule such as

If RANGE *is* AT START *and* LINEUP *is* LARGE LEFT *and* DRIFT RATE *is not* LARGE RIGHT *or* MEDIUM RIGHT, *then* ROLL *is* LARGE POSITIVE

The pulse can be ended by a rule such as

If RANGE *is* AT START *and* LINEUP *is* LARGE LEFT *and* DRIFT RATE *is* LARGE RIGHT, *then* ROLL *is* ZERO

The pulse also can be moderated by rules such as

If RANGE *is* AT START *and* LINEUP *is* LARGE LEFT *and* DRIFT RATE *is* MEDIUM RIGHT *and* DRIFT ACCELERATION *is* LARGE RIGHT, *then* ROLL *is* MEDIUM POSITIVE

Lineup rules also change as a function of range. The AT START rules are mainly used to correct large errors as fast as possible. The lineup rules for IN MIDDLE, are more concerned with correcting drift rate errors than lineup errors. This mirrors the pilot being more concerned with trends than small static errors. Thus, in the IN MIDDLE range, the rule base will not correct a trivial static lineup error. For example

If LINEUP *is* LITTLE LEFT *and* DRIFT RATE *is* NEAR ZERO *and* DRIFT ACCELERATION *is* NEAR ZERO, *then* ROLL *is* ZERO

Finally, IN CLOSE and AT RAMP, the lineup rules level the wings with the final rule

If RANGE *is* AT RAMP *or* IN CLOSE, *then* ROLL *is* ZERO

The next rule base to be examined is glideslope. Like the lineup rules, the glideslope rules are a combination between human and automatic strategy. Vertical Glideslope control is also the most important task in carrier landing and is the most likely place for a catastrophic error to occur. Like the lineup rules, large errors must be corrected quickly in the AT START range. By the IN MIDDLE range, small static errors are not as important as trends in errors. However, these rules are not symmetrical. For example, for a high error the following rule applies:

If RANGE *is* IN MIDDLE *or* IN CLOSE *and* SLOPE *is* LITTLE HIGH *and* SINK RATE *is* OK *and* SINK ACCELERATION *is* NEAR ZERO, *then* SINK RATE COMMAND *Is* NOMINAL

However, even small static low errors are dealt with due to the possibility of a ramp strike from a low condition. An example is the rule:

If RANGE *is* IN MIDDLE *or* IN CLOSE *and* SLOPE *is* LITTLE LOW *and* SINK RATE *is not* VERY SLOW, *then* sink RATE COMMAND *is* LITTLE SLOW

Finally, like the lineup rules, there is a rule to maintain a constant safe sink rate close to the carrier.

The final rule base is the close-in rules. This rule base uses a different set of membership functions for range. The membership functions are FAR FROM BURBLE, NEAR BURBLE, IN BURBLE, and PAST BURBLE. The change in rule base is necessary because the close-in rules are used to deal with the carrier air wake and carrier motion(the rest of the rule bases use a stabilized glide slope). An example of a rule dealing with a normal reaction to the carrier air wake is:

If RANGE *is* IN BURBLE *and* SLOPE *is* NEAR ZERO *and* ANGLE-OF-ATTACK *is* NEAR NOMINAL *and* SINK RATE *is* OK, *then* SINK RATE COMMAND *is* POSITIVE MEDIUM SMALL

There are several possible responses to the carrier air wake, depending on the flight situation. For example, if the aircraft is a little high before the burble with an OK sink rate, the close in rule base allows the air wake to reduce the high error, although it carefully watches sink rate to prevent the aircraft from going low. There are also a number of possible responses to carrier motion within this rule base. The outputs of this rule base are then combined with outputs from the stabilized glideslope rule base.

SIMULATION RESULTS

The performance of the Fuzzy logic Automatic Carrier Landing System(FACLS) was compared with the conventional F/A-18 ACLS on the simulation model briefly described in Sec. II. Before presenting the results, there are several advantages each system had that should be noted. First, the FACLS has an advantage because it was tuned for the simulation model, whereas the conventional ACLS was tuned for the real aircraft. The FACLS also has an advantage because the performance criterion was created by the same person that created the FACLS. The conventional ACLS had an advantage because it has a specially designed inner loop, whereas the FACLS uses a modified version of the manual inner loops, which are not ideal for automatic control.

To classify the performance of the systems, landings were considered excellent, acceptable, or poor. Excellent landings were those in which all states were close to desired on touchdown. This includes landing with proper pitch attitude, wings level, speed and sink rate close to nominal, lineup close to desired, and catching the number three wire. It also meant there was not a wave-off(aborted landing due to exceeding safe bounds during approach), bolter(missed catching any wires), or ramp strike(hook strikes carrier ramp). Acceptable landings were similar to excellent ones except that the requirements were loosened so all parameters only had to be within safe limits. All other landings were considered poor.

Table 1 shows the percentage change over the conventional ACLS in excellent and acceptable landings for simulations in severe, moderate and calm wind and sea-state conditions. The calm condition had no deck motion and the severe condition had up to 12 ft. of deck motion at the ramp from pitch and heave. In all three conditions, a variety of initial position errors were used. The FACLS improves the number of excellent and acceptable landings over the conventional ACLS in all environments, but it clearly has the largest improvements in severe weather, since the conventional F/A-18 ACLS is a very well designed system for less extreme environments.

There are several reasons for the FACLS improvements. One reason is the improvement in open loop step response over the F/A-18 autopilot by the alpha, sink rate, and roll rule bases. Table 2 shows a comparison in rise time, settling time, and percent overshoot for open loop step sink rate commands. The FACLS significantly decreases overshoot and settling time while maintaining slightly faster rise time. A time response comparison of the FACLS and the conventional ACLS is shown in Fig. 6 for a 5 ft/sec sink rate correction. These improvements in open loop response are due largely to the FACLS having an effective nonlinear control strategy in which it acts as a high gain proportional controller for large errors and

Table 1 Change in Number of Excellent and Acceptable Landings

ENVIRONMENT	EXCELLENT	ACCEPTABLE
CALM	12%	7%
MODERATE	32%	21%
SEVERE	47%	28%

Table 2 Open Loop Sink Rate Step Responses

STEPS (FT/S.)	RISE TIME(S.)		PERCT. OVRSHT.		SETT. TIME(S.)	
	ACLS	FACLS	ACLS	FACLS	ACLS	FACLS
5	1.04	0.87	18.9	0.92	3.52	1.94
10	1.32	1.07	16.5	0.83	4.41	2.58
15	1.79	1.42	14.2	0.69	4.79	3.21

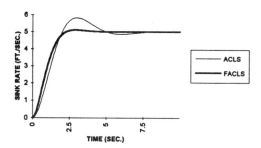

Fig. 5 Sink Rate Step Comparison

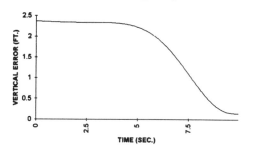

Fig. 6 FACLS Correction in Air Wake

phases in damping after the error starts decreasing.

There were also performance improvements due to the "intelligence" in the lineup, sink rate, and close-in rule bases. An example of "intelligence" in the Glideslope rule base is shown in Fig. 7. In this example, the aircraft is a little high before flying into the carrier air wake, which has a settling effect on the aircraft. The FACLS makes no attempt to correct the error and allows the settling effect of the air wake to reduce the high error. The FACLS also carefully monitors and adjusts sink rate throughout the settling to insure that the aircraft does not go low.

Conclusions

The results given in this paper demonstrate that fuzzy logic control has the potential to combine sound automatic control techniques with skilled pilot "intelligent" behavior to create a sophisticated outer loop aircraft control system. The excellent time responses of this nonlinear controller, combined with the use of some pilot heuristics clearly demonstrated substantial performance benefits in both guidance and control tasks. However, one thing that the simulation results do not demonstrate is whether the FACLS would be acceptable to Navy pilots. This is impossible to determine without real-time simulation and flight testing. Further, since the pilot study completed as part of this work was very limited in scope, it is uncertain if the pilot "intelligence" and performance criteria used in this work would be optimal for the average fleet pilot. Nonetheless, the results do show the ability of fuzzy logic control to implement a variety of pilot specified heuristics and qualitative performance criterion. Despite this, the improvements from the use of fuzzy logic are probably not significant enough to justify the use of such a radical technology for carrier landing. However, future more advanced outer loop systems may need this type of technology to be possible.

References

[1] Steinberg, M., "Role of Neural Networks and Fuzzy Logic in Flight Control Design and Development," *AIAA Aerospace Design Conference*, AIAA Paper No. 92-0999, 1992.

[2] Kosko, B., *Neural Networks and Fuzzy Systems*, Englewood Cliffs, NJ, Prentice Hall, 1990.

[3] Lee, C., "Fuzzy Logic in Control Systems, Fuzzy Logic Controller," *IEEE Transactions on Systems, Man, and Cybernetics*, Vol. 20, No. 2, Mar., 1990, pp. 404-430.

[4] Sugeno, M., *Industrial Applications of Fuzzy Control*, Elsevier Science Publishers, 1985.

[5] Chiu, S., Chand, S., Moore, D., Chaudhary, A., "Fuzzy Logic for Control of Roll and Moment for a Flexible Wing Aircraft," *IEEE Control Systems Magazine*, Vol. 11, No. 4, June, 1991, pp. 42-48.

[6] Larkin, L., "A Fuzzy Logic Controller for Aircraft Flight Control," *Industrial Applications of Fuzzy Control*, Elsevier Science Publishers, 1985.

[7] Steinberg, M., "A Fuzzy Logic Based F/A-18 Automatic Carrier Landing System," *Proceedings of the AIAA Guidance, Navigation, and Control Conference*, Aug., 1992.

[8] Heffley, R., "Outer-Loop Control Factors for Carrier Aircraft," NAVAIR Report RHE-NAV-90-TR-1, Dec., 1990.

[9] Durand, T., Wasicko, R., "Factors Influencing Glide Path Control in Carrier Landing," *AIAA Journal of Aircraft*, Vol. 4, No. 2, Mar-Apr, 1967, pp. 146-158.

[10] Durand, T., Teper, G., "An Analysis of Terminal Flight Path Control in Carrier Landing," STI Technical Report No. 137-1, Aug., 1964.

[11] Smith, R., "LSO-Pilot Interviews on Carrier Approach," U.S. Navy Report No. NADC-AM-TM-1681, Apr., 1973.

[12] Durand, T., "Theory and Simulation of Piloted Longitudinal Control in Carrier Approach," STI Technical Report No. 130-1, Mar., 1965.

[13] Jex, H., Bimal, A., Clement, W., Ashkenas, I., "Dynamic Model for Carrier Approach Lateral Lineup", STI Technical Report No. 1272-1, April 1991.

[14] "NATOPS Flight Manual, Navy Model F/A-18A/B/C/D," U.S. Navy Report No. A1-F18AC-NFM-000, 1988.

[15] Automatic Carrier Landing System, Airborne Subsystem, General Requirements for," Naval Air Systems Command, AR-40A, May, 1975.

[16] Urnes, J., Hess, R., "Development of the F/A-18A Automatic Carrier Landing System," *AIAA Journal of Guidance, Control, and Dynamics*, Vol. 8, No. 3, May, 1985, pp. 289-295.

[17] Craig, S., Ringland, R., Ashkenas, I., "An Analysis of Navy Approach Power Compensator Problems and Requirements," STI Technical Report No. 197-1, March, 1971.

[18] Kessler, G., Huff, R., "A-6F Airplane Power Approach Systems Development and Analysis," U.S. Navy Report No. SA-118R-86.

Chapter 10: Communications

AN RLS FUZZY ADAPTIVE FILTER, WITH APPLICATION TO NONLINEAR CHANNEL EQUALIZATION

Li-Xin Wang
Department of Electrical Engineering and Computer Science
University of California at Berkeley
Berkeley, CA 94720

and

Jerry M. Mendel
Department of Electrical Engineering-Systems
University of Southern California
Los Angeles, CA 90089-2564

Abstract—

A fuzzy adaptive filter is constructed from a set of fuzzy IF-THEN rules which change adaptively to minimize some criterion function as new information becomes available. In this paper, we develop a fuzzy adaptive filter which uses a recursive least squares (RLS) adaptation algorithm. The RLS fuzzy adaptive filter is constructed through the following four steps: 1) define fuzzy sets in the filter input space $U \subset R^n$ whose membership functions cover U; 2) construct a set of fuzzy IF-THEN rules which either come from human experts or are determined during the adaptation procedure by matching input-output data pairs; 3) construct a filter based on the set of rules; and, 4) update the free parameters of the filter using the RLS algorithm. The most important advantage of the fuzzy adaptive filter is that linguistic information (in the form of fuzzy IF-THEN rules) and numerical information (in the form of input-output pairs) can be combined into the filter in a uniform fashion. Finally, this fuzzy adaptive filter is applied to nonlinear communication channel equalization problems; the simulation results show that: 1) without using any linguistic information, the RLS fuzzy adaptive filter is a well-performing nonlinear adaptive filter (similar to polynomial and neural-net adaptive filters); 2) by incorporating some linguistic description (in fuzzy terms) about the channel into the fuzzy adaptive filter, the adaptation speed is greatly improved; and, 3) the bit error rate of the fuzzy equalizer is very close to that of the optimal equalizer.

I. INTRODUCTION

Filters are information processors. In practice, information usually comes from two sources: sensors which provide numerical data associated with a problem, and human experts who provide linguistic descriptions (often in the form of fuzzy IF-THEN rules) about the problem. Existing filters can only process numerical data, whereas existing expert systems can only make use of linguistic information; therefore, their successful applications are limited to problems (or portions of problems) where either linguistic rules or numerical data do not play a critical role. There are, however, a large number of practical problems in economics, seismology, management, etc., where both linguistic and numerical information are critical. At present, when we are faced with such problems, we use linguistic information, consciously or unconsciously, in the: choice among different filters, evaluation of filter performance, choice of filter orders, interpretation of filtering results, etc. There are serious limitations to using linguistic information in this way, because for most practical problems the linguistic information (in its natural form) is not about which kind of filter should be chosen or what the order of the filter should be, etc., but is in the form of IF-THEN rules concerning fuzzy concepts like "small", "hot", "not very fast", "very large but not very very large", etc. The purpose of this paper is to develop a new kind of nonlinear adaptive filter, which we refer to as *fuzzy adaptive filter*, that makes use of both linguistic and numerical information in their natural form, i.e., as fuzzy IF-THEN rules and input-output data pairs.

II. RLS FUZZY ADAPTIVE FILTER

Our RLS fuzzy adaptive filter solves the following problem.

Problem: Consider a real-valued vector sequence $[\underline{x}(k)]$ and a real-valued scalar sequence $[d(k)]$, where $k = 0, 1, 2, ...$ is the time index, and $\underline{x}(k) \in U \equiv [C_1^-, C_1^+] \times [C_2^-, C_2^+] \times \cdots \times [C_n^-, C_n^+] \subset R^n$ (we call U and R the input and output spaces of the filter, respectively). At each time point k, we are given the values of $\underline{x}(k)$ and $d(k)$. The problem is: at each time point $k = 0, 1, 2, ...$, determine an adaptive filter $f_k : U \subset R^n \to R$ such that

$$J(k) = \sum_{i=0}^{k} \lambda^{k-i}[d(i) - f_k(\underline{x}(i))]^2 \quad (1)$$

is minimized, where $\lambda \in (0, 1]$ is a forgetting factor.

The above problem is quite general. If we constrain the f_k's to be linear functions, the problem becomes an FIR adaptive filter design problem [4]. If the f_k's are Volterra series expansions, we have an adaptive polynomial filter design problem [6]. If the f_k's are multi-layer perceptrons or radial basis function expansions, the problem becomes the neural nets adaptive filter design problem [2,3].

Design Procedure of the RLS Fuzzy Adaptive Filter:

Step 1: Define m_i fuzzy sets in each interval $[C_i^-, C_i^+]$ of the input space U, which are labeled as F_i^{ji} ($i = 1, 2, ..., n$; $ji = 1, 2, ..., m_i$), in the following way: the m_i membership functions $\mu_{F_i^{ji}}$ cover the interval $[C_i^-, C_i^+]$ in the sense that for each $x_i \in [C_i^-, C_i^+]$ there exists at least one $\mu_{F_i^{ji}}(x_i) \neq 0$. These membership functions are fixed and will not change during the adaptation procedure of Step 4.

Step 2: Construct a set of $\prod_{i=1}^{n} m_i$ fuzzy IF-THEN rules in the following form:

$$R^{(j1,...,jn)} : \text{IF } x_1 \text{ is } F_1^{j1} \text{ and } \cdots \text{ and } x_n \text{ is } F_n^{jn},$$
$$\text{THEN } d \text{ is } G^{(j1,...,jn)}, \quad (2)$$

where $\underline{x} = (x_1, ..., x_n)^T \in U$ (the filter input), $d \in R$ (the filter output), $ji = 1, 2, ..., m_i$ with $i = 1, 2, ..., n$, F_i^{ji}'s are the same labels of the fuzzy sets defined in Step 1, and the $G^{(j1,...,jn)}$'s are labels of fuzzy sets defined in the output space which are determined in the following way: if there are linguistic rules from human experts in the form of (2), set $G^{(j1,...,jn)}$ to be the corresponding linguistic terms of these rules; otherwise, set $\mu_{G^{(j1,...,jn)}}$ to be an arbitrary membership function over the output space R. *It is in this way that we incorporate linguistic rules into the fuzzy adaptive filter, i.e., we use linguistic rules to construct the initial filter.*

Step 3: Construct the filter f_k based on the $\prod_{i=1}^{n} m_i$ rules in Step 2 as follow:

$$f_k(\underline{x}) = \frac{\sum_{j1=1}^{m_1} \cdots \sum_{jn=1}^{m_n} \theta^{(j1,...,jn)}(\mu_{F_1^{j1}}(x_1) \cdots \mu_{F_n^{jn}}(x_n))}{\sum_{j1=1}^{m_1} \cdots \sum_{jn=1}^{m_n} (\mu_{F_1^{j1}}(x_1) \cdots \mu_{F_n^{jn}}(x_n))}, \quad (3)$$

where $\underline{x} = (x_1, ..., x_n)^T \in U$, $\mu_{F_i^{ji}}$'s are membership functions defined in Step 1, and $\theta^{(j1,...,jn)} \in R$ is the point at which $\mu_{G^{(j1,...,jn)}}$ achieves its maximum value. Due to the way in which we defined the $\mu_{F_i^{ji}}$'s in Step 1, the denominator of (3) is nonzero for all the points of U, therefore the filter f_k of (3) is well-defined. Equation (3) is obtained by combining the $\prod_{i=1}^{n} m_i$ rules of Step 2 using product inference and centroid defuzzification [8,9].

In (3), the weights $\mu_{F_1^{j1}}(x_1) \cdots \mu_{F_n^{jn}}(x_n)$ are fixed functions of $\underline{x}$; therefore, the free design parameters of the fuzzy adaptive filter are the $\theta^{(j1,...,jn)}$'s which are now collected as a $\prod_{i=1}^{n} m_i$-dimensional vector

$$\begin{aligned}\underline{\theta} &= (\theta^{(1,1,...,1)}, ..., \theta^{(m_1,1,...,1)}; \theta^{(1,2,1,...,1)}, ..., \theta^{(m_1,2,1,...,1)}, \\ &\quad ..., \theta^{(1,m_2,1,...,1)}, ..., \theta^{(m_1,m_2,1,...,1)}; ...; \theta^{(1,m_2,...,m_n)}, \\ &\quad ..., \theta^{(m_1,m_2,...,m_n)})^T. \end{aligned} \quad (4)$$

Define the *fuzzy basis functions* [9]

$$p^{(j1,...,jn)}(\underline{x}) = \frac{\mu_{F_1^{j1}}(x_1) \cdots \mu_{F_n^{jn}}(x_n)}{\sum_{j1=1}^{m_1} \cdots \sum_{jn=1}^{m_n}(\mu_{F_1^{j1}}(x_1) \cdots \mu_{F_n^{jn}}(x_n))}, \quad (5)$$

and collect them as a $\prod_{i=1}^{n} m_i$-dimensional vector $\underline{p}(\underline{x})$ in the same ordering as the $\underline{\theta}$ of (4), i.e.,

$$\begin{aligned}\underline{p}(\underline{x}) &= (p^{(1,1,...,1)}(\underline{x}), ..., p^{(m_1,1,...,1)}(\underline{x}); p^{(1,2,1,...,1)}(\underline{x}), \\ &\quad ..., p^{(m_1,2,1,...,1)}(\underline{x}); ...; p^{(1,m_2,1,...,1)}(\underline{x}), ..., \\ &\quad p^{(m_1,m_2,1,...,1)}(\underline{x}); \cdots; p^{(1,m_2,...,m_n)}(\underline{x}), ..., \\ &\quad p^{(m_1,m_2,...,m_n)}(\underline{x}))^T. \end{aligned} \quad (6)$$

Based on (4) and (6) we can now rewrite (3) as

$$f_k(\underline{x}) = \underline{p}^T(\underline{x})\underline{\theta}. \quad (7)$$

Step 4: Use the following RLS algorithm [4] to update $\underline{\theta}$: let the initial estimate of $\underline{\theta}$, $\underline{\theta}(0)$, be determined as in Step 2, and $P(0) = \sigma I$, where σ is a small positive constant, and I is the $\prod_{i=1}^{n} m_i - by - \prod_{i=1}^{n} m_i$ identity matrix; at each time point $k = 1, 2, ...$, do the following:

$$\underline{\phi}(k) = \underline{p}(\underline{x}(k)), \quad (8)$$

$$P(k) = \frac{1}{\lambda}[P(k-1) - P(k-1)\underline{\phi}(k)(\lambda + \underline{\phi}^T(k)P(k-1)\underline{\phi}(k))^{-1}\underline{\phi}^T(k)P(k-1)]. \quad (9)$$

$$K(k) = P(k-1)\underline{\phi}(k)[\lambda + \underline{\phi}^T(k)P(k-1)\underline{\phi}(k)]^{-1}, \quad (10)$$

$$\underline{\theta}(k) = \underline{\theta}(k-1) + K(k)(d(k) - \underline{\phi}^T(k)\underline{\theta}(k-1)), \quad (11)$$

where $[\underline{x}(k)]$ and $[d(k)]$ are the sequences defined above in the Problem, $p(*)$ is defined in (6), and λ is the forgetting factor in (1).

Remark 1: The RLS algorithm (9)-(11) is obtained by minimizing $J(k)$ of (1) with f_k constrained to be the form of (7). Because f_k of (7) is linear in the parameter, the derivation of (9)-(11) is the same as that for the FIR linear adaptive filter [4]; therefore, we omit the details.

Remark 2: It was proven in [8,9] that functions in the form of (3) are universal approximators, i.e., for any real continuous function g on the compact set U, there exists a function in the form of (3) such that it can uniformly approximate g over U to arbitrary accuracy. Consequently, our fuzzy adaptive filter is a powerful nonlinear adaptive filter in the sense that it has the capability of performing any nonlinear filtering operation.

Remark 3: Linguistic information (in the form of the fuzzy IF-THEN rules of (2)) and numerical information (in the form of desired input-output pairs $(\underline{x}(k), d(k))$) are combined into the filter in the following way: due to Steps 2-4, linguistic IF-THEN rules are directly incorporated into the filter (3) by constructing the initial filter based on the linguistic rules; and, due to the adaptation Step 4, numerical pairs $(\underline{x}(k), d(k))$ are incorporated into the filter by updating the filter parameters such that the filter output "matches" the pairs in the sense of minimizing (1). It is natural and reasonable to assume that linguistic information from human experts is provided in the form of (2) because the rules of (2) state what the filter outputs should be in some input situations, where "what should be" and "some situations" are represented by linguistic terms which are characterized by fuzzy membership functions. On the other hand, it is obvious that the most natural form of numerical information is provided in the form of input-output pairs $(\underline{x}(k), d(k))$.

III. APPLICATION TO NONLINEAR CHANNEL EQUALIZATION

Nonlinear distortion over a communication channel is now a significant factor hindering further increase in the attainable data rate in high-speed data transmission [1,5]. Because the received signal over a nonlinear channel is a nonlinear function of the past values of the transmitted symbols, and the nonlinear distortion varies with time and from place to place, effective equalizers for nonlinear channels should be nonlinear and adaptive. In this section, we use our RLS fuzzy adaptive filter (which is clearly nonlinear and adaptive) as equalizers for a nonlinear channel.

The digital communication system considered in this paper is shown in Fig. 1, where the "channel" includes the effects of the transmitter filter, the transmission medium, the receiver matched filter, and other components. The transmitted data sequence $s(k)$ is assumed to be an independent sequence taking values from $\{-1, 1\}$ with equal probability. The inputs to the equalizer, $x(k), x(k-1), \cdots, x(k-n+1)$, are the channel outputs corrupted by an additive noise $e(k)$. The task of the equalizer at the sampling instant k is to produce an estimate of the transmitted symbol $s(k-d)$ using the information contained in $x(k), x(k-1), \cdots, x(k-n+1)$, where the integers n and d are known as the order and the lag of the equalizer, respectively.

We use the geometric formulation of the equalization problem due to [2,3]. Using similar notation to that in [2,3], define

$$P_{n,d}(1) = \{\underline{\hat{x}}(k) \in R^n | s(k-d) = 1\}, \quad (12)$$
$$P_{n,d}(-1) = \{\underline{\hat{x}}(k) \in R^n | s(k-d) = -1\}, \quad (13)$$

where

$$\underline{\hat{x}}(k) = [\hat{x}(k), \hat{x}(k-1), \cdots, \hat{x}(k-n+1)]^T, \quad (14)$$

$\hat{x}(k)$ is the noise-free output of the channel (see Fig. 1), and $P_{n,d}(1)$ and $P_{n,d}(-1)$ represent the two sets of possible channel noise-free output vectors $\underline{\hat{x}}(k)$ that can be produced from sequences of the channel inputs containing $s(k-d) = 1$ and $s(k-d) = -1$, respectively. The equalizer can be characterized by the function

$$g_k : R^n \longrightarrow \{-1, 1\} \quad (15)$$

with

$$\hat{s}(k-d) = g_k(\underline{x}(k)), \quad (16)$$

where

$$\underline{x}(k) = [x(k), x(k-1), \cdots, x(k-n+1)]^T \quad (17)$$

is the observed channel output vector. Let $p_1[\underline{x}(k)|\underline{\hat{x}}(k) \in P_{n,d}(1)]$ and $p_{-1}[\underline{x}(k)|\underline{\hat{x}}(k) \in P_{n,d}(-1)]$ be the conditional probability density functions of $\underline{x}(k)$ given $\underline{\hat{x}}(k) \in P_{n,d}(1)$ and $\underline{\hat{x}}(k) \in P_{n,d}(-1)$, respectively. It was shown in [2,3] that the equalizer which is defined by

$$f_{opt}(\underline{x}(k)) = sgn[p_1(\underline{x}(k)|\underline{\hat{x}}(k) \in P_{n,d}(1)) - p_{-1}(\underline{x}(k)|\underline{\hat{x}}(k) \in P_{n,d}(-1))] \quad (18)$$

achieves the minimum bit error rate for the given order n and lag d, where $sgn(y) = 1(-1)$ if $y \geq 0$ ($y < 0$). If the noise $e(k)$ is zero-mean and Gaussian with covariance matrix

$$Q = E[(e(k),...,e(k-n+1))(e(k),...,e(k-n+1))^T], \quad (19)$$

then from $x(k) = \hat{x}(k) + e(k)$ we have that

$$\begin{aligned}&p_1[\underline{x}(k)|\underline{\hat{x}}(k) \in P_{n,d}(1)] - p_{-1}[\underline{x}(k)|\underline{\hat{x}}(k) \in P_{n,d}(-1)] \\ &= \sum exp[-\frac{1}{2}(\underline{x}(k) - \underline{\dot{x}}_+)^T Q^{-1}(\underline{x}(k) - \underline{\dot{x}}_+)] \\ &- \sum exp[-\frac{1}{2}(\underline{x}(k) - \underline{\dot{x}}_-)^T Q^{-1}(\underline{x}(k) - \underline{\dot{x}}_-)], \quad (20)\end{aligned}$$

where the first (second) sum is over all the points $\underline{\dot{x}}_+ \in P_{n,d}(1)$ ($\underline{\dot{x}}_- \in P_{n,d}(-1)$).

Now consider the nonlinear channel

$$\hat{x}(k) = s(k) + 0.5s(k-1) - 0.9[s(k) + 0.5s(k-1)]^3, \quad (21)$$

and white Gaussian noise $e(k)$ with $E[e^2(k)] = 0.2$. For this case, the optimal decision region for $n = 2$ and $d = 0$,

$$\begin{aligned}[\underline{x}(k) \in R^2 | p_1[\underline{x}(k)|\underline{x}(k) \in P_{2,0}(1)] \\ -p_{-1}[\underline{x}(k)|\underline{x}(k) \in P_{2,0}(-1)] \geq 0], \quad (22)\end{aligned}$$

is shown in Fig. 2 as the shaded area. The elements of the sets $P_{2,0}(1)$ and $P_{2,0}(-1)$ are illustrated in Fig. 2 by the "o" and "*", respectively. From Fig. 2 we see that the optimal decision boundary for this case is severely nonlinear. We now use the RLS fuzzy adaptive filter to solve this specific equalization problem (channel (28), $e(k)$ white Gaussian with variance 0.2, equalizer order $n = 2$ and lag $d = 0$).

Example 1: Here we used the RLS fuzzy adaptive filter without any linguistic information. We chose $\lambda = 0.999, \sigma = 0.1, m_1 = m_2 = 9$, and $\mu_{F_i^j}(x_i) = exp[-\frac{1}{2}(\frac{x_i - \bar{x}_i^j}{0.3})^2]$ with $\bar{x}_i^j = -2, -1.5, -1, -0.5, 0, 0.5, 1, 1.5, 2$ for $j = 1, 2, ..., 9$, respectively, where $i = 1, 2$, $x_1 = x(k)$ and $x_2 = x(k-1)$. For the same realization of the sequence $s(k)$ and the same randomly chosen initial parameters $\underline{\theta}(0)$ (within [-0.5, 0.5]), we simulated the cases when the adaptation algorithm (9)-(11) stopped at: (i) $k = 30$, and (ii) $k = 100$. The final decision regions, $[\underline{x}(k) \in R^2 | f_k(\underline{x}(k)) \geq 0]$, for the two cases are shown in Figs. 3 and 4, respectively. From Figs. 3 and 4 we see that the decision regions obtained from the RLS fuzzy adaptive filter tended to converge towards the optimal decision region.

Example 2: Next, we used the RLS fuzzy adaptive filter and incorporated the following linguistic information about the decision region. From the geometric formulation we see that the equalization problem is equivalent to determining a decision boundary in the input space of the equalizer. Suppose that there are human experts who are very familiar with the specific situation, such that although they cannot draw the specific decision boundary in the input space of the equalizer, they can assign degrees to different regions in the input space which reflect their belief that the regions should belong to 1-catalog or -1-catalog. Take Fig. 2 as an example. We see from Fig. 2 that the difficulty is to determine which catalog the middle portion should belong to; in other words, as we move away from the middle portion, we have less and less uncertainty about which catalog the region should belong to. For example, for the left-most region in Fig. 2, we have more confidence that it should belong to the 1-catalog rather than the -1-catalog. Similarly, for the right-most region in Fig. 2, we have more confidence that it should belong to the -1-catalog rather than the 1-catalog. Also, we assume that the human experts know that a portion of the boundary is somewhere around $x(k) = -1.2$ for $x(k-1)$ less than 1 and around $x(k) = 1.2$ for $x(k-1)$ greater than 1. To make these observations more specific, we have the fuzzy rules shown in Fig. 5 where the membership functions N3, N2, etc. are the $\mu_{F_i^j}$'s defined in Example 1. We have 48 rules in Fig. 5, corresponding to the boxes with numbers; for example, the bottom-left box corresponds to the rule: "IF $x(k)$ is N4 and $x(k-1)$ is N4, THEN f_k is G," where f_k is the filter output, and the center of μ_G is 0.6. Because the filter output f_k is a weighted average of these centers (see (3)), the numbers 0.6, 0.4, -0.4, -0.6 in Fig. 5 reflect our belief that the regions should correspond to the 1-catalog or the -1-catalog. For example, if the input point $[x(k), x(k-1)]$ falls in the left-most region of Fig. 5, then we have more confidence that the transmitted $s(k)$ should be 1 rather than -1, and, we represent this confidence by assigning the center of the fuzzy term in the corresponding THEN part to be 0.6.

It should be emphasized that the rules in Fig. 5 provide very fuzzy information about the decision region, because: 1) the regions are fuzzy, i.e., there are no clear boundaries between the regions, and 2) the numbers 0.6, 0.4, -0.4, -0.6 are conservative, i.e., they are away from the real transmitted values 1 or -1. We now show that although these rules are fuzzy, the speed of adaptation is greatly improved by incorporating them into the RLS fuzzy adaptive equalizer (filter). Figure 6 shows the final decision region determined by the RLS fuzzy adaptive filter, $[\underline{x}(k) \in R^2 | f_k(\underline{x}(k)) \geq 0]$ (shaded area), when the adaptation stopped at $k = 30$ after the rules in Fig. 5 were incorporated, where the $\mu_{F_i^j}$'s and the sequence $s(k)$ were the same as those in Example 1. Comparing Figs. 6 and 3 we see that the adaptation speed was greatly improved by incorporating these fuzzy rules.

Example 3: In this final example, we compared the bit error rates achieved by the optimal equalizer (18) and the fuzzy adaptive equalizer for different signal-to-noise ratios, for the channel (21) with equalizer order $n = 2$ and lag $d = 1$. The optimal bit error rate was computed by applying the optimal equalizer (18) to a realization of 10^6 points of the sequences $s(k)$ and $e(k)$. We chose the filter parameters to be the same as in Example 1. We ran the RLS fuzzy adaptive filters for the first 1000 points in the same 10^6 point realization of $s(k)$ and $e(k)$ as for the optimal equalizer, and then used the trained fuzzy equalizers to compute the bit error rate for the same 10^6 point realization. Figure 7 shows the bit error rates of the optimal equalizer and the RLS fuzzy equalizer for different signal-to-noise ratios. We see from Fig. 7 that the bit error rate of the fuzzy equalizer is very close to the optimal one.

IV. CONCLUSIONS

In this paper, we developed a new nonlinear adaptive filter, namely: RLS fuzzy adaptive filter. The key elements of the fuzzy adaptive filter are a fuzzy logic system, which is constructed from a set of fuzzy IF-THEN rules, and an adaptive algorithm for updating the parameters in the fuzzy logic system which is an RLS type of algorithm. The most important advantage of the fuzzy adaptive filter is that linguistic information from human experts (in the form of fuzzy IF-THEN rules) can be directly incorporated into the filters. If no linguistic information is available, the fuzzy adaptive filter becomes a well-defined nonlinear adaptive filter, similar to the polynomial, neural nets, or radial basis function adaptive filters. We applied the RLS fuzzy adaptive filter to nonlinear channel equalization problems. Simulation results showed that: 1) the fuzzy adaptive filter worked quite well without using any linguistic information; 2) by incorporating some linguistic rules into the fuzzy adaptive filter, the adaptation speed was greatly improved; and, 3) the bit error rate of the fuzzy equalizer was close to that of the optimal equalizer.

V. REFERENCES

[1] Biglieri, E., A. Gersho, R. D. Gitlin and T. L. Lim, "Adaptive cancellation of nonlinear intersymbol interference for voiceband data transmission," *IEEE J. on Selected Areas in Communications*, Vol. SAC-2, No. 5, pp. 765-777, 1984.

[2] Chen, S., G. J. Gibson, C. F. N. Cowan and P. M. Grant, "Adaptive equalization of finite non-linear channels using multilayer perceptrons," *Signal Processing*, Vol. 20, pp. 107-119, 1990.

[3] Chen, S., G. J. Gibson, C. F. N. Cowan and P. M. Grant, "Reconstruction of binary signals using an adaptive radial-basis-function equalizer," *Signal Processing*, Vol. 22, pp. 77-93, 1991.

[4] Cowan, C. F. N. and P. M. Grant (ed.), "Adaptive Filters," Prentice-Hall, Inc., New Jersey, 1985.

[5] Falconer, D. D., "Adaptive equalization of channel nonlinearities in QAM data Transmission systems," *The Bell System Technical Journal*, Vol. 57, No. 7, pp. 2589-2611, 1978.

[6] Mathews, V. J., "Adaptive ploynomial filters," *IEEE Signal Processing Magazine*, pp. 10-26, July, 1991.

[7] Mulgrew, B. and C. F. N. Cowan, "Adaptive filters and Equalisers," Kluwer Academic Publishers, 1988.

[8] Wang, L. X., "Fuzzy systems are universal approximators," *Proc. IEEE Conf. on Fuzzy Systems*, pp. 1163-1170, San Diego, 1992.

[9] Wang, L. X. and J. M. Mendel, "Fuzzy basis functions, universal approximation, and orthogonal least squares learning," *IEEE Trans. on Neural Networks*, Vol. 3, No. 5, pp. 807-814, 1992.

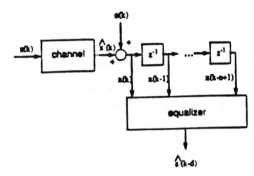

Figure 1: Schematic of data transmission system.

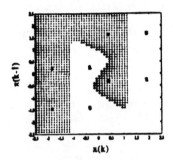

Figure 2: Optimal decision region for the channel (21). Gaussian white noise with variance $\sigma_e^2 = 0.2$, and equalizer order $n = 2$ and lag $d = 0$.

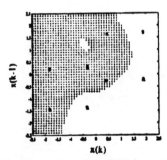

Figure 3: Decision region of the RLS fuzzy adaptive filter without using any linguistic information and when the adaptation stopped at $k = 30$.

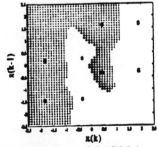

Figure 4: Decision region of the RLS fuzzy adaptive filter without using any linguistic information and when the adaptation stopped at $k = 100$.

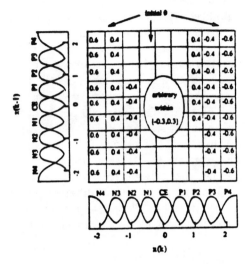

Figure 5: Illustration of some fuzzy rules about the decision region.

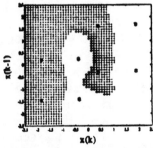

Figure 6: Decision region of the RLS fuzzy adaptive filter after incorporating the fuzzy rules illustrated in Fig. 5 and when the adaptation stopped at $k = 30$.

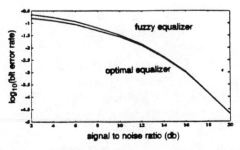

Figure 7: Comparison of bit error rates achieved by the optimal and fuzzy equalizers.

Model-Reference Neural Color Correction for HDTV Systems based on Fuzzy Information Criteria

Po-Rong Chang, and C. C. Tai
National Chiao-Tung University,
Hsin-Chu, Taiwan, R.O.C.

Abstract—This paper presents a new adaptive color correction process for High Definition Television (HDTV) system based on human color perception model. To achieve the high-fidelity color reproduction, it requires eliminating the color error sources resulted from camera, monitor, intermediate color image processing and their mutual interaction. The intermediate processing may include the coding/decoding, quantization, and modulation/demodulation. And, the mutual interaction denotes the color gamut mismatch and resolution conversion between camera and monitor. It can be shown that those color error sources of unknown form are highly nonlinear. Hence, a cost-effective indirect adaptive control scheme consisting of both back-propagation neural net controller and forward model is proposed to overcome the difficulty of dealing with both nonlinearity and model-identification of the HDTV systems. The forward model is used to convert the plant output error (color difference) into control signal error for training the back-propagation controller. Furthermore, the values of derivative-related parameters of the forward model are determined by Hornik's multi-layer neural net identification process. In addition, it is noted that the measure of the color difference (plant output error) in human color perception space should be quantified by fuzzy information in order to perform the system control. Finally, the effectiveness of our method is being verified by a number of experiments.

I. INTRODUCTION

The colorimetry of current 525 line NTSC television system is vaguely defined in terms of its display–i.e., the receiver/monitor. The standard simply states that the signals should be suitable for an NTSC display. Future television systems should be designed to take advantage of future display technologies, which may have different, perhaps improved color characteristics than current CRT based displays. High Definition Television (HDTV) offers the potential for a vastly improved television service. In addition to higher resolution and a wide aspect ratio, it provides an opportunity to improve color reproduction. Its color improvements include complete definitions of color characteristics and a large color gamut which is addressed by the SMPTE Ad-Hoc Group on HDTV production[1][2]. Meanwhile, there are several color error sources which will degrade the quality of color reproduction. The color errors may be caused by the deviation from the ideal image taking characteristics and the nonlinearity of the color camera. Because of the nonlinear gamma-correction used with all color camera, the division of luminance and chrominance in the transmitted signals is imperfect. Therefore, the chrominance signal, which should transmit only chrominance, carries a considerable amount of luminance information too. In other words, the constant luminance principle does not work precisely. As a result, luminance errors that occur are greater in an area with high saturation figures. Therefore, contour in the high saturated image area is somewhat blurred because of the addition of the converted narrow-band chrominance signal to the luminance signal. In contrast to camera, the second kind of color errors is caused by deviation of the primary colors of monitor due to such factors as inadequate luminous efficiency and light residues. Obviously, the noisy channel and the nonlinearity involved in the coder and decoder would also degrade the reproduction quality. This kind of color errors caused by transmission is identified as the third color error source. Additionally, the color gamut mismatch and resolution conversion between the camera and monitor may also lead to be a the fourth color error source to the entire system. Finally, the last color error source comes from the deterioration of the color-rendering properties of illuminating light. It is desired to render correct colors under a varity of lighting conditions.

Figure 1, shows how a neural-based color correction processor can be used to eliminate those undesired color errors based on human perceptual color space. In order for colors to be reproduced desirably, it is necessary to analyze what colors people feel or sense to be desirable. Usually, it is lack of a quantitative definition of subjective image quality and human color perception. Nevertheless, the psychological color difference perceived by human can be characterized by the fuzzy information measure[8]. The details of our adaptive fuzzy neural-based correction will be described in the next two section.

II. Fuzzy Neural based Control for Color Correction

Recently, Jordan[3], Psaltis, Sideris, and Yamanura[4] have proposed a promising adaptive control by using the back-propagation neural network as the plant's controller. Meanwhile, it requires propagating the errors between actual and desired plant outputs back to produce the error in the control signal which can be used to train the back-propagation controller. To achieve the goal, their idea is essentially to treat the plant as an additional back-propagation network which is connected to the back-propagation controller directly. This additional network is called the forward model. In Figure 2, this back-propagation process is illustrated by the dashed line passing back through the forward model and continuing back through a second layered network that uses it to learn a control rule. In additon, Jordan[3] showed that the Jacobian or derivative of plant is required in performing the back-propagation process. Next, we will describe how the traditional error back-propagation rule can be applied to convert the plant output error into the control signal error.

For the traditional multilayered back-propagation network shown in Figure 3, error back-propagation rule would modify the weight, ω_{ab}^m between neurons a and b of the m-th and $(m-1)$-th layers, in order to minimize the summed squared output error, $\|\vec{\epsilon}\|^2$, respectively, as follows:

$$\Delta \omega_{ab}^m = -\eta \frac{\partial \|\vec{\epsilon}\|^2}{\partial \omega_{ab}^m} \qquad (1)$$
$$= \eta \delta_a^m q_b^{m-1} \qquad (2)$$

where $\vec{\epsilon}$ is the error vector between the actual output vector $\mathbf{u}(=\mathbf{q}^n)$ and the desired output vector $\mathbf{u}^*(=\mathbf{q}^{n*})$, and η is an acceleration constant. In equation(2), the output of the a-th neuron is q_a, its input is p_a, and δ_a is the back-propagation error given for the output layer, n, by

$$\delta_a^n = G_n'(p_a^n)(u_a - u_a^*) \qquad (3)$$

and for all other layers, $1 \le m \le n-1$, by

$$\delta_a^m = G_m'(p_a^m) \sum_i \delta_i^{m+1} \omega_{ia}^{m+1} \qquad (4)$$

where u_a and u_a^* are the a-th component of $\mathbf{u}$ and $\mathbf{u}^*$ respectively, and $G_i'(p_a^i)$ is the derivative of the sigmodal activation function for the a-th neuron at the i-th layer.

For a cascade system shown in Figure 4 consisting of the back-propagation controller and the feedforward model, we can apply error back-propagation rule directly to the system with the following modifications. For simplicity, the feedforward model is thought of as an additional, although unmodifiable layer. Obviously, the i-th plant output y_i is identical to the i-th neuron output q_i^{n+1} located at the output layer of the $(n+1)$ layer cascade system. Therefore, the squared error of the cascade system can be given by

$$E = \frac{1}{2} \sum_i (y_i^* - y_i)^2 = \frac{1}{2} \mathbf{e}^T \mathbf{e} \qquad (5)$$

where y_i^* is the desired i-th component of the desired reference output vector $\mathbf{y}^*$, and $\mathbf{e}$ is the error vector between $\mathbf{y}^*$ and $\mathbf{y}$.

Since the weights of the forward model or the output layer are unmodified, our objective is to adjust the weights of the preceeding n-layers in order to minimize E. Next, we would like to derive the relationship between the back-propagated errors for the output layer and the n-layer denoted by δ_i^{n+1} and δ_i^n respectively. According to equation (1), the weight increment between neurons a and b of the n-th and $(n-1)$-th layer is given by

$$\Delta \omega_{ab}^n = -\eta \frac{\partial E}{\partial \omega_{ab}^n} \qquad (6)$$

Substituting (5) into (6) and by chain rule, one may obtain

$$\Delta \omega_{ab}^n = -\eta \frac{\partial E}{\partial u_a} \frac{\partial u_a}{\partial \omega_{ab}} \qquad (7)$$
$$= \eta \{ \sum_i (y_i^* - y_i) \frac{\partial y_i}{\partial u_a} \} \frac{\partial u_a}{\partial \omega_{ab}}$$

Furthermore, by using the fact that $u_a = q_a^n$, $\partial q_a^n / \partial p_a^n = G_n'(p_a^n)$, $p_a^n = \sum_j \omega_{aj} q_j^{n-1}$ and chain rule, (7) becomes as follows

$$\omega_{ab}^n = \eta q_b^{n-1} \{ G_n'(p_a^n) [\sum_i (y_i^* - y_i) \frac{\partial y_i}{\partial u_a}] \} \qquad (8)$$

Compared to (2), δ_i^{n+1} and δ_i^n may be defined as

$$\delta_i^{n+1} = y_i^* - y_i \qquad (9)$$

and

$$\delta_a^n = G_n'(p_a^n) \sum_i \delta_i^{n+1} \frac{\partial y_i}{\partial u_a} \qquad (10)$$

As shown above, the error propagation rule requires computing the partial derivatives or Jacobian of the plant. If the plant is a function of unkown form, the evaluation of its Jacobian becomes a difficult job. Fortunately,

Hornik, Stnchcombe, and White [5][6] proposed a systematic method to identify the Jacobian of the plant. They showed that their multilayer back-propagation-like networks with appropriately smooth hidden layer activation functions are capable of arbitrary accurate approximation to an arbitrary function and its derivatives. As a result, the Jacobian of a plant can be identified by Horink's method. This determined plant Jacobian is then applied to the model-reference indirect adaptive control when the plant is a plant of unknown form. The system architecture of this adaptive control is shown in Figure 5.

It is known that the controller placed in front of plant acts as a pre-equalizer for the plant and forces the plant output be equal to the desired output. The proposed model-reference indirect adaptive control (MIAC) with back-propagation controller is not restricted in performing the control process described in Figure 5. Since the back-propagation neural network [7] has the capability to equalize the second plant placed in front of itself, one may easily show that the MIAC can be extended to perform the control process shown in Figure 6.

Clearly, the generalized MIAC can be applied to the color correction for HDTV system directly. The detailed function diagram is shown in Figure 7. It should be noted that plant 1 includes the HDTV monitor and human evaluation model. Usually, it requires quantifying human evaluation model in order to perform system control. Therefore, we use the Xie's fuzzy information measure [8] in our system and given by

$$\begin{aligned} H_T &= H_s(P_1^R, P_0^R) + H_f(\mu_R) + H_s(P_1^G, P_0^G) \\ &\quad + H_f(\mu_G) + H_s(P_1^B, P_0^B) + H_f(\mu_B) \end{aligned} \quad (11)$$

As mentioned previously, μ_R, μ_G, and μ_B denote the membership functions for R (Red), G (Green) and B (Blue) respectively. And P_1^X and P_0^X represent the X having grade of membership 1 and membership 0, respectively, where X = R or G or B. $H_s(\cdot)$ and $H_f(\cdot)$ are called the shannon information and fuzzy information, respectively. In contrast to plant 1, plant 2 consists of the HDTV camera and communication channel model. To verify the effectiveness of our model, a number of experiments is being tested.

III. Conclusion

In this paper, we have developed a cost-effective color correction algorithm based on a generalized neural-based indirect adaptive control which consists of both back-propagation controller and forward model. The back-propagation controller is placed between plant 1 and plant 2 which denote a combination of HDTV monitor and human color perception model and a combination of HDTV camera and communication channel respectively. It has been pointed out that this particular controller has two different functions. In other words, it acts as a equalizer to plant 2 and also the pre-equalizer (or controller) to plant 1 simultaneously. To be a controller (or pre-equalizer) for plant 1, it requires propagating the output error of plant 1 throughth a converter called the forward model to control signal error for training the back-propagation controller. In addition, the color difference (or the plant output error) evaluated in human perception space is characterized by the Xie's fuzzy information measure.

References

[1] L. E. DeMarsh "HDTV production colorimetry," *SMPTE J.*, pp.796–805 Oct. 1991.

[2] LeRoy DeMarsh "Colorimetry for HDTV," *IEEE Trans. on Consumer Electronics*, pp.1–6, vol. 37, No. 1, 1991.

[3] Jordan, M. " Generic constraints on underspecified target trajectories," In *Proceedings of the 1989 International Joint Conference on Neural Networks*, pp. 217–225 1989 New York: IEEE Press.

[4] D. Psaltis, A. Sideris, and A. A. Yamamura, "A Multilayered Neural Network Controller," *IEEE Contr. Syst. Mag.*, vol 4, pp.17–21 1988

[5] Horink, K., Stinchcombe, M., and White, H. "Multilayer feedforward nteworks are universal approximators," *Neural Networks* vol 2, pp.359–366. 1989

[6] Horink, K., Stinchcombe, M., and White, H. "Universal approximation of an unknown mapping and its derivatives using multilayer feedforward networks." *Neural Networks*, vol 3, pp.551–560. 1990

[7] S. Siu, G. J. Gibson, and C. F. N. Cowan "Decision feedback equalisation using neural network structures and performance comparison with stand architecture," *IEE Proceedings*, vol 137, No 4, 1990.

[8] XIE Wei-Xin "An information measure for color space," *Fuzzy sets and system*, pp157–165 1990 Elsever Science Publishers

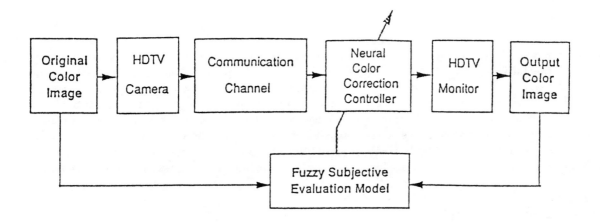

Fig. 1 Color reproduction system

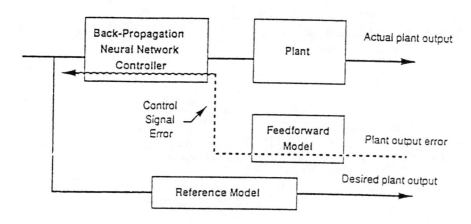

Fig. 2 Model-reference neural network control with backpropagating through a forward model of the plant to determine control signal error

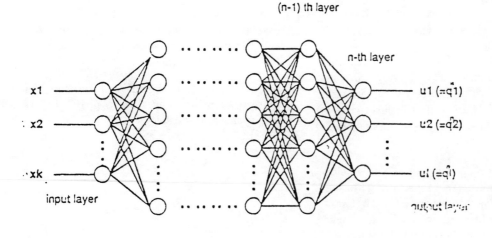

Fig. 3 Traditional n-layer back-propagation

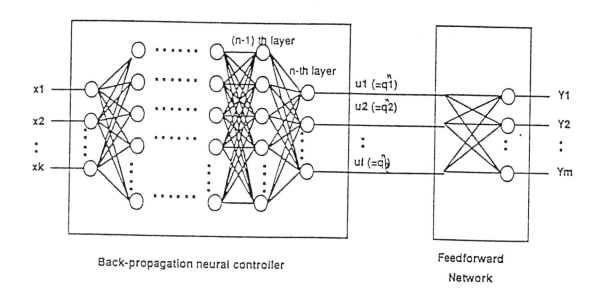

Fig. 4 A cascade system as a (n+1)-layer

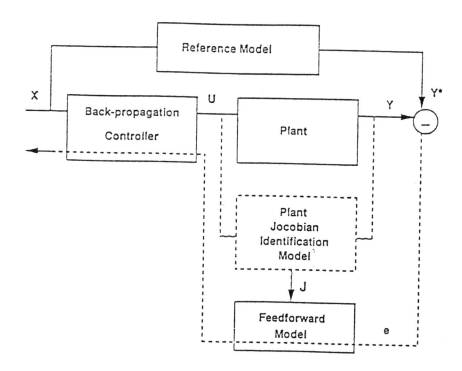

Fig. 5 The system architecture when the plant is unknown

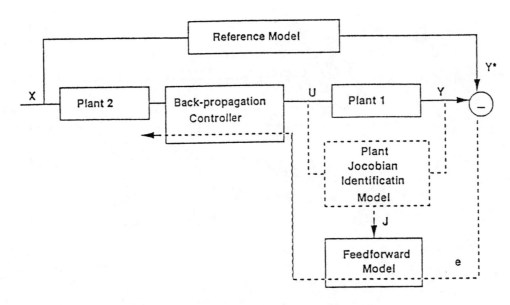

Fig. 6 Generalized MIAC with back-propagation controller

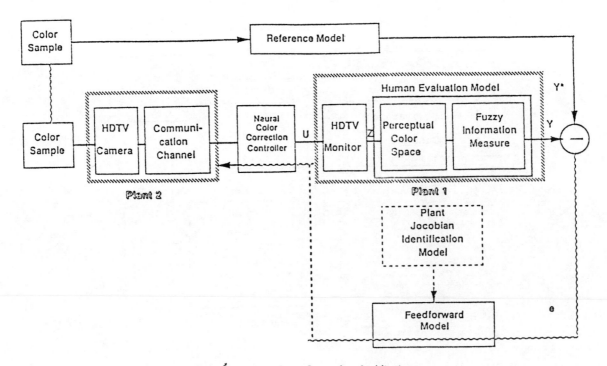

Fig. 7 Model-Reference Color Correction Architecture

CLASSIFIED VECTOR QUANTIZATION USING FUZZY THEORY

Ferran MARQUÉS
Dept. Teoría de la Señal y Comunicaciones
E.T.S.E.T.B. - U.P.C. Apdo. 30002
08080 Barcelona, SPAIN

C.-C. JAY KUO
Signal and Image Processing Institute
Dep. of Elec. Engineering-Systems U.S.C.
Los Angeles, California 90089-2564

ABSTRACT

Classified Vector Quantization (CVQ) is a technique for coding images that achieves good perceptual results while reducing the computational load of the process. However, some aspects of this technique can be improved by facing them from the fuzzy theory viewpoint. In this research, a method that uses fuzzy theory to perform the classification involved in the CVQ technique is presented. In order to formulate the membership function of different categories, the concept of surface representation by triangular elements is introduced. This method yields a more natural classification scheme. Moreover, the problem of determining the size of each subcodebook is also solved. The subcodebooks are built using fuzzy classifiers and the degree of accuracy of each codeword is tested by computing their fuzzy entropy. The codewords with larger fuzzy entropy are further split until an optimal point is reached.

1.- Introduction

Vector quantization (VQ) is a technique that has been widely applied for coding speech as well as image signals. Since image signals present some special characteristics, several methods have been proposed that utilize these characteristics to achieve better coding performance [1]. One way to improve the coding performance is by employing the fact that edges contain a large part of the perceptual information of the image. However, as edges are not very likely within an image, they get few codewords when applying ordinary VQ, which leads to an edge degradation. Thus, a technique able to overcome this degradation should be sought.

Towards this goal, Gersho and Ramamurthi introduced a modified VQ scheme known as the Classified Vector Quantization (CVQ) [2]. In this method, classification of each block within the image (input vectors) into several categories is firstly carried out. After this classification, a separate codebook is designed for each class. In the initial work [2], the input vectors were classified into two different categories: *shade* and *edge*. By means of this classification, the edges in the original image are better preserved compared to the ordinary VQ.

In a later work [3], Ramamurthi and Gersho extended the previous CVQ by using more (mainly four) classes: *shade* (no significant gradient), *midrange* (moderate gradient, no edge), *edge* (four different orientations: 0º, 45º, 90º, and 135º), and *mixed* (no definite single edge, but significant gradient). The main reason for this extension was to further improve the fidelity in representing edges. Furthermore, different coding processes were applied to different classes to take advantage of the special characteristics of each category (for instance, Mean/Shape Vector Quantizers (M/SVQ) [4]). In order to determine the size of each subcodebook, several combinations of sizes were experimented [3]. This trial and error procedure was based on a first estimate of the optimal sizes obtained assuming an asymptotically large number of training vectors [5]. The above CVQ yields perceptually better results while reducing the algorithm complexity over the CVQ proposed in [2].

Despite these good results, some aspects of the CVQ technique may be reexamined and naturally simplified by introducing the concept of fuzziness [6]. The first aspect that can be handled in a more natural way from the fuzzy viewpoint is the classification stage. As above commented, in the original paper [2] the classification stage dealt with only two different categories, based on some a priori knowledge of the problem. Afterwards, more categories were added owing to two reasons: the lack of accuracy when classifying all vectors in just two categories, and the seeking of simpler coding structures. Both, the introduction of some a priori knowledge and the expansion of the number of classes, motivated us to cope with the problem by using fuzzy clustering algorithms in the classification stage. That is, each one of the categories may be defined as a fuzzy set and the a priori knowledge of the problem may be utilized to formulate their membership functions. Moreover, the values of these membership functions, once the classification has been performed, give us an idea of the necessity of creating new categories.

Another aspect that can be easily handled by means of fuzzy concepts is the problem of choosing the size of each subcodebook. Rather than applying a trial and error procedure and the LBG algorithm [7], a fuzzy classifier may be used. In this way, the partial distortion introduced by each single codeword can be computed by means of its fuzzy entropy [8]. In order to obtain a fair set of subcodebook sizes, the fuzzy entropies of different codewords are compared and the codewords that happen to have larger fuzzy entropies are further split. As in the ordinary CVQ, a first estimate of the expected sizes is initially used, and later refined by the above procedure.

After this introduction, the organization of the paper is as follows. In Section 2, the representation of an image by triangular elements is described, and its use to formulate the fuzzy membership function is explained. Section 3 is devoted to the formulation of the fuzzy membership functions, and to the use of the information provide by them. The way to determine the size of subcodebooks by means of fuzzy theory is discussed in Section 4. Finally, in Section 5, some results and conclusions are presented.

2.- Image representation with triangular surface patches

In this work, three different categories have been initially used to classify input vectors. These categories are: *plane* (no significant change in gradient), *edge* (a main change in gradient) and *texture* (rapid variation of gradient). A convenient tool to discriminate among these three categories is the analysis of the surface normals of triangular elements that are contained in each block. Consider the representation of the image intensity function as the union of triangular surface patches. This representation is often used in the context of the Finite Element Method [9], and has been applied before, in the image processing framework, to handle the problem of Shape from Shading [10].

Given a rectangular lattice Ω, it can be divided into a set of nonoverlapping triangles T_i, as shown in Figure 1. Each one of these triangles contains three points from the original lattice and, the image grey levels at these three points define a unique plane. The equation that describes this plane can be written as follows

$$p \cdot x + q \cdot y + r = z, \qquad (1)$$

where the pair (p, q) is the gradient of the surface z(x, y), and the parameter r is just an offset of the grey level over the triangular domain T_i [10].

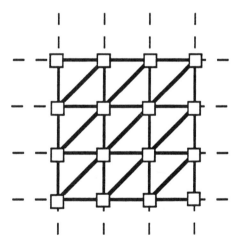

Figure 1.- Division of a lattice into triangular domains.

In our case, the parameter r is not of interest, since the set of offsets in a block is a useless feature in order to classify the block into the three categories above mentioned. On the other hand, the analysis of the set of pairs (p, q) that a block contains gives us enough information to classify it. In the sequel, the size of the blocks is assumed to be 4x4 pixels, so that each block is divided into 18 triangular domains.

A block belonging to the *plane* class should yield 18 samples very close to each other (a clear cluster) and located at any position in the (p, q) plane (the location depends on the main gradient of the block). On its turn, a block classifiable as an *edge* should form two or three compact clusters located also at any position in the (p, q) plane (one cluster for each side of the edge, which may have the same gradient, and one cluster containing the samples for the edge). Finally, the samples from the *texture* blocks do not gather in a clear cluster, as the grey level values of their pixels change rapidly. Due to these characteristics, the samples in the (p, q) plane give us two different features that combined can discriminate among the three categories: the number of clusters formed by 18 samples in the (p, q) plane and the compactness of these clusters.

3.- Formulation of the membership function

The way to utilize the above two features to formulate the fuzzy membership function is described as follows. Firstly, the underlying structure of the data in each block is analyzed. That is, the (p, q) samples of each block are classified by means of the fuzzy c-means (FCM) algorithm [11], generating a sequence of classifications containing 1, 2, 3 and 4 clusters. To each one of the performed classifications, a measure of compactness is applied. Although the computation of the fuzzy entropy of the classification would provide for a possible measure of compactness, the variance of the clusters has been utilized instead. The reason for this choice is that the use of the variance leads also to a convenient measure while avoiding the problem that arises when trying to define the fuzzy entropy of a single category classification. The variance is computed as

$$\sigma^2 = \sum_j \sigma_j^2 \quad \text{where} \quad \sigma_j^2 = \frac{\sum_i u_{ji}^m \|x_i - v_j\|^2}{\sum_i u_{ji}^m}, \quad (2)$$

and where x_i is a vector, u_{ji} stands for the grade of membership of the vector x_i to the fuzzy set u_j, m is a constant such that m > 1 (m = 1.4 has been used), and v_j is the center of the jth cluster calculated as

$$v_j = \frac{\sum_i u_{ji}^m x_i}{\sum_i u_{ji}^m}. \quad (3)$$

To allow a fair comparison among the four different variances obtained from each block, they have been weighted by the amount of clusters that they represent. That is, a block is said to contain mainly m clusters if

$$m\sigma_m^2 < n\sigma_n^2 \quad \text{with} \quad m \neq n \quad \text{and} \quad m, n \in \{1...4\}. \quad (4)$$

In Figure 2, the chosen values for each block in Lenna image are shown. The grey level values filling the blocks have been set ranging from 0, for the single category classification, to 255, for the 4 categories classification. As it was expected, the largest set is the single category blocks (87% in this image). Blocks within this set may belong to the *plane* class as well as to the *texture* class, and more unlikely to the *edge* class. An analogous reasoning could be made for each one of the different sets.

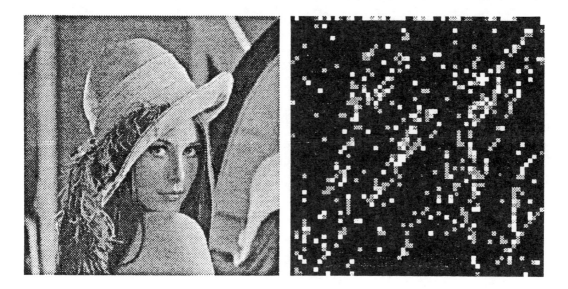

Figure 2.- Original Lenna image and its blocks classified as single category (3561), two categories (288), three categories (112) and four categories (135).

Taking into account these reasonings, a first approximation of the membership functions could be something close to the functions sketched in Figure 3. However, by using this kind of membership functions, the information contained in the variance of the chosen classification is withdrawn. Hence, a better way to formulate the membership function is by using the variance in order to either increase or decrease the value of different membership functions. That is, the greater the variance, the smaller the membership function of the classes *plane* and *edge*, and the opposite for the class *texture*.

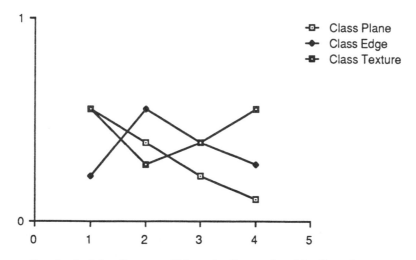

Figure 3.- A sketch of a possible set of membership functions

That has been performed by taking the first estimate of the membership function $\{\tilde{\mu}\}$ to the power of a function of the variance. Since $\mu_{ji} \in [0, 1]$ and the variance of a classification can get large values, the final membership functions assume the following form:

$$\mu_p(n) = k \, [\tilde{\mu}_p(n)]^\alpha$$
$$\mu_e(n) = k \, [\tilde{\mu}_e(n)]^\alpha \qquad (5)$$
$$\mu_t(n) = k \, [\tilde{\mu}_t(n)]^{10/\alpha}$$

where the subscripts p, e, and t denote *plane*, *edge* and *texture*, respectively, k is a normalizing parameter ($\Sigma \mu_j = 1$), $\alpha = \log(\sigma^2)$, and n is the number of clusters in the block. Different values for the initial membership functions have been tested, achieving very similar results for a large range. For the results here presented, the values shown in Table 1 were used.

n	1	2	3	4
$\tilde{\mu}_p$	0.9	0.7	0.33	0.25
$\tilde{\mu}_e$	0.33	0.9	0.80	0.5
$\tilde{\mu}_t$	0.5	0.4	0.43	0.5

Table 1.- Values used in defining the membership functions in (5).

Once the $\{\mu_j(n)\}$ have been calculated, the blocks can be classified into 7 different classes, depending on the values of different u_j for each vector. These classes are *plane* (C_1), *edge* (C_2), *texture* (C_3), *edge - texture* (C_4), *texture - plane* (C_5), *plane - edge* (C_6) and *plane - edge - texture* (C_7). The rules to obtain these classes and the order to apply them are:

1) if $\mu_{ji} > 0.7$ then block $\mathbf{x}_i \in C_j$
2) if $\mu_{ji} < 0.2$ then block $\mathbf{x}_i \in C_{j+3}$
3) if $\mu_{ji} < 0.4 \; \forall j$ then block $\mathbf{x}_i \in C_7$

As in the case of $\{\tilde{\mu}(n)\}$, different values for thresholds have been tested, yielding similar results in all cases. The set of values given above is the one that has been used for the results here shown.

When analyzing the data obtained after classification, it can be seen that C_7 very seldom has elements (less than 0.01%). Therefore, the vectors assigned to this class are very unlike to appear in an image, and may not be used when designing the codebooks. Moreover, by using a fuzzy classifier, new classes appear as the intersection between two expected classes (C_1, C_2 or C_3); that is, a vector is assigned to a new class when it does not wholly fulfill the characteristics of any of the expected classes. This is a more natural way to produce intermediate classes than the one used in the ordinary CVQ algorithm, in which these classes are set before hand.

4.- Refining the subcodebook sizes with fuzzy entropy

The main advantage of using CVQ is that, since classification is performed in the first stage, a separate codebook can be designed for each class. This separate design allows to accelerate the coding process (just a subcodebook has to be checked for each input vector), and to improve the coding quality (each class can be coded taking into account its special features). On the other hand, since each subcodebook is to be designed separately, their sizes have to be controlled. The two approaches proposed in [3] to control these sizes consist of a first estimate of the sizes and a later refinement by means of using the LBG algorithm in a trial and error procedure. That is, different combinations of subcodebook sizes are experimented until some conditions are fulfilled.

Instead of using a trial and error procedure, the refinement can be achieved easily by comparing the degree of accuracy of different codewords, and further splitting those with lower degree. This approach makes the design of subcodebooks non-fully independent, but since the comparison is performed in the refinement stage, the first estimate of the subcodebooks is still carried out separately. In order to compare the partial distortion introduced by each codeword, a fuzzy classifier is used to obtain the subcodebooks, and a measure of the degree of fuzziness is computed on each codeword. Among the existing measures (for a complete discussion see [12]), the normalized entropy, as defined in [8], has been chosen. That is,

$$h_f(u_j^{(n)}) = -\frac{1}{|C_n|} \sum_{x_i^{(n)}} u_{ji}^{(n)} \log(u_{ji}^{(n)}) \qquad (6)$$

where $|C_n|$ stands for the cardinality of the class C_n, $u_j^{(n)}$ is the membership function of the jth set of the class C_n, $x_i^{(n)}$ is the ith vector of the class C_n, and $u_{ji}^{(n)}$ is the grade of membership of the vector x_i to the fuzzy set $u_j^{(n)}$ of the class C_n.

5.- Results and conclusions.

In Figure 4 the result obtained using ordinary VQ is compared with the result obtained by means of the fuzzy CVQ algorithm (both with the same compression ratio). As it can be seen, the quality of the reproduction obtained by using the fuzzy CVQ algorithm is better than the one obtained with the ordinary VQ, since the edge distortion has been almost completely removed.

The work that has been presented is in its first stage, and further research is being carried out. The future work has to deal with several aspects. First, a study of different techniques that have been proposed to design subcodebooks has to be performed. Furthermore, improvements achieved by introducing fuzzy theory in these techniques have to be analyzed. Besides, a comparison of the performance of different measures of degree of fuzziness should be made.

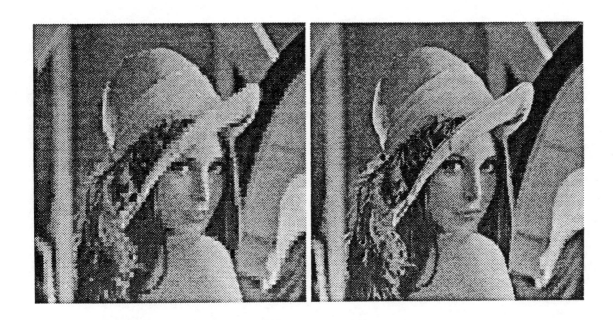

Figure 4.- Results obtained by using ordinary VQ (left) and fuzzy CVQ (right).

REFERENCES

[1] N. M Nasrabadi and R. A. King, "Image coding using Vector Quantization: A review," IEEE Trans. Commun., vol. COM-36, pp. 957-971, Aug. 1988.
[2] A. Gersho and B. Ramamurthi, "Image coding using Vector Quantization," in Proc. IEEE Int. Conf. Acoust., Speech, Signal Proc. pp. 428-431, Apr. 1982.
[3] B. Ramamurthi and A. Gersho, "Classified Vector Quantization of Images," IEEE Trans. Commun., vol. COM-34, pp. 1105-1115, Nov. 1986.
[4] R. L. Baker and R. M. Gray, "Image compression using non-adaptive spatial vector quantization," in Proc. 16th Asilomar Conf. Circuits, Sys., Comput., Oct. 1982.
[5] Y. Yamada, S. Tazaki and R. M. Gray, "Asymptotic performance of block quantizers with difference distorsion measures," IEEE Trans. Inform. Theory,vol IT-26, pp. 6-14, Jan 1985.
[6] L. A. Zadeh, "Fuzzy sets," Inform. Contr., vol. 8, pp. 338-353, 1965.
[7] Y. Linde, A. Buzo and R. M. Gray, "An algorithm for vector quantizer design," IEEE Trans. Commun., vol. COM-28, pp. 84-95 Jan. 1980.
[8] A. De Luca and S. Termini, "A definition of a nonprobabilistic entropy in the setting of Fuzzy Sets Theory," Information and Control, vol 20, pp. 301-312, 1972.
[9] H. Schwartz, "Finite Element methods," Academic Press, 1988.
[10] Kyoung Mu Lee and C.-C. Jay Kuo, "Shape from Shading with a linear triangular element surface model," USC - SIPI, Report num. 172, Feb. 1991.
[11] J. C. Bezdek, "Pattern Recognition with Fuzzy Objective Function algorithms," New York: Plenum 1981.
[12] S. K. Pal and D. K. D. Majumder, "Fuzzy mathematical approach to Pattern Recognition," John Wiley & Sons, 1986.

Chapter 11: Bioengineering

Fuzzy Control of Blood Pressure During Anesthesia with Isoflurane

R. Meier, J. Nieuwland, S. Hacisalihzade*, D. Steck, A. Zbinden**
Institute of Automatic Control, Swiss Federal Institute of Technology (ETH), CH-8092 Zürich
*) Landis & Gyr, Corporate R&D (4737), CH-6301 Zug, Switzerland
**) Institute of Anesthesiology and Intensive Care, Inselspital, CH-3010 Bern, Switzerland

Abstract

A fuzzy controller which controls the depth of anesthesia during surgery with isoflurane was designed and implemented on a personal computer. The mean arterial pressure (MAP) was taken as a measure for the depth of anesthesia. The design process was iterative and the reference points of the membership functions as well as the linguistic rules were determined by trial and error. The control rules made use of the error between the desired and the actual values of MAP as well as the integral of the error. The controller was tested in several surgical operations and it was observed that the anesthetists supervising the controller never had to intervene or override it. Therefore, one might conclude that such a controller might be routinely used during surgery calling for anesthesia with a single agent.

Introduction

One of the anesthetist's main tasks during surgery is to control the depth of anesthesia, which however, is not readily measurable. In clinical practice depth of anesthesia is evaluated by measuring blood pressure, heart rate, and clinical signs such as pupil size, motor activity, etc. (*Cullen et al., 1972*). The depth of anesthesia has to be such that the patient's Mean Arterial Blood Pressure (MAP), lies within a predefined range. Furthermore, it is desirable to minimize anesthetic agent consumption for both economical and ecological reasons. The main reason for automating the control of the depth of anesthesia is to release the anesthetist so that s/he can devote her/his attention to other tasks - such as controlling the fluid balance, ventilation, and drug application - which cannot be adequately automated, thus increasing the patient's safety.

The control of depth of anesthesia is achieved by using a mixture of drugs which are injected intravenously or inhaled as gases. Most of these agents decrease MAP. Among the inhaled gases isoflurane is widely used, most often in a mixture of 0-2 vol % of isoflurane in oxygen and/or nitrous oxide or nitrogen. The isoflurane concentration in the inspired air is adjusted by the anesthetist depending on the patient's physiological condition, surgery, MAP, and other clinically relevant parameters. To deliver the anesthetic agent to the patient, a semi-closed circuit is used. It allows reuse of the exhaled anesthetic gases as shown in Figure 1.

The isoflurane concentration at the end of each expiratory cycle is called endexpired or endtidal concentration. It mirrors the arterial and thus the brain concentration of the anesthetic agent. The semi-closed circuit makes the inspired concentration more difficult to control, because of an additional delay and a variable dilution of the inflowing gas by the exhaled gas. Thus, a variable gradient exists between the inflowing, inspired and expired isoflurane concentrations.

First experiments in the automatic control of the depth of anesthesia started in the late sixties with constant gain PID-controllers. Then came controllers with one-step adaptation of the gains, which were unable to cope with the time varying parameters of the patient during surgery. It was thought that this problem could be solved by using adaptive controllers, for instance, as described by (*Vishnoi and Roy, 1991*). Such controllers show robustness with respect to variations in the patient's parameters. More fashionable studies use rule-based and fuzzy controllers, to cope also with the time varying structure of biological systems (*Linkens and Mahfouf, 1988, Ying and Sheppard, 1990, Linkens and Hasnain, 1991*). For a review of this subject see (*Chilcoat, 1980, Linkens and Hacisalihzade, 1990*).

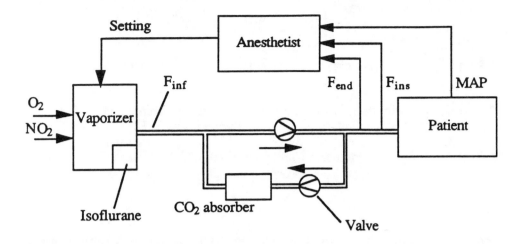

Figure 1: Flow and signal diagram of anesthesia. F_{inf}: Inflowing isoflurane concentration, F_{ins}: Inspired isoflurane concentration, F_{end}: Endtidal isoflurane concentration

Controller Design and Simulations

The controller should mimic the control actions of the anesthetist (and not act, for instance, like a 'bang bang' controller). This way a supervising anesthetist can easily ensure proper functioning of the controller. Furthermore, modelling a biological process like anesthesia is very complex, because it has a nonlinear, time varying structure with time varying parameters. The facts suggest the use of rule based controllers such as fuzzy controllers which are suitable for the control of such systems.

The control loop studied has the structure shown in Figure 2:

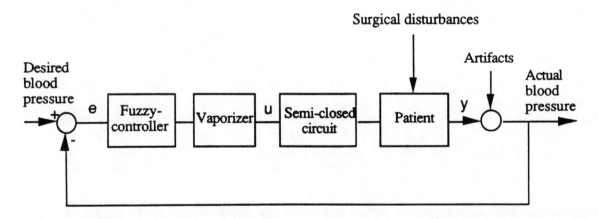

Figure 2: The block diagram of the control loop for the control of depth of anesthesia.

There are two different kind of disturbances: a) System noise due to pain caused by surgery, cardiovascular disease, concomitantly applied other drugs etc. (for example, a skin incision can lead to rapid changes in blood pressure of more than 10 mmHg) b) Measurement noise in the blood pressure and artifacts caused by calibration, electrocautery etc.

For simulation and controller design purposes the relationship between inflow concentration of isoflurane u and blood pressure y is modelled as a second order low pass with two pure time delays (the model includes the patient and also the semiclosed circuit). The corresponding step response is

$$h(t)=K_1\left(1 - e^{-(t-\tau_1)}\right) + K_2\left(1 - e^{-(t-\tau_2)}\right) \qquad (1)$$

z-transforming (1) leads to

$$y(k) = -a_1 y(k-1) - a_2 y(k-2) + b_1 u(k-\tau_1) + b_2 u(k-\tau_1-1)$$
$$+ b_3 u(k-\tau_2) + b_4 u(k-\tau_2-1) \qquad (2)$$

The parameters used in the design process were identified off-line from the data collected during surgery:

$\tau_1 = 23$ [s] $\tau_2 = 101$ [s]
$a_1 = -1.331$ $a_2 = 0.335$
$b_1 = 0.030$ $b_2 = -0.048$ $b_3 = 0.017$ $b_4 = -0.041$

For the simulation study a random noise signal was added to *y*. The artifacts and the low frequency noise were not included in the model.

The first simple linguistic rules that describe the anesthetist's actions were tested and systematically extended in several simulation runs. During the first design phase the error *e*, change of error *ce* and integral of error *ie* were used for the computation of the control variable *u*. The point in using the integral part was to eliminate the steady state error; the point in using the derivative part was to speed up the controller. The membership functions and the reference points were chosen by using the data recorded during several operations where blood pressure was controlled by an anesthetist. The bell shaped membership functions used can be described by the equation

$$y = e^{-k(x-a)^2} \qquad (3)$$

where *x* is the input value and *a* the shifting of the function in relation to zero. The factor *k* determines the "width" of the bell. The evaluation of the linguistic rules was performed using the max-min composition and the center of gravity method (*Zimmermann, 1990*).

The control characteristics and behavior under different disturbances were tested in simulations with different noise amplitudes and parameters. The best results were achieved with the linguistic rules and parameters of the membership functions shown in Table 1. Figure 4 depicts a simulation run with the reference points and linguistic rules shown in Table 1.

Some conclusions drawn from the simulation study are: a) More rules do not necessarily result in better control characteristics. Conflicting rules can even lead to unstable behavior. Furthermore, computational requirements increase with an increasing number of rules; b) The contribution of the derivative part is minor. Even after smoothing, relatively fast changes still remain in the signal. Thus, a controller using the derivative part shows a lot of switching of the control variable without improving the quality of control; c) The controller is robust with respect to variations in process parameters.

The results of the simulations encouraged us to apply the controller during surgery under "real life" operation room conditions.

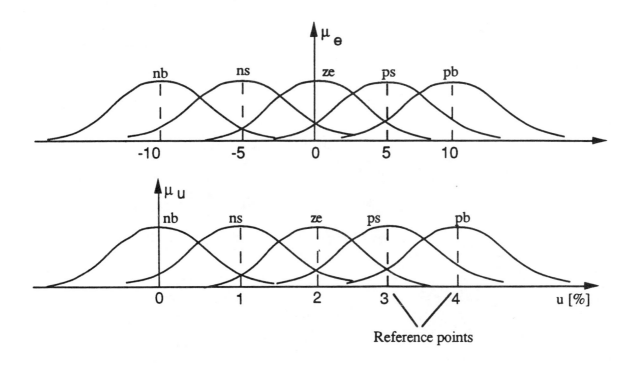

Figure 3: The membership functions and their reference points. **nb**: negative big, **ns**: negative small, **ze**: zero, **ps**: positive small; **pb**: positive big.

	INPUTS		OUTPUT
	e	ie	u
nb	-10	-160	0
ns	-5	-90	1
ze	0	0	2
ps	5	90	3
pb	10	160	4

(a)

	INPUTS		OUTPUT
	e	ie	u
1	ns	-	ps
2	ps	-	ns
3	nb	-	pb
4	pb	-	nb
5	ze	ze	ze
6	ze	ps	ns
7	ze	ns	ps
8	-	nb	pb
9	-	pb	nb

(b)

Table 1: a) The reference points of the membership functions and b) the linguistic rules which result in the best control and noise rejection characteristics.

Ensuring the patient's safety during surgery has the highest priority. In addition, the following points are important for the anesthetist: s/he has to guarantee that the patient's hemodynamics (MAP, heart rate, etc.) remain stable, that the patient remains sufficiently anesthetized. Moreover, s/he has to recognize malfunctioning of monitors and other devices. The safety concept of an automatic depth of anesthesia control should also fulfill these requirements. Therefore, the implemented safety concept supervises both the MAP and the endtidal

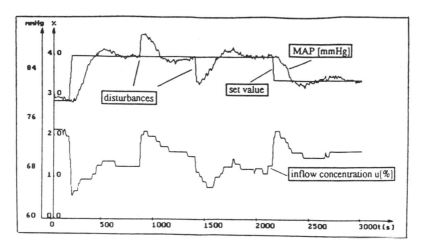

Figure 4: At 200 seconds the set value is raised from 80 to 88 mmHg. At 800 and 1600 seconds two disturbances are applied. The controller is fast and only a small overshoot is observed. At 2400 seconds the set value is reduced to 84 mmHg.

concentration. The data for monitoring were collected separately from the data used for the control of MAP. An acoustic alarm is sounded if the difference between the desired and the actual value exceeds 10mmHg although the controller keeps on working normally (the alarm is suppressed during 3 minutes after a change of the set-value). If the endtidal concentration sinks below 0.4% (danger of awakening), the inflow concentration is increased by 0.1% and held there for 3 minutes. If the endtidal concentration is still under 0.4% the process is repeated. If the endtidal concentration exceeds 2.5% (toxical limit), it is also signaled by an acoustic alarm, the inflow concentration is decreased by 0.2% and the controller is switched off. When the endtidal concentration returns to a level between 0.4% and 2.5%, the controller is switched on again. Also, the anesthetist can either override or switch the controller off at any time without having to worry about any adverse consequences.

Implementation

A Toshiba T5100 personal computer was used for collection, display, control and storage of data as well as feedback control. The analog input and output data were converted to/from digital data by using a Burr-Brown PCI-20041C-A carrier board in connection with the analog input module PCI-20019M-1A and the analog output module PCI-20003M. Additional digital communication was done via the RS 232 serial port. The desired inflow isoflurane concentration was obtained using a servo motor driven Dräger Vapor 19.3 vaporizer. The servo-motor was driven by a PID-controller-amplifier, which in turn was controlled by a conventional electronic interface, a D/A converter The servo-motor had a switch enabling toggling between manual and automatic control at any time. After determining the curve relating the D/A-output voltage (0..10 V) to the resulting isoflurane concentration (0..5 Vol%), the computer software calculated the D/A-value required for a specified inflow isoflurane concentration, taking into account the actual ambient temperature and barometric pressure. The precise concentration was reassured by a Dräger Irina gas analyzer with an analog output voltage for the A/D-converter.

Automatic control was not started before an acceptably deep anesthesia was achieved. A catheter was inserted into a radial artery and connected to a transducer. The obtained electrical signal was then preamplified and displayed by a Hellige Servonic STDV monitor. Its analog output voltage was transmitted to the A/D converter to measure (MAP).

The gas concentrations (O_2, CO_2 and isoflurane) were measured at the mouthpiece of the patient by a DATEX Capnomac gas analyzer. It transformed the gas concentrations to a data string that was sent to the RS232 serial port of the computer. Update rate of the data was 0.1 Hz.

Computer programs (modules) written in Modula-2 were developed to perform the tasks of data acquisition, display, control and storage. The main program can be started at any instance. The anesthetist is first asked to enter the ambient temperature, barometric pressure and the desired MAP. An internal clock and the variables are then initialized. Thereafter, the controller starts working. During automatic control a loop, consisting of MAP-updating (A/D-conversion), filtering, control calculations, control signal (i.e. isoflurane concentration) updating (D/A-conversion), data display and data storage on file is repeated every 10 seconds. The computer displays the actual and the desired blood pressure, as well as the inspiratory and endtidal isoflurane concentrations. The desired MAP may be changed at any time without interrupting automatic control. A filtering procedure for MAP was developed, in order both to smooth the incoming signal and to detect disturbances during data acquisition. Therefore, the incoming MAP signal is processed by a median-filter with a moving window size of 21 samples. The advantage of the median-filter is its good performance in the presence of disturbances having a wide probability density function. Then, the signal is checked for consistency, i.e. it is compared with a number of conditions that have to be satisfied if the data acquisition works properly. Otherwise the MAP data are not updated and, if the malfunction continues, a warning is displayed. All data of interest are stored in files for later examination.

Results

The controller was first tested during surgery in the so called "consultative mode" which means that the controller suggested a value for the inflow concentration and the anesthetist could then decide to set set this value or not. This way the patient's safety was assured and the controller could be tested under realistic operation room circumstances. Subsequent analysis of data collected during surgery showed that the correlation between the inflow concentration suggested by the controller and set by the anesthetist was sufficiently high, so that a fully automatic run was permissible during which the controller actually set the inflow concentration administered to the patient. Figure 5 shows the results of that operation.

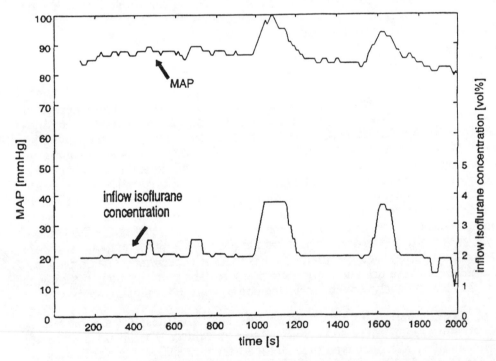

Figure 5: The variation of MAP and the inflow isoflurane concentration as adjusted by the controller during the first 2000 seconds of an abdominal operation shows that the controller can adjust the blood pressure adequately.

Conclusions

A fuzzy controller which controls the depth of anesthesia during surgery with isoflurane was designed and implemented on a personal computer. The controller was tested in several surgical operations. The anesthetists supervising the controller never had to intervene or over ride it. Therefore, it is concluded that such a controller might be routinely used during surgery calling for anesthesia with a single agent. It is, however, necessary to test the controller in many more operations with a wide range of patients to see its limitations. Also, the controller has to be enhanced to account for the use of more than one anesthetic agent.

References

Chilcoat RT (1980), "A review of the control of depth of anesthesia", *Transactions of the Measurement and Control*, 2, 38-45.

Cullen DJ, et al. (1972), "Clinical signs of anesthesia", *Anesthesiology*, 36: 21-36.

Linkens DA, Hacisalihzade SS (1990), "Computer control systems and pharmacological drug administration: a survey", *Journal of Medical Engineering and Technology*, 14: 41-54.

Linkens DA, Hasnain SB (1991), "Self-organising fuzzy logic control and application to muscle relaxant anesthesia", *IEE Proceedings-D*, 138: 274-284.

Linkens DA, Mahfouf M (1988), "Fuzzy logic knowledge-based control for muscle relaxant anesthesia", *1st IFAC Symposium on Modeling and Control of Biomedical Systems* (Venice), 185-190.

Vishnoi R, Roy RJ (1991), "Adaptive control of closed-circuit anesthesia", *IEEE Transactions on Biomedical Engineering*, 38: 39-46.

Ying H, Sheppard LC (1990), "Real-time expert-system-based fuzzy control of mean arterial pressure in pigs with sodium nitroprusside infusion", *Medical Progress through Technology*, 16: 69-76.

Zimmermann HJ (1990), *Fuzzy Sets Theory - and its Applications*, Kluwer-Nijhoff Publishing.

Real-time Fuzzy Control of Mean Arterial Pressure in Postsurgical Patients in an Intensive Care Unit

Hao Ying[1,2,3], Michael McEachern[4], Donald W. Eddleman[4] and Louis C. Sheppard[1,2]

1) Department of Physiology and Biophysics
2) Biomedical Engineering Center
3) Office of Academic Computing
University of Texas Medical Branch, Galveston, TX 77555, USA
4) Carraway Methodist Medical Center, Birmingham, AL 35234, USA

Abstract — We have developed a fuzzy control system to provide closed-loop control of mean arterial pressure (MAP) in postsurgical patients in a cardiac surgical intensive care unit setting, by regulating sodium nitroprusside infusion. The core of the control algorithms was a nonlinear proportional-integral (PI) controller, which precisely represents a simplest fuzzy controller. The proportional-gain and integral-gain adjusted continuously according to error and rate change of error of MAP. Twelve postoperative patients who exhibited elevated MAP following coronary artery bypass grafting procedures took part in the study. The length of time that 12 patients were on the fuzzy control system ranged from 1 hour 45 minutes to 18 hours 7 minutes. The total fuzzy-controller-run time was 95 hours 13 minutes. Clinical results showed that the average percentage of time in which MAP stayed between 90% and 110% of the MAP setpoint was 89.31%, with a standard deviation of 4.96%.

I. INTRODUCTION

The fast-acting vasodilator drug sodium nitroprusside (SNP) is used to treat patients who demonstrate elevated systemic arterial blood pressure after open-heart surgery. The rapid and powerful action of SNP imposes upon nursing personnel the task of frequent monitoring of mean arterial pressure (MAP) followed by adjustment of SNP infusion rate. Because nurses have many other duties, inappropriate or infrequent control actions on SNP adjustment may occur, which may lead to poor system performance.

To improve the quality of patient care, automatic closed-loop control SNP delivery systems have been developed. A nonlinear proportional-integral-derivative (PID) control system was first built and used clinically in the mid-1970s [7]. Various control algorithms including nonlinear adaptive control, multiple-model adaptive control and adaptive multivariable control were developed and tested [1][2][3][4][5][6][8][9]. The success of developing the above-mentioned drug delivery control systems, especially the adaptive control systems, heavily depended on mathematical models of patients. However, accurately identifying a mathematical model of patients is a very difficult task, due to the complexity of the human body. Alternatively, fuzzy control may be used.

We previously developed a generalized expert-system-shell-based fuzzy controller [10], which we then utilized to control MAP by regulating SNP infusion, in both digital computer simulation and real-time in pigs [11][12]. In this paper, we report the results of fuzzy control of MAP in postsurgical patients clinically (see also [14]).

II. NONLINEAR FUZZY CONTROL ALGORITHMS

The generalized expert-system-shell-based fuzzy controller had two input fuzzy sets, three output fuzzy sets, four fuzzy control rules, fuzzy logic AND and OR and a center of gravity defuzzification algorithm. To greatly reduce execution time and reveal the structure of the fuzzy controller, we analytically converted the expert-system-shell-based fuzzy controller into nonfuzzy controller which turned out to be a nonlinear PI controller [13]. The inputs of the fuzzy controller at sampling time nT are the scaled error and rate change of error of MAP:

$$GE \cdot e(nT) = GE[\text{ setpoint} - MAP(nT)], \quad (2.1)$$

$$GR \cdot r(nT) = GR[\text{ } MAP(nT-T) - MAP(nT) \text{ }]/T \quad (2.2)$$

where GE and GR are the input scalars, T is sampling period and the setpoint is the desired MAP level. The nonlinear PI controller is described as:

$$GI \cdot \delta SNP(nT) = K_i \cdot e(nT) + K_p \cdot r(nT) \qquad (2.3)$$

where the nonlinear proportional-gain and integral-gain are

$$K_p = \frac{L \cdot GI \cdot GR}{3L - input} \qquad (2.4)$$

$$K_i = \frac{L \cdot GI \cdot GE}{3L - input} \qquad (2.5)$$

and

$$input = \begin{cases} GE|e(nT)|, & GR \cdot |r(nT)| \le GE \cdot |e(nT)| \le L \\ GR|r(nT)|, & GE \cdot |e(nT)| \le GR \cdot |r(nT)| \le L \end{cases}.$$
(2.6)

$\delta SNP(nT)$ is the incremental SNP infusion rate. GI is the output scalar for $\delta SNP(nT)$. It is obvious that K_p and K_i change with $e(nT)$ and $r(nT)$. The larger the absolute value of $e(nT)$ or $r(nT)$, the larger K_p and K_i, which help reduce $e(nT)$ and $r(nT)$ quickly. On the other hand, the smaller the absolute value of $e(nT)$ and $r(nT)$, the smaller K_p and K_i, which drive SNP gradually to the setpoint in a stable manner. Hence, the fuzzy controller is a nonlinear adaptive PI controller.

A new SNP infusion rate, $SNP(nT)$, is computed as

$$SNP(nT) = SNP(nT-T) + GI \cdot \delta SNP(nT) \cdot T. \qquad (2.7)$$

It should be noted that above nonlinear PI control algorithms are the analytical description of the expert-system-shell-based fuzzy controller, precisely representing the fuzzy controller.

III. CLINICAL SETTING

The fuzzy controller was used to maintain desired MAP in patients in the Cardiac Surgical Intensive Care Unit (CICU) of the Carraway Methodist Medical Center. Fig. 1 is a block diagram of the implemented fuzzy control SNP delivery system. A Hewlett-Packard 78534 Monitor/Terminal was used to collect, process and display MAP, systolic pressure, diastolic pressure, left atrial pressure, right atrial pressure, heart rate and the electrocardiogram. A Puritan-Bennett 7200a Microprocessor Ventilator was connected to the patients to maintain respiration. MAP values were fed from the Hewlett-Packard Monitor into an IBM PS/2 Model 70 computer, ran the fuzzy controller in the form of the nonlinear control algorithms encoded in C programming language. SNP infusion rate calculated by the fuzzy controller was sent to an Abbott/Shaw LifeCare™ Pump Model 4. The pump infused SNP to patients.

Twelve postoperative patients who exhibited elevated MAP following coronary artery bypass grafting procedures took part in the study. Typically the trials began within one to two hours after the patients arrived in CICU. The typical MAP setpoint, determined by the attending medical doctors or the nurses, was 80 mm Hg. The fuzzy control system was started by technical personnel when the attending nurses thought SNP was needed for a patient. The fuzzy control system was always initiated at a SNP infusion rate of zero.

IV. CLINICAL PERFORMANCE OF THE FUZZY CONTROL SNP DELIVERY SYSTEM

During the trials, all normal patient care duties were performed by the nurses. The duties included sampling patient blood, suctioning the patient to clear his/her airway, bathing the patient, changing bed linen, injecting drugs other than SNP, infusing blood, and so on. MAP in the patient frequently fluctuated considerably when the above-mentioned duties were being carried out. Besides these situations, other factors also affected MAP. Substantial fluctuation of MAP took place as body temperature of the patients was changing or if the patients were in pain. Spontaneous fluctuation of MAP occurred as well. In addition to these, sensitivity of the patients to SNP changed with time. The response delay to SNP varied among the patients.

Fig. 2 shows a typical trend plot of both MAP and the corresponding SNP infusion rate obtained from a fuzzy controller controlled patient. For this specific patient, blood was sampled at 12:57, 13:42, 15:56 and 17:50. Suctioning the patient began at 13:04, 17:00 and 19:17. The patient was bathed between 15:36 to 15:50. Changing bed linen started at 19:45 and lasted for several minutes. Injection of Vallium took place at 13:09, 14:41 and 17:57. The drugs Pavulon and Morphine were injected into the patient at 14:50 and 17:10, respectively. As the result shows, the fuzzy control SNP delivery system regulated MAP satisfactorily even with the fluctuation of MAP caused by the various factors stated above. For this patient, the percentage of time in which MAP stayed within the band between 90% and 110% of the MAP setpoint was 86.5%. The trial lasted 7 hours and 49 minutes.

The length of time that 12 patients were on the fuzzy control system ranged from 1 hour 45 minutes to 18 hours 7 minutes. The total fuzzy-controller-run time was 95 hours 13 minutes. For the sampling period T=10 seconds, 34,278 MAP samples were collected from the patients. The overall performance of the fuzzy control SNP delivery system in 12 patient trials is summarized in Table 1. The table exhibits that MAP is tightly controlled around the desired MAP level.

V. DISCUSSION

A wide variation of patient sensitivity to SNP was experienced during the clinical trials. The fuzzy controller could cope with different sensitivity by continuously adjusting its nonlinear proportional-gain K_d and integral-gain K_i. Fig. 3 shows simulated results using a patient model [8], which indicate that the fuzzy control SNP system could adapt to a wide range of patient sensitivity, from the sensitive patients (K=-2.88) to the insensitive patients (K=-0.18), a ratio of 16:1. This range of sensitivity covers that of most patients.

VI. CONCLUSION

Results of the clinical trials on 12 patients revealed that the performance of the fuzzy control SNP delivery system was clinically acceptable. Based on the clinical results and the simulated results, we expect the fuzzy control SNP delivery system to perform well for most patients.

REFERENCES

[1] R. A. de Asla, A. M. Benis, R. A. Jurado, and R. S. Litwak, "Management of postcardiotomy hypertension by microcomputer-controlled administration of sodium nitroprusside," *J Thorac Cardiovasc Surg*; 89: 115-120, 1985.

[2] J. J. Hammond, W. M. Kirkendall, and R. V. Calfee, "Hypertensive crisis managed by computer controlled infusion of sodium nitroprusside: a model for the closed loop administration of short acting vasoactive agents," *Comput Biomed Res*; 12: 97-108, 1979.

[3] J. F. Martin, A. M. Schneider, and N. T. Smith, "Multiple-model adaptive control of blood pressure using sodium nitroprusside," *IEEE Trans Biomed Engineering*; 34: 603-611, 1987.

[4] L. J. Meline, D. R. Westenskow, N. L. Pace, and M. N. Bodily, "Computer-controlled regulation of sodium nitroprusside infusion," *Anesth Analg*; 64: 38-42, 1985.

[5] J. S. Packer, D. G. Mason, J. F. Cade, and S. M. Mckinley, "An adaptive controller for closed-loop management of blood pressure in seriously ill patients," *IEEE Trans Biomed Engineering*; 34: 612-616, 1987.

[6] J. H. Petre, D. M. Cosgrove, and F. G. Estafanous, "Closed loop computerized control of sodium nitroprusside," *Trans Am Soc Artif Intern Organs*; XXXIX: 501-505, 1983.

[7] L. C. Sheppard, "Computer control of the infusion of vasoactive drugs," *Ann of Biomed Engineering*; 8: 431-444, 1980.

[8] J. Slate, and L. C. Sheppard, "Automatic control of blood pressure by drug infusion," *IEE Proc*; 129, Pt. A, No. 9, 1982.

[9] G. I. Voss, P. G. Katona, and H. J. Chizeck, "Adaptive multivariable drug delivery: control of arterial pressure and cardiac output in anesthetized dogs," *IEEE Trans Biomed Engineering*; 34: 617-623, 1987.

[10] H. Ying, W. M. Siler, and D. M. Tucker, "A new type of fuzzy controller based on fuzzy expert system shell FLOPS," Proceedings of IEEE International Workshop on Expert Systems and Their Application in Industry, Japan, May, 1988.

[11] H. Ying, L. C. Sheppard, and D. M. Tucker, "Expert-system-based fuzzy control of arterial pressure by drug infusion," *Med Prog thr Technol*; 13: 202-215, 1988.

[12] H. Ying, and L. C. Sheppard, "Real-time expert-system-based fuzzy control of mean arterial pressure in pigs with sodium nitroprusside infusion," *Med Prog thr Technol*; 16: 69-76, 1989.

[13] H. Ying, W. M. Siler, and J. J. Buckley, "Fuzzy control theory: a nonlinear case," *Automatica*; 26: 513-520, 1990.

[14] H. Ying, M. McEachern, D. W. Eddleman, and L. C. Sheppard, "Fuzzy control of mean arterial pressure in postsurgical patients with sodium nitroprusside infusion," *IEEE Trans Biomed Engineering*; 39(10): 1060-1070, 1992.

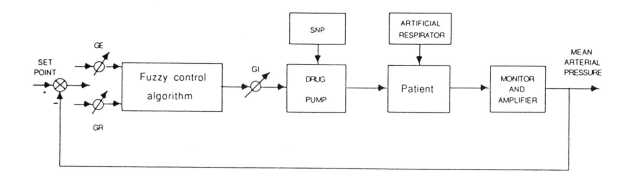

Fig. 1. Block diagram of the fuzzy control SNP delivery system.

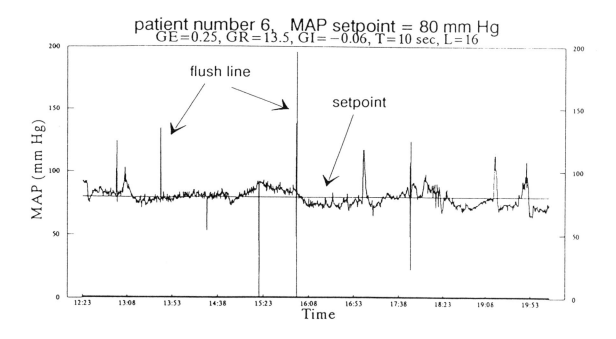

Fig. 2(a). MAP response for a single patient obtained by using the fuzzy control SNP delivery system clinically.

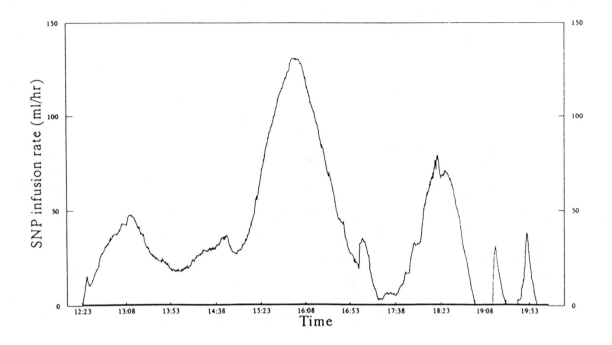

Fig. 2(b). The corresponding SNP infusion rate.

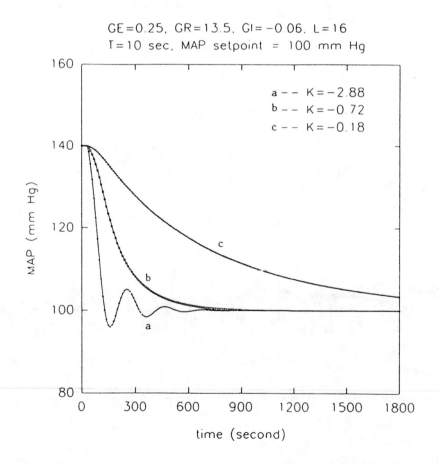

Fig. 3. Simulated MAP for the sensitive patients (K = -2.88), the normal patients (K = -0.72) and the insensitive patients (K = -0.18).

Table 1. Mean (μ) percent of total fuzzy-controller-run time (and standard deviation σ) for different MAP intervals. The calculation is based on 12 patient trials. MAP_d is the desired MAP.

	$< 0.8 MAP_d$	$(0.8-0.9)MAP_d$	$(0.9-1.1)MAP_d$	$(1.1-1.2)MAP_d$	$> 1.2 MAP_d$
μ	1.00	3.92	89.31	3.85	1.92
σ	1.09	2.72	4.96	1.84	1.14

Fuzzy ARTMAP Neural Network Compared to Linear Discriminant Analysis Prediction of the Length of Hospital Stay in Patients with Pneumonia

Philip H. Goodman, Vassilis G. Kaburlasos, Dwight D. Egbert
Departments of Medicine and Electrical Engineering, University of Nevada
Reno, NV 89557

Gail A. Carpenter, Stephen Grossberg, John H. Reynolds, David B. Rosen
Department of Cognitive and Neural Systems, Boston University
Boston, MA 02215

Arthur J. Hartz
Division of Biostatistics/Clinical Epidemiology, Medical College of Wisconsin
Milwaukee, WI 53226

Abstract — Health care databases may comprise hundreds of predictive variables, thousands of cases, and complex outcomes. Artificial neural networks may provide an alternative to established predictive algorithms for analyzing massive health care databases, potentially overcoming obstacles arising from the number of cases, missing data, variable selection, multicollinearity, specification of important interactions, and sensitivity to erroneous values. On an actual database derived from patients hospitalized with pneumonia, we compared the cross-validated predictions of linear discriminant analysis (LDA) to a new, supervised adaptive resonance theory network called ARTMAP. Unbiased proportionate reduction in error using ARTMAP was 50% greater than LDA. Under conditions of simulated noise and increasing-proportion learning, ARTMAP demonstrated further advantages over LDA. The promising performance of ARTMAP warrants further evaluation on larger health care databases.

I. INTRODUCTION

A major national effort is underway to determine patterns of medical practice that most effectively result in favorable health outcomes [1], [2]. Databases arising from medical effectiveness research may contain tens of thousands of cases and hundreds of variables intended to predict outcome status. Established statistical prediction algorithms may be suboptimal for such tasks because of obstacles arising from massive number of cases, missing data, variable selection, multicollinearity, specification of important interactions, and sensitivity to erroneous values. Artificial neural networks may offer an alternative method that overcomes some of the aforementioned analytic problems.

To address these inadequacies, we developed a new self-organizing supervised neural network that incorporates fuzzy set logic into adaptive resonance theory mapping (ARTMAP) to simultaneously predict outcome and define category patterns within outcomes. In voting fuzzy ARTMAP, multiple subnetworks resulting from random permutations of a learning set are created until a stable "voting" consensus (predictive score) is achieved.

The purpose of this study was to determine whether such a self-organizing neural network could accurately predict the length of stay of patients admitted to a community hospital with a diagnosis of pneumonia. Comparison was made to the performance of linear discriminant analysis, the most suitable of the established predictive methodologies.

II. METHODS

A. Clinical Database

1) Predictive Measures: The database was generated by one of us (AJH) through abstraction of 239 charts carrying a principal diagnosis of pneumonia on patients hospitalized between 1988 and 1990 at the Medical College of Wisconsin (MCW) affiliated hospitals. Stepwise linear regression, performed on about 200 clinically tenable measurements, resulted in 16 factors documented within the first 2 days of admission on the 214 patients whose charts had no missing data. The Reno and Boston researchers were blinded to the relative importance of the 16 variables. Risk factors included continuous as well as binary measurements (displayed as part of Fig. 5).

2) Outcome Measure: The outcome measure chosen for this preliminary study was length of stay (LOS), because it is related to both the severity of illness and the process of care, and is a major determinant of the cost of medical care. We broke the LOS into 3 intervals, based on inspection of its distribution (Fig. 1). The distribution is skewed rightward, with a mean LOS of 6.9 days and a median LOS of 5 days.

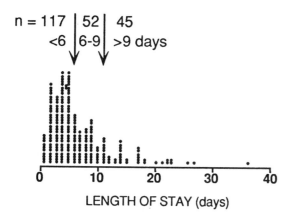

Fig. 1. Distribution of the length of stay for 214 patients hospitalized with pneumonia. Arrows indicate empiric cut-points for trichotomous classification of the length of stay.

Short LOS reflected either relatively healthy patients who responded briskly to therapy, or those who were very ill and died despite treatment. Long LOS reflected relatively sick patients, often requiring intensive care, who usually survived after prolonged hospitalization. There were only 10 deaths, distributed across all 3 LOS categories.

B. Accuracy, Cross-Validation, and Simulations

1) Accuracy: In order to apply algorithmic prediction in a clinical setting, cut-points or thresholds must be used to categorized the predicted outcomes. An ideal predictive algorithm would create probabilities clustered near 0 and 1 for each predicted outcome, so that the accuracy (as judged by the fraction of true positives and negatives) would be insensitive to the choice of a particular cut-point. For both fuzzy ARTMAP and linear discriminant analysis (LDA), we used the predicted outcome with the greatest probability. In the case of LDA, the predictive formula was adjusted for the prior probabilities of each length of stay category. In order to adjust accuracy for chance correctness of classification, we made use of the proportionate reduction in error (PRE) statistic [3],

$$PRE = \frac{(total\ correct - expected\ correct\ by\ chance)}{(total\ number\ of\ cases - expected\ correct\ by\ chance)}$$

The PRE is zero, therefore, when the diagonal sum in a confusion matrix is equal to that expected by chance alone, and 1 when classification is perfect.

2) Cross-Validation: In the analysis of large datasets, validity is threatened more by bias than variance. If a dataset in trained then tested on the same cases (i.e., resubstituted), the predictive accuracy is favorably biased [4]. We obtained nearly unbiased estimates for our dataset by the appropriate use of the k-fold cross-validation technique. In k-fold cross-validation, the data set is randomly divided into k partitions of approximately equal size; the cases not found in each partition are used to train the classifier, and the partitioned cases are used as a testing set. This is performed on all k partitions and the overall predicted class assignments are tallied, from which average accuracy is computed. We varied k from 2 to 100, and found no substantial loss of accuracy with k=10. This is supported by empirical evidence that the partition fraction can approximate the prevalence of each class without significant loss of accuracy [4].

3) Simulations: In the first simulation, we assessed the robustness of the techniques to uncorrelated noise by adding 16 uniform random noise variables to the dataset of 16 existing pneumonia variables, for a total of 32 predictive variables. In the second simulation, we created a larger but correlated data set by replicating each patient record 3 times, distorting the continuous variables by a random positive or negative amount (within one standard deviation). On 10 random permutations of this new 642-record database, we then performed an increasing-proportion comparison of the cross-validated predictive abilities of fuzzy ARTMAP and LDA, using a progressively larger portion of a randomly permuted data set for training, and the remainder for testing.

C. Linear Discriminant Analysis (LDA)

Most of the published studies on predicting outcome from severity-of-illness and treatment utilized linear regression, analysis of variance, LDA, or logistic regression. Each of these models makes distributional assumptions. For instance, dichotomous outcomes like mortality are better modeled by logistic regression if the predictor variables are binary or not normally distributed, whereas LDA performs better if normality holds [5], [6]. Continuous measurements (e.g., LOS) can be broken into intervals for analysis of variance or discriminant analysis, or predicted directly with linear regression. Because LOS was broken into 3 levels, we employed LDA. Analysis was performed using SYSTAT version 5 on an Apple Macintosh platform.

D. Fuzzy ARTMAP

Adaptive resonance theory (ART) neural networks use feedback and competition (analogous to interneuronal and recurrent neural circuits) to self-organize stable recognition codes in real time in response to arbitrary sequences of input patterns. Within the ART architecture, the process of adaptive pattern recognition is a special case of the more general cognitive process of hypothesis discovery, testing, search, classification, and learning. This property opens up the possibility of applying ART systems to the more general problem of adaptively processing large abstract information sources and databases. The development stems from the formulation of synaptic learning as compartmental interactions characterized by differential equations. Fortunately, competitive models can be formulated as Liapunov functions, so the system is asymptotically stable. Since a parallel architecture described by differential equations can be modeled on a von Neumann computer, we can experiment with neural networks on a multipurpose computer and reserve parallel hardware development for specific applications.

The original ART paradigm of Carpenter and Grossberg [7], called ART1, clustered only binary variables. Subsequently, Carpenter and Grossberg developed ART2 [8] and ART3 [9] for analog variables wherein the similarity of new input vectors to existing category vectors was determined Euclidean dot-products. This scheme required the addition of substantial circuitry to automatically scale input vectors. Recently, these Boston University Center for Adaptive Systems researchers proposed an alternative way to represent nonbinary variables using fuzzy set membership theory [10]. Minimal modification of the ART1 was required, as nonbinary variables could be normalized to the 0-1 range, and a fuzzy subsethood membership function [11] substituted for the ART1 matching formula (which reduces to ART1 if the vectors contain only binary elements). The key difference is that the choice and vigilance equations use the logical "AND" function in ART1 but the "MIN" in fuzzy ART. For example, if a newly input vector (1,1,0,1) is being tested for degree of match to an existing pattern vector (1,0,1,1), the AND operator results in the vector (1,0,0,1). The MIN operator selects the minimum of the 2 values for each variable in the vector, which would result in the same vector as the AND function when only binary data is used. For the case of analog data, consider the vector (1, .8, .2, .7) being tested for goodness-of-category match with the existing pattern (1,0,1,1); the MIN operator produces the vector (1, 0, .2, .7). Both the choice function, T_j (j is the index for the F2 nodes), and the vigilance, ρ, use the L1 norm (absolute sum) of this resultant vector. For either ART1 or fuzzy ART, T_j is maximal when the intersection of a newly input F1 vector with an F2 category vector is identical to the F2 vector norm. To deal with ties, the parameter a exerts a normalizing effect using bottom-up weights (w_{ij}), so that T_j will be maximal for the F2 category with the greatest absolute norm.

To enable ART to learn from experience, or map from input to outcome data vectors, a supervised architecture called adaptive resonance theory mapping, or ARTMAP, was proposed by the Boston researchers, first incorporating binary vectors [12], and later generalized to analog vectors using fuzzy ART [13]. As shown in Fig. 2, ARTMAP utilizes 2 ART modules, one to cluster input variables (ART_a) and one to cluster outcome variables (ART_b), with linkage by a map field of nodes (F_{ab}). While ART_b operates like a typical ART module, clustering multiple patterns of outcome if necessary before presentation to F_{ab}, ART_a function is modified to predictively optimize the formation of its F2 category patterns (which have mutually exclusive assignments, or expectations, for outcome classification). This is accomplished during training by elevating ρ_a on a vector-by-vector basis until the best F2 choice (T_j maximum) for that vector is assigned to the outcome class expected by ART_b. Upon successful assignment, the winning F2 category pattern is modified, or updated, to reflect the impact of the new vector. If the vector is allowed to have maximal impact, the learning is called "fast". Slow learning simply takes a weighted average of the new vector (b) with the pre-existing category vector (1-b). Training and prediction can occur in real time, since ARTMAP makes an outcome prediction upon presentation of each new input vector; only if it learns the correctness of the prediction does it change the F2 coding, which will affect the subsequent input vector's prediction, and so on. If the data records are presented in a relatively unbiased

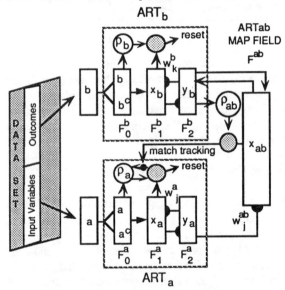

Fig. 2. ARTMAP Architecture (Supervised Learning).

TRAINING: Measurement vectors (a) are preprocessed into (a,a^c) by 0-1 normalization and creation of a complement for each variable. Likewise, (b) are preprocessed into (b,b^c). (a,a^c) is distributed by F1 to all existing F2 category nodes, which feed back learned category weights to F1. The outcome class assignment of the best F2 category match meeting the current vigilance (ρ_a) setting is sent to Fab; if the prediction is confirmed, the F2 category is modified as it learns (a,a^c). If the prediction by ART_a is disconfirmed, MAP FIELD activation induces the match tracking process, raising ρ_a just above the F_1 to F_0 match ratio $|x_a|/|(a,a^c)|$. This triggers another ART_a search for the next best F2 category match, and so on, which leads ultimately to match with an existing ART_a category pattern that correctly predicts (b), or, if none exist, to the assignment of pattern (a) to a previously uncommitted ART_a F2 node. In voting ARTMAP, permutations of (a,a^c) are simultaneously processed in multiple ARTa strata, which send predictions into ARTab, where the consensus is judged.

TESTING: Each new (a,a^c) input to the trained ARTa finds its best F2 match; the outcome class assigned to that F2 category node is the prediction. In voting ARTMAP, the consensus of multiple ARTa strata is used as the final prediction.

fashion, as would be the case in the real temporal sequence of hospital admissions (which we simulate by random permutation), then most of the learning occurs in a single pass through the data, i.e., an ARTMAP program could be used on-line as a real-time "self-taught" expert system. However, a modest increase in accuracy (several percent) is achieved by allowing the input vectors to cycle through several times, until there are no "re-learning" effects on the early-formed F2 categories due to late case presentation. In our experience, only a few cycles, or epochs, of training occur until all input training vectors are learned 100% correctly. This is in contrast to the backpropagation model, which typically must cycle through a data set thousands of times before converging.

In consultation with the Boston Center, the Reno group fully implemented voting fuzzy ARTMAP using C language during the summer of 1991. The Reno ARTMAP was compiled and run on DOS, Apple Macintosh, and UNIX platforms, including the Cray YMP/2 platform at the University of Nevada National Supercomputing Center for Energy and the Environment.

III. RESULTS

A. Accuracy

Under resubstitution, LDA correctly categorized LOS in 67% of cases when trained and tested on the same data set, whereas fuzzy ARTMAP learned to completely discriminate 100% of cases (Table I). Under 10-fold cross-validation, LDA correctly predicted the LOS category in 52% (PRE 0.20), a single ARTMAP network correctly predicted in 56% of cases (PRE 0.26), and a consensus of 30 voting fuzzy ARTMAP networks correctly predicted in 59% of cases (PRE of 0.30) (Table I).

TABLE I
PREDICTIVE ACCURACY OF MODELS

METHOD	Resubstitution	10-fold Cross-Validation	
	Accuracy	Accuracy	PRE[b]
Linear Discriminant	0.67	0.52	0.20
Single ARTMAP	1.00	0.53	0.22
30-Vote ARTMAP	1.00	0.59	0.30

[a]Resubstitution refers to testing after training on all cases.
[b]PRE is the proportionate reduction in error (see text for explanation).

B. Simulations

With the addition of 16 predictive variables consisting of uncorrelated uniform random noise, the PRE of voting ARTMAP actually improved from 0.257 to 0.265 (a 3% increase), whereas the PRE of LDA decreased from 0.203 to 0.164 (19% decrease) on the same data set.

Under the increasing-proportion simulation (Fig. 3), training on only the first 42 cases resulted in about 55% accuracy (PRE 0.33) on the remaining 600 cases for both ARTMAP and LDA. However, as an increasing portion of the database was used to train (with the remainder used as the independent testing set), the predictive accuracy increased only marginally for LDA to about 62% (the covariance matrix changes minimally), but ARTMAP accuracy increased rapidly to an asymptote at about 90% (PRE 85%) with about 300 records or fewer used for testing. Training with 442 and testing on 200 cases increased the PRE for LDA almost two-fold, but more than tripled the PRE for ARTMAP. These findings demonstrate the superior capacity of ARTMAP over LDA to learn from data regionally clustered in data space (i.e., to recognize similar patterns under noisy or systematically biased conditions).

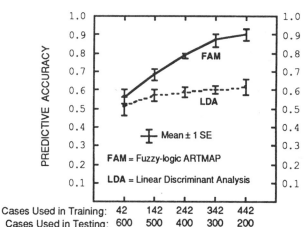

Figure 3. Increasing-Proportion Prediction by the Models. The original 214 cases were replicated twice, and the continuous variables were randomly varied by up to 1 standard deviation. FAM and LDA were trained and tested on an increasing proportion of the new 642-case dataset. The process was repeated 10 times to establish mean accuracies and standard errors.

C. Pattern Identification

One reason to prefer ARTMAP or other locally clustering algorithms over global activation models (such as backpropagation) for the analysis of health care data is to be able to "explain" the partitioning of input feature-space. As an example, during resubstitution trial with a single network, ARTMAP created a total of 44 long-term memory patterns for the 214 MCW pneumonia patients, representing a 5-fold reduction in data categories. The sequence of formation and distribution of patients' records by F2 memory pattern is shown in Figure 4. Figure 4 demonstrates that populous, more "generalized" memory patterns tend to be recruited early in the neural network training process, consistent with real-time learning capability. The reason for this is that early memory patterns are those most subject to change by subsequently presented cases; later patterns emerge to classify the more atypical case clusters.

Fig. 5 displays the structure of the 9 memory patterns from Figure 4 that each clustered at least 10 patients. Among the analog features, there is a trend towards longer hospitalization with increasing age, lower hemoglobin (i.e.,

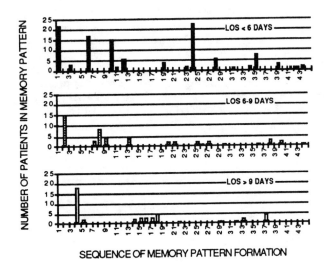

Figure 4.

worsening anemia), and greater time spent in an intensive care unit. Among binary features, short stays demonstrate consistent absence of some risk factors (open rectangles) and inconsistent mixture of others (shaded rectangles). Intermediate stays also cluster the absence of many risk factors, as well as non-white racial background, those without insurance, and arrival by ambulance (filled rectangles). Long stays include categories clustering the blind, those not able to control their urine, not able to walk, and with consistently abnormal laboratory results.

ARTMAP's clustering within outcomes is clinically tenable. For instance, increasing age, worsening anemia, and the presence of risk factors associated with concurrent diseases would reasonably explain, in part, why such patients would require more prolonged hospital treatment. It is conceivable that displays similar to Fig. 5 could enable ARTMAP to "explain itself", alleviating the concern of many physicians and hospital administrators about the use of "black box" predictive algorithms.

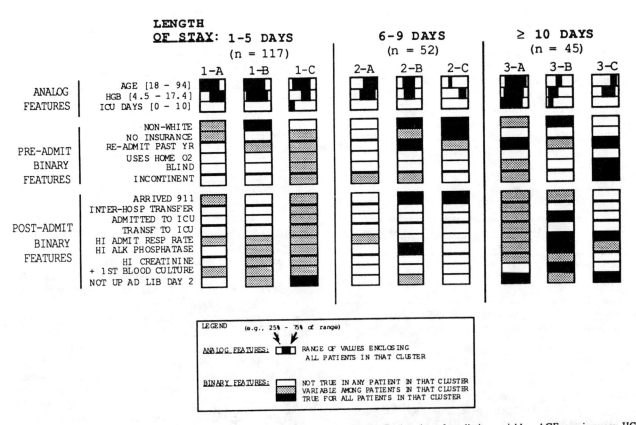

Figure 5. ARTMAP Internal Coding of Categories within Length of Stay Predictions. Explanation of predictive variables: AGE, age in years; HGB, hemoglobin, gm/dl; ICU DAYS, days spent in an intensive care unit; NON-WHITE, race other than white; NO INSURANCE, no private or public source of health insurance; RE-ADMIT PAST YR, admitted at least once to hospital for any problem within the preceding year; USES HOME O2, used continuous home oxygen prior to admission; BLIND, legally blind; INCONTINENT, incontinent of urine; ARRIVED 911, arrived by ambulance; INTER-HOSP TRANSFER, transferred from another hospital; TRANSF to ICU, transferred into an ICU during hospital stay; HI ADMIT RESP RATE, admission respiratory rate > 24; HI ALK PHOSPHATASE, elevated serum alkaline phosphatase; HI CREATININE, serum creatinine > 2; + 1ST BLOOD CULTURE, growth on blood culture from admission; NOT UP AD LIB DAY 2, not able to walk spontaneously by second hospital day.

IV. CONCLUSIONS

The fuzzy ARTMAP consensus of 30 parallel voting networks outperformed the linear discriminant function in predicting length of stay in 214 patients hospitalized for pneumonia (Table I). On this data set, PRE by voting ARTMAP was 50% greater than by LDA (0.30 vs. 0.20). Each ARTMAP layer reached steady state after only 5 to 10 cycles through the dataset.

While the addition of uncorrelated noise variables degraded the predictive function of LDA, it actually improved the accuracy of ARTMAP. It is possible that the ARTMAP system learned to recognize and ignore noise, but this behavior needs to be replicated and further characterized.

Each voting fuzzy ARTMAP network generated outcome-specific multivariate memory categories distinguished by simultaneous ranges within variables (Fig. 5). Memory categories capturing many (populous) and few (sparse) input vectors may both be clinically important. Populous patterns reflect consistent associations among input variables (in our setting, these are severity-of-illness groupings), and thereby contribute to good generalization on future input patterns not previously encountered. In addition, these populous patterns facilitate an "explanation" by the network of those interactions accounting for the predictions. On the other hand, sparse patterns reflect multivariate outliers, resulting from either statistical variation, systematic error, aberrant care (e.g., by hospitals or individual practitioners), or emerging trends (e.g., unrecognized adverse events or new disease entities). In the ARTMAP algorithm, these sparse, outlying memory patterns are not degraded by continued training, reflecting the local nature of the learning paradigm. In our study, inspection of the internal memory patterns revealed that ARTMAP clustered clinically meaningful interactions among the severity-of-illness variables. We are presently exploring the partitioning of variable space by the multiple voting networks in order to improve the "self-explanatory" ability of the system.

Further work should involve larger datasets and missing data. Consideration should also be given to improving predictive performance by tandem or hybrid networks of competitive, locally clustering models like ARTMAP with global activation models like backpropagation.

ACKNOWLEDGMENTS

The authors thank Bahram Nassersharif, Michael Ekedahl, and Sam West of the University of Nevada National Supercomputing Center for Energy and the Environment for their assistance in compiling our software and providing computer time. Supported in part by Washoe Health System, DARPA (AFOSR 90-0083), the National Science Foundation (NSF IRI 90-00530), and the Office of Naval Research (ONR N00014-91-4-4100).

REFERENCES

[1] Goodman PH. The Agency for Health Care Policy and Research: opportunities for research and guidelines leadership. SGIM Newsletter 1990; 13(4):4.

[2] Institute of Medicine. Effectiveness and outcomes in health care. Washington, DC: National Academy Press, 1990.

[3] Liebentrau AM. Measures of association. Beverly Hills: Sage Publ., 1983:16-30.

[4] Weiss SM Kulikowski CA. Computer systems that learn. San Mateo, CA: Morgan Kaufmann, 1991:108-110.

[5] Kandel A. Fuzzy mathematical techniques with applications. Reading, MA: Addison-Wesley, 1986:17-18.

[6] Efron B. The efficiency of logistic regression compared to normal discriminant analysis. J Amer Statistical Assoc 1975;70:892-98.

[7] Carpenter GA, Grossberg S. A massively parallel architecture for a self-organizing neural pattern recognition machine. Computer Vision, Graphics, and Image Processing 1987;37:54-115.

[8] Carpenter GA, Grossberg S. ART 2: self-organization of stable category recognition codes for analog input patterns. Applied Optics 1987;26(23):4919-30.

[9] Carpenter GA, Grossberg S. ART 3: hierarchical search using chemical transmitters in self-organizing pattern recognition architectures. Neural Networks 1990;3:129-52.

[10] Carpenter GA, Grossberg S, Rosen DB. Fuzzy ART: an adaptive resonance algorithm for rapid, stable classification of analog patterns. Neural Networks 1991;6:759-71.

[11] Kosko B. Fuzzy entropy and conditioning. Information Sciences 1986; 40:165-74.

[12] Carpenter GA, Grossberg S, Reynolds JH. ARTMAP: supervised real-time learning and classification of nonstationary data by a self-organizing neural network. Neural Networks 1991;4:503-44.

[13] Carpenter GA, Grossberg S, Markuzon N, Reynolds JH, Rosen DB. Fuzzy ARTMAP: a neural network architecture for incremental supervised learning of analog multidimensional maps. IEEE Transactions on Neural Networks, in press, 1992.

FUZZY CLASSIFICATION OF HEART RATE TRENDS AND ARTIFACTS

Dean F. Sittig [1], Ph.D., Kei-Hoi Cheung [2], M.S., Lewis Berman [2,3], M.D.

[1] Department of Biomedical Engineering and Center for Biomedical Informatics
Vanderbilt University, 1500 21st Avenue South, Suite 2000 Nashville, TN 37232-8143

[2] Yale University Center for Medical Informatics, and
[3] Section of Pulmonary and Critical Care Medicine, Yale University School of Medicine
333 Cedar Street New Haven, CT 06510

Abstract

Fuzzy set theory allows one to map inexact data, concepts, and events to fuzzy sets via user-defined membership functions. This paper describes a method for 1) robustly estimating the mean and slope of an arbitrary number of data points, 2) developing a set of fuzzy membership functions to classify various properties of heart rate trends, and 3) finding the longest consecutive sequence of heart rate data that fit a particular fuzzy membership function. Preliminary results indicate that fuzzy set theory has significant potential in the development of a clinically robust method for classifying heart rate data, trends, and artifacts.

Introduction

Many investigators have experimented with various signal processing methodologies for the identification and classification of trends from physiologic signals [Avent, 1990]. Most of these attempts at trend detection have concentrated on the rapid identification of relatively short term trends (i.e., ones that occur for less than 10 minutes). Once such a short term trend is identified, an alarm is sent to the clinician. While we are interested in being able to generate alarms, we are also interested in the identification of longer term trends (i.e., on the order of 20 - 60 minutes) which may be more representative of underlying changes in the patient's physiological state. Once these longer term trends have been detected, we need a method for classifying them according to 1) severity, 2) duration, 3) stability, and 4) our confidence in these estimates. As with most physiologic data, there are no clear cut demarcations between normal and abnormal states. This paper describes early experiments in the use of fuzzy set theory for the classification of trends and artifacts from the heart rate.

Background

Fuzzy Set Theory

Fuzzy set theory, originally developed by Zadeh [1965], allows one to map inexact physiologic data, concepts, and events to fuzzy sets via user-defined membership functions. For example, a systolic blood pressure of 100 mmHg could be considered low or normal depending upon the clinical situation. This reading might correspond to the fuzzy set: {low: 0.6, normal: 0.4, high: 0.0}. Once these fuzzy membership functions have been defined and a particular data point (or trend) classified accordingly, fuzzy logic provides powerful reasoning mechanisms for combining information from multiple fuzzy sets. Whereas many investigators have attempted to model various aspects of the medical domain using fuzzy set theory, few have applied this methodology to the classification of time-varying physiologic data, trends, or artifacts.

Linear Regression

Linear regression addresses the question of how to fit a set of N data points to a linear model of the form: $y = mx + b$ (where m represents the slope and b represents the y-intercept of the line). Truly useful answers to this question must include 1) estimates of the parameters m and b, 2) estimates of the errors associated with each of these parameters, and 3) a statistical measure of the goodness of fit of the model to the data. If the goodness of fit is poor, suggesting that the chosen data model is not representative of the underlying data set, then the estimates of the model's parameters and their associated errors are unlikely to be correct or even of use. Many different statistical techniques have been developed in an attempt to find the "best fit" to the linear model. Each of these techniques is based on minimizing (or maximizing) a particular merit function. Examples of merit functions that have been developed include: 1) least squares, 2) chi-square, and 3) absolute deviation. Of these, the absolute deviation is the most "robust".

Robust Estimation

In referring to statistical estimators, the term "robust" was first used in 1953 by G.E.P. Box to mean "insensitive to small departures from the idealized assumptions for which the estimator is optimized." These small departures may be either proportionally small deviations in a large majority of the data points or proportionally large deviations in a small minority of the data points [Press, 1988]. The second case, which are commonly referred to as outliers in statistics and artifacts in hemodynamic monitoring, is the most difficult for the majority of statistical procedures. By minimizing the mean absolute deviation (MAD) between the data and the linear estimate rather than the square of these differences (as one does in a least squares fit), one obtains an estimate of the data's slope that is not as sensitive to outliers, or artifacts in the data.

Figure 1 shows the difference between a least squares fit and a mean absolute deviation fit on a sample of heart rate data collected once every 15 seconds during surgery. Notice how the least squares fit overcompensates for a relatively small number (4/30) of data points, while the mean absolute deviation fit more closely represents the true underlying slope of the data.

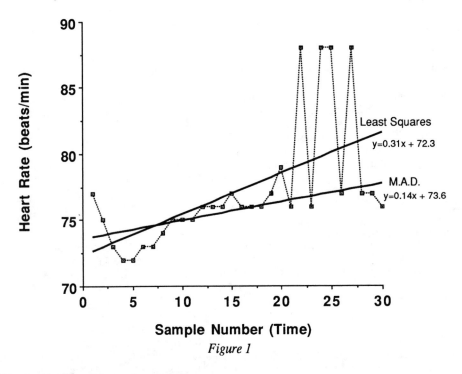

Figure 1

Design Considerations

Our overall goals in these experiments were to 1) find a robust method for estimating the mean and slope of an arbitrary number of data points, 2) develop a set of fuzzy membership functions to classify various properties of hemodynamic trends, and 3) find the longest consecutive sequence of hemodynamic data that fit a particular fuzzy membership function.

Methods

To focus our initial experiments we used heart rate data routinely collected once every 15 seconds over the course of two surgical cases (approx. 5 hours of data). We began by developing mathematical functions (described in more detail in the following sections) to characterize various aspects of the heart rate signal. For example, we calculate the rate of change of the heart rate based on the slope of the best fit line to recent data. We go on to calculate our confidence in this rate of change based on a measure of the goodness-of-fit of the line to the data. We estimate our confidence in the current data point by comparing our prediction of where the data point should be (found by substituting the current x-value into the equation of the best fit line) to the current data point. In an effort to measure the stability of the system, we keep track of the previous slope estimates and compare them to the current slope estimate. Finally, we send these quantitative values to our fuzzy classification scheme. The following sections describe each of our mathematical formulations and fuzzy classification methodology in more detail.

Identification of Longest Linear Data Trend

To establish a robust quantitative value for the slope of the heart rate data, we used an algorithm which minimizes the mean absolute deviation (MAD) of the difference between the data and the estimated line [Press, 1988]. This technique provides one with an estimate of the slope and intercept of the data along with a measure of the degree of fit of the model to the data.

We use an iterative process to find the longest consecutive sequence of hemodynamic data that fits a particular fuzzy membership function. Each iteration is initiated by calculating the slope, intercept and MAD of all data points acquired over the last 60 minutes. We pass these parameters to our fuzzy membership function which classifies the data's fit to a linear function over a range from too loose to too close. If the fit falls within this range then processing is complete. If the fit is too restrictive, we hypothesize that a longer range of data might still fit the model. In this case, we increase the length of the test interval by 50% and restart the process. If the fit is too loose, we decrease the length of the interval by 50% and reiterate. The iterative process continues until 1) we find a segment that fits within our predefined criteria (neither too loose nor too close), or 2) we have decreased the length of the data set through repeatedly halving it to a length of 3, or 3) we have increased the length of the data set by 50%. If condition one is satisfied then that particular segment is chosen as the longest linear data segment.

To handle cases which do not meet our criteria of a good fit (Figure 2), we developed an empiric function (*fit_length*) that produces a metric which takes into account both the length of the data set and the fit (see equation 1).

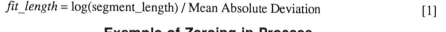

$$\mathit{fit\_length} = \log(\mathit{segment\_length}) \, / \, \text{Mean Absolute Deviation} \qquad [1]$$

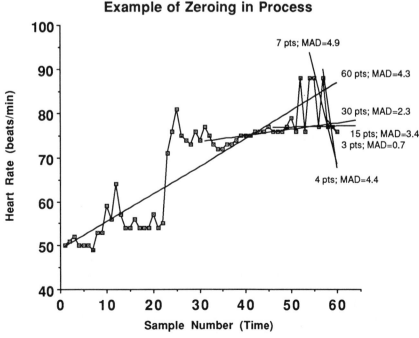

Figure 2

We developed this function to account for the fact that while fit is important we also needed to factor in the length of the data set. We believe, in general, that a less exact fit over a longer data set is preferable to a better fit over a shorter data segment (the short extreme, in which any two data points always lie on a line helps illustrate our reasoning). Therefore, we divide the log of the length of the data set (in number of samples) by the mean absolute deviation of the fit to this data set. Any line with a mean absolute deviation of less than 1 is arbitrarily assigned a *fit_length* of 0.75. This eliminates those segments that are extremely short and have too close, or restrictive, of a fit. Each of the line segments are then ranked according to *fit_length* in descending order. In the example shown in figure 3, the line containing 30 data points has a *fit_length* of 1.48 while its closest competitors are the 60 points line (0.95) and the 15 points line (0.78).

Estimation of Confidence in Current Data Point

We use one minus the ratio of the difference between the current data point and the predicted data point (based on the best fit line segment) to that of the predicted data point, to calculate the closeness of the current data point to the previous data points.

$$Confidence = 1 - (\text{Absolute value (current pt - predict pt)}/ \text{predict pt}) \quad [2]$$

Fuzzy classification of data

We used fuzzy membership functions based on those presented by Adlassnig [1980], in which a sigmoid-shaped function, f_1 (shown in figure 3) can be characterized by parameters $a, b,$ and g such that:

$$f1(x; a, b, g) = \begin{cases} 0 & \text{for } x \leq a \\ 2 * \frac{(x-a)^2}{(g-a)^2} & \text{for } a < x \leq b \\ 1 - 2 * \frac{(x-a)^2}{(g-a)^2} & \text{for } b < x \leq g \\ 1 & \text{for } x > g \end{cases}$$

To obtain a function that has both a leading and trailing edge (i.e., hump-shaped), one can use the following equations:

$$f2(x; a, b) = \begin{cases} f1(x; b-a, b-a/2, b) & \text{for } x \leq b \\ 1 - f1(x; b, b+a/2, b+a) & \text{for } x > b \end{cases}$$

Fuzzy membership functions (see Appendix) were developed to characterize the slope, goodness-of-fit, stability, and length of an arbitrary number of data points. We determined these fuzzy membership functions by first questioning several clinicians regarding their classifications of these parameters. Next, we assigned fuzzy boundaries to their estimates to account for the range of opinions regarding the exact cutoffs for each of the categories. Each of these descriptors of the line segments (i.e., slope, fit, segment length, confidence of current point, and magnitude of current point) are passed to a fuzzy classifier.

Figure 3

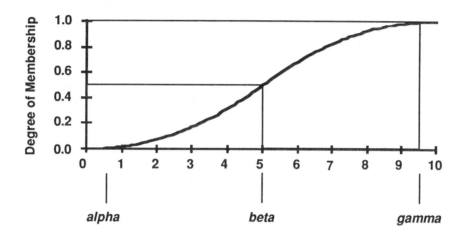

Figure 3. Sample fuzzy membership function. Important fiducial points are delineated by alpha (0.5), beta (5.0), and gamma (9.6) which correspond to fuzzy membership values of 0.0, 0.5, and 1.0 respectively.

Determination of Stability

To obtain an estimate of the stability of the underlying signal, we keep track of the fuzzy classification of the previous best fit line. At each point in time, if the current best fit line's slope is in the same fuzzy category as the previous line's slope we increase our stability measure by one. The stability measure is then sent to the fuzzy classifier.

Identification and Elimination of Artifacts

Any data point that falls within the "artifact" range of the confidence assessment is replaced by a point that is one half of the distance between the actual data point and the predicted data point.

New Data Point = Old Data Point - (Old Data Point - Predicted Data Point) / 2   [3]

Results and Discussion

Preliminary results indicate that fuzzy set theory has significant potential for the development of clinically robust methods for classifying heart rate data, trends, and artifacts.

Figure 4 shows the data collected during one surgical case (total time approximately 2.5 hours). Each of the 11 circles (out of 625 data points) denotes a point that we determined to be an artifact. Although, several potentially artifactual points were missed, and we cannot be sure that the artifacts identified were actually incorrect data points, most clinicians (on retrospective review, with all data available) would consider the points identified to be unrepresentative of the patient's true physiological state at that point in time. We made these determinations in simulated real-time (i.e., without looking "forward" in time, or beyond the current data point).

Using the same data set shown in figure 4, we measured the stability of our slope estimates and found that there were 43 changes from negative to positive or vice versa. Of these 43 changes 7 fell into the low classification (see Appendix), 4 into the medium classification and 1 into the high classification. Although at first sight these results suggest that the majority of the time the data were very unstable, upon closer inspection it turned out that the highly stable segment was 442 units long (out of a total of 625 units). In other words, over 70% of the time our system determined that the heart rate's slope (and hence the patient) was very stable.

Using the *fit_len* criteria to determine the "best" segment, our system chose a segment 120 units or longer (i.e., 30 minutes of data) 55% of the entire time (343/625) and 68% (343/505) of the time when there actually was a data segment at least 120 units long. In addition, the system chose a segment 240 units long 31% of the total time and 51% of the time that there was a segment at least 240 units long.

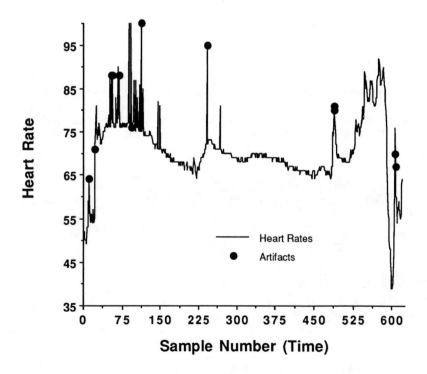

Figure 4

Conclusion and Future Direction

Based on these preliminary results, we hypothesize that fuzzy classification of trends and artifacts from other physiologic data streams will yield similar results. We are now in the process of developing fuzzy classification criteria for the arterial blood pressure signal. Once these classifications have been developed, we will develop a means of combining the information from these two signals to make clinically meaningful physiologic state determinations.

References

Adlassnig KP. A fuzzy logical model of computer-assisted medical diagnosis. *Methods of Information in Medicine* 19(3):141-148; 1980.

Avent R. K. A critical review of trend-detection methodologies for biomedical monitoring systems. *Critical Reviews in Biomedical Engineering* 17(6); 621-659; 1990.

Press WH, Flannery BP, Teukolsky SA, Vetterling WT. Numerical Recipes in C. Cambridge University Press, New York, NY. 1988.

Sanchez E. Medical applications with fuzzy sets. from Fuzzy Sets Theory and Applications, A. Jones et al (eds.) D. Reidel Publishing Company: 331-347; 1986.

Zadeh LA. Fuzzy sets. *Information and Control* 8:338-353; 1965.

Acknowledgements

This research was supported in part by NIH grants R29 LM05284 and T15 LM07056 from the National Library of Medicine and by a grant from the Whitaker Foundation.

Appendix

The following tables describe the key fiduciary points in the definition of our fuzzy classification scheme. Note that the *a*, *b*, and *g* correspond to the alpha, beta, and gamma shown in figure 2 and referred to again in the Methods section entitled fuzzy classification of data. The CLASSIFICATION column of the tables contains the text description of the fuzzy category. The RANGE column gives a rough estimate of most of the values that will fall in that particular category. The *a*, *b*, and *g* under the "rising edge" heading describe a one-sided function much like that shown in figure 2. In some instances only an *a* and *b* are listed. In these instances, they refer to the values for f_2, the hump-shaped described in the Methods section. The values for *a*, *b*, and *g* listed under the "falling edge" heading describe a sigmoid-shaped function that starts high and goes down to zero (i.e., 1 - (the graph shown in figure 3)).

Heart rate magnitude (beats/min)

CLASSIFICATION	RANGE	Rising edge a	b	g	Falling edge a	b	g
LOW THREATENING	< 50				44	50	55
LOW MARGINAL	50 - 60	10	55				
SATISFACTORY	61 - 125	55	60	65	120	125	130
HIGH MARGINAL	126 - 140	120	125	130	135	140	145
HIGH THREATENING	141 - 250	135	140	145	200	250	300
ARTIFACT	> 250	200	250	300			

Heart rate slopes (delta beats/min) (absolute value of change is classified)
(sign of change is appended to fuzzy classification to get e.g., slow rise, fast fall, etc.

CLASSIFICATION	RANGE	Rising edge a	b	g	Falling edge a	b	g
ZERO	< 0.4				0.25	0.4	0.5
STEADY	0.5 - 2	0.25	0.4	0.5	1.8	2.0	2.2
SLOW	2.1 - 8	1.75	1.9	2.1	7.0	8.0	9.0
FAST	> 8	7.0	8.0	10.0			

Heart rate term (time interval in minutes)

CLASSIFICATION	RANGE	Rising edge a	b	g	Falling edge a	b	g
INSTANTANEOUS	< 1				0.8	1.0	1.2
VERY SHORT	1 - 5	1.0	1.2	1.4	4.0	5.0	6.0
SHORT	6 - 20	5.0	6.0	7.0	15	20	25
MEDIUM	21 - 60	20	25	30	50	60	70
LONG	> 60	50	60	80			

Heart rate confidence   (equation 2)

CLASSIFICATION	RANGE	Rising edge a	b	g	Falling edge a	b	g
ARTIFACT	< 0.8				0.7	0.8	0.85
MINIMAL	0.8 - 0.84	0.05	0.825				
LOW	0.85 - 0.89	0.05	0.875				
MEDIUM	0.9 - 0.95	0.05	0.925				
HIGH	> 0.95	0.92	0.95	0.975			

Heart Rate Slope Stability   (number of segments with no change in slope classification)

CLASSIFICATION	RANGE	Rising edge a	b	g	Falling edge a	b	g
UNSTABLE	< 5				3	5	7
LOW	5 - 10	3	7.5				
MEDIUM	11 - 25	8	10	12	20	25	30
HIGH	> 25	20	25	30			

Example shows how one particular data point might be classified.

ITEM	VALUES	FUZZY CLASSIFICATION
Heart rate	128 beats/min	high marginal
slope	0.8 delta beats/min	steady positive
term	18 min	short
confidence	0.91	medium
stability	8	low

Chapter 12: Image Processing and Recognition

Application of the Extended Fuzzy Pointing Set to Coin Grading

P. A. Laplante – Fairleigh Dickinson University,
D. Sinha – City University of New York
C. R. Giardina – City University of New York

July 30, 1992

Abstract

Recent work has focused on the application of an extension to the fuzzy pointing set in artificial intelligence [6], [7].

One of the earliest applications of the extended fuzzy pointing set involved the processing of images where star values are used in representing pixels with no grey-value assigned [6]. This approach is useful in robotics where a damaged or spurious sensor may be producing data without enough reliability to be considered in the fuzzy range. While an automated coin grading system could utilize extended fuzzy pointing sets in the optics – image processing and pattern recognition, we will use these sets and induced algebraic properties in the associated expert system.

I. The EXTENDED FUZZY POINTING SET

Consider the set $[0,1]$ with the non-real numbers $*_1, *_2, \ldots *_n$ appended. We call this set the **extended fuzzy pointing set** and denote it S_n. The elements $*_1, *_2, \ldots *_n$ represent levels of uncertainty in the algebraic system and differ significantly from fuzzy values. For example, in an image processing framework $*_n$ can be thought of as representing a background value or missing information while $*_1, \ldots, *_n$ can be thought of as "uncertain" information where in some sense, $*_i$ is "more uncertain" than $*_j$ when $i > j$. The Cayley tables describing special algebraic operations on S_n should agree with this interpretation.

II. ALGEBRAIC OPERATIONS

Consider the binary operations of $(\vee)$, $(\wedge)$, $(+)$ and $(\cdot)$ defined on elements from S_n by Cayley tables 1, 2, 3 4 respectively. Within the tables the operations on $a, b \in [0, 1]$ are the usual ones.

Next, consider the partial ordering defined by the relation $\leq$ on the set S_n defined as follows: if $a, b \in [0, 1]$ then the relation is that usually defined on fuzzy numbers, otherwise

$$*_i \leq a \quad a \in [0,1]$$
$$*_n \leq *_m \quad n \geq m$$

Moreover, the algebraic system $(S_n, \leq)$ enjoys the properties of a lattice. Since if $a \leq b$ then

$$a(\wedge)b = a$$

and

$$b(\vee)a = b$$

are greatest lower an least upper bounds respectively for any $a, b \in S_n$. From an application standpoint, the partial ordering relation, $\leq$ is used to render value judgments when comparing two items with associated values from S_n.

However we are more interested in the algebraic structure of

$$A = (S_n, (+), (\cdot))$$

for use in image processing, signal processing and information fusion. In addition, the properties of A will be the same for algebra

$$B = (S_n, (\vee), (\wedge))$$

since the table structure such as $(\vee)$ and $(\wedge)$ are similar to those for $(+)$ and $(\cdot)$ respectively.

Specifically, it can be shown [6] that $(S_n, (+))$ and $(S_n, (\cdot))$ are commutative semigroups with pseudo-inverses and that $(\cdot)$ distributes over $(+)$. For example the fact that both $(S_n, (+))$ and $(S_n, (\cdot))$ are commutative follows directly from their Cayley tables.

Next, it can be easily shown [6] that $(S_n, (+))$ and $(S_n, (\cdot))$ are associative. Similarly by inspecting the tables, it can be seen that $*_n$ represents the identity for $(+)$ while 1 is an identity for $(\cdot)$.

III. COIN GRADING USING THE EXTENDED FUZZY POINTING SET

All grading systems utilize organized rules and provide standards mainly for specifying the degree of preservation of a coin. Buying and selling prices of coins are highly influenced by the grade: a slight difference in the grade often results in great price fluctuations. This is particularly true for mint state grading of non-circulated coins. Here the surface impairment (such as hair-line scratches and nicks), the luster, the strike, and the eye appeal are important.

Professional coin grading services usually encapsulate each graded coin along with an accompanying grade. Traditionally, the grade is given on the Sheldon integer scale ranging from 1 through 70, with the higher numbers indicating better grades. These numbers can of course be normalized as fuzzy values. However, when a coin is tampered with, many grading services refuse to specify a grade. This is particularly true for potentially higher graded coins in the 60-70 range, where the luster is often covered up or created using artificial means.

These "fakes" include whizzed coins, where mechanical wire-brushes are used to simulate original mint luster on circulated coins, and cleaned coins which are often identified by parallel hairline on the surface as well as luster disruptions. Several star values can be used to account for artificially enhanced coins and one can now grade any coin.

In practice one would require that a coin be graded by two or more grading services, each using the same valuation space, to assign value judgments to the coin. The judgments are then fused to obtain the overall judgment using an average operation that we now introduce.

Suppose that N grading services are to grade a coin, using the rating values in S_{n_2} each rendering grades g_i. Then the average coin grade $\hat{G}_{avg}$ is given by

$$\hat{G}_{avg} = (g_1(\cdot)1/N) \, (+) \, (g_2(\cdot)1/N) \, (+) \cdots (+) \, (g_N(\cdot)1/N) \quad (1)$$

IV. EXAMPLE 1

A coin is to be graded by 5 grading services. The grading services can issue a grade from the Sheldon scale, $G = \{1, 2, \ldots, 70\}$. These grades are mapped into $[0,1]$ by dividing the grade by 70. The grading services also have the option of issuing a grade from the set $\{*_1, *_2, *_3, *_4\}$. Thus the valuation space for the coins is S_4.

As far as ranking goes, the $*_i$ follow the relations given by Cayley tables 1 and 2. However for computational purposes, each $*_i$ behaves like a "0" in accordance with Cayley table 3. Suppose that the grades obtained from the services are

$$G_5 = \{*_1, 58, *_2, 69, 63\}$$

Then

$$\begin{aligned}\hat{G}_{avg} &= (*_1(\cdot)1/5) \, (+) (58/70(\cdot)1/5) \, (+) \, (*_2(\cdot)1/5) \\ &(+) \, (69/70(\cdot)1/5) \, (+) \, (63/70(\cdot)1/5) = .5428\end{aligned}$$

For ease of interpretation, we can convert a score in $\hat{G}$ to the Sheldon scale using the transformation (note how the $*_i$ values map to a "0" or undefined in the Sheldon scale).

$$\begin{aligned}T : S_4 &\longrightarrow G \cup \{0\} \\ s &\longrightarrow s * 70\end{aligned}$$

Hence, for our coin we get

$$G_{avg} = .5428 * 70 = 38$$

The very low score of 38 reflects the fact that two of the services thought that the coin was fake. In the interpretation of the final grade, the very high grades issued by the other three services; 58, 69, and 63 are offset by relative likelihood that the coin is a forgery.

The computational value assigned to the $*_i$ were chosen to reflect a pessimistic view of forgeries. A more aggressive buyer may have assigned other computational values to the $*_i$, which would result in a higher G_{avg}. This would necessitate a reconstruction of the Cayley tables to preserve the rankings, especially those for $(+)$ and $(\cdot)$.

V. EXAMPLE 2

Suppose a more aggressive buyer wishes to interpret $*_1$ computationally as 61.5 and $*_2$ as 34.5. $*_3$ and $*_4$ are both to be interpreted as 0.

The Cayley tables for $(+)$ and $(\cdot)$ would have to be modified to account for these new computational values. The addition operation would be interpreted as in Cayley table 5. Multiplication would be defined similarly, as in Cayley table 6. This buyer would interpret G_{avg} as

$$\begin{aligned}\hat{G}_{avg} &= (61.5/70(\cdot)1/5) \, (+)(58/70(\cdot)1/5) \, (+) \, (34.5(\cdot)1/5) \\ &(+) \, (69/70(\cdot)1/5) \, (+) \, (63/70(\cdot)1/5) = .81715\end{aligned}$$

Finally normalizing to the Sheldon scale gives:

$$G_{avg} = .81715 * 70 = 57.2$$

VI. CONCLUSIONS

The coin grading example takes advantage of the fact that S_n has been equipped with the special operators with certain induced properties. In addition, we exploit the fact that the valuation space is endowed with a lattice structure.

Finally, we believe that coin grading is simply a paradigm for more important applications. For example this technique could be used in grading semiconductor wafers before implantation or for the interpretation of data over noisy or damaged channels such as in space communications.

(∨)	a	$*_1$	$*_2$	·	·	$*_n$
b	$b \vee a$	b	b	·	·	b
$*_1$	a	$*_1$	$*_1$	·	·	$*_1$
$*_2$	a	$*_1$	$*_2$	·	·	$*_2$
·	·	·	·			·
·	·	·	·			·
$*_n$	a	$*_1$	$*_2$	·	·	$*_n$

Table 1: Binary Operator (∨) on $S_n \times S_n$

(∧)	a	$*_1$	$*_2$	·	·	$*_n$
b	$b \wedge a$	$*_1$	$*_2$	·	·	$*_n$
$*_1$	$*_1$	$*_1$	$*_2$	·	·	$*_n$
$*_2$	$*_2$	$*_2$	$*_2$	·	·	$*_n$
·	·	·	·			·
·	·	·	·			·
$*_n$	$*_n$	$*_n$	$*_n$	·	·	$*_n$

Table 2: Binary Operator (∧) on $S_n \times S_n$

(+)	a	$*_1$	$*_2$	·	·	$*_n$
b	$min[b+a,1]$	b	b	·	·	b
$*_1$	a	$*_1$	$*_1$	·	·	$*_1$
$*_2$	a	$*_1$	$*_2$	·	·	$*_2$
·	·	·	·			·
·	·	·	·			·
$*_n$	a	$*_1$	$*_2$	·	·	$*_n$

Table 3: Binary Operator (+) on $S_n \times S_n$

(·)	a	$*_1$	$*_2$	·	·	$*_n$
b	$b \cdot a$	$*_1$	$*_2$	·	·	$*_n$
$*_1$	$*_1$	$*_1$	$*_2$	·	·	$*_n$
$*_2$	$*_2$	$*_1$	$*_2$	·	·	$*_n$
·	·	·	·			·
·	·	·	·			·
$*_n$	$*_n$	$*_n$	$*_n$	·	·	$*_n$

Table 4: Binary Operator (·) on $S_n \times S_n$

(+)	a	$*_1$	$*_2$	$*_3$	$*_4$
b	$min[b+a,1]$	$min[61.5/70+b,1]$	$min[34.5/70+b,1]$	b	b
$*_1$	$min[61.5/70+a,1]$	1	$*_1$	$*_1$	$*_1$
$*_2$	$min[34.5/70+a,1]$	$*_1$	$69/70$	$*_2$	$*_2$
$*_3$	a	$*_1$	$*_2$	$*_3$	$*_4$
$*_4$	a	$*_1$	$*_2$	$*_3$	$*_4$

Table 5: Modified Cayley table for (+) on $S_4 \times S_4$

References

[1] Birkoff, G. and J.D. Lipson, "Heterogeneous Algebras", *Journal of Combinatorial Theory*, Vol. 8, pp. 113-115.

[2] Giardina, C.R., "Morphology in a Wraparound Image Algebra", *Proceedings of the SPIE Conference*, 1988.

[3] Giardina, C.R. and D. Sinha "Image Processing Using Pointed Fuzzy Sets", *Proceedings of the SPIE Conference*, 1989.

[4] Gougen, J.A., "L-fuzzy Sets", *Journal of Math. Anal. Appl.*, Vol. 18, pp. 145-174.

[5] Kandel, Abraham, <u>Fuzzy Mathematical Techniques with Applications</u>, Addison-Wesely, 1986.

[6] Laplante, Phillip, "On the Volterra Series and Morphological Operations", Ph.D. Dissertation, Stevens Institute of Technology, May 1990.

(·)	a	$*_1$	$*_2$	$*_3$	$*_4$
b	$b * a$	$61.5/70 * b$	$34.5/70 * b$	$*_3$	$*_4$
$*_1$	$61.5/70 * a$	.77188	.43301	$*_3$	$*_4$
$*_2$	$34.5/70 * a$	.43301	.24290	$*_3$	$*_4$
$*_3$	$*_3$	$*_3$	$*_3$	$*_3$	$*_4$
$*_4$	$*_4$	$*_4$	$*_4$	$*_4$	$*_4$

Table 6: Modified Cayley table for (·) on $S_4 \times S_4$

[7] Giardina, C.R. and P.A. Laplante, "Extensions to the Fuzzy Pointing Set and Its Application to Artificial Intelligence", *Proceedings of 1990 IEEE Conference on Systems, Man and Cybernetics*, November 1990.

[8] Sinha,D., "Fuzzy Sets, Possibility Distributions and their Application to Image Processing", Ph.D. dissertation, Stevens Institute of Technology, 1987.

[9] Zadeh, L.A., "Fuzzy Sets", *Journal of Information Control*, Vol. 8, pp. 338-353.

Region Extraction for Real Image based on Fuzzy Reasoning

Koji MIYAJIMA, and Toshio NORITA

Laboratory for International Fuzzy Engineering Research
Siber Hegner Bld. 3Fl. 89-1 Yamashita-Cho Naka-Ku Yokohama-Shi 231,
Japan

Abstract

In this paper, an approach to extract regions of a natural object by binarization using fuzzy reasoning is proposed. An advantage of this approach is that the threshold is obtained by fuzzy reasoning based on the shape instead of the user having to determine the threshold by trial and error. This approach was applied to extracting the image of a real flower as an example of a natural object.

In experiments, we use the image of real flowers whose color is same as the one of background. And we have demonstrated that our approach is more effective than statistical methods for extracting objects whose shape is almost known.

1. Introduction

Understanding what is printed in a natural image is very important in image recognition. In image analysis, segmentation, which can group pixels together to construct object regions, is very important. The most basic approach to extract objects in an image is the region segmentation by binarization. Where there is sufficient contrast between the object and background, the object region may be easily extracted by binarization. However, the threshold is difficult to determine due to subtle changes of gray level. Some approaches that can obtain the stable threshold automatically have been proposed[1]-[4]. Because these are statistical approaches, it is not necessary to extract object regions with subtle changes, such as shadow, reflectance and noise in the image.

On the other hand, a large number of studies have been advanced on image understanding for artificial objects, although only a few attempts at image understanding for natural objects have been advanced so far. There are several reasons why image recognition for natural objects is difficult. One is that an algorithm for segmentation that is based on bottom-up approach cannot extract regions of the object, because image features are changed by conditions such as light, shade and object definition. Therefore, the extraction of image features with not only a bottom-up approach but also a top-down approach is needed using knowledge of image processing and objects[5]. Another reason is that it is more difficult to describe models of natural objects than artificial ones. Although it is difficult to describe the shape of natural objects as object knowledge, the model should approximate geometrical shape in order to handle it easily. As mentioned above, it is effective to extract regions of natural objects in an image using the model as knowledge of objects.

The aims of our study are as follows:
(1) To enable some natural objects in an image to be extracted using an approximate geometrical model.
(2) To obtain the threshold of binarization with fuzzy reasoning, in order to best extract the objects.

In this paper, natural object regions are extracted by binarization determined with fuzzy reasoning. The advantage of this approach is that the threshold is obtained using fuzzy reasoning based on the shape instead of the user having to determine the threshold by trial and error. The approach is applied to extracting flower image features as an example of a natural object.

2. Basic assumptions

In this paper, we deal with the image printed flowers and background which are same color. The proposed approach is applied to images with regions that are not able to be extracted using statistical approaches.

We assume that the rough shape and scale of the extracted object are known by global interpretation. The rough shape and scale of an object are assumed by observation around the object. For example, where there is a narrow green color area, a flower region is assumed to exist at the end of the green color region and the scale of the flower region is restricted. In this way, the rough shape and scale are estimated and an object is extracted by the top-down approach. Figure 1 shows an overview of image processing using top-down approach. The flow diagram is described below.

Image features, such as a region and edge, are extracted with initial threshold and parameters of model and are matched to the model. If the result of matching is not good, the threshold and parameters of model are changed according to feedback from the results of matching and the image features are extracted again with modified threshold and model. If the result of matching is good, then extraction is completed. It is possible to extract image features that are suitable even if the objects have been changed because the models are described with only simple approximate geometrical shapes. The goal of this paper is to extract regions of objects that include complex condition, such as shadow and reflection, that cannot be extracted by statistical approaches.

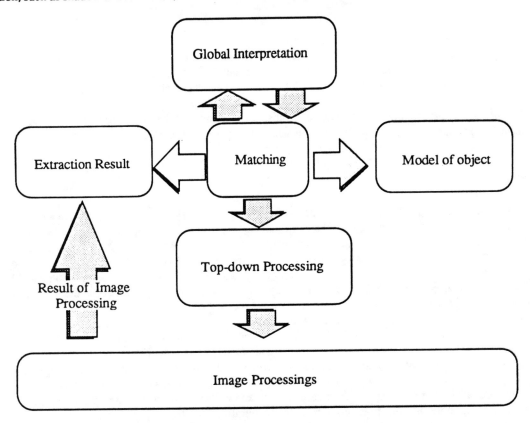

Figure 1: Overview of Top-Down processing

3. Object's Region Extraction Approach

Figure 2 shows an outline of the region extraction approach. The procedure is as follows:
First, by 'Global Interpretation' the information of regions of an object such as shape, scale and location are given. Initial regions are extracted by binarization with an initial threshold based on such conditions as shape, scale and location. Secondly, parameters of the geometrical shape model are obtained from the extracted region using the least squares method. The extracted region is then matched to the model. The model is expressed by a geometrical filled shape. The matching of the model with the extracted region is given by:

$$M = 1 - \frac{\sum_{i=0}^{N} \sum_{j=0}^{M} (F(x_i, y_j) \text{ EXOR } I(x_i, y_j))}{\sum_{i=0}^{N} \sum_{j=0}^{M} F(x_i, y_j)} \qquad (1)$$

where F is the binary image of the model, I is the binary image of the extracted region, N is vertical length of image and M is horizontal length of image. Figure 3 shows matching between the binary image of a model and the extracted region. The threshold is changed by the result of the fuzzy reasoning. The reasoning is as follows:

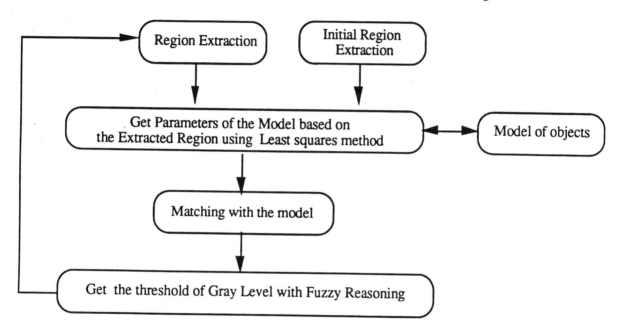

Figure 2: Outline of the object's region extraction approach

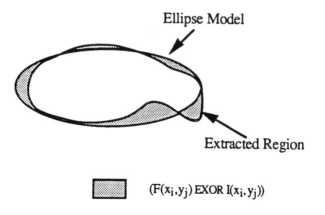

$(F(x_i, y_j) \text{ EXOR } I(x_i, y_j))$

Figure 3 : An example of matching between model and extracted region

The difference between the result of this matching and the previous matching is described in conditional parts of production rules. The rate of change of the threshold is described in the action parts of the rules. The thresholds are then determined using fuzzy reasoning based Mamdani's approach. The advantage of this approach is that the thresholds are easily obtained, because the user's decision is based on the result. Then, using the threshold, a region of the object is extracted again. When the result of matching is good, the procedure is completed. A detailed account of the approach is given below.

The rules are as follows:

IF $M_k - M_{k-1}$ is PB THEN $rd_{low\,k}$ is PB
IF $M_k - M_{k-1}$ is PS THEN $rd_{low\,k}$ is PS
IF $M_k - M_{k-1}$ is ZE THEN $rd_{low\,k}$ is ZE
IF $M_k - M_{k-1}$ is NS THEN $rd_{low\,k}$ is NS
IF $M_k - M_{k-1}$ is NB THEN $rd_{low\,k}$ is NB

IF $M_k - M_{k-1}$ is PB THEN $rd_{high\,k}$ is PB
IF $M_k - M_{k-1}$ is PS THEN $rd_{high\,k}$ is PS
IF $M_k - M_{k-1}$ is ZE THEN $rd_{high\,k}$ is ZE
IF $M_k - M_{k-1}$ is NS THEN $rd_{high\,k}$ is NS
IF $M_k - M_{k-1}$ is NB THEN $rd_{high\,k}$ is NB

where k is the number of times for matching, rd_{low} is rate of change of low threshold, rd_{high} is rate of change of high threshold. And the thresholds are given by

$$d_{low\,k} = d_{low\,k-1} \times rd_{low\,k} \tag{2-1}$$

$$d_{high\,k} = d_{high\,k-1} \times rd_{high\,k} \tag{2-2}$$

$$th_{low\,k} = th_{low\,k-1} + d_{low\,k} \tag{3-1}$$

$$th_{high\,k} = th_{high\,k-1} + d_{high\,k} \tag{3-2}$$

where $d_{low\,k}$ is difference between the low threshold of k times and k-1 times, $d_{high\,k}$ is difference between the high threshold of k times and k-1 times, $th_{low\,k}$ is the low threshold, and $th_{high\,k}$ is the high threshold. In this way, the thresholds can be obtained depending on the match.

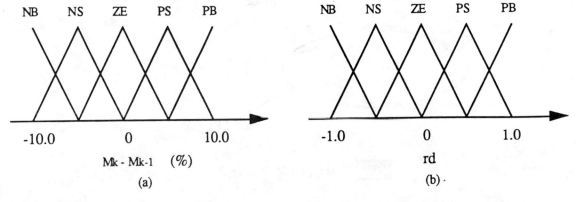

Figure 4 : (a) Membership Functions of the difference between the the result of this matching and previous matching,
(b) Membership Functions of the rate of change of thresholds

4. Experimental results

The algorithm was tested on the image printing a white flower in front of white wall. The image used in the experiment was composed of 300 × 300 pixels and 8bit of depth. Figure 5 shows the input image used in the experiments. It is very difficult to recognize flower regions, because both the color of flowers and the one of background are white. The contrast between the flower and the background is very delicate. The model of flower shape, which approximates to ellipse, is assumed in 'Global Interpretation'. Figure 6 shows the approximate shape model of flower.

Figure 7 shows the result of the discriminant analysis method. The threshold of the discriminant analysis method is obtained from variances of two clusters. Figure 8 shows the results of the differential histogram approach. The threshold of the differential histogram approach is the gray level which is correspond to the peak of differential histogram. The flower is not extracted suitably as an object, because the approaches are statistical and do not have knowledge of the shape.

Figure 9 shows the results of our approach. The flower is extracted more suitably than the approaches above. In spite of the influence of shadow and delicate contrast, the flower region is extracted suitably by binarization using fuzzy reasoning.

Figure 5: Input image

Figure 6: Ellipse model for white flower

Figure 7: Binarization result for discriminant analysis approach.

Figure 8: Binarization result for differential histogram approach

Figure 9: Binarization result using the threshold obtained fuzzy reasoning approach

5. Conclusion

We have described a method for region extraction which can obtain an adaptive threshold using fuzzy reasoning. In the experiments, we have shown that our approach is better than statistical approaches for extracting objects whose shapes are already known, in spite of the complex condition, the flower region is extracted suitably by binarization using fuzzy reasoning.

The implementation and results of region extraction by binarization using a threshold determined by fuzzy reasoning are available for image recognition.

References

[1] N.Otsu, A threshold selection method from gray-level histograms., IEEE Trans.,vol. SMC-9, pp.62-66(1979)

[2] S.Watanabe et al.,An automated apparatus for cancer prescreening: CYBEST.,CGIP, vol.3, pp.350-358(1974)

[3] J.S.Weszka,et al., A threshold selection techniques, IEEE Trans. on computer, vol. C-23, No.12, pp.1322-1326,(1974)

[4] J.Kittler and J. Illingworth, Minimum error thresholding, Pattern Recognition, vol.19, No.1, pp.41-47,(1986)

[5] V.S.Hwang,L.S.Davis,and T.Matsuyama, Hypothesis Integration in Image Understanding Systems, Computer Vision, Graphics, and Image Processing, 36, 321-371(1986)

A Fuzzy Approach To Scene Understanding

Weijing Zhang
Aptronix, Inc., 2150 North First Street
San Jose, CA 95131, U.S.A.

Michio Sugeno
Tokyo Institute of Technology, 4259 Nagatsuta,
Midori-ku, Yokohama 227, Japan

Abstract: In this paper, we suggest an approach to scene understanding by proposing a memory model that contains the necessary knowledge to understand a scene. Fuzzy sets theory is used for knowledge representation and reasoning. A system called SEE was developed based on the proposed approach and examples of natural color scenes are provided.

1. INTRODUCTION

The purpose of computer vision is to enable a computer to understand its environment from visual information. Objects of computer vision may be outdoor scenes, indoor scenes, machine parts, characters and so on. Many image processing techniques [8,9,10,11] have been developed to extract features from input image data for recognition. However, aiming at natural scene understanding, computer vision has been proved to be one of the most difficult fields for scientists. Because the input image data is very sensitive to the environment, a scene in the real world may produce completely different data under different situations.

We can understand a scene with little difficulty no matter in what a situation we are. Why can we do that? Knowledge about the scene is a key point to this question [1,2,3,4,6,7]. We see things with our eyes, but we understand them with our brain. A baby may see a same scene of the real world, and do the same image processing with eyes as adults do, but certainly cannot understand what he/she is seeing because he/she knows very little about the real world.

If we want to build a system that is able to understand a scene in the same way as we human do, we must give the system the knowledge to understand the scene. In other words, in addition to conventional procedures to preprocess an input image and to extract a set of features, such as enhance an image or extract edges from the image, procedures of how to use these image processing results should be provided for to a vision system.

In the following sections, a memory model for scene understanding is introduced at first. This model contains knowledge we need to understand a scene. Fuzzy sets theory is employed for knowledge representation and processing. Recognitive processes using the memory model is provided. At last, system **SEE**, which has been developed based on the proposed approach, is introduced. Examples of natural color scenes are provided.

2. KNOWLEDGE BASED APPROACH TO SCENE UNDERSTANDING

2.1 *A Memory Model*

To understand a natural scene, the first thing we need is a plan for the process *see*. Then we need to know the objects in the scene, not only the features of the objects but also the relations among the objects. Also, we need to know how different a scene looks in different situations. From this observation, we proposed a memory model as a knowledge structure for scene understanding [3,4], which consists of following five sub-memories:

- Individual-Memory (IDM) is used to store particular knowledge about an individual object. In IDM we can find information like *my car is of red color*.

- Category-Memory (CM) contains general features of a class of objects. For instance, *a car has four wheels*. Information in CM is shared by a group of objects.

- Scene-Memory (SNM) provides information about relations among objects in scenes, such as *a car on a road*. We may define a scene as a set of relations among objects.

- Situational-Memory (STM) tells us how SNM and CM should be revised according to different situations. For instance, we know that *things look big when they are near* and *trees may turn yellow in Autumn*.

- Intentional-Memory (IM) plays a role of controlling the whole process of scene understanding. If we had no intention to find something, nothing would come to our mind even though we are seeing. When we are looking for something, we have a goal, and we try to create a plan to achieve the goal. For example, when we want to find a car, we try to find if there is a road; we do not look at sky to search for a car although this is what computers usually do. IM is a guidance when we are looking for something.

These five memories correspond to five different kinds of knowledge concerning a scene. All these memories are necessary when we want to understand a scene. Fig. 1 illustrates the basic idea of this memory model.

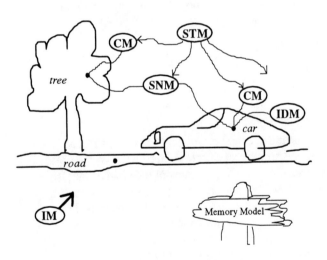

Fig. 1. Memory Model

2.2 The Memory Model In System SEE

This section shows how to build a proposed memory model. The examples shown here are used in system **SEE**.

2.2.1 Intentional Memory(IM)

IM gives a procedure to achieve a goal. Table 1 shows an example of IM in system **SEE**. **vb, b, vps** are abbreviations of **very big, big,** and **vertical position** respectively. An IM record contains the items to be confirmed and the weights assigned to them. An item can be a feature stored in CM, such as color, vps, or a scene ID number in SNM, such as 10002, 10003, 10014, etc.

Table 1: IM-sky

Object	Type	Weight	Item
sky	1	vb	color
sky	1	b	vps
sky	2	b	10002
sky	2	b	10003
sky	2	b	10014
sky	2	b	10015
sky	2	b	10016

2.2.2 Category Memory(CM)

Table 2 gives an example of CM records in system **SEE**, where **cf** means certainty factor, **lblue, dblue, z** are abbreviations of **light blue, dark blue** and **zero**. Values of vps, **t, c, b** are abbreviations of **top, center** and **bottom** respectively; **tc** means the area between top and center, while **cb** means the area between center and bottom. **cf=zero** indicates the value is impossible for the feature of the object. For example, sky cannot have a **green** color as shown in Table 2.

Table 2: CM-sky

Object	Feature	Type	Value	CF	STM-ID
sky	color	2	blue	vb	00000
sky	color	2	lblue	vb	00000
sky	color	2	dblue	b	00000
sky	color	2	white	vb	00000
sky	color	2	green	z	00000
sky	color	2	dgreen	z	00000
sky	color	2	lgreen	z	00000
sky	color	2	yellow	z	00000
sky	vps	2	t	vb	00000
sky	vps	2	tc	b	00000
sky	vps	2	c	m	00000
sky	vps	2	cb	s	00000
sky	vps	2	b	vs	00000
sky	size	2	vb	vb	00000
sky	size	2	b	b	00000
sky	size	2	m	m	00000
sky	size	2	s	s	00000
sky	size	2	vs	vs	00000

Each color name like **blue, lblue** corresponds to a color distribution and represented by fuzzy sets[1].

2.2.3 Situational Memory(STM)

A lot of information can be stored as situation items in STM, such as weather, time, season, distance, angle, etc. Table 3 shows an example of records about weather. Although STM has not been used yet in current **SEE**, it is obvious that knowledge in STM is very useful. For instance, Table 4 gives some examples of how a feature value in CM varies in accordance with different situations. From the table we can read

In Situation No.00011 (Weather=**Clear**)
Possibility(is(X, sky) | color(X)=**blue**)=**very_big**

In Situation No.00012 (Weather=**Cloudy**)
Possibility(is(X, sky) | color(X)=**blue**)=**small**

Table 3: Examples of STM records

STM-ID	Type	Situation	Value
00000	0	weather	unknown
00011	1	weather	clear
00012	1	weather	cloudy
00013	1	weather	rainy
00014	1	weather	snowing

Table 4: CM-sky: STM effect

Object	Feature	Type	Value	CF	STM-ID
sky	color	2	blue	vb	00011
sky	color	2	blue	s	00012
sky	color	2	blue	vs	00013
sky	color	2	blue	vs	00014

2.2.4 Scene Memory(SNM)

Table 5 (see last page) shows SNM records about object sky. A SNM record gives a piece of information about the certainty of a region to be the object when a relation is satisfied. For example, SNM No.10011 says that

If there exists a region Ax, provided
   color(Ax) = **darkblue**,
and there exists a region Ay, such that
   above(Ax, Ay),
   connect(Ax, Ay) and
   color(Ay) = **blue**
then possibility(is(Ax, sky)) = **big**

2.3 Recognitive Processes

We can say a system understands a scene if the system can give reasonable solutions to following two types of problems. One is to find a particular object in a given scene, i.e. to answer whether the object is in the scene and give its location; another is to tell something about the whole scene, i.e., to give a linguistic description of the input image, such as *a red car on a road, trees around a park*, etc. The first one is a search problem while the second is an interpretation one. Recognitive process is different according to the problem type.

Procedure 1: Search for an object in a scene
  1) Input scene and object name
  2) Implement image segmentation
  3) Set initial matching table
  4) [Procedure-IM]
  5) Output position of the object
     if the answer is YES

Procedure 2: Interpretation of a scene
  1) Input scene and scene type
  2) Implement image segmentation
  3) Set initial matching table
  4) Read objects decided by scene type
    4.1) Search for current object
    4.2) [Procedure-IM]
    4.3) Go to next object if any
  5) Output scene description

[Procedure-IM]
  a) Read IM of the object
  b) Check current IM item
    b1) CM type
    b2) SNM type
  c) Renew matching table
  d) Go to next IM item if any

3. FUZZY KNOWLEDGE

3.1 Linguistic Values

Knowledge in the memory model is represented by linguistic values[5], as we have seen in Section 2.2. Linguistic values, characterized by fuzzy sets, make it easy to denote vague and/or uncertain data such as **vb** (very big), **s** (small), for certainty factor; position **t** (top), color **dblue** (dark blue) for features of an object; **und** (under), **cnt** (connect) for relationship among objects in a scene. Definitions of this kind of values in **SEE** can be found in [1].

3.2 Fuzzy Rules

Knowledge for a vision system can be basically represented by following two types of implications,

(1) If $f(Ax) = v$ then $Ax = Obj$: CF

(2) If $r(Ax, Ay)$ and $f(Ay) = v$
        then $Ax = Obj$: CF

where f is a feature, such as color, v is a value of f, such as **green**; Ax and Ay are regions in an image; r is a relation defined among regions, such as **above**; Obj is an object name in a natural scene of the real world, such as sky; CF means certainty factor. v is a linguistic value and r is a fuzzy relation, both can be characterized by fuzzy sets. As we can see from the examples shown above, knowledge in CM is of type (1), and knowledge in SNM type (2). Each rule of either type is assigned to a certainty factor CF. When a value is impossible for a feature of an object, CF is set to be **z(zero)**.

3.3 Overall Evaluation of Matching Degrees

Matching degrees of each region in an input image with the model data change whenever a new item in IM is matched. Suppose current matching degree of a region with the model data is m and the matching degree of a new IM item is m_i then the new value of m, denoted by m_n, is a function with two variables m and m_i. Usually a product or minimum is taken in this case if the IM item is a CM type. If the IM item is an SNM, we cannot simply use product to calculate m_n. Because if there exists the relation then the whole matching degree will certainly get bigger, however if the relations can not be satisfied, we can say little about how should change. In **SEE** we use following to calculate new matching degree when an IM item is of SNM type.

$$m_n = \begin{cases} m + 4(1-m)(m-0.5)(m_i - 0.5) & m > 0.5, m_i > 0.5 \\ m + 0.2(m_i - 0.5) & m = 0.5, m_i > 0.5 \\ m + 0.4m(m_i - 0.5) & m < 0.5, m_i > 0.5 \\ m & m_i = 0.5 \\ 2mm_i & m_i < 0.5 \end{cases}$$

4. Outline Of System SEE

SEE is a system we developed based on the approach proposed above, which runs on X windows at UNIX workstations (Fig.4, last page). Fig.2 shows the system diagram.

4.1 Input Of SEE

Inputs to **SEE** are color pictures of natural scene read from an image scanner. Each input scene can be written as a 256x256 matrix whose each element can take one of eight different values that represent eight different colors.

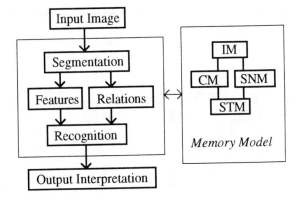

Fig.2. System Diagram of **SEE**

4.2 Scene Type

Usually we use a great deal of commonsense knowledge to understand a scene. It is almost impossible to input all commonsense knowledge into a computer so some assumptions are necessary. Suppose we know the picture is an outdoor scene, we would expect there is a blue or gray sky at the top of the scene, the areas in green color are trees, two parallel long lines make a road, small blocks on the road are cars, and so on.

Scenes can be divided into some groups, such as indoor scenes, outdoor scenes, etc. Scene type may be automatically decided by a computer, but in current **SEE**, we simply give system the scene type of an image as an input.

4.3 Image Segmentation

As we human see a scene, we grasp the rough impression of the scene from our first glance at it. If we have

Scene

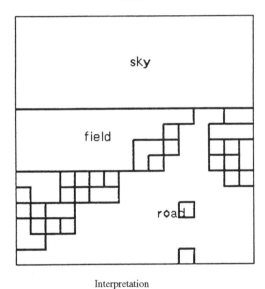
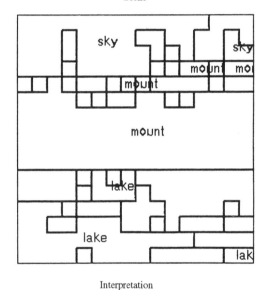

Interpretation

Fig. 3a. Example 1

Fig.3b. Example 2

something interested in the image, we pay attention to that for details. This suggests to divide the whole image into some suitable sized regions at first. We think a 8x8 or 16x16 division of the original image is suitable. These correspond to the 3rd and 4th layers in a quadtree structure, which is often used to represent an image. Segmentation is carried out based the color distributions, denoted by fuzzy vectors, of the 8x8 and 16x16 partition of the original image[1].

4.4 *Examples*

Two types of natural scenes are tested in system **SEE**. One includes sky, field and road (Fig.3a) and another includes sky, mountain and lake (Fig.3b). With only little knowledge in the system, the outputs show satisfactory results.

5. CONCLUSIONS

Knowledge plays a critical role in an intelligent vision system. More attention should be paid to knowledge structure and knowledge processing. Conventional image processing techniques are necessary, but not enough. In the field related to knowledge processing, fuzzy sets theory can provide us with a powerful tool.

REFERENCES

[1] W.Zhang and M.Sugeno, "SEE: a knowledge based scene understanding system using fuzzy reasoning," unpublished.

[2] W.Zhang and M.Sugeno, "Knowledge based scene understanding," *Proc. 7th Fuzzy System Symposium*, 1991.

[3] M.Sugeno and W.Zhang, "An approach to scene understanding using fuzzy case based reasoning," *Proc. Sino-Japan Joint Meeting on Fuzzy Sets and Systems*, 1990.

[4] W.Zhang and M.Sugeno, "Dynamic memory for scene understanding," *Proc. 6th Fuzzy System Symposium*, 1990.

[5] L.A.Zadeh, "The Concept of a Linguistic Variable and Its Application to Approximate Reasoning," I, II, III, *Information Sciences*, vol. 8, pp. 199-249, pp. 301-357, 1975, vol. 9, pp. 43-80, 1975.

[6] R.C.Schank, *Conceptual Information Processing*, North-Holland, 1975.

[7] R.C.Schank, *Dynamic Memory*, Cambridge University Press, 1982 (in Japanese translation).

[8] M. Ejiri, *Industrial Image Processing*, Shoukodo, 1988 (in Japanese).

[9] M. Nagao, *Image and Language Recognitions*, Corona Publishing Co. Ltd., 1989 (in Japanese).

[10] Y. Shirai, *Three-Dimensional Computer Vision*, Springer-Verlag, 1987.

[11] *Image Processing Handbook*, Shoukodo, 1987 (in Japanese).

Table 5 : SNM-sky(T:type abv:above und:under cnt:connect)

ID	Obj	CF	STM	T								
10001	sky	b	00000	1	0			1	cnt		1	color	white
10002	sky	vs	00000	1	0			1	und		1	color	dgreen
10003	sky	s	00000	1	0			1	und		1	color	dgrey
10004	sky	b	00000	0	0			1	cnt		1	cloud	
10011	sky	b	00000	6	1	color	dblue	2	abv	cnt	1	color	blue
10012	sky	b	00000	6	1	color	dblue	2	abv	cnt	1	color	lblue
10013	sky	b	00000	6	1	color	blue	2	abv	cnt	1	color	lblue
10014	sky	s	00000	6	1	color	dblue	2	und	cnt	1	color	blue
10015	sky	s	00000	6	1	color	dblue	2	und	cnt	1	color	lblue
10016	sky	s	00000	6	1	color	blue	2	und	cnt	1	color	lblue

Fig. 4. Scene Understanding System **SEE**

Chapter 13: Pattern Recognition

Recognition of Facial Expressions using Conceptual Fuzzy Sets

Hirohide USHIDA*, Tomohiro TAKAGI*, and Toru YAMAGUCHI**

*Laboratory for International Fuzzy Engineering Research,
SiberHegner Buil 3F, 89-1 Yamamashita-cho,
Naka-ku, Yokohama-shi, 231, JAPAN
(Tel.)+81-45-212-8227, (Fax.)+81-45-212-8255

**Systems & Software Engineering Laboratory, Research & Development Center, Toshiba Corp.
70 Yanagi-cho, Saiwai-ku, Kawasaki-shi, 230, JAPAN
(Tel.)+81-44-548-5637, (Fax.)+81-44-533-3593

Abstract

A facial expression is a vague concept that is difficult to explicitly describe. Conceptual Fuzzy Sets (CFS) have the ability to explicitly represent vague concepts. CFS are realized by using bi-directional associative memories, and a multi-layer structured CFS represents the meaning of a concept by various expressions in each layer. Multi-layered Reasoning in CFS has the following features: 1. Capability of simultaneous abstract and concrete representation 2. Capability of simultaneous top-down and bottom-up processing. In this paper, we apply CFS to the recognition of facial expressions and show that it can achieve context sensitive recognition.

Key Words: Facial Expression, Fuzzy Sets, Bidirectional Associative Memory, Context Sensitive, Multi-layered Reasoning

1. Introduction

Fuzzy sets theory provides an effective means to combine the real world (consisting of a very large number of instances of events and continuous numeric values) with human logical knowledge (consisting of abstracted concepts and symbols). It is difficult to delineate clear boundaries among human concepts because most of them are vague. Fuzzy sets theory is expected to provide knowledge processing in real-world applications, because it can support both symbolic processing and numeric processing, by connecting the logic-based world and the real world.

However, it is difficult to represent the meaning of a concept by using simple fuzzy sets because the meaning depends on the context. For example, a human facial expression consists of hazy and weak patterns on different parts of the face rather than one conspicuous pattern on one part of the face[1]. Although eyes look angry, for instance, the whole face may seem to be laughing depending on the expression of the mouth or other parts. Also it is difficult to explicitly represent the concepts of facial expressions by using ordinary fuzzy sets. For example, in the case of an angry face, it is easy to tell the degree of anger but not easy to explain why the face looks angry. Therefore, in the definition of a fuzzy set, a denotative description is generally used by means of the membership function, which associates each element with a grade of membership in the interval [0,1]. However the definition of the fuzzy set, described by instances "A, B, C", cannot determine the membership value of "D", which is a new instance. This means that a fuzzy set is not able to generalize knowledge from instances.

All these problems relate to the representation of the meaning of a concept. According to Wittgenstein[2], the meaning of a concept is represented by the totality of its uses. In this spirit we proposed the notion of Conceptual Fuzzy Sets (CFS)[3]. In the CFS the meaning of a concept is represented by the distribution of activations of labels that have concepts. Since the distribution changes, depending on the activated labels, to indicate a situation, CFS can represent context dependent meanings.

CFS also carry out Multi-Layered Reasoning (called MLR in this paper) based on association that is driven by propagation of activation of labels[4]. MLR by means of CFS has the following features:

1. Capability of simultaneous symbolic and quantitative processing (semantic guideline)
2. Capability of simultaneous top-down and bottom-up processing (context sensitive processing)

In this paper, we apply MLR by means of CFS to the recognition of human facial expressions. In section 2, we discuss the general characteristics of CFS. In section 3, we propose a CFS network for recognizing human facial expressions and discuss the results of recognition experiments.

2. Conceptual Fuzzy Sets

2.1. Conceptual Fuzzy Sets for Concept Representation

A label of a fuzzy set represents the name of a concept and the fuzzy set represents the meaning of the concept. Therefore, the shape of a fuzzy set should be determined from the meaning of the label depending on various situations. According to the theory of meaning representation from use proposed by Wittgenstein, the various meanings of a label (word) may be represented by other labels (words). We can assign grades of activation showing compatibility degrees between different labels in CFS, so that the distribution of activation can represent the meaning of a label depending on context. In CFS, once a label is activated, the activation is propagated to other labels. The distribution determined by activations agrees with the region of thought corresponding to the word expressing its meaning and the grade of the activation expresses the degree of the relation among labels. Since situations are also indicated by activations, the meaning is expressed by overlapping the region of thought determined by these activations. Figure 2-1 illustrates the different meanings of the same label L1 in different situations S1 and S2.

CFS are realized using bi-directional associative memories, in which a node represents a concept and the strength of a link is determined by the strength of the relation between two connected concepts. Activations of nodes produce reverberation and system energy is stabilized to a local minimum. As a result, corresponding concepts are recollected. In this paper, the recollections are realized by means of Bidirectional Associative Memories (BAMs)[5]. During the association in BAMs reverberation is carried out according to:

$$Y(t) = \phi(M \cdot X(t)), \quad X(t+1) = \phi(M^T \cdot Y(t)) \quad (1)$$

where, $X(t) = [x1, x2, ..., xm]^T$, $Y(t) = [y1, y2, ..., yn]^T$ are activation vectors on x and y layers at the reverberation step t, and $\phi(\cdot)$ is a sigmoid function of each node. BAMs memorize corresponding pairs of elements at each layer in terms of a synaptic weight matrix M to memorize CFS. When the pairs of patterns $(X_1, Y_1), (X_2, Y_2), ...,$ and (X_p, Y_p) are given, M is calculated from these pairs with coefficient β:

$$M = \sum_{i=1}^{p} \beta Y_i X_i^T, \quad M^T = \sum_{i=1}^{p} \beta X_i Y_i^T \quad (2)$$

2.2. Construction of CFS by Learning

CFS can be inductively constructed using neural network learning laws[6]. It means that the construction is carried out in terms of instances.

In this paper, a CFS is inductively constructed using the Hebbian learning law. The CFS is realized using associative memories in which a link represents the strength of the relation between two concepts. In the Hebbian learning law the strength m_{ij} of a link is modified by the product of the activation of two nodes x_i and y_j according to:

$$\dot{m}_{ij} = -m_{ij} + x_i y_j \quad (3)$$

A complex CFS is realized by combining several pieces of associative memory structured individually. If $C_1, C_2, ..., C_n$ denote individual CFSs and $M_1, M_2, ..., M_n$ are their corresponding correlation matrices then we can combine them to obtain a CFS, whose correlation matrix M is given by:

$$M = M_1 + ... + M_n \quad (4)$$

Such a combination of pieces of knowledge enables the CFS to realize context dependent representation.

2.3. Multi-Layered Reasoning by means of CFS [4]

Generally in image processing, recognition is carried out using characteristic values that are already obtained by low level image processing. However, when the model or context of an object is known, image recognition is more efficient. In fact, human mechanisms simultaneously realize both image processing and recognition, by means of the effective fusion of bottom-up and top-down processing supported by simultaneous information exchange and parallel processing. CFS can realize parallel processing to support the fusion of bottom-up and top-down processing by combining the semantic information processing in the upper layer and local processing in the lower layer. For example, in image recognition, the upper layer describes the knowledge about a context while the lower layer describes primitive concepts or instances. The concepts in the upper

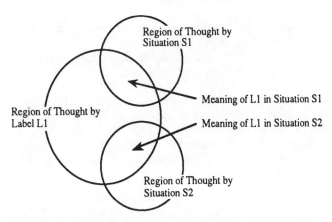

Fig. 2-1 Different meanings in different situations

layer are explained by the instances in the lower layer. The characteristic values extracted from an original image activate the corresponding nodes in the lower layer. This results in the activation of the concept in the upper layer. At the same time the context described in the upper layer depresses the contradictory patterns of distribution of activation and promotes the meaningful patterns of activation in the lower layer. Thus the nodes denoting instances become active so as to satisfy both the characteristic values and the context. This context sensitive processing provides us with an accurate result. It uses the context to eliminate vagueness which may come from noisy and vague data and which would otherwise cause misunderstandings.

3. Recognition of Facial Expressions using CFS networks

Recently there has been much research in which human facial expression is regarded as a means of communicating human intention or emotion and recognition of these facial expressions is applied to human-machine interfaces[7]. However, it is difficult to explicitly describe a facial expression because of its vagueness. In this section, we apply MLR by means of CFS to the recognition of facial expressions.

3.1. Network Design

There are generally six types of basic facial expressions: surprise, fear, disgust, anger, happiness, and sadness. These expressions mostly consist of hazy and weak patterns on different parts of the face but do not very often strongly appear at one part of the face[1]. People judge other people's facial expressions by the whole face rather than one part of it. In psychology, various rules for facial expressions, AUs (Action Units), have been developed in the FACS (Facial Action Coding System) and most AUs are characteristics of eyebrows, eyes, or mouth[8]. Therefore characteristics of these three parts of the face are required for recognition of facial expressions. Also people usually judge by using knowledge obtained from experience. For example, if one has knowledge about angry faces and he look at an angry face, he can recognize that the face is angry by comparing the facial expression with his knowledge. Thus he can judge the expression of a new face by estimating similarities between it and his knowledge about several other facial expressions.

In this paper, we limit facial expressions to three categories: anger, happiness, and sadness, and propose a network model that recognizes these facial expressions. The network has links containing the knowledge that is constructed of characteristics extracted from instances (Fig. 3-1). The network consists of three stages and the lowest stage is composed of input layers. The network between the lowest stage and the middle stage is like an LVQ (Learning Vector Quantization) network. This network compares the knowledge in links with inputs by using the Kohonen algorithm[9] so that activations of nodes in the middle stage increase according to the degree of agreement. The network between the middle stage and the highest stage is composed of CFS networks and the distribution of activations of nodes in the network converges to satisfy contexts after the reverberation.

The details of the network design are as follows. The lowest stage of the network has three layers and each layer corresponds to an eyebrow, an eye, or a mouth. The characteristic vector extracted from each part of a face is input to the nodes of the corresponding layer. The middle stage has nine layers that consist of nodes denoting instances of people. The layers in the middle stage are divided to three parts: an eyebrow part, an eye part, and a mouth part; moreover, each part has three layers that

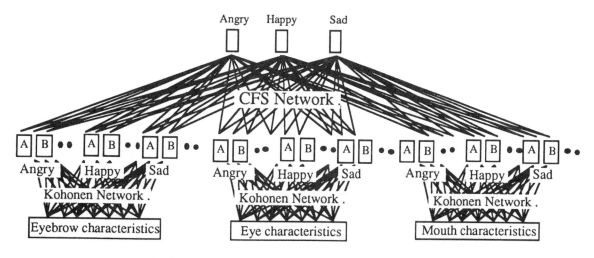

Fig. 3-1 CFS network for recognition of facial expressions

Table 1 Equations to obtain input values

Eyebrow Data	Eye Data	Mouth Data
org_x = (x1 + x2) / 2	e1 = y3 - y1	org_x = (x17 + x18) / 2
org_y = (y1 + y2) / 2	e2 = y4 - y1	org_y = (y17 + y18) / 2
b1 = x9 - org_x	e3 = y5 - y1	m1 = x15 - org_x
b2 = y9 - org_y	e4 = y7 - y1	m2 = y15 - org_y
b3 = x10 - org_x	e5 = y8 - y1	m3 = x16 - org_x
b4 = y10 - org_y	e6 = y6 - y4	m4 = y16 - org_y
b5 = x11 - org_x	e7 = y3 - y9	m5 = x17 - org_x
b6 = y11 - org_y		m6 = y17 - org_y
b7 = x12 - org_x		m7 = x18 - org_x
b8 = y12 - org_y		m8 = y18 - org_y
b9 = x13 - org_x		
b10 = y13 - org_y		
b11 = x14 - org_x		
b12 = y14 - org_y		* ai = (xi, yi)

correspond to three facial expressions. There are nodes of instances of people in each layer, and the node denoting the person who has the nearest characteristics is activated by the largest value when the values of the characteristics are inputted to the network. For Example, if the eyebrow characteristics of "Angry A" are the nearest to those of the input, the node denoting "Angry A" has the largest activation within the eyebrow layer. Also if the eye characters of "Sad B" are the nearest to those of the input, the node denoting "Sad B" has the largest activation within the mouth layer. Kohonen's network, i.e. LVQ network, is available in order to get activation that is proportional to the nearness between the characteristic vectors of people denoted by nodes and those from the input. Kohonen's network is composed of two layers: the input and output layers. The nodes in the output layer connect with all of the nodes in the input layer and the activation η_j of the output node is proportional to the nearness between the node's vector (reference vector) m_j and input vector x. Equation (5) shows this:

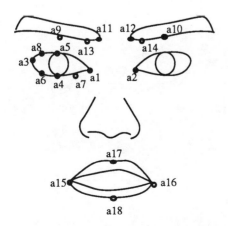

Fig. 3-2 Facial characteristic points

$$\eta_j = \Phi\left(\sum_{i=1}^{n} \mu_{ji} \cdot \xi_i\right)$$

$$m_j = \{\mu_{j1}, \ldots, \mu_{jn}\} \quad (5)$$

$$x = \{\xi_1, \ldots, \xi_n\}$$

where $\phi(\cdot)$ is a monotone increasing function.

In this paper, the coordinates of each facial part illustrated in Fig. 3-2 are used as the characteristic values and transformed into relative coordinates for inputting to the network. The input nodes of each facial part connect with the nodes of the corresponding part in the middle stage as shown in Fig. 3-1. The weight vector of the links between the input nodes and the instance node in the middle stage is the vector whose components are the inputted relative coordinates. Thus if the input nodes are activated by a new pattern, the activations are propagated to the middle stage and the instance nodes have activations proportional to the nearness between the link vector and the input vector. Therefore these activations represent the instance of the middle stage that is the nearest to the input pattern in each layer. The distribution of the activations of the instance nodes can represent the facial expression nearest to the input pattern in each part but cannot do this for the whole face. For example, when the input vector of the eyebrow is near to "Angry A" but those of the mouth are near to "Sad B", we cannot explain the facial expression of the input. CFS networks between the middle stage and the highest stage are bi-directional associative memories in order to solve such problems. The activation in the middle stage is propagated to the upper stage. The highest stage has three nodes and each node denotes a facial expression: anger, happiness, or sadness. These nodes connect with all of the nodes in the middle stage. The Hebbian learning law is available to get the weight of the links in the CFS network. The network memorizes facial expressions as fuzzy sets. If a CFS network receives signals from the lower layers, reverberation occurs in the network and the distribution of activations converges in order to match the contexts. Thus the final distribution of activations in the highest stage represents the facial expression of the input. In this network, the contexts are standard patterns so a combination of characteristics similar to that of the standard patterns is promoted and the mismatch with the context is inhibited. Therefore the CFS network realizes context sensitive recognition. The result of recognition is represented by the activation of three nodes in the highest stage.

3.2. Facial Expression Recognition Experiment

In our experiment, 56 faces were used. These included three types of expressions: anger, happiness, and sadness. The data were the same as Kobayashi's[1]. Three facial expressions for each person, A, B, and C, were used to construct the knowledge of the network. The data were

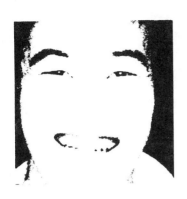

Fig. 3-3 Subject image

transformed into relative coordinates and normalized to make the length of the vector equal to 1. Table 1 shows equations to obtain the relative coordinates from the coordinates on the face. The Numbers of elements in the characteristic vectors of each facial part are 12 (eyebrows), 7(right eye), and 8(mouth). We used data only from the right eye, on the assumption that the face is symmetrical.

After training the network with 9 data, the network scored 100% correct faces about 9 training data. About 47 generalization testing data, the score was 78.7%. An example of a subject image is shown in Fig. 3-3 and the results of recognition are illustrated in Figs. 3-4 and 3-5. The face in Fig. 3-3 is a happy face and was not taught to the network. In Fig. 3-4, the activations of the nodes in the lowest stage denote values of characteristics of the face. The middle stage in the network shows the initial activation state (before reverberation). The state of the eyebrow part is near to "Angry" or "Sad" but the state of the eye is near to "Happy" or "Sad". The mouth part clearly shows nearness to "Happy". Thus we cannot judge the facial expression of the object from the output of the middle stage. Figure 3-5 illustrates the state of the network after reverberation. Fuzzy entropies have decreased during cycles of the reverberation and the distribution of activation of nodes converged to the state representing "Happy". The experimental results show that the distribution of activations converges to a state that is meaningful to the context after reverberation though the initial state of activation is too vague to judge the facial expression.

4. Conclusion

We described a network model in which Multi-Layered Reasoning using CFS was applied to recognition of vague facial expressions. MLR by means of CFS can simultaneously process both numeric values and abstract concepts, and achieve recognition depending on context by bi-directional parallel processing.

A facial expression is represented by vague and weak patterns at different parts of the face instead of one clear pattern at one part. We showed that the proposed network matches characteristics of each part with the knowledge constructed from instances and achieves context sensitive recognition by decreasing fuzzy entropies by means of bi-directional processing.

Acknowledgements

We express our deepest gratitude to Dr. Hara and Mr. Kobayashi, Science University of Tokyo, for useful advice and providing us with important data.

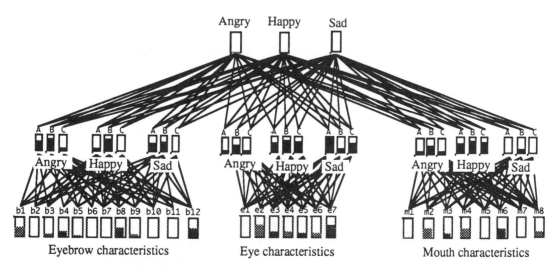

Fig. 3-4 Recognition of a facial expression (Initial State)

References

[1] P.Ekman and W.V.Friesen: Unmasking The Face, Prentice-Hall, Inc., Englewood Cliffs, New Jersey (1975)
[2] Wittgenstein: Philosophical Investigations, Basil Blackwell, Oxford (1953)
[3] T.Takagi, T.Yamaguchi and M.Sugeno: Conceptual Fuzzy Sets, International Fuzzy Engineering Symposium '91, PART II, pp. 261-272 (1991)
[4] T.Takagi, A.Imura, H.Ushida and T.Yamaguchi: Multi-layered Reasoning by Means of Conceptual Fuzzy Sets, Third International Workshop on Neural Networks and Fuzzy Logic '92, NASA Johnson Space Center (1992)
[5] B.Kosko: Adaptive Bidirectional Associative Memories, Applied Optics, Vol.26, No.23, pp. 4947-4960 (1987)
[6] T.Takagi, A.Imura, H.Ushida and T.Yamaguchi: Inductive learning of Conceptual Fuzzy Sets, 2nd International Conference on Fuzzy Logic and Neural Networks IIZUKA '92 (1992)
[7] H.Kobayashi and F.Hara: The Recognition of Basic Facial Expressions by Neural Network, Proc. of IJCNN '91, pp. 460-466 (1991)
[8] P.Ekman and W.V.Friesen: The Facial Action Coding System, Consulting Psychologists Press, Inc., San Francisco, CA (1975)
[9] T.Kohonen: The neural phonetic typewriter, IEEE Computer, 21, 3, pp. 11-22 (1988)

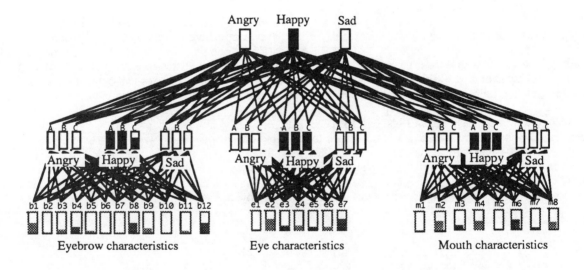

Fig. 3-5 Recognition of a facial expression (After reverberation)

Qualitative/Fuzzy Approach to Document Recognition

Hiroko Fujihara and Elmamoun Babiker

Department of Computer Science, Texas A&M University
hiroko@cs.tamu.edu, babiker@cs.tamu.edu

Abstract

It is becoming significantly important to develop an image processing technology which efficiently analyzes images and interprets the printed document as the human eye captures images and processes them. Tasks involved in document recognition are image processing and reasoning about the spatial layout of the various blocks in the document image. Reasoning uses information such as shape, relative size, and position. In the past, several document recognition applications have been built, however, they lack abstraction and data compression capability, completeness, efficiency, or flexibility.

The approach in this paper uses qualitative spatial reasoning and fuzzy logic [15][16]. The model is based on a new point-interval representation which retains both topological and geometrical attributes of the block regions within a qualitative non-numeric framework, and a grade of membership vector for each block. We show how this model is used to generate a description that covers a wide range of documents, and parse several technical documents based on a single generic model.

AI Topic: Document analysis and recognition, Image processing, Qualitative spatial reasoning, Fuzzy logic
Domain area: Automating information retrieval
Language: C, Lisp, and KEE (expert system shell)
Status: A prototype was implemented.
Effort: One person - six months
Impact: It is more robust and flexible document recognition system in a wider class of documents.

1 Introduction

Paper documents are still a common medium to transmit information in our society. However, as the technology advances, the electronic document has been increasing its share in the communication medium. It is becoming significantly important to develop an image processing technology which efficiently analyzes images and interprets the printed document as the human eye captures images and process them. To be able to recognize, interpret, and effectively utilize images, the machine must be provided with considerable knowledge of the object included in the image, the environmental situation in which the image is taken, and also the methods of processing images [1].

One of the analytic image processing methods is called *pattern recognition*, which outputs the category names of the patterns by analyzing the images. This pattern recognition can be used to analyze images of printed documents in the classification of segmented blocks such as title block, author's name block, etc. Reasoning and recognizing document layouts is an important problem in automating information retrieval. Document layouts deal with rectangular regions or blocks in the image. A subtask involves reasoning about the spatial layout of the various blocks in the document image using information such as shape, relative size, and position. One of the first steps in identifying document images is to create a model that reflects the logical structures behind each block in the model. Many characteristics of the layout relate directly to logical content such that the title is almost always above the author's name. Other significant attributes in defining the block are its size and its position.

Our interest is to model these generic spatial characteristics, to evolve a representation language that meets several criteria for usage, and to make the system robust.

1.1 Document structure and ODA

How is one to describe the functional aspects of various blocks in the document? The Office Document Architecture (ODA) is a standard description for document processing tools such as editors and formatters in multi-vendor office systems. It is an object-oriented architecture for describing both functional and geometric aspects of document modeling [6]. Previous works in document recognition [7][2] have often chosen the ODA standard as an intermediate representation for document structure.

In the ODA model, a document is hierarchical and object-oriented, and is described by two structures: *logical structure* and *layout structure*. The *logical structure* is a tree which consists of hierarchically organized blocks: each block represents the content of the document that means something to the human reader or writer. The *layout structure* is concerned with the positions of each block; it divides the largest rectangular region, page frame, into subdivided rectangular regions within pages such as columns and sections. These rectangular regions core tools of our system. We begin by describing a point-interval algebra that we have formalized for this purpose.

3.1 Interval logic

In order to model the spatial relations between blocks in a document layout, we can consider the projections of these blocks along two orthogonal directions and model these separately as one dimensional intervals [9, 10].

3.1.1 Point-interval algebra

We developed a set of qualitative tools for modeling relative positions, size, and aspect ratios [5]. If we consider a point a and an interval B then a can be before, inside or after B. In addition, there are two more cases of interest - coincidence with either the front or the back boundary of B, which constitute the *tangency* cases. Altogether there are five regions of interest: $+, f, i, b, -$ (ahead, front, interior, back, and posterior respectively). The relation between two intervals A *and* B can then be expressed for example, as $A(++) B$, which would mean that *A is after B*. The detailed discussion is in [5].

Sometimes, the relation between two objects may not be known directly. In such instances, we use the transitivity property. Compiling all possible transitive relations is NP-complete [11][12], but is fortunately unnecessary in any type of document. Transitive relations are used only when an input relation cannot be unified directly with one of the predicates in a rule.

3.1.2 Multi-dimensional representation for spatial relations

Consider the first page of a typical technical document. After initial processing, the document is decomposed into blocks. For each block, we have available the pixel data regarding size and position. This data is fundamentally quantitative in nature, and we would like to specify: 1) a non-subjective qualitative description of binary relations between adjacent blocks, 2) between the current block and a global reference (layout structure) such as the pageframe-block or the column-block, or 3) fuzzy attributes. Consider as an example, the three page fragments shown in Figure 1. In Figure 1 (a), the block B_1's position with respect to B_2 can be represented by

$$B_1 \begin{pmatrix} bi \\ ++ \end{pmatrix} B_2$$

where the relation $\begin{pmatrix} bi \\ ++ \end{pmatrix}$ is a vector of the x and the y qualitative interval relations: {bi} in x represents the fact that the back or minimum x of B_1 is at the same position as the back *(b)* of B_2 ($B_1.x_1 = B_2.x_1$), and that the front of B_1 is within the interior *(i)* of B_2 ($B_1.x_2 < B_2.x_2$); i.e., the left side of B_1 and B_2 are flush on the same vertical line, and B_1 is shorter than B_2. Similarly, in the y direction, B_1 is {++} with respect to B_2, i.e. it is above.

Sometimes we would like to reference the position with respect to a layout block such as the page frame. For example, in Figure 1 (b) B_3 and B_4 can be represented with respect to the page frame:

$$B_3 \begin{pmatrix} cs \\ if \end{pmatrix} Frame \wedge B_4 \begin{pmatrix} bi \\ ii \end{pmatrix} Frame$$

i.e., B_3 is centered or {cs} on the frame in x and starts inside the frame in y but is flush with it at the top {if}. B_4 is flush with the right margin of the page {bi}, and is contained inside it in y{ii}.

Similarly, the positional relation between B_5 and B_6 in Figure 1 (c) can be represented as

$$B_6 \begin{pmatrix} cl \\ -- \end{pmatrix} B_5 \iff B_5 \begin{pmatrix} cs \\ ++ \end{pmatrix} B_6$$

where the two relations are complimentary, given the position of B_6 with respect to B_5, the position of B_5 with

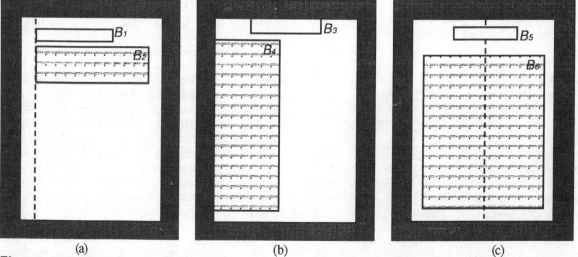

Figure 1. Examples of typical block layouts in documents. Left justified (a), not justified (b), and centered (c) where (c) is disjunction of center (s) and center (l). (y-axis is from bottom to top.)

can define nested rectangular regions within them. The lowest-level rectangular regions are defined as blocks and only blocks have contents by definition [6][7]. A frame might be used to represent a column of text, for example, with nested blocks representing the regions used for sections, which contain a section heading, a section body, a figure and its caption.

This paper presents an approach for document recognition using qualitative spatial reasoning and fuzzy logic. This is a substantial extension of our previous work, which was implemented based on a qualitative spatial reasoning [5]. The approach consists of three phases: image processing, reasoning the blocks based on qualitative spatial reasoning, and refining them using fuzzy logic. Qualitative spatial reasoning allows us to recognize the logical structures based on the corresponding layout structures as defined in ODA model. The additional use of fuzzy logic allows us to recognize a wider class of documents and make the qualitative spatial reasoning approach more robust and complete. The main contribution of our model is to define spatial relations between layout components and relating these to the logical structure. We also present an implementation based on the spatial attributes alone, and show this capability of recognizing a large class of technical documents.

2 Related works in document recognition

Several applications [1][2][3][4] have modeled rectangular tessellations in the past. These works used;
a) quantitative approach
b) image resolution and texture features approach
c) formalism for document layout description and modeling approach.

Qualitative approach

This approach is to build quantitative models specific to a certain fixed format, often based on the numerical description of the block in terms of two opposite corners for each block. Yashiro, et al. [2] presented a model which analyzes technical documents using generic layout rules for document images. A formal language named Form Definition Language (FDL) is used to describe the generic layout rules of a document, i.e., how the document page is organized in terms of rectangular blocks. FDL provides many functions for programming with the steps which determine the value of variables. These steps are: a) extract rectangles, b) sort the rectangles, c) group the rectangles, d) calculate area and join small rectangles, and e) resolve the variables. This kind of quantitative approach lacks the power of abstraction and data compression needed to refer to generic document classes and be independent of scale. Also, quantitative approaches often make it difficult to distinguish the logical structure underlying the layout from the layout itself, since there is no separate instantiation for the logical structure.

Image resolution and texture feature approach

Image resolution and texture approach is used to analyze and classify segmented blocks into categories such as half-tone photographs, text with large letter, text with small letters, line drawings, etc. [3]. Considerable work has been done [3] on modeling image texture; such models have been used with great efficiency on recognition of objects with specific layouts, such as newspaper. Texture models are known to be sensitive to font type and spacing used, while it is quite difficult to abstract such information across generic object classes. At the same time, a large amount of human knowledge of documents is encoded in terms of the relative shapes, sizes, and positions of blocks in the image. In a sense, this information is more universal; we know the title usually appears somewhere near the top of the page, irrespective of font type or its size.

Formalism for document layout description and modeling approach

Dangel [4] used this approach to implement the knowledge based document analysis system. The system uses a new formalism for document layout description and modelling. In this approach, the model is realized by a tree structure, which describes the layout of a document page in different layout abstraction levels. First, a given document page is captured by a scanner. Then, the system automatically segments the binary image into components like characters, text-lines and text-blocks, and graphics. Using projection profile cut [8] a document image are segmented into horizontal and vertical axis. As a result of the segmentation procedure the identified layout components are mapped into a hierarchical data structure. The syntactical structure of the document is established by stepwise refinement. This system was tested for business letters with different layouts and compositions. However, it is not efficient or practical if wide variation of styles in a class of documents exist. This would force the system to have a large database. It is because the system uses a breadth-first search in its hierarchical database that search time and size of database increases significantly.

We have studied and analyzed three different approaches in document recognition. Each approach has its advantages and disadvantages, however none of them has logical abstraction, completeness, and efficiency together in a model. In the following section, we present an approach combining qualitative spatial reasoning and fuzzy logic. A qualitative spatial reasoning is based on a new point-interval representation which retains both topology and geometrical attributes of the block regions within a qualitative non-numeric framework, and fuzzy logic uses membership functions and fuzzy reasoning.

3 Background knowledge

As we mentioned in previous sections, a qualitative spatial reasoning using a new point-interval logic and fuzzy reasoning using fuzzy membership functions are the respect to B_6 can be inferred directly, e.g., in x, B_6 is *centered-smaller* {cs} on B_5, or equivalently, B_5 is *centered-larger* {cl} on B_6. Sometimes in describing the logical structure between two blocks the relation may be a disjunction of several basic interval primitives; e.g. the author-affiliation and author-name blocks have the relation {cl} V {cs} V {bf}, which provides a mechanism for expressing the hierarchical abstraction "centered" (c). Together, these positional representations enable the system to determine the tangencies and singularities in the logical structure effectively and efficiently.

3.2 Fuzzy logic

Fuzzy logic was developed by Lotfi A. Zadeh of the University of California at Berkeley, and it is firmly grounded in mathematical theory. Combining multi-valued logic, probability theory, artificial intelligence (AI), and neural networks, it is a digital control methodology that simulates human thinking by incorporating the imprecision inherent in all physical systems [13]. The low specificity of fuzzy logic makes it useful; it allows a more flexible response to a given input similar to how a human behaves. Fuzzy systems base their decisions on inputs in the form of linguistic variables, such as above, close, and near. If the system is very complex, or cannot be easily represented by a series of IF-THEN rules, fuzzy approach becomes useful.

An approach to document recognition should have the capability for several levels of abstraction, so that recognition may be performed over a wide range of object classes [14]. The use of fuzzy logic enables the intermediate decision, which was obtained through a qualitative spatial reasoning, to be more precise and concrete. Fuzzy logic allows the system to flexibly determine what logical block the recognized block can be based on fuzzy functions and fuzzy rules. This is an important aspect in recognizing blocks in a technical document, due to the wide variety of writing styles in the document class.

Membership functions

In order to assign a grade of membership vector for each block, we define a membership function for each of the fuzzy sets *left*, *top*, *center*, and *small*. All functions are defined on the set B of blocks into [0,1] and are based on the x and y coordinates points for each block. In the following, x_p and y_p denote the maximum x, y coordinates for page frame (maximum among all x, y coordinates for all blocks). The membership function for the fuzzy set *left* is defined as:

$L(b_i) = 1 - x\_min_i/x_p,$

where x_min_i is the minimum x-coordinate for block b_i. The value obtained using L for a given block b_i reflects how much left justified block b_i is. Similarly, we define the membership function for the fuzzy set *top* as follows:

$T(b_i) = 1 - y\_min_i/y_p,$

where y_min_i is the minimum y-coordinate for block b_i. Furthermore, the membership function for fuzzy set *small* is defined as:

$S(b_i) = 1 - area\_b_i/max\_area,$

where $area\_b_i$ is the area of the block i, and max_area is the area of the biggest block. Finally, the membership function for the fuzzy set *center* is defined as:

$C(bi) = \min\{(x_p - x\_max_i)/x\_min_i, x\_min_i/(x_p - x\_max_i)\}$

where x_max_i, x_min_i are the maximum and minimum x-coordinates for block $b_i, x\_max_i <> x_p$, and $x\_min_i <> 0$. If $x\_min_i = 0$ and $x\_max_i = x_p$, then $C(bi) = 1$. If $x\_min_i = 0$ or $x\_max_i = x_p$, then $C(bi) = 0$.

4 DOCREC-2: A flexible document recognition system

DOCREC-2, consists of three phases: image processing (segmentation into blocks), reasoning blocks (analysis of block spatial position using the qualitative model), and refining the results using fuzzy functions. DOCREC-2 was implemented in *C* and *KEE-Lisp* as was DOCREC (the first version). The significant improvement to DOCREC-2 over the first version is its refining phase using fuzzy logic. The set of rules instantiating the logical structure for technical documents is concise and powerful compared to other systems. However, all the possible components which any technical document might use may not be in the scope of the rules. The rules were constructed to be robust and complete, however, flexibility was not incorporated, which humans use in making decisions when they do not have clear knowledge. We look first at the model construction process, and then the filters for block segmentation, and recognition and refinement procedures.

4.1 Constructing a document model: The generic technical paper

The predicates for qualitative spatial reasoning are defined only between two objects. In the case of a technical paper, page frame and columns are the layout objects, with no instance in the logical structure. The textual elements of a document page are laid out as rectangular blocks and the element orientation is along the X and Y axis directions only. Thus, we can define the qualitative spatial relationship by referring the minimum and maximum positions of the two objects. The block of interest is referenced with respect to its corresponding layout object (block) such as the page frame, or logical object which is normally its adjacent block.

The outermost layout block of the nested blocks is a page frame, which is established from the minimum and maximum corners among all the image blocks.

Similarly, the column is defined between white long rectangles along the y direction. These layout objects are useful in defining the relative positions of the blocks in a page. Particularly, blocks which normally appear at distinctive positions such as the top of a page (column), corner of a page, or bottom of a page (column) are defined with a small rule base using this layout concept. Every block can be defined with respect to the other blocks because of the transitive property of the qualitative spatial algebra. The underlying logical structure may be defined in terms of the notation used in the ODA model (Figure 2).

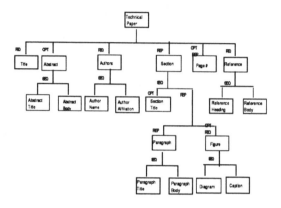

Figure 2. Logical structure based on the ODA Notation OPT: Optional (0 or 1 occurrence), REQ: Required (exactly 1 occurrence) REP: Repetitive (1 or more occurrence) SEQ: A sequence (occurring in the order specified)The complete set of rules instantiating this logical structure for technical documents is given in [5].

4.2 Image processing: segmentation into blocks - phase 1

Given a scanned image of a document, two filters - Run Length Smoothing Algorithm (RLSA), and Recursive X-Y Cuts (RXYC) are used to segment the image into blocks which may bind with the various classes of logical objects. Under the RLSA, any two black pixels (1's), which are separated by at most a certain threshold t, are merged into a continuous stream of black pixels. The RLSA is first applied row-by-row and then column-by-column, yielding two distinct bit maps. The two results are then combined by applying a logical AND to each pixel location [3]. The resulting RLSA image displays covered rectangular blocks with uneven edges. In this study, we used the threshold for x $t_x=15$ for one-column technical paper, and $t_x=10$ for two-column documents. In reality, the decision of which pixel to use is a complex one, and one must analyze the fonts and overall line spacings in order to do this well.

Subsequent to the RLSA processing, the recursive projection profile cuts (X-Y cut) were applied recursively, first only for the rows and then for the columns. The objective here is to identify the spaces between columns and to recognize the higher levels of the layout structure. A projection along a line parallel to the X or Y-axes is a sum of all the pixels (1's). Subdivision along the two directions is determined by comparing the sum and a predetermined threshold. If the sum does not reach the threshold, the row or column of pixels is treated as a blank. We used different thresholds in the vertical and horizontal directions - for x $t_x=15$ and for y $t_y=50$. Because of the orientation of a page frame, the threshold for y needs to be larger than x - total number of pixels for y direction is to be larger. In the recursive processes, the higher blocks in the logical hierarchy are decomposed into smaller blocks by reducing the threshold for y until it reaches the basic block level.

The positional information of the layout structure of a page obtained through RLSA and RXYC algorithms is the input for our spatial reasoning. Before any blocks can be labeled, their relative positions must be described in the qualitative form and these can then be unified against the rules to obtain a match for the block labels. Blocks are identified starting from the top of the page.

A reader of technical paper can differentiate among logical blocks in the paper by using attributes such as location and size of each block. For example, a title block has different size and location than a section block. Similarly, a page number block and an abstract block can easily be differentiated. Therefore, we incorporate these attributes in the refinement phase. For each block identified earlier in the intermediate phase, we assign a vector based on a membership functions to be described latter. The components in the vector represent the grade of membership in the fuzzy sets *left*, *top*, *small*, and *center*. For instance, a block with a vector (0.9, 0.8, 0.7, 0.1) indicates the block is almost left justified and almost top of the page frame, small size, and not centered almost at all.

4.3 Using the rule structure in logical structure identification - phase 2

The title block (if present) is generally centered and at the top of the frame, and is the first block that is identified. Subsequent blocks refer to a neighboring block in the positive y-direction (close to the top of the page). The qualitative relationships are first defined from the blocks which belong to the higher level of the layout structure. Thus, the page frame is recursively divided into smaller frames and each block is defined in terms of the logical frame of which it is a part and/or one adjacent block. As shown in figure 2, the logical blocks can occur in the order specified or for a certain number of times. As an example, consider the author-block which contains the author's name and affiliation, is adjacent to the title block (SEQ). Inside the author-block, the author's names always comes before the author's affiliations (SEQ). Thus, the author-block is defined with respect to the title based on the logical structure, and the authors' names and affiliations are defined with respect to the logical structure. The flexibility of the number of appearance of the blocks as well as the position of the block can be represented by disjunctions. When more than one relationship needs to be defined, the representations are logically "'OR'ed" in the rule-base.

As discussed before, this mechanism does not require a large rule-base. The blocks such as Title and Author are defined only by one pair of relationships. Column is defined inside the page frame for abstract, section, and references blocks. When the paper is organized into two columns, this column concept becomes significantly useful. The use of the hierarchical representation mechanism and adjacent layout reference enabled the rule base to remain simple and efficient. Generally two or at most three conjunct forms can represent one element of the paper sufficiently. Our discussion concentrates on the first page of the technical document, however our rule-base works for any kind of page layout.

4.4 Refinement using fuzzy logic - phase 3

In this phase, the system examines the intermediate results which are produced by the reasoning phase. We define a set of rules for such refinement. Each rule is applied specifically to qualified blocks. For instance, one rule can be used to examine whether a particular block is a title block. Each rule compares membership vectors for a generic logical block to the membership vector for the qualified block. If a candidate block is matched (based on threshold) with the generic block, then the candidate block is accepted, otherwise, it is rejected. In the rejection case, the system will attempt to match other generic blocks with the candidate block. The procedure for matching can be summarized as follows:

1. The absolute difference vector is computed by taking the absolute values from subtracting the components of the generic block vector from their corresponding components in the intermediate block vector.
2. The absolute difference vector is multiplied (component wise) by a generic weighted vector (each generic membership vector is multiplied by a certain weight to reflect the importance of each component).
3. The average of the components of the product vector is computed.
4. The sum is compared to the threshold to determine the degree of matching.

Example

Suppose a candidate block for a title with membership vector (0.7, 0.2, 0.5, 0.08) is passed to the refinement phase. This block is compared with the generic block membership vector (0.3, 0.95, 0.8, 0.85) for the title as follows. The absolute difference vector is computed; in this case it is (0.4, 0.75, 0.3, 0.77). This vector is multiplied by the weighted vector for the title (0.1, 0.4, 0.1, 0.3); the resulted vector in this case is (0.04, 0.3, 0.03, 0.23). The average of the sum is computed to be (0.04 + 0.3 + 0.03 + 0.23)/4 = 0.15. This results shows that the candidate block is less likely to be a title based on a threshold of 0.9.

Block/Function	left	top	small	center
Title	0.30	0.95	0.80	0.85
	0.10	0.40	0.10	0.30
Author-Name	0.10	0.80	0.90	0.85
	0.10	0.30	0.40	0.30
Author-Affiliation.	0.10	0.75	0.85	0.85
	0.10	0.25	0.30	0.30
Abstract-Title	0.50	0.70	0.95	0.75
	0.15	0.20	0.40	0.25
Abstract-Body	0.80	0.65	0.50	0.85
	0.10	0.15	0.15	0.30
Section-Title	0.90	0.50	0.90	0.20
	0.40	0.15	0.40	0.10
Section-Body	1.00	0.45	0.20	0.00
	0.75	0.10	0.10	0.00
Page	0.10	0.30	0.99	0.50
	0.10	0.15	0.70	0.10

Table 1. Fuzzy Functions: The following are the generic membership vectors and their corresponding weighted vectors for generic blocks used in our system.

5. Implementation and output

Lisp programs in KEE create the units corresponding to the blocks which were recognized in the previous step, and put the qualitative representations into slots. The rules for the logical structure of the technical papers were defined in KEE (Figure 3). The details of the rule base are discussed in [5]. When a rule fires and the block is recognized, the block is plugged into the defined structure of the technical paper (Figure 4). DOCREC-2 displays how many blocks are recognized through the previous steps and what their logical structure is.

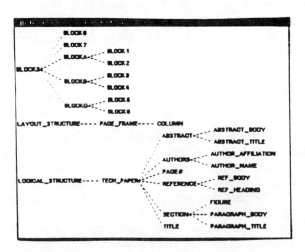

Figure 3. KEE units corresponding to the recognized blocks.

As figure 3 shows, the blocks have been identified at several levels of resolution and organized into hierarchical groups - e.g. Block B in the figure will be labelled as the

Abstract-Block after running the rules, and it has two child blocks 3 and 4, which will be labelled Abstract-title and Abstract-body.

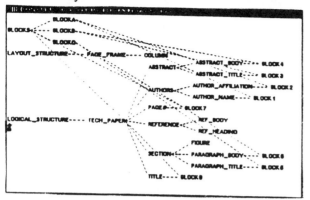

Figure 4. The result of firing rules. When rule base is fired, the recognized blocks are plugged into the logical structure display.

6. Conclusion

Both a qualitative and a fuzzy representations have been presented for modeling the geometric and topological attributes of orthogonal image blocks in a document layout. Qualitative models provide a compact representation since there is no absolute metric; at the same time it preserves much of the important information such as tangency relations, and symmetry. Fuzzy logic approach allows some degree of uncertainty in recognizing document blocks and modeling human perception. We have shown that using qualitative reasoning in conjunction with fuzzy logic enables us to produce a model that is capable of handling a wider class of documents despite variable locations and sizes of the blocks in the document.

Acknowledgement

We would like to thank Dr. Amit Mukerjee for his guidance and encouragement to accomplish this work.

References

[1] M. Ejiri, "Knowledge-based approaches to practical image processing," *Int'l Workshop on Industrial Applications of Machine Intelligence and Vision* (MIV-89), Tokyo, April 10-12, 1989, pp. 1-8.

[2] H. Yashiro, T. Murakami, Y Shima, Y. Nakano, and H. Fujisawa, "A new method of document structure extraction using generic layout knowledge," *Int'l Workshop on Industrial Applications of Machine Intelligence and Vision* (MIV-89), Tokyo, April 10-12, 1989, pp. 282-287.

[3] D. Wang and S Srihari, "Classification of newspaper image blocks using texture analysis," *Computer Vision, Graphics and Image Processing*, Vol.47, 1989, pp. 327-352.

[4] A. Dengel, "Automatic visual classification of printed documents," *Int'l Workshop on Industrial Applications of Machine Intelligence and Vision* (MIV-89), Tokyo, April 10-12, 1989, pp. 276-281.

[5] H. Fujihara and A. Mukerjee, "Qualitative reasoning about document structures," *Texas A&M University Technical Report*, TR91-010, February, 1991.

[6] W. Horak, "Office document architecture and office document interchange formats: Current status of international standardization," *Computer*, October 1985, pp.50-60.

[7] J. Rosenberg, M. Sherman, A. Marks, and J. Akkerhuis, *Multimedia Document Translation - ODA and the EXPRES Project*, Spring-Verlag, New York, 1991.

[8] G. Nagy, S. Seth, and S. Stoddard, "Document analysis with an expert system," *Pattern Recognition in Practice II*, Elsevior Science Publ. B.V. 1986.

[9] J. Allen, "Maintaining knowledge about temporal intervals," *Communications of the ACM*, Vol.26 (11), November 1983, pp.832-843.

[10] J. Malik and T. Binford, "Reasoning in Time and Space," *Proceeding of the Eighth I JCAI* 1983, pp.343-345.

[11] P. Van Beek, "Exact and approximate reasoning about qualitative temporal relations," Ph.D. thesis, University of Alberta TR90-29, August 1990.

[12] P. Van Beek "Reasoning about qualitative temporal information," *AAAI-90*, Boston, July 29-August 3, 1990, pp.728-734.

[13] K Self, "Designing with fuzzy logic," *IEEE Spectrum*, November 1990, pp.42-44.

[14] H. Samet, "The design and analysis of spatial data structures," *Reading*, Mass.: Addison-Wesley, 1990 (Addison-Wesley series in computer science).

[15] A. Zadeh, "Fuzzy Sets," *Inform. Control*, 8, 1965, pp.338-353.

[16] A. Zadeh, "Fuzzy Logic," *IEEE Computer,* April 1988, pp.83-93.

Fuzzy Artificial Network and its Application to a Command Spelling Corrector

N. Imasaki,* T. Yamaguchi,** D. Montgomery,*** and T. Endo**

*,** Systems & Software Engineering Laboratory, Research & Development Center, TOSHIBA Corporation
70, Yanagi-cho, Saiwai-ku, Kawasaki-shi 210, JAPAN
*** Faculty of Engineering, University of Victoria, CANADA
(* Currently a visiting researcher at University of California, Berkeley, U.S.A.)

Abstract - This paper proposes a Fuzzy Artificial Network (FAN) which utilizes associative memories and is constructed by a method which makes it easy to represent and to modify fuzzy rule sets. While conventional fuzzy inference methods induce much fuzziness on multi-layered fuzzy rule sets, the associative memory based FAN results in inferences which fit human sense better. We call this type of fuzzy inference "associative inference." For memorizing fuzzy rule sets, the proposed FAN system employs a correlation matrix which is constructed from a nominal correlation matrix, a bias matrix, and a scale parameter, so that it is easy to carry out refinement and cut-and-paste operations for rule sets. Using a FAN development system, we compose a command spelling corrector which uses a multi-layered fuzzy rule set. The spelling corrector application shows the eligibility of associative inference for multi-layered fuzzy rule sets.

1. INTRODUCTION

In order to compose intelligent systems, not only linguistic (conceptual) value processing but also physical (numerical) value processing is necessary [1],[2]. Fuzzy set theory [3],[4] is known as a suitable theory for an interface between linguistic value and physical value, and has been applied to many actual intelligent systems [5],[6]. In this movement, we are aware that fuzzy systems should advance so that they achieve the following two points: 1) inference which fits human sense better, and 2) easier refinement of rule sets. Toward the realization of these points, many trials to combine fuzzy set theory and neural networks have been performed. We call, in this paper, neural network systems which represent fuzzy rules "Fuzzy Artificial Network (FAN)." FAN systems using feed forward neural network learning techniques are known to be able to refine fuzzy rules automatically (one area of "inductive learning"). On the other hand, FAN systems which employ associative memory networks, i.e., Bidirectional Associative Memories (BAM) [7], result in inferences which fit human sense reasonably well [8],[9], whereas conventional fuzzy inference methods induce results with much fuzziness when the inference is made on a multi-layered fuzzy rule set. We call such an associative memory based fuzzy inference "associative inference" [10].

This paper proposes a FAN system which inherits both the inductive learning and the associative inference. For memorizing fuzzy rule sets, the FAN system employs a correlation matrix which is constructed from a nominal correlation matrix, a bias matrix and a scale parameter. This employment makes it easy to carry out both refinement and cut-and-paste operations for fuzzy rule sets, which is difficult for conventional FAN systems. The FAN system stores the nominal correlation matrix (given by an actual relationship on the fuzzy rule set) instead of a conventional correlation matrix (calculated from input-output data converted to bipolar elements on the fuzzy rule set). As the nominal correlation matrix shows the actual relationships on the fuzzy rule set, it is easy to carry out both refinement and cut-and-paste operations for the fuzzy rule set, and easy to check the memorizing ability of associative memories.

For the purpose to apply the FAN system to various fields, we have developed a design tool called "FAN Development System (FANDeS)." Using the FANDeS package, we composed a command spelling corrector which employs a multi-layered fuzzy rule set. In this application, we mainly show the eligibility of the associative inference for multi-layered fuzzy rule sets. The FAN based spelling corrector introduces two types of recognition: 1) recognizing a correct character in the neighborhood of a keyboard-input character using fuzzy membership functions [11], and 2) recognizing a command (word) as an element of the command line (sentence). The FAN system combines these two types of recognition by simultaneous bottom-up and top-down associative inference processing.

2. ASSOCIATIVE MEMORY BASED FUZZY ARTIFICIAL NETWORK

The FAN system which this paper proposes uses associative memories so that it represents a fuzzy rule set as shown in Fig.1. It consists of three elements: 1) if-part, 2) if-then-rule-part, and 3) then-part. The fuzzy rule set represents expert knowledge describing the relationships between "conditions" and "operation models." The if-part has membership functions which abstract and characterize the conditions. The then-part has membership functions or input-output functions which represent the operation models. Associative memories (a kind of BAM) store the relationships between the if-part's conditions and the then-part's operation models.

BAM, which has two layers, memorizes pattern (vector) pairs in terms of a correlation matrix and its transpose matrix. Assume that a BAM memorizes x-y pair (each element $\in$ [0,1]) and a noisy x is given to the x layer, the BAM recalls

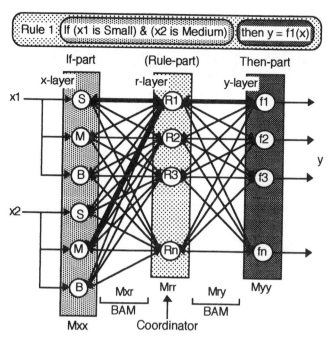

Fig.1 FUZZY RULES REPRESENTATION BY FAN

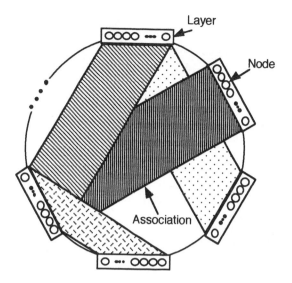

Fig.2 MULTI-LAYERED FAN

the x-y pair on its two layers. BAM recalls a pair from its memory by reverberation which is performed as follows:

$$Y_t = \phi(M \cdot X_t), \quad X_{t+1} = \phi(M^T \cdot Y_t). \tag{1}$$

In (1), $X_t=[ax_1, ax_2, ..., ax_m]^T$, $Y_t=[ay_1, ay_2, ..., ay_n]^T$ are activation vectors on the x, y layer respectively at reverberation step t, and $\phi(\cdot)$ is a sigmoidal function of each unit. The correlation matrix M and M^T are given as follows:

$$M = \sum_{i=1}^{N} \beta_i \cdot y_i \cdot x_i^T, \quad M^T = \sum_{i=1}^{N} \beta_i \cdot x_i \cdot y_i^T. \tag{2}$$

In (2), x_i, y_i ($i=1\sim N$) are memorized pairs, β_i is an association parameter.

In conventional associative memory systems, each element of vector x_i, y_i ($\in\{0,1\}$) is often converted to bipolar form $\{-1,1\}$ before the calculation of (2) in consideration of an energy function [7]. We denote this conventional correlation matrix using bipolar elements "M_b."

BAM recollection by means of reverberation has a feature that the most suitable pair for input conditions appears as a result of decreasing fuzzy entropy which shows fuzziness of inference output.

The FAN system introduces this BAM's feature to control fuzziness at fuzzy inference. A single BAM has not so high memorizing capacity that the system uses two BAM-like associative memories as shown in Fig.1 to represent fuzzy rules. Each unit in x layer represents the if-part membership function given by the fuzzy rule set as conditions. Each unit in y layer represents the then-part operation model. Each unit in r layer represents a fuzzy rule. Two BAMs connect the x layer to the r layer and the r layer to the y layer. The correlation matrices are M_{xr} and M_{ry}. The r layer has mutual negative weighted connections between the units in order to recall the most suitable rule for input conditions, and minor positive weighted recessive connections on each unit in order to assist its activation. These connections are represented by the r layer's correlation matrix M_{rr}, called the coordinator, which stores fuzzy rules. There are also minor positive weighted recessive connections at x and y layers in order to assist activation. The x and y layers' self correlation matrices are denoted as M_{xx} and M_{yy}, respectively.

A reverberation with those correlation matrices (M_{xx}, M_{xr}, M_{rr}, M_{ry}, M_{yy}) and their transpose matrices carries out the associative inference. The associative inference comes up with a weak point of the conventional fuzzy inference, that is fuzziness increasing with respect to the inference output, because the association can regulate the fuzziness. Extraction of activation values before the sufficient regulation provides some imitated fuzzy inference methods, free from worrying about excessive induced fuzziness. We use some activated fuzzy rules to synthesize each output. Suppose that γ_i is the normalized activation value of the i-th unit in the y layer, the output of fuzzy rule set (which has n output nodes) is given as follows:

$$y^* = \sum_{i=1}^{n} \gamma_i \cdot f_i(x). \tag{3}$$

FAN systems using the conventional correlation matrix M_b, are weak in that it is difficult to carry out fuzzy rule cut-and-paste operations and difficult to check the memorizing ability after setting M_b. In order to improve this point, we introduce an another correlation matrix M_e which is composed from a normalized version of the nominal correlation matrix M, a bias matrix B, and a scale parameter a as follows:

$$M_e = a(M + B), \tag{4}$$

where

$$a = 1/(3c), \quad B = \begin{bmatrix} -c\ -c\ \cdots\ -c \\ \cdots \\ -c\ -c\ \cdots\ -c \end{bmatrix}, \quad c = \frac{1}{2}max\ (\ element\ of\ M\ ).$$

(5)

The usage of the nominal correlation matrix M which represents actual relationships in the rule set, makes it easy to carry out cut-and-paste operations for fuzzy rules and to check the memorizing ability (which is calculated from the Hamming distance between the memorized pairs) even after setting the correlation matrix M_e. The cut-and-paste operations change the dimension of the matrix M according to the number of rules.

If we need a multi-layered fuzzy rule set, the FAN system connects necessary number of layers as shown in Fig.2 [10]. For example, the 1st rule output layer connects to the 2nd rule input layer. The usage of the nominal correlation matrix M also makes it easy to build a multi-layered rule system.

3. FAN DEVELOPMENT SYSTEM

A software package which allows us to design associative memories including the FAN system, add as many associations (or layers) as desired, and then load the network into an application program has been developed.

The system, called Fuzzy Artificial Network Development System (FANDeS), treats an associative memory network as a group of connected "layers," each containing neural nodes. The following declaration defines the data structure used for each layer.

```
LAYER {
  char layername[LAYERNAME_LENGTH];
  char numofnodes;
  MATRIX *oldactivations;
  MATRIX *newactivations;
  double inputweight;
  double lambda;
  double coordinator;
  NODE *nodes;
  CONNECTION *connectedlayers;
  LAYER *nextlayer;
};
```

The parameter `layername` is simply the name of the layer in string form. Similarly `numofnodes` describes the number of nodes the layer contains. The pointers `*oldactivations` and `*newactivations` point to matrices (pointer based matrix implementations) of size (1, numofnodes) which store the activations of each node in the layer. These pointers are updated during each iteration of the reverberation calculations. The parameter `inputweight` simply defines the weight assigned to the input vector (which is stored as part of the structure `NODE`). This parameter comes into play during the reverberation calculations. The next parameter is `lambda`, which is a parameter used in the sigmoidal function during reverberation calculations. Next is `coordinator`, which defines the layer's self-coordinator matrix. The coordinator can be either positive-negative (positive weighted connections from each node to itself and equal magnitude connections from each node to each other node in the layer), positive-zero (just positive weighted connections from each node to itself), or zero (no coordinator connections). The next parameter is the pointer `*nodes`, of type `NODE`, which is defined as follows.

```
NODE {
  char nodename[NODENAME_LENGTH];
  double input;
  NODE *nextnode;
};
```

As can be seen, the structure `NODE` contains each each node's name, its input value (which is used when the layer's inputweight is nonzero), and a pointer to the next node in the linked list.

The next parameter in the `LAYER` structure is the pointer `*connectedlayers`, which describes the connections between layers. This parameter is also a linked list, so each layer can be connected to as many other layers as the user wants. The final parameter is a pointer to the next layer in the linked list.

The structure `CONNECTION` is defined as follows.

```
CONNECTION {
  LAYER *connectedlayer;
  MATRIX *connectionweights;
  CONNECTION *nextconnection;
};
```

This structure is used as a linked list describing the connections of each layer. When two layers are connected, each layer has a "connection" added to its connection list. Each connection points to the other layer, and they both point to the same connection matrix. The matrix is defined by adding associations between the two layers and then is normalized by (4) and (5) to improve network energy function, thus improving recollections of the memorized pair.

The connection associations are defined by another linked list structure.

```
ASSOCIATION {
  CONNECTION *C1, *C2;
  MATRIX *V1, *V2;
  double beta;
  char flag;
  ASSOCIATION *nextassoc;
};
```

This structure describes everything about an individual association. The user can define as many associations as desired.

After the network has been built, the user can save it to a file and then write a simple C program to use it. There are functions in the software module which can be linked into a program which allow one to load the network, set input vectors, and make reverberation calculations. The reverberation algorithm is as follows:

```
if first-iteration {
    set all old activations based on input vectors;
}
for (all layers) {
    if (input weight) {
        set new activations based on input;
    } else clear new activations;
    if (coordinator) {
        add (old activations) * (coordinator matrix)
            to new activations;
    }
    for (all connected layers) {
        add (connected layer old activations)
            * (connection matrix) to new activations;
    }
    new activations = φ( new activations );
}
for (all layers) {
    old activations = new activations;
}
```

Essentially this is the algorithm described in (1).

EXAMPLE:

Consider the following fuzzy rules.

R1: if (X1 is BIG) and (X2 is BIG)
            then (Y is NEGATIVE)
R2: if (X1 is BIG) and (X2 is SMALL)
            then (Y is ZERO)
R3: if (X1 is SMALL) and (X2 is SMALL)
            then (Y is POSITIVE)

The network in Fig.3 is created using FANDeS. The associations are made with the following vectors:

R1: [1 0 1 0] ↔ [1 0 0] ↔ [0 0 1]
R2: [1 0 0 1] ↔ [0 1 0] ↔ [0 1 0]
R3: [0 1 0 1] ↔ [0 0 1] ↔ [1 0 0].

A simple C program is written to test this network. After loading the network, the program executes a loop which takes as input two real numbers, X1 and X2, assigns layer X input based on these numbers, and then lets the network relax. The layer X input is set by the fuzzy membership functions shown in Fig.4.

Table 1 shows the result of this experiment (after 4 iterations) with various input values.

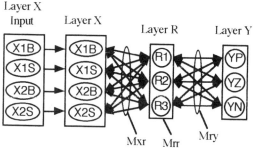

Fig.3. FAN EXAMPLE

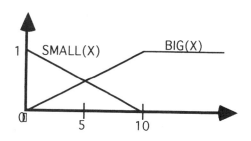

Fig.4. MEMBERSHIP FUNCTIONS FOR SMALL AND BIG

Table 1 EXAMPLE RESULT

X1	X2	X1B-in	X1S-in	X2B-in	X2S-in	YP	YZ	YN
10	10	1.0	0.0	1.0	0.0	0.170	0.224	0.988
10	0	1.0	0.0	0.0	1.0	0.182	0.989	0.182
7	3	0.7	0.3	0.3	0.7	0.183	0.989	0.183
0	0	0.0	1.0	0.0	1.0	0.988	0.224	0.170

4. AN APPLICATION: COMMAND SPELLING CORRECTOR

A common problem that occurs when typing commands on computers is that when typing quickly, one often makes simple errors that go unnoticed. A common source of these errors is the proximity of keys on the keyboard. For example, the key 's' is very close (i.e., next) to 'd' and 'a.' When typing quickly, sometimes one will press 'd' instead of the intended 's.'

We propose a command spelling corrector based on the fuzzy "closeness" of keys. This concept stems from Araki and Kaguei [11], where a command corrector was constructed based on fuzzy Hamming distance. In this paper, inputted commands are recognized by two factors: 1) the actual inputted character sequence, and 2) the number of arguments (or "words") in the command line (or "sentence").

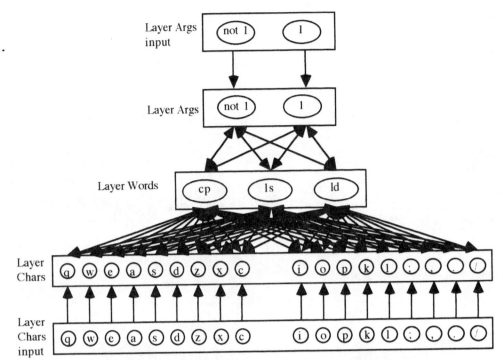

Fig.5 FAN BASED COMMAND SPELLING CORRECTOR

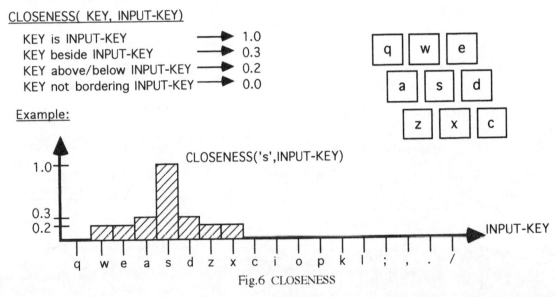

Fig.6 CLOSENESS

A FAN system for this application has two input layer so that rules are multi-layered. The system employs a simultaneous top-down and bottom-up inference. Figure 5 shows the FAN (constructed using FANDeS) for a simple (18 keys, 3 commands) prototype version of the system. A program was written which loads the corrector FAN, sets the input vectors according to keyboard input, and then reverberates the network to make an associative inference as to what the input probably was intended to be. The program sets the input to layer "Args" by counting the "words" in the command line. The input to layer "Chars" is derived from the fuzzy "closeness" of each key with each inputted character in the first "word" of the command line. The membership function for "closeness" is shown in Fig.6. The program was executed for the input strings shown in Table 2. Table 3 shows the results. We built the networks to recognize some UNIX commands. The input "xp a b" is clearly incorrect input - there is no command "xp." The command "ld" exists, but has to have arguments, so one can assume that "ld" by itself was intended to be "ls." The command "la" does not exist, so the input "la a" was probably meant to be "ls a."

As can be seen, the association network successfully recalls the intended command in each case. Successful recollection is assisted by the normalized correlation matrix M_e, which improves the energy function of associative memory networks.

Some future enhancements are recommended and are currently being worked on: 1) expansion of the network to include more commands and the full keyboard, and 2) introduction of a Hebbian learning algorithm for the fuzzy membership functions, such that a commonly made error will make a stronger association to the intended key.

It is of note that the network architecture used for this application (simultaneous top-down and bottom-up inference) is also useful in many other applications including image

Table 2 TRIAL INPUT STRINGS

Correct Input	Incorrect Input
cp a b	xp a b
ls	ld
ls a	la
ld a	la a

Table 3 SPELLING CORRECTOR RESULTS

Input line	cp	ls	ld	Probable command
cp a b	1.000	0.025	0.034	cp a b
xp a b	0.999	0.077	0.117	cp a b
ls	0.001	1.000	0.028	ls
ld	0.000	0.997	0.754	ls
ls a	0.002	0.995	0.861	ls a
la	0.002	1.000	0.068	ls
ld a	0.003	0.476	0.997	ld a
la a	0.005	0.982	0.948	ls a

processing (facial pattern recognition) and conceptual fuzzy sets [12]. The multi-layered FAN has shown itself to be a useful and versatile inference technique.

5. CONCLUSION

This paper proposed a Fuzzy Artificial Network (FAN) which uses associative memory networks and is constructed by a method which allows easy representation and modification of fuzzy rules. The proposed FAN system brought inference results which fit human sense reasonably well compared to conventional fuzzy inference, which induces much fuzziness when making inference on a multi-layered fuzzy rule set. The FAN system uses a correlation matrix which is derived from the normalization of a nominal correlation matrix and thus it is easy to carry out not only refinement but also cut-and-paste operations on fuzzy rules. These operations are difficult for the conventional FAN systems. Using a FAN development system, we composed a command spelling corrector which used a multi-layered fuzzy rule set. In this spelling corrector application, we mainly explained the usefulness of associative inference for the multi-layered fuzzy rule set using simultaneous bottom-up and top-down inference processing. The corrector corrected each command using information on keyboard input closeness and information on the entire command line (number of arguments).

REFERENCES

[1] J. Russmussen: Skills, Rules, and Knowledge, "Signal, signs, and symbols, and other distinctions in human performance models," *IEEE Trans. on System, Man and Cybernetics*, vol.13, no.3, 1983, pp. 257-266.

[2] T. Sawaragi, T. Norita and T. Takagi, "Fuzzy theory," *Journal of the Robotics Society of Japan*, vol.9, no.2, 1991, pp. 238-255 (*in Japanese*).

[3] L. A. Zadeh, "Fuzzy algorithms," *Information and Control*, vol.12, 1968, pp. 94-102.

[4] T. Terano, *Fuzzy System Theory and Its Applications*, Ohmu-sya, 1987 (*in Japanese*).

[5] T. Yamaguchi, K. Goto, T, Takagi, K. Doya and T. Mita, "Intelligent control of a flying vehicle using fuzzy associative memory system," *IEEE Int. Conf. on Fuzzy Systems 1992*, vol.2, pp. 1139-1149.

[6] T. Yamaguchi, T. Takagi and T. Mita, "Self-organizing control using fuzzy neural networks," *Int. J. Control*, 1992, *in press*.

[7] B. Kosko, "Adaptive bidirectional associative memories," *Applied Optics*, vol.26, no.23, 1987, pp. 4947-4960.

[8] T. Yamaguchi, N. Imasaki and K. Haruki, "Fuzzy rule realization on associative memory system," *IJCNN-90-WASH-DC Conf. Proc.*, vol.2, 1990, pp. 720-723.

[9] T. Takagi, T. Yamaguchi and M. Sugeno, "Conceptual fuzzy sets," *International Fuzzy Engineering Symposium: IFES'91 Conf. Proc.*, vol.2, 1991, pp. 261-272

[10] S. Yamamoto, T. Yamaguchi and T. Takagi, "Fuzzy associative inference system and its features," *IEEE Int. Conf. on System Engineering*, 1992, *in press*.

[11] H. Araki and S. Kaguei, "Error correction of key-inputted instructions using fuzzy Hamming distance," *Proc. of the 2nd Int. Conf. on Fuzzy Logic & Neural Networks*, 1992, pp. 701-704.

[12] T. Takagi, A. Imura, H. Ushida and T. Yamaguchi, "Multilayered reasoning by means of conceptual fuzzy sets," *Third International Workshop on Neural Networks and Fuzzy Logic'92*, NASA Johnson Space Center, 1992, *in press*.

A New Similarity Measurement Method for Fuzzy-Attribute Graph Matching and Its Application to Handwritten Character Recognition

G.M.T. MAN, J.C.H. POON

Department of Electronic Engineering Hong Kong Polytechnic, Hung Hom, Kowloon, Hong Kong.

ABSTRACT - Attributed graph is widely used as a straightforward representation of structural patterns. The vertices of the graph represent pattern primitives describing the pattern while the arc is the relation between these primitives. For the problems of pattern recognition and classification procedures, the concept of fuzziness is usually applied when patterns are mapped into the feature space. To handle such fuzzy concepts, we extend the attributed graph to fuzzy-attribute graph (FAG) by making the attributes fuzzy. However, such extension will cause new problems in graph matching as we can no longer check the equality of attributes. In this paper, we will give a formal definition of FAGs and introduce a new method of similarity measurement for FAGs to solve the problem of graph matching. A handwritten numeral recognizer has been developed to evaluate the effectiveness of the proposed algorithm and it gives an accuracy of 93 percent in correct classification.

1. INTRODUCTION

During the past three decades, there has been a considerable growth of interest in problems of pattern recognition leading to a variety of mathematical methods. Traditionally, these methods can be grouped into two major categories : (a) statistical, and (b) structural.

a. Statistical methods

Statistical methods are applied primarily for the purpose of pattern classification. The concept of pattern classification may be expressed in terms of partitioning the pattern feature space. A pattern is transformed into a feature vector by taking characteristic measurements. So each pattern can be considered as a point in the feature space. The problem of classification is to assign each feature vector to a proper pattern class. The statistical methods have a long tradition in pattern recognition and are based on a well-founded mathematical theory such as statistical or decision theory. They have been proven useful in numerous applications. However, the statistical methods are basically classification oriented without any pattern description. Certain pattern structures are also difficult to describe with numerical features and cause ineffectiveness in feature extraction [1].

b. Structural methods

In the structural methods, pattern structures are expressed as compositions of structural units, called primitives, and a pattern is recognized by matching its structural representation with that of a reference pattern or by parsing the representation according to a set of syntax rules. The structural methods allow not only pattern classification but also the inference of a structural description of an unknown input pattern. They are also based on a well-founded mathematical theory such as the theory of formal languages and automata. However, the structural methods are difficult to handle the majority of real world images which are corrupted by noise and distortions. Time-consuming pattern matching or parsing is required when the pattern representations and the syntax rules for describing complicated pattern structures are complex [2,3].

The previously mentioned weak points of the conventional structural method are partially due to the discreteness implicitly carried by the symbols and the production rules. Such discreteness implies that only discrete primitives, subpatterns, and pattern structures can be analyzed. Continuous types of numerical features extracted from the primitives must be discretized first and then transformed into symbols for use in matching or parsing. This increases rounding or truncation errors and reduces recognition accuracy. On the other hand, only exact matching or parsing is allowed in the conventional method. Thus structural variations which come from pattern noise or distortion and result in the changes of symbols cannot be handled easily. We can remove the above two weak points, discretness and exactness, of the conventional graph analysis method by introducing numerical features called attributes into the symbolic representations for the primitives and their relations.

Attributed graph is widely used as a hybrid approach to pattern representation and recognition. It is introduced by Fu and Tsai [4]. In this representation, a vertex with attributed values is used to represent a pattern primitive while the attributed arc is used to represent the relation between these primitives. When an unknown input pattern is represented by an attributed graph, the recognition process is to find an isomorphic reference pattern graph. For isomorphism between two attributed graphs means the corresponding matched pairs of vertices and arcs must have the same attribute values for each attribute [5]. However, such graph matching can not be able to handle real world images. A number of algorithms have been proposed in the literature by using a similarity criterion or graph distance measure [4,6-9].

Fuzzy sets theory [10], introduced by L.A.Zadeh, can be used to cope with many natural properties and relations which are fuzzy in nature. The development of fuzzy sets has a strong impact on techniques of pattern recognition [11-13]. It acts as a reasonable tool for modeling and imitating cognitive processes of the human being, especially those concerning recognition aspects. Besides, it offers a lot of novel algorithms which are useful for designing of classification procedures [13]. Therefore, we try to extend the attributed graphs to fuzzy-attributed graph (FAGs) by making the attributes fuzzy in order to enhance the recognition and classification of real world images. In this paper, we will give a formal definition of FAGs in section 2 and introduce a new method of similarity measurement for FAGs in section 3. Finally, an example of applying the proposed method to the recognition of handwritten numerals is described in section 4.

2. Definitions of Attributed Graphs and Fuzzy-Attributed Graphs

The basic definitions of attributed graphs given below are adapted from [4].

a. Attributed Graphs

Each vertex may take attributes from the set $Z=\{z_i| i=1,2,...,I\}$. For each attribute z_i it will have possible values $S_i=\{s_{ij}|j=1,2,...,J\}$. The set of possible attribute value pairs of the vertices is denoted by the set $L_v=\{(z_i,s_{ij})| i=1,...,I; j=1,...,J_i\}$. A valid pattern primitive is just a subset of L_v in which each attribute appears only once and Π denotes the set of all those valid pattern primitives. Thus, each node will be represented by an element of Π.

Similarly, for the arcs, we have the attribute set $F=\{f_i| i=1,...,I'\}$ in which each f_i may take values $T_i=\{t_{ij}|j=1,...,J_i'\}$. $L_a=\{(f_i,t_{ij})|i=1,...,I'; j=1,...,J_i'\}$ denotes the set of possible relational attribute value pairs. A valid relation is just a subset of L_a in which each attribute appears once. The set of all those valid relation is Θ.

Definition 1: An attributed graph G over $L=(L_v,L_a)$, with an underlying graph structure $H=(N,E)$, is defined to be a pair (V,A) where $V=(N,\mu)$ is called an attributed vertex set and $A=(E,\delta)$ is called an attributed arc set. The mappings $\mu:N\rightarrow\Pi$ and $\delta:E\rightarrow\Theta$ are called vertex interpreter and arc interpreter, respectively.

Definition 2: Two attributed graphs $G_1=(V_1,A_1)$ and $G_2=(V_2,A_2)$ are said to be structurally isomorphic if there exists an isomorphism $T:H_1\rightarrow H_2$ where $H_1=(N_1,E_1)$ and $H_2=(N_2,E_2)$ represent the structural aspects of G_1 and G_2, respectively. G_1 and G_2 are said to be completely isomorphic, written $G_1=G_2$, if there exists an attribute value preserving structural isomorphism T between G_1 and G_2.

b. Fuzzy-Attributed Graphs (FAGs)

A fuzzy set (A) in a space of points $X=\{x\}$ is a class of events with a continuum of grades of membership and is characterised by a membership function $\mu_A(x)$ which associates with each point in X a real number in the interval [0,1] with the value of $\mu_A(x)$ at x representing the grade of membership of x in A. The membership function reflects the ambiguity in a set, and as it approaches unity, the grade of membership of an event in A becomes higher. For example, $\mu_A(x)=1$ indicates strictly the containment of the event x in A. If on the other hand x does not belong to A, $\mu_A(x)=0$. Any intermediate value would represent the degree to which x could be a member of A.

Because of this generalisation, we can modify our definition of attributed graph as follows [16].

The set of all possible fuzzy attribute value pair is $L_v=\{(z_i,A_{S_i})| i=1,2,...,I\}$ where A_{S_i} is a fuzzy set on the attribute-value set S_i. A pattern primitive Π is a subset of L_v and let Ω denote the set of all possible pattern primitives.

Similarly, we can assume the adjacency relations to be fuzzy, and then it can be generalized that the set of possible attribute-value pairs of the arcs denoted by $L_a=\{(f_i,B_{T_i})| i=1,2,...,I'\}$ where B_{T_i} is a fuzzy set on the relational attribute-value set T_i. Φ denotes the set of all relations.

Definition 3: A fuzzy-attribute graph G over $L=(L_v,L_a)$ with an underlying graph structure $H=(N,E)$ is defined to be an ordered pair (V,A), where $V=(N,\sigma)$ is called a fuzzy vertex set and $A=(E,\delta)$ is called a fuzzy arc set. The mapping $\sigma:N\rightarrow\Omega$ and $\delta:E\rightarrow\Phi$ are called fuzzy vertex interpreter and fuzzy arc interpreter respectively.

3. A New Method of Similarity Measurement for FAGs

Formally, two graphs G_1 and G_2 are isomorphic if there exists a one-to-one and onto function $f: V(G_1)\rightarrow V(G_2)$ such that x and y belong to $V(G_1)$ if and only if $f(x)$ and $f(y)$ belong to $V(G_2)$. Similarly, two attributed graphs $G_1=(V_1,A_1)$ and $G_2=(V_2,A_2)$ are said to to isomorphic, denoted by $G_1=G_2$, if there exists an attributed value preserving isomorphism f between G_1 and G_2. However, in the case of fuzzy set, equality is too strict a condition for such a matching. Hence, we try to relax the requirement and define a similarity measurement between FAGs.

Definition 4: Let $F(U)$ denotes all the fuzzy subsets in the universe U and $A,B\in F(U)$. A fuzzy membership function α is defined as a relation between the Cartesian product, $F(U)XF(U)$ and the real number interval [0,1], which is denoted by $\alpha:F(U)XF(U)\rightarrow[0,1]$, $(A,B)\mapsto\alpha(A,B)$, if α satisfies

(i) $\alpha(A,B)=1$ iff $A=B$;
(ii) $\alpha(A,B)=\alpha(B,A)$;
(iii) $A\subseteq B\subseteq C \Rightarrow \alpha(A,C)\leq\alpha(A,B)\wedge\alpha(B,C)$;

it denotes the similarity function of $F(U)$. $\alpha(A,B)$ represents the degree of similarity between A and B. We proposed a similarity function as follows:

Let $F_{1xn}=\{(x_1,...,x_n):0\leq x_i\leq 1, i=1,...,n\}$,

and $a,b\in F_{1xn}$, $\alpha(a,b)=\dfrac{\sum_{i=1}^{n}(a_i\wedge b_i)}{\sum_{i=1}^{n}(a_i\vee b_i)}$ satisfies the above conditions and $\alpha(a,b)=1 \Rightarrow a=b$.

The above similarity function can be adopted as a similarity measurement of two fuzzy attributed graphs and is defined as follows.

Definition 5: Let α_1 be the similarity function between vertices v_1 and v_2 of two FAGs G_1 and G_2. Let A_{1Si} be the fuzzy subset that gives the attribute value for z_i of v_1 and A_{2Si} be that of z_i of v_2.

$$\alpha_1(v_1,v_2) = \frac{\sum_{i=1}^{I}\sum_{j=1}^{J_i}(\mu_{A_{1Si}}(s_{ij}) \wedge \mu_{A_{2Si}}(s_{ij}))}{\sum_{i=1}^{I}\sum_{j=1}^{J_i}(\mu_{A_{1Si}}(s_{ij}) \vee \mu_{A_{2Si}}(s_{ij}))}$$

where $\mu(s_{ij})$ is the membership grade of s_{ij} in the fuzzy subset A_{ksi}, $k=1,2$.

Definition 6: Let α_2 be the similarity function between arcs a_1 and a_2 of two FAGs G_1 and G_2. Let B_{1Ti} be the fuzzy set that gives the attribute value of t_i of a_1 and B_{2Ti} be the attribute value of t_i of a_2.

$$\alpha_2(a_1,a_2) = \frac{\sum_{i=1}^{I'}\sum_{j=1}^{J_i'}(\mu_{B_{1Ti}}(t_{ij}) \wedge \mu_{B_{1Ti}}(t_{ij}))}{\sum_{i=1}^{I'}\sum_{j=1}^{J_i'}(\mu_{B_{2Ti}}(t_{ij}) \vee \mu_{B_{2Ti}}(t_{ij}))}$$

where $\mu(t_{ij})$ is the membership value of t_{ij} in the fuzzy set B_{kti}, $k=1,2$.

We combine the above two definitions to formulate a similarity function of two fuzzy-attributed graphs.

Definition 7: Let α be the similarity function of two FAGs G_1 and G_2 such that the underlying graph H_1 of G_1 is isomorphic to the underlying graph H_2 of G_2.

$$\alpha(G_1,G_2) = \frac{\sum_p \alpha_1(p,h(p)) \wedge_{q \geq p} \alpha_2(a(p,q),a(h(p),h(q)))}{\sum_p \alpha_1(p,h(p)) \vee_{q \geq p} \alpha_2(a(p,q),a(h(p),h(q)))}$$

where p,q denote the vertices of G_1 and h(p),h(q) denote the corresponding vertices of G_2 under the isomorphism h. The value of $\alpha(G_1,G_2)$ represents the degree of similarity between G_1 and G_2.

In recognition, let $B \in F(X)$ be the unknown input pattern to be classified and $A_i \in F(X)$, $i=1,2,...,n$ be the reference patterns by the above similarity function. A selection threshold δ is defined as follows:

Definition 8: Let G_1 and G_2 be two FAGs. G_1 is δ-isomorphic to G_2 if the degree of similarity $\alpha(G_1,G_2) \geq \delta$.

Then the unknown pattern B can be classified as reference pattern A_i if

$$\alpha(B,A_i) = \max_{1 \leq j \leq n} \alpha(B,A_j) \geq \delta.$$

It is well-known that conventional graph isomorphism can be solved by tree-search methods [5]. The time complexity of these algorithms are known to be NP-complete. Various attempts have been tried to reduce the time requirement [5,14]. In this paper, we propose an algorithm for finding the optimal δ-isomorphism between two attributed graphs by adopting the branch-and-bound algorithm. The proposed similarity measurement is used to act as the evaluation function in the branch-and-bound algorithm. The optimal isomorphism is the best matching in which the degree of similarity is maximized. This search algorithm expands fewer number of nodes and then requires less CPU time. Its pseudo-code is given below:

Algorithm : Similarity Measurement
Begin
for all reference pattern FAGs do
{
    initialize the node_list with the root node;
    set the first node in the list be the current node,p;
    while (the list is not empty) do
    {
      - compute $\alpha(p,h(p))$;
      - set the upper bound = MAX($\alpha(p,h(p))$);
      - insert in the list in decending order according to the value of $\alpha(p,h(p))$;
      - remove the current node and any node in the list whose $\alpha(p,h(p)) \leq$ upper bound;
      - add the upper bound to the total degree of similarity;
    }
    output the optimal_solution with the total similarity score
}
assign the unknown FAG into the reference pattern FAG with the greatest degree of similarity;
End.

During the search, the degree of similarity of all combination of vertice pairs on the same level are computed. Only the most similar vertice pair is extended one level deeper and all the other possible paths are pruned. The procedure repeats until all the vertices of the input numeral are considered. Since the path with maximum degree of similarity is chosen for extension, the path first reaches the last vertex of the input numeral is certain to be optimal. For example, Fig.1 shows the FAGs of an input numeral and a reference prototype. The complete search tree for the input FAG and the reference FAG based on the above algorithm is shown in Fig.2.

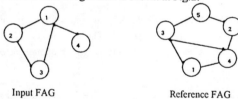

Input FAG   Reference FAG

Fig.1 The FAG of an input numeral and a reference prototype.

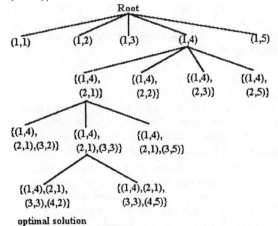

Fig.2 The search tree for the input FAG and the reference FAG.

4. Application to Handwritten Numeral Recognition

The numerals are inputted via a camera. The pattern obtained is a grey-scale one and it transforms into a binary pattern by thresholding. Thresholding is achieved by setting a threshold to the grey scale pattern. The binary pattern is represented by its skeleton via the Hilditch's approach [15]. Then some dominant points such as tips, corners and juctions are extracted from the skeleton, and the line segments connecting a pair of adjacent dominant points are measured in term of straightness and orientation. Each of the line segments are then classified into one of the primitive classes such as straight line, portions of circles or circles according to its membership grade on each class. In this application, a sufficient set of features [11] is used and shown in Fig.3.

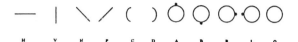

Fig.3 The sufficient set of features for handwritten numeral.

Based on these information, the FAG of the unknown numeral are constructed and classified into one of the reference prototypes by the proposed method. The overall block diagram of the handwritten numeral recognition system is shown in Fig.4.

Fig.4 The block diagram of the numeral recognition system.

In the training stage, a set of training samples from different writers are used to build up a knowledge database. The database includes each reference pattern FAGs with their membership grade on each primitive classes. Automatic supervised training is used to compute these membership grades for each reference pattern class. A set of 10 samples for each numeral class is used as training samples. On the other hand, 500 numerals from different writers are inputted to the recognition system and the percentage of correct classification is 93 percent. Some of the numerals are shown in Fig.5.

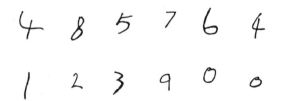

Fig.5 Examples of numerals.

5. Conclusion

In this paper, the adoption of fuzzy-attributed graph is suitable in coping with real world images which are always corrupted by noise and distortion. The fuzzy set theory provides a well-founded mathematical theory to encounter the uncertainty problems in pattern representation and classification. However, such extension will cause new problems in graph matching as we no longer check the equality of the fuzzy attributes. A new method of similarity measurement for FAGs is introduced to solve the problem of graph matching. This algorithm reduces the time complexity for graph matching by using the branch-and-bound approach. Besides, it is a general approach applicable to the recognition of various types of patterns. Finally, the proposed algorithm is applied to the recognition of handwritten numerals. The recognizer reduces as much as possible the written constraints while maintaining overall recognition rates to those existing methods.

REFERENCES

1. K.Fukunaga, "Introduction to Statistical Pattern Recognition ", New York : Academic, 1972.
2. K.S.Fu, "Syntactic Pattern Recognition and Applications." Englewood Cliffs, NJ : Prentice-Hall, 1982.
3. T.Pavlidis, "Structural Pattern Recognition.", New York : SpringerVerlag, 1977.
4. Wen-Hsiang Tsai and King-Sun Fu, "Error-Correcting Isomorphisms of Attributed Relational Graphs for Pattern Analysis", IEEE Trans. on SMC-9, No.12, Dec. 1979.
5. J.R.Ullmann,"An Algorithm for Subgraph Isomorphism.", J. ACM, Vol.23, No.1, 1976, pp. 31-42.
6. A.Sanfeliu and K.S.Fu,"A Distance Measure Between Attributed Relational Graphs for Pattern Recognition." IEEE Trans. on SMC-13, 1983, pp. 353-362.
7. A.K.C.Wong and D.E.Ghahraman,"Random graphs : Structural Contextual Dichotomy." IEEE Trans. on PAMI-2, 1980, pp. 341-348.
8. A.K.C.Wong and Manlai You,"Entropy and Distance of Random Graphs with Application to Structural Pattern Recognition", IEEE Trans. on PAMI-7, No.5, Sept. 1985, pp. 599 - 609.
9. M.A.Eshera and K.S.Fu, "An Image Understanding System Using Attributed Symbolic Representation and Inexact Graph Matching", IEEE Trans. on PAMI-8, No.5, Sept 1986.
10. Sankar K.Pal and Dwijesh K.Dutta Majumder,"Fuzzy Mathematical Approach to Pattern Recognition", John Wiley & Sons, 1986.
11. Pepe Siy and C.S.Chen,"Fuzzy Logic for Handwritten Numeral Character Recognition", IEEE Trans. on SMC-4, Nov. 1974, pp.570-575.
12. Sankar K.Pal, Robert A.King and Abdullah A. Hashim, "Image Description and Primitive Extraction Using Fuzzy Sets", IEEE Trans. on SMC-13, No.1, 1983.
13. W.Pedrycz,"Fuzzy Sets in Pattern Recognition : Methodology and Methods", Pattern Recognition, Vol.23, No.1/2, 1990, pp. 121-146.
14. A.K.C.Wong, M.You and S.C.Chan,"An Algorithm for Graph Optimal Monomorphism", IEEE Trans. on SMC-20 No.3, May/June 1990, pp. 628-636.
15. Nabil Jean Naccache and Rajjan Shinghal, " An Investigation into the skeletonization approach of Hilditch", Pattern Recognition, Vol.17, No.3, 1984, pp. 279-284.
16. K.P.Chan and Y.S.Cheung, "Fuzzy-Attribute Graph and Its Application to Chinese Character Recognition".

Automatic Target Recognition Fuzzy System for Thermal Infrared Images

Christiaan Perneel[*], Michel de Mathelin[†], and Marc Acheroy[*]

[*] Electrical Engineering Department, Royal Military Academy, 30 Renaissance avenue, 1040 Brussels, Belgium.
[†] Electrical & Computer Engineering Department, Carnegie Mellon University, Pittsburgh, PA 15213-3890.

Abstract

An expert system for automatic recognition of armored vehicles from short distance infrared images is presented in this paper. This expert system uses fuzzy logic to combine information coming from different heuristics and to guide the search of the solution to the recognition problem.

1 Introduction

The problem consists in identifying armored vehicles based on short distance 2-D infrared images. The major difficulty lies in the lack of knowledge of the position and orientation of the vehicle with respect to the camera. Indeed, it is completely unpractical to realize a close matching of the image of the vehicle with a template given all the possible positions and orientations of the vehicle, and also given all the possible vehicles. Therefore, the problem is divided in two subproblems where the first subproblem is the computation of the orientation and position of the vehicle based on a crude model of the vehicle and the second subproblem is the identification of the vehicle in a reference position and orientation.

A blackboard type expert system (*cf.* Nii [1, 2] or Hayes-Roth [3] for a definition and Murdock & Hayes-Roth [4] for an example of application) was designed to accomplish these two tasks. The first task, position and orientation detection, consists in putting a system of three axes, $\{X, Y, Z\}$, on the image of the vehicle according to predefined conventions. In our application, the conventions are the following:

- X **axis**: the line on the side of the vehicle between the wheel train and the ground.

- Y **axis**: the line on the front or on the rear of the vehicle between, respectively, the front or the rear wheels on each side of the vehicle and the ground.

- **origin**: the origin of the system of axes is, by convention, at the extremity of the X axis on the engine side if the image gives a side view of the vehicle, on the left of the Y axis if the image gives a front or rear view of the vehicle but the engine is on the front of the image, or on the right of the Y axis if the image gives a front or rear view of the vehicle but the engine is on the rear of the image.

- Z **axis**: the vertical direction of the vehicle, normalized on the wheel train height if the image gives a side view of the vehicle or on the distance between the floor of the vehicle body and the ground if the image gives a front or rear view of the vehicle.

An illustration of these conventions can be found in Fig. 1. The position and orientation detection is then

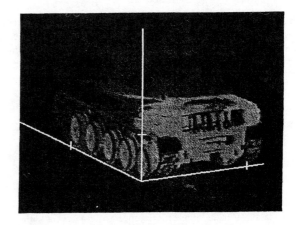

Figure 1: Axes conventions

divided into $N = 5$ subtasks. These 5 subtasks correspond to 5 different levels of increasing knowledge of the position and orientation of the vehicle.

- **Level 1**: Determination of one principal direction, either the direction of the X axis if the image gives a side view, or the direction of the Y axis if the image gives a front or rear view.

- **Level 2**: Determination of the line of the X axis if it is a side view or the line of the Y axis if it is a front or rear view.

- **Level 3**: Determination of the position of the origin.

- **Level 4**: Determination of the Z axis.

- **Level 5**: Determination of the third axis, either the Y axis if it is a side view, or the X axis if it is a front or rear view.

The following assumptions are made about the expert system considered in this paper

- Level 1 is the initial level of knowledge of the solution in the search process. When the expert system reaches the next level in the search process, some more information about a candidate solution is generated. A candidate solution consists in values for the N predefined characteristics or sets of characteristics (one per level). A candidate partial solution up to level i consists in values for i predefined characteristics or sets of characteristics. Level N is the last level of knowledge in the search process where a possible solution to the problem can be proposed.

- At each level i, at most n_i new possible partial solutions or, equivalently, new branches in the search tree are proposed. The n_i new candidate partial solutions consist in a candidate partial solution up to level $(i-1)$ combined with some incremental knowledge under the form of a possible value for one or more predefined characteristics of the solution. Consequently, the size of the candidate solution space $S \leq \prod_{i=1}^{N} n_i$. For our application, $\{n_1 = 8, n_2 = 30, n_3 = 2, n_4 = 8, n_5 = 8\}$. Therefore, the size of the solution space, $S = 30720$.

- At each level i, m_i different heuristics give a rating to the n_i new candidate partial solutions of level i. These ratings must be combined to give a global rating to the partial solutions. For our application, the number of heuristics is $\{m_1 = 3, m_2 = 12, m_3 = 2, m_4 = 3, m_5 = 6\}$.

The image processing algorithms involved in the computation of the different candidates and in the different heuristics are too time consuming to make possible an exhaustive exploration of the solution space. Therefore, a heuristic search strategy is implemented. Typically, the knowledge of an expert is required in order to guide the search toward the solution in a more efficient manner. This knowledge takes the form of heuristics and a search procedure using information of this sort is called heuristic search. At any given time in the search process all the possible paths are evaluated with the help of the heuristics and the most promising path according to the heuristics is selected. The heuristics are designed based on the knowledge and the experience of the expert. The main difficulty in this approach resides in finding an appropriate way to combine the knowledge coming from the heuristics into a unique evaluation function. Furthermore, as the search of the solution progresses, new paths must be compared with paths abandoned earlier. However, new paths are usually much more advanced in the building of a solution than old paths and it might become very difficult to find a common ground for comparison. In the section to follow, we will see how we can design a global rating function to evaluate the different candidates using fuzzy logic.

Once the first task of detecting the position and orientation of the vehicle is accomplished, it remains to identify the type of vehicle. This is done by defining characteristic details, such as, e.g., number of wheels, engine position, tracks size, exhaust system, for each vehicle to be recognized. The location and the shape of these characteristic details is known *a priori*. Therefore, templates can be created with their location on the vehicle specified. Since the position and orientation of the vehicle is known, it is sufficient to verify that the templates match the corresponding areas on the image. This is done by using pattern recognition techniques as cross-correlation, or neural networks (*cf.* Duda & Hart [5] and Carpenter & Grossberg [6]).

2 Fuzzy reasoning

The heuristics rate the candidate solutions based on some measurements or observations. Very often, the range of the measured variable is more important than its exact value. The measurement can be often classified in simple categories describing how well it fits the hypothesis that the candidate is the solution to the problem. For example, **Good**, **Average**, and **Bad** could be the categories or linguistic terms describing how the observation fits the hypothesis. Furthermore, the transitions between these categories (*i.e.*, **Good** and **Average** or **Average** and **Bad**) are often blurred. Therefore, there must be also a smooth transition in the rating of the different candidate solutions.

2.1 Modelization

The following assumptions are made to model uncertainty and fuzziness (based on the terminology used in, *e.g.*, Bellman & Zadeh [7], or Mamdani [8])

- The values which define uniquely a solution (*cf.* Section 1) belong to nonfuzzy support sets of universes of discourse like, for example, the k-th characteristic or set of characteristics of the solution. Let X_k be the set of possible values for the characteristics of the solution receiving a value at level k. Then, the X_k's are the nonfuzzy support sets, having a finite number, S_k where $S_k \leq \prod_{i=1}^{k} n_i$, of elements, of the universes of discourse "k-th characteristic or set of characteristics of the solution". Based on these definitions, the characteristics defining a candidate partial solution up to level k will be represented by the nonfuzzy support set

$$Y_k = (\ X_1, \ \ldots, \ X_k\ )$$

Therefore, a partial solution up to level i will be an element of Y_i and a complete solution will be an element of Y_N.

- Suppose that the n_i values $\{x_{ik}(y_{i-1})\}$, $k = 1, \ldots, n_i$, are all the elements depending on y_{i-1} which belongs to the nonfuzzy support set X_i, where y_{i-1} is an element selected in the previous levels nonfuzzy support set, Y_{i-1}. Suppose that there exist M different categories or linguistic terms (*e.g.*, **Good**, **Average** and **Bad**, $M = 3$), for describing to what degree x_{ik} fits the characteristics of the solution represented by X_i. These linguistic terms define fuzzy subsets of the universes of discourse "i-th characteristic or set of characteristics of the solution". Then, for each level i, there exists M membership functions (one per linguistic term)

$$f_i^l(x_{ik}, y_{i-1}) : X_i \times Y_{i-1} \to [0,1] \ k = 1, \ldots, n_i$$
$$x_{ik} = x_{ik}(y_{i-1})$$

which define fuzzy subsets associated to the linguistic terms, l, describing to what degree the values $\{x_{ik}(y_{i-1})\}$ belongs or fits *a priori* the i-th characteristic or set of characteristics of the solution represented by X_i. Let's define the vector f_i as the combination of these M membership functions, $\{f_i^l\}$, in one vector. For example, if the linguistic terms are (**Good**, **Average**, **Bad**), the fuzzy subset $[x_1, x_2, x_3] = f_i(x_{ik})$ expresses that x_{ik} belongs *a priori* to the **Good** fit category with a factor x_1, to the **Average** fit category with a factor x_2, and to the **Bad** fit category with a factor x_3, for the characteristics of the solution represented by X_i.

Without loss of generality, it can be assumed that the membership functions, f_i^l, are normalized, so that

Normality rule: $\max_x f_i^l(x) = 1$

Furthermore, to minimize the number of linguistic terms and avoid redundant categories, it is assumed that the M linguistic terms describe different (but possibly overlapping) degrees of fit. Consequently, if the degree of fit of one category is 1, then the degree of fit of the others must be 0. This translates into the following rule

Economy rule: $\sum_l f_i^l(x_{ik}) \leq 1 \quad \forall i, k$

Furthermore, if it is assumed that the linguistic terms span all the possible degrees of fit, then the following rule applies

Completeness rule: $\sum_l f_i^l(x_{ik}) \geq 1 \quad \forall i, k$

Consequently, if the three rules apply

$$\sum_l f_i^l(x_{ik}) = 1 \quad \forall i, k$$

Finally, by default if the *a priori* degree of fit is unknown, a constant membership function will be assumed, *i.e.*,

$$f_i^{\mathbf{Good}}(x_{ik}, y_{i-1}) = 1$$
$$f_i^{\mathbf{Average}}(x_{ik}, y_{i-1}) = 0$$
$$f_i^{\mathbf{Bad}}(x_{ik}, y_{i-1}) = 0$$
$$f_i(x_{ik}, y_{i-1}) = [\ 1,\ 0,\ 0,\ ]$$

- There are m_i different rating functions at level i (one per heuristic). Let $h_{ij}^l(p_{ij})$ be the j-th membership function of level i associated to the linguistic term l, where p_{ij} is an observation depending on the characteristics described by Y_i. Let $h_{ij}(p_{ij})$ be the vector of M membership functions $h_{ij}^l(p_{ij})$. Given a candidate partial solution up to level i, $y_i \in Y_i$, an observation $p_{ij}(y_i)$ is made and the j-th heuristic returns a fuzzy subset, $h_{ij}(p_{ij})$, describing how the observation $p_{ij}(y_i)$ fits the hypothesis that y_i is the solution up to level i.

Without loss of generality, it can be assumed that the membership functions, h_{ij}^l, are normalized, so that $\max_p h_{ij}^l(p) = 1$. Furthermore, it is assumed that the Economy rule applies, *i.e.*, $\sum_l h_{ij}^l(p) \leq 1$, $\forall i, j$ and $\forall p$.

Given a measurement p and assuming that $M = 3$, with linguistic terms (**Good**, **Average**, **Bad**), a typical heuristic membership function, $h(p)$, with an overlapping factor of 50%, may look like in Fig.2. The observation p belongs to the **Good** fit category

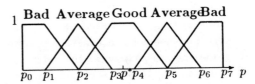

Figure 2: Heuristic fuzzy membership function, $h(p)$

with a factor 1 around a reference value, p^*, then slowly starts to belong to the **Average** category and finally to the **Bad** category outside this area. Finally, note that the overlapping factor between the different categories should be proportional to the fuzziness of the boundaries between categories. In his book, Kosko [9] recommends an overlapping factor of 25%. We use ourselves an overlapping factor of 50% in our application.

2.2 Rating function

A global rating function must be defined to follow an opportunistic search strategy. This global rating function should logically reinforce candidate partial solutions whose heuristics give mostly a **Good** rating and should disadvantage candidate partial solutions whose heuristics give mostly a **Bad** rating.

2.2.1 Heuristics rating functions

Each heuristics return a fuzzy vector of size M, the number of different linguistic terms. This fuzzy vector is made of the membership values for each of the M different linguistic terms. Now, these M different values must be combined in one unique value which is the rating given by the heuristic. This operation of transforming the fuzzy vector $h(p)$ in a unique nonfuzzy rating value is called defuzzification. This is usually done by assigning to each linguistic term, l, a rating membership function, $g^l(r)$, where r is the nonfuzzy rating value, obeying to the following rules

- $g^l: \Re \to [0, 1]$.

- $\forall\, r$, the rating must increase when the linguistic term expresses an improvement and must decrease when the linguistic term expresses a worsening. For example,

$$g^{\mathbf{Good}}(r) \geq g^{\mathbf{Average}}(r) \geq g^{\mathbf{Bad}}(r) \quad \forall\, r$$

- Assuming that the rating values must belong to the interval $[K_{\min}, K_{\max}]$ then $g^l(K_{\min}) = 1$ if l is the "worst degree of fit" linguistic term and $g^l(K_{\max}) = 1$ if l is the "best degree of fit" linguistic term.

For example, suppose that $M = 3$ and the linguistics terms are (**Good**, **Average**, **Bad**) and assume that the minimum rating is $K_{\min}$ and the maximum rating $K_{\max}$, then typical rating membership functions g^l are like in Fig.3. Once the rating member-

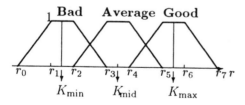

Figure 3: Rating membership functions

ship function have been defined, given an observation, p, given a heuristic with heuristic membership functions, $h^l(p)$, where l spans all the possible linguistic terms, we can define the cumulated rating membership function, $g_c(r)$, as

$$g_c(r) = \sum_l \min(g^l(r), h^l(p)) \quad \text{minimum inference}$$

For example, suppose that the the linguistics terms are (**Good**, **Average**, **Bad**), that $h(p)$ is defined as in Fig.2, that the observation p is such that $h(p) = [4/5, 1/5, 0]$, and that the rating membership functions are defined as in Fig.3, then $g_c(r)$ will look like in Fig.4. Now, the simplest defuzzification scheme is

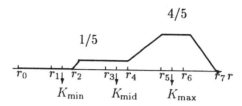

Figure 4: Cumulated rating membership functions

the *maximum-membership defuzzification* scheme defined as

$$R = \arg\,\max_r g_c(r) = \arg\,\max_r (\sum_l \min(g^l(r), h^l(p)))$$

However, this scheme has two important imperfections. First, R is not always unique (see for example, Fig.4). Second, most of the information in the shape of the rating membership functions, g^l, is lost. A better alternative is the well known *fuzzy centroid defuzzification* scheme defined as

$$R = \frac{\int_{-\infty}^{+\infty} r g_c(r)\, dr}{\int_{-\infty}^{+\infty} g_c(r)\, dr} = \frac{\sum_l \int_{-\infty}^{+\infty} r \min(g^l(r), h^l(p))\, dr}{\sum_l \int_{-\infty}^{+\infty} \min(g^l(r), h^l(p))\, dr}$$

For the example in Fig. 4, the fuzzy centroid defuzzifier returns

$$R = \frac{K_{\mathrm{mid}} \int_{-\infty}^{+\infty} \min(g^{\mathbf{Average}}(r), 0.2)\, dr + K_{\max} \int_{-\infty}^{+\infty} \min(g^{\mathbf{Good}}(r), 0.8)\, dr}{\int_{-\infty}^{+\infty} \min(g^{\mathbf{Average}}(r), 0.2)\, dr + \int_{-\infty}^{+\infty} \min(g^{\mathbf{Good}}(r), 0.8)\, dr}$$

2.2.2 Level rating function

Now, the individual heuristic ratings must be combined for each levels. Suppose that y_i is a candidate solution up to level i, the level rating function is the combination of the m_i heuristic rating functions h_{ij} and of the *a priori* rating function f_i where all these rating functions represents fuzzy subsets. A classical approach in fuzzy logic (see, *e.g.*, Bellman & Zadeh [7] and Mamdani [8]) consists in taking the minimum of the different membership functions, *i.e.*, if the fuzzy centroid defuzzification scheme is adopted

$$\tilde{R}_i(y_i) = \frac{\sum_l \int_{-\infty}^{+\infty} r \min(\min_j(h^l_{ij}(p_{ij}(y_i))), f^l_i(y_i), g^l(r))\, dr}{\sum_l \int_{-\infty}^{+\infty} \min(\min_j(h^l_{ij}(p_{ij}(y_i))), f^l_i(y_i), g^l(r))\, dr}$$

Unfortunately, as it was pointed out by Kosko in [10], this type of combination rule is often too pessimistic, since the rating is always based on the heuristic giving the worst rating, even when all the other heuristics give a good rating. Much better, in our opinion, is the weighted sum approach where a weight, w_{ij}, is given to each heuristics according to its reliability and where a weighted sum of all the different membership functions is done

$$\tilde{R}_i(y_i) = \frac{\sum_l \int_{-\infty}^{+\infty} r(w_i \min(f_i^l(y_i), g^l(r)) + \sum_{j=1}^{m_i} w_{ij} \min(h_{ij}^l(p_{ij}(y_i)), g^l(r))) \, dr}{\sum_l \int_{-\infty}^{+\infty} (w_i \min(f_i^l(y_i), g^l(r)) + \sum_{j=1}^{m_i} w_{ij} \min(h_{ij}^l(p_{ij}(y_i)), g^l(r))) \, dr}$$

This rating function does not take into account the rating obtained at the previous levels ($< i$). Only the heuristics of the current level are combined in this manner. We propose different methods to take into account the ratings from the previous levels: either a weighted sum of the different level ratings, or a global centroid of all the different levels, or even the minimum of the different level ratings.

1. Weighted sum level combination

$$R_i(y_i) = \sum_{k=1}^{i} \tilde{R}_k(y_k)$$

$$= \frac{K_k \sum_l \int_{-\infty}^{+\infty} r(w_k \min(f_k^l(y_k), g^l(r)) + \sum_{j=1}^{m_k} w_{kj} \min(h_{kj}^l(p_{kj}(y_k)), g^l(r))) \, dr}{\sum_{k=1}^{i}(\sum_{k=1}^{i} K_k) \sum_l \int_{-\infty}^{+\infty} (w_k \min(f_k^l(y_k), g^l(r)) + \sum_{j=1}^{m_k} w_{kj} \min(h_{kj}^l(p_{kj}(y_k)), g^l(r))) \, dr}$$

2. Centroid level combination

$$R_i(y_i) =$$

$$\frac{\sum_l \sum_{k=1}^{i} \int_{-\infty}^{+\infty} r(w_k \min(f_k^l(y_k), g^l(r)) + \sum_{j=1}^{m_k} w_{kj} \min(h_{kj}^l(p_{kj}(y_k)), g^l(r))) \, dr}{\sum_l \sum_{k=1}^{i} \int_{-\infty}^{+\infty} (w_k \min(f_k^l(y_k), g^l(r)) + \sum_{j=1}^{m_k} w_{kj} \min(h_{kj}^l(p_{kj}(y_k)), g^l(r))) \, dr}$$

3. Minimum level combination

$$R_i(y_i) = \min_{1 \leq k \leq i}(\tilde{R}_k(y_k))$$

$$= \min_{1 \leq k \leq i} \left(\frac{\sum_l \int_{-\infty}^{+\infty} r(w_k \min(f_k^l(y_k), g^l(r)) + \sum_{j=1}^{m_k} w_{kj} \min(h_{kj}^l(p_{kj}(y_k)), g^l(r))) \, dr}{\sum_l \int_{-\infty}^{+\infty} (w_k \min(f_k^l(y_k), g^l(r)) + \sum_{j=1}^{m_k} w_{kj} \min(h_{kj}^l(p_{kj}(y_k)), g^l(r))) \, dr} \right)$$

Which method should be selected depends on the application. For example, if it is critical to find the exact solution at each level and if the level have the same importance, the minimum combination is the most suitable method. However, if the different levels have different importance and if the different level characteristics X_i, $i = 1, \ldots, N$, are poorly related, it is much better to use the weighted sum method. Finally, if the different level characteristics X_i, $i = 1, \ldots, N$ are very interdependent, then the centroid combination method is better. In the case of our application the weighted sum level combination gives the best results. Finally, these level rating functions are unable to compare candidate partial solutions at different levels. To do so, a global rating function must be defined.

2.2.3 Global rating function

To compare two candidate partial solutions at different levels, the rating of the candidate at the lowest level should be extrapolated up to the highest level. In other words, an estimate of the rating that a candidate could possibly obtained at a higher level must be found. Therefore, to compare candidate solutions at every possible levels the rating of every candidates should be estimated up to the highest possible level, N. Assume that y_i is a candidate solution up to level i, then an estimate of

$$h_{kj}(p_{kj}(y_k)) \quad \text{and} \quad f_k(y_k) \quad k = i+1, \ldots, N$$

$$y_k = \begin{bmatrix} y_i^T & (\text{unknown subvector})^T \end{bmatrix}^T$$

must be found. Let's define $E\{h_{kj}\}(y_i)$ and $E\{f_k\}(y_i)$, $k = i+1, \ldots, N$ as these estimates. Then, they could be computed different ways depending on the type of behavior which is desired for the expert system.

1. **Best case approach:** For each of the heuristics and *a priori* ratings to be estimated, the maximum value is always assumed. For example, if the linguistic terms are (**Good, Average, Bad**)

$$E\{h_{kj}\}(y_i) = \begin{bmatrix} 1, & 0, & 0, \end{bmatrix}$$
$$E\{f_k\}(y_i) = \begin{bmatrix} 1, & 0, & 0, \end{bmatrix}$$

This is a cautious approach. Indeed, the candidate partial solutions at the lower levels might be advantaged with respect to the candidates at the higer levels.

2. **Arbitrary value approach:** If a less cautious approach is desired, expected values should be used. Estimates are selected arbitrarily and then later tuned for the application. For example, if the linguistic terms are (**Good, Average, Bad**)

$$E\{h_{kj}\}(y_i) = \begin{bmatrix} \bar{h}_{kj}^{\textbf{Good}}, & \bar{h}_{kj}^{\textbf{Average}}, & 0, \end{bmatrix}$$
$$E\{f_k\}(y_i) = \begin{bmatrix} \bar{f}_k^{\textbf{Good}}, & \bar{f}_k^{\textbf{Average}}, & 0, \end{bmatrix}$$

3. **Expected value approach:** Ofcourse, if the probability distributions of the observations p_{kj}, $k = i+1, \ldots, N$, are known for a partial solution equal to y_i, then the expected values could be directly computed. So, let $d_{kj}(p)_{|y_i}$, $k = i+1, \ldots, N$, be the corresponding density of probability functions, then

$$E\{h_{kj}\}(y_i) =$$
$$\begin{bmatrix} \int_{-\infty}^{+\infty} h_{kj}^{\textbf{Good}}(p) d_{kj}(p)_{|y_i} dp, \\ \int_{-\infty}^{+\infty} h_{kj}^{\textbf{Average}}(p) d_{kj}(p)_{|y_i} dp, \\ \int_{-\infty}^{+\infty} h_{kj}^{\textbf{Bad}}(p) d_{kj}(p)_{|y_i} dp \end{bmatrix}$$

This is basically equivalent to the computation of the probability of the fuzzy event: "observation p_{kj} is **Good**", as defined in Bellman & Zadeh [7].

Based on these estimates, the global rating function is defined the following way. Suppose that y_i is a candidate solution up to level i

Weighted sum level combination

$R(y_i) =$

$$\sum_{k=1}^{i} \frac{K_k}{\sum_{k=1}^{N} K_k} \frac{\sum_l \int_{-\infty}^{+\infty} r(w_k \min(f_k^l(y_k), g^l(r)) + \sum_{j=1}^{m_k} w_{kj} \min(h_{kj}^l(p_{kj}(y_k)), g^l(r))) \, dr}{\sum_l \int_{-\infty}^{+\infty} (w_k \min(f_k^l(y_k), g^l(r)) + \sum_{j=1}^{m_k} w_{kj} \min(h_{kj}^l(p_{kj}(y_k)), g^l(r))) \, dr} +$$

$$\sum_{k=i+1}^{N} \frac{K_k}{\sum_{k=1}^{N} K_k} \frac{\sum_l \int_{-\infty}^{+\infty} r(w_k \min(E\{f_k^l\}(y_i), g^l(r)) + \sum_{j=1}^{m_k} w_{kj} \min(E\{h_{kj}^l\}(y_i), g^l(r))) \, dr}{\sum_l \int_{-\infty}^{+\infty} (w_k \min(E\{f_k^l\}(y_i), g^l(r)) + \sum_{j=1}^{m_k} w_{kj} \min(E\{h_{kj}^l\}(y_i), g^l(r))) \, dr}$$

3 Experimental results

Trapezoidal type rating functions, as in Fig.2, are used as rating functions. They are 3 different linguistic terms selected: **Good, Average**, and **Bad**. The overlapping factor is 50% for the heuristics and the rating membership functions. Finally, the parameters of the rating functions are tuned to maximize the quality of the results. The minimum inference is used, with the centroid defuzzifier for the level ratings, and the weighted sum level combination with the arbitrary value approach (for the estimation of the unknown ratings) for the global rating.

The results obtained for a database containing infrared images of 8 vehicles in 16 different positions are shown in Table 1. Perfect detection consists in finding the exact system of 3 axes. Almost perfect detection consists in finding the 3 axes directions, with small normalization or position errors. Sufficient detection consists in finding at least 2 axes (this is, in fact, sufficient for the vehicle identification). Partial detection consists in finding at least the direction of 2 axes. Even if this is not enough to identify directly the vehicle, the correct two axes are among the limited number, (< 240), of candidates having these same 2 directions. Once the first task of detecting the position and orientation of the vehicle is accomplished, it remains to identify the type of vehicle. This is done by defining characteristic details, such as, *e.g.*, number of wheels, engine position, tracks size, exhaust system, for each vehicle to be recognized. The location and the shape of these characteristic details is known *a priori*. Therefore, templates can be created with their location on the vehicle specified. Since the position and orientation of the vehicle is known, it is sufficient to verify that the templates match the corresponding areas on the image. Then, a final rating is given to the proposed solution based on the number of matching details and on the quality of this match. If the final rating is not judged sufficiently large, the expert system backtracks to the position and orientation task to propose another candidate solution.

Quality of detection	Percentage
Perfect	83.6 %
Almost perfect	88.8 %
Sufficient	94.0 %
Partial	94.8 %

Table 1: Position and orientation detection results

References

[1] H. Penny Nii. Blackboard systems: the blackboard model of problem solving and the evolution of blackboard architectures. *The AI Magazine*, pages 38–53, Summer 1986.

[2] H. Penny Nii. Blackboard systems: blackboard application systems, blackboard systems from a knowledge engineering perspective. *The AI Magazine*, pages 82–91, August 1986.

[3] B. Hayes-Roth. A blackboard architecture for control. *Artificial Intelligence*, 26(3):251–321, 1985.

[4] J. L. Murdock and B. Hayes-Roth. Intelligent monitoring and control of semiconductor manufacturing equipment. *IEEE Expert*, 6(6):19–31, 1991.

[5] R. O. Duda and P. E. Hart. *Pattern classification and scene analysis*. John Wiley & Sons, New-York, NY, 1973.

[6] G. A. Carpenter and S. Grossberg. The art of adaptive pattern recognition by a self-organizing neural network. *Computer*, 21:77–88, 1988.

[7] R. E. Bellman and L. A. Zadeh. Decision making in a fuzzy environment. *Management Science*, 17:141–164, 1970.

[8] E. H. Mamdani. Application of fuzzy logic to approximate reasoning using linguistic synthesis. *IEEE Trans. on Computers*, C-26:1182–1191, 1977.

[9] B. Kosko. *Neural networks and fuzzy systems*. Prentice-Hall, Englewood Cliffs, NJ, 1992.

[10] B. Kosko. Fuzzy knowledge combination. *Int. J. of Intelligent Systems*, I:293–320, 1986.

A vowel recognition using adjusted fuzzy membership functions

Sung-Soon Choi and Kyung-Whan Oh

Artificial Intelligence Research Lab., Department of Computer Science,
SoGang Univ., C.P.O.BOX 1142 Seoul 100-611, Korea

ABSTRACT

In this paper, we propsed a two stage recognition procedure using adjusted fuzzy membership functions. In the first stage, recognition procedure inspects the rough feature in unknown patterns, and generates alpha-cut set by the adusted fuzzy membership function. If ambiguity is found in the first stage, the second stage procedure is invoked. Also weight values for calculating the distnace are generated at each stage. An adjusted fuzzy membership function consists of a conventional fuzzy membership function and a fuzzy membership function changed by objective conditions.

I. Introduction

In conventional pattern matching techniques for voice recognition, it is difficult to expect the correct recognition because voice patterns include ambiguous features such as noise, natural characteristics of the speaker, and inaccurate articulation. As an approach solving this problem, researches using fuzzy logic have been studied. Sankar K. Pal[1] proposed vowel and speaker recognition using fuzzy membership functions representing distance measure, Fujimoto[2] has developed a word recognition system through fuzzy membership function about formant locations, and De Mori[3] proposed phonetic and phonemic labeling of continuous speech thru verification of hypothesis using a fuzzy algorithm, Francisco Casacuberta[4] attempted to extract phoneme using fuzzy automata and hierachical structure of knowledge representation. Michel Lamotte[5] introduced a phonemerecognition method using uncertainty degrees and fuzzy relations.

The method proposed in this paper has a two stage recognition procedure. The transition of the first stage to the second stage is determined by alpha-cut set of adjusted fuzzy membership functions. In the second stage, we calculate weight values according to alpha-cut set generated in the first stage, in order to process the detailed pattern matching. In general, when pattern matching techniques are applied using in one-stage, it occasionally fails to recognize unknown patterns producing ambiguous results. To solve it generally, recognition procedure by multi-stage is recommended. In muti-stage recognition, we must find the processing method at each stage and the criteria to decide whether to transit from the current stage to another or not. First-stage recognition procedure classifies reference patterns according to rough features. As the number of transitions becomes greater, recognition procedure inspects more important features of reference patterns in detail. Therefore, the recognition rate is enhanced.

II. Feature Extraction

A feature vector used in our approach is a smoothed log spectrum. To extract the feature vector, the processing mechanisms such as FFT, normalization, and smoothing is needed. In this section, we discuss normalization and smoothing methods process.

1. Normalization.

In a general normalization method, the element of feature vector is a real value which has 0 thru 1 according to dividing the value of each element by maximum value in feature vector. But this method has not only overheads like floating point calculation but also the drawback that the distance value becomes higher. Therefore we get the feature vector which has integer value using quantization techniques in Fig-1. Formally, we define the normalization formula as

$$SMIN = \min_i (S_i) \quad 0 <= i < N$$

$$SMAX = \max_i (S_i) \quad 0 <= i < N$$

$$FV_i = \frac{(LV - 1)(S_i - SMIN)}{(SMAX - SMIN)} \quad 0 <= i < N$$

where FV is the feature vector
LV is the number of quantization level
N is the half of FFT length

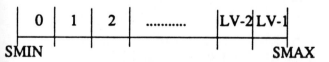

Ffg-1 The normalization using quantization

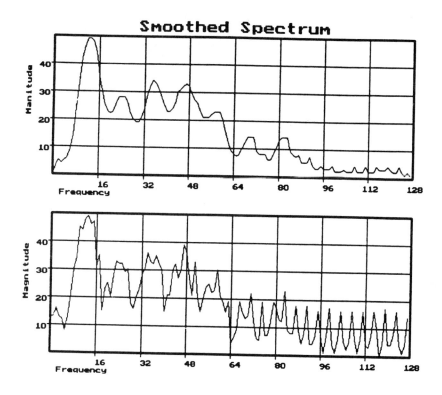

Fig-2. The approximate smoothed spectrum

2. Smoothing

In general, smoothed spectrum can be calculated through cepstral analysis. But this method has an overhead required to process FFT and inverse FFT procedures. To avoid it, we get the approximate smoothed spectrum using the following algorithm by linear interpolation techniques.

```
SMOOTHED_SPECTRUM( FV )
BEGIN
  WHILE FV is modified
    FOR i = 0 TO N-1
      IF FV_i <= FV_{i-1} AND FV_i <= FV_{i-1} AND
         ( (FV_{i-1} - FV_i) > th OR
           (FV_{i+1} - FV_i) > th )
      THEN
           FV_i = ( FV_{i-1} + FV_{i+1} ) / 2
      ENDIF
    END
  END
END
```

The above algorithm repeats linear interpolation until the given condition is satisfied. And it stops after finite iterations until FV value is not changed. Fig-3 shows the approximate smoothed spectrum resulted from the above algorithm.

III. Adjusted fuzzy membership function

Fuzzy membership functions that are determined subjectively have some problems. For example, consider a fuzzy membership function labled as 'cold'. This concept 'cold' must be changed by season and region, because 'the concept in autumn differs from one in winter, and the concept in Alaska differs from one in Florida.

In fuzzy pattern matching method, the same problem exists. When fuzzy membership function defined as
$$\mu(x) = ((1 + x/E)^F)^{-1}$$
we consider distances in Fig-3 that are calculated by three unknown patterns with the same attributes and five reference patterns. If we use MIN-MAX operation of fuzzy membership function, the result is the 5th pattern. If we use MAX-MAX operation of it, the result is the 4th pattern. But we know that the 1st pattern has a very high possibility.

reference patterns

	1	2	3	4	5
1	9	20	16	18	13
2	8	12	17	3	10
3	11	23	20	13	15

unknown patterns

fig-3 distance table

To solve this problem, fuzzy membership functions must be adjusted using objective conditions. We adjust fuzzy membership function using statistical approach. Adjusted fuzzy membership functions defined as

$$xmax = MAX(X)$$
$$xmin = MIN(X)$$
$$amf(x) = ((1 + x/E)^F)^{-1}$$
$$rmf(x) = 1 - (x - xmin) / ((xmax - xmin) * 2)$$
$$\mu_{ad}(x) = MIN(amf(x), rmf(x))$$
$$\text{where } x \in X$$

Amf(x) is a kind of conventional fuzzy membership function. It is a kind of threshold. Rmf(s) is adjusted by xmax and xmin value. A set X consist of distances with each reference pattern. Xmax and xmin is a maximum and minimum value in set X.

IV. Recognition

The vowel recognition system in this paper has a two stage recognition procedure that uses adjusted fuzzy membership function of distance measure. First-stage recognition procedure generate the crisp set by alpha-cut of fuzzy membership degree that calculate by distance of each reference pattern. If the number of elements in crisp set is greater than one, it transits to the second stage recognition because an unknown pattern is ambiguous. In this section, we refer to the reference pattern and the weight generation.

1. Reference pattern

Generation of a reference pattern is very important for correct recognition. It would be idealif each vowel had an unique reference pattern. But approximate smoothed spectrum that is the reference pattern in this system includes not only feature of vowel but also natural characteristics of speaker and noise. It is difficult to remove these components from the spectrum. Fig-4 shows that reference patterns in this system consist of m vowels that are articulated by n speakers. Therefore each reference pattern has the relation of column and row. We generate weight values using these relations.

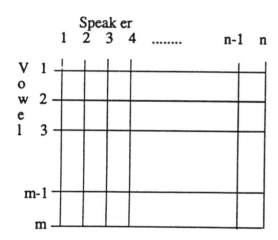

Fig-4. The structure of reference patterns

2. Weight generation for distance measure

We use the following distance measure that Sankar K. Pal[1] used.

$$d(UP, RPm) = \bigwedge_{n} \left(\sum_{i} (W_i (up_i - rp_{imn})^2)^{0.5} \right)$$

where UP is an unknown pattern and
RP is a reference pattern of mth vowel

The ambiguity of reference patterns referred to in the above can be reduced according of generating weight values of the above distance measure. We calculate the weight value according to each relation to reference pattern using statistical approach. For example, in case of m vowels by the same speaker, lets consider the meaning of variance of each frequency component in vowel spectrum. If the variance is greater than others, it means that the frequency component of variance varys frequently. Therefore features of each pattern may exist in frequency components of variance that is greater than others. We extend this scheme for n speakers as

$$VAR_i = \bigwedge_{n} (\sigma_{m} (RP_{in}))$$

$$W_i = VAR_i / \max_{i}(VAR_i)$$

In fig-5 generated by the above scheme, lower frequency components have a small weight value, because most patterns are similar. Also higher frequency components are eliminated. It means that weight values are enhanced.

In our method, weight values are calculated at each stage to improve the recognition rate. The second stage recognition procedure calculates weight values by vowels included in the alpha-cut set.

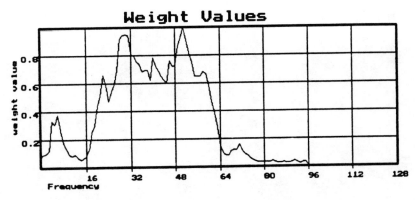

Fig-5. The result of the weight generation

3. Recognition procedure

Recognition procedure consists of Step I and Step II. First stage recognition procedure calculates distance values with an average value for unknown patterns. This means that it classifies reference patterns. The second stage recognition procedure calculates all distance values by each unknown pattern in case that the first stage recognition has ambiguity.

a. recognition step I

1: $AFV = \sum_j FV_j / NFR$ /* NFR is the number of frames */
2: $DST_m = d(AFV, RP_m)$
3: $mf1_m = \mu_{ad}(DST_m)$
4: alpha-cut by TH
5: IF the number of elements in alpha-cut set > 1
    THEN CALL recognition step II
6: MAX ($mf1_m$) is result of this system

b. recognition step II

1: weight generation according to alpha-cut set
2: $IDST_{jm} = d(FV_j, RP_m)$
3: $DST_m = \sum_j IDST_j / NFR$ /* NFR is the number of frames */
4: $mf2_m = \min(\max_j(\mu_{ad}(IDST_{jm})), \mu_{ad}(DST_m))$
5: $mf1_m = \min(mf1_m, mf2_m)$
6: return

V. Experimental results

Fig-6 shows the structure of this system. We consider each step roughly. In the recording step, a voice signal is recorded using 20kHz sampling rate. The segmentation step determines the start and the

end points using log energy of the input signal and extracts significant frames.
The preprocessing step emphasizes high frequency components of each frame and put a hamming window on it. The log FFT step gets log magnitude spectrums of each frame using a 256 FFT algorithm. In the normalization step, each spectrum is normalized using quantization techniques. The smoothing step gets approximate smoothed spectrum
using simple algorithm by linear interpolation techniques. Recognition step I and II are processed by distance measure and threshold value is 0.75. This system was implemented on IBM/PC 386 with DSP board using TMS320C25.

Reference pattern was generated by /a/,/e/,/i/,/o/, and /u/ that was articulated by female and male speakers. About 98 percent of /a/ patterns were recognized in recognition step I because these differ from the other patterns explicitly. More than 70 percent of /i/, /e/,
and /o/ were recognized in recognition step I, and about 90 percent in recognition step II. But /u/ was recognized only about 30 percent in recognition step I and 80 percent in step II, because the /u/ pattern was very similar to /o/.

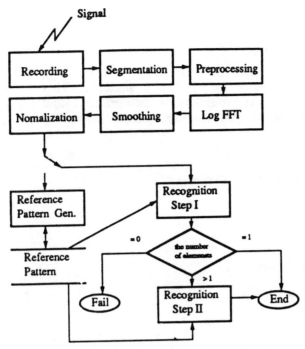

Fig-6. The structure of the vowel recognition

VI. Conclusions

In ambiguous patterns where we can not extract features without difficulty, we proposed the two stage recognition procedure using adjusted fuzzy membership functions. In the first stage, recognition procedure inspects the rough feature in unknown patterns, and generates alpha-cut set by the adusted fuzzy membership function. If ambiguity is found in the first stage, the second stage procedure is invoked. Also, we proposed the generation of the weight values and an adusted fuzzy membership function. To improve system performance, weight values are generated at each stage. An adjusted fuzzy membership function consists of a conventional fuzzy membership function and a fuzzy membership function changed by objective conditions. This method could be applied in other field. For the futher research, we are considering approaches extended n stage procedure.

VII. Reference

[1] Sankar K. Pal and Dwijesh Dutta Majumder, "Fuzzy Sets and Decisionmaking Approaches in Vowel and Speaker Recognition," IEEE Transactions on Systems, Man, and Cybernetics, pp 625-629, August 1977
[2] Jun-ichiroh Fujimoto, Tomofumi Nakatani and Masahide Yoneyama, "Speaker-independent Word Recognition Using Fuzzy Pattern Matching," Fuzzy Sets and Systems, Vol 32, pp 181-191, 1989
[3] Renato de Mori and Pietro Laface, "Use of Fuzzy Algorithms for Phonetic and Phonemic Labeling of Continuous Speech," IEEE Transactions on Pattern Analysis and Machine Intelligence, Vol. PAMI-2, No2, pp 136-148, March 1980
[4] Francisco Casacuberta and Enrique Vidal, "Interpretation of Fuzzy Data by Means of Fuzzy Rules with Applications to Speech Recognition," Fuzzy Sets and Systems, Vol 23, pp 137-389, 1987
[5] Michel Lamotte, Lucien Bour, Gerard Hirsch, "Fuzzy Phoneme Recognition," Fuzzy Sets and Systems, Vol 28, pp 363-374, 1988

Chapter 14: Management

DYNAMICS AND FUZZY CONTROL OF A GROUP

Kenji Kurosu*, Tadayoshi Furuya**, Masaaki Nakamura*, Hiroshi Utsunomiya* and Mitsuru Soeda**

* Kyushu Institute of Technology
Kawazu, Iizuka-shi, Fukuoka-ken, 820, JAPAN
** Kitakyushu College of Technology
5-20-1, Shii, Kokura-minami-ku, Kitakyushu, 803, JAPAN

Abstract

Some investigations about group behaviors are studied to obtain more efficient rules of guidance, some improvements of policies in the panics, etc. This paper proposes a simulation method for analyzing group movements by assuming dynamic equations of individual members, and a guidance strategy for controlling the members of the group. The mathematical models and the fuzzy controller which describe the group's actions are presented.

Some simulations of group behaviors, assumed that there are 10 members in the group with one leader, were carried out with some initial states. Some individuals became unable to follow the path of the leader because the group inertia makes difficult for them to follow the group. It also proved that the member's positions at the initial stage affected the results. It is found that the guidance failed unless the leader's path is carefully planed. It is concluded that the behaviors of the group can be simulated by the proposed mathematical models.

1. Introduction

There have been some investigations about group behaviors, or some experimental trials concerning how individuals behave in a certain situation. These studies, based on dynamic characteristics of the members in some groups, aim to obtain more efficient rules of guidance, some improvements of policies in the panics, or necessary leader's requirements.

In the cases where flocks of sheep, flows of cars, or crowds of people have to be guided towards some designated directions, the behaviors of members are too complex to be predicted or analyzed. Simplifying each member's movements into one group dynamic equation makes it easier to analyze such as the group's moving directions, speeds, etc. But it is impossible to find each member's behavior, or the behaviors of strayed members left from the group. Therefore, it is necessary to focus on each member of the group for determination of the group movements. Some studies, analyzed by solving the dynamic equations of the individual members about pedestrians[1],people in panic[2],marine traffic[3][4],and car's traffic control[5], are reported. But non of them discussed completely about the important roles of leaders.

This paper proposes a simulation method for analyzing group movements by assuming dynamic equations of individual members, and a guidance strategy for controlling the members of a group. The dynamic equations of each member are assumed on the situations, which include the relationship between the members and the combined forces affected on the individuals. The simulation provides an analyzing method for the group behavior, which leads to controlling the members belonged to the group. Also Fuzzy controller assigned to the leader characteristics in the group are simulated and studied for the guidance strategy of the groups with the rules of the leadership stated as the results of simulation.

2. Models for a Group

2.1 A Model for Individuals in a Group

Assumed that a leader exits to guide the individuals in a group toward a designated path, the individual members move as determined by some dynamic equations shown in the paper[1],[2]. The paper [2] has presented two cases: One is about mice, and the other is about human beings, where nothing but the coefficients differ in the same dynamic equations. Here, the case about mice is simulated because the group of men can be too easily controlled by the leader due to the coefficients.

The movements of the individual are governed by the following dynamic equations.

$$m_i \cdot \ddot{x}_i + v_i \cdot \dot{x}_i = F_i \qquad (1)$$

,where i corresponds to the i-th individual member of the N member group. m_i represents the mass of the i-th member, and v_i, x_i, F_i represents viscous resistances, the vector of the positions, respectively. F_i is the force applied to the mass caused by three different kinds of components. The resultant force F_i is determined by the nonlinear functions and each component has different types of causes, i.e the different forcing functions as the followings.

The force acted on the i-th member is divided into three kinds as follows.

(a) the force F_1 which makes the members stick together and form a forward moving group.
(b) the force F_2 influenced by the environment
(c) the disturbance F_3

Furthermore, the force F_1 are composed of the following detailed forces of four kinds.

(a1) the force F_a which drives the member move forward.
(a2) the force F_b which makes the member stick together.

(a3) the force F_c which makes every member move to the same direction.
(a4) the force F_d which help the members avoid collisions.

The force F_2 is caused by two kinds of forces
F_w: repulsion force from the wall.
F_m: Force resulted by the past influences like the history between the position of the leader and the i-th member.
F_3 is the random disturbance.

The forces are represented as follows,

$$F_{ai} = a \frac{\dot{x}_i}{|\dot{x}_i|} \quad (2)$$

where a is a positive constant., and $\dot{x}_i$ is the i-th member's velocity.

$$F_{bi} = \sum_j \frac{c(r_{ij}, \phi_{ij}) \cdot (x_j - x_i)}{r_{ij}} \quad (3)$$

where r_{ij}: the distance between i-th and j-th.
ϕ_{ij} : the angle formed as shown in Fig.1.
$c(r_{ij}, \phi_{ij})$: the nonlinear function.

$$F_{ci} = \sum_j \frac{h(r_{ij}, \phi_{ij}) \cdot (\dot{x}_j - \dot{x}_i)}{M_c} \quad (4)$$

where r_{ij}, ϕ_{ij}'s are the same as Eq.3
$h(r_{ij}, \phi_{ij})$: the nonlinear function
M_c: the number of the members influenced on the i-th members.

The force which makes the members to avoid collisions are

$$F_{di} = \frac{1}{M_d} \sum_j \frac{k \cdot l_{ij} \cdot e_{Pij}}{t_{Pij}} \quad (5)$$

as shown in Fig.2.

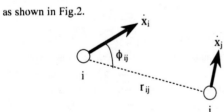

Fig.1 The i-th and j-th members

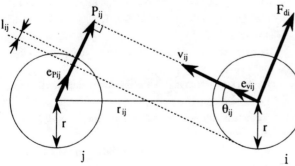

Fig.2 The force F_{di}

2.2 A model for the leader.

It is assumed that the leader moves by the same dynamic equation, but forces, being different from the member's ones, are determined by a fuzzy controller with the leader's rules. The model for the leader is shown in Fig.3. The leader detects the positions and velocities of the group to determine its own courses. The dynamic equation is represented as

$$\dot{x}_{k+1} = \frac{\Delta t \cdot F + m \cdot \dot{x}_k}{m + \Delta t \cdot v} \quad (6)$$

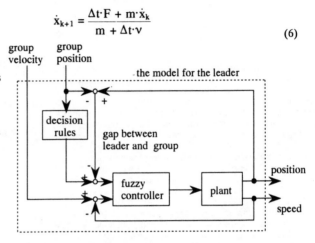

Fig.3 The model for the leader

The rules of behaviors give the reader's own desired behaviors, which the leader sets its targets of behaviors. To simplify these rules, the leader's behaviors are restricted to two actions; one is moving forward and another is moving backward. It allows the leader to move on a straight line called leader 's path. On this leader's path, the leader decides the gap between the top of the group and its own position. With these rules where the leader maintains its speed as fast as the group follows, the leader behaves as determined by the gap and the group velocity.

The objects such as the gap and its speed are controlled by the fuzzy controller, which gets errors and determines the output by some fuzzy rules. Their membership functions and rules are shown in Fig.4 and Table 1 . The desired values of the gap are determined as

$$D_{cont} = \begin{cases} 3.5 & (0.0 < E \leq 1.5) \\ 3.0 & (1.5 < E \leq 9.0) \\ -1.5 & (9.0 < E \leq 20.0) \end{cases} \quad (7)$$

where E is the distance between the leader and the center of the group. The value 20 in Eq.7 comes from 40, the width of the passage. So, the rules of behaviors are summarized as
(1) The gap between the leader and the center of the group is controlled.
(2) The speed of the leader is maintained as fast as the speed of the group.

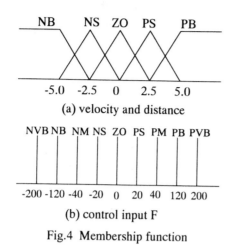

(a) velocity and distance

(b) control input F

Fig.4 Membership function

Table 1 Fuzzy rule

		distance				
		NB	NS	ZO	PS	PB
speed	NB	PVB	PB	PM	PS	ZO
	NS	PB	PM	PS	ZO	NS
	ZO	PM	PS	ZO	NS	NM
	PS	PS	ZO	NS	NM	NB
	PB	ZO	NS	NM	NB	NVB

3. Simulation

By using the model of proposed in the previous section, some simulations on guidance of group were experimented, assumed that there are 10 members in the group with one leader. In these experiments, the simulations were carried out with some initial states, where the members of group were scattered at random and the leader was placed in front of the group. The experiments tried to find whether the leader guided the individuals towards a designated path in a certain passage. Three kinds of passages, a straight path and two different ones with corners, were used in the experiments. Changing the initial positions of individual members, simulations were performed ten times on each case.

3.1 Straight Passage

On the straight passage in Fig.5, the results of simulations are shown in Tab 2 No.A. The guidance is called successful when no members are strayed from the guided direction. The failure in guidance is shown in Fig.9. In these figures, a line through the center represents a path of the leader, and the circles on the path represent the position of each individual member at some sampling time. As shown in Tab 2 No.A, the overall successful rate is 70%.

In Fig.9, an individual member, who is apart from group at the initial stage, becomes gradually strayed from others, because other members of the group have accumulated forces from each other to accelerate them, but on the other hand, a solitary member has no forces to accelerate itself.

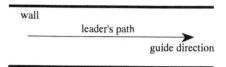

Fig.5 Straight passage

3.2 Passage with a gentle angle corner

A simulation case on a passage with a 45 degree angle corner is performed, where a leader guides the group to the direction of the arrow as shown in Fig.6. The results of ten examples are shown in Tab 2 No.B. The failure of guidance is shown in Fig.10.

The experiment results show that a corner with a gentle angle does not affect on the failure rate in guidance of the leader.

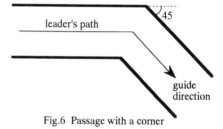

Fig.6 Passage with a corner

3.3 Passage with two corners of 90° angle (1)

On the passage with two turns of a 90 degree angle, simulations on guidance of group in the direction of the arrow were experimented. The results are shown in Table 2 No.C. The failure of guidance are shown in Fig.11.

The experiment results show that the successful cases decreased extremely in comparison with the previous experiments. The reasons are discussed as follows.

First, each individual became unable to follow the path of the leader at first corner, so the paths of individual went beyond the path of the leader. This is caused by the leader's sudden change in its direction. When the leader changes its direction suddenly, the force from the leader affects and makes some members to turn, but all the other members are influenced from each other; the forces produce the group inertia and overshoot damping.

Second, as in Fig.11, the group turned at the first corner without strayed members, but at the second turns, the group could not follow the leader. The typical example of leader and a member who turned into left direction at the second corner is shown in Fig.7. Some strayed member, positioned inside the path of the leader before the leader changed his course, passed the leader. Then the strayed member lost sight of the leader completely. Those strayed members' courses influenced on the entire members of the group. These results show that the failure in guiding the group depends on the course of the leader.

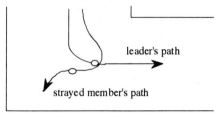

Fig.7 A strayed member

3.4 Passage with two corners of 90° angle (2)

Fig.8 shows that the leader turned at an angle of 45°, twice at each corner. The results in these cases are shown in Table 2 No.D as the successful percentage up to 60%. In comparison with No.C, the leader guided the members better in this way. Some examples of the results are shown in Fig.12.

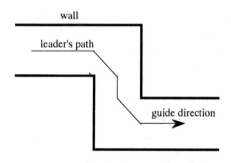

Fig.8 Passage with two corners

The leader failed to guide in the experiment No.D, though it succeeded in guidance No.A, No.B, and No. C. It is because a strayed individual passed the leader, lost his direction and became unable to catch the group. The failure cases of simulation No.1, No.2 and No.6 are not caused by the shape difference of passages since they failed even in the straight passage. Examining closely on the individual member who separated from the group leads to the comment that the initial state of each member causes outcomes of the strayed members.

The initial states of the individual members on the straight passage are shown in Fig.13, where ● represents the position of each member, ■ represents the position of the leader, and × represents the center of the group. Fig.13(a) is the failed case, and Fig.13(b) is the successful case. The lines on each dot represent the vectors of the velocity of each member. Every vector has the same magnitude but different directions.

Fig.13 indicates that the individual in the success cases gathered around the center of the group, but some members in the failed case were placed far from the center. The member marked by the arrow apart from the center became strayed in this simulation. It may be concluded that a member apart from the center of the group at the initial stage becomes strayed.

4. Conclusion

A proposed method is studied for analyzing group behaviors described by the dynamic equations of the individual members guided by a leader. The experimental consideration for the mathematical method leads to the following conclusion.

(1) Simple mathematical model for the members and the leader was introduced.
(2) The behaviors described by the fuzzy rules was studied to check whether the leader can guide the individual members.
(3) In some cases, some individuals of the group become strayed.
(4) The group inertia was proved to be responsible for the failures of guidance.
(5) The member's positions at the initial stage affected the results. It is difficult for the individual placed far from the center to follow the group.
(6) There are failures of guidance unless the leader's path was carefully planed.
(7) The group behaviors can be analyzed by obtaining the mathematical models and their fuzzy controller.

References

[1] Helbing,D: A mathematical model for the behavior of pedestrians, Behav. Sci.,36, 4, pp.298-310 (1991)
[2] Hirai, K.,Tarui,K.: A simulation of the behavior of a crowd in panic, System and Control(Japan), 21, 6, pp.331-338(1977)
[3] Jin,Y., Koyama,T. and Zhang,Z: The marine traffic control system as a distributed solving network, IEEE Int. Con on Sys. Man and Cyb., pp.876-881, 929 (1990)
[4] Sugisaki, A.M.: On a marine traffic flow simulator, Navigation(Japan), 91, MARCH (1987)
[5] Gantz, D.T., Mekemson, J.R.: Flow profile comparison of a microscopic car-following model and a macroscopic platoon dispersion model for traffic simulation.

Table 2 Experimental result

No.	1	2	3	4	5	6	7	8	9	10	success
A	×	×	○	○	○	×	○	○	○	○	70.0%
B	×	×	○	○	○	×	○	○	○	○	70.0%
C	×	×	○	×	○	×	○	×	×	×	30.0%
D	×	×	○	○	×	×	○	○	○	○	60.0%

(○: success   ×: failure with some strayed members)

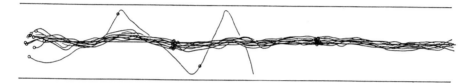

Fig.9 Failure Example (straight passage)

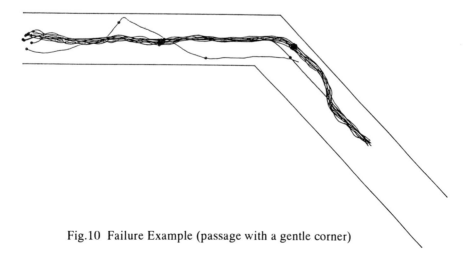

Fig.10 Failure Example (passage with a gentle corner)

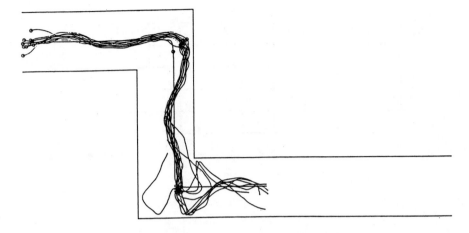

Fig.11 Failure Example (passage with two corners)

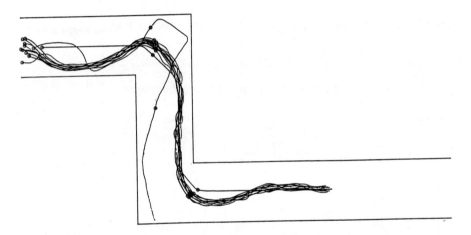

Fig.12 Failure Example 2 (passage with 2 corners)

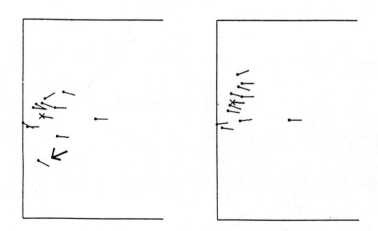

Fig. 13 Initial states of the individuals

The Fuzziness Index for Examining Human Statistical Decision-Making

Sumiko Takayanagi
Norman Cliff
Department of Psychology
University of Southern California
Los Angeles, California 90089-1061

Abstract - The fuzziness index is described as one of the most efficient and useful experimental tools to examine human decision-making in the area of statistical judgment. This study investigated how statistical significance levels were treated and interpreted by various researchers by using the fuzziness index to measure the degree of fuzziness in hypothesis-testing. The fuzziness index showed a significant association between the magnitude of the fuzziness and the critical point of the human statistical decisions. The research demonstrated the possibility of measuring the human imprecision of the decision boundary, and the effectiveness of the fuzziness index in assessing underlying psychological mechanisms in decision-making.

1. INTRODUCTION

Human decision-making enters into just about every conceivable task in our daily lives ranging from weather forecasting to formal scientific inquiry and has become one of the fastest growing fields in the behavioral sciences over the past 20 years [1]-[2]. Accordingly, integrative research in fuzzy set theory and human decision-making has come to be one of the most extensively investigated areas to explore imprecise human cognitive boundaries and judgmental behaviors [3]-[4]. As one of the areas in decision-making, statistical hypothesis-testing has gained a normative popularity among researchers from various disciplines and is widely used as the major foundation of data assessment techniques. The technique of hypothesis-testing seems to allow many researchers to perform a more fail-free objective assessment of their research outcomes, and such assessment would be believed as relatively consistent among different researchers. However, the conclusion drawn from hypothesis-testing actually is reflected by a large amount of human subjectivity and vagueness [5]-[6]. In other words, the significance level obtained in hypothesis-testing may hold the essential responsibility for the research conclusion, but the final process in interpreting the significance level is largely dependent on our subjective estimation [6].

Therefore, the present study attempted to explain the human imprecision of the choice boundary in hypothesis-testing. The rationale of fuzzy set theory and the fuzziness index was used as a measurement technique in the study in searching for the unclear boundary of human inferential judgments. In particular, the study attempted to determine how membership functions and the fuzziness index clearly showed the human decision ambiguities that are reflected in formal scientific investigations.

II. RESEARCH ISSUES

There were three key issues that this study attempted to investigate. They were; 1) issues in "Thresholders" or "Estimators" [7], 2) the alpha level controversy and fuzziness differences among research version types, scenarios and sample sizes, and 3) the fuzziness difference between verbal and numerical statistical expressions.

The issues in "Thresholders" or "Estimators" came from the idea that there are two types of researchers; one believes statistical judgments are a dichotomous judgment and the other a continuous evaluation [8]. "Thresholders," take the firm position of "Hypothesis-Testing," as a clear dichotomous decision of rejecting or not rejecting the null hypothesis by using a pre-determined alpha level, and not taking into account how far the probability levels are into the critical region [6], [8]-[10]. On the other hand, "Estimators" interpret statistical significance on a continuum with significance levels justifying disbelief of the null hypothesis linearly and inversely related to the probability of its null hypothesis [11]-[12]. Therefore, this study attempted to examine what actual proportions of the subjects who would be the "Thresholders - crisp decision-makers" or "Estimators - fuzzy decision-makers" by using the logic of the fuzzy set theory.

In terms of the alpha level controversy, it is shown that the vast majority of researchers generally placed the critical alpha levels at 0.05 or 0.01 [6], [8], [13], [14]. These universally accepted critical alpha levels are, in practice, treated as magical numbers which dictate publication potentiality [15] and decide research worthiness even though they were originally assigned just as conventional alpha levels [16]. Thus, the study tried to investigate whether the fuzziness is precisely related to these critical points of decisions. In particular, it examined the difference in the degree of fuzziness among different significant levels nested within different experimental conditions including versions, scenario type, and the sample sizes.

The last issue that the study investigated was the differences in fuzziness between verbal and numerical statistical expressions. Many decision-makers recommend the use of less vague, numerical probability terms for probabilistic expressions rather than verbal expressions [2]. However, in reality, many decisions are verbally expressed for their convenience and some psychological desirability [17]. Therefore, it attempted to investigate whether consistent results between verbal and numerical statistical expressions in terms of fuzziness would be obtained.

III. FUZZINESS INDEX

Researchers in the field of fuzzy set theory have developed useful indices to measure the degrees of fuzziness in-between boundaries [4], [18]-[19]. One central issue of this study was to measure these in-between boundaries in hypothesis-testing by using the fuzziness index with an assumption that the critical regions would be hypothesized as the two mutually exclusive sets.

The indices were based on an assumption that a set must be "fuzzier" when its membership responses tend to lie between "in a set" and "not in a set," and a set is less fuzzy as it approaches closer to "in the set" or "not in the set." Fuzziness indices were created for calculating the degree of fuzziness, using standard values between 0 and 1, with 0 indicating "no fuzziness" - a response is either in or outside of the set, while 1 is "most fuzzy" - at the middle location between in or outside of a set.

The fuzziness measure was defined as a positive function [f] of a set of [A] which must meet several conditions where $m_A(x)$ is a rating response between 0 and 1: 1) $f(A)=0$ if a set A is a crisp set; 2) $f(A)=1$ if a set A's fuzziness is maximum and, therefore, $m_A(x)=0.5$ for all x in [A], where $m_A(x)$ is the possibility rating from 0 to 1; 3) $f(A) \leq f(B)$ if for all x in either $m_A(x) \leq m_B(x)$ when $m_B(x) \leq 0.5$, or $m_A(x) \geq m_B(x)$ when $m_B(x) \geq 0.5$; and 4) $f(A)=f(A')$ [4], [18]. Based on the above assumptions, the most common and simple formula which measures the linear distance of a fuzzy set from its counterpart was developed by Kaufmann [19],

$$f_k(A) = \sum_{i=1}^{N} |m_{Ai} - m_{A*i}|, \qquad (1)$$

where "A*i" is the unfuzzy part of the set A and $m_{A*i}=1$ when $m_{Ai} \geq 0.5$, and $m_{A*i}=0$ otherwise. The maximum fuzziness is located at the raw score value of 0.5. Smithson [4] kept this above concept and created the formula,

$$FK = \sum_{i=1}^{N} (1 - |(1-2(X_i))|) / N, \qquad (2)$$

which made the middle raw score ($X_i=0.5$) the maximum fuzziness value as 1, and reduces fuzziness as the raw score value approaches to 0 or 1. N is the number of observations.

IV. EXPERIMENT

The experiment used three different questionnaires, Questionnaires 1 (Reject Model), 2 (Significance Model), and 3 (Verbal Model). All questionnaires had hypothetical research narratives with a response section followed by each narrative. Two different prototype scenarios and sample sizes, N=300 and N=40, were created. Scenario 1 was designed as a less serious research scenario and Scenario 2 was considered as more serious one. The Reject Model (RM) had two judgmental versions "Rejected" or "Not Rejected" in a narrative, whereas the Significance Model (SM) had three versions, "Highly Significant," "Significant," and "Not Significant." For these two models, a numerical response section followed by each narrative was presented. Unlike the above two models, the Verbal Model (VM) contained the three numerical versions, "P=.001," "P=.05," or "P=.06," and a verbal response section followed by each narrative.

In each response section for RM and SM, ten numerical p-values (p=.0005,.001, .007, .01, .03, .04, .05, .06, .07, .1) with 0-10 rating scales were created, whereas for VM, ten verbal expressions, Very Highly Significant, Highly Significant, Quite Significant, Significant, Slightly Significant, Marginally Significant, Nearly Significant, Borderline Significance, Not Significant, Not At All Significant, and 0-10 rating scales were given. One example is as follows:

<u>Reject Model, Scenario 2, Rejected, N=300</u>

A researcher in a Cancer Research Institute was about to use his new drug, Drug A, as a skin cancer treatment. However, the drug also could produce some dangerous side effects in people who are allergic to it. Thus, the researcher decided not to consider the drug for actual use unless it showed an outstanding effect for treating skin cancer. He randomly assigned <u>300</u> skin cancer patients into two groups. One group was given the new drug at the minimum level of danger. The other group, which served a control group, did not receive Drug A. After six months, the researcher performed a t-test on the prognosis scores for the two groups. As a result, he REJECTED his null hypothesis, on the basis of his obtained p-value. Indicate to what extent you agree with him, if his obtained p-value was the value specified in each statement below.

DEFINITELY DISAGREE										DEFINITELY AGREE

p=.07
0 1 2 3 4 5 6 7 8 9 10

p=.05
0 1 2 3 4 5 6 7 8 9 10

p=.001
0 1 2 3 4 5 6 7 8 9 10

p=.1
0 1 2 3 4 5 6 7 8 9 10

p=.0005
0 1 2 3 4 5 6 7 8 9 10

p=.03
0 1 2 3 4 5 6 7 8 9 10

p=.01
0 1 2 3 4 5 6 7 8 9 10

p=.06
0 1 2 3 4 5 6 7 8 9 10

p=.007
0 1 2 3 4 5 6 7 8 9 10

p=.04
0 1 2 3 4 5 6 7 8 9 10

A. Method

Subjects. The subjects were 34 psychology graduate students at the University of Southern California who had a basic knowledge of inferential statistics. They participated in this experiment on a voluntary basis.

Apparatus. The materials included questionnaires 1-3 (RM, SM, & VM). Questionnaire 1 had a total of eight, whereas Questionnaires 2 and 3 had 12 different research narratives subjects had to choose for the ten p-values or verbal expressions.

Procedure. Three questionnaires were given to the subjects. The subjects' task was to choose the most appropriate numerical rating value, expressing to what extent they agreed with the hypothetical experimenter if the experimenter's obtained p-values or verbal expressions were the values or statements given in the response section. Most of the subjects reported that they took approximately one and a half hours to complete these questionnaires.

V. RESULTS AND DISCUSSION

A. Graphical Presentation

The membership function for each subjects for each experimental condition was constructed as in Fig. 1. There was a substantial variability among the subjects. The results clearly revealed that a majority of the subjects were <u>Estimators</u>, regardless of the experimental condition types (1 Thresholder, 31 Estimators, & 2 missing subjects). This finding confirmed that many subjects do not perceive hypothesis-testing as a clear binary decision-making. Therefore, further assessment was made to measure the degree of fuzziness for this unclear boundary by using the fuzziness index.

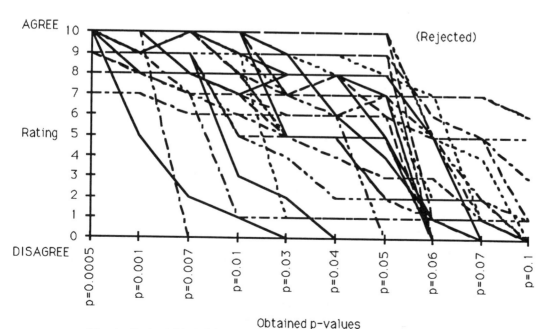

Fig. 1. Reject Model (Quest 1), Scenario-2, N=300, Psych St. (n=32)

B. Analysis by Fuzziness Index

The graphical representation of the membership function for each individual showed a tendency for the majority of subjects as being Estimators. This result invites prospects for investigating the degree of uncertainty which seems to exist in the area of statistical judgments. The rating scores, 0 to 10, were converted into the interval between 0.0 to 1.0 (e.g., raw score 5 was assigned to the value of 0.5). Then, the fuzziness index was calculated by the equation (2) as Xi for the raw transformed scores between 0.0 to 1.0, and each fuzziness index was used to obtain the average fuzziness for each condition.

The result for the individual differences in fuzziness (to what extent, the subjects were Thresholders or Estimators) was significant for all models, RM [$F(32,230)=28.79$, $p<.0001$: $w^2=.771$], SM [$F(33,373)=29.69$, $p<.0001$: $w^2=.699$], and VM [$F(32,362)=34.79$, $p<.0001$: $w^2=.732$], indicating that there was a significant difference in terms of the degree of fuzziness among the subjects.

A second analysis showed whether the fuzziness was different between the two different versions, scenarios, and N sizes across the subjects. For the Reject Model, a repeated measures analysis, version (2) X scenario (2) X N size (2), indicated no main effect of version, but there were main effects of scenario [$F(1,31)=7.00$, $p<.05$: $r^2=.014$] and N [$F(1,31)=5.25$, $p<.05$: $r^2=.004$]. Both variables, scenario and N, showed a higher fuzziness in the Scenario 2 and N2 than Scenario 1 and N1, indicating that fuzzier decision trends in more serious scenario and a smaller sample size conditions were present. In the Significance Model, fuzziness was the highest in the version of "Highly Significant," [$F(2,64)=4.06$, $p<.05$: $r^2=.004$]. The Verbal Model also showed a main effect for the numerical versions [$F(2,62)=12.59$, $p<.0001$: $r^2=.038$], showing the version where "P=.05" as the most fuzzy (mean fuzziness MF=.34) followed by "P=.06" (MF=.30) and "P=.001" (MF=.24).

These findings showed that the fuzziness was significantly different among individuals as well as context dependent (e.g., scenarios and sample sizes). These results suggest that the fuzziness is maximized when the condition is most vague (Highly Significant) or the place where the category is laid on a transition point of the decision (P=.05). Accordingly, the fuzziness of each p-value and the verbal expression was calculated to confirm this fuzziness-critical point association as described in the next section.

The fuzziness of each p-value (Reject and Significance Models) and verbal expressions (Verbal Model) were analyzed. The first analysis used a one-way ANOVA to determine the differences in fuzziness among the ten p-values across subjects, scenarios, and N sizes. The result showed a significant effect of p-values or verbal expressions for each model, RM [$F(9,70)=39.22$, $p<.0001$: $w^2=.810$], SM [$F(9,110)=21.79$, $p<.0001$: $w^2=.609$], and VM [$F(9,110)=10.21$, $p<.0001$: $w^2=.077$]. In Fig. 2, the fuzziness was maximized when p-values were at .04, .05, and .06, and gradually declined to the both ends for the RM and SM, whereas VM showed "Slightly Significant" at the highest and showed a similar decline at the both ends with some spikes. These results clearly indicated a correlation between the degree of fuzziness and the commonly assigned critical point of hypothesis-testing.

In addition, three, two-way ANOVAs were performed in order to clarify the above findings. These analyses [p-value (10) X version(2); p-value (10) X scenario (2); and p-value (10) X N-size (2)] were designed to examine fuzziness differences between p-values and the three variables, version, scenario type, and sample size, in RM and SM. RM showed a significant "p-value" main effect in these three variables, [F-version$(9,60)=34.99$, $p<.001$; $w^2=.808$: F-scenario$(9,60)=106.25$, $p<.001$; $w^2=.825$: and F-N-size$(9,60)=37.17$, $p<.001$; $w^2=.808$], indicating that the p-value fuzziness differences were prominent. In the p-value X scenario, an interaction effect was significant [$F(9,60)=9.776$, $p<.001$: $w^2=.068$], showing that the fuzziness in Scenario 2 was larger than in Scenario 1, when the given p-values were between .0005 and .04, and this feature was reversed when p-values became larger than .04 (Fig. 3). This significant interaction effect showed a unique point of maximum fuzziness between Scenario 1 (maximum fuzziness at P=.05) and 2 (maximum fuzziness at P=.04), implying a more stringent choice of the critical alpha was made by the subjects for more serious scenario, Scenario 2. Likewise, a theoretically consistent result was obtained in SM. In Fig. 4, an interaction between p-values and versions, indicated clear relationships between the highest points of fuzziness and the critical point of the decision, such as P=.01 to .05 for HS whereas P=.06 and .05 for Significant and Not Significant versions, respectively.

Unlike the first two models, VM showed a unique result, indicating the verbal expression, "Slightly Significant" for "P=.001" and "P=.05," and "Borderline Significance" for "P=.06" version as most fuzzy. This model showed some unique features that the fuzziness is related to the familiarity and frequency of the terms that are commonly used to express statistical findings. This result implies not only uncertainty of the decision in terms of statistical significance, but verbal preference of expressions. People are generally fuzzy in their perceptions and judgments, but at the same time, they are reluctant and wary of using fuzzy terms to express formal statistical findings. Therefore, the fuzziness was increased to express the statistical result by the unfamiliar terms.

In summary, these findings indicated that fuzziness is strongly correlated with the critical position of the decision. The analyses strongly confirmed the effect that p=.04, p=.05, and p=.06 were fuzzier than the other p-values. Similarly, the interaction in the p-values and scenarios showed differences in the maximum point of fuzziness for different scenarios. This study confirmed the effectiveness of the fuzziness index in showing systematic differences in the range of subtle human decision-making.

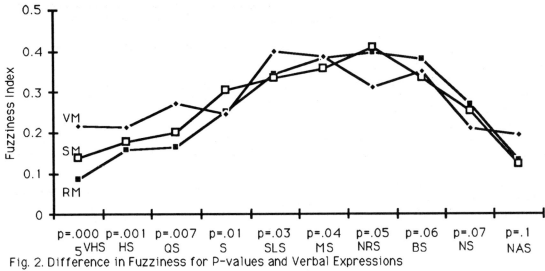

Fig. 2. Difference in Fuzziness for P-values and Verbal Expressions

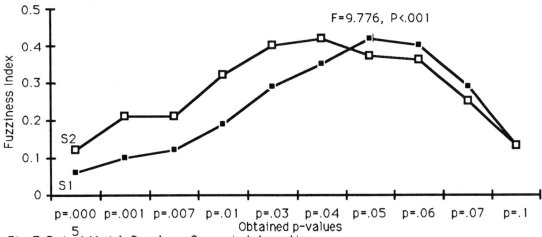

Fig. 3. Reject Model, P-value x Scenario Interaction

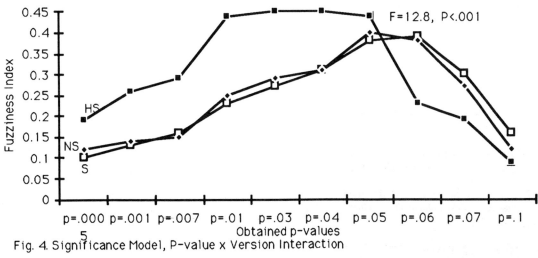

Fig. 4. Significance Model, P-value x Version Interaction

VI. CONCLUSIONS

The integration of statistical significance levels and the method developed by the idea of fuzzy set theory revealed several aspects of statistical judgments. First of all, it revealed that decisions are naturally fuzzy despite the dichotomous judgment that we have been instructed to follow. However, as shown in the Verbal Model, it is also true that many people still preferred to state their decision in formal statistical terms, such as "Significant" and "Not Significant," rather than expressions which are not often used.

Furthermore, the fuzziness index provided a useful way to assess the relationship between decision-making difficulty and the magnitude of the fuzziness. Overall, decisions which were on the borderline were the most difficult for subjects to make and these were also the fuzziest points on the fuzziness index. This result was present throughout the experiments. The study well documented that in its simplest form, the fuzziness index worked as a precise measurement tool to assess the underlying state of human psychological mechanisms in decision-making.

REFERENCES

[1] A. Tversky and D. Kahneman, "Judgment under uncertainty: Heuristics and biases," *Science*, 185, pp. 1124-1131, 1974.

[2] D. von Winterfeldt and W. Edwards, *Decision Analysis and Behavioral Research*. Cambridge: Cambridge University Press, 1986.

[3] H. J. Zimmermann, *Fuzzy Sets, Decision Making and Expert Systems*. Boston: Kluwer, 1985.

[4] M. Smithson, *Fuzzy Set Analysis for Behavioral and Social Science*. New York: Springer-Verlag, Inc., 1987.

[5] W. W. Rozeboom, "The fallacy of the null hypothesis significance test," *Psychological Bulletin*, 57, pp. 416-428, 1960.

[6] D. Bakan, *On method: Toward a Reconstruction of Psychological Investigation*. San Francisco: Jossey-Bass Inc., Publishers, 1967.

[7] M. Kochen, "Applications of fuzzy sets in psychology," in *Fuzzy Sets and Their Applications to Cognitive and Decision Processes*, L. A. Zadeh, K. Fu, K. Tanaka, and M. Shimura, Eds., New York: Academic Press, Inc., 1975.

[8] B. D. Franks and S. W. Huck, "Why does everyone use the .05 significance level?" *Research Quarterly*, 57, 3, pp. 245-249, 1986.

[9] J. Neyman and E. S. Pearson, "On the problem of the most efficient tests of statistical hypotheses," *Philosophical Transactions of the Royal Society-A*, 231, pp. 289-337, 1933.

[10] E. R. Harcum, "The highly inappropriate calibrations of statistical significance," *American Psychologist*, 44, 6, p. 964. 1989.

[11] D. Kempthorne and J. L. Folks, *Probability, Statistics, and Data Analysis*. Ames, Iowa: State University Press, 1971.

[12] S. Kanekar, "Statistical significance as a continuum," *American Psychologist*, 45, 2, p. 296, 1990.

[13] M. Cowles and C. Davis, "On the origins of the .05 level of statistical significance," *American Psychologist*, 37, 5, pp. 553-558, 1982.

[14] G. R. Loftus, "On the tyranny of hypothesis testing in the social sciences," *Contemporary Psychology*, 36, 2, pp. 102-105, 1991.

[15] D. R. Atkinson, M. J. Furlong, and B. E. Wampold, "Statistical significance, reviewer evaluations and the scientific process: Is there a (statistically) significant relationship?" *Journal of Counseling Psychology*, 29, pp. 189-194, 1982.

[16] D. K. Beale, "What's so significant about .05?" *American Psychologist*, 27, 11, pp. 1079-1080, 1972.

[17] T. S. Wallsten, "The costs and benefits of vague information," in *Insights in Decision Making: Theory and Applications. A Tribute to the late Hillel Einhorn*, R. Hogarth, Ed., Chicago: University of Chicago Press, 1989.

[18] A. DeLuca and S. Termini, "A definition of a non-probabilistic entropy in the setting of fuzzy sets theory," *Information and Control*, 20, pp. 301-312, 1972.

[19] A. Kaufmann, *Introduction to the Theory of Fuzzy Subsets, 1*. New York: Academic Press, 1975. Cited in M. Smithson, *Fuzzy Set Analysis for Behavioral and Social Sciences*. New York: Springer-Verlag, 1987.

Linking the fuzzy set theory to organizational routines: a study in personnel evaluation in a large company

Alessandro Cannavacciuolo
Fiat Research Center, Strada Torino 50, 10043 Orbassano (Torino), Italy

Guido Capaldo
Via S.Lucia 133, 80132 Napoli, Italy

Aldo Ventre
Mathematics Institute, Faculty of Architecture, University of Naples "Federico II"
Via Monteoliveto 3, 80134 Napoli, Italy

Giuseppe Zollo
ODISSEO, Dept. of Computer Science and Systems, University of Naples "Federico II"
Via Diocleziano 328, 80124 Napoli, Italy

Abstract–This paper is concerned with the application of the fuzzy set theory to an personnel evaluation procedure. The effectiveness of the fuzzy concepts and methods depends on the approach used for the analysis of organizational issues. The fuzzy set theory allows us to model the weak signals existing in evaluation processes, and highlights part of the tacit knowledge involved in individual judgments.

I. Introduction

Usually researchers, consultants and managers have a rather qualitative approach to organizational problems. The natural language is the preferred instrument for describing the organizational situations because the shades of meaning and the ambiguity of the verbal statements allows the company actors to manage diverging opinions, tensions and conflicts [1], [2], [3], [4], [5].

On the other hand, the logical-mathematical models tend to represent a world of certainty and coherence where doubts, contradictions, divergences, polysemy, conflicts, and ambiguities are usually typified, dissolved, degraded and linearized. Within this same conceptual framework mathematicians, computer scientists, A.I. researchers and engineers, in search of formal coherence, quantifiable variables and efficient algorithms, usually tend to use the fuzzy set theory without considering the complexity and the ambiguity of organizational situations [6], [7], [8], [9], [10], [11] [12], [13], [14], [15]. Consequently, they grasp only secondary aspects of company problems.

II. Individual Skills and Organizational Routines

The experiment illustrated regards the use of fuzzy concepts in redefining a procedure for evaluating personnel.

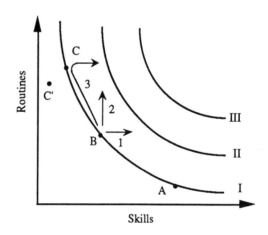

Fig. 1. Combination of skills and routines in the evaluation process

Putting together such a procedure means, above all, modelling human judgment capacity and singling out the social system which enables the principal actors of the organization to converge, in some form of agreement.

The problem of company skills has been widely debated in organizational studies [1], [16], [17]. In most cases skills and routines are combined to produce a determinate level of performance as illustrated in Fig. 1. On the abscissas we have individual skills, on the ordinates we have organizational routines. The points showing a given combination of skills and routines define the iso-performance curves. The curves that are farther from the origin represent the most efficient combinations of skills and routines.

This personnel evaluation contains two extreme cases, defined by points A and C. In point A the contribution of the individual skills prevails. By interpreting company needs the evaluator expresses his own subjective evaluation. The contribution of routines is reduced to a minimum: it is a matter of drawing up documents and motivating the appraisal in some way.

In point C personnel evaluation is carried out by giving priority to the contribution of procedures, while the role played by the individual is less important. The procedure is rather rigid: the company has defined an ideal professional profile, and singled out measuring scales, criteria for scores, etc. Each evaluator is asked to follow the procedure and supply his evaluations on pre-established criteria. In order to minimize the variability of individual judgments, meetings of evaluators are organized in order to check and compare the single evaluations. The evaluation process is the result of a social rather than individual activity.

Point B represents a middle course between the extreme cases: the evaluation of personnel is not executed by a single person but by several persons who meet and discuss - there is a minimum agreement on how the evaluation will be expressed, on the attributes to consider most important, etc.

In order to improve the capability for personnel evaluation the company must pass from curve I to curves II and III. The company could do this in various ways:

1) by improving evaluation skills (horizontal trajectory 1). In this case the company would depend even more on the evaluative capability of each individual;

2) by using better procedures (vertical trajectory 2), but this presupposes that the company cannot fully exploit the existing skills using the new developed procedures.

Actually, the way to pass over to a higher performance curve is a harder task. In fact the company, which is usually at point A or B, plans to move to C by transforming individual evaluation skills into procedures, and to rest assured that the new evaluation system does not reduce the performance level. At the same time it tries to better utilize individual skills to pass onto a higher curve (trajectory 3). Adopting this strategy the company achieves various goals: it translates individual skills into organizational capabilities; it frees individual skills that it can use to improve company performance.

The success of this strategy lies mostly in the ability to model individual evaluative capabilities. In fact, if the routines that are set up simplify the evaluation process too much or reduce its flexibility needlessly, the result of this strategy will no longer reach point C, but point C', which lies on a curve of lower performance. In this case the company is forced to use individual skills only to get back to the initial performance. This situation gives rise to intolerance of procedures and frustration at individual level, as well as a waste of resources at company level.

The traditional approach to defining organizational routines often produces this result because it overlooks facts that cannot be modelled by the current techniques and does not take into account how the evaluation process develops at an individual level as well as at social level. The fuzzy set theory constitutes an effective response to this problem, as long as it is used together with organizational methods and techniques that allow us to recognize and model the phenomena that evade traditional analysis.

III. THE EVALUATION PROCESS

A. The Existing Procedure

The current evaluation process of the potential candidates for higher positions in the company hierarchy is managed by a procedure which comprehends three important aspects: organizational relationships, rating sheets, and operating rules.

Organizational relationships. The evaluation process involves the CEO, the Direct Evaluator (DE), usually the hierarchical superior of the candidate, and the Personnel Manager (PM). The DE interviews the candidate and draws up a first draft of his profile. The PM, together with the DEs, checks each individual profile and compares the different profiles of all the candidates to define a definitive profile. At last, the PM draws up a final graded scale among the candidates. The role of interpreting the ratings is performed by the CEO.

Rating sheet. The candidate's profile is compared with the ideal profile of the position to be filled. The evaluation is made rating about 15 items with respect to three different areas:
 - professional skills (from 3 to 5 items);
 - managerial skills (10 items);
 - personal characteristics (4 items).

Each item is given a value ranking from 1 to 5.

Operating rules. In order to attribute a value to the items, the evaluator uses a document called "Operating Notes" in which he finds for each item:
 - the definition of the items;
 - analytic references for the characteristics of the items.

Each evaluator is supposed to use the analytic references as a frame to express the judgement.

The purpose of the procedure, as it emerges from interviews to company executives, is: "to get homogeneous, clear and reliable judgements - based on explicit evaluation rules that are shared by the evaluators - that can grasp the individual knowledge in the broadest possible sense and, contemporaneously, filter the subjective elements that might have a distortive effect on the overall judgement". Moreover the quantitative approach is needed to keep the evaluation time at the minimum.

B. Critical Aspects

The goal of the present research was neither to modify the organizational aspects nor the rating sheet, but to focus only on the meaning of the ratings. In this way the company is not forced to make organizational changes which require more time than was expected.

Three critical points are singled out: a) the items' meanings; b) the items' ratings; c) the rating-aggregation.

(a) *The items' meanings*. The evaluators attribute subjective meanings to the items. In the Operating Notes, for example, the item "decision" is explained as follows: "To adopt the appropriate decisions, according to the received delegations, and take on the responsibility of carrying them out". What does "the appropriate decisions" mean ? Does it depend on how the decision was taken? Or, whether the collaborators are satisfied? Or, whether the outcomes are successful? Each evaluator interprets "appropriate" in different ways. On the other hand, the consultant who defined the items perhaps meant all of the above meanings, along with still others. Each definition hides these ambiguities and the evaluator is not expected to specify which circumstances he considered significant.

(b) *The items' ratings*. Many dangers are hidden in the rating scale. The consultant who designed it foresaw a value ranging from a minimum of 1 to a maximum of 5. For example, the minimum of the item "decision" is defined as follows: "He is not capable of making a decision, he procrastinates, postpones, constantly asks for approval , keeps reconsidering his decision even when there are no new elements involved". Whereas the maximum is defined as follows: "He takes risky decisions without manifesting excessive stress and with considerable realism, is timely without being hasty and does not reconsider his decision unless he is faced with elements that were unknown to him earlier." Consequently, the minimum and maximum values are extreme values that will actually never be reached. According to the point of view of the evaluator, the minimum value is considered unrealistic, whereas the maximum value is interpreted as nothing to do with his own experience. The conclusion is that, while the theoretical model suggests anchoring his judgment to the extreme values of the scale, the evaluator anchors his judgment to the average value of the scale.

(c) *The ratings aggregations*. The problems arise from the fact that the evaluator aggregates the results in a totally arbitrary way, and compensates between various items in order to reach his final judgement. This compensation is hidden in the evaluation process, because the evaluator tries to make his overall judgement of the candidate convergent with the result coming from the analytical ratings.

IV. THE USE OF FUZZY SET CONCEPTS AND OPERATORS

The fuzzy concepts and operators are used to interpret the meanings attributed to the items by each evaluator and to propose a procedure of rating aggregation.

A. The "ideal candidate" fuzzy sets

The field interviews confirm that the evaluator usually uses three concepts of *ideal candidate*:

Ideal Candidate A: "The candidate is ideal when his profile is absolutely excellent".

Ideal Candidate B : "The candidate is ideal when his profile is coincident with the profile of the position requested by the company";

Ideal Candidate C: "The candidate is ideal when his profile is equal to or exceeding the profile of the position requested by the company";

These concepts of ideal candidate have been defined as normal fuzzy sets A, B, C, characterized by the fuzzy membership functions: μ_A, μ_B, μ_C.

The membership functions could be interpreted as operators transforming mathematical distance into perceived distance within different cognitive frames [18]. Consequently, we define the support sets of the fuzzy sets B and C as crisp sets defined by the mathematical distance $d(x,y)$ between the rating R_x of the candidate x and the rating R_y of the (requested) position y, while the support set of the fuzzy set A depends only by the distance d_x between the rating R_x of the candidate x and average value of the scale.

The membership functions that seem suitable to represent the fuzzy sets are:

$$\mu_B(x) = 1 - \exp\left(-\left(\frac{a}{|d(x,y)|}\right)^b\right) \quad -4 \leq d(x,y) \leq 4 \quad (1)$$

representing the membership function of the fuzzy set *ideal candidate B*, and

$$\mu_C(x) = \begin{cases} 1 - \exp\left(-\left(\frac{a}{|d(x,y)|}\right)^b\right) & -4 \leq d(x,y) < 0 \\ 1 & 0 \leq d(x,y) \leq 4 \end{cases} \quad (2)$$

representing the membership function of the fuzzy set *ideal candidate C*.

During this first phase of the research the values of the parameters are:

$a = 1$ and $b = 1.5$

These values represent only the first hypothesis on the average attitudes of the evaluators.

The first improvement was the representation of the sensibility of the evaluator to the absolute value of the position rating R_y (that is the rating of the position demanded by the company). To represent this behaviour we choose to define the parameters a and b as function of R_y, as follows:

$a = 0.25\, R_y$ and $b = 0.1\, R_y + 1.5$

The fuzzy sets associated with the membership functions with these calculated parameters are called respectively Bp and Cp.

For the fuzzy set A, *the* membership function was defined as follows:

$$\mu_A(x) = \frac{1}{1+\exp(-d_x)} \qquad -2 \leq d_x \leq 2 \qquad (3)$$

It is important to notice that the purpose was to design the whole procedure, that is: the field research to render explicit the tacit assumptions of the evaluators, the definition of the fuzzy sets of the Ideal Candidate, the aggregation of methods of the analytic ratings and the comparison of the candidates. Consequently, minor attention was dedicated to the definition of the best membership functions and, as we will see later, to the best way to aggregate the results. For this reason, the membership functions could be considered as hypothesis, partially tested on the basis of the field research, and the whole procedure as a laboratory prototype. By the use of this prototype we test the effectiveness of the five fuzzy sets previously defined (A, B, Ap, Bp, C) in representing the different points of view of the evaluators and present them to the decision-maker.

B. The aggregation of the analytic judgments

The evaluator uses about 15 rating factors for each candidate. The rating are grouped into three intermediate sets: Professional Skills, Managerial Skills and Personal Characteristics. Finally, these aggregate evaluations are grouped in a synthetic evaluation of the candidate.

The problem is the formulation of a synthesis procedure on the basis of field studies, through fuzzy composition operators. For the purpose of the aggregation two diverse grouping criteria were used: the implicit quantifier *most* and the γ-model.

The mental processing of common-sense knowledge uses implicit quantifiers. An important function of the quantifiers is represented by the analysis and grouping of data to devise a general meaning of real problems. As has emerged from the interviews, the raters commonly used the quantifier *most* to justify the value of the intermediate concepts. For example, they usually said: "The rating of the Professional Skills is high because most characteristics are high".

The implicit quantifier *most* has been implemented using the algebraic consensory method [19].

The γ-model is an attractive operator to aggregate partial judgments, because this model operationalize the compensatory behaviour of the evaluator between low and high ratings. The simplest algebraic representation of the model is to set the parameter γ as a weight of the non compensatory connective *and* and the fully compensatory *or* [13].

The problem consists in determining empirically the value of the parameter γ. We found that the γ-model interprets the compensatory behaviour of the evaluator in the case of Ideal Candidate B, where the Ideal Candidate is equal to or exceeding the demanded position.

TABLE 1
PHASES AND METHODS OF A QUANTITATIVE DECISION-MAKING PROCEDURE

Phases	Methods
intelligence	interviews with interested individuals. translation of verbal statements into numbers.
design	criteria for classifying and weighting information, determination of alter-natives: use of reference models, often implicit; use of typologies and weights.
choice	synthesis of the criteria and hierarchy of the alternatives: sums, averages.

The rater uses the exceeding ratings to compensate the failing ratings. To modelize the behaviour of the evaluator we introduce for each rating factor the concepts of Demanded Compensation C^d and Supplied Compensation C^s.

V. RESULTS

The main result is a multiple ranking (Fig. 2). The decision-maker can use the multiple ranking of the candidates to choose the most appropriate criterion according to the strategy of the company and to the specific management issues. In this way the decision maker can easily adapt the criterion of ideal candidate to different environmental situations. For example, when the environment is stable and the position profile is known, the decision-maker can use the criterion of choosing a candidate fitting the requested profile (Ideal Candidate A), raising in this way the operational efficiency of the company. Otherwise, when the environment becomes more turbulent or is engaged in a growth strategy, it is necessary to have candidates with exceeding abilities with respect to the position profile. In this case the decision-maker can use the second or the third concept of Ideal Candidate.

Moreover, the decision-maker can assign his own criterion for each rating factor and modify, through the fuzzy membership functions, the original ratings. In this way the decision-maker can explore the situation from diverse viewpoints and eventually he forms a more precise idea of the characteristics of the candidate.

The reliability of the evaluation procedure increases when the evaluator makes explicit the criterion used for each rating factor, reducing in this way the halo effect, that is a favourable rating given to all rating factors, based on the impressive performance of just one job factor.

VI. DISCUSSION

If we analyze any decision-making procedure we will easily recognize the three phases of the classic decision-making process: intelligence, design, choice. The mediation between simplicity and efficiency is made by quantifying the verbal judgments and using mathematical and statistical techniques, capable of dealing with an enormous amount of data (table 1).

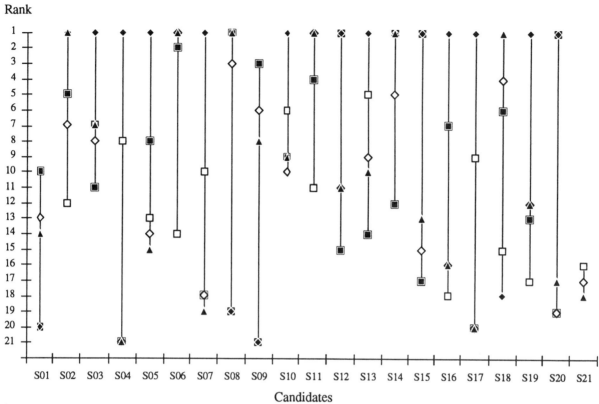

Fig. 2. The multiple classification of candidates

Essentially, the complex and contradictory knowledge the company possesses is filtered and traced back to a manageable sub-set. The result is that the company uses and controls only a small part of the skills it possesses to make choices.

The fuzzy set theory could avoid the oversimplification of the company's capacities and use its organizational complexity (with all its attributes of variety, diversity, ambiguity, uncertainty) as a resource to create better company procedures. In fact using the fuzzy set theory in the area of organizing management could:

1) help avoid the rigidity of the traditional models: the transitions from one judgment to another are progressive and there is no need to fix precise thresholds between one category and another which would make it artificial to manage;

2) simplify the translation from human reasoning into procedures, thus overcoming the fictional precision of decision-making procedures based on numerical values, scores, sums and averages;

3) provide concepts and techniques for an easy modelling of complex systems, because the complexity is not represented by complicated algorithms but by a more exact representation of the events;

4) allow for the use of situational rules in decision-making procedures: it is possible to model very specific rules without undermining the general evaluation model;

5) make it possible to grasp common sense: the evaluator easily recognizes the results of the procedure because it does not represent a leap in his reasoning;

6) make it possible to represent contrasting objectives and judgements that can be resolved by the membership of elements to be evaluated in various fuzzy sets at different degrees;

7) make it possible to easily represent the interrelation between elements that make up the decision-making process through the fuzzy operators;

8) provide for representation of the ambiguities present in the decision-making process more accurately with regard to the information content of imprecise reasonings.

The fuzzy concepts - in order to be truly effective - must not simply take existing procedures as a starting point but become an integral part of a new approach to organizational issues, built with the aim of highlighting and representing the part of the organizational situation that is normally overlooked by traditional models. In the experiment carried out, and still in progress, the procedure followed is listed below:

1) identification of the weak signals present in the organization (shades of meaning, concurrent processes, contradictions, divergent opinions);

2) identification of the role that the weak signals have in evaluating and judging (what is gained in terms of efficiency and efficacy by modifying the existing models);

3) building an evaluation model that incorporates the weak signals;

4) formalizing the model by using concepts and operators of the fuzzy set theory;

5) introducing the model into the organization to correct the current procedure.

VI. Conclusions

The experiment carried out so far has highlighted the difficulty of creating a reliable model of the company evaluation process. The elements to be analyzed are incorporated in individuals and in the organizational procedures. The most important issue is making clear the evaluative capacity of the various actors without reducing variety to uniformity, differences to averages, contradictions to a fictitious coherence. The challenge to management in the '90s is to build procedures that allows differences to interact, and to use them as a resource for innovation and competitiveness. Fuzzy techniques could play an important role in assisting companies to take on this challenge.

References

[1] C. Argyris, D. A. Schon, *Organizational Learning. A Theory of Action Perspective*, Addison-Wesley, Reading (Mass), 1978.

[2] K.S. Cameron, "Effectiveness as paradox: consensus and conflict in conceptions of organizational effectiveness", *Management Science*, vol. 32, 5, 1986.

[3] D.J. Orton, K.E. Weik, "Loosely coupled system: A reconceptualization", *Academy of Management Review*, 1990, vol. 15, 2, pp. 203-223.

[4] C.R. Schwenk, "Cognitive simplification processes in strategic decision-making", *Strategic Management Journal*, vol. 5, 1984, pp. 111-128.

[5] K.E. Weick, "Educational organizations as loosely coupled systems", *Administrative Science Quarterly*, vol. 21, 1976, pp. 1-19.

[6] R.E. Bellman and L.A. Zadeh, "Decision making in a fuzzy environment", *Management Science*, vol. 17, 1970, pp. 141-164.

[7] C. Carlsson, "On the relevance of fuzzy sets in management science methodology", in H.J. Zimmermann et al. (eds.), *Fuzzy sets and Decision Analysis*, North-Holland, Amsterdam, 1984.

[8] B.R. Gaines, "Fundamental of decision: probabilistic, possibilistic and other forms of uncertainty in decision analysis", in H.J. Zimmermann et al. (eds.), *Fuzzy sets and Decision Analysis*, North-Holland, Amsterdam, 1984.)

[9] B.R. Gaines, L.A. Zadeh and H.J. Zimmermann, "Fuzzy sets and decision analysis: A perspective", in H.J. Zimmermann et al. (eds.), *Fuzzy sets and Decision Analysis*, North-Holland, Amsterdam, 1984.

[10] R.M. Tong, P. P. Bonissone, "Linguistic solutions to fuzzy decision problems", in H.J. Zimmermann et al. (eds.), *Fuzzy sets and Decision Analysis*, North-Holland, Amsterdam, 1984.

[11] R.R. Yager, "Satisfaction and fuzzy decision functions", in P.P. Wang and S.K. Chang, *Fuzzy Sets: Theory and Applications to Policy Analysis and information Systems*, Plenum Press, New York, 1980.

[12] P.L. Yu, "Dissolution of fuzziness for better decision-perspective and techniques", in H.J. Zimmermann et al. (eds.), *Fuzzy sets and Decision Analysis*, North-Holland, Amsterdam, 1984.

[13] L.A. Zadeh, "Making computers think like people", *IEEE Spectrum*, August 1984, pp. 26-32.

[14] S. French, "Fuzzy decision analysis: Some criticism", in H.J. Zimmermann et al. (eds.), *Fuzzy sets and Decision Analysis*, North-Holland, Amsterdam, 1984.

[15] L.A. Zadeh, "Outline of a new approach of the analysis of complex systems and decision processes", *IEEE Trans. on System, Man and Cyb.*, vol.3, 1973, pp. 28-44.

[16] G. Morgan, *Images of Organization*, Sage Publications, Beverly Hills, 1986

[17] R.R. Nelson and S. G. Winter, *An Evolutionary Theory of Economic Change*, Harvard Business Press, Cambridge, Mass., 1982

[18] H.J. Zimmermann, *Fuzzy Sets Decision Making and Expert Systems*, Boston, Kluwer, 1987.

[19] J. Kacprzyk, "A 'down-to-earth' managerial decision making via a fuzzy-logic-based representation of commonsense knowledge", in L.F. Pau, (editor), *Artificial Intelligence in Economics and Management*, Elsevier Science Publishers, 1986.

Chapter 15: General & Multi-Discipline

AN ELECTRONIC VIDEO CAMERA IMAGE STABILIZER OPERATED ON FUZZY THEORY

Yo Egusa, Hiroshi Akahori, Atsushi Morimura and Noboru Wakami

Central Research Laboratories,
Matsushita Electric Industrial Co., Ltd.
3-15, Yakumo-nakamachi, Moriguchi, Osaka 570, JAPAN

ABSTRACT

A new electronic video camera image stabilizer has been developed, which eliminates a substantial part of the image instability caused by the involuntary movement of camera holders. The discrimination between image movement caused by unstable hand-holding and that of moving objects becomes possible by developing the following process: 1) dividing the image taken by the camera into four regions 2) providing two signals to discriminate between the causes of image instability 3) evaluating these two discriminating signals after they are transformed into reliability values by membership functions and 4) tuning the membership functions using a simplex method. This image stabilizer has been incorporated into a new compact video camera and its substantially improved field performance has been confirmed.

1. INTRODUCTION

While the development and popularization of compact and light weight video cameras have been appreciated by consumers, the problem of shaky and unstable images produced by unstable camera holding has been left unresolved until recently.

Although an effective electronic image stabilizer which automatically compensates for the movement of images by extracting motion vectors and counter-shifting the image according to the magnitude of the motion vector in each image has been developed, it suffers from inherent instabilities due to difficulties in discriminating between movement caused by unstable camera holding and the movement of objects[1],[2].

We have developed a new electronic image stabilizer using fuzzy theory to discriminate between the two kinds of image movement. In the process, the image is divided into four regions, and motion vectors corresponding to these four regions are derived. At the same time, two signals for discriminating the causes of image movement are selected for each of the regions. Since the discriminating signals greatly depend on the character of the image, these signals are transformed into reliability factors by employing membership functions. The parameters of the membership functions are then tuned by a simplex method which is an optimization technique.

The effectiveness of this stabilizing process has been verified by the high capability of the procedure to discriminate between the two kinds of image movement.

2. PRINCIPLE OF THE IMAGE STABILIZER

Fig. 1 shows the system configuration for the electronic video camera image stabilizer which divides the image into four regions. The motion vector v_i and the discriminating signal w_{ij} ($i=1,2,3,4$, $j=1,2$) for each region are determined. The motion vector is derived from the correlation between the movement of the images. The discriminating signal is calculated from the mean and minimum values for the correlation factors.

The two discriminating signals selected from each region are converted into two values (e_{i1}, e_{i2}) in the closed interval [0,1] by using membership functions. Hereafter these are referred to as reliability factors. When the reliability factor approaches 1, it indicates the moving image is caused by unstable camera holding. Movement toward 0 indicates that the moving image is caused by movement of the object being filmed.

These two reliability factors are averaged to yield a result E_i for the discrimination. Based on this E_i, the motion vector v_i produced by unstable camera holding is selected for determining the motion vector V for the image. Thus, a stable image can be obtained by shifting the image in an equal and opposite direction to the motion vector V.

Tuning of the membership function is conducted at the design stage of the system. A function to evaluate the discrimination performance is introduced on the basis of the result E_i. Then, the membership function is tuned using a simplex method.

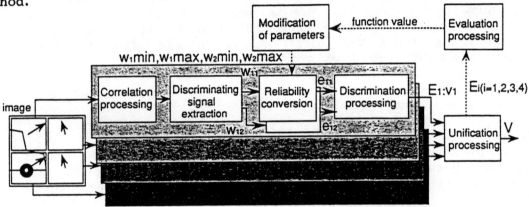

Fig.1 System configuration of electronic image stabilizer

2.1 Determination of The Motion Vector

Fig. 2 shows the division of the image into four regions, each of which is further divided into 30 sub-regions, and each of these has a representative point. The correlation value R_i is derived from the difference between the present image data for the representative point and the preceding image data for the point shifted by the amounts (p,q). A shift to give a minimum correlation value defines the motion vector v_i.

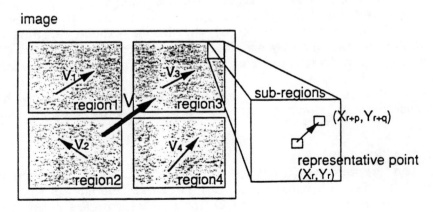

Fig.2 Determination of motion vector from image data

$$R_i(p,q) = \sum_{r=1}^{N} | S(t-1, X_r, Y_r) - S(t, X_{r+p}, Y_{r+q}) | \quad (1)$$

$v_i = (p,q)$ for min $R_i(p,q)$

where $R_i(p,q)$: correlation value for the position shifted from the representative point by (p,q), i : i-th divided region, N : the number of representative points in one region, $S(t-1, X_r, Y_r)$: image data at the position of r-th representative point (X_r, Y_r) in the preceding image, t : field number, $S(t, X_{r+p}, Y_{r+q})$: image data at the position shifted from r-th representative point by (p,q) in the present image, and v_i : motion vector in each region.

2.2 Extraction of Discriminating Signals

Fig. 3 shows a change in the correlation value expressed by Eq. 1 caused by the shift of (p,q). The correlation value takes the minimum value shown in curve(a) for normal conditions. The position of the minimum value determines the corresponding motion vector. However, when a moving object comes into the region or a different motion vector comes to exist within the region, a decrease in the difference between the minimum and the average correlation value takes place as shown in curve(b).

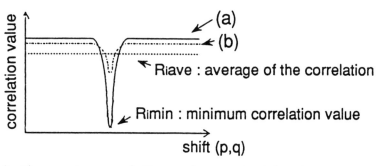

Fig.3 Change in correlation value against (p,q)

This reduces accuracy in detecting the motion vector because of a lowered correlation value between the images. In order to discriminate between such conditions, the discriminating signals, based on the minimum, average and differentiated minimum correlation values, are developed as shown below.

$$w_{i1} = R_i\text{ave} - R_i\text{min} * \beta \quad (2)$$
$$w_{i2} = d(R_i\text{min})/dt \quad (3)$$

where w_{i1} and w_{i2} : discriminating signals, $R_i$ave : average of the correlation values, $R_i$min : minimum correlation value, β : constant.

The lower the value of w_{i1} or the higher the value of w_{i2}, the higher the possibility that conflicting motion vectors exist.

Fig. 4 shows comparisons of time-changes (field-changes) between respective signals for a moving object and for moving images caused by unstable camera holding in one region. Figs. 4(a) and 4(b) show time-change for the motion vector for an invading object and time-changes for the signals w_{i1} and w_{i2} respectively. Figs. 4(c) and 4(d) show time-change for the motion vector for the unstable image and time-changes for w_{i1} and w_{i2} respectively. Fig. 4(a) shows the detection of the motion vector for an invading object for the intervals a' and a". Fig. 4(b) shows a decrease in w_{i1} and an increase in w_{i2} at respective times prior to each of these intervals. Such changes in the signals w_{i1} and w_{i2} indicate invading objects.

The pattern shown in Fig. 4(c) has to be consistently interpreted as unstable camera holding, but as shown in Fig. 4(d), other cases of decrease in w_{i1} or increase in w_{i2} are possible since the values of w_{i1} and w_{i2} are subject to change with a variation in image. Therefore, a considerably high possibility of erratic discrimination is to be expected despite the two-valued logic which provides a threshold value for each of the signals.

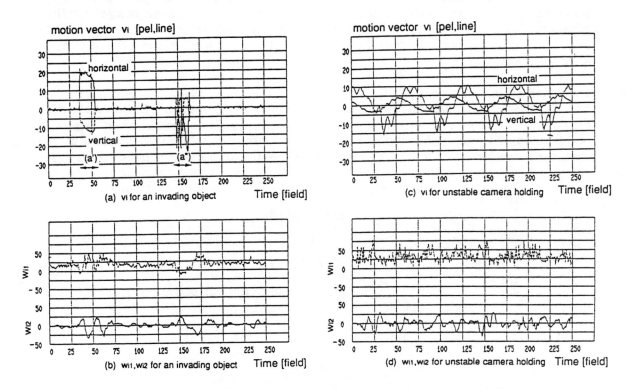

Fig.4 Motion vector v_i, and discrimination signals w_{i1} and w_{i2} for one region

2.3 Discrimination Based on Membership Functions

As mentioned before, since the value of discriminating signal w_{ij} depends on the image character, it is transformed into reliability factor e_{ij} within a closed interval [0,1] for evaluation. Fig. 5 shows the membership functions for the discriminating signals w_{i1} and w_{i2}.

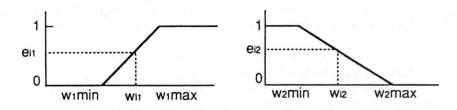

Fig.5 Membership functions

In this case, the higher the value of signal w_{i1}, the higher the probability of unstable camera holding, while the higher the value of signal w_{i2}, the higher the probability of a moving object. The membership functions for both cases are expressed by the two patterns shown in Fig. 5. The respective reliability factor e_{ij} for each signal w_{ij} is derived from Eq. 4 using these membership functions.

$$e_{ij} = \mu_j(w_{ij}) \quad (i=1,2,3,4), (j=1,2) \tag{4}$$

where μ_j is the membership function.

The accuracy of the discrimination is improved by deriving an average E_i for the reliability factors.

$$E_i = (e_{i1} + e_{i2})/2 \tag{5}$$

The determination of the cause of regional image movement is made according to the conditional equations given below.

$$\begin{array}{ll} \text{Unstable camera holding} & \text{if } E_i > 0.5 \text{ or} \\ \text{Movement of object} & \text{if } E_i \leq 0.5 \end{array} \tag{6}$$

However, even if a moving object can be detected using Eq. 6, when a moving object completely fills the region, patterns for the discriminating signals similar to that for unstable camera holding will be produced. In order to eliminate this problem, if the motion vector is attributed to a moving object, and afterward attributed to unstable camera holding by Eq. 6, then the motion vector is regarded as belonging to a moving object unless it coincides with the other motion vectors attributed to unstable holding.

Moreover, even when the motion vectors are attributed to unstable camera holding, their maximum and minimum values should be attributed to movement different from that of the background. Therefore, as a final process to unify the four motion vectors, a median of motion vectors attributed to the unstable camera holding is calculated in order to determine the motion vector V for the image.

2.4 Automatic Tuning of Membership Functions

The membership function has to be optimized in order to derive a precise result for the discrimination. There are four real numbers for break points represented in the shape of membership functions. They are $w_j\text{min}$ and $w_j\text{max}$ ($j=1,2$) as shown in Fig. 5. These four real numbers have to be optimized by using a simplex method[3].

The tuning of fuzzy controllers using the simplex method has been experimented with before. It is considered effective when it is impossible to differentiate the tuning model[4]. The simplex method is a trial and error method by which an n-dimensional geometrical figure called "simplex" (having points of more than n+1, each of which corresponds to function values) is constructed for deriving a minimum solution for the function values. An n-dimensional space X and its constraint condition are expressed by the following.

$$X = (x_1, x_2, ..., x_n) \tag{7}$$

where $a_\ell < x_\ell < b_\ell$, a_ℓ and b_ℓ are constants, and $\ell = 1, 2, ..., n$.

Consider a simplex having m points, each of whose coordinates is expressed by:

$$X_k = (x_{1k}, x_{2k}, ..., x_{nk}) \tag{8}$$

where $k = 1, 2, ..., n+1, ..., m$.

The function value at each point X_k is expressed below.

$$P_k = P(X_k) \qquad (9)$$

A point taking the maximum value from function values $P_1,..., P_m$ of m points is called Xw. A reflecting point with Xg as the center is determined to be X_w'.

$$X_w' = X_G + \alpha (X_G - X_w) \qquad (10)$$
$$X_G = \frac{1}{m-1} \sum_{k=1, k \neq w}^{m} X_k$$

where α is a weighting coefficient in determining the reflection point.

Fig. 6 shows, for n = 3 and m = 4, the relationship between X_w and X_w' where X_G is the center between points X_1, X_2 and X_3. X_w' is the point obtained by reflecting Xw by α times about Xg. This means that the point X_w' is an exterior division point at which the line segment $X_w X_G$ is divided by a ratio of $(1+\alpha) : \alpha$.

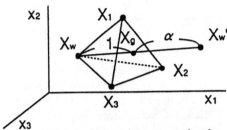

Fig.6 Searching method

The weighting coefficient α is adjusted until the function value at the reflecting point Xw' determined by Eq. 10 becomes lower than the function value at Xw. Thereafter, trial and error operation is repeated, and a minimum value at which all of the values for m-points converge is obtained.

This operation is ended when the standard deviation of the function value for each point becomes less than ε, or the condition shown below is satisfied.

$$\left[\frac{1}{m} \sum_{k=1}^{m} (P_k - Pave)^2 \right]^{1/2} < \varepsilon \qquad (11)$$

where Pave : average of P_k.

In the simplex method, Eq. 7 gives real numbers for the break points of the membership functions expressed by ($w_1$min, $w_1$max, $w_2$min, $w_2$max). The function values given by Eq. 9 for each set of image data are determined by the following process:
  1)Invading Object : The minimum E_{in}min for the function value is determined in a field right before the motion vector v_i for a moving object appears.
  2)Unstable Camera Holding : The ratio between the interval attributed to the inappropriate movement of an object and the total interval of data is determined to be S_{in}.

The sum of these two values is defined as the function P_k as shown below, and its minimum can be obtained.

$$P_k = \sum_{n=1}^{M} \sum_{i=1}^{4} (E_{in}min + S_{in}) \qquad (12)$$

where M is the number of sets of data available for evaluation.

3. EXPERIMENTAL RESULTS

Tuning experiments are conducted by preparing two sets of image data, one caused by unstable holding and the other by an invading object. The data in this experiment includes that in Fig. 4.

The results of tuning made on the four break points are $w_1 \max = 14.6$, $w_1 \min = 2.7$, $w_2 \max = 30.2$, $w_2 \min = 16.7$.

Figs. 7(a) and (b) show the motion vectors for the over-all image and the respective results of discriminations made for each region. Although the discrimination conditions for the image data shown in Fig. 4 presents one of the most difficult situations, the results in Fig. 7(a) show a successful elimination of the motion vector produced by a moving object. Fig. 7(b) shows consistent discrimination for the motion vector attributed to unstable camera holding for almost all intervals.

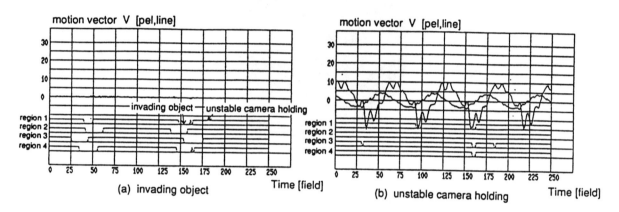

Fig.7 Motion vector after tuning and results of discrimination

An experiment to evaluate the discrimination capability of this system was conducted by constructing the device shown in Fig. 8, wherein a stationary video camera films a rectangular object moving at a constant speed in front of a stationary background.

The error rate in mistaking moving objects for unstable camera holding was measured for every twenty runs of the moving object made at various speeds. The results of this experiment are summarized in Fig. 9, which shows satisfactory results for the higher speeds but a slightly higher error rate at lower speeds.

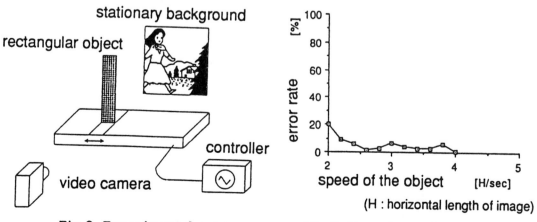

Fig.8 Experimental setup   Fig.9 Experimental results

This is attributed to the smaller difference between the average and the minimum for correlation values. In other words, this is caused by the smaller differences between the motion vectors for moving objects and the background. This is compensated for by suppressing the amount of compensation for the total image in actual cameras.

4. CONCLUSION

A method to discriminate between moving images caused by unstable camera holding and a moving object has been developed based on fuzzy theory. The results of this have been applied to an electronic image stabilizer incorporated in the recently developed video camera shown in Fig. 10.

Fig.10 Video camera with electronic image stabilizer

The discriminating capability of this image stabilizer has been substantially improved by providing a membership function for each of the discriminating signals, and evaluating the signals in terms of their reliability. The parameter of the membership function can be optimized by using a simplex method.

An experiment for determining the relationship between the speed of invading objects and the correct discrimination showed satisfactory results except for a slightly increased error rate in the low speed range.

References:
1) M.Oshima, et al. : "VHS Camcorder with Electronic Image Stabilizer", IEEE Trans. Consumer Electronics, Vol. 35, No. 4, pp. 749-758, Nov.1989.
2) K. Uomori, et al. : "Automatic Image Stabilizing System by Full-Digital Signal Processing", IEEE Trans. Consumer Electronics, Vol. 36, No. 3, pp. 510-519, Aug. 1990.
3) Nelder.J.A and R.Mead : "A Simplex Method for Function Minimization", Computer Journal, Vol.7, pp. 308-313, 1965
4) H. Akahori and Y. Egusa : "Auto-Tuning of Fuzzy Inference by Optimization Method and Its Application to a Running Cart", Resume of 5-th Fuzzy Symposium, pp. 77-82, 1989 (in Japanese).

Fuzzy Logic Based Banknote Transfer Control

Masayasu Sato[†], Tohru Kitagawa[†], Takehito Sekiguchi[†], Keisuke Watanabe[†]
and Masao Goto[††]

[†]Human Interface Laboratory, Oki Electric Industry Co., Ltd.
550-5, Higashiasakawa, Hachioji, Tokyo 193, Japan
[††]Human Interface Technology R&D Center, Oki Electric Industry Co., Ltd.
3-1, Futaba, Takasaki, Gunma Prefecture 370, Japan

Abstract-A fuzzy logic based banknote transfer control mechanism is proposed. In this mechanism, linear stepping motors are used for adjusting the gap between a feed roller and a gate roller. The gap is automatically adjusted to an optimal value through fuzzy reasoning, based on the statistical mean value of the feeding pitch and on the statistical mean value of the skew of the banknotes that have been fed.

Experimental results show that the mean value of the feeding pitch and the mean value of the skew are successfully adjusted automatically to error of ±0.28 msec and ±0.25° or less, respectively, by performing control within one to four times. We have confirmed that fuzzy logic is applied effectively to a banknote separation and feeding mechanism.

I. INTRODUCTION

A large number of automatic teller machines (ATMs) and cash dispensers (CDs) have been put into use to streamline and improve the efficiency of the teller window business at banks. In recent years, substantial liberalization of financial markets has occurred in Japan, and ATMs and CDs have come to be installed not only in bank premises but also in train stations, department stores, offices, convenience stores and the like, to be used much more widely. In addition, their operating hours have been extended and they are now available for certain time of periods at night and on holidays. These factors - installation in more varied places, longer operating hours and automated operation on holidays - have brought about an even greater demand for highly reliable machines that are free from troubles or failure. The separation and feeding technology is the most important for obtaining high reliability in ATMs and CDs.

Friction separation methods are commonly used at present for the separation and feeding of the banknotes, as they enable high speed separation and feeding as well as downsizing of machines. In these methods, the banknotes are separated and fed one by one from the gap between the feed roller and the gate roller. This separation and feeding capability varies by assembly error and changes in the operating environment and the passage of time. For example, the interval between one banknote and another may become abnormally small or a large skew may occur, thus affecting reliability of ATMs and CDs. Therefore, initial adjustment of the gap and subsequent adjustment must be performed frequently. And yet it is delicate to adjust the gap properly, as it requires skill.

In this paper, a fuzzy logic based banknote transfer control mechanism is proposed that automatically adjusts the gap to an optimal value through fuzzy reasoning based on the statistical mean value of the feeding pitch and the statistical mean value of the skew obtained from the banknotes that have been fed, and the results of experiments are described.

II. FRICTION SEPARATION MECHANISM

Friction separation mechanisms commonly used at present consist of two pickup rollers, two feed rollers and two gate rollers (see Fig. 1, but ignore linear stepping motors and the control system). The surfaces of these rollers are covered with a rubber material. Banknotes are pressed by spring force onto the pickup rollers. As the pickup rollers rotate, the top banknote is separated from the other banknotes and moved in the direction of rotation of the pickup rollers. The banknote then passes through the left and right gaps (X_L and X_R) between the feed rollers and the gate rollers (which are installed face to face against these feed rollers to prevent the feeding of more than one banknote at a time). The gaps (X_L and X_R) can be varied independently by the gap adjustment mechanism. In the friction separation methods, the conditions of feeding banknotes such as the feeding pitch, the skew, and the duplicate feeding are substantially affected by the coefficient of friction of the rubber on each roller and the gaps. The gaps have an especially substantial effect on the characteristics of the separation and feeding, depending on how they are adjusted. Thus adjusting the gaps demands skill.

III. PRINCIPLE OF FUZZY LOGIC BASED CONTROL MECHANISM

Fig. 1 shows the banknote separation and feeding mechanism based on fuzzy logic proposed in this paper.

For the following reasons, we have analyzed a system for automatically adjusting the gaps of a banknote separation and feeding mechanism using fuzzy logic.
(1) It is difficult to know the quantitative characteristics of the controlled object because of a large number of parameters.
(2) It is quite difficult to physically model the controlled object and the control system.
(3) The manual adjustments performed by an expert through understanding the qualitative characteristics of the controlled object and the control system can be partly expressed linguistically and turned into rules.

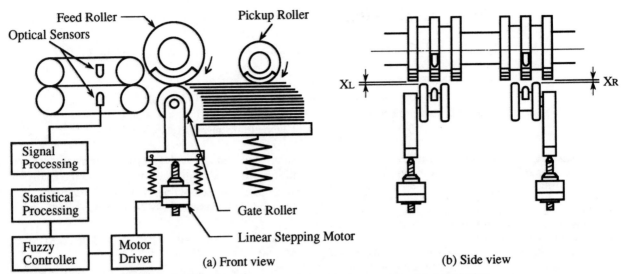

Fig. 1. Schematic diagram of fuzzy logic based banknote transfer control mechanism.

A. Control System

In the mechanism proposed, conventional manual adjustment mechanism is replaced by two linear stepping motors (25 μm/step). Two pairs of optical sensors in the banknote feed path detect the feeding pitch and the skew of the banknotes, and a feedback loop is structured so that the gaps (X_L and X_R) are controlled through fuzzy reasoning based on these data.

This is a two-input, two-output control system. The input fuzzy variables are the statistical mean value of the feeding pitch, and the statistical mean value of the skew of the banknotes that have been fed. The output fuzzy variables are the manipulating variable for the left motor and the manipulating variable for the right motor.

It is difficult to know the correlation between a banknote that has just passed through the optical sensors and the banknote to be separated and fed next. Therefore it would make the control system unstable if the gaps were controlled for each banknote that had been fed. This could cause repetitive fluctuations on the feeding pitch and the skew. In our system, therefore, the mean value of the feeding pitch and the mean value of the skew are used for control. This reduces the effect of the differences in the characteristics of the banknotes.

The feeding pitch and the skew of a banknote fed into the feed path through the gaps are detected by the optical sensors. As each banknote is fed into the feed path, these data detected are accumulated successively. Then each mean value is obtained. The difference between the mean value (p_m) of the feeding pitch (p_i) and the target value (p_r) is represented by Δp (= $p_m - p_r$), while the difference between the mean value (θ_m) of the skew (θ_i) and the target value (θ_r) is represented by $\Delta \theta$ (= $\theta_m - \theta_r$). Through fuzzy reasoning based on Δp and $\Delta \theta$, drive signals to the left and right motors are calculated independently. In response to these signals, the left and right gaps are adjusted independently by the motors so that the left and right gaps become optimal.

B. Fuzzy Control

Fuzzy control is studied on the basis of the specifications of fuzzy chip set developed by our company. Its specifications are shown in Table I. The Oki fuzzy chip set has two chips, a fuzzification chip (MSM91U044) and a defuzzification chip (MSM91U045). The max-min-gravity method proposed by Mamdani [1] is used. Up to 16 input variables and up to 2 output variables are available. For each input variable, up to 14 membership functions (up to 56 membership functions for all of the input variables) can be set. For the output variables, up to 15 membership functions can be set for each variable. The universe of discourse of the membership functions in the if-part is fixed at 64 elements. The universe of discourse of the membership functions in the then-part is 32 elements with a single output variable and 16 elements with double output variables. There are 16 grades in both the if-part and the then-part.

TABLE I
BASIC SPECIFICATIONS OF OKI FUZZY INFERENCE CHIP

Parameter	Condition	Standard Value			Unit
		MIN	TYP	MAX	
Inference Speed				7.5	MFLIPS
Number of Input Valuables		1		16	
Number of Output Valuables		1		2	
Number of Rules		1		960	Rule
Number of Membership Function	IF-part			56	Kind
	THEN-part			30	
Universe of Discourse	IF-part		64		Element
	THEN-part	16		32	
Grade			16		Step
Clock Frequency			20	30	MHz

An example of the basic characteristics of the gap and the feeding pitch is shown in Fig. 2. The horizontal axis represents the relative value of the gap, while the vertical axis represents Δp and $\Delta \theta$. As the gap is made larger, the feeding pitch becomes shorter than the target value. Conversely, as the gaps are made smaller, the feeding pitch becomes longer. The linear area is small, exhibiting nonlinear characteristics.

Table II shows the rules that are prepared for our system in consideration of nonlinear characteristics. The meanings of the labels given to Δp in Table II are as follows. ZR means that the mean value of the feeding pitch is almost equal to the target value, with a negligible difference ($\Delta p=0$); N means that the mean value of the feeding pitch is shorter than the target value ($\Delta p<0$), that is, the interval between one banknote and the next one is shorter than the target value, and P means that the mean value of the feeding pitch is longer than the target value ($\Delta p>0$), that is, the interval between one banknote and the next one is longer than the target value. As for the meanings of the labels given to $\Delta \theta$, ZR means that there is little skewing of banknotes ($\Delta \theta=0$); N means that the left side of the banknotes is ahead of the right side (left ahead) ($\Delta \theta<0$) and P means that the right side of the banknotes is ahead of the left side (right ahead) ($\Delta \theta>0$). For both Δp and $\Delta \theta$, L, M and S respectively mean "the difference from the target value is LARGE," "the difference is MEDIUM" and "the difference is SMALL". With respect to the meanings of the labels given to the output for the motor, ZR means that the output is made so as to hardly drive the motor; N means that the output is made so as to drive the motor in the direction of reducing the gap, and P means that the output is made so as to drive the motor in the direction of widening the gap. In addition, L, M, S and VS respectively indicate "drive to a LARGE degree," "drive to a MEDIUM degree" "drive to a SMALL degree" and "drive to a VERY SMALL degree". There are nine labels for the output to the motors.

Fig. 3 shows the membership functions. In Fig. 3, (a) is the membership function for the feeding pitch; (b) is the membership function for the skew, and (c) is the membership function for the output to the motor. The shapes of the membership functions for the skew are steep, to ensure more sensitive control . Moreover, for the membership functions for output to the motor, two functions, NVS and PVS, have been added to the seven general functions, for a total of nine functions, so that fine control can be achieved in the neighborhood of $\Delta p=0$ and $\Delta \theta=0$.

In Fig. 4, the results of calculation of the outputs to the left and right motors by the max-min-gravity method with Δp and $\Delta \theta$ as parameters are shown three-dimensionally using the rules and the membership functions described above. The (a) and (b) graphs are the results of calculation of the outputs to the left and right motors respectively. The X axis represents the skew. The center is where $\Delta \theta=0°$. Moving right from the center, the right-ahead skew becomes larger, while when moving left from the center, the left-ahead skew becomes larger. The Y axis represents the feeding pitch. The center of the Y axis is where $\Delta p=0$ msec. Moving forward, the feeding pitch becomes shorter than the target value, while when moving backward, the feeding pitch becomes longer than the target value. The Z axis represents the output to the motor. The center of the Z axis is where the output to the motor is zero. Moving up from the center, the output is such that the motor is driven in the direction of widening the gap. Moving down from the center, the output is such that the motor is driven in the direction of reducing the gap.

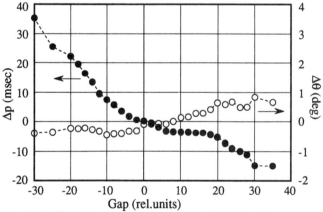

Fig. 2. Feeding pitch and skew characteristics. The left and right gaps are varied by a certain quantity at the same time.

TABLE II
RULE TABLES FOR THE MOTOR CONTROL

(a) Rules for controlling the left motor

		Left			$\Delta \theta$			Right
		NL	NM	NS	ZR	PS	PM	PL
Short	NL	NL	NL	NL	NL	NM	NS	ZR
	NM	NM	NM	NM	NL	NS	ZR	PS
	NS	NS	NS	NS	NS	ZR	PS	PM
Δp	ZR	NM	NS	NVS	ZR	PVS	PS	PM
	PS	NM	NS	ZR	PS	PS	PS	PS
Long	PM	NS	ZR	PS	PM	PM	PM	PM
	PL	ZR	PS	PM	PL	PL	PL	PL

(b) Rules for controlling the rigth motor

		Left			$\Delta \theta$			Right
		NL	NM	NS	ZR	PS	PM	PL
Short	NL	ZR	NS	NM	NL	NL	NL	NL
	NM	PS	ZR	NS	NL	NM	NM	NM
	NS	PM	PS	ZR	NS	NS	NS	NS
Δp	ZR	PM	PS	PVS	ZR	NVS	NS	NM
	PS	PS	PS	PS	PS	ZR	NS	NM
Long	PM	PM	PM	PM	PS	ZR	NS
	PL	PL	PL	PL	PL	PM	PS	ZR

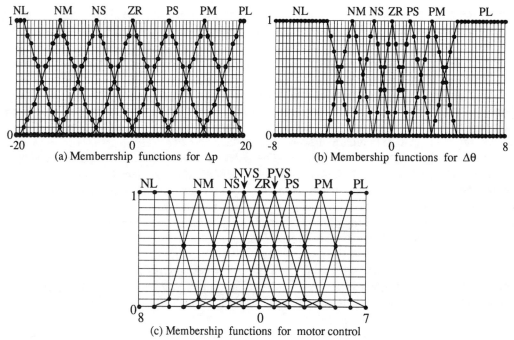

Fig. 3. Membership functions for the motor control

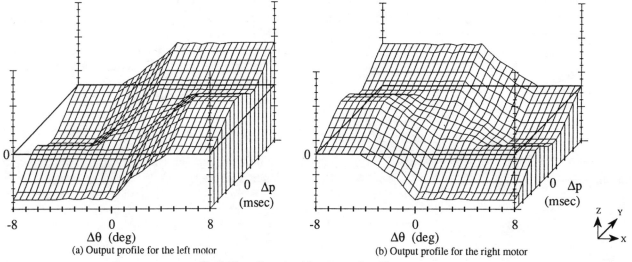

Fig. 4. Three dimensional fuzzy control profile.

IV. EXPERIMENTAL RESULTS

Experiments are performed using 500 in-circulation banknotes. They vary widely in paper quality, extent of damage, and other conditions. They are not arranged uniformly in terms of either direction or faces. The pickup roller, the feed roller, and other parts are not new, which means that they have been already used for some time. Therefore, they have some abrasion.

The experiments are performed according to the following steps:

1) The 500 in-circulation banknotes are used. The mean value of the feeding pitch (pm) and the mean value of the skew (θm) of the 500 banknotes that have passed through are obtained.

2) The differences from the respective target values (Δp and $\Delta \theta$ respectively) are obtained. Fuzzy control is performed based on those values.

3) Operations 1) and 2) are repeated until the outputs to both the left and the right motors become zero.

The results of the experiment are shown in Fig. 5, Fig. 6 and Fig. 7.

Fig. 5 shows the results in histogram form. The upper graphs are histograms of the feeding pitch, while the lower graphs are histograms of the skew. In the upper graphs, the horizontal axis represents the difference between the feeding pitch of the banknote that have been fed and the target value. In the lower graphs, the horizontal axis represents the difference between the skew and the target value. In both the upper and lower graphs, the vertical axis represents frequency. The (a) graphs are histograms of the feeding pitch and the skew for the 500 banknotes passed through without the fuzzy control. The (b) graphs indicate the results of the fuzzy control performed on (a) (first control). And the (c) graphs indicate the results of the fuzzy control performed on (b) (second control). In the initial state, or in (a), Δp=23.6 msec and $\Delta\theta$=-2.4°. After the first control, Δp=2.1 msec and $\Delta\theta$=0.05°. That is, the skew had been already controlled to within the target value. After the second control, or in (b), Δp=0.23 msec and $\Delta\theta$=0.09°. That is, the feeding pitch was also controlled to within the target value.

Fig. 6 shows the process of the feeding pitch and the skew converging on the respective target values as a result of the fuzzy control. These are data for a set of 100 banknotes (1

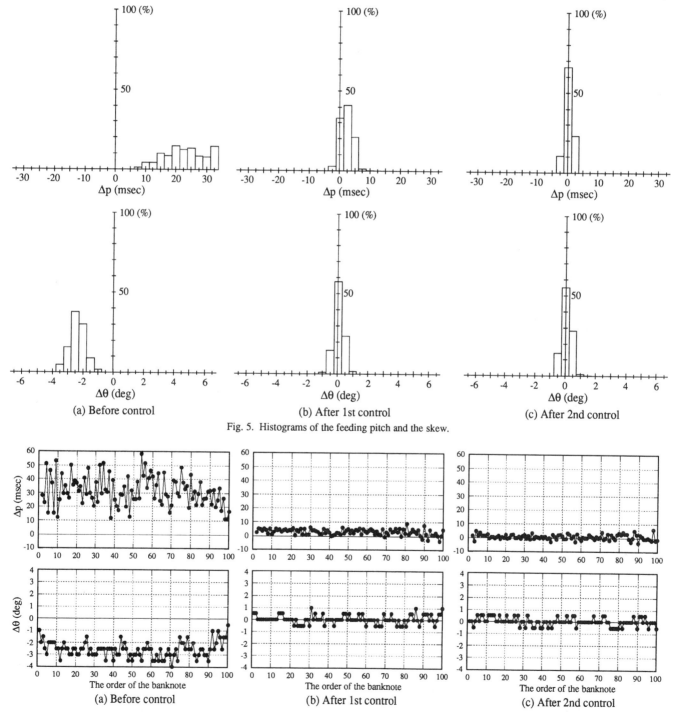

Fig. 5. Histograms of the feeding pitch and the skew.

Fig. 6. Variations of the feeding pitch and the skew measurement when a series of banknote feeding by fuzzy logic control.

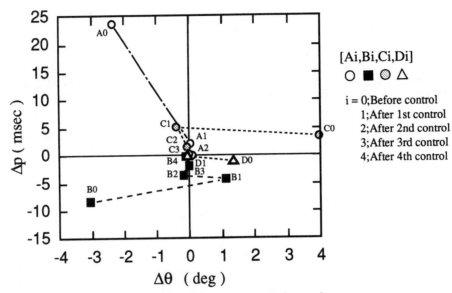

Fig. 7. Feeding pitch and skew trajectory by fuzzy logic control

unit) among the data on the feeding of the banknotes shown in Fig. 5. The (a), (b) and (c) graphs in Fig. 6 indicate "(a) before control," "(b) first control" and "(c) second control," respectively as in Fig. 5. The horizontal axis represents the order of banknote separated and fed. The vertical axis in the upper graphs indicates the difference between the feeding pitch that have been fed and the target value, while the vertical axis in the lower graphs indicates the difference of the skew and the target value. These graphs show the process by which the uneven feeding pitch and skew in the initial state converge on the respective target values through the repetition of control. In the initial state, the standard deviation of the feeding pitch is 10.7 and that of the skew is 0.66. After the second control, these values are 1.81 and 0.32 respectively.

Fig. 7 shows the results of examining the effect of fuzzy control by changing the initial state in various ways. The vertical axis represents the feeding pitch and the horizontal axis represents the skew. The point of intersection of the vertical axis and the horizontal axis is a target point ($\Delta p=0$ and $\Delta\theta=0$). The each marked points (○, ■, ◉ and △) represent the mean values of the feeding pitch and the skew obtained from the experiment on the 500 in-circulation banknotes. As a result of the experiment, the mean values of the feeding pitch and the skew are successfully controlled within ±0.28 msec and ±0.25°, respectively, by performing control once or twice, at most four times. Even when control had to be repeated four times in order to reduce the differences to ±0.28 msec and ±0.25° or less, differences of ±2.8 msec and ±0.5° or less were achieved in the third control. Moreover, when the initial state is in the neighborhood of the target values, the convergence is achieved by performing control only once.

V. CONCLUSION

The adjustment of a banknote separation and feeding mechanism, which is manually performed by experts, has been successfully automated by applying fuzzy control. Through the performance of control from one to four times, the mean value of the feeding pitch and the mean value for skew are automatically adjusted to differences of less than ±0.28 msec and ±0.25°, respectively, from the target values. When the differences are ±2.8 msec or less and ±0.5° or less, control are required only once. In short, a highly precise adjustment mechanism with a simple structure is realized by grasping the basic characteristics of a friction separation mechanism. Thus, we have confirmed that fuzzy control can be applied effectively to a friction separation mechanism.

ACKNOWLEDGMENTS

We express our deep gratitude to Mr. Tuji, Mr. Koshida, Mr. Ishidate and Mr. Ozawa, who helped us prepare rules and offer the experimental mechanism, and Mr. Kuroe who helped us design control system and further gave us precious advice in our execution of the experiment.

REFERENCE

[1] E. H. Mamdani : "Application of Fuzzy Algorithms for Control of Simple Dynamic Plant ", Proc. IEE, vol. 121, no. 12, pp. 1585-1588, 1974

Electrophotography Process Control Method based on Neural Network and Fuzzy Theory

Tetsuya MORITA†, Mitsuhisa KANAYA†, Tatsuya INAGAKI†,
Hisao MURAYAMA‡, Shinji KATO‡

Ricoh Co., Ltd. R&D Center†. RP development center‡ E-mail: morita@ipe.rdc.ricoh.co.jp

Keywords: Electrophotography, Latent image, Toner supply, Fuzzy control, Neural Network

Abstract

We present an electrophotography process(EP) control method using fuzzy theory and a neural network to control the toner supply and maintain the latent-image[1]. Although the accuracy and stability of the voltage control system is satisfiable, a response delay to control the copy density is not compensated by the toner supply control. In this paper, a developing bias control is applied to stabilize the copy density against the fluctuation of the toner concentration.

1 Introduction

In the electrophotography process, an electrostatic image on a photoconductor is first produced by photoelectric formation, then a visible image is developed by transferring electrically charged toner particles to the latent image by electrostatic power. The primary architecture of electrophotography process is composed of (1)Photosensitive drum, (2)Charging unit, (3)Exposure unit, (4)Developing unit, (5)Transferring unit, (6)Fixing unit, (7)Discharging unit, and (8)Cleaning unit, as shown in Figure 1.

The first step of electrophotography(EP) is to create a uniform charge on the surface of the photosensitive drum and to maintain the surface voltage at a required potential. The scorotron sensitizing method, which uses a corona discharge, is well known as a charging method. The second step is to create a latent image by exposure using a halogenous lamp or a laser diode. The third step is a development to make the powdered toner stuck to the latent image by electrostatic power.

The characteristics of corona discharge in the air are strongly affected by temperature or humidity. Also, the sensitivity and the charge retentivity of a photoconductor, which is made of selenium or organic materials, are also affected by environmental conditions. In addition, change in toner consumption or fatigue by long term use sometimes cause problems by fluctuation of charging and exposure voltages.

We applied neural networks to the electrophotography process to control the surface voltage of latent image and also applied fuzzy theory to control the toner supply quantity. As a result, the process had more stabilized image density even in a large volume

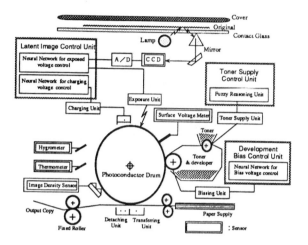

Figure 1: Diagram of the Electrophotography Process Control System.

photocopy and less toner blots on the paper ground even in environmental conditions changes.

2 Control of the electrostatic latent image

In order to maintain the latent image conditions at a necessary potential, we need to know the relationship between the charging grid voltage and the charge potential on the surface of photoconductor drum, as well as the relationship between the exposing light power and exposed potential. These relations are, however, hard to describe as a physical model to control, because they are readily affected by many environmental parameters, the values of which might vary in a non-linear and non-orthogonal way. Furthermore, some parameters such as sensitivity or fatigue of the photoconductor are unmeasurable.

Figure 2 illustrates the conventional feed back loop for latent image control. In this system, a special latent pattern for measuring a charged voltage and exposed voltage are written on the drum. Then, these voltage values are returned to the latent image controller to determine the next charging voltage and exposing power. The latent image controller calculates the differential values between the target voltages and the measured voltages to determine the next values to reduce errors. Such a feed back system, however, needs to repeat measurements and calculations of the surface voltages until these values satisfy the target condition. This process consumes time and wastes toner. It may also put a burden on the cleaning unit.

posing light power and exposed potential. Thus the neural networks allow us to estimate the surface voltages without the actual charging or exposure of the photoconductor. Forming a pseudo feed back loop with the previous networks, the system is able to determine the appropriate charging voltage and exposing power to produce the necessary surface voltage. This process saves time and toner, and prevents the drum and the cleaning unit from excessive abrasion.

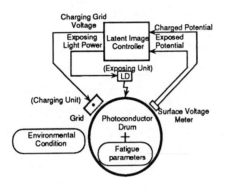

Figure 2: Conventional feed back loop for EP control.

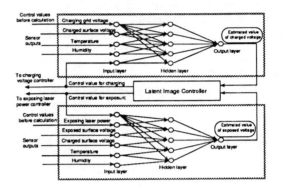

Figure 4: Network architecture which decide the charging grid and the LD voltage.

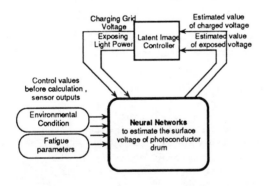

Figure 3: Pseudo feed back loop with Neural Network which estimates the OPC surface voltage for EP control.

Figure 3 illustrates the pseudo feed back loop with neural networks for controlling the latent image voltage. In detail, figure 4 shows the neural networks which estimates the OPC surface voltages. The neural networks learn the charging and exposing characteristics of the photoconductor. This means, after learning, the networks know a relationship between a charging grid voltage and charged potential on the surface of the drum and a relationship between an ex-

In one learning cycle, two sets of parameters are sampled in the consecutive two photocopies are used. Some of the input parameters to the networks describe the environmental conditions such as temperature and humidity, while the others are related to the fatigue condition of the photoconductor such as a charging grid voltage, exposed power, and the surface voltages which were measured before. On the other hand, the surface voltages which are measured during the present photocopy are used for the output teaching values. To sample the various combinations of the input and output data, we conducted experiments under the three different temperature and humidity conditions, using five photoconductors with different thicknesses, five charging grid voltages and exposing powers for each parameter. The backpropagation learning cycle was carried out 500,000 iterations on a SUN Sparc Station.

For the evaluation of the voltage control system, we employed five drums which are used in the learning process and two drums which are not used in the learning. The conventional fluctuation of the surface voltage was 50 to 60V by the feed back method using a standard organic photo conductor(OPC). As a result of the new method, both of the surface voltages can be controlled to within ±10V of the target voltage. The errors in controlling the charging voltage is shown in Figure 5. Even in the case of two

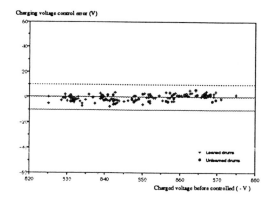

Figure 5: Charged surface voltage before control v.s. control errors.

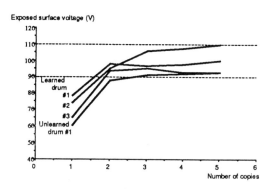

Figure 6: Exposed surface voltage before and after control.

unlearned drums, the precision of control was same as the learned drums. That means the generalization capability of the Neural Network is enough to estimate the surface voltage of unlearned drums. Figure 6 shows the fluctuation of the exposed surface voltage, which is controlled in the target range of 100V±10V.

3 Image density compensation by developing bias control

3.1 Fluctuation of toner concentration and tone reproduction

The purpose of conventional developing bias control is to reduce toner blots, moreover to reproduce tone in highlight area. In our previous system, the latent image is sufficiently maintained against the environmental conditions and fatigue parameters. Consequently, the developing bias control unit for clearing those problems is not always necessary. In this section, a new developing bias control unit which compensates the fluctuations in tone reproduction caused by that of toner concentration. The reproduction process of an original image with electrophotography is shown in figure 7. This process is composed of leading a plate voltage from an original density in a second quadrant, leading amounts of toner deposition from a plate voltage in a third quadrant and leading a copy density from amounts of toner deposition in a fourth quadrant. These processes which are represented in second to fourth quadrants are called as "ILLUMINATION & EXPOSURE", "IMAGE DEVELOPMENT", and "TRANSFER & FIX", respectively. These are well matched with the primary processes in the electrophotography method. As a result of these processes, the variations of tone reproduction are drawn in the first quadrant. Figure 7 shows the relations between toner concentration(TC) and tone rendition in a EP process. For example, when a TC is high, the reproduced tone is saturated. Conversely, when a TC is low, the shadow areas are not well reproduced and what is worse, the copy densities are washy as a whole area.

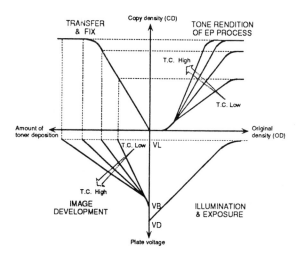

Figure 7: EP subsystems for describing the relation between TC and tone rendition

These problems are compensated by controlling a developing bias voltage. Figure 8 shows the characteristics between the original density and the copy density. When the developing bias(VB) is low, such as VB' and VB", the copy densities(CD) are shifted to washy. Conversely, when VB is high, CDs are made darker. Thus, the compensation of the image density is achieved by using these characteristics; high VB

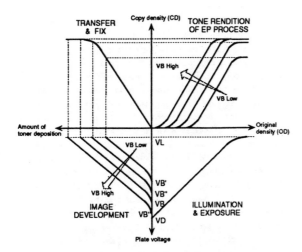

Figure 8: EP subsystems for describing the relation between VB and tone rendition

should be used in low TC and low VB should be used in high TC.

3.2 Neural network for VB control

The neural network in figure 9 requires the optimal VB by using data from the image density sensor, thermometer, hygrometer and copy counter. During the learning process, a number of environmental conditions and fatigue parameters are input on the input-layer, and the optimal VBs are used as the teaching value. The optimal VB means the VB value with which the ID value of the standard pattern exposed by 11th level is reproduced to "1.2". Figure 10 represents the printer gammas which are sampled with five kinds of VBs and eight kinds of TCs in constant temperature and humidity. The optimal VBs for teaching values are selected from these data.

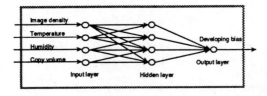

Figure 9: Network architecture which decide the developing bias voltage.

The gamma curves after or before compensation by using the new developing bias control are shown

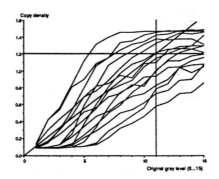

Figure 10: Uncontrolled printer gamma curves.

in figure 11. When the TC is maintained to the target value and the VB is set to the default value, the gamma curve which is labeled as "Normal" is almost linear and passes near the cross point of the lines of 11th level on the original gray and 1.2 on the copy density. On the other hand, if the TC is higher or lower than the target TC and the VB is set to the default value, the gamma curves which are labeled as "Saturated" and "Lean" are not linear and far from the cross point which is mentioned above. The two lines labeled as "Controlled" represent the gamma curve controlled by the developing bias control unit which is shown in figure 9. High VB or low VB is applied to compensate for the low TC or high TC respectively. As a result, MSE(Mean of Square Errors) between the normal curve and the controlled one are reduced to 70% to 50% as shown in table 1, and MSE between the linear and the controlled are shown is table 2.

	Low TC	Target TC	High TC
Low VB	–	–	0.092
Default VB	0.223	0.128	0.202
High VB	0.160	–	–

Table 1: MSE between the normal curve and the controlled one.

	Low TC	Target TC	High TC
Low VB	–	–	0.170
Default VB	0.327	0.128	0.195
High VB	0.139	–	–

Table 2: MSE between the linear curve and the controlled one.

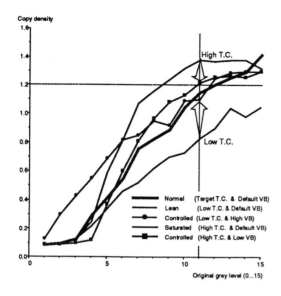

Figure 11: Gamma curves before or after compensation using the developing bias control.

4 Toner supply control using Fuzzy Theory

In the two-component development, toner concentration might vary compounded ratio of toner particle and developer. Decreasing the compounded ratio debases the image density, and too high ratio leads some toner blots on the paper ground. To maintain the ratio, consumed toner by development on each photocopy should be supplied quickly, but the toner supply unit retains some toner in it so that the transient control of the compounded ratio is difficult. Furthermore a toner particle is attached with a developer using the phenomenon of triboelectrification, so the variance of the temperature and humidity affect the compounded ratio. Thus it is tough to form a physical model for the control system. Fuzzy control is commonly used to control such systems which are difficult to model but could be described with some linguistic rules.

In the toner supply control, we applied fuzzy reasoning to calculate a coefficient of the quantity of toner, called GAIN, using the number of dots in a original image, the image density value and its differential. The number of dots means the total number of black dots in the original image which is sensed by the CCD scanner. The image density sensor is composed of a semiconductor-laser part and a photo diode, which measures the amount of toner particles by optical reflection. The following equation was employed to specify the quantity of toner.

$$Toner\_Supply = GAIN \times Num\_Dots \times Toner\_1dot \quad (1)$$

where Toner_Supply is the quantity of toner to be supplied, GAIN is the proportional coefficient, Num_Dots is the number of dots in an original image and Toner_1dot is the quantity of toner for printing one dot.

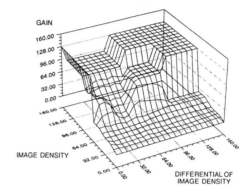

Figure 12: GAIN function with fuzzy reasoning.

Figure 12 illustrates the input and output function of the fuzzy reasoning, and its fuzzy rules and membership functions are adjusted for requiring the GAIN from the image density value and its differential. Two fuzzy input variables that are the image density and its differential are applied to each "IF" clause of 49 rules, and one fuzzy output variable that is GAIN value is led from "THEN" clause by the max-min composition.

For the toner supply experiment, we measured the changes in the image density values during 500 copies with several charts which include 7%, 50% or 80% of black area in each chart. The fluctuation of the image densities by a conventional crisp control is shown in Figure 13, and by the fuzzy control is shown in Figure 14. Five lines in each figure designate the cases of different exposure levels. Table 3 shows the standard deviations and the peek-to-peek errors of image density values for each chart.

Chart	Crisp control			Fuzzy control		
	7%	50%	80%	7%	50%	80%
Std. dev.	0.083	0.126	0.162	0.010	0.016	0.018
Max. err.	0.28	0.39	0.40	0.04	0.06	0.05

Table 3: Standard deviation and peek-to-peek errors. The variable number of dots are used in both methods.

As a result, even in the terms of copies of high percentage charts, such as 50% or 80% chart, the fluctuation of image densities are reduced to 1/6 to 1/9.

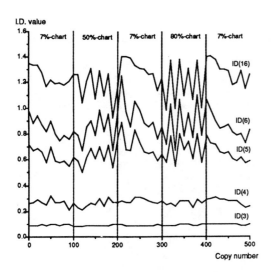

Figure 13: Changes in I.D. using conventional crisp control.

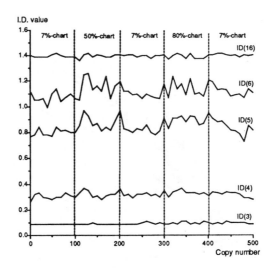

Figure 14: Changes in I.D. using Fuzzy control.

The above mentioned toner supply method is only available for the digital EP process, because the number of dots in an original image is employed in the formula (1). A digital EP process include the CCD scanner to digitally produce the area data in according to an original image, while an analog EP process does not include such means. However, the number of dots can't be known in the analog EP process, so we conducted the next experiments with a constant Num_Dots value in formula (1).

	Crisp control			Fuzzy control		
Chart	7%	50%	80%	7%	50%	80%
Std. dev.	0.051	0.094	0.100	0.046	0.053	0.023
Max. err.	0.20	0.27	0.34	0.14	0.15	0.05

Table 4: Standard deviation and peek-to-peek errors. Constant number of dots are used in both methods.

Table 4 shows the results of the second experiment. Using the constant number of dots value in the fuzzy control, the fluctuation became slightly worse than using the variable number of dots value. However, as well as in the case of variable number of dots, both of the fluctuations and peek-to-peek errors are reduced to 1/4 to 1/7 by the fuzzy control. From the above results, the new toner supply control system is available for both the digital and the analog EP process.

5 Conclusion

We introduced a new electrophotography process control method which includes the latent image voltage control system using a Neural Network, as well as a toner supply control system using fuzzy theory. Furthermore, a developing bias control is applied to stabilize the copy density against the fluctuation of the toner concentration.

As a result, the charged and the exposed potentials were controlled to within ±10V of the target voltage. This fluctuation is about 1/2 to 1/3 of that of the conventional voltage control system. Furthermore, by applying fuzzy reasoning for the toner supply control system, fluctuation of image density was reduced from 1/6 to 1/9. The new toner supply control system was useful for both the digital and the analog EP process. Using the developing bias control, the MSE between the normal gamma curve and the controlled one was reduced from 70% to 50%.

References

[1] T.Morita, et.al, "Electrophotography Process Control Method based on Neural Network and Fuzzy Theory." Proc.of the 2nd Intl. Conf. on Fuzzy Logic and Neural Networks, 1992, will be published.

[2] R.M.Schaffert, Electrophotography. London: The Focal Press, 1965.

[3] M.Kawato, et.al, "Hierarchical Neural Network Model for Voluntary Movement with Application to Robotics.", IEEE Control Systems Magazine, 8, 2, PP.8-16, 1988.

Fuzzy Logic Implementation of Intent Amplification in Virtual Reality

John Dockery
Center for Excellence in C3I[1]
George Mason University, Fairfax, Virginia 22030
and
David Littman
Computer Science Department
George Mason University, Fairfax, Virginia 22030

ABSTRACT--This note continues exploration of a concept, which the authors have proposed, called intelligent virtual reality. One of its features is detection and amplification of user intent. A first order implementation of that concept, which takes the form of a model, is introduced using the idea of possible worlds discussed by Ruspini in a paper on the semantics of fuzzy logic. The authors have previously proposed intelligent virtual reality as a universal interface for the handicapped; and, moreover, surveyed the possibility of extensive implementation in fuzzy set logic.

BACKGROUND

The idea of combining artificial intelligence with virtual reality in a construct called intelligent virtual reality has been introduced by Dockery and Littman [1,2]. The idea first arose in the context of a universal computer interface for the handicapped, which would be capable of detecting schematic user intent. What is evolving is a construct, which amplifies the barest outline of user intent into a full fledged scenario capable of execution by "smart", external robotic agents. A first order survey of possible implementation strategies revealed distinct advantages accruing to the use of fuzzy set concepts.

In Figure 1, following, the relationship of intelligence and reality is schematically portrayed. In the lower left corner we have what we call the scene shop or prop room. In the corner of which is predicate calculus. This quadrant expresses the backdrop or backcloth to reality analogous to a stage setting. Moving left one enters virtual reality while moving up brings one into normal physical reality. The entry in parenthesis along each side is the proposed primary mechanism for crossing the boundary. Also shown are schematic axes indicating the relative importance of fuzzy data logic and approximate reasoning in general to navigating in each quadrant.

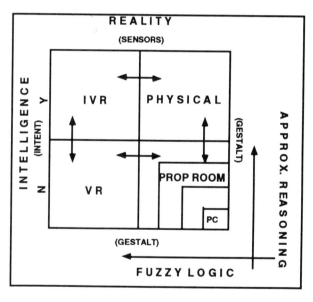

Fig. 1: Schematic interpretation of the relationships between intelligent virtual reality and other realities.

INTENT AMPLIFICATION

Space does not permit us to dwell further on the overall IVR concept. Rather, in this discussion we focus on what we consider to be a key aspect. This is the notion of detecting and amplifying the user's intent, which we now sketch. The process is illustrated in Figure 2. We seek therefore to model the relation $m: I \to I'$ where I, I' express intent. (See endnote for sample scenario).

0-7803-0614-7/93$03.00 ©1993 IEEE

The user begins by signalling intent, which for a handicapped person may be from a very limited repertoire. As a consequence the controller must work from a very fuzzy input command combining same with a fuzzy description of context and user profile to produce a trial construct of a virtual world in which the action intended action(s) may be carried out. A refinement process then ensues as the controller populates the trial virtual world construct until it is sufficiently detailed to both permit the action(s) and to plan for execution by external agents These agents are "smart" and may negotiate among themselves. The cycle ends with the user signalling satisfactory accomplishment.

POSSIBLE WORLDS

A general scheme for implementation of the process illustrated in Figure 2 is suggested in an article by Ruspini, which treats the consideration of evidence in worlds connected to each other by a relation $\Re$ called the *accessibility* relation [3]. Briefly Ruspini's model has these elements:
• A non-empty set of possible worlds U which contain evidence E;
• A function for assigning truth values to proposition in worlds written, when the proposition is true as $w \mapsto \varnothing$; and
• Two operators of the relation $\Re$, which are $\mathbf{N}$ and Π standing for necessity and possibility. $\mathbf{N}$ requires that true proposition in w be true in all possible worlds w' accessible via $\Re$ while Π relaxes the requirement so that there need only exist at least one w' in which truth is possibly maintained.

We have adapted this model to the process in Figure 2 by identifying the trial virtual worlds together with the real world as constituting the set U. To initiate the process we require that the first cut at virtual world w' be reachable from the real world by the

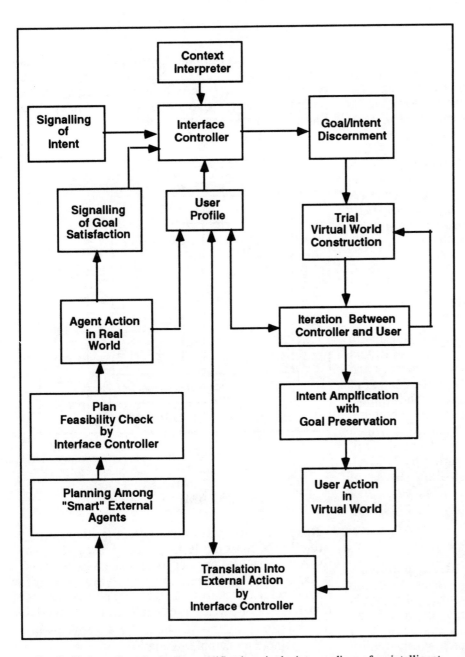

Fig. 2: Process flow for intent amplification via the intermediary of an intelligent virtual reality controller interface.

relation Π. Subsequent iterations must preserve the developing set of propositions. Ergo, to that sequence we assign the operator **N**. Eventually an IVR is constructed which permits both the accomplishment of the real and virtual action(s), which the user intends. Abduction is suggested as the process by which the evidence is weighed in successive trials. Conventionally abduction reasons backward to the antecedent from an observation of the consequent. Thus, $p \to q$, with q observed. Trouble is that there may exist several p_j which usually imply q. The problem has been treated by Whalen and Schott and by Whalen. [4-6] A general overview of evidential reasoning may be found in Ruspini, et al. [7]

After the action(s) are completed by the user, there is at least one proposition p which is no longer true in the altered virtual world. Consistency is maintained by the requirements of goal preservation which we shall shortly treat. It is also required that the virtual world, altered by internal action(s), still be an accessible one from the real world.

The problem is now to cause action(s) in the real world. First the IVR must plan; and then relay the instructions to the external agents followed by a feasibility check. The subject of team agents acting under uncertainty is treated by Cohen and Day in a general overview. [8]. We have also had recourse to Whalen and Berenji. [9]

Figure 3 following illustrates the steps outlined above. The first operator Π_1 establishes the initial trial world. Π_2 *establishes the possible truth that things are altered as required under the intended action(s)*. **A** is the action in the real world which changes at least one relation indicated by the prime.

PRESERVING THE INTENT

The foregoing was addressed to the problem of structuring a virtual world adequate to the task of building and populating an environment in which the user could carry out his intended action(s). We turn to the much more fuzzy concept of goal preservation. Again we have recourse to Ruspini's model for a possible formalism. [3] Formally stated we require the following: That there exist no assertion in the post-action condition which implies the negation of goals deduced from context (or explicitly stated) in the pre-action condition.

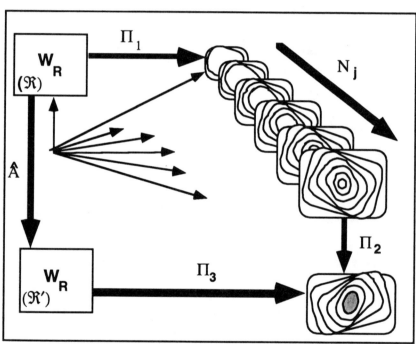

Fig. 3: Schematic interpretation of intent amplification using Ruspini's model of possible worlds. [3]

Ruspini introduces a similarity relation to express the similarity between accessible worlds. He defines it as follows: $S: U \times U \mapsto [0,1]$. Our own goal is to maintain a value as close as possible to unity for $S(w,w')$. S is a transitive relation satisfying the following relationship.

$S(w,w") \geq S(w,w') \otimes S(w',w")$ where $\otimes : [0,1] \times [0,1] \mapsto [0,1]$

As Ruspini says, the real complexity comes from comparing similarity relations between pairs of subsets of possible worlds.

He addresses the problem by defining additional operators. Additional development concludes that S is the $\otimes-transitive$ dual of the Hausdorff distance.

To attack the problem of partial implication an operator defining the degree of implication is introduced into the model as follows:

$$I(p|q) = \inf_{w' \mapsto q} \sup_{w \mapsto p} S(w,w') \quad \text{where } q \Rightarrow \Pi_\beta p \text{ if } \beta \leq I(p,q)$$

The interpretation given **I** is that it is a measure of *minimal amount of stretching* necessary to reach a p-world from an q-world. $[S(w,w') \geq a]$. Such an interpretation is in accord with our concept of intent amplification.

To bridge the gap between the virtual world in which the action is to take place and the virtual world altered by the action we have recourse to the degree of consistency *of the goals* between p and q defined as

$$C(p|q) = \sup_{w' \mapsto q} \sup_{w \mapsto p} S(w,w')$$

In Figure 4 following the application of these model operators to the preservation of goal-intent is schematically illustrated.

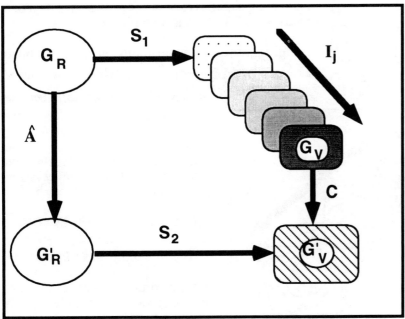

Fig. 4: Schematic interpretation of goal preservation under intent amplification using the intermediary of IVR according to the model of Ruspini. [3]

DISCUSSION

In order to produce practical model implementation, the definitions of Ruspini (above) need to be expanded to a basis which includes the entire set of evidence (E). For this purpose we have recourse to conditional possibility and necessity distributions defined in terms of a pseudo-inverse for triangular norms $\otimes$ as $a \oslash b = \sup\{c : b \otimes c \leq a\}$. They may then be written as follows:

$$Nec(q|p) \leq \inf_{w \mapsto E}[I(q|w) \oslash I(p|w)] \quad \text{and} \quad Pos(p|w) \geq \sup_{w \mapsto E}[I(q|w) \oslash I(p|w)]$$

The possibility of using T-norm based computations in the model under discussion means that a hierarchy of decision processes can be experimented with ranging from risk averse to risk taking according to user profile and context. It can easily develop that the achievement of conflicting goals can put the user at risk if the "wrong" possible world is constructed. To take an absurd example consider three goals out of n possible, to wit: g_1-stay on feet, g_2-own a pen, and g_3-stay alive. We move now to two possible worlds, w_1 and w_2, which are post action worlds with consequent actions a_1, a_2, and a_3 respectively. Now a_1 implies g_3 while the other two imply g_1 and g_2. In most cases the choice is obvious.

What is not covered in this first stage of implementation is any development of algorithms. We have treated only model formalism. Stage two will concentrate on efficient search algorithms as the search space of possible worlds is probably infinite. The metric introduced for goal preservation is intended as part of an efficient search strategy.

ADDITIONAL OBSERVATIONS

Modelling the user intent so as to produce a virtual world in which the handicapped can achieve barrier free access seems to us to be a perfect example of "soft" computing requirements. It is complex reasoning under uncertainty. Both judgment (assessment) skills and action skills are necessary.

Look at the computational environment. Intent is quintessentially a fuzzy concept; likewise, goal satisfaction. Sensor data and subsequent context interpretation will be fuzzy at best. Fuzzy interpolation seems the order of the day. However, as the construction of the virtual world converges to some satisfactory exemplar, conventional computing techniques will be relied upon for efficiency. For example, the task of refining the trial virtual worlds has been considered briefly. It appears to be amenable to the application of genetic algorithms.

As regards various modalities for reasoning about the evidence, we examined Whalen and Schott. [6] They discuss six types of inference of which two, presumption and prejudice, are considered an aberration. In this type of reasoning the truth value (believability) of the implication itself places limits on the truth values of antecedent and consequent. Thus, in prejudice the truth value of the consequent is always less than the antecedent. With presumption the consequent carries the roughly the same truth value as the implication. While perhaps a nuisance in standard applications, both prejudice and presumption could set default values on quirky sensors when context is being determined

SUMMARY

A model developed from consideration of the semantics of fuzzy logic has been advanced as a formalism for a first order implementation of intent amplification aspects of IVR as a universal computer interface. Goal preservation is the proposed metric by which to test the process. This article has introduced a model formalism. The next step is development of algorithms.

REFERENCES

1. Dockery, J .and Littman, D., "Intelligent Virtual Reality as a Universal Interface for the Handicapped", in *Proceedings of a Conference .on Interface to Real and Virtual Worlds*, Montpelier, France, March 23-27, 1992, available through Informatique, EC2, 269, rue de la Garene, 92024 Nanterre, Cedex, France

2. Dockery, J .and Littman, D. "Intelligent Virtual Reality in the Setting of Fuzzy Sets ", to be presented at the annual conference of NAFIPS [North American Fuzzy Information Processing Society], to be held in Puerta Vallarta, Mexico, December 14-17, 1992

3. Ruspini, E. H., "On the Semantics of Fuzzy Logic", *International Journal of Approximate Reasoning*, **5** (1), pp 45-88 (1991).

4. Whalen, T., and Schott, B., "Usuality, Regularity, and Fuzzy Set Logic", *International Journal of Approximate Reasoning*, **6** (4), pp 481-504 (1992).

5. Whalen, T., Private communication (1992).

6. Whalen, T. and Schott, B., "Presumption and Prejudice in Logical Inference", *International Journal of Approximate Reasoning*, **3** (5). pp 359-382 (1989).

7. Ruspini, E.H., Lowrance, J.D. and Strat, T.R., "Understanding Evidential Reasoning", *International Journal of Approximate Reasoning*, **6** (3), pp 401-424 (1992).

8. Cohen, P.R. and Day, D.S., "The Centrality of Autonomous Agents in Theories of Action Under Uncertainty", *International Journal of Approximate Reasoning*, **2** (3), pp 303-326 (1988).

9. Whalen, T., and Berenji, H.R., "Actions as Evidence: Multiple Epistemic Agents Acting Under Uncertainty", IEEE Conference on AI, Evaluation and Planning, Florida (1991).

ENDNOTE

Consider reaching for a book. Some generic hand motion begins the signalling. A trial virtual world is constructed, which is populated with objects within visual range. A subsequent external motion is translated so that a virtual hand "reaches" toward a particular field of objects in the trial virtual world. Subsequent refinement permits selection, grasping and removal. When completed the IVR assesses it external agents and attempts to plan the analogous action in the external world while at the same time checking feasibility.

---

[1] On an Intergovernmental Personnel Act assignment from the Defense Information Systems Agency at GMU on a part time basis.

On-line Analysis of Music Conductor's Two-Dimensional Motion

Zeungnam Bien and Jong-Sung Kim

Department of Electrical Engineering ,KAIST
P.O.Box 150, Cheongyang, Seoul, 130-650, KOREA

Abstract

This paper proposes an on-line method of understanding human's conducting action for chorus or orchestra observed through a vision sensor. The vision system captures images of conducting action and extracts image coordinates of each endpoint of the baton every 1/15 second. A fuzzy-logic-based inference algorithm is then applied for the information involved in conducting action. Complementary algorithms are also proposed for identifying first beat, static point, and dynamics of playing. Through a sequence of experiments organized around a system including IBM PC-386, a special purpose vision system and sound module, the algorithm is found to faithfully follow up the conductor's motion in playing a given song.

I. Introduction

An image grabbed through a vision sensor may be classified into static image and dynamic image. Note, however, that a dynamic image is in actuality a sequence of static images that are related by a time function defining orders and time interval of the sequences[1]. While the analysis of a static image is not affected by the computer processing time, the analysis of dynamic images is generally constrained by the processing time for relevant information should be extracted not only from each frame, but also from the sequences of the frames. That is, the details derived from each image must be integrated into a coherent whole.

Among various analysis methods for dynamic image that have been developed so far, three important approaches are the differencing technique, the temporal-spatial gradient analysis, and the matching analysis[2]. First, the differencing technique[3] handles point-by-point intensity changes in images. This can often be done efficiently by subtracting one image frame from another and thresholding the result. Second, the temporal-spatial gradient analysis(or optical flow analysis)[4] uses the intensity changes of corresponding points of subsequent images over both time and space to estimate the rate of translation of a surface. Third, in the matching analysis a set of structures in one image frame is used(as a reference), and an organized search for the corresponding structures is performed in subsequent frames.

The methods described above, however, show many practical difficulties in real-time analysis of moving objects because of inherent complexities of the algorithms and the lack of robustness to signal distortion[5]. Those difficulties may be eliminated by using special purpose hardwares and by simplifying image data(binarization) or the algorithm itself.

Examples of on-line pattern recognition can be found in handwriting recognition[6], gesture recognition[7], voice recognition[8], and music score recognition[9]. In these examples, one can see that human can extract features of input images amazingly fast. One of the most distinguished aspects of human knowledge processing is its non-numerical inferencing mechanism.

In this paper, we propose a system which extracts relevant information from motions of the conductor's baton and plays a music according to the intention of the conductor. For this, a fuzzy logic algorithm is developed for analyzing the motions of the baton and a special purpose hardware is designed for real-time processing of image data. The analysis of the motions of the baton is closely related with determination of upper and lower corners. Fuzzy logic is applied in this determination, and some examples are given to compare the proposed algorithm with numerical methods.

II. Recognition of Conducting Diagram

1. Conducting and conducting diagram

Conducting is a pantomine of gestures, poses, facial expressions, manual motions and bodily attitudes by means of which a conductor transmits silently his individual interpretation of a musical composition to an assemblage of performers-sometimes singers, sometimes instrumentalists, sometimes both[10]. In the beginning, a conductor indicates the start of playing and afterwards he forwards many orders related with tempo, rhythm, speed, accent, and so on. The conducting is performed primarily by drawing so called "conducting diagram". The diagram has been changed and modified along with time. In the early stage, a conductor used to indicate the start of music and keep the tempo by drawing simple circles. Conducting skill, howerever, has been developed to represent more detail information.(See Fig 1.)

Basic principles related to the conducting diagram are as follow. The upper and lower corners in the diagram are used to indicate division of beats. The first beat represents the start of each measure, which is indicated by the motion from top to down about the center of the conducting diagram. Also, the dynamics of music is represented by variation of the overall size of the conducting diagram. For examples, when the conductor sees the symbol f (or p)-mark on the way of playing, he enlarges(or reduces) the size of conducting diagram while speeding up(or down) the conducting not to cause changes of tempo. When he meets a *fermata*-mark, the conductor executes delay by stopping a baton for a while.

2. Recognition of upper corner and lower corner using fuzzy logic

In general, conductors perform various musics by using both hands. All the explicit and the implicit information of general form of conducting may be, however, very difficult to represent, and hence the following are assumed.

1) The conductor conducts using only one hand.
2) Only the principal conducting diagram is inputed.
3) The conducting motion remains in the predescribed visual window for efficient camera sensing.

The motion of baton is then tracked by a high-speed vision system while the following features are extracted according to the principles of conducting.

1) Every upper and lower corner in the conducting diagram.
2) First beat followed by normal rhythm (first of each measure).
3) Static point denoted as *fermata*.
4) Size of conducting diagram when dynamics($mf, f, mp, p...$) exists in the music.

The upper corner points and the lower corner points can be extracted by detecting direction change in the conducting diagram, and the first beat can be detected by finding the highest upper corner. The static points are also identified by comparing y-axis coordinates.

The upper corner and the lower corner that divide beats are the most important features in conducting motion. To identify these corners quickly, one can use direction changes as features[7]. In this case, however, difficulties may arise by noise factors such as conductor's hand-trembling or inherent noise of processing hardware. To overcome these difficulties, we have adopted a fuzzy logic inference method which takes direction changes along x & y axes as input, and the possibility of being a corner point as output. We have performed many experimens with a high speed vision system[11] sending x,y-coodinate data every 1/15 second. After examining the results, we have summarized rules for identifying corner points in the forms of linguistic rules given in Table 1. The input fuzzy variables are assigned to NB, NS, PS, and PB as shown in Fig.2. One of the rules, for example, reads as follows :

IF dy1 is positive big AND dy2 is negative big THEN the possibility of being a corner is completely high.

The output is inferred by the compositional rule of inference with the two inputs as direction changes along x and y coordinates. We use the "center-of-area" method(COA)[13] for defuzzification described as ;

$$u_o = \frac{\sum_{j=1}^{n} u(u_j) \cdot u_j}{\sum_{j=1}^{n} u(u_j)} ,$$

where u is the membership function, n is the number of quantization levels of the output. The output fuzzy variables are assigned to NB, NS, PS, and PB as shown in Fig.3. With two inputs dy1 and dy2 representing the direction changes of the latest three points, the defuzzified output indicates the possibility of being the upper corner or being the lower corner. The upper corners and the lower corners are determined by thresholding the resultant possibility : if the defuzzified output is greater than th1(threshold), the second point of the latest three points is a upper corner or lower corner, and if the defuzzified output is less than th1 and greater than th2(threshold; th1 > th2), we can't decide it at the current stage. Indecision may arise when the second point of the latest three points is possibly contaminated by noise or distorted by slow motion, and in this case the checking procedure resumes at the next stage. For reducing the processing time, we don't apply the inference method when dy1 and dy2 are the same sign.

3. Recognition of first beat and static point

The first beat of every measure is indicated by a downward motion of the conductor's right arm. This is "downbeat," or the count of ONE in each measure. This beat is checked by finding an upper corner above y-axis reference level. The existence of *fermata* also can be easily checked by examining the rate of changes in the motions of the baton.

4. Recognition of dynamics

Unlike tempo, which can be exactly determined by metronome, dynamics of conductor's motion is rather subjective and conveys only relative values. There is no absolute level for *piano* or *forte*, for example[12]. In the paper, we propose to adjust the volume of sound by comparing the lengths of links connecting upper corner and lower corner with respect to a reference length. Fig.4 shows two different sizes of conducting diagrams.

III. Experiments

A sequence of experiments are performed with the system composed of a special purpose vision system and a main computer equipped with sound module.

1. System Configuration

The vision system involves a MVME-110 unit operated under XINU and is developed at KAIST for high-speed image processing[11]. The center of object can be tracked every 1/15 second. The vision system is connected to main computer(IBM PC-386) via RS-232C. One image frame is digitized by the size of 256 × 256. The sound module can handle 12 channels independently.(Fig.5)

2. Image processing and recognition of conducting diagram

For real-time image processing, the vision system is equipped with lookup table for binarization, and binary image projection processor. The sequence of operation is as follow. "Yesterday"(sung by Beatles) is stored in sound module. A conductor conducts in such way that his motion resides in a prespecified window. The trajectory of baton calculated by the vision system is transmitted to IBM PC-386 via RS-232C every 1/15 sec. The received data is processed by fuzzy-logic-based algorithm and the song is played via the music card. An example of conducting motion trajectory and resultant possibility of inference is shown in Fig.6, where "*" marks sampling point and each sampling point is connected by a straight line. The numbers in parenthesis represent tempo in M.M.(Metronome by Malzel), which indicates the number of beats per minute[12].

3. Sound module

Given information of scores, the format of music data has three elements; note-key, steptime, and first beat. Each element represents pitch, interval between notes, and start of measure, respectively. Fig.7 is an example of format of music data. For the example of "62 1 1"(the first part of "Yesterday"), the music card makes sound of 'RE'(62) at the first lower corner indicated by last "1" of conducting diagram and continues till a next upper corner represented by the second "1" appears. Fig.8 is the procedure of the total system operation.

IV. Concluding Remarks

In this paper, we have proposed an on-line method of understanding human's conducting action observed through a high-speed image processing system. We showed the feasibility of the proposed algorithm by performing a sequence of experiments in which a popular song was played by a synthesizer according to human conducting. To speed up image processing, a special purpose hardware was adopted, and to make up for noisy data, a fuzzy-logic-based inference algorithm is developed. Experiments showed that the proposed algorithm faithfully followed up such intentions of conductor as speed and accent related with the dynamics of playing. The proposed algorithm, however, should be extended to more general cases such that analysis of the motion image of a conductor's body may convey various subtle information.

References

[1] W. N. Martin and J. K. Aggarwal, "Survey - Dynamic Scene Analysis," *Computer Graphics and Image Processing*, vol. 7, pp. 356-374, 1978.

[2] W. B. Thompson and S. T. Barbard, "Lower-level Estimation and Interpretation of visual motion," *IEEE Trans. Computer*, Aug. 1981.

[3] S. Yalamanchili, W. N. Martin, and J. K. Aggarwal, "Extraction of Moving Object Descriptions via Differencing," *Computer Graphics and Image Processing*, vol. 18, pp. 188-201, 1982.

[4] B. K. P. Horn and B. G. Schunck, "Determining Optical Flow," *Artificial Intelligence*, vol. 17, pp. 185-203, 1981.

[5] T. Uno, M. Ejiri and T. Tokunaga, "A Method of Real-Time Recognition of Moving Objects and its Application," *Pattern Recognition Pergamon Press*, vol. 8, pp. 201-208, 1976.

[6] S. Impedovo, B. Marangelli, and V. L. Plantamura, "Real-Time Recognition of Handwritten Numerals," *IEEE Trans. Syst., Man, Cybern.*, pp. 145-148, Feb. 1976.

[7] Joonki Kim, "On-line Gesture Recognition by Feature Analysis," *Computer Vision and Shape Recognition*, World Scientific, 1989.

[8] R. D. Mori and P. Laface, "Use of Fuzzy Algorithms for Phonetic and Phonemic Labeling of Continuous Speech," *IEEE Trans. Pattern Anal. Mach. Intell.*, vol. PAMI-2, No. 2, pp. 136-148, Mar. 1980.

[9] T. Toyo and H. Aoyama, "Automatic Recognition of Music Score," *IEEE Conference on Pattern Recognition*, pp. 1223, 1982.

[10] Robert Sabin, *The International Cyclopedia of Music and Musicians*, DODD, MEAD & Company, 1964.

[11] Z. Bien, S. R. Oh, J. Won, B. J. You, D. Han, and J. O. Kim, "Development of a Well-Structured Industrial Vision System," *16th Annual Conference of IEEE Industrial Electronics Society*, pp. 501-506, Nov. 1990.

[12] M. M. Miller, *Introduction to Music*, Barnes & Noble Books, 1958.
[13] C. C. Lee, "Fuzzy Logic in Control Systems: Fuzzy Logic Controller, Part II," *IEEE Trans. Syst., Man. Cybern.*, vol. SMC-20, No. 2, pp. 419-435, Apr. 1990.

Table 1. Rule table

dy1\dy2	NB	NS	PS	PB
NB			HI	CO
NS			LO	HI
PS	HI	LO		
PB	CO	HI		

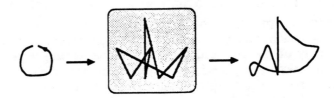

Fig 1. The transition of conduction diagram

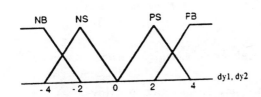

Fig 2. Input fuzzy variable

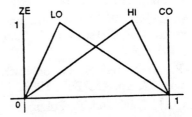

Fig 3. Output fuzzy variable

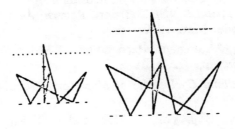

Fig. 4. The size of conduction diagram

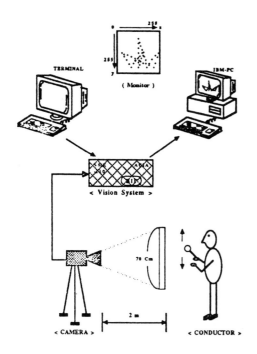

Fig 5. Total system configuration

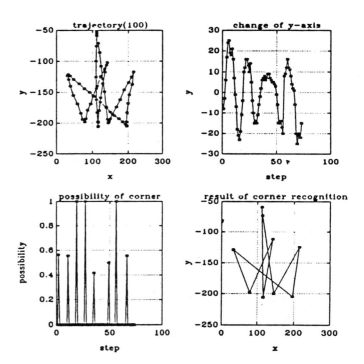

Fig 6. The result of recognition

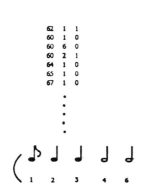

Fig 7. The format of music data and length of note

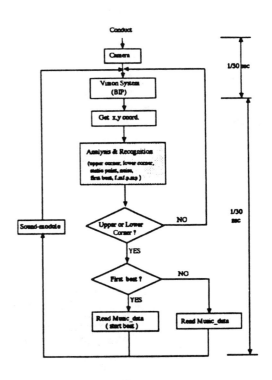

Fig 8. Operation procedure

555

Multilevel Database Security Using Information Clouding

Sujeet Shenoi *
Department of Mathematical and Computer Sciences
Keplinger Hall, University of Tulsa
Tulsa, Oklahoma 74104, USA

Abstract—Database security models typically employ information hiding together with polyinstantiation which helps create fictitious – and misleading – versions of confidential information. Although strict security-control is maintained, a high price is paid in terms of user convenience and data accuracy. Authorized users are often prevented from accessing necessary information which may not compromise security. Moreover, users accessing and unknowingly making decisions based on the misleading information used as cover stories can lead to disastrous consequences.

In this paper we employ fuzzy sets in a multilevel model for general-purpose database security. Sensitive information in database relations is meaningfully clouded by fuzzy sets; this is accomplished by broadening the possibility distributions constraining the values of sensitive attributes. The technique promotes the use of data and also maintains database security. Clouding with fuzzy sets is the middle ground between information release and information hiding/falsification. It nicely supplements the two security techniques and helps strike the right balance between user convenience and database security.

I. INTRODUCTION

The efficient implementation of strict database security is a problem of prime concern. A database must release information to support legitimate activities; at the same time, it must secure sensitive information from unauthorized users [1]. The simplest strategy is to hide information in views which present non-sensitive information to database users [1,3]. Although it is easily implemented, pure information hiding has several limitations. Authorized users can be prevented from accessing needed information which may not compromise database security.

Moreover, in a multilevel database environment where information is tagged with different classification levels and is accessed by users with different clearance levels, information hiding can give rise to covert channels which permit the flow of high-level information to low-level users [3-7].

The SeaView multilevel database security model introduced polyinstantiation as a mechanism for eliminating covert channels [6,7]. It allows a database to hold different versions – or cover stories – for an entity in the real-world [4]. A cover story conveys "sanitized" information about an entity that can be securely released to users at a lower clearance level.

As an example, consider a relation containing the tuple:

Name	Occupation
$Max\ Smart\ (c)$	$Spy\ (s)$

The entire tuple is visible to users with clearance levels secret (s), or higher. A confidential (c) user would know of the existence of *Mr. Smart*; but his occupation at the confidential level would appear as a *null* value. The entire tuple is invisible to an unclassified (u) user. Now suppose a confidential user attempts to replace the *null* value listed as *Max Smart*'s occupation with the new value *Salesman*. Notifying the user about the conflict could lead to the disclosure of high-level information, or at the very least signal the existence of high-level information. Therefore, a "polyinstantiated element" or cover story is created for *Mr. Smart* as shown below.

Name	Occupation
$Max\ Smart\ (c)$	$Spy\ \ \ \ (s)$
$Max\ Smart\ (c)$	$Salesman\ (c)$

Users with secret clearance levels or higher see the real and the fictitious tuple. Confidential users are blissfully unaware of *Max Smart*'s real occupation. They are under the impression that he is a salesman.

---

*Research supported by NSF Grants IRI-9110709 and IRI-9244550 and by OCAST Grants AR2-002 and AR9-010.

Polyinstantiation is a powerful mechanism for securing database data. However, in attempting to hide the existence of sensitive information it releases misleading information to low-level users. Due to their very nature, military databases and certain corporate databases must necessarily release false information, e.g., to hide and/or cover the fact that an individual is an undercover agent or that an employee is involved in a secret company project. Nevertheless, implementing security-control by falsifying information is needlessly restrictive and is often unacceptable in general-purpose database environments.

Consider the case of a secretary in an investment firm who is asked to send a mailing to "middle-income" investors. In a polyinstantiation environment, since a cover story is not feasible, the secretary would have to be assigned a higher clearance level to access the correct salary information. Assigning the higher clearance level can compromise database security. It might permit the access of sensitive information that would otherwise not be available.

It is important to recognize that the availability of exact sensitive information is not essential to many database activities, e.g., the investment firm mailing. Rather, users can perform their tasks adequately using abstracted or clouded versions of sensitive information. Sensitive values with underlying unordered (i.e., scalar) or ordered (e.g., numeric) domains can be conveniently secured by information clouding. For example, the exact salaries of investors can be securely replaced by fuzzy sets. The "granularity" of the fuzzy sets should be large enough to ensure data security and small enough so as not to dilute the notion "middle-income". Such clouding with fuzzy sets captures natural data semantics and preserves confidentiality while supporting normal database activities.

In this paper we employ fuzzy sets in a model for database security. Although fuzzy sets have been extensively applied in databases geared for AI and expert system applications [2,9,10,13], security-control remains a novel and relatively unexplored area. Sensitive information in database relations can be meaningfully clouded using fuzzy sets. This technique promotes data usage and also maintains access- and inference-control. Clouding with fuzzy sets is the middle ground between information release and information hiding/falsification. More importantly, it can be used in combination with information hiding and polyinstantiation. It nicely supplements the two techniques and helps strike the right balance between user convenience and database security.

II. FUZZY SETS AND CLOUDED INFORMATION

Sensitive information in a multilevel relation is clouded by simply replacing it with a meaningful fuzzy set. For example, *Mr. Smart*'s salary of $50K$ in the relation below can be clouded using the fuzzy set "moderate." The classification level (s^*) means that secret (s) information is clouded and is visible to users with the appropriate clearance level.

Name	Occupation	Salary
Max Smart (c)	Spy (s)	50K (s)
Max Smart (c)	Spy (s)	moderate (s^*)

In terms of possibility theory [12], information clouding relaxes the possibility distribution constraining the value of a sensitive attribute. The precise fact, "*Salary* is $50K$," implies that

$$Poss\{Salary = x\} = \mu_{50K}(x), \quad x \epsilon D,$$

where μ_{50K} is the characteristic function of the classical set $50K$, and $Poss\{Salary = x\}$ is the possibility that *Salary* has the value x. On the other hand, the piece of clouded information "*Salary* is *moderate*," implies that

$$Poss\{Salary = x\} = \mu_{moderate}(x), \quad x \epsilon D.$$

Here *moderate* is a *normal* fuzzy set which properly contains the classical set $50K$ and which has several elements $x \epsilon X$ for which $\mu_{moderate}(x) = 1$.

Exact information is clouded using normal fuzzy sets whose $\alpha_{1.0}$-cuts are supersets of the original singleton sets. A fuzzy set with a larger $\alpha_{1.0}$-cut broadens the possibility distribution constraining the attribute value. This secures the exact information from unauthorized users who are presented with <u>multiple</u> elements with possibility values of 1.0. Note that normal "triangular" fuzzy sets cannot be used to cloud information. The cardinality of the $\alpha_{1.0}$-cut of a triangular fuzzy set is 1. ¿From the security point of view it is equivalent to exact information.

Since security considerations require only that the $\alpha_{1.0}$-cuts of clouded information have cardinalities greater than 1, it is possible to cloud information using classical sets and intervals. However, there are many advantages to using normal fuzzy sets in clouding sensitive information. Fuzzy sets are more general in that they include classical sets and intervals. More importantly, they better capture natural semantics. This helps release secure yet meaningful information to database users and promotes flexible (fuzzy) querying of the secured database.

A singleton classical set defined on a linearly-ordered domain and having a "spiked" characteristic function can be clouded by a "trapezoidal" fuzzy set. A trapezoidal fuzzy set is denoted by (l_0, l_1, r_1, r_0). The support is the open interval (l_0, r_0) and its $\alpha_{1.0}$-cut set is the closed interval $[l_1, r_1]$. Using this notation *Mr. Smart*'s *moderate* salary can be expressed as $(20K, 30K, 60K, 70K)$.

We now introduce the notion of a "context" [11] to formalize the treatment of clouded information and to enforce the compatibility of classical relations embodying exact information and fuzzy relations containing clouded information.

Definition:

A *context* C is a partition defined by an equivalence relation ρ on a set of elements $\hat{D}$. The set of elements $\hat{D}$ participating in ρ is called its *restricted domain*; $\hat{D}$ is a subset of an underlying scalar database domain D. The set of all contexts capable of being constructed using restricted domains on D is denoted by $\mathcal{C}_D$.

Contexts in $\mathcal{C}_D$ are ordered by the quantity of equivalences captured by the generating equivalence relations. Each equivalence class in a finer context is a subset of an equivalence class in a coarser context.

Definition:

Let C and C' be contexts in $\mathcal{C}_D$ induced by the equivalence relations ρ and ρ', respectively. Then, $C' \sqsubseteq_C C$, i.e., C' is *coarser than* C, whenever $\rho \subseteq \rho'$.

A context acts as a *sieve* — with equivalence classes as its openings — for controlling the size of fuzzy sets released as tuple components.

Definition:

Let t_i be normal fuzzy set. Then, t_i is a *consistent* tuple component with respect to a context C_i if and only if the $\alpha_{1.0}$-cut of t_i, denoted by $(t_i)_{1.0}$, is a non-empty subset of an equivalence class in C_i.

Since the $\alpha_{1.0}$-cut of a released fuzzy set can be no larger than an equivalence class, a context specifies the "high water mark" for clouded tuple components. "Precise" contexts with singleton equivalence classes release exact (classical) information, i.e., singleton classical sets with spiked characteristic functions. Contexts with larger equivalence classes permit the release of clouded information expressed as trapezoidal fuzzy sets. The coarsest context $\{D\}$ releases the most clouded information. Such clouding is equivalent to pure information hiding. Note that the null set is not a valid tuple component as the maximal chunk D denotes an unknown, but defined value. Since contexts are derived from equivalences based on data semantics, the equivalence classes consist of closely related elements. This ensures that the clouded information chunks released as tuple components are always meaningful.

III. MULTILEVEL CLOUDED RELATIONS

The first step in extending classical multilevel relations to embody clouded information is to define the access classes used for exact information and those used for clouded information. We consider the following set of linearly-ordered exact information access classes:

$$S = \{unclassified(u), confidential(c), secret(s), topsecret(ts)\}.$$

In addition to the access classes in S, multiple clouded access classes of varying degrees are employed for each $a \in S - \{u\}$. A clouded access class is denoted by a^*, e.g., ts^* is a clouded class corresponding to top secret (ts). Note that clouded classes are not required for the unclassified (u) access class because unclassified information is nonsensitive by definition and is never clouded.

The set comprising an exact information access class a and its clouded classes is denoted by $\mathcal{A}$. The elements in $\mathcal{A}$ are partially ordered by $\sqsubseteq_*$ ("more clouded than"). The exact access class a dominates all the elements in $\mathcal{A}$ because it is associated with precise information. The expression $a^{**} \sqsubseteq_* a^*$ means that a^{**} is associated with more clouded information than a^*.

One of the main advantages of the context formalism is that it provides a uniform mechanism for viewing exact and clouded information in database relations. A relational attribute A_i with domain D_i is assigned a separate "view context" C_i (on D_i) for each distinct exact/clouded data classification level used in the extended relation. Within each access class set $\mathcal{A}_i$ used for attribute A_i, the assigned contexts must respect the natural ordering ($\sqsubseteq_*$) on the access classes. The dominant (top) element a_i of each $\mathcal{A}_i$ (e.g., s_i) is assigned the finest context generated on the domain D_i. This context comprises singleton equivalence classes and only releases exact information. The more clouded the access class a_i^* in $\mathcal{A}_i$ ($s_i^{**} \sqsubseteq_* s_i^*$), the coarser is the context assigned to it ($C_i^{**} \sqsubseteq_C C_i^*$), and the larger or more clouded are the information chunks released as tuple components.

Consider the following extended multilevel relation comprising exact and clouded information at various data classification levels.

Name	Occupation	Salary	
Smart (c)	Spy (s)	50K	(s)
Smart (c)	Spy (s)	(20K, 30K, 60K, 70K)	(s*)
Smart (c)	Spy (s)	(0K, 0K, 100K, 100K)	(s**)
Smart (c)	Salesman (c)	24K	(c)
Smart (c)	Salesman (c)	(0K, 20K, 25K, 45K)	(c*)

Since the secret (s) and confidential (c) classifications convey exact information, the *Salary* attribute is associated with the precise context on the *Salary* domain for each of these classifications. (For simplicity we denote exact information as an atomic value rather than as a singleton set). The coarser context { [0K, 29K], [30K, 64K], [65K, 100K], [101K+] } is appropriate for the secret clouded classification s^*. ($[l_i, h_i]$ denotes an interval on the set of integers.) The context used for s^{**}, say { $[0K, 100K]$, $[101K+]$ }, is even coarser, as s^{**} conveys more clouded information than s^*. The s^{**} salary value is actually the classical set $[0K, 100K]$. Thus, exact information (singleton sets), classical sets, and more generally, fuzzy sets, can be uniformly stored/viewed in multilevel relations using the context mechanism.

The *Salary* context used for clouded confidential (c^*) information, say { $[0K, 50K]$, $[51K+]$ }, is independent of the contexts used for clouded secret (s^* and s^{**}) information; however, it must be coarser than the precise context used for exact confidential (c) information.

Multilevel information clouding is implemented by providing a user with a clearance level $a \in S$ which permits access to exact information, and a clouded clearance level for each relational attribute A_i which is to be clouded. To eliminate, or at least to reduce, covert channels, the clouded levels assigned to a user must correspond to his/her clearance level for exact information. Thus, a secret (s) user can only be provided with secret clouded clearance levels, say s^* for one relational attribute and s^{**} for another. The issue of clouded information flow through covert channels is discussed in a later section.

We are now in a position to define the actual information seen by users at various clearance levels. As in the classical multilevel model, users who have exact clearance levels would see information whose classifications do not dominate their clearance levels. Since such users see exact information conveyed at all (lower) levels, it is not necessary for them to see any clouded information. Secret (s) users thus see (exact) tuples 1 and 4 in their views of the multilevel relation above; confidential (c) users only see tuple 4. Likewise, users with clouded clearance levels for a certain attribute, say s^*, would only see s^* information, and not s^{**} information even if it exists. Thus, secret (s) users with s^* clearance for *Salary* see tuples 2 and 4 (tuple 4 is visible because secret users are cleared for confidential information). Similarly, users with s^{**} clearance for *Salary* see tuples 3 and 4. Those with c^* clearance only see tuple 5.

The final issue to be disussed deals with updates to clouded databases. It is clear that update operations are crucial to achieving the desired levels of security in classical multilevel database systems; they are equally important in clouded multilevel database systems. Unfortunately, there is no clear consensus in the classical security community about the precise update semantics to be adopted in multilevel databases. Since the original SeaView model [6] several versions of an update semantics have been proposed for multilevel relations (see e.g., [5,7]. They differ in the number of core integrity properties employed and in the nature of these properties (e.g., whether they are state or transition properties). When discussing updates in the clouded model it is important to observe that clouding is designed to supplement rather than compete with the classical multilevel formulation. Two approaches to update handling are feasible as they support this view and would naturally integrate with any of the existing update semantics proposals. The simplest approach is to consider clouding purely as a mechanism for viewing database data. Thus, only users with exact clearance levels for relational attributes may update their values; users with clouded clearance levels are banned from performing updates. A better strategy is to allow updates by users with exact or clouded clearance levels as long as they input exact information which is subsequently tagged with the corresponding exact classification levels. For example, a user with an s^* clearance level for *Salary* could enter an exact value, say $55K$ in place of $50K$ in the relation above, which would be tagged as secret (s). The authorization to make such updates would of course require a certain level of trust on the part of the user.

IV. QUERYING CLOUDED RELATIONS

Contexts and their equivalences play an important role in query specification. The notion of equality used with exact information is generalized to a context-based notion of equivalence for uniformly querying exact and clouded/fuzzy information.

<u>Definition:</u>

The normal fuzzy sets, t and t', are formally *equivalent* with respect to a context C, denoted by $t \sim_C t'$, whenever their $\alpha_{1.0}$-cuts are subsets of a single equivalence class in C.

Equivalence classes comprise indistinguishable elements. Therefore, all non-empty subsets of an equivalence class (sieve opening) in a context are equivalent. Exact information (spiked fuzzy sets) and clouded information chunks (e.g., trapezoidal fuzzy sets, or normal fuzzy sets for unordered scalar domains) which pass through a given sieve opening are equivalent. Note that $\sim_C$ reduces to equality ($=_C$) for a precise context C.

A relation containing clouded information is queried by attaching contexts to values in query specifications. These "query contexts" act as sieves controlling the quality of the information recovered. A query specifies certain sieve openings (equivalence classes containing the specified values) and all information passing through these openings is retrieved. Classical query languages employ precise contexts. They formulate "precise queries" based on the equality of atomic values (equivalence in precise contexts). An imprecise query employs coarser contexts. The weaker equivalence permits the extraction of exact and clouded information which "approximately" matches the specifications.

Querying multilevel clouded relations actually involves the interaction of two types of contexts on the relational attributes. The first is the user-selected query context C_q which determines the precision of the "submitted" query. The second is the "view context" C_v assigned to a relational attribute for each exact/clouded data classification level in a multilevel relation. It determines the precision of the information released to the user. The two contexts are combined during query processing to produce an "effective context" C_e used for information retrieval. The combination must ensure strict access-control: Users should not obtain less clouded versions of the information than they are cleared to receive by adjusting the precision of their query contexts or by carefully selecting the target values in their query specifications.

Definition:

An *effective context* C_e is computed as $glb(C_v, C_q)$. If ρ_v and ρ_q are the equivalence relations generating C_v and C_q, respectively, then C_e is the context generated by $\rho_e = eq(\rho_v \cup \rho_q)$ (eq is the equivalence relation closure operator).

C_e is the finest context coarser than both C_v and C_q. Therefore, regardless of how users phrase their queries, the effective precision is never finer than their view contexts. Since a multilevel view is already secured by information clouding, "creative querying" by snoopers only causes additional information to match their targets. The transformation coarsens precise queries to secure fuzzy queries and converts somewhat fuzzy queries into fuzzier and even more secure queries.

Proposition:

A query evaluated with effective contexts C_e is *secure* with respect to access-control.

To understand the notion of an effective context consider a secret (s) user provided with an s^* clearance for *Salary*. By definition, he/she also sees exact confidential (c) information. Assume that the coarse view context C_v used for s^* is { [0K, 29K], [30K, 64K], [65K, 100K], [101K+] }; a precise view context is used for exact confidential (c) information. Suppose the user issues the precise query ($C_q = \{ \{60K\} \}$):

$$Select(r) \text{ where } Salary = 60K.$$

The effective context C_e at the confidential classification is the precise *Salary* context. Thus, the user can query confidential (c) information precisely:

$$Select(r; c) \text{ where } Salary = 60K.$$

On the other hand, since C_e for s^* information is { [0K, 29K], [30K, 64K], [65K, 100K], [101K+] }, the query is transformed and evaluated as:

$$Select(r; s^*) \text{ where } Salary \sim_{C_e} 60K,$$

i.e., $Salary = moderate$. The view context for s^* does not allow the user to distinguish between "moderate" salaries in [30K, 64K]. He/she can however differentiate between the fuzzy concepts "low," "medium," "high," and "very-high" salaries. Note that using coarser contexts C_q for s^* or c information (e.g., those "spanning" two equivalence classes in C_v) only results in fuzzier queries.

Clouding eliminates inference attacks based on deduction [4,8] as users can only reason approximately with fuzzy information presented in their views. Consider the multilevel relation $R_1(Name, Occupation, Salary)$ above, and the unclassified tax-table $R_2(Salary, Tax)$ embodying the precise dependency between *Salary* and *Tax*. Since a secret user with s^* clearance for *Salary* sees clouded salaries at the secret level, he/she can only obtain clouded tax assessments. A context-based *join* automatically produces clouded tax assessments based on s^* information because it uses the glb of the *Salary* contexts of the two relations.

V. CLOUDING AND COVER STORIES

Polyinstantiation plugs covert channels by eliminating the flow of high-level information to low-level users. E.g., updating *Mr. Smart*'s salary at the secret (s) level does not affect the value seen by confidential (c) users. Clouded multilevel relations are also "leakproof," e.g., s to c, or s to c^*. However, information flow is possible between an exact class and its clouded counterparts, e.g, s to s^* to s^{**}. This arises from the need to maintain the veracity of clouded information. A clouded cover story is useful because it does not lie; but it should reduce or perhaps even eliminate information flow.

Suppose the fuzzy set *moderate* is defined with respect to the s^* context { [0K, 29K], [30K, 64K], [65K, 100K], [101K+] }. The veracity of the clouded cover story is maintained until *Mr. Smart*'s salary at the secret (s) level is updated to a value residing in a different equivalence class in the s^* context. The s^* salary must then be updated to "high" or "low" to preserve truth. Information flow to the s^* level due to the update can be eliminated by using the coarsest possible context for s^* salaries; but this is equivalent to providing no information at all.

Name	Occupation	Salary
Max Smart (c)	Spy (s)	50K (s)
Max Smart (c)	Spy (s)	moderate (s^*)
Max Smart (c)	Salesman (c)	24K (c)
Max Smart (c)	Salesman (c)	low (c^*)

Nevertheless, information flow can still be eliminated, or at least reduced, if the clouding context is selected based on likely future salaries. Appropriately selecting the context and the clouded salary can maintain the veracity of the clouded cover story without ever changing its value. However, its granularity should be sufficiently small to be useful to database users. It is interesting to note that the requirements for good clouded cover stories are similar to those placed on the depositions of well-coached defense witnesses!

The final point deals with the connection between clouded cover stories at different exact access classes: How should *Mr. Smart*'s clouded salary at the c^* level relate to its value at the s^* level? Although the clouded contexts and values are assigned independently, it is helpful if the clouded values are the same, e.g., *moderate* for both s^* and c^* above. This in turn implies that the exact values for different access classes should be semantically "close" to each other and the "fuzzy" dependencies between attributes should be preserved at the clouded levels. The present model based on fuzzy sets can help address this issue. A good cover story is sensible, but it must also be corroborated by other information sources. Otherwise it will soon be ridden with holes. Perhaps the best way to fabricate a cover story is to first ascertain that it is reconciled with the original information at the clouded level.

VI. CONCLUSIONS

Multilevel database security is a novel and unexplored area for applying fuzzy set theory. Sensitive information in multilevel relations is meaningfully clouded using fuzzy sets. This is accomplished by broadening the possibility distributions constraining the values of sensitive attributes. Clouding is appealing because it maintains access-control and also eliminates inference attacks by enforcing fuzzy reasoning with sensitive information. It is the middle ground between information release and information hiding/cover stories. Moreover, unlike other multilevel security techniques, information clouding preserves truth. This promotes data usage and helps strike the right balance between user convenience and database security.

REFERENCES

[1] N.R. Adam and J.C. Wortmann, "Security-control methods for statistical databases," *ACM Computing Surveys*, vol. 21, pp. 515-556, 1989.

[2] B.P. Buckles and F.E. Petry, "A fuzzy representation of data for relational databases," *Fuzzy Sets and Systems*, vol. 7, pp. 213-226, 1982.

[3] D.E. Denning, S.G. Akl, M. Heckman, T.F. Lunt, M. Morgenstern, P.G. Neumann and R.R. Schell, "Views for multilevel database security," *IEEE Transactions on Software Engineering*, vol. 13, pp. 129-140, 1987.

[4] T.D. Garvey and T.F. Lunt, "Cover stories for database security," *Proceedings of the Fifth IFIP WG11.3 Workshop on Database Security*, Shepherdstown, West Virginia, 1991.

[5] S. Jajodia and R. Sandhu, "Toward a multilevel secure relational data model," *Proceedings of ACM SIGMOD*, ACM Press, New York, pp. 50-59, 1991.

[6] T.F. Lunt, D.E. Denning, R.R. Schell, M. Heckman and W.R. Shockley, "The SeaView security model," *IEEE Transactions on Software Engineering*, vol. 16, pp. 593-607, 1991.

[7] T.F. Lunt and D. Hsieh, "Update semantics for a multilevel relational database system," in *Database Security IV: Status and Prospects*, S. Jajodia and C.E. Landwehr, Eds. New York: Elsevier Science, pp. 281-296, 1991.

[8] M. Morgenstern, "Security and inference in multilevel database and knowledge base systems," *Proceedings of ACM SIGMOD*, ACM Press, New York, pp. 357-373, 1987.

[9] K.V. Raju and A.K. Majumdar, "Fuzzy functional dependencies and lossless join decomposition of fuzzy relational database systems," *ACM Transactions on Database Systems*, vol. 13, pp. 129-166, 1988.

[10] S. Shenoi, A. Melton and L.T. Fan, "An equivalence classes model of fuzzy relational databases," *Fuzzy Sets and Systems*, vol. 38, pp. 153-170, 1990.

[11] S. Shenoi, K. Shenoi and A. Melton, "Contexts and abstract information processing," *Proceedings of the 4th International Conference on Industrial and Engineering Applications of AI and Expert Systems*, Kauai, Hawaii, pp. 44-50, 1991.

[12] L.A. Zadeh, "Fuzzy sets as a basis for a theory of possibility," *Fuzzy Sets and Systems*, vol. 1, pp. 3-28, 1978.

[13] M. Zemankova and A. Kandel, *Fuzzy Relational Databases: A Key to Expert Systems*. Cologne: Verlag TUV Rheinland, 1984.

Active Control of Broadband Noise Using Fuzzy Logic

Oscar Kipersztok
Research & Technology
Boeing Computer Services
P.O. Box 24346, MS: 7L-64
Seattle, WA, 98124

Abstract - This paper explores the use of fuzzy logic for the active control of broadband noise, as an alternative to more conventional linear filter approaches. A simple example is chosen in which a compact random noise source radiating into free space is used. The simulated control is provided by a cancelling noise source in the form of a loudspeaker placed between the primary noise source and a location in the far-field where the noise reduction is desired. The peak cross correlation between signals from an "early" microphone placed near the noise source and an error microphone in the far-field is used as the error function. The results obtained demonstrate that fuzzy logic performs well in actively controlling broadband noise for the cases shown. The fuzzy controller achieves significant noise reduction levels in a reasonably fast and stable manner, suggesting the continued investigation to evaluate its performance in more complicated applications of this type.

I. INTRODUCTION

In active noise and vibration control (ANVC) secondary noise sources are used to create a sound field which is equal and opposite to the undesired, primary sound field. For noise control the secondary sources are usually loudspeakers while for vibration control shakers are used as vibration sources. The physical mechanism by which the vibrations are reduced, whether sound vibrations in air or structural vibrations, is that of reducing the resistance to motion of the medium close to the sources, thereby reducing their ability to do work on the medium. Another control mechanism, not used here, is to increase the resistance to motion, thereby reducing the ability of forces to do work on the medium.

Potential applications of ANVC span the range from noise control in ducts and rooms to vibration control on space structures, from automotive mufflers to chatter-free machine tools.

With the advent of affordable computer power in the early 80's, active noise control has increasingly become a practical choice as a method for noise reduction. A number of survey papers have summarized the state of the art in active noise control including work by Eghtesadi, et al [1], Ffowcs-Williams [2], Warnaka [3] and more recently Stevens and Ahuja [4].

By implementing a fuzzy logic controller the essence of the required control can be captured in a small number of model-free heuristic rules [5,6]. This approach is particularly appealing for applications of a time-critical nature and even more so when the system displays non-linear behavior. The rules used in the fuzzy controller may depend on subjective noise measures which are non-linear functions of the acoustic pressure signal. These subjective measures cannot easily be handled by linear controllers.

The intent of this paper is to demonstrate the feasibility of using a fuzzy-logic control approach to the ANVC problem by means of a simple example.

II. DESCRIPTION OF THE PROBLEM

The selected problem is described in Figure 1 and is composed of a compact monopole used as a noise source radiating into free space. The aim of the active noise-control system is to provide a cancelling noise source in the form of a loudspeaker placed between the primary noise source and a location in the far-field where the noise reduction is desired.

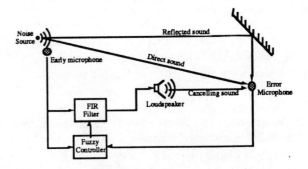

Figure 1. Description of the physical plant

An "early" microphone close to the source picks up the noise signal which is then amplified, inverted and sent to the loudspeaker. Due to a number of factors complete cancellation will not take place unless the signal is further modified. This modification will take place in a *nonrecursive* or finite impulse response (FIR) filter. The coefficients of this filter are set by the fuzzy controller based on inputs from both the early microphone and an error microphone, placed in the far-field.

Factors which require consideration for complete cancellation to take place include the response of the speaker, the separation between speaker and the source, and reflections from solid boundaries.

The FIR filter contains a number of adjustable parameters which are the filter coefficients and constitute outputs of the fuzzy controller. Another output of the controller is the change to the parameter used in the exponential averaging applied to the source broadband signal. The inputs to the controller are the error defined by the peak value of the cross-correlation function between the early and error microphones, and the corresponding change in error defined by the difference between two error signals at a particular interval of time. An additional input to the controller is the random-error estimate of the cross-correlation function.

III. SIMULATION OF THE PHYSICAL PLANT

The fuzzy controller is evaluated in a simulated acoustic environment. Both simulation and control are performed in the time domain. The compact noise source is simulated with a band-limited random number generator and the wave propagation is modelled as a time delay. The radiation of the sound is taken to be that of a monopole. This defines the sound radiated from the source to the error microphone.

The early microphone signal is modeled as the direct noise source. The error microphone signal is modeled as the sum of the source noise delayed by the distance from the source, with some uncorrelated noise, and the cancelling noise radiated from the loudspeaker.

The loudspeaker radiation is also assumed to emanate from a compact monopole source. It is modeled as the output of an FIR filter [7] applied to the early microphone signal and is defined in Equation 1.

$$y(i) = \sum_{k=0}^{n} h(k)x(i-k) \qquad (1)$$

Here, $x(i)$ is the signal from the early microphone and $h(k)$ is the filter coefficient that corresponds to the time delay k. A simple model of a loudspeaker transfer function from applied voltage to sound pressure can be incorporated in order to model the modification of the early microphone signal due to the presence of the loudspeaker in the circuit. In the simulation cases discussed in section 5, the loudspeaker transfer function is ignored, i.e., set to unity.

IV. THE CONTROL PRINCIPLES

In this section, a definition will be given of the parameters used in the fuzzy control, followed by a description of the principles underlying the control.

A. *Definition of Parameters*

The peak value of the cross correlation between early and error microphones is used as the error function for the controller. The cross-correlation function is estimated by the sum

$$r_{xz}(s) = \frac{1}{n}\sum_{i=0}^{n} x(i)z(i+s) \qquad (2)$$

where $x(i)$ and $z(i)$ are the early- and error-microphone signals respectively, s is a particular time delay, and n is the number of samples used.

In a real-time application where the cross correlation needs to be computed continuously, the Kalman predictor can be used to compute an exponential average. This is given by the following recurrence equation:

$$r_i(s) = \varepsilon\, x(i)\, z(i+s) + (1 - \varepsilon)\, r_{i-1}(s) \qquad (3)$$

where ε is the exponential averaging parameter which for large n in Equation 2, is equivalent to $1/n$.

Dividing the cross-correlation by the product of the individual auto-correlations evaluated at time delay zero, results in the cross-correlation coefficient $\rho_{xz}(s)$, which has the property that $|\rho| \leq 1$.

$$\rho_{xz}(s) = \frac{r_{xz}(s)}{\sqrt{r_{xx}(0)\, r_{zz}(0)}} \qquad (4)$$

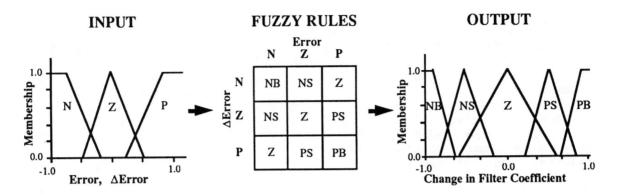

Figure 2. Fuzzy Control of the Filter Coefficient

The auto-correlation at time delay zero is the variance or the random error of the signal [8].

In the simulation cases that will be shown, the parameters controlled by the fuzzy controller are the coefficient h of the FIR filter and the exponential averaging parameter ε. The inputs for the control of h are the peak value of the cross-correlation coefficient ρ_{peak}, used as the error function, and the change in error approximated by $\Delta\rho/\Delta t$. The input for the control of ε is the estimate of the cross-correlation random error $E_r[r_{xz}(s)]$, defined [8] as

$$E_r[r_{xz}(s)] = \sqrt{\frac{1+\rho_{xz}^{-2}(s)}{n}} \quad (5)$$

which for large n can be approximated by

$\sqrt{\varepsilon(1+\rho_{xz}^{-2}(s))}$. This quantity represents the ratio of the variance of the cross-correlation estimate (i.e., the random noise introduced by finite sampling limitations) to the magnitude of its peak value.

The initial step, before applying the fuzzy control, is to identify the time delay corresponding to the difference in propagation time between the source and loudspeaker from the error microphone. This value will set the delay for the filter coefficient.

The identification of this time delay is achieved by first setting the filter delay to zero, resulting in a cross-correlation function with two peaks, one corresponding to the signal from the loudspeaker, the other to the signal from the source. The time difference between the two peaks is the desired time delay. The magnitude of the filter coefficient is then set arbitrarily to some initial value.

B. The Fuzzy Control

The fuzzy control structure for the nonrecursive filter coefficient h is shown in Figure 2. The peak-value of the cross-correlation is defined as the error function. A different correlation peak occurs at a different time delay for each sound propagation path. The same control algorithm applies for each correlation peak.

The inputs to the controller are values of the error and error-change functions. Three membership functions are defined for each input; negative (N), zero (Z) and positive (P), resulting in the nine fuzzy rules shown. The five output membership functions range from negative big (NB) to positive big (PB).

The heuristic principle captured in the rules is to continuously change the magnitude of the filter coefficient until the peak correlation value is zero. An increase in the magnitude of the filter coefficient causes the correlation peak value to decrease; and conversely, a decrease in the magnitude of the coefficient results in a correlation increase. Adding the change in error to the rules helps to minimize "overshoots" as the cross-correlation function decreases.

Figure 3 shows the fuzzy control structure for the exponential averaging parameter ε. The estimate of the cross-correlation random error, defined in Equation 5, is used to indicate how well the features of the cross-correlation (i.e., the peak values) can be detected from the random noise introduced by sampling limitations.

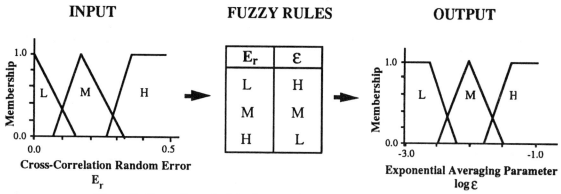

Figure 3. Fuzzy Control of the Exponential Averaging Parameter

The parameter ε varies from 0 to 1 and while applying exponential averaging, as the value of ε decreases the corresponding number of samples averaged, increases; with more weight given to the most recent samples when using exponential averaging. The heuristic principle captured in the rules is that as long as the random error E_r is low (L), that is, when the correlation features are well resolved above the random noise, we can afford to select a high (H) value for ε and therefore, reduce the number of averaging samples. Conversely, if the correlation features are not well resolved above the random noise, we should increase the number of averages by selecting a lower value of ε to achieve better feature resolution. Note that for the output membership functions $\log \varepsilon$ is used to linearize the scale of ε.

In both control cases the max-dot fuzzy inference mechanism [6] is used.

V. SIMULATION RESULTS AND DISCUSSION

Following is a description of the three simulation cases shown. They were run on an i486, 33 MHz PC.

A. Case 1

In this case a single propagation path is simulated between the early and error microphones. In the absence of any control this appears as a single peak in the cross-correlation, as shown in Figure 4. The time delay of the peak corresponds to the difference in propagation time between the source and loudspeaker from the error microphone. Since no propagation losses are assumed other than the inverse square law of spherical divergence, the magnitude of the correlation coefficient is near 1. The random fluctuations in the correlation are due to sampling limitations and depend on the parameter ε (i.e., the inverse of the number of averages) and the sampling rate.

After the control is applied, a level of about 35 dB in noise reduction is achieved quite rapidly. The noise reduction is defined as ten times the 10-based logarithm of the ratio of the uncontrolled to controlled error-microphone autocorrelations, both evaluated at time delays of zero. The autocorrelation function evaluated at zero, corresponds to the variance of the signal. The amount of maximum noise reduction obtained depends on the level of external, uncorrelated, random noise added in the model of the error-microphone signal.

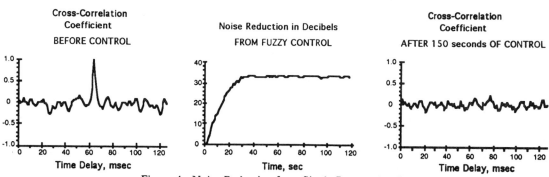

Figure 4. Noise Reduction for a Single Propagation Path

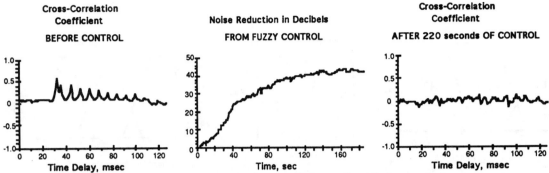

Figure 5. Noise Reduction for Multiple, Different-Sized Reflection Paths

Note that after the first 10 seconds, more than 12 dB reduction is obtained. That amount of reduction is already quite effective considering it corresponds to a reduction of the sound intensity by a factor of 16, and of the sound pressure by a factor of 4.

After 150 seconds of control, the maximum noise reduction level has remained stable, and the cross correlation peak is virtually gone.

B. Case 2

In case 2, ten reflection paths are modelled, each of a distinctly different distance, and therefore in Figure 5, the ten peaks in the cross-correlation coefficient are spread out. Note that the individual peaks have much smaller magnitudes than the single peak of case 1. The reason is that for all the different propagation paths, the cross-correlation coefficient obeys the following rule

$$\sum_{k=0}^{n} \rho^2(s_k) = 1 \qquad (6)$$

where s_k is the time delay associated with each path.

Here, the noise reduction curve rises at a slower rate than in case 1, nevertheless after about 30 seconds more than 12 dB of noise has already been reduced. The controller applies the same control principle to each correlation peak, working on the highest peak at any given time. Because of that, compared to case 1, it takes longer to reach the maximum noise reduction level.

After 220 seconds the noise reduction curve levels off at about 40 dB and, the correlation peaks are virtually gone.

C. Case 3

Case 3 is shown in Figure 6, where a number of reflection paths of similar size are modelled. In the absence of any control the cross-correlation shows all the peaks occuring in a "haystack" within a narrow range of time delays. The noise reduction curve has a similar shape to that of case 2; about 12 dB reduction after the first 30 seconds, with a faster rise at first and then a slower increase before leveling off at a maximum reduction of about 40 dB.

Note that the noise reduction curve in this case is much more jagged than in case 2. The reason has to

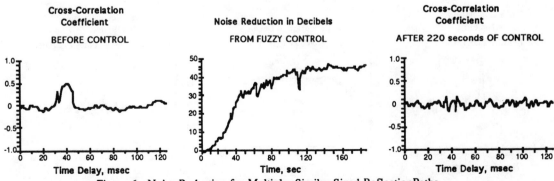

Figure 6. Noise Reduction for Multiple, Similar Sized ReflectionPaths

566

do with the closeness of the reflection peaks in the correlation function. In this case, it is more difficult for the controller to detect the highest peak on which to apply the fuzzy control. Because of random fluctuations inherent in the correlation function the controller will ocationally select the wrong peak, i.e., one which does not correspond to a reflection path; therefore, causing a temporary increase in noise and a valley in the noise reduction curve.

D. Discussion

The simulation results show that good control of broadband noise can be achieved using fuzzy logic for the simple cases demonstrated herein. The performance of the fuzzy controller is reasonably fast and stable in all cases shown. In situations where "dips" occur in the noise reduction curves, the recovery time is very good, as in case 3. Typically, those dips appear after significant attenuation has already taken place. In situations with a large number of reflections present, since the same control applies equally to each reflection path, one could apply the control in parallel to improve the noise reduction time.

In more complicated cases, where several secondary sources and sensors may interact, it is difficult to predict how the fuzzy control will do. It is possible that in such situations a conventional, linear-filter approach may be computationally more efficient. In such cases the fuzzy logic can be used for tuning the parameters of the linear-filter algorithm. Furthermore, in cases where noise reverberation is present the use of recursive filter coefficients may be required, which include non-linear calculations for which fuzzy logic might also be useful.

VI. CONCLUSIONS

Fuzzy logic performs well in actively controlling broadband noise for the simple, simulated cases shown in this work. The fuzzy controller is effective in achieving significant noise reduction levels in a reasonably fast and stable manner.

The results shown demonstrate the feasibility of using fuzzy logic for the active control of broadband noise, and render possible the continued investigation to evaluate the performance of fuzzy control in more complicated and realistic applications of this type.

ACKNOWLEDGEMENT

Very special thanks to Andy Andersson from Noise Engineering in the Boeing Commercial Airplane Group for sharing his sound knowledge of acoustics, without which this work would have not been possible.

Special thanks also to Kish Sharma, Wendell Miller, Ron Hammond and Bob Schneble for their continued support and encouragement through the peaks and troughs of this project.

REFERENCES

[1] Eghtesadi, Kh., Hong, W. K. W., and Leventhall, H. G., "Evaluation of Active Noise Attenuator Systems", Vol. 1, Published by Noise Control Foundation, 1986, pp. 577-582.

[2] Ffowcs-Williams, J. E., "Anti-Sound", *Proceedings of the Royal Society of London*, Series A, Vol 395, 1984, pp. 63-88.

[3] Warnaka, G. E. and Zalas J. M., "Active Attenuation of Noise in a Closed Structure", U.S. Patent # 4,562,589, December 31, 1985.

[4] Stevens J. and Ahuja K., "The State-of-the-Art in Active Noise Control", *AIAA 13th Aeroacoustics Conference*, AIAA-90-3924, October 22-24, 1990, Tallahassee, FL.

[5] Chuen C. L, "Fuzzy Logic in Control Systems: Fuzzy Logic Controller-Part I", *IEEE Transactions on Systems, Man and Cybernetics*, Vol. 20, No. 2, March/April 1990.

[6] Chuen C. L, "Fuzzy Logic in Control Systems: Fuzzy Logic Controller-Part II", *IEEE Transactions on Systems, Man and Cybernetics*, Vol. 20, No. 2, March/April 1990.

[7] Press W. H., Flannery B. P., Tuey D., Vetterling W. T., "Numerical Recipes in C: The Art of Scientific Computing", Cambridge Univ. Press, 1988.

[8] Bendat J. S., Piersol A. G., "Engineering Applications of Correlation and Spectral Analysis", John Wiley & Sons, 1980.

Author Index

A

Acheroy, M., 486
Akahori, H., 563
Alzamora, G., 312
Arai, F., 168, 347

B

Babiker, E., 469
Bay J. S., 123
Berenji, H. R., 363
Bergstrom, R. A., 123
Berman, L., 430
Bezdek, J. C., 3
Bien, Z., 549
Borges da Silva, L. E., 196, 221
Bosacchi, B., 13
Brown, J. D., 274

C

Cannavacciuolo, A., 515
Capaldo, G., 515
Chand, S., 101
Chang, P.-R., 395
Chen, R. L., 267
Chen, J.-L., 229
Cheung, K.-H., 430
Chiu, S., 101
Cho, K.-B., 212
Choi, S.-S., 493
Cliff, N., 509
Consoli, A., 177
Curruto, E., 177

D

Davidson, J., 285
Davis, Jr., L. I., 68
de Mathelin, M., 486
de Sam Lazaro, A., 353
Dockery, J., 543
Doya, K., 369

E

Eddleman, D. W., 418
Egbert, D. D., 424
Egusa, Y., 523
Elbuluk, M. E., 204
Endo, T., 476

F

Feldkamp, L. A., 43, 68
Ferreira da Silva, V., 196
Fie, J., 298
Franssila, J., 325
Fujihara, H., 469
Fujino, M., 293
Fujisawa, Y., 168
Fukuda, T., 129, 168, 347
Fukushima, T., 293
Furuya, T., 503

G

Giardina, C. R., 443
Goodman, P. H., 424
Goto, K., 369
Goto, M., 531
Goulter, I., 285
Graf, F., 88
Greiss, H., 221
Gurocak, H. B., 353

H

Hacisalihzade, S., 411
Haga, T., 94
Hans, V., 345
Heller, J., 50
Herron, L. H., 233
Hilloowala, R. M., 304
Hoblit, J., 154, 403
Hofmann, W., 190
Hollinger, J. G., 123
Hoshino, H., 168
Hou, L., 107
Hou, S.-L., 253
Huang, P.-H., 239

I

Imasaki, N., 476
Inagaki, T., 537

J

Jani, Y., 154, 363
Jawahir, I. S., 298

K

Kaburlasos, V. G., 424
Kalam, A., 233
Kamada, N., 293
Kanaya, M., 537
Kato, S., 537
Keller, J. M., 25
Khedkar, P., 363
Kim, J.-S., 549
Kim, D.-II, 184
Kim, S., 184
Kimura, G., 56
Kipersztok, O., 562
Kitagawa, T., 571
Klir, G. J., 274
Koivo, H. N., 325
Konolige, K., 148
Kosuge, K., 168
Krause, M., 190
Krishnapuram, R., 25
Kubota, T., 312
Kurosu, K., 503

L

Lambert-Torres, H. G., 196, 221
Laplante, P. A., 443
Lea, R. N., 154, 363
Lee, J.-W., 184
Lee, K.-C., 212
Lei, W., 345
Lin, R.-B., 333
Littman, D., 543
Liu, M.-H., 160

M

Madau, D. P., 68
Maeda, Y., 135
Magdalena, L., 117
Malkani, A., 363
Mamdani, E. H., 19
Man, G. M. T., 482
Marks II, R.J., xviv, xxi, 573
Marques, F., 401
Marudarajan, A. R., 337
Masaki, I., 13
Matsuura, H., 347
McEachern, M., 418
Meier, R., 411
Mendel, J. M., 389
Min, S.-S., 212
Mir, S. A., 204
Mita, T., 369
Miyajima, K., 447
Miyazaki, T., 168
Monasterio, F., 117
Montgomery, D., 476

Morimura, A., 523
Morita, T., 537
Mukhedkar, D., 221
Muller, R., 74
Murakoshi, S., 247
Murayama, H., 537
Muro, E., 168

N

Naito, N., 293
Nakamura, M., 503
Nakashima, K., 196
Neuber, S., 50
Nieuwland, J., 411
Nijhuis, J., 50
Nocker, G., 74
Norita, T., 447

O

Oh, K.-W., 493
Ohtsubo, K., 168

P

Pal, S. K., 33
Pattay, R. S., 319
Pedryczk, W., 285
Perneel, C., 486
Pin, F. G., 319
Poon, J. C. H., 482
Probst, G., 87
Puskorius, G. V., 43

R

Raciti, A., 177
Rattan, K. S., 81
Ricker, N. L., 253
Ruspini, E. H., 148

S

Sabharwal, D., 81
Saffiotto, A., 148
Sakaguchi, S., 94
Sakai, I., 94
Sameda, Y., 280
Sano, M., 261
Sato, M., 531
Sato, N., 293
Sato, T., 247
Sekiguchi, T., 531
Sharaf, A. M., 304
Shenoi, S., 556
Sheppard, L. C., 418
Shi, J., 233
Shibata, T., 129
Shimojima, K., 347

Shioya, M., 56
Shragowitz, E., 331
Sinha, D., 443
Sittig, D. F., 430
Slator, T. R., 337
Song, J.-H., 141
Song, K.-T., 212
Spanos, C. J., 267
Sponnemann, J., 50
Steck, D., 411
Steinberg, M., 380
Sugeno, M., 455
Symon, J., 319

T

Tai, J.-C., 141
Tai, C. C., 395
Takagi, T., 135, 369, 463
Takagi, H., 8
Takayanagi, S., 509
Tanabe, M., 135
Tanaka, K., 247, 261
Tani, T., 247
Testa, A., 177
Tran, R. T., 229
Tsai, R., 229
Tsao, Y. J., 109
Tzeng, H.-W., 229

U

Uehara, K., 168
Umano M., 247
Ushida, H., 463
Utsunomiya, H., 503

V

Valiquette, B., 221
Ventre, A., 515
Vijeh, N., 62

W

Wakami, N., 523
Wang, L.-X., 389
Wang, Z., 107
Wangemann, R.T., vii
Wantanabe, H., 319
Watanabe, K., 531
Watanabe, H., 261
Weil, H.-G., 88

Y

Yamaguchi, T., 369
Yamagugchi, T., 463, 476
Yamane, S., 312
Yeh, E., 109

Ying, H., 418
Yuan, F., 68
Yuasa, N., 56
Yuta, M., 135

Z

Zadeh, L., xvii
Zbinden, A., 411
Zollo, G., 515
Zhang, W., 455
Zinger, D. S., 204

Subject Index

A

Advanced distance control, 78
AC Drives, 190
AC Servo motors, 181, 184
Adaptive,
 algorithm, 17
 filters, 389
Adaptive Resonance Theory (ART), 444
Aerospace, 363, 369, 380
Aircraft carrier, 380
Air conditioners, xxiv, 9
Air pollution, 261
Alpha cut, 493
Ambiguity, 34
Anesthesia, 411
Anti-lock braking (see Braking)
Arbitration protocol, 338
Arterial pressure control, 418
Artifact identification, 435
Artificial intelligence, 19, 129, 443
Associative,
 inference, 476
 memory, 369, 476
Attribution, 483
Attitude control, 364
Automated guided vehicle (AGV), 50, 56
Automatic,
 teller machines, 531
 transmissions, 88, 94
Autonomous,
 navigation, 319
 robots, 129, 135, 148

B

Banknote transfer, 531
Behavior control, 503
Bi-directional associative memory, 463
Biped, 117
Bioengineering, 411, 418, 424, 543
Blending, 148
Blood pressure, 411
Books (on fuzzy systems), xxiii
Boundary, 27
Braking, 14, 68
Broadband noise, 562
Built-in-tests, 274
Bus artribration, 337
Bus load forecasting, 221

C

Cameras, 312, 395, 523, 557
Carbon monoxide concentration, 261
Carburization, 107
Carpet, 9
Cascade type, 11
Cash dispensers, 551
Cement plants, 196
Channel equalization, 391
Character recognition, 482
Chassis control, 15
Characteristic function, 3
Chemical vapor deposition, 107, 267, 319
Chip breaker, 298
Claptrap, xviv
Classification, 300, 401, 430, 443
Clothes dryer, 9
Clouded relations, 548
Coin grading, 443
Collision avoidance, 15, 51
Color, 396, 447, 455
Color corrections, 395
Compliant wrist, 354
Conceptual fuzzy sets, 463
Context sensitive reasoning, 463
Control,
 algorithm, 163
 structure, 118, 148
Consumer products, 8
Commutation, 184
Computer vision, 25, 455
Cooperation, man robot, 169
Corrective control, 162
Corrector type, 10
Cult of analyticity, 21
Current control, 187, 212, 214
Cutting conditions, 302
Cruise control, 15, 74

D

Database security, 546
Deblurring, 160
Decision,
 making, 509
 model, 285
Decisions, 514
Defuzzification (see Fuzzification)
Detection, 27
Development tools, 10
Direct self control, 206
Distance control, 56, 76, 77
Drive train, 89
Document recognition, 469
Drives, 177, 190, 196, 212

E

Electric,
 heater, 9
 thermo pot, 9
Electronics, 319, 325, 331, 337
Electrophotography, 537
Energy conversion, 304
Engine,
 diagnosis, 15
 model, 44
Entropy, 34, 401, 407
Environment, 253, 261, 347
Equalization, 389
Estimation, 431
Evaluation,
 matching, 458
 personnel, 515
Expert system, 53, 88, 269, 418, 486

F

Facial expression, 463
Fairness, 338
Fans, 9
Fault isolation, 229
Feature extraction, 447, 494
Feed rate, 312
Filters, 389
Flight control, 380
FLn, xvii
FLw, xvii
Flux control, 177, 190
Flying vehicle, 369
Force control, 123, 162
Forecasting, 221
Friction separation, 531
Fuel control, 366
FUZZ-IEEE (see IEEE International Conference on Fuzzy Systems)
Fuzzification, xxiii, 5, 63, 84, 89, 191
Fuzziness index, 510
Fuzzy,
 artificial networks, 476
 control, 19, 212, 294, 307, 313
 entropy, 8, 34
 logic control, 325
 logic technology, 13
 membership functions, 496
 models, 3, 6, 369
 neural networks, 371
 pointing set, 443
 regulation, 178
 rule, 89, 102, 171, 192, 206, 214, 282, 295, 300, 307
 set image processing, 33
 sets and systems, xxi, 7, 177

G

Gas system, 285
Gait synthesis, 119
Gauge control, 293
Gear shift control, 15
Genetic algorithms, 19, 37
Global,
  directedness, 148
  rating, 491
Goal directedness, 148
Grading, 443
Graph matching, 482
Group,
  behavior, 508
  control, 503
Gulf War, 380

H

Handwriting recognition, 482
Hard contact, 123
High definition television (HDTV), 395
Heart rate, 430
Heuristics rating, 489
Hierarchical control, 118, 131, 135, 162
High Performance Motion Systems, 177
Highway (see Intelligent highway system)
History (of fuzzy systems), 19
Hot strip mill, 293
Hover flight control, 372
Hospitals, 424
Human interface, 129
Hybrid control, 249, 293
Hydraulic system, 70
Hysteresis band control, 212

I

Idemitsu Chiba Refinery, 251
Identification algorithm, 262
Idle speed control (ISL)
  (see Speed control)
IEEE Educational Activities Board, xxii
IEEE International Conference on
  Fuzzy Systems, xxiv
IEEE,
  Neural Networks Council, xxiii
  Transactions on Fuzzy Systems,
    xxii, 3
  Transactions on Neural Networks,
    xxiii, xxiv, 3
  Transactions on Systems, Man &
    Cybernetics, xxiii

Image,
  coding, 411
  definitions, 34
  density, 539
  processing, 33, 395, 401, 443,
    447, 457
  recognition, 463, 469, 486
  stabilizer, 523
Impedance control, 169
Independent type, 10
Indicator function, 3
Induction,
  heating, 9
  motors, 196, 204, 212
Industry applications, 129, 247, 253,
  261, 267, 274, 280, 285, 293, 298,
  304, 312
Inference, 90, 96, 225, 269, 281, 348
Information clouding, 556
Intelligent,
  control, 117
  highway system, 13
  vehicle, 15
Intensive care, 418
Intent amplification, 543
Intentional inference, 98
Interface, 206
International,
  Fuzzy Systems Association,
    (IFSA), 6
  Journal of Approximate
    Reasoning, xxii

J

Johnson Space Center, 154

K

Kerosene heater, 9
Knowledge, 95

L

Laboratory of Industrial Fuzzy
  Engineering (LIFE), xxiv
Landing control, 380
Lasers, 280
Learning control, 369
Level control, 62, 247
Linear,
  discriminant, 425
  interpolation, 495
  quadratic (LQ) control, 15, 43
  regression, 431
Linguistic,
  inputs, 391
  variables, 268, 332
  rules, 109, 200
Load forecasting, 222
Low level vision, 26

M

Machining, 160, 298
Manipulator, 168
Mean arterial pressure, 411
Membership function, xxi, 3, 36, 214,
  496
Memory model, 455
Mercedes-Benz, 74
Meta-knowledge, 131
Microwave ovens, 9
Microprocessor, 184
Mobile robot, 148, 319
Modelization, 89, 261, 487
Motion,
  analysis, 589
  control, 177
Motors, 177, 184, 196, 204, 212, 571
MITI, xxiv
Multilayer reasoning, 464
Multipardigm, 19
Multiprocessor systems, 338
Multivariable systems, 183
Music, 589

N

Natural language, 513
Natural gas systems, 285
Navigation, 144
Neural networks, xxiii-xxiv, 8, 36, 50,
  107, 129, 247, 285, 349, 369, 389,
  396, 424, 465, 476, 577
Neuromorphic control, 133
Noise control, 562
Non-analytic, 22
Non-consumer applications, 10
Nonlinear,
  algorithms, 418
  filtering, 391
  function, 61
North American Fuzzy Information
  Processing Society (NAFIPS),
  xxiv
Numerical recognition, 485

O

Object recognition, 28, 35
Obstacle avoidance, 145, 148
Operator experience, 225
Optimization, 253

P

Parameter identification, 294
Patents, xviv
Pattern recognition, 148, 427, 443
  469, 476, 482, 486, 492, 493
Personal preference, 11

Personnel evaluation, 515
Photo-voltaics, 304
Pitch axis, 81
PI and PID Control, 43, 56, 74, 107, 123, 328, 411, 418
Placement, 331
Pneumonia, 424
Pointing set, 443
Pollution, 253
Position,
    control, 177, 162
    error, 356
    sensor, 184
Possible worlds, 544
Potential fields, 152
Power forecasting, 221
Power systems, 221, 229, 233, 239
Power systems stabilizer, 233, 239
Pre-cooling, 12
Prediction, 193, 250, 424
Probability (contrasted to fuzzy), xxi-xxii
Process control, 107, 267, 537
Proportional force control, 123
Psychology, 503, 509, 515
Pulse width modulation, 190, 212, 306

R

Ramp comparison controller, 212
Range test, 274
Reactive control, 149
Real-time embedded systems, 253, 274
Recognition, 27, 33, 497, 499
    processes, 28, 457
Refrigerators, 9
Region extraction, 448
Regression, 431
Region extraction, 28, 447
Regulator, 178
Reinforcement learning, 363
Relationships, 29
Rice cooker, 9
Robotic deburring, 160
Robotics, 117, 123, 129, 135, 141, 148, 154, 160, 168, 312, 325, 347, 353
Robust estimation, 431
Rule based controller, 12, 89, 307

S

Scaling factor, 59
Scene understanding, 455
Seam tracking, 280
Seattle Metro, 253
Segmentation, 26, 35, 458
Self-learning, 270
Self-organizing, 63, 270
Self-tuning, 270, 282
Semiconductor processing, 267
Sensors, 50, 88, 131, 184, 345, 347
Servo control, 129, 184
Sewers, 253
Sheet feed, 294
Similarity measure, 483
Simulator, 213
Slip-recovery drive control, 196
Smoothing, 495
Society of Fuzzy Theories, The, (SOFT), xxii
Sociology, 503
Solar energy, 304
Space Shuttle, 363
Speed,
    control, 43, 76, 177, 196
    measurement, 187
Spell checking, 476
Stability, 21, 233, 239
Statistical,
    component, 250
    decisions, 509
    judgments, 514
Steam engine control, 19
Steel mills, 293
Steering control, 15
Stepper,
    motor control, 123
    motor robot, 123
Strip mills, 293
Surgery, 418
Suspension design, 109
Symbolic control, 132
Synchronous,
    generators, 233
    motor, 177, 183

T

Taiwan Power Company, 229
Tank level control, 96, 247
Target recognition, 486
Technology evolution, xvii, xxiv
Testing, real-time, 274
Theory limits, 20
Thickness control, 293
Torque control, 89, 177, 190, 204, 236
Train control, xxiv
Training, 43
Traffic signal control, 15, 101
Transputers, 325
Transmissions, xxiv, 88, 94

U

User environment, 12

V

Valuation space, 445

Vasodilator control, 419
Vector quantization, 401, 466
Virtual reality, 543
Vision, 15, 25, 26, 353
    fuzzy rules, 458
Vacuum cleaners, 9
Vehicular technology, xxi, 13, 43, 50, 56, 62, 68, 74, 81, 88, 94, 107, 109, 168, 319
Videos, xxii
Voice recognition, 493
Voting consensus, 424
Vowel recognition, 493
VLSI, 319, 331
Vlsi placement, 331

W

Walking, 117
Washing machines, 9
Water pumping, 196
Welding, 280, 312, 353

Editor's Biography

Robert J. Marks II

Robert J. Marks II is a Professor in the Department of Electrical Engineering at the University of Washington, Seattle. Prof. Marks was awarded the *Outstanding Branch Councilor* award in 1982 by IEEE and, in 1984, was presented with an *IEEE Centennial Medal*. He was named a Distinguished Young Alumnus of *Rose-Hulman Institute of Technology* in 1992 and, in 1993, was inducted into the *Texas Tech Electrical Engineering Academy*.

Dr. Marks was Chair of *IEEE Neural Networks Committee* (1989) and served as the first President of the *IEEE Neural Networks Council* (1990-91). In 1992, he was given the honorary title of *Charter President*. He is an IEEE Fellow. Dr. Marks was named an *IEEE Distinguished Lecturer* in 1992. He is a Fellow of the *Optical Society of America*. Dr. Marks was the co-founder and first President of the *Puget Sound Section of the Optical Society of America* and was elected that organization's first *Honorary Member*. He is co-founder and current President of **Multidimensional Systems Corporation** in Lynnwood, Washington and is a founder of *Financial Neural Networks, Inc.* in Kirkland, WA.

Prof. Marks is the Editor-in-Chief of the ***IEEE Transactions on Neural Networks*** (1992-present) and serves as an Associate Editor of the ***IEEE Transactions on Fuzzy Systems*** (1993-present). He serves on the Editorial Board of the ***Journal on Intelligent Control, Neurocomputing and Fuzzy Logic*** (1992-present). He was also the topical editor for *Optical Signal Processing and Image Science* for the ***Journal of the Optical Society on America - A*** (1989-91) and a member of the Editorial Board for ***The International Journal of Neurocomputing*** (1989-92). Dr. Marks served as *North American Liaison* for the 1991 *Singapore International Joint Conference on Neural Networks* (IJCNN), *International Chair* of the 1992 *RNNS/IEEE Symposium on Neuroinformatics and Neurocomputing* (Rostov-on-Don, USSR) and Organizational Chair for both the 1993 IEEE Virtual Reality Annual International Symposium (VRAIS) in Seattle and the *IEEE-SP International Symposium on Time-Frequency and Time-Scale Analysis* (Victoria, BC, 1992). He also served as the *Program and Tutorials Chair* for the *First International Forum on Applications of Neural Networks to Power Systems* (Seattle, 1991).

Dr. Marks was elected to the Board of Governors of the *IEEE Circuits and Systems Society* (1993-96) and was the co-founder and first Chair of the *IEEE Circuits & Systems Society Technical Committee on Neural Systems & Applications*. He is the *General Chair* of the *1995 International Symposium on Circuits and Systems*, Seattle. Dr. Marks is also the *Technical Program Director* for the first *IEEE World Congress on Computational Intelligence*, Orlando, July 1994. Five of his papers have been reproduced in volumes of collections of outstanding papers. He has two US patents in the field of artificial neural networks.

Dr. Marks is the author of the book *Introduction to Shannon Sampling and Interpolation Theory* (Springer Verlag, 1991) and is editor of the companion volume, ***Advanced Topics in Shannon Sampling and Interpolation Theory*** (Springer Verlag, 1993). Dr. Marks is a co-founder of the *Christian Faculty Fellowship* at the University of Washington and serves as the faculty advisor to the University of Washington's chapter of *Campus Crusade for Christ*.